The History of Mathematics: A Reader

This reader is one part of an Open University integrated teaching system and the selection is therefore related to other material available to students. It is designed to evoke the critical understanding of students. Opinions expressed in it are not necessarily those of the course team or of the University.

The editors wish to record their gratitude to *The Leverhulme Trustees* for their grant which enabled the Open University to appoint Dr Cynthia Hay as Research Fellow in the History of Mathematics, and thus for assisting the editors with basic research involved in the preparation of this book.

The History of Mathematics: A Reader

Edited by John Fauvel and Jeremy Gray
at the Open University

in association with

The Open University

First published 1987 by
MACMILLAN EDUCATION LTD
This edition now published by
THE MACMILLAN PRESS LTD
Houndmills, Basingstoke, Hampshire RG21 2XS
and London
in association with
THE OPEN UNIVERSITY
Walton Hall
Milton Keynes MK7 6AA

ISBN 0–333–42790–4 hardcover
ISBN 0–333–42791–2 paperback

A catalogue record for this book is available from the British Library.

Printed in Hong Kong

Reprinted 1988 (twice), 1991, 1992, 1993

Contents

Acknowledgements

The authors and publishers wish to thank the following who have kindly given permission for the use of copyright material:

American Oriental Society for extracts from 'Schooldays: a Sumerian Composition Relating to the Education of a Scribe' by S. N. Kramer in *Journal of the American Oriental Society*, **65** (1949).

British Journal for the History of Science for extracts from 'The Geometer and the Archaeoastronomers' in *BJHS*, **18** (1985).

Cambridge University Press for extracts from *The Works of Archimedes* by T. L. Heath (1910); *The Thirteen Books of Euclid's Elements* translated by T. L. Heath (1925); *Diophantus of Alexandria* by T. L. Heath (1910); *The Mathematical Papers of Isaac Newton* (edited and translated by D. T. Whiteside, 1967); and *Philebus*, translated by J. Hackforth.

Jonathan Cape Ltd and Harper and Row (Publishers) Inc for extracts from *One Hundred Years of Solitude* by G. Márquez, translated by G. Rabassa. English translation copyright © 1970 by Harper & Row.

W. H. Freeman and Co for extracts from 'Ishango' by J. de Heinzelin in *Scientific American*, June (1962).

Griffith Institute for extracts from *Egyptian Grammar* by A. Gardiner, Oxford University Press.

Grafton Books for extracts from *Science and Society in Prehistoric Britain* by E. W. MacKie (1977).

Harvard University Press for extracts from *Lore and Science in Ancient Pythagoreanism* by W. Burkert, translated by E. L. Minar, Jr, copyright © 1972 by the President and Fellows of Harvard College; *A Sourcebook in Greek Science* by M. R. Cohen and U. E. Drabkin, copyright © 1948 by the President and Fellows of Harvard College and copyright © 1976 by Leonora C. Rosenfield; and *A Sourcebook in Mathematics 1200–1800* edited by D. J. Struik, copyright © 1969 by the President and Fellows of Harvard College.

Heinemann Books and Harvard University Press for extracts from Loeb Classical

Library, *Selections Illustrating the History of Greek Mathematics*, Vols I & II, translated by I. Thomas, © 1939 by I. Thomas.

Hodder and Stoughton Ltd for extracts from *The Growth of Mathematical and Scientific Concepts in Children* by K. Lovell (1961).

Barbara Hooper Sude for extracts from her PhD Thesis 'Ibn-al-Haytham's Commentary on the Premises of Euclid's Elements' (1975).

Isis for extracts from 'History of Ancient Mathematics' by S. Unguru in *Isis*, **70** (1979).

Jarrold Colour Publications for the extract from *Discovery*, June (1945).

Librairie Droz S. A. for extracts from *The Italian Renaissance of Mathematics* by P. L. Rose.

Macmillan Publishing Co for extracts from *Discourse on Method* by Descartes, translated by L. Lafleur, © 1951 by Macmillan Publishing Co., copyright renewed 1979, and *Preliminary Discourse to the Encyclopedia of Diderot of 1751* by D'Alembert, translated by R. N. Schwab with W. E. Rex, © 1963 by Macmillan Publishing Co.

A. Marshack for extracts from his book *The Roots of Civilisation*, McGraw-Hill.

The Mathematical Association of America for extracts from 'Sherlock Holmes in Babylon' by R. Creighton Buck in *American Mathematical Monthly*, **87** (1980); and from 'The Rhind Mathematical Papyrus' by A. B. Chace (1927).

The MIT Press for extracts from *Number Words and Number Symbols* by K. Menninger (1969), *Philosophy of Mathematics and Deductive Structure in Euclid's Elements* by I. Mueller (1981) and *Origins of Cauchy's Rigorous Calculus* by J. Grabiner (1981).

Nieuw Archiv voor Wiskunde for extracts from 'On the Rectification of Curves' by van Heurat, translated by A. W. Grootendorst and J. A. van Maanen, in *Nieuw Archiv voor Wiskunde*, **25**, No. 3 (1982).

The Open University for translations of items 7.A1, 7.B3, 8.A1, 8.B2, 11.A2, 11.A3, 11.A5, 11.A6, 14.A4 and 15.A2, copyright © 1986 The Open University Press.

Oxford University Press for extracts from *Megalithic Sites in Britain* by A. Thom (1967); 'Mathematics and Astronomy' by G. J. Toomer in *The Legacy of Egypt* edited by J. R. Harris (2nd edn, 1971); *Plato's Theaetetus*, translated and edited by J. McDowell (1974); *Plato: Thirteen Epistles*, translated by L. A. Post (1925); *Graph Theory 1736–1936* by N. L. Biggs *et al.* (1976); *Ideas of Space* by J. J. Gray; *A History of Greek Mathematics* by Sir T. L. Heath (1921); *Mathematics in Aristotle* by Sir T. L. Heath (1949); and with Princeton University Press for extracts from *Charles Babbage: Pioneer of the Computer* by A. Hyman (1982).

Penguin Books Ltd for extracts from *The Birds and Other Plays* by Aristophanes, translated by D. Barrett and A. Sommerstein, © by D. Barrett and A. Sommerstein (1978); *Letters on England* by Voltaire, translated by L. Tancock, © by L. Tancock (1980); *Makers of Rome* by Plutarch, translated by I. Scott-Kilvert, © by I. Scott-Kilvert (1965); *Protagoras and Meno* by Plato, translated by W. K. C. Guthrie, © W. K. C. Guthrie (1956); *The Republic* by Plato, translated by H. D. P. Lee, © H. D. P. Lee (1974); *Timaeus and Critias* by Plato, translated by H. D. P. Lee, ©

H. D. P. Lee (1965); and *Brothers Karamazov* by F. Dostoevsky, translated by D. Magarshack, © by D. Magarshack (1958).

Princeton University Press for extracts from *Proclus: A Commentary on The First Book of Euclid's Elements*, translated by G. R. Morrow copyright © 1970 by Princeton University Press.

Random House Inc. for extracts from *Episodes from the Early History of Mathematics* by A. Aaboe.

D. Reidel Publishing Co for *Sophie Germain: An Essay* by I. L. Bucciarelli and N. Dworsky (1980).

The Royal Economic Society for extracts from 'Newton, the Man' by J. M. Keynes in *Collected Writings*, Vol. 10: *Essays in Biography*, pp. 363–4, 364–5, 366–7 and 371 (1972).

The Royal Society for extracts from *The Correspondence of Isaac Newton*, edited by H. W. Turnbull (1960).

St Martin's Press for extracts from *Critique of Pure Reason* by I. Kant, translated by N. Kemp Smith (1970).

Secker and Warburg Ltd for extracts from *Aubrey's Brief Lives*, edited by O. Lawson Dick.

Springer-Verlag for extracts from 'Greek and Arabic Constructions of the Regular Heptagon' by J. P. Hogendijk in *Archive for History of Exact Sciences*, Vol. 30, © 1984 Springer-Verlag, Berlin, Heidelberg, New York, London, Paris and Tokyo; 'On the Representation of Curves in Descartes Geometrie' by H. J. M. Bos in *Archive for History of Exact Sciences* Vol. 24, © 1981 Springer-Verlag; *The Geometrical Work of Girard Desargues*, by J. V. Field and J. J. Gray (1987); and *Fermat's Last Theorem* by H. M. Edwards (1979).

John Stillwell for extracts from his translation of 'Essay on the Interpretation of Non-Euclidean Geometry' by E. Beltrami in *Geometry*, Paper No. 4, Monash University (1982).

The University of California Press for extracts from *Mathematical Principles of Natural Philosophy and His System of the World*, by Isaac Newton, translated by A. Motte (1729), revised by F. Cajori (1934).

University of Wisconsin Press for extracts from *Algebra* by abu-Kamil, edited by M. Levey (1966).

D. T. Whiteside for extracts from his article 'Newton, the Mathematician' in *Contemporary Newtonian Research,* Reidel (1982).

Yale University Press for extracts from *Disquisitiones Arithmeticae* by Carl Friedrich Gauss, translated by A. Arthur and S. J. Clarke (1966).

The British Library for permission to use the front-cover illustration based on Johannes Kepler's *Mysterium Cosmographicum*, and the photographs in sections 14.D1 and 17.B4.

Every effort has been made to trace all the copyright holders but if any have been inadvertently overlooked the publishers will be pleased to make the necessary arrangement at the first opportunity.

Introduction

This selection of readings has been made for students of the Open University course MA290 *Topics in the History of Mathematics*, but it is hoped that others may find it interesting and useful too, whether students of mathematics interested in the work of past mathematicians or students of the past intrigued by George Sarton's claim that 'the history of mathematics should really be the kernel of the history of culture'.

'History of mathematics' is an ambiguous phrase. The reader may expect mathematical works written in the past, or historical studies by present-day historians of mathematics. These are, indeed, but two extremes of the range of material presented here. While the bulk of the entries are selected from the generally agreed main directions of Western mathematical thought and practice, we did not wish to encourage passive acceptance of the view of a source book as a collection of mathematical classics. It seemed right that a selection of readings made at a time when the history of mathematics, as a historical discipline, is more lively and vigorous than ever before should reflect something of the energy and excitement of that enterprise.

Thus, examples of modern historical discussion are included, as well as some of the fruits of recent scholarship in the form of newly discovered important ancient texts. Activities of these kinds are almost as old as Greek mathematics itself, as the extracts from works of Hellenistic and Islamic scholars testify. We have sought to make available, too, examples of other source material for the history of mathematics and mathematicians: letters, extracts from diaries, and other evidence of past mathematical practice, including its perception in literature. It is interesting and surprising, for example, that our earliest evidence for the investigation of the classical problem of 'squaring the circle' should be found in the work of an Athenian comic dramatist. And it will interest today's student of mathematics to discover that the greatest mathematicians of the past have frequently had the keenest knowledge of *their* predecessors.

The arrangement of such varied material has presented problems which any anthology editor will recognize. Within a broadly chronological framework, the extracts are given as seemed best in the light of content and theme. The chapter titles are flexible gestures, not binding constraints. The material of history is always richer than any categorization of it can reveal.

In order to make available as full a selection as possible for the general reader, we have purposefully discarded the apparatus of scholarship and have kept editorial commentary to a minimum. In the accompanying Open University course, students are both given a fuller background and encouraged to evaluate arguments and query historical texts. Note that inclusion of a historical study, argument or view in this Reader implies only that the editors find it interesting or important, not that we necessarily agree with it.

Those who care to study the extracts in their contexts may wish to consult the original sources listed at the end. Others may wish to complement the extracts with a general account of the history of mathematics such as is widely available in such standard texts as those of Carl Boyer, Morris Kline or Dirk Struik (details of these books are given at the end of this introduction). Biographies of many mathematicians appear in reference books such as *Chambers's Biographical Dictionary*, and most fully in the *Dictionary of Scientific Biography*. A valuable guide to other writings on the history of mathematics is the annotated bibliography edited by Joseph Dauben. Note that, for the period 1200–1800, we have endeavoured not to overlap more than is essential with the excellent source book of Dirk Struik.

In assembling this book we have been especially aided by Cynthia Hay, who not only provided substantial editorial input for Chapters 6–8, but also led with enthusiasm our team of translators – Judith Field, Gale Jagger and Fenny Rankin – to whom thanks are also due. Graham Flegg has been a constant source of helpful advice. We thank, too, Steven Engelsman, Jesper Lützen, Elinor Schuess and Trista Selous for specific proposals. John Taylor has been characteristically calm and supportive; and preparation of this book for the press would have been impossible without the generous and painstaking endeavours of Roger Lowry. John Fauvel was responsible for editing Chapters 1–5, 9 and 19; the editors with Cynthia Hay for Chapters 6–8; and Jeremy Gray was responsible for Chapters 10–18, as well as for the rest of the new translations.

John Fauvel
Milton Keynes, 1986 Jeremy Gray

Further Reading

Carl B. Boyer, *A History of Mathematics*, John Wiley, 1968.

Joseph W. Dauben (ed.), *The History of Mathematics from Antiquity to the Present: A Selective Bibliography*, Garland, 1985.

Charles Coulston Gillespie (ed.), *Dictionary of Scientific Biography*, 16 vols, Scribners, 1970–80.

Morris Kline, *Mathematical Thought from Ancient to Modern Times*, Oxford University Press, 1972.

Dirk J. Struik, *A Concise History of Mathematics*, Dover, 1967.

Dirk J. Struik, *A Source Book in Mathematics 1200–1800*, Harvard University Press, 1969; Princeton University Press, 1986.

1 Origins

1.A On the Origins of Number and Counting

Since at least the time of the ancient Greeks, there has been interest in trying to understand how and why mathematics began. While few have doubted that counting was one of the early activities that gave rise to mathematics (appreciation of shape, form and decoration was perhaps another), how counting itself started is harder to answer, if indeed it is a meaningful question. The extracts in this section are chosen not for being necessarily influential—still less, 'true'—but rather to illustrate the range of views that have been held. Aristotle (1.A1) wrote in the fourth century BC, Sir John Leslie (1.A2) wrote at the beginning of the nineteenth century (AD), and 1.A3–1.A5 are a selection of twentieth-century views. It is noticeable how the accounts reflect the period in which they were written, both in attitudes towards early societies and their life conditions and in the kind of historical and mathematical analysis exemplified.

1.A1 Aristotle

Why do all men, whether barbarians or Greeks, count up to ten and not to some other number, such as two, three, four or five, so that they do not go on to repeat one of these and say, for example, 'one-five', 'two-five', as they say 'one-ten' [i.e. eleven], 'two-ten' [twelve]? Or why, again, do they not stop at some number beyond ten and then repeat from that point? For every number consists of the preceding number [taken as base] plus one or two, etc., which gives some different number; nevertheless ten has been fixed as the base and people count up to that. Chance cannot account for the fact that, apparently, everyone does this always: what happens always and in all cases is not the result of chance but is in the nature of things. Is it because ten is a perfect number, seeing that it comprises all kinds of number, even and odd, square and cube, linear and plane, prime and composite? Or is it because ten is the beginning of number, since ten is produced by adding one, two, three, and four? Or is it because the moving bodies [in the heaven] are nine in number? Or because within ten [compounded] ratios four cubic numbers are completed, of which numbers the Pythagoreans hold that the universe is constructed? Or is it because all men had ten fingers, so that having, as it were, counters for the appropriate number, they employed this number for counting other things as well? A certain race among the Thracians alone count up to four, because like that of children, their memory is not able to take in more, and they never use any large number.

1.A2 Sir John Leslie

The idea of number, though not the most easily acquired, remounts to the earliest epochs of society, and must be nearly coëval with the formation of language. The very savage, who draws from the practice of fishing or hunting a precarious support for himself and family, is eager, on his return home, to count over the produce of his toilsome exertions. But the leader of a troop is obliged to carry farther his skill in numeration. He prepares to attack a rival tribe, by marshalling his followers; and, after the bloody conflict is over, he reckons up the slain, and marks his unhappy and devoted captives. If the numbers were small, they could easily be represented by very portable emblems, by round pebbles, by dwarf-shells, by fine nuts, by hard grains, by small beans, or by knots tied on a string. But to express the larger numbers, it became necessary, for the sake of distinctness, to place those little objects or counters in regular rows, which the eye could comprehend at a single glance; as, in the actual telling of money, it would soon have become customary to dispose the rude counters, in two, three, four, or more ranks, according as circumstances might suggest. The attention of the reckoner would then be less distracted, resting chiefly on the number of marks presented by each separate row. [...]

These simple arrangements would, on their first application, carry the power of reckoning but a very little way. To express larger numbers, it became necessary to renew the process of classification; and the ordinary steps by which language ascends from particular to general objects, might point out the right path of proceeding. A collection of *individuals* forms a *species*; a cluster of *species* makes a *genus*; a bundle of *genera* composes an *order*; a group of *orders* constitutes a *class*; and an aggregation of *classes* may complete a *kingdom*. Such is the method indispensably required in framing the successive distribution of the almost unbounded subjects of Natural History.

In following out the classification of numbers, it seemed easy and natural, after the first step had been made, to repeat the same procedure. If a heap of pebbles were disposed in certain rows, it would evidently facilitate their enumeration, to break down each of those rows into similar parcels, and thus carry forward the successive subdivision till it stopped. The heap, so analysed by a series of partitions, might then be expressed with a very few low numbers, capable of being distinctly retained. The particular system adopted for this decomposition would soon become clothed in terms borrowed from the vernacular idiom.

1.A3 K. Lovell

When early man returned to his abode he may have wanted to tell his family of his experiences and to describe the animals he encountered. He may well have used terms corresponding to our 'many' and 'few', a large group for example, being described as 'many, many'. Out of these and other experiences there arose a need for quantitative exactness. The use of names for animals or objects helped at this stage, providing a man's possessions were few. If he had, say, three sheep, he would have names for each, and he could tell if they were all present. On the other hand, if he had no names for the sheep in his flock he had only a vague idea whether or not they were all present. Later, he hit upon the device of matching the objects of one group with those of another group. For example, if flint axes were handed round to a group of men, it would soon be clear if there were enough axes or too many or too few. This one-to-one correspondence was important for the later concept of number, and it also led, no doubt, to terms like 'more', 'less', 'as many as'. This matching was doubtless accidental at first, but later it was deliberate in the sense that the group of objects was matched against a model group, e.g. wings of a bird, paws of the lion, fingers of the hand. Then when he spoke about his group, he said that he had seen as many bears as there were toes on his feet.

But to deal with large groups he eventually resorted to tallying. A notch was cut in a stick for each object, or a pebble put aside for each animal, so that the herdsman could check his sheep against a number of stones. The stones and sheep were quite dissimilar, but each sheep and each stone represented a unit and there was a one-to-one correspondence between them. Tallying was very useful and a marked step forward, but even when primitive man had 'tallied', he still could not think of, or name, the number.

The uses of model groups and tallying were limited. Early man's first concept of fiveness would be in terms of the number of fingers on the hand, not the abstract 'five'. It was a big step, intellectually, to move from words that stand for model groups to words that stand for abstract numbers. Exactly how this took place we do not know. The associated sensory impressions — the perceptions — of a one-to-one correspondence, and/or the actions involved in setting up this correspondence, were most helpful in helping man to arrive at the concept of the natural numbers; but they may not have been in themselves sufficient. These perceptions and actions, perhaps, did not in themselves bring in the idea of numbers, but they increased greatly the chance of the concept being formed. There had to be in early man, as in every child today, an intellectual jump, to the idea of an abstract 'twoness' and 'threeness'.

How, when objects are matched in front of the child, or primitive adult, is insight gained into the idea of the natural numbers? How does the child recognize the quantity of 'threeness' in two or more groups of three (three apples, three marbles, etc.)? No definite answer can be given to these questions, as mathematicians are not agreed among themselves. Some, for example the French mathematician and philosopher H. Poincaré, believe that at this point the idea of a series of natural numbers becomes clear to everyone. Such people believe that the concept of the natural numbers is the result of a primitive intuition at this stage.

1.A4 Karl Menninger

Haven't people always counted as we do today?

We shall find the answer to this question if we descend the ladder of culture down to the very lowest steps, scarcely above the level where mind could not rise above its environment. Early man counted, too, whenever he merely gathered fruits or hunted, whether he grew his own food by more or less primitive methods of cultivation or drove his herds from pasture to pasture or whether, like many tribes living near the coast, he sought to earn his living by trade. His way of life taught him to count, the nature of his economy determining the extent of his number sequence. Why should a pygmy people, living in isolation in the primeval forest, need to count beyond 2? Anything over that is considered 'many'. But the cattlebreeder must count his herd, head by head, up to 100 or even more. For him 'many' is something far greater, something that no longer has economic meaning to him. Thus early man's environment determined his thinking and actions, and also his counting.

So that we can understand his number sequences, which we shall now consider, let us dwell for a moment on the way primitive man perceives the world around him. It still impinges on him directly, in all its myriads of colors and forms. Things have not yet been 'cooled off' for him by his intellect, which sifts them and orders them and separates them, filing their elements away in the gray, colorless pigeon-holes of concepts. On the contrary, in their immediate, hot-blooded, many-colored uniqueness they touch his innermost heart. Thus they are not objects to him, things which are alien to him and stand 'outside' himself—*here* am I and *there* is the world—rather they are completely absorbed in his own life. He is a part of them, as they are of him. He is woven into the very fabric of the universe by powerful strands of religion; he does not, like 'modern' man, like ourselves, stand before it in wonderment, in calculation, or indifference.

Yet some remaining fragments of that early perception of the world still loom up in our own. Many a superstition, many an oddity, survives unrecognized in the midst of the intense consciousness of our own culture. Who today knows or cares much about the number 7? Though for us it has lost its supernatural content, though it offers not the slightest advantage in measuring and reckoning time, the seven-day week still governs our whole external life. From that early interpenetration of man with the world arises the infusion of objects and numbers with mystic significance, and hence the 'holiness' of the numbers 3 and 7 and the auspiciousness or bad luck of the number 13. It is the task of mythology to uncover the concepts that led early man to impart supernatural significance to certain numbers.

Our own purpose, however, is to understand how early man gave expression to things and events, with all their profusion of kaleidoscopic detail, in his own primitive language.

1.A5 Abraham Seidenberg

If we agree that counting is not a simple process invented in one stroke, we may examine it for elements it contains that could have had an independent existence and out of which it might have grown. There is no particular call to count the fingers, and we may safely suppose that they have no bearing on the invention. The basic things needed for counting are a definite sequence of words and a familiar activity in which they are employed. The creation ritual offers us precisely such sequence and activity. Processions of couples in ritual are well known. 'Male and female he created them.' 'There went in two and two unto Noah into the ark, the male and the female, as God had commanded Noah.' Presumably, a couple is announced (in ritual the Word always accompanies the Deed) and then they make their appearance on the ritual scene. The sequence of words so used might have come to be used as the initial number-words.

The above is in accordance with, and explains, the phenomena: first, the association of numbers with deities, in fact, according to the above, numbers did not exist prior to their ritual application, viz, for the purpose of invoking 'deities', that is, participants in ritual; second, 'the marked preference not infrequently observed among savages for counting by pairs'—this is *our* explanation of the 2 of the 2-system; third, the ritual division of numbers into even and odd, known also to the Egyptians, Babylonians, Indians, Chinese, and Incas, as well as to savages in Africa, Sumatra, the Philippines, Polynesia, and North America.

Our conjecture is, then, first, that the names of participants in ritual, or the words announcing them, were the initial number words. The ritual suffuses itself into all sorts of activities—according to Raglan 'all rites were originally creation rites, and in essence are so still'. A collection of objects, say a catch of fish, must undergo a rite. The 'serializing' is the rite, and the 'counting' is the myth. The original intention is to mimic a portion of the Creation ritual. It is in this way that we envisage 'counting' to have become detached from the ritual and to have acquired its abstract or general character.

Second, we think that the higher counting may have started as a method of taking care of longer and longer processions (not with the idea of counting them, however). The base (which is not logically inherent in counting) corresponds to the number of persons in the basic ritual, and the higher counting derives from the continued repetition, with slight modifications, of this basic ritual.

1.B Evidence of Bone Artefacts

Among the earliest evidence for proto-mathematical activity is a bone dated at 9000–6500 BC, dug up in the 1950s at Ishango in what is now Zaire. The bone has what appear to be tallying marks engraved on it, notches carved in groups. To interpret aright the meaning and intention of these marks would be highly revealing, but the difficulty of doing so is illustrated by the contradictory interpretations advanced in the two extracts below: 1.B1 describes the conclusions reached by the bone's discoverer, Jean de Heinzelin; 1.B2, the view of Alexander Marshack, who restudied the bone and considers it to be a tallying of lunar phases.

1.B1 Jean de Heinzelin on the Ishango bone as evidence of early interest in number

The most fascinating and most suggestive of all the artifacts at Ishango is not a harpoon point but a bone tool handle with a small fragment of quartz still fixed in a narrow cavity at its head. In the first place, its shape and the sharp stone in its head suggest that it may have been used for engraving or tattooing, or even for writing of

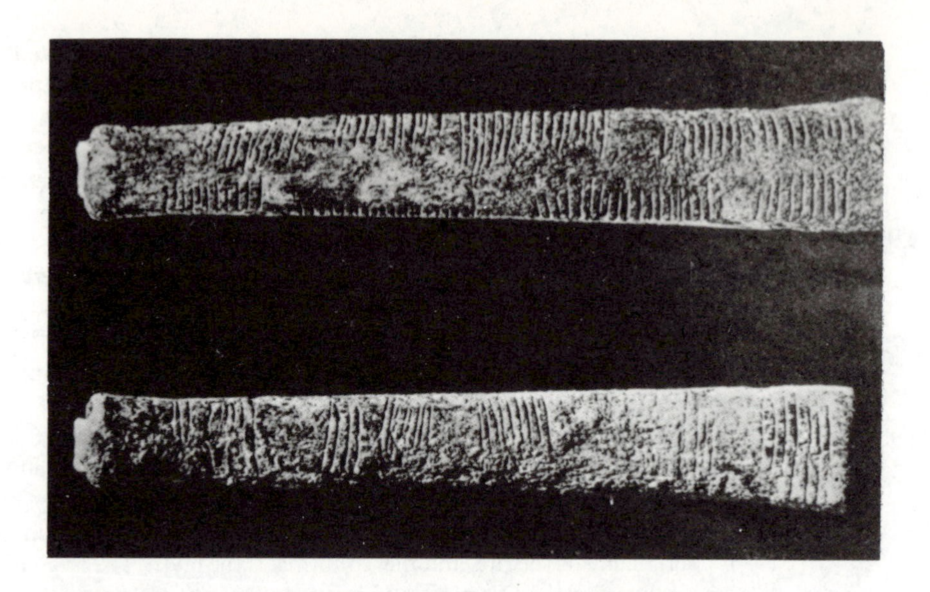

some kind. Even more interesting, however, are its markings: groups of notches arranged in three distinct columns. The pattern of these notches leads me to suspect that they represent more than pure decoration. When one counts them, a series of number sequences emerges. In one of the columns they are arranged in four groups composed of 11, 13, 17 and 19 individual notches. In the next they are arranged in eight groups containing 3, 6, 4, 8, 10, 5, 5 and 7 notches. In the third they are arranged in four groups of 11, 21, 19 and 9.

I find it difficult to believe that these sequences are nothing more than a random selection of numbers. The groupings in each column are quite different from one another and each column contains internal relationships unlike those found in either of the others. Take the first column, for example: 11, 13, 17 and 19 are all prime numbers (divisible only by themselves and by one) in ascending order, and they are the only prime numbers between 10 and 20. Or consider the third: 11, 21, 19 and 9 represent the digits 10 plus one, 20 plus one, 20 minus one and 10 minus one. The middle column shows a less cohesive set of relations. Nevertheless, it too follows a pattern of a sort. The groups of 3 and 6 notches are fairly close together. Then there is a space, after which the 4 and 8 appear—also close together. Then, again after a space, comes the 10, after which are the two 5s, quite close. This arrangement strongly suggests appreciation of the concept of duplication, or multiplying by two.

It is of course possible that all the patterns are fortuitous. But it seems probable that they were deliberately planned. If so, they may represent an arithmetical game of some sort, devised by a people who had a number system based on 10 as well as a knowledge of duplication and of prime numbers.

1.B2 Alexander Marshack on the Ishango bone as early lunar phase count

Looking at the Ishango bone I therefore asked: Why should an adult Mesolithic man—or woman—before history and agriculture, be playing arithmetical games?

Certainly not as the child plays with water, mud, sticks, or stones, testing, enjoying, and developing his manipulative skills and knowledge of these things. Adult intellectual games, including arithmetical games, all have some kind of storied meaning. This holds whether the number games are used in magic or divination, in astrology or gambling, or whether the numbering is used as a storytelling and remembering device, say to recall an event, or to indicate a tattoo or a ritual pattern. This 'storied' content for number thinking existed in the early arithmetic and geometric skills and games of the Egyptians, Babylonians, and Greek Pythagoreans, and it exists today for all branches of mathematics. Considered as an abstract skill, numbering may perhaps be thought of as 'pure' and unrelated to storied meaning, but psychologically it is, nonetheless, a symbolizing and cognitive process. [...] With these general thoughts I assumed it was far more than a mere game of numbers that was being played in the Mesolithic.

It seemed to me that the bone from Ishango, instead of representing a simple arithmetical game, was more significantly some sort of *notation*—arithmetic or not—and that it had been intentionally made for some purpose, whether playful or useful, and that it contained some storied meaning.

Martin P. Nilsson, the Danish historian, in his work *Primitive Time Reckoning* (1920), had indicated that most primitive peoples of whom there is record had some knowledge of time reckoning and made use of either the stars or moon or both. His 'primitives', though, were almost all of our time, those studied by the anthropologist or those whose records were left us in writing.

My first assumption, then, was to ask if this series of odd and different 'counts' on the prehistoric Ishango bone could be related somehow to a time count and perhaps to a lunar count. The lunar count would be the simplest possibility for an early system of time reckoning and might, for instance, include a recognition or count from full moon to full moon or from invisibility (new moon) to invisibility or from crescent to full moon.

[After a detailed analysis of the bone markings against lunar phases, the text continues as follows.]

Without any certainty about the results, I had taken a number of new steps. I had shown that what had seemed to be a Mesolithic number game or decimal system to the excavator could instead be the result of a sequential notation, *perhaps* notating the lunar phases and periods. Whether my answers and results were entirely right or wrong was only partially important. The questions and techniques were new and might lead to new insights and findings regarding prehistoric thought. By my questions I had been forced to devise tests for this prehistoric material that had inherent controls, and similar tests could now be used on other examples with comparable markings. Perhaps, though I could not be sure, I had stumbled on a technique and result that could help 'crack the code', not of a grammar or writing but nevertheless of some of man's earliest intellectual activities. 'Intellectual' because it was not, apparently, art, but rather notation—though one can see implied tendencies toward decoration in such accumulations; 'intellectual' because it was not toolmaking, though to the extent that notation may have been a useful procedure, it was an intellectual or cultural 'tool'.

1.C Megalithic Evidence and Comment

The previous section concerned the possible mathematical meaning of prehistoric bone markings. This one concerns another potential route of discovering prehistoric mathematical attainments—that of trying to infer from surviving constructs the particular mathematical knowledge presupposed in them. Careful surveying and analysis of the large stone remains, of which Stonehenge is the best known, in north-western Europe has suggested that the mathematical activity in their period of construction (c. 3500–2000 BC) was greater and more sophisticated than has previously been realized. Alexander Thom (1.C1), in particular, has concluded from statistical analysis of measurements of the sites that there was both a standard unit of measurement (the 'megalithic yard') held constant over a wide area and a comparatively advanced geometrical knowledge. Although this conclusion has not been universally accepted by prehistorians (1.C2), it has contributed to a growing re-evaluation of our knowledge of megalithic societies (1.C3). B. L. van der Waerden (1.C4) has conjoined to Thom's thesis evidence from elsewhere, to argue for an early and widespread mathematical science. But difficulties with the interpretation of the evidence remain, as is pointed out by the historian of mathematics Wilbur Knorr in his review of van der Waerden's book (1.C5).

1.C1 Alexander Thom on the megalithic unit of length

It is fortunate for us that Megalithic man liked, for some reason or another, to get as many as possible of the dimensions of his constructions to be multiples of his basic unit. We are thereby enabled to determine unequivocally the exact size of this unit. In fact probably no linear unit of antiquity is at present known with a precision approaching our knowledge of the Megalithic yard. The reason for his obsession with integers is not entirely clear, but undoubtedly the unit was universally used, perhaps universally sacred. It may have been that in the absence of paper and pen he found it necessary to record in stone his geometrical and perhaps also his arithmetical discoveries. Such of these as are known to us are of no mean order and there is no reason to suppose that our knowledge of what he knew is by any means complete. When it is recalled that our knowledge of his achievements in this field is only a decade or so old it is obvious that we have no right to imagine that it is complete. This mistake has indeed been made too often.

It is remarkable that 1000 years before the earliest mathematicians of classical Greece, people in these islands not only had a practical knowledge of geometry and were capable of setting out elaborate geometrical designs but could also set out ellipses based on Pythagorean triangles. [. . .] They concentrated on geometrical figures which had as many dimensions as possible arranged to be integral multiples of their units of length. They abhorred 'incommensurable' lengths. This is fortunate for us because once we have established their unit of length we can very often unravel designs which would otherwise be meaningless. These people also measured along curves and so it is necessary to devote some space to the methods of calculating the perimeters of the various rings which they developed.

The basic figure of their geometry, as of ours, is the triangle. Today everyone knows the Pythagorean theorem which states that the square on the hypotenuse of a right-angled triangle is equal to the sum of the squares on the other two sides. We do not know if Megalithic man knew the theorem. Perhaps not, but he was feeling his way towards it. One can almost say that he was obsessed by the desire to discover and record in stone as many triangles as possible which were right-angled and yet had all three sides integers. The most famous of the so-called Pythagorean triangles is the 3, 4, 5—right-angled because $3^2 + 4^2 = 5^2$. He used this triangle so often that he may well have noticed the relation. Limiting the hypotenuse to 40 there are six true Pythagorean triangles. These are:

(1) 3,4,5. (4) 7,24,25.
(2) 5,12,13. (5) 20,21,29.
(3) 8,15,17. (6) 12,35,37.

Megalithic man knew at least three of these. He may have known all six and we simply have not yet found the sites where they were used, but we shall see later that there were other conditions to be fulfilled and these certainly restricted the use of some of these triangles. The remarkable thing is that the largest, the 12, 35, 37, was known and exploited more than any other with the exception of the 3, 4, 5.

[...] It is one of the objects [here] to demonstrate unequivocally the existence of a common unit of length throughout Megalithic Britain and to show that its value was accurately 2·72 ft.

1.C2 Stuart Piggott on seeing ourselves in the past

This afternoon we paid tribute[...]to the long, patient, accurate and modestly pursued work of Professor Alexander Thom, on which he has based a thesis which if accepted demands the recognition of considerable mathematical skills among the non-literate societies of northwestern Europe from the fourth to the second millennia BC; mute inglorious Newtons who somehow managed to command the labor and organization necessary to construct stone circles or alinements from the Bay of Biscay to the Arctic Ocean. Here, as subsequent discussion showed, however cogent his reasoning may be on purely mathematical grounds, many archaeologists, including myself, would feel that a great number of difficulties have not yet been faced in an evaluation of this hypothesis. [...]

There is always the danger of seeing ourselves in the past, of becoming victims of the fallacy whereby 'ideas are imported from present-day experience, and ancient man is anachronistically saddled with views he would have found at best strangely unfamiliar', as Ian Richmond put it, or of the unconscious tendency 'to project the axioms, habits of thought and norms of the present day into the past', in Henri Frankfort's phrase. God-like, we try to make ancient man in our own image, and the preferred image varies with the changes of taste and preference of our society. We desire to find admired qualities in the past, and mathematical and scientific qualities are admired today. If ecstasy and shamanism were more highly regarded than these, this is what we might be looking for—and doubtless finding—in prehistory.

1.C3 Euan MacKie on the social implications of the megalithic yard

One aspect of the detailed conclusions made by Thom about the megalithic yard, fathom and rod in the stone circles cannot pass without comment. This is his deduction that the evidence plainly shows that the megalithic yard was carried in the form of actual rods of standard length made of wood or bone from one end of England to the opposite end of Scotland (Ireland has not yet been investigated), and without varying in length more than 3/100ths of an inch. In other words, the metrology and geometry that the circle builders used in their projects was highly organised to the extent that the lengths of the measuring rods were standardised at a single centre: if the rods had been copied from one region to the next errors should have accumulated and the actual variation between the rods of different areas should have been much greater. If it is correct, this deduction must have very important implications for the social organisation of Late Neolithic and Early Bronze Age Britain since it is scarcely conceivable that such a situation could have come about unless there was one major training centre for the wise men of that period where the appropriate knowledge and skills were taught by the wisest of the order and from which the 'graduate' astronomer priests and magicians were sent out all over the country.

Leaving aside for the moment the question of the astronomical skills possessed, one may conclude that if it were not for this particular inference about the absolute uniformity of the rods, it would be possible to assume a lower level of organisation among the 'priesthood' and perhaps to draw an analogy with the Early Christian missionaries and monks in Scotland in the middle of the first millennium AD, who were in general united by their faith but who owed allegiance primarily to local religious leaders, who lacked (with a few exceptions) great intellect and had no rigidly uniform body of doctrine and ritual. A centralised but widespread prehistoric order would be more analogous to the highly organised medieval Roman church with its training colleges and monasteries, uniform doctrine and ritual, and rigid hierarchical organisation.

Pursuing the Early Christian analogy, one might imagine that at 2000 BC there was a body of practical knowledge of, and a religious tradition about, geometry and measurement, perhaps originating among some earlier inspired group, which was carried gradually throughout the country by 'missionaries' of some kind and handed down in local versions or sects through several generations. It would not be necessary to suppose that the specialist class of each generation was trained at a single centre nor that its members were thus closely integrated over the whole country, nor even that the practitioners were a full-time priestly class: part-time shamans and medicine men might suffice. However, the concept of an exactly standardised yardstick implies that there was such an integrated class, and it presupposes an altogether more advanced and better organised class of wise men and priests. The evidence for it therefore needs to be carefully examined.

1.C4 B. L. van der Waerden on neolithic mathematical science

Until quite recently, we all thought that the history of mathematics begins with Babylonian and Egyptian arithmetic, algebra, and geometry. However, three recent discoveries have changed the picture entirely.

The first of these discoveries was made by A. Seidenberg. He studied the altar constructions in the Indian Śulvasūtras and found that in these relatively ancient texts the 'Theorem of Pythagoras' was used to construct a square equal in area to a given rectangle, and that this construction is just that of Euclid. From this and other facts he concluded that Babylonian algebra and geometry and Greek 'geometrical algebra' and Hindu geometry are all derived from a common origin, in which altar constructions and the 'Theorem of Pythagoras' played a central rôle.

Secondly I have compared the ancient Chinese collection 'Nine Chapters of the Arithmetical Art' with Babylonian collections of mathematical problems and found so many similarities that the conclusion of a common pre-Babylonian source seemed unavoidable. In this source, the 'Theorem of Pythagoras' must have played a central rôle as well.

The third discovery was made by A. Thom and A. S. Thom, who found that in the construction of megalithic monuments in Southern England and Scotland 'Pythagorean Triangles' have been used, that is, right-angled triangles whose sides are integral multiples of a fundamental unit of length. It is well-known that a list of 'Pythagorean Triples' like (3, 4, 5) is found in an ancient Babylonian text, and the Greek and Hindu and Chinese mathematicians also knew how to find such triples.

Combining these three discoveries, I have ventured a tentative reconstruction of a mathematical science which must have existed in the Neolithic Age, say between 3000 and 2500 BC, and spread from Central Europe to Great Britain, to the Near East, to India, and to China. [...]

We have seen that there are so many similarities between the mathematical and religious ideas current in England in the Neolithic Age, in Greece, in India, and in China in the Han-period, that we are bound to postulate the existence of a common mathematical doctrine from which these ideas were derived.

Can we make a reasonable conjecture about the place of origin of this mathematical doctrine? [...]

The Beaker people, who built Stonehenge II, Woodhenge and other henges, came to England from the continent shortly after 2500 BC. It is possible that they already spoke an Indo-European language. In any case, they lived in a region where Indo-European languages were spoken at an early date.

The Indo-European languages are connected with a perfect decimal counting system, including a method of designating fractions. The English expression 'the fifth part' corresponds to the Greek *τὸ πέμπτον μέροσ*. This number system is an important cultural achievement and an excellent basis for teaching arithmetic and algebra. Also, the religions of ancient Indo-European populations have so much in common that the existence of an Indo-European religion can hardly be doubted. Hence, if we find quite similar ideas about the ritual importance of geometrical constructions in Greece and India, and the same set of Pythagorean triangles with ritual applications in England and India, and the same geometrical constructions in Greece and India, the conclusion that these religious and mathematical ideas have a common Indo-European origin is highly probable.

1.C5 Wilbur Knorr's critique of the interpretation of neolithic evidence

The strengths of van der Waerden's book—the perceptive mathematical analysis of results and techniques, the far-ranging coverage of fields and traditions, and the inclusion of materials (as on Hindu and Chinese) not yet easily accessible in standard discussions—are overwhelmed by the weaknesses of his interpretive framework. These weaknesses stem from his failure to scrutinize the implications of his hypothesis of neolithic origins. All his evidence is compatible with the generally accepted and far more plausible view of independent Egyptian and Sumerian origins in the third millennium BC, followed by their development and transmission to the Hindu, Greek and Chinese, through an intricate pattern of cultural interactions extending over several thousand years.

The same limitations mark even more strongly the discussion of the neolithic evidence, the veritable linchpin of van der Waerden's thesis. Speculations on the *astronomical* significance of neolithic sightlines have received considerable attention in recent years, from the time that G. S. Hawkins began to announce his findings on Stonehenge two decades ago to the present rise of the professional group of 'archaeoastronomers' representing a wide spectrum of particular hypotheses, but a general consensus on the validity of Hawkins' approach. The related speculations on neolithic geometry are less familiar, however, and I had not encountered them before reading van der Waerden's book. From his own account and those he cites, one derives the impression that the painstaking effort of measuring and reconstructing the neolithic sites has accorded virtual certainty to ambitious conclusions about the mathematical expertise of their ancient designers. But reading as an interested nonspecialist, I could discern strong cause for scepticism.

[...] If one *assumes* an advanced neolithic technique, then one is predisposed to discern complicated patterns in the megalithic rings; with ingenuity and diligence, interesting patterns will be found. But if one questions that assumption, the same rings will as easily be conceived as roughly laid circles and more complicated schemes will be received with scepticism. This *assumption* of the level of expertise, however, is the fundamental issue. Thus, the effort to measure and reconstruct the rings cannot provide the evidence by which this assumption could be confirmed or convincingly argued.

It seems to me that the sensational impact of finding such complex geometric patterns in these ruins has obscured for analysts like van der Waerden and the Thoms the phenomenal difficulties implicit in their views. A comparable insensitivity to deeper implications bears on the Thoms' hypothesis of the 'megalithic yard', the common measuring unit inferred from measurements of the diameters and circumferences of hundreds of neolithic rings. Surely, the notion that the neolithic builders adopted such a standard unit must be considered intrinsically implausible. How could such a unit be kept standard over more than a millennium and a geographical area of thousands of square miles? What would the standard be made of (wood? stone?) and where would it be kept? How could a population of dispersed, migratory agricultural tribes maintain standardized measures, or even want to? In modern industry and science, careful measurement is typically crucial for the successful operation of precision instruments and machines. But at the more basic level of manual crafts, distances can be laid out in a more rough-and-ready manner without detriment to the product. Even today, dealers in yard goods, for instance, can measure out lengths of fabric by the distance from nose to outstretched hand and stay within an acceptable range of variation. Neolithic life is far more likely to have followed this more basic pattern, than to have instituted norms of measurement comparable to those of modern science and technology.

Common sense, I think, should lead one to balk at the implications of such a hypothesis about neolithic technique and to reexamine the data in the search for simpler hypotheses. The Thoms' data for the 'MY', for instance, do not reveal sharply marked preferences for certain distances (namely, multiples of a hypothetical unit), nor even pronounced statistical distributions around those distances. Their expression of the MY as 2·72 feet gives the misleading impression that megalithic distances were typically surveyed to high degrees of accuracy (e.g., 1 part in 272 or better); further, it does not state the degree of uncertainty in the result. [...] One infers from their data that only one fourth of their sample (40 of 163 items) lies within $\pm 0{\cdot}1$ MY from an integral multiple. If this is claimed to signify a preference for integral values of this unit, the preference can hardly be considered a strong one. But even if we accept that something other than a random distribution is at issue here, we could hardly suppose that a pattern very different from this would obtain for the variation in forearm lengths or pacing distances among adult neolithic males. Surely, these God-given and eminently portable units of measure would serve the purposes of the neolithic builders far better than an artificial standardized 'megalithic yard'.

By accepting these claims of the archaeogeometers and incorporating them into his thesis of mathematical origins, van der Waerden has, in effect, produced a *reductio ad absurdum* of their claims. For he has displayed more fully what they seem only dimly to have perceived, that the appearance of configurations like Pythagorean numerical triangles must point to the presence of a highly elaborated system of number theory and geometry. Neolithic life can hardly have posed the demands or furnished the resources for developing such sophisticated mathematical theory. In view of this, one ought to return to the data in search of simpler ways to describe the plans of the megaliths, rather than persist in the discredited hypothesis.

The frame of mind in which scholars like van der Waerden and the Thoms can even countenance such theses of neolithic expertise, let alone presume to argue them on the basis of the available evidence, is utterly alien to my own intuitions as a historian of ancient mathematics. [...] The professional mathematicians, scientists and engineers who dominate this field of scholarship, by their ready acceptance of such notions and by their reluctance to seek simpler alternatives, more in line with the technical and social level of the neolithic tribes, strike me as having abandoned the basic canons of scientific method and rational inquiry. I fear that the lure of notoriety for sensational 'discoveries', with the resulting rewards of public attention, may have clouded their professional judgment. I fear even more the regrettable impact on credulous nonspecialists, who may not know to distinguish between the general enterprise of scientific research and the reckless notions of some scientists.

1.D Egyptian Mathematics

For a literate civilization extending over some 4000 years, that of the ancient Egyptians has left disappointingly little evidence of its mathematical attainments. Even though the classical Greeks believed mathematics to have been invented in Egypt (1.D4)—though their accounts are far from unanimous on how this happened—there are now but a handful of papyri and other objects to convey a sense of Egyptian mathematical activity. The largest and best preserved of these is the Rhind papyrus (1.D1, 1.D2), now in the British Museum, a copy made in about 1650 BC of a text from two centuries earlier. A lively picture of one of the contexts in which mathematics was used is provided by a satirical letter (1.D3) from later that millennium (perhaps 1500–1200 BC); the writer adopts a jocular attitude towards his colleague's attempts at quantity surveying. 1.D5–1.D7 are modern commentaries. In 1.D5 the Egyptologist Sir Alan Gardiner explains an initially puzzling feature of Egyptian arithmetic, the Egyptian concept of fraction or part. 1.D6 and 1.D7 are contrasting perceptions of Egyptian mathematics, from the translator of the Rhind papyrus and from a historian of mathematics.

1.D1 Two problems from the Rhind papyrus

(*a*) *Problem 24*

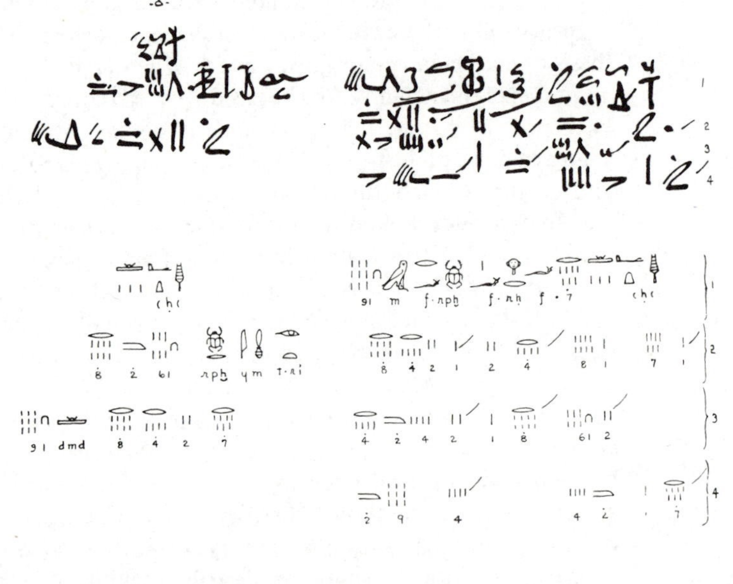

‘ḥ‘ 7̇ · f ḥr · f ḫpr · fm 19.
A quantity, $\frac{1}{7}$ of it added to it, becomes it: 19.

\1 7
\7̇ 1

1	8
\2	16
$\dot{2}$	4
\\$\dot{4}$	2
\\$\dot{8}$	1
\1	2 $\dot{4}$ $\dot{8}$
\2	4 $\dot{2}$ $\dot{4}$
\4	9 $\dot{2}$

ỉr·t my ḫpr
The doing as it occurs.

‘ḥ‘	16 $\dot{2}$ $\dot{8}$
The quantity	
$\dot{7}$	2 $\dot{4}$ $\dot{8}$
dmd	19.
Total	

(*b*) *Problem 40*

tꜣ·w 100 n s 5 $\dot{7}$ n 3 ḥry·w n s 2 ẖry·w pty twnw
Loaves 100 for man 5, $\frac{1}{7}$ of the 3 above to man 2 those below. What is the difference of share?

ỉr·t my ḫpr twnw 5 $\dot{2}$ ir·ḫ[r]·k wꜣḥ–tp m 1 $\ddot{3}$
The doing as it occurs. The difference of share being 5 $\frac{1}{2}$, Make thou the multiplication: 1 $\frac{2}{3}$

\1	23
\1	17 $\dot{2}$
\1	12
\1	6 $\dot{2}$
\1	1
dmd	60
Total	
\1	60
\\$\ddot{3}$	40

r	sp	23	ḫpr·ḫr·f m	38 $\dot{3}$
up to times			becomes it:	
“		17 $\dot{2}$	“	29 $\dot{6}$
“		12	“	20
“		6 $\dot{2}$	“	10 $\ddot{3}$ $\dot{6}$
“		1	“	1 $\ddot{3}$
dmd		60	dmd	100.
Total			Total	

1.D2 More problems from the Rhind papyrus

(a) Problem 25

A quantity and its $\frac{1}{2}$ added together become 16. What is the quantity?
Assume 2.

\1	2
\\$\frac{1}{2}$	1
Total	3.

As many times as 3 must be multiplied to give 16, so many times 2 must be multiplied to give the required number.

\1	3
2	6
\4	12
$\frac{2}{3}$	2
\\$\frac{1}{3}$	1
Total $5\frac{1}{3}$.	
1	$5\frac{1}{3}$
\2	$10\frac{2}{3}$

Do it thus:

The quantity is	$10\frac{2}{3}$
$\frac{1}{2}$	$5\frac{1}{3}$
Total	16.

(b) Problem 31

A quantity, its $\frac{2}{3}$, its $\frac{1}{2}$, and its $\frac{1}{7}$, added together, become 33. What is the quantity?
Multiply $1\ \frac{2}{3}\ \frac{1}{2}\ \frac{1}{7}$ so as to get 33.

1	$1\ \frac{2}{3}\ \frac{1}{2}\ \frac{1}{7}$
\2	$4\ \frac{1}{3}\ \frac{1}{4}\ \frac{1}{28}$
\4	$9\ \frac{1}{6}\ \frac{1}{14}$
\8	$18\ \frac{1}{3}\ \frac{1}{7}$
$\frac{1}{2}$	$\frac{1}{2}\ \frac{1}{3}\ \frac{1}{4}\ \frac{1}{14}$
\\$\frac{1}{4}$	$\frac{1}{4}\ \frac{1}{6}\ \frac{1}{8}\ \frac{1}{28}$.

Total $14\frac{1}{4}$. $14\frac{1}{4}$ times $1\ \frac{2}{3}\ \frac{1}{2}\ \frac{1}{7}$ makes $32\frac{1}{2}$ plus the small fractions $\frac{1}{7}\ \frac{1}{8}\ \frac{1}{14}\ \frac{1}{28}\ \frac{1}{28}$. $32\frac{1}{2}$ from 33 leaves the remainder $\frac{1}{2}$ to be made up by these fractions and a further product by a number yet to be determined.

$$\tfrac{1}{7}\quad \tfrac{1}{8}\quad \tfrac{1}{14}\quad \tfrac{1}{28}\quad \tfrac{1}{28}$$

taken as parts of 42 are

$$6\quad 5\tfrac{1}{4}\quad 3\quad 1\tfrac{1}{2}\quad 1\tfrac{1}{2},$$

making in all $17\frac{1}{4}$, and requiring $3\,\frac{1}{2}\,\frac{1}{4}$ more to make 21, $\frac{1}{2}$ of 42.

Take $1\,\frac{2}{3}\,\frac{1}{2}\,\frac{1}{7}$ as applying to 42:

$\backslash 1$	42
$\backslash \frac{2}{3}$	28
$\backslash \frac{1}{2}$	21
$\backslash \frac{1}{7}$	6
Total	97.

That is, $1\,\frac{2}{3}\,\frac{1}{2}\,\frac{1}{7}$ applied to 42 gives 97 in all. $\frac{1}{42}$ of 42, or 1, will be $\frac{1}{97}$ of this, and $3\,\frac{1}{2}\,\frac{1}{4}$ will be $3\,\frac{1}{2}\,\frac{1}{4}$ times as much. Therefore we multiply $\frac{1}{97}$ by $3\,\frac{1}{2}\,\frac{1}{4}$.

$\backslash \frac{1}{97}$	$\frac{1}{42}$ or 1 as a part of 42
$\backslash \frac{1}{56}\,\frac{1}{679}\,\frac{1}{776}$	$\frac{1}{21}$ or 2 as a part of 42
$\backslash \frac{1}{194}$	$\frac{1}{84}$ or $\frac{1}{2}$ as a part of 42
$\backslash \frac{1}{388}$	$\frac{1}{168}$ or $\frac{1}{4}$ as a part of 42.

The total is $14\,\frac{1}{4}\,\frac{1}{56}\,\frac{1}{97}\,\frac{1}{194}\,\frac{1}{388}\,\frac{1}{679}\,\frac{1}{776}$, which multiplied by $1\,\frac{2}{3}\,\frac{1}{2}\,\frac{1}{7}$ makes 33.

(c) *Problem 41*

Find the volume of a cylindrical granary of diameter 9 and height 10.

Take away $\frac{1}{9}$ of 9, namely, 1; the remainder is 8. Multiply 8 times 8; it makes 64. Multiply 64 times 10; it makes 640 cubed cubits. Add $\frac{1}{2}$ of it to it; it makes 960, its contents in *khar*. Take $\frac{1}{20}$ of 960, namely 48. 4800 *hekat* of grain will go into it.

Method of working out:

1	8
2	16
4	32
$\backslash 8$	64.

1	64
$\backslash 10$	640
$\backslash \frac{1}{2}$	320
Total	960
$\frac{1}{10}$	96
$\backslash \frac{1}{20}$	48.

(*d*) *Problem 42*

Find the volume of a cylindrical granary of diameter 10 and height 10.

Take away $\frac{1}{9}$ of 10, namely $1\,\frac{1}{9}$; the remainder is $8\,\frac{2}{3}\,\frac{1}{6}\,\frac{1}{18}$. Multiply $8\,\frac{2}{3}\,\frac{1}{6}\,\frac{1}{18}$ times $8\,\frac{2}{3}\,\frac{1}{6}\,\frac{1}{18}$; it makes $79\,\frac{1}{108}\,\frac{1}{324}$. Multiply $79\,\frac{1}{108}\,\frac{1}{324}$ times 10; it makes $790\,\frac{1}{18}\,\frac{1}{27}\,\frac{1}{54}\,\frac{1}{81}$ cubed cubits. Add $\frac{1}{2}$ of it to it; it makes $1185\,\frac{1}{6}\,\frac{1}{54}$, its contents in *khar*. $\frac{1}{20}$ of this is $59\,\frac{1}{4}\,\frac{1}{108}$. $59\,\frac{1}{4}\,\frac{1}{108}$ times 100 *hekat* of grain will go into it.

Method of working out:

1	$8\,\frac{2}{3}\,\frac{1}{6}\,\frac{1}{18}$
2	$17\,\frac{2}{3}\,\frac{1}{9}$
4	$35\,\frac{1}{2}\,\frac{1}{18}$
\8	$71\,\frac{1}{9}$
\\$\frac{2}{3}$	$5\,\frac{2}{3}\,\frac{1}{6}\,\frac{1}{18}\,\frac{1}{27}$
$\frac{1}{3}$	$2\,\frac{2}{3}\,\frac{1}{6}\,\frac{1}{12}\,\frac{1}{36}\,\frac{1}{54}$
\\$\frac{1}{6}$	$1\,\frac{1}{3}\,\frac{1}{12}\,\frac{1}{24}\,\frac{1}{72}\,\frac{1}{108}$
\\$\frac{1}{18}$	$\frac{1}{3}\,\frac{1}{9}\,\frac{1}{27}\,\frac{1}{108}\,\frac{1}{324}$
Total	$79\,\frac{1}{108}\,\frac{1}{324}$.

1	$79\,\frac{1}{108}\,\frac{1}{324}$
10	$790\,\frac{1}{18}\,\frac{1}{27}\,\frac{1}{54}\,\frac{1}{81}$
$\frac{1}{2}$	$395\,\frac{1}{36}\,\frac{1}{54}\,\frac{1}{108}\,\frac{1}{162}$
Total	$1185\,\frac{1}{6}\,\frac{1}{54}$
$\frac{1}{10}$	$118\,\frac{1}{2}\,\frac{1}{54}$
\\$\frac{1}{20}$	$59\,\frac{1}{4}\,\frac{1}{108}$.

(*e*) *Problem 48*

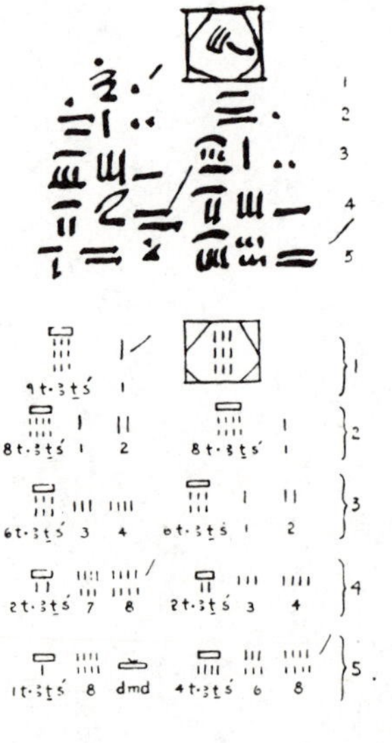

Compare the area of a circle and of its circumscribing square.

The circle of diameter 9.	
1	8 *setat*
2	16 *setat*
4	32 *setat*
\8	64 *setat*

The square of side 9.	
\1	9 *setat*
2	18 *setat*
4	36 *setat*
\8	72 *setat*
Total	81 *setat*

(*f*) *Problem 51*

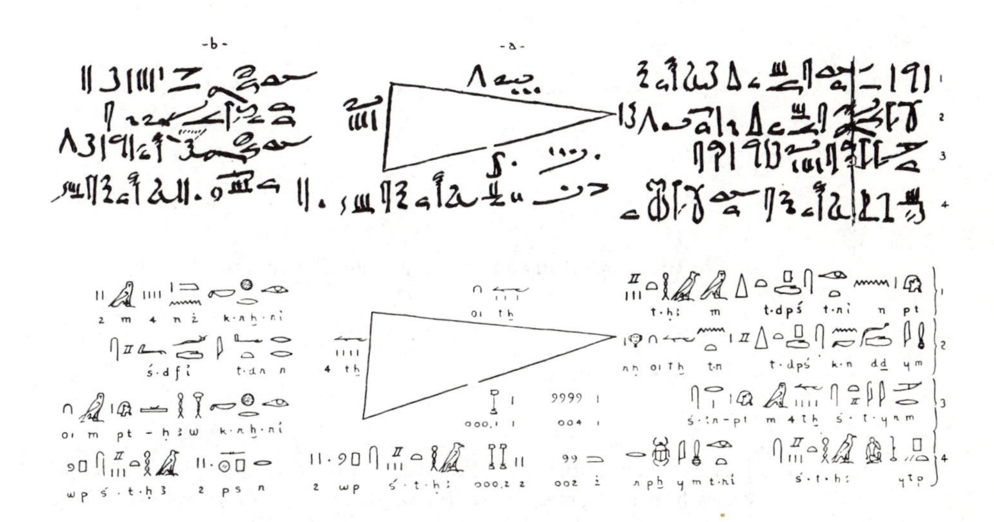

Example of a triangle of land. Suppose it is said to thee, What is the area of a triangle of side 10 khet *and of base 4* khet?

Do it thus:

1	400
$\frac{1}{2}$	200
1	1000
2	2000.

Its area is 20 *setat*.

Take $\frac{1}{2}$ of 4, in order to get its rectangle. Multiply 10 times 2; this is its area.

(*g*) *Problem 65*

Example of dividing 100 loaves among 10 men, including a boatman, a foreman, and a door-keeper, who receive double portions. What is the share of each?

The working out. Add to the number of the men 3 for those with double portions; it makes 13. Multiply 13 so as to get 100; the result is $7\,\frac{2}{3}\,\frac{1}{39}$. This then is the ration for seven of the men, the boatman, the foreman, and the door-keeper receiving double portions.

For proof we add $7\,\frac{2}{3}\,\frac{1}{39}$ taken 7 times and $15\,\frac{1}{3}\,\frac{1}{26}\,\frac{1}{78}$ taken 3 times for the boatman, the foreman, and the door-keeper. The total is 100.

(*h*) *Problem 66*

If 10 hekat *of fat is given out for a year, what is the amount used in a day?*

The working out. Reduce the 10 *hekat* to *ro*; it makes 3200. Reduce the year to days; it makes 365. Get 3200 by operating on 365. The result is $8\,\frac{2}{3}\,\frac{1}{10}\,\frac{1}{2190}$. This makes for a day $\frac{1}{64}$ *hekat* $3\,\frac{2}{3}\,\frac{1}{10}\,\frac{1}{2190}$ *ro*.

Do it thus:

1	365
2	730
4	1460
8	2920
$\frac{2}{3}$	$243\,\frac{1}{3}$
$\frac{1}{10}$	$36\,\frac{1}{2}$
$\frac{1}{2190}$	$\frac{1}{6}$

Total $8\,\frac{2}{3}\,\frac{1}{10}\,\frac{1}{2190}$.

Do the same thing in any example like this.

1.D3 A scribe's letter

[...] Another topic. Behold, you come and fill me with your office. I will cause you to know how matters stand with you, when you say 'I am the scribe who issues commands to the army'.

You are given a lake to dig. You come to me to inquire concerning the rations for the soldiers, and you say 'reckon it out'. You are deserting your office, and the task of teaching you to perform it falls on my shoulders.

Come, that I may tell you more than you have said: I cause you to be abashed [?] when I disclose to you a command of your lord, you, who are his Royal Scribe, when you are led beneath the window [of the palace, where the king issues orders] in respect of any goodly [?] work, when the mountains are disgorging great monuments for Horus [the king], the lord of the Two Lands [Upper and Lower Egypt]. For see, you are the clever scribe who is at the head of the troops. A [building-] ramp is to be constructed, 730 cubits long, 55 cubits wide, containing 120 compartments, and filled with reeds and beams; 60 cubits high at its summit, 30 cubits in the middle, with a batter of twice 15 cubits and its pavement 5 cubits. The quantity of bricks needed for it is asked of the generals, and the scribes are all asked together, without one of them knowing anything. They all put their trust in you and say, 'You are the clever scribe, my friend! Decide for us quickly! Behold your name is famous; let none be found in this place to magnify the other thirty! Do not let it be said of you that there are things which even you do not know. Answer us how many bricks are needed for it?'

See, its measurements [?] are before you. Each one of its compartments is 30 cubits and is 7 cubits broad.

1.D4 Greek views on the Egyptian origin of mathematics

(*a*) *Herodotus* (*mid fifth century BC*)

The king moreover (so they say) divided the country among all the Egyptians by giving each an equal square parcel of land, and made this his source of revenue, appointing the payment of a yearly tax. And any man who was robbed by the river of a part of his land would come to Sesostris and declare what had befallen him; then the king would send men to look into it and measure the space by which the land was diminished, so that thereafter it should pay in proportion to the tax originally imposed. From this, to my thinking, the Greeks learned the art of geometry; the sun-clock and the sundial, and the twelve divisions of the day, came to Hellas not from Egypt but from Babylonia.

(*b*) *Plato* (*early fourth century BC*)

SOCRATES: I have heard that at Naucratis, in Egypt, there was one of the ancient gods of that country, to whom was consecrated the bird, which they call Ibis; but the name of the deity himself was Theuth. That he was the first to invent numbers and arithmetic, and geometry and astronomy, and moreover draughts and dice, and especially letters, at the time when Thamus was king of all Egypt, and dwelt in the great city of the upper region which the Greeks call Egyptian Thebes, but the god they call Ammon; to him Theuth went and showed him his arts, and told him that they ought to be distributed amongst the rest of the Egyptians. [...]
PHAEDRUS: Socrates, you easily make Egyptian and any other country's tales you please.

(*c*) *Aristotle* (*mid fourth century BC*)

Hence it was after all such inventions [the practical arts] were already established that those of the sciences which are not directed to the attainment of pleasure or the necessities of life were discovered; and this happened in the places where men had leisure. This is why the mathematical arts were first set up in Egypt; for there the priestly caste were allowed to enjoy leisure.

(*d*) *Proclus* (*fifth century AD*)

According to most accounts geometry was first discovered among the Egyptians, taking its origin from the measurement of areas. For they found it necessary by reason of the rising of the Nile, which wiped out everybody's proper boundaries. Nor is there anything surprising in that the discovery both of this and of the other sciences should have had its origin in a practical need, since everything which is in process of becoming progresses from the imperfect to the perfect. Thus the transition from perception to reasoning and from reasoning to understanding is natural. Just as exact knowledge of numbers received its origin among the Phoenicians by reason of trade and contracts, even so geometry was discovered among the Egyptians for the aforesaid reason.

1.D5 Sir Alan Gardiner on the Egyptian concept of part

The commonest method of expressing fractions in Egyptian was by the use of the word *r* 'part', below which (or partly below it in the case of the higher numbers) was written the number described in English as the denominator. Thus *r-5* 'part 5' is equivalent to our $\frac{1}{5}$, *r-276* 'part 276' to our $\frac{1}{276}$.

For the Egyptian the number following the word *r* had ordinal meaning; *r-5* means 'part 5', i.e. 'the fifth part' which concludes a row of equal parts together constituting a single set of five. As being the part which completed the row into one series of the number indicated, the Egyptian *r*-fraction was necessarily a fraction with, as we should say, unity as the numerator. To the Egyptian mind it would have seemed nonsense and self-contradictory to write *r-7 4* or the like for $\frac{4}{7}$; in any series of seven, only one part could be the seventh, namely that which occupied the seventh place in the row of seven equal parts laid out for inspection. Nor would it have helped matters from the Egyptian point of view to have written *r-7(+)r-7(+)r-7(+)r-7*, a writing which would likewise have assumed that there could be more than one actual 'seventh'. Consequently, the Egyptian was reduced to expressing (e.g.) $\frac{4}{7}$ by $\frac{1}{2}(+)\frac{1}{14}$. For more complex fractions even as many as five terms, all representing fractions with 1 as the numerator and with increasing denominators, might be needed; thus the Rhind mathematical papyrus, dating from the Hyksos period, gives as equivalent of our $\frac{2}{61}$ the following complex writing: *r-40 r-244 r-488 r-610* '$\frac{1}{40}+\frac{1}{244}+\frac{1}{488}+\frac{1}{610}$'. It is not generally known that the same cumbrous methods of expression were in common use with the Greeks and Romans. It would seem also that a relic of them survives in the use of English ordinals in the names of our fractions, though we speak of 'one-third' and 'three-fifths' without any qualms. [. . .]

Though the Egyptians were unable to say 'three-sevenths' or 'nine-sixteenths', yet they made a restricted use of certain fractions which appear, at first sight, to stand on the same footing: a great rôle is played in Egyptian arithmetic by the fraction *rwy* 'the two parts' (out of three), i.e. $\frac{2}{3}$, and a very rare sign *r-3* (perhaps to be read *ḫmt rw*) can be quoted for 'the three parts' (out of four), i.e. $\frac{3}{4}$. These 'complementary fractions' represent the parts remaining over when 'the third' or 'the fourth' is taken away from a set of three or four, and indeed their existence is practically postulated by the terms *r-3*, *r-4*. But we must be careful to note that in *r-3* $= \frac{3}{4}$ the numeral is a cardinal, not an ordinal, and that the expression means 'the three parts' and was not construed, as with ourselves, as meaning 'three *fourths*'. In ordinary arithmetic the only complementary fraction used was $\frac{2}{3}$. Compare in English 'two parts full', i.e. two-thirds full, doubtless a survival of the old Egyptian way of regarding the same fraction.

1.D6 Arnold Buffum Chace on Egyptian mathematics as pure science

A careful study of the Rhind papyrus convinced me several years ago that this work is not a mere selection of practical problems especially useful to determine land values, and that the Egyptians were not a nation of shopkeepers, interested only in that which they could use. Rather I believe that they studied mathematics and other subjects for their own sakes. In the Rhind papyrus there are problems of area and problems of volume that might be of use to the farmer who owns land and raises grain. There are pyramid problems that might furnish specifications to the builders, or enable an interested observer to determine the dimensions of a pyramid before him. Many of the arithmetical problems concern a division of loaves or of a quantity of grain among a certain number of men, or the relative values of different amounts of food or drink. But when we come to examine the conditions laid down and the numbers involved in these various problems as well as the purely numerical ones, we see that they are more like theoretical problems put in concrete form. In one (Problem 63) 700 loaves are divided among four men in shares that are proportional to the four fractions $\frac{2}{3}$, $\frac{1}{2}$, $\frac{1}{3}$ and $\frac{1}{4}$, the first four terms of their fraction-series. In two (Problems 40 and 64) there is a dividing into shares that form an arithmetical progression, in Problem 67 the tribute for cattle is determined as $\frac{1}{6}\,\frac{1}{18}$ of the herd and the problem asks for the number of the herd when the number of tribute cattle is given, and Problem 31 is a problem whose answer is

$$14\,\frac{1}{4}\,\frac{1}{56}\,\frac{1}{97}\,\frac{1}{194}\,\frac{1}{388}\,\frac{1}{679}\,\frac{1}{776}.$$

Such problems and such quantities were not likely to occur in the daily life of the Egyptians. Thus we can say that the Rhind papyrus, while very useful to the Egyptian, was also 'an example of the cultivation of mathematics as a pure science, even in its first beginnings' [H. Wieleitner, 'Zur ägyptischen Mathematik', *Zeitschrift für mathematischen und naturwissenschaftlichen Unterricht*, **56** (1925) pp. 129–137].

1.D7 G. J. Toomer on Egyptian mathematics as strictly practical

The Rhind and Moscow papyri are handbooks for the scribe, giving model examples of how to do things which were a part of his everyday tasks. This is confirmed, if confirmation were needed, by a papyrus in the form of a satirical letter [see 1.D3] in which a scribe ridicules a colleague for his inability to do his job, and cites among other examples of his failures calculations of the rations of soldiers and of the number of bricks required for building a ramp of given dimensions. A further indication of the origin of these texts is the kind of expression used to introduce problems, for instance: 'If a scribe says to you . . . , let him hear . . .'. The texts are in one respect similar to the Babylonian mathematical texts, in that these too are in the form not of treatises but of specific problems with solutions. But there the similarity ends: the cuneiform texts have a claim to be called mathematical in a fully scientific sense. The problems are only formally about the measurement of areas, determination of lengths, etc. Many of them are not of a kind which could conceivably ever occur in actual mensuration, and the whole point of them is the algebraic procedure involved. They are really 'pure' mathematics. However, this difference from the Egyptian texts is not the important one; mathematics can be applied to practical ends without losing any of its scientific quality. What really distinguishes Babylonian mathematics is the systematic development of intricate algebraic techniques which we can deduce from the working of the problems. These techniques could never have been created by mere empiricism, and we must posit an order of mathematical reasoning of which there is no trace in the Egyptian sources.

To illustrate the elementary and practical nature of Egyptian mathematics, we set out in full Problem 42 of the Rhind papyrus [see 1.D2].

The problem is to determine the cubic content of a cylinder of diameter (D) 10 cubits and height (h) 10 cubits. This is complicated by the fact that for the Egyptian cubic content means how much it will hold of some specific thing, so an answer in cubic cubits is not satisfactory. It is therefore necessary to convert to hundreds of quadruple-*hekat* of corn by way of the equivalences:

$$1 \text{ cubic cubit} = 1\tfrac{1}{2}\ \textit{khar}.$$

$$1\ \textit{khar} = 20 \text{ hundreds of quadruple-}\textit{hekat}.$$

In the working, some of the steps, which would require the use of auxiliary fractions, have certainly been omitted. But what is set down is enough to show that the real difficulty for the Egyptian scribe was the mastering of elementary arithmetical calculations; we can see how hemmed in he was by his numerical system, his crude methods, and his concrete mode of thought. [...]

The truth is that Egyptian mathematics remained at much too low a level to be able to contribute anything of value. The sheer difficulties of calculation with such a crude numeral system and primitive methods effectively prevented any advance or interest in developing the science for its own sake. It served the needs of everyday life (it is only a relatively advanced technology, such as was never achieved in the ancient world, which demands more than the most elementary mathematics), and that was enough. Its interest for us lies in its primitive character, and in what it reveals about the minds of its creators and users, rather than in its historical influence.

1.E Babylonian Mathematics

The most startling revaluation this century in the history of mathematics has been the realization of the scope and sophistication of mathematical activity in ancient Mesopotamia, thanks to the work of scholars in deciphering the cuneiform texts. Extracts from a selection of these are given in 1.E1, illustrating the range and achievement of scribes of the Old Babylonian period (c. 1800–1600 BC). 1.E2 is an enthralling investigation by the mathematician R. Creighton Buck of possibly the most remarkable of all Old Babylonian tablets, the so-called Plimpton 322. As Buck describes, various views have been advanced on what this tablet was for, and the most recent considered judgement, that of the historian Jöran Friberg, forms 1.E3. The remaining items bear mainly on the preceding Sumerian period. Something of the educational context and self-conscious professionalism of scribal activity is revealed in 1.E4, two popular compositions about the training of the scribe. The second of these was written in Sumerian, perhaps about 2000 BC. 1.E5 is a discussion of two very short Sumerian mathematical texts of half a millennium earlier (c. 2500 BC), part of the historian Marvin Powell's argument that the origins of Babylonian mathematics are even earlier than had been realized. 1.E6 ends our chapter on 'Origins' by sketching an argument that mathematics itself—as a coherent, systematic body of knowledge—was the product of the Sumerian scribal school.

1.E1 Some Babylonian problem texts

(a) YBC 4652

Transcription

19 [1]na$_4$ ʾ-pá ki-lá nu-na-tag 6-bi ì-lá 2 gín [bí-daḫ-ma]
[2]igi-3-gál igi-7-gál a-rá-24-kam tab bí-daḫ-ma
[3]ì-lá 1 ma-na sag na$_4$ en-nam sag na$_4$ $4\frac{1}{3}$ gín

20 [4]na$_4$ ì-pà ki-lá nu-na-tag 8-bi ì-lá 3 gín bí-daḫ-ma
[5]igi-3-gál igi-13-gál a-rá 21 e-tab bí-daḫ-ma
[6]ì-lá 1 ma-na sag na$_4$ en-nam sag na$_4$ $4\frac{1}{2}$ gín

21 [7]na$_4$ ì-pà ki-lá nu-na-tag igi-6-gál ba-zi
[8]igi-3-gál igi-8-gál bí-daḫ-ma ì-lá 1 ma-na
[9]sag na$_4$ en-nam sag na$_4$ 1 ma-na 9 gín $21\frac{1}{2}$ še
[10]ù ⟨igi-⟩ 10-gál še kam

Translation

---

19 [1]I found a stone, (but) did not weigh it; (after) I weighed (out) 6 times (its weight), [added] 2 gín, (and)
[2]added one-third of one-seventh multiplied by 24,
[3]I weighed (it): 1 ma-na. What was the origin(al weight) of the stone? The origin(al weight) of the stone was $4\frac{1}{3}$ gín.

---

20 [4]I found a stone, (but) did not weigh it; (after) I weighed (out) 8 times (its weight), added 3 gín,
[5]one-third of one-thirteenth I multiplied by 21, added (it), and then
[6]I weighed (it): 1 ma-na. What was the origin(al weight) of the stone? The origin(al weight) of the stone was $4\frac{1}{2}$ gín.

---

21 [7]I found a stone, (but) did not weigh it; (after) I subtracted one-sixth (and)
[8]added one-third of one-eighth, I weighed (it): 1 ma-na.
[9]What was the origin(al weight) of the stone? The origin(al weight) of the stone was 1 ma-na, 9 gín, $21\frac{1}{2}$ še,
[10]and one-tenth of a še.

Commentary
According to the colophon, the tablet contained twenty-two examples when complete, but only eleven are even partly preserved; of these, six can be fully restored. All the problems of the tablet are obviously of the same type, resulting in a linear equation for one unknown quantity, the 'original' weight of a stone.

[Three of] the preserved problems are the following:

19 $$(6x + 2) + \frac{1}{3}\cdot\frac{1}{7}\cdot 24(6x + 2) = 1,0$$

Solution: $x = 4\frac{1}{3}$ gín.

Indeed, $6x + 2 = 28 \qquad \frac{24}{21}(6x + 2) = 32.$

20 $$(8x + 3) + \frac{1}{3}\cdot\frac{1}{13}\cdot 21(8x + 3) = 1,0$$

Solution: $x = 4\frac{1}{2}$ gín.

Indeed, $8x + 3 = 39 \qquad \frac{21}{39}(8x + 3) = 21.$

21 $$\left(x - \frac{x}{6}\right) + \frac{1}{3}\cdot\frac{1}{8}\left(x - \frac{x}{6}\right) = 1,0$$

Solution: $x = 1$ ma-na $+ 9$ gín $+ 21\frac{1}{2}$ še $+ \frac{1}{10}$ še $= 1,9;7,12$ gín.

Indeed, $\frac{x}{6} = 11;31,12 \qquad x - \frac{x}{6} = 57;36 \qquad \frac{1}{24}\left(x - \frac{x}{6}\right) = 2;24.$

The remaining problems are completely broken away or too badly preserved to be restored with certainty.

The main difficulty encountered in interpreting the text of the problem consists in placing the parentheses correctly. The terminology alone is in itself inadequate; only experience with analogous problems, when combined with the given solution, indicates the correct interpretation. The ancient scribes of course had the oral interpretation of their teachers at their disposal.

(*b*) *YBC 4186*

---

[1]A cistern was 10 GAR square, 10 GAR deep.

[2-3]I emptied out(?) its water; with its water how much field did I irrigate to a depth of 1 šu-si?

[4]Put (aside) 10 and 10 which formed the square.

[5]Put (aside) 10, the depth of the cistern.

[6-6a]And put (aside) 0; 0, 10, the depth of the water which irrigated the field.

[7-8]Take the reciprocal of 0; 0, 10, the depth of the water which [irri]gated the field, and (the resulting) 6, 0 [mul]tiply by 10, the depth of the cistern, (and the result is) 1, 0, 0.

[9]1, 0, 0 ke[ep] in your head.

[10][Square (?)] 10, which formed the square, [and (the result is)] 1, 40.

[11-12]Multiply 1, 40 by 1, 0, 0, which you are ke[eping] in your head. I irrigated 1, 40, 0, 0 (SAR) field.

Commentary

The text assumes a cistern (túl) in the shape of a cube, such that its length l, width b, and depth h are 10 GAR each. The problem which is posed requires the calculation of the area A of a field irrigated to a depth h_A of 1 šu-si by the water contained in the cistern. After the transformation of $h_A = 1$ šu-si to 0; 0, 10 GAR, which is necessary because h is expressed in units of GAR, is made, the actual computation is carried out according to the formula

$$\frac{h}{h_A} \cdot l \cdot b = A.$$

The transformation of the final answer 1, 40, 0, 0 (SAR) to the standard 3 šár 2 bur'u is not made in the text.

The situation described in the text is strongly idealized in that the water is required to be spread to a uniform depth of one finger's breadth over a field which is approximately $3\frac{1}{2}$ kilometers square.

(*c*) *YBC 4608*

---

[13]A triangle. 6, 30 is the length, [11, 22], 30 the area; I did not know [its (?)] width.

[14]6 brothers divided it. One brother('s share) exceeded the other('s), but

[15]how much he exceeded I did not know.
[16]How much did one brother exceed the other?
[17]When you perform (the operations), multiply the area by two, (and the result is) 22, 45, 0.
[18-19]The reciprocal of 6, 30 is not obtainable. What should I put to 6, 30 which will give me 22, 45, 0? Put 3, 30, (which is)
[20]the upper width. Take the reciprocal of 6, the brothers, (and) [multiply] the (resulting) 0; 10 by 6, 30, and (the resulting)
[21]1, 5 (is) the length which each too[k ....]
[22]35 GAR is the breadth. 35 [from 3, 30....................................]
[23]35 from 2, 5[5 ....]
[24]35 from 2, 2[0 ....]
[25]35 from 1, 4[5 ....]
[26]subtr[act] 35 from 1, 10 [....................................]
[27]subtract 35, and the width (?) [....................................]

Commentary
We have here one of the 'inheritance' problems which require the partition of property to be distributed among a given number of brothers. The field in question is of triangular shape with length l and area A:

$$l = 6, 30 \qquad A = 11, 22, 30.$$

This area is divided among 6 brothers by equidistant lines parallel to the base of the triangle. The question asked by the text concerns the difference between the allotments of the brothers.

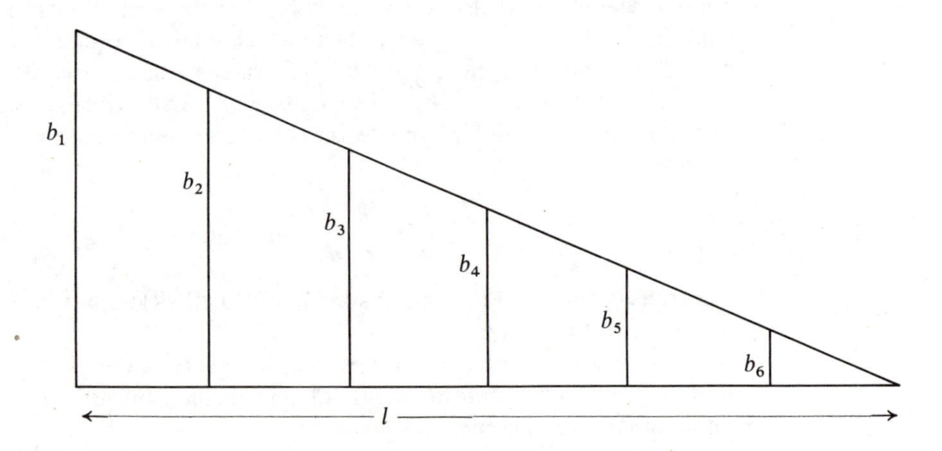

(*d*) *YBC 6967*

[1][The *igib*]*ūm* exceeded the *igūm* by 7.
[2]What are [the *igūm* and] the *igibūm*?

[3-5]As for you—halve 7, by which the *igibūm* exceeded the *igūm*, and (the result is) 3; 30.
[6-7]Multiply together 3; 30 with 3; 30, and (the result is) 12; 15.
[8]To 12; 15, which resulted for you,
[9]add [1, 0, the produ]ct, and (the result is) 1, 12; 15.
[10]What is [the square root of 1], 12; 15? (Answer:) 8; 30.
[11]Lay down [8; 30 and] 8; 30, its equal, and then
(*Reverse*)
[1-2]subtract 3; 30, the *takīltum*, from the one,
[3]add (it) to the other.
[4]One is 12, the other 5.
[5]12 is the *igibūm*, 5 the *igūm*.

Commentary
The problem treated here belongs to a well-known class of quadratic equations characterized by the terms igi and igi-bi (in Akkadian, *igūm* and *igibūm*, respectively). These terms refer to a pair of numbers which stand in the relation to one another of a number and its reciprocal, to be understood in the most general sense as numbers whose product is a power of 60. We must here assume the product

$$xy = 1,0 \tag{1}$$

as the first condition to which the unknowns x and y are subject. The second condition is explicitly given as

$$x - y = 7. \tag{2}$$

From these two equations it follows that x and y can be found from

$$\left.\begin{matrix} x \\ y \end{matrix}\right\} = \sqrt{\left(\frac{7}{2}\right)^2 + 1,0} \pm \frac{7}{2},$$

a formula which is followed exactly by the text, leading to

$$\left.\begin{matrix} x \\ y \end{matrix}\right\} = \sqrt{1,12;15} \pm 3;30 = 8;30 \pm 3;30 = \left\{\begin{matrix} 12 \\ 5 \end{matrix}\right.$$

(*e*) *YBC 4662*

---

21 [24]A ki-lá. $7\frac{1}{2}$ SAR is the area, 45 SAR the volume; one-seventh
[25]of that by which the length exceeded the width is its depth. What are the length, the width, and its depth?
[26]When you perform (the operations), take the reciprocal of $7\frac{1}{2}$ SAR, the area, [multiply by] 45, [the volume, (and)]
[27]you will get its depth. Halve the one-seventh which has been assumed, (and)
[28]you will get 3; 30. Take the reciprocal of its depth, (and) you will get 0; 10;
[29]multiply 0; 10 by 45 (SAR), the volume, (and) you will get 7; 30.
[30-31]Halve 3; 30, (and) you will get 1; 45; multiply together 1; 45 times 1; 45, (and) you will get 3; 3, 45; add 7; 30 to 3; 3, 45, (and)

[32]you will get 10; 33, 45; as for 10; 33, 45, [take] its square root, (and)
[33]you will get 3; 15; operate with 3; 15 ⟨twice⟩:
[34]add 1; 45 to one, subtract 1; 45 from the other, (and)
[35]you will get the length and the width. 5 GAR is the length; [$1\frac{1}{2}$ GAR is the width].

---

22 [36]A ki[-lá. 5 GAR is the length, $1\frac{1}{2}$ GAR the width], $\frac{1}{2}$ GAR its depth, 10 [gín (volume) the assignment].
[37][How much length did one man take? When] you perform (the operations),
[38][multiply together the width and its depth, (and) you] will get 9;
[39-40][take the reciprocal of 9, (and)] you will get [0; 6, 40; multiply] 0; 6, 40 times the assignment, (and) you will get [0; 1, 6, 40]. 0; 1, 6, 40 (GAR) is the taking of one man.

---

23 [41][A ki-lá. 5 GAR is the length, $1\frac{1}{2}$ GAR the width, $\frac{1}{2}$ GAR] its depth, 10 gín (volume) the assignment.
[42][How much length did 30 workers take?] When you [perform (the operations)],
... (three or four

(*Reverse*)

lines missing) ...

---

24 [3][A ki-lá. 5 GAR is the length, $1\frac{1}{2}$ GAR the width, $\frac{1}{2}$ GAR] its [depth], 10 gín (volume) [the assignment].
[4]In how many [days] did [30 workers] finish?
[5]When you perform (the operations), multiply together the length and the width, (and)
[6]you will get [7;] 30; multiply 7; 30 by its depth, (and) you will get 45.
[7]Take the reciprocal of the assignment, (and) you will get 6; multiply 45 by 6, (and)
[8]you will get 4, 30. Take the reciprocal of 30 workers, (and) you will get 0; 2;
[9]multiply by 4, 30, (and) you will get 9.
[10]30 workers finished on the 9th day.

---

25 [11]A ki-lá. $1\frac{1}{2}$ GAR is the width, $\frac{1}{2}$ GAR its depth, 10 gín (volume) the assignment;
[12]30 workers finished on the 9th day.
[13]What is its length? When you perform (the operations),
[14]multiply together the width and its depth, (and) you will get 9. Take the reciprocal of the assignment, [(and) you will get 6];
[15]multiply 6 by 9, (and) you will get 54; take the reciprocal of 54, (and) you will get 0; 1, 6, 40.
[16-17]Multiply together 30 and 9, (and) you will get 4, 30; multiply 4, 30 by 0; 1, 6, 40, (and) you will get the length. 5 GAR is the length.

---

26 [18]A ki-lá. 5 GAR is the length, $\frac{1}{2}$ GAR its depth, 10 gín (volume) the assignment; 30 workers
[19]finished on the 9th day. What is its width?

[20]When you perform (the operations), multiply together the length and its depth, (and)
[21]you will get [3]0. Take the reciprocal of the assignment, (and) you will get 6;
[22]multiply [30] by 6, (and) ⟨you will get 3, 0⟩; take ⟨the reciprocal⟩ of 3, 0, (and) you will see 0; 0, 20. 30 workers and 9
[23][multiply] together, (and) you will get 4, 30; multiply 4, 30 by 0; 0, 20, and
[24]you will get the width. $1\frac{1}{2}$ GAR is the width.

---

27 [25]A ki[-lá. 5 GAR is the length, $1\frac{1}{2}$ GAR] the width, 10 gín (volume) the assignment; ⟨30 workers finished on the 9th day⟩.
[26]What is its depth? When [you] perform (the operations),
[27]multiply together the length and the width, (and) you will get [7; 30]. Take the reciprocal of the assignment, ⟨multiply by 7; 30⟩, (and)
[28]you will get 45; take the reciprocal of 45, (and) you will get 0; 1, 20.
[29]Multiply [together] 30 workers (and) the 9th day, (and) you will get [4, 3]0;
[30]multiply 4, 30 by 0; 1, 20, [(and) you will get 6. $\frac{1}{2}$ GAR is its depth].

---

28 [31]A ki-lá. 5 GAR is the length, [$1\frac{1}{2}$ GAR the width], $\frac{1}{2}$ GAR its depth; ⟨30 workers finished on the 9th day.⟩ What is the assignment?
[32]When you perform (the operations), multiply together the length and the width, (and)
[33]you will see 7; 30; multiply 7; 30 by its depth, (and) you will see 45.
[34]Multiply together 30 workers and the 9th day, (and) you will see 4, 30;
[35]take the reciprocal of 4, 30, (and) you will see 0; 0, 13, 20; [multiply] 0; 0, 13, 20 by [45], (and)
[36]you will get the assignment. 10 gín (volume) is the as[signment].

(*f*) *BM 13901*

---

1 I have added up the area and the side of my square: 0; 45. You write down 1, the coefficient. You break off half of 1. 0; 30 and 0; 30 you multiply: 0; 15. You add 0; 15 to 0; 45: 1. This is the square of 1. From 1 you subtract 0; 30, which you multiplied. 0; 30 is the side of the square.

---

2 I have subtracted the side of my square from the area: 14, 30. You write down 1, the coefficient. You break off half of 1. 0; 30 and 0; 30 you multiply. You add 0; 15 to 14, 30. Result 14, 30; 15. This is the square of 29; 30. You add 0; 30, which you multiplied, to 29; 30. Result 30, the side of the square.

---

7 I have added up seven times the side of my square and eleven times the area: 6; 15. You write down 7 and 11. You multiply 6; 15 by 11: 1, 8; 45. You break off half of 7. 3; 30 and 3; 30 you multiply. 12; 15 you add to 1, 8; 45. Result 1, 21. This is the square of 9. You subtract 3; 30, which you multiplied, from 9. Result 5; 30. The reciprocal of 11 cannot be found. By what must I multiply 11 to obtain 5; 30? 0; 30, the side of the square is 0; 30.

---

10 The surfaces of my two square figures I have taken together: 21; 15. The side of one is a seventh less than the other. You write down 7 and 6. 7 and 7 you multiply: 49. 6 and 6 you multiply. 36 and 49 you add: 1; 25. The reciprocal of 1; 25 cannot be found. By what must I multiply 1; 25 to give me 21; 15? 0; 15. 0; 30 the side. 0; 30 to 7 you raise: 3; 30 the first side. 0; 30 to 6 you raise: 3 the second side.

---

12 The surfaces of my two square figures I have taken together: 21; 40. The sides of my two square figures I have multiplied: 10, 0. You break off half of 21; 40. 10; 50 and 10; 50 you multiply: 1, 57, 21, 40. 10, 0 and 10, 0 you multiply, 1, 40, 0, 0 inside 1, 57, 21, 40 you tear off: 17, 21, 40. 4, 10 the side. 4, 10 to the first 10, 50 you add: 15, 0. 30 the side. 30 the first square figure. 4, 10 inside the second 10, 50 you tear off: 6, 40. 20 the side. 20 the second square figure.

(*g*) *YBC 7289*

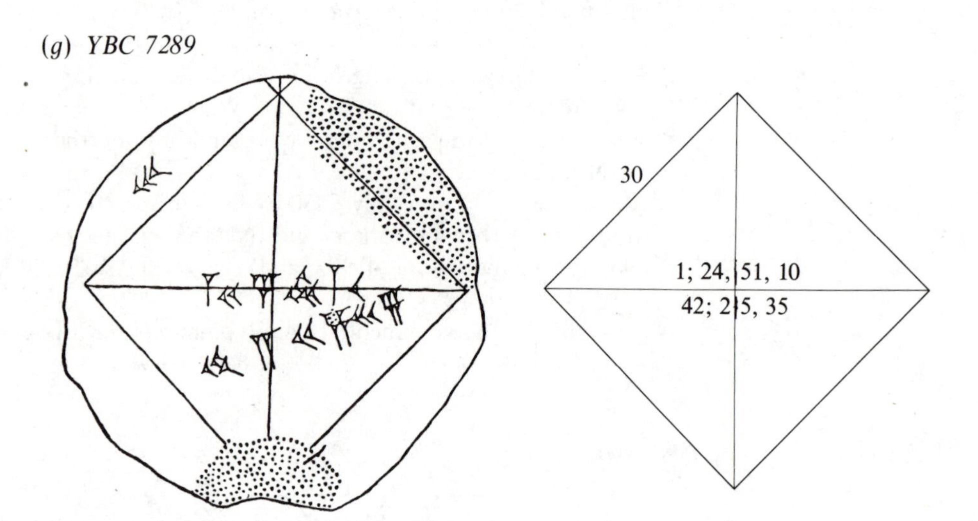

The number 30 indicates the side a of the square, and 1, 24, 51, 10 means

$$1; 24, 51, 10 \approx \sqrt{2}; \tag{1}$$

we therefore find

$$d = a\sqrt{2} = 42; 25, 35$$

for the diagonal. The value (1) for $\sqrt{2}$ is very good, as can be seen from

$$(1; 24, 51, 10)^2 = 1; 59, 59, 59, 38, 1, 40.$$

1.E2 Sherlock Holmes in Babylon: an investigation by R. Creighton Buck

With this brief introduction to the arithmetic of the Babylonians, we turn to another tablet whose mathematical nature had been overlooked until the work of Neugebauer and Sachs [see 1.E1]. It is in the George A. Plimpton Collection, Rare Book and Manuscript Library, at Columbia University, and usually called Plimpton 322. The left side of this tablet has some erosion: traces of modern glue on the left edge suggest that a portion that had originally been attached there has since been lost or stolen. Since it was bought in a marketplace, one may only conjecture about its true origin and date, although the style suggests about −1600 for the latter. As with most such tablets, this had been assumed to be a commercial account or inventory report. We will attempt to show why one can be led to believe otherwise.

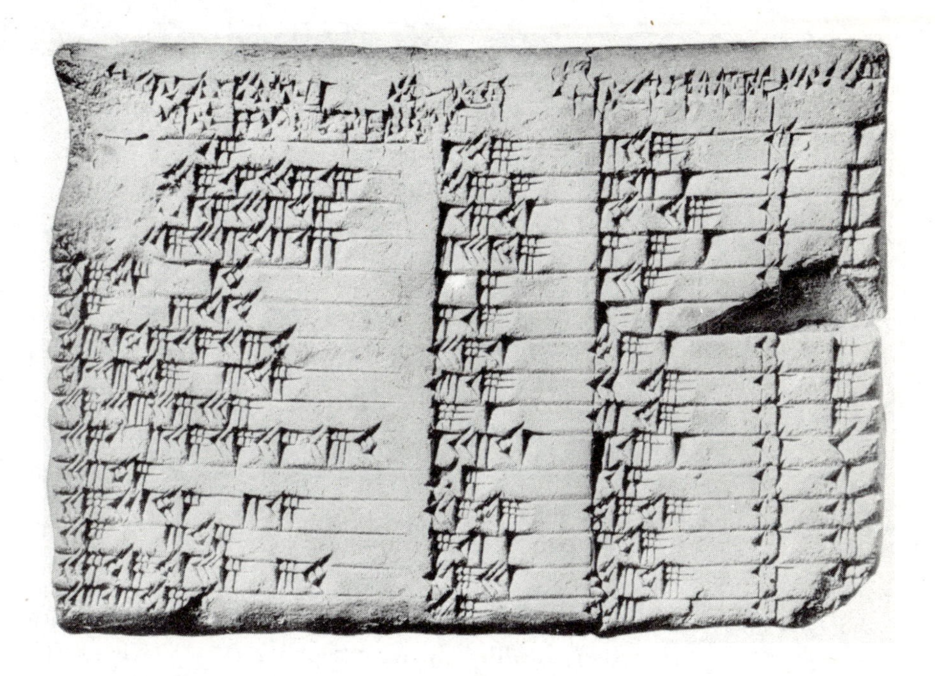

Column *A*	Column *B*	Column *C*
15	1/59	2/49
58/14/50/6/15	56/7	3/12/1
1/15/33/45	1/16/41	1/50/49
5 29/32/52/16	3/31/49	5/9/1
48/54/ 1/40	1/5	1/37
47/ 6/41/40	5/19	8/1
43/11/56/28/26/40	38/11	59/1
41/33/59/ 3/45	13/19	20/49
38/33/36/36	9/1	12/49
35/10/2/28/27/24/26/40	1/22/41	2/16/1
33/45	45	1/15
29/24/54/ 2/15	27/9	48/49
27/ 3/45	7/12/1	4/49
25/48/51/35/6/40	29/31	53/49
23/13/46/40	56	53

Figure 1

First, let us transcribe it into the slash notation, as seen in Figure 1. We have reproduced the three main columns, which we have labelled A, B, and C. We note that there are gaps in column A, due to the erosion. However, it seems apparent that the numbers there are steadily decreasing. We note that some of the numerals there are short and some long, apparently at random. In contrast with this, all the numerals in columns B and C are rather short, and we do not see any evidence of general monotonicity.

B	C
119	169
3367	11521
4601	6649
12709	18541
65	97
319	481
2291	3541
799	1249
541	769
4961	8161
45	75
1679	2929
25921	289
1771	3229
56	53

Figure 2

$C + B$	$C - B$
288	50
14888	8154
11250	2048
31250	5832
162	32
800	162
5832	1250
2048	450
1310	228
13132	3200
120	30
4608	1250
26210	−25632
5000	1458
109	−3

Figure 3

Since it is easier for us to work with Arabic numerals, let us translate columns B and C into these numerals and look for patterns. (See Figure 2.) We see at once that B is smaller than C, with only two exceptions. Also, playing with these numbers, we find that column B contains exactly one prime, namely, 541, while column C contains eight numbers that are prime.

In the first 20,000 integers, there are about 2,300 primes, which is about 10 per cent; among 15 integers, selected at random from this interval, we might, then, expect to see one or two primes, but certainly not eight! This at once tells us that the tablet is mathematical and not merely arithmetical. (Imagine your feelings if you were to find a Babylonian tablet with a list of the orders of the first few sporadic simple groups.)

Encouraged, one attempts to find further visible patterns, for example, by combining the entries in columns B and C in various ways. One of the earliest tries is immediately successful. In Figure 3, we show the results of calculating $C + B$ and $C - B$. If you are sensitive to arithmetic you will note that, in almost every case, the numbers are each twice a perfect square.

If $C + B = 2a^2$ and $C - B = 2b^2$, then $B = a^2 - b^2$ and $C = a^2 + b^2$. Thus the entries in these columns could have been generated from integer pairs (a,b). In passing, we note that B, being $(a - b)(a + b)$, is not apt to be prime; on the other hand, when a and b are relatively prime, every prime of the form $4N + 1$ can be expressed as $a^2 + b^2$.

In Figure 4, we have recopied columns B and C, together with the appropriate pairs (a, b) in the cases where this representation is possible. As a further confirmation that

B	C	(a, b)
119	169	12, 5
3367	11521	?
4601	6649	75, 32
12709	18541	125, 54
65	97	9, 4
319	481	20, 9
2291	3541	54, 25
799	1249	32, 15
541	769	?
4961	8161	81, 40
45	75	?
1679	2929	48, 25
25921	289	?
1771	3229	50, 27
56	53	?

Figure 4

Corrected Version

B	C	(a, b)
119	169	12, 5
3367	4825	64, 27
4601	6649	75, 32
12709	18541	125, 54
65	97	9, 4
319	481	20, 9
2291	3541	54, 25
799	1249	32, 15
481	769	25, 12
4961	8161	81, 40
45	75	$1, \frac{1}{2} = 30$
1679	2929	48, 25
161	289	15, 8
1771	3229	50, 27
56	106	9, 5

Figure 5

we are on the right track, we note that in every such pair the numbers a and b are both 'nice', that is, factorable in terms of 2, 3, and 5. In five cases, the pattern breaks down and no pair exists. It will be a further confirmation if we can explain these discrepancies as errors made by the scribe who produced the tablet. We make a simple hypothesis and assume that B and C were each computed independently from the pair (a, b) and that a few errors were made but each affected only one number in each row. Thus in each vacant place we will assume that either B or C is correct and the other wrong, and attempt to restore the correct entry. Since we do not know the correct pair (a, b) we must find it; because of the evidence in the rest of the table, we insist that an acceptable pair must be composed of 'nice' sexagesimals.

We start with line 9; here, $B = 541$, which happens to be the only prime in Column B. We therefore assume B is wrong and C is correct, and thus write $C = 769 = a^2 + b^2$. This has a single solution, the pair (25, 12). (We also note that both happen to be nice sexagesimals.) If this is correct, then B should have been $(25)^2 - (12)^2 = 481$, instead of 541 as given. Is there an obvious explanation for this mistake? Yes, for in slash notation, $541 = 9/1$ and $481 = 8/1$. The anomaly in line 9 seems to be merely a copy error.

Turn now to line 13; here, B is far larger than C, which is contrary to the pattern. Assume that B is in error and C is correct, and again try $C = 289 = a^2 + b^2$. There is a 'nice' unique solution, (15, 8), and using these, we are led to conjecture that the correct value of B is $(15)^2 - (8)^2 = 161$. Again, we ask if there is an obvious explanation for arriving at the incorrect value given, 25921. A partial answer is immediate: $(161)^2 = 25921$; so that for some reason the scribe recorded the *square* of the correct value for B.

Continuing, consider line 15. Since $B = 56$ and $C = 53$, we have $B > C$, which does not match the general pattern. However, it is not clear whether B is too large or C too small. Trying the first, we assume C is correct and solve $53 = a^2 + b^2$, obtaining the unique answer (7, 2). We reject this, since 7 is not a nice sexagesimal. Now assume that

B is correct, and write $56 = a^2 - b^2 = (a + b)(a - b)$. This has two solutions, (15, 13) and (9, 5). We reject the first and use the second, obtaining $9^2 + 5^2 = 106$ as the correct value of C. Seeking an explanation, we note that the value given by the scribe, 53, is exactly *half* of the correct value.

Turning now to line 2 of Figure 4; we have $B = 3367$ and $C = 11521$, either of which might be correct. Assume that $C = a^2 + b^2$ and find two solutions (100, 39) and (89, 60). While 100 and 60 are nice, 39 and 89 are not, so we reject both pairs and assume that B is correct. Writing $3367 = (a - b)(a + b)$ and factoring 3367 in all ways, we find four pairs: (1684, 1683), (244, 237), (136, 123), (64, 27), of which we can accept only the last. This yields $(64)^2 + (27)^2 = 4825$ as the correct C. Comparing this with the number 11521 that appeared on the tablet, we see no immediate naive explanation for the error. For example, since $4825 = 1/20/25$ and $11521 = 3/12/1$, it does not seem to be a copy error. Without an explanation, we may have a little less confidence in this reconstruction of the entries in line 2.

The last misfit in the table is line 11, where we have $B = 45$ and $C = 75$. This is unusual also because this is the only case where B and C have a common factor. The sums-and-differences-of-squares pattern failed because neither $C + B = 120$ nor $C - B = 30$ is twice a square. However, everything becomes clearer if we go back to base 60 notation and remember that we use floating point; for $120 = 2/0$, which is twice $1/0$ and which we can also write as 1, clearly a perfect square. In the same way, 30 is twice 15, which is also 4^R and which is the square of 2^R. The pattern is preserved and no corrections need be made in the entries: with $a = 1 = 1/0$ and $b = \frac{1}{2} = 2^R = 30 = 0/30$, we have $a^2 = 1/0$ and $b^2 = 0/15$, and

$$C = a^2 + b^2 = 1/0 + 0/15 = 1/15 = 75$$

$$B = a^2 - b^2 = 1/0 - 0/15 = 0/45 = 45.$$

(Another aspect of the line 11 entries will appear later.)

With this, we have completed the work of editing the original tablet. In Figure 5, we give a corrected table for columns B and C, together with the appropriate pairs (a, b) from which they can be calculated.

It is now the time to raise the second canonical question: What was the purpose behind this tablet? Speculation in this direction is less restricted, since the road is not as well marked. We can begin by asking if numbers of the form $a^2 - b^2$ and $a^2 + b^2$ have any special properties. In doing so, we run the risk of looking at ancient Babylonia from the twentieth century, rather than trying to adopt an autochthonous viewpoint. Nevertheless, one relation is extremely suggestive, involving both algebra and geometry. For any numbers (integers) a and b,

$$(a^2 - b^2)^2 + (2ab)^2 = (a^2 + b^2)^2. \qquad (*)$$

In addition, if we introduce $D = 2ab$, then B, C, and D can form a right-angled triangle with $B^2 + D^2 = C^2$. And finally, these formulas generate *all* Pythagorean triplets (triangles) from the integer parameters (a, b). (See Figure 6.)

There is no independent information showing that these facts were known to the Babylonians at the time we conjecture that this tablet was inscribed, although, as will appear later, their algebra had already mastered the solution of quadratic equations. If the tablet indeed is connected with this observation, then the unknown column A numbers ought to be connected in some way with the same triangle. The next step is,

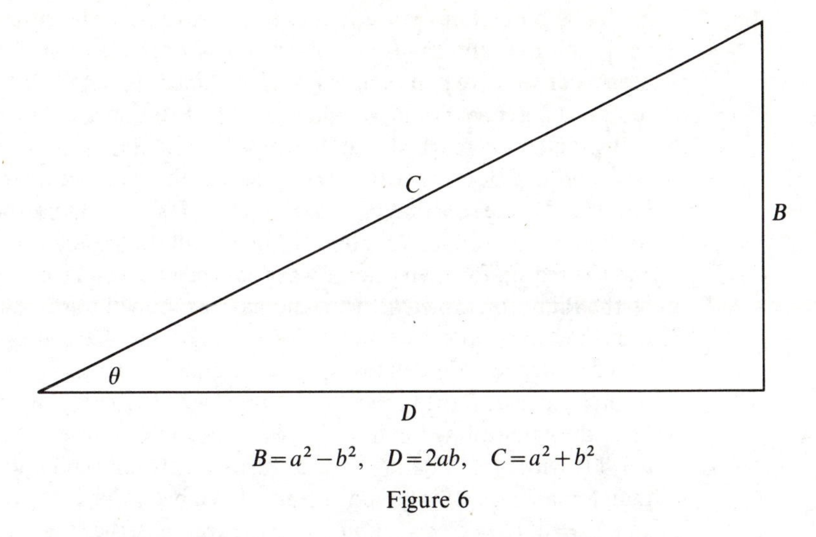

$B = a^2 - b^2, \quad D = 2ab, \quad C = a^2 + b^2$

Figure 6

then, to proceed as before and try many different combinations of B, C, and D, in hopes that one of these will approximate the entries in column A. Slopes and ratios are an obvious starting point, so one calculates $C \div B$, $C \div D$, $B \div D$, etc. After discarding many failures, one arrives at the combination $(B \div D)^2$. In Figure 7, we give the values of this expression, calculated from the corrected values of B and using the hypothetical values of (a, b) to find D. (We remark that it was very helpful to have a programmable pocket calculator that could be trained to work in sexagesimal arithmetic!)

If we now return to Figure 1 and compare the numerals given there in column A with those that appear in Figure 7, we see that there is almost total agreement. For example, in line 10 we have exact duplication of an eight-digit sexagesimal! On probabilistic grounds alone, this is an overwhelming confirmation. Of course, at the top of the tablet where there were gaps due to erosion, Figures 1 and 7 are not the same, but it is evident that the calculated data in Figure 7 can be regarded as filling in the gaps. There are two minor disagreements in the two tables. In line 13, the tablet does not show an internal

line	Calculated Values of $(B \div D)^2$
1	59/0/15
2	56/56/58/14/50/6/15
3	55/7/41/15/33/45
4	53/10/29/32/52/16
5	48/54/1/40
6	47/6/41/40
7	43/11/56/28/26/40
8	41/33/45/14/3/45
9	38/33/36/36
10	35/10/2/28/27/24/26/40
11	33/45
12	29/21/54/2/15
13	27/0/3/45
14	25/48/51/35/6/40
15	23/13/46/40

Figure 7

'0' that is present in Figure 7. This could have been the custom of the scribe in dealing with such an event. In line 8, the scribe has written a digit '59' where there should have been a consecutive pair of digits, '45/14'. Since $59 = 45 + 14$, it is not difficult to invent several different ways in which an error of this sort could have been made.

It should be remarked that Neugebauer and Sachs did not use $(B \div D)^2$ as a source for column A but rather $(C \div D)^2$. Because of the relationship between B and C, and formula (*), one sees that $(C \div D)^2 = (B \div D)^2 + 1$. Thus, the only effect of the change would be to introduce an initial '1/' before all the sexagesimals that appear in Figure 7, and the reason for their choice was that they believed that this was true for column A on the Plimpton tablet. Others who have examined the tablet do not agree. (I have not seen the tablet, and I do not believe it matters which alternative is used.)

We now know the relationship of columns A, B, and C. Referring to Figure 6, C is the hypotenuse, B the vertical side, and A is the square of the slope of the triangle; thus, in modern notation $A = \tan^2 \theta$. It is interesting to observe that the anomalous case of line 11, with $B = 45$ and $C = 75$, turns out to be the familiar 3, 4, 5 triangle; in the Babylonian case, this would seem to have been the $\frac{3}{4}, 1, \frac{5}{4}$ triangle, since $45 = 3 \times 4^R$ and $75 = 1/15 = 5 \times 4^R$. Of course the triangle, the side D, and the parameters (a, b) are all constructs of ours and not immediately visible in the original tablet. All that we can assert without controversy is that $A = B^2 \div (C^2 - B^2)$.

Let us reexamine some of our reasoning. In lines 2, 9, 13, and 15, the scribe recorded correct values for A but incorrect values for C, B, B, and C, respectively. This suggests strongly that A was not calculated directly from the values of B and C, but that A, B, and C were all calculated independently from data that do not appear on the tablet; our hypothetical pair (a, b) gains life. (Of course there is the possibility that the tablet before us is merely a copy from another master tablet.) In either case, it seems odd that column A should be error free while columns B and C, involving simpler numbers, should have four errors.

Other questions can be raised. If, as argued by Neugebauer (*The Exact Sciences in Antiquity*, Dover, 1969), the purpose of the tablet was to record a collection of integral-sided Pythagorean triangles (triplets), why do we not see the value of D, or at least the useful parameters (a, b)? And why would one want the values in column A which are squares of the slope? And why should the entries be arranged in an order that makes the numbers A decrease monotonically?

Variants of this explanation have been proposed. If one computes the values of the angle θ for each line of the tablet, they are seen to decrease steadily from about 45° to about 30°, in steps of about 1°. Is this an accident? Could this tablet be a primitive trigonometric table, intended for engineering or astronomic use? But again, why is $\tan^2 \theta$ useful?

Additional confirmation of such a hypothesis could be given by an outline of a computation procedure leading to the tablet, which makes all of the errors plausible and also shows why they would have occurred preferentially in columns B and C.

Building upon an earlier suggestion of Bruins, an intriguing explanation has been recently proposed by Voils. In Nippur, a large number of 'school texts' have been found, many containing arithmetic exercises. Among these, a standard puzzle problem is quite common. The student is given the difference (or sum) of an unknown number and its reciprocal and asked to find the number. If x is the number (called 'igi') and x^R is its reciprocal (called 'igibi'), then the student is to solve the equation $x - x^R = d$. Thus, the 'igi and igibi' problems are quadratic equations of a standard variety.

The school texts teach a specific solution algorithm: 'Find half of d, square it, add 1, take the square root, and then add and subtract half of d'. This is easily seen to be nothing more than a version of the quadratic formula, tailored to the 'igi and igibi' problems. Voils connects this class of problems, and the algorithm above, with the Plimpton tablet as follows.

First, assume with Bruins that the tablet was computed not from the pair (a, b) but from a single parameter, the number $x = a \div b$. Since a and b are both 'nice', the number x and its reciprocal x^R can each be calculated easily. Indeed, $x = a \times b^R$ and $x^R = b \times a^R$, and a^R and b^R, each appear in a standard reciprocal table. Next observe that

$$B = a^2 - b^2 = (ab)(x - x^R)$$

$$C = a^2 + b^2 = (ab)(x + x^R)$$

$$A = \left(\frac{B}{D}\right)^2 = \left\{\frac{1}{2}(x - x^R)\right\}^2.$$

This shows that the entries A, B, C in the Plimpton tablet could have been easily calculated from a special reciprocal table that listed the paired values x and x^R. Indeed, the numbers B and C can be obtained from $x \pm x^R$ merely by multiplying these by integers chosen to simplify the result and shorten the digit representation.

Voils adds to this suggestion of Bruins the observation that the numbers A are exactly the results obtained at the end of the second step in the solution algorithm, $(d/2)^2$, applied to an igi-igibi problem whose solution is x and x^R. Furthermore, the numbers B and C can be used to produce other problems of the same type but having the same intermediate results in the solution algorithm. Thus Voils proposes that the Plimpton tablet has nothing to do with Pythagorean triplets or trigonometry but, instead, is a pedagogical tool intended to help a mathematics teacher of the period make up a large number of igi-igibi quadratic equation exercises having known solutions and intermediate solution steps that are easily checked.

It is possible to point to another weak confirmation of this last approach. Suppose that we want a graduated table of numbers x and their reciprocals x^R. We start with the class of *all* pairs (a, b) of relatively prime integers such that $b < a < 100$ and each integer a and b is 'nice', factorable into powers of 2, 3, and 5. It is then easy to find the terminating Babylonian representation for both $x = a \div b$ and for $x^R = b \div a$. Make a table of these, arranged with x decreasing. Impose one further restriction:

$$\sqrt{3} < x < 1 + \sqrt{2}.$$

(This corresponds to the limitation $30° < \theta < 45°$, where θ is the base angle in the triangle in Figure 6.)

Then, the resulting list of pairs will coincide with that given in Figure 5, the corrected Plimpton table, except for three minor points. The pair (16, 9) does not appear, the pair (125, 54) does appear, and instead of the pair (2, 1) we have the pair $(1, \frac{1}{2})$; in passing, we recall that the last pair yields the standard 3, 4, 5 Pythagorean triangle.

Unlike Doyle's stories, this has no final resolution. Any of these reconstructions, if correct, throws light upon the degree of sophistication of the Babylonian mathematicians and breathes life into what was otherwise dull arithmetic. [...] I can do no better than to close with an analogy used by Neugebauer (*The Exact Sciences in Antiquity*, p. 177):

> In the 'Cloisters' of the Metropolitan Museum in New York there hangs a magnificent tapestry which tells the tale of the Unicorn. At the end we see the miraculous animal captured, gracefully resigned to his fate, standing in an enclosure surrounded by a neat little fence. This picture may serve as a simile for what we have attempted here. We have artfully erected from small bits of evidence the fence inside which we hope to have enclosed what may appear as a possible living creature. Reality, however, may be vastly different from the product of our imagination; perhaps it is vain to hope for anything more than a picture which is pleasing to the constructive mind, when we try to restore the past.

1.E3 Jöran Friberg on the purpose of Plimpton 322

We start by making the following simple but extremely important observation. With very few exceptions all Babylonian mathematical problem texts contain problems whose solutions are rational numbers or, more precisely, semiregular numbers which can be expressed by use of the Babylonian sexagesimal notation. It is evident that the authors of these Babylonian mathematical texts must have devoted a lot of work and ingenuity in *choosing* the right kind of *data* in their formulation of the problems, and in *devising problems* they knew would possess solutions of the indicated kind. For brevity, I call such problems *solvable*. For example, the problem of finding the third side of a right triangle when two of the sides are given becomes 'solvable' only if the sides of the given triangle are multiples of the sides of a primitive Pythagorean triangle with one of the shorter sides regular.

Thus it appears that the reason for the construction of the tables on the Plimpton tablet was not an interest in number-theoretical questions, but rather the need to *find data for a 'solvable' mathematical problem*. More precisely, it is my belief that the purpose of the author of Plimpton 322 was to write a *'teacher's aid' for setting up and solving problems involving right triangles*. In fact, a typical Babylonian problem text contains not only the formulation of the problem but also the details of its numerical solution for the given data. Hence the contents of the table on the (intact) Plimpton table would have given a teacher the opportunity to set up a large number of solved problems involving right triangles, with full numerical details, as well as to formulate a series of exercises for his students where only the necessary data were given, although the teacher *knew* that the problem was solvable, and where he could *check* the numerical details of the students' solutions by using the numbers in the table.

1.E4 The scribal art

(a) The scribal art is the mother of orators, the father of masters,
The scribal art is delightful, it never satiates you,
The scribal art is not (easily) learned, (but) he who has learned it need no longer
be anxious about it.

Strive to (master) the scribal art and it will enrich you,
Be industrious in the scribal art and it will provide you with wealth and abundance,
Do not be careless concerning the scribal art, do not neglect it,
The scribal art is a 'house of richness', the secret of Amanki,
Work ceaselessly with the scribal art and it will reveal its secret to you,
If you neglect it, they will make malicious remarks about you.
The scribal art is a good lot, richness and abundance.
Since you were a child it causes you grief, since you have grown up [it . . .],
The scribal art is the 'bond' of all . . . [. . .] . . .
Work hard for it [and it will . . . you] its beautiful prosperity,
To have superior knowledge in Sumerian, to learn . . . , [to learn] Emesal,
To write a stele, to draw a field, to settle accounts, [. . .],
. . . the palace . . . ,
The scribe may be its (of the scribal art) servant, he calls for the corvée basket, [. . .].

(b) 'I neglected the scribal art, [I forsook] the scribal art,
My teacher did not . . . ,
. . . d me his skill in the scribal art.
The . . . of words, the art of being a young scribe,
the . . . of the art of being a big brother, let no one . . . to school.'
'Give me his gift, let him direct the way to you,
let him put aside counting and accounting;
the current school affairs
the schoolboys will . . . , verily they will . . . me.'
To that which the schoolboy said, his father gave heed.
The teacher was brought from school;
having entered the house, he was seated in the seat of honor.
The schoolboy took the . . . , sat down before him;
whatever he had learned of the scribal art,
he unfolded to his father.
His father, with joyful heart
says joyfully to his 'school-father':
'You "open the hand" of my young one, you make of him an expert,
show him all the fine points of the scribal art.
You have shown him all the more obvious details of the tablet-craft, of counting and accounting, [*or* 'Of the mathematical tablets, of counting and accounting, you explain their solution to him,']
you have clarified for him all the more recondite details of the . . .'
'Pour out for him . . . like good wine, bring him a stand,
make flow the good oil in his . . . -vessel like water,
I will dress him in a (new) garment, present him a gift, put a band about his hand.'
They pour out for him . . . like good date-wine, brought him a stand,
made flow the good oil in his . . . -vessel like water,
he dressed him in a (new) garment, gave him a gift, put a band about his hand.
The teacher with joyful heart gave speech to him:
'Young man, because you did not neglect my word, did not forsake it,

May you reach the pinnacle of the scribal art, achieve it completely.
Because you gave me that which you were by no means obliged (to give),
you presented me with a gift over and above my earnings, have shown me great honor,
may Nidaba, the queen of the guardian deities, be your guardian deity,
may she show favor to your fashioned reed,
may she take all evil from your hand copies.
Of your brothers, may you be their leader,
of your companions, may you be their chief,
may you rank the highest of (all) the schoolboys.

1.E5 Marvin Powell on two Sumerian texts

Two texts which call for comment here represent two versions of the same exercise, and they are significant, not only because of the abstract interest in numerical relations indicated by the enormous numbers involved, but also because they concern the problem of irregular numbers.

One of these texts (Jestin Number 50) was treated by Geneviève Guitel (mistakenly, I believe) as a problem in division, and she posits a method of solution 'absolutely analogous to modern practice'. It is, however, precisely this close correspondence to modern practice that makes the solution suspect. If modern long division had been used in the Fara period, it is virtually certain that it would appear somewhere in Old Babylonian mathematical texts, which is not the case. Moreover, the most significant text pertaining to the problem of how the calculation was performed is not Jestin's Number 50 at all. It is rather a text (Jestin Number 671) written by a bungler who did not know the front from the back of his tablet, did not know the difference between standard numerical notation and area notation, and succeeded in making half a dozen writing errors in as many lines, but nevertheless was not without a modicum of ability and probably finished school with a low passing grade, took a post with the government and became a bureaucrat. The writer of Number 50 no doubt became a scholar and died penniless. However probable these postulated eventualities may be, the modern scholar may well be more grateful to our third millennium bungler than to his competent classmate. The reason for this will, I believe, be apparent if we compare the two texts.

Jestin Number 50	*Jestin Number 671*	*Translation of Number 50*
še guru$_7$: 1	(rev) *še sila*$_3$ 7$^!$ *guru*$_7$	The grain (is) 1 silo.
sila$_3$ 7		7 sila (seven liters)
lú: 1 *šu ba-ti*	*lú*: 1 *šu ba-ti*	each man received.
lú-bi	(obv) *guruš*	Its men:
45, 42, 51	45, 36, 0 (written on three lines)	45, 42, 51.
še sila$_3$: 3 *šu*$^?$-*tag*$_4$$^?$		3 sila of grain (remaining.)

As one can see, the two problems are identical in type and form. Number 671, in addition to the handwriting errors, which I have not shown, also has *guruš* (man = Latin *vir*) instead of *lu* (man = *homo*) and omits the verb form at the end, because, as we shall see shortly, his solution did not require a remainder. A silo (*guru*) in this period contained 40, 0 *gur*, each of which contained 8, 0 *sila*. Thus, the number being 'divided' by 7 is 5, 20, 0, 0. Seven is the only integer between 1 and 10 that will not produce an even result, therefore, given this fact and the fact that two exercises dealing with the same problem have survived, the choice of 7 can hardly be coincidental. Moreover, the choice of 7 has no material explanation, because the seven-day week played no role in Sumerian accounting procedures, and, having read thousands of Sumerian and Akkadian texts from the third millennium, I cannot recall a single case where 7 functions as a divisor. Thus, the choice of 7 can hardly be motivated by any other cause than that it is an irregular number. Moreover, the two different answers suggest that the object is an exercise in using the reciprocal of an irregular number.

This deduction flows from the following considerations: (1) multiplication of the 'dividend' by the reciprocal of the 'divisor' is the only means of 'dividing' attested in the Babylonian mathematical tradition, except when the 'divisor' is 2 (halving); (2) the two answers obtained to the problem are explicable by a single hypothesis, but only if one assumes multiplication by the reciprocal. The correct answer seems to have been obtained by the following process:

(1) $5, 20, 0, 0 \cdot 0; 8, 34, 17, 8 = 45, 42, 51; 22, 40$

(2) $45, 42, 51 \cdot 7 = 5, 19, 59, 57$

(3) $5, 20, 0, 0 - 5, 19, 59, 57 = 3$

In (1), the number of *sila* is multiplied by the reciprocal of 7 calculated to the fourth place (a three-place reciprocal will not work, unless one assumes the use of rounding). In (2), the fractional number of men is discarded and the whole number multiplied by 7 to obtain the number of *sila* passed out on a seven-each basis. In (3), the product of (2) is subtracted from the original number of *sila*, giving the remainder 3. In the text containing the wrong answer (Number 671), the pupil has apparently used 0; 8, 33 as the reciprocal of 7, for $5, 20, 0, 0 \cdot 0; 8, 33 = 45, 36, 0$ which is the answer contained in the text. How the pupil arrived at the choice of 0; 8, 33 for the reciprocal of 7, I have no idea, but perhaps someone else will see the solution where I have not.

It may seem rather startling to suggest that the Sumerians were working problems involving the use of reciprocals calculated to the fourth place in the middle of the third millennium, but I must confess that I find it difficult to believe that the relationships exhibited in the two problems are merely the result of coincidence. Also, although up to the present time there has been no definite evidence for this sort of thing, the Sumerians had been using a type of sexagesimal notation and dealing with very large numbers for several hundred years. A context out of which the use of reciprocals could have emerged was, therefore, clearly in existence by the middle of the third millennium.

1.E6 Jens Høyrup on the Sumerian origin of mathematics

As so many other elements of our modern culture, mathematics came into being for the first time in Sumer, in Southern Mesopotamia. This happened in connection with the development of writing, around 3000 BC.

By claiming that mathematics came into being in Sumer and in exactly this epoch I do not want to deny that Sumerian mathematics has its roots back in the Stone Age societies of the Near and Middle East, nor that these and other Stone Age societies were in possession of elements of mathematical thought. Many Stone Age peasant peoples have applied geometrical principles in construction techniques and for decorative purposes, and even geometrical play can be found. In the Near and Middle East a system for arithmetical accounting related to the principles of the abacus (and to later Sumerian notation) was known as early as 8000 BC. Outlines of pre-Sumerian temple buildings were laid out in advance by strings and thus by use of geometry before the development of the earliest script; metrological systems for lengths and probably even for capacities, used seemingly in connection with arithmetical calculations, were employed before the rise of Sumerian civilization. What I wanted to express by my introductory phrase was, that only in the late fourth millennium BC, when the first primitive writing was born, were all these different *elements* of mathematical thought moulded *into one coherent system*: *MATHEMATICS*.

It happened in Sumer. The social context of the 'event' (which of course was a process) was the incipient formation of the *state*, where a social elite concentrated around the temple used its key position in a number of important functions (the construction of irrigation systems; trade; exchange between the various groups of producers of the products of agriculture, herding, fowling, fishing and handicrafts; genuine ritual functions; and probably even more) as a base from where it would gradually secure for itself a politically ruling position, and simultaneously ensure for its own mouth the lion's share of that social surplus which was secured precisely through some of the social functions of the priestly elite. The context was also that of the 'urban' revolution, where the city rose to the position of a dynamic centre from where development was determined, even though the city-dwellers remained in most cases a demographic minority.

The urban revolution and the rise of the incipient class-state do not constitute two mutually independent developments. On the contrary, they must be viewed as different aspects of the same social development, conditioned by the concrete natural, technological and social conditions prevailing in late fourth millennium Mesopotamia. In the concrete shape which they took on because of these conditions, they necessitated and furthered together the transformation of the above-mentioned token-based notation into a genuine, primitive script, including a numeral notation.

Until now I have not mentioned the school. However, the use of the proto-Sumerian script consisting of perhaps c. 1000 basic signs was hardly learned by the future temple official just while following the footsteps of elder colleagues. From the earliest proto-Sumerian times there is ample evidence that organized teaching has taken place, presumably in the temple, and that the teaching methods current in later Mesopotamian schools were already in use. The same evidence proves that the school was a place where knowledge was organized *systematically*, and that the schools of the single independent city-states were in mutual contact (since the development of the script and of numerical and metrological notations was the same even in far-separated cities). So, the school was the organizer of knowledge in general and not just of writing abilities (and probably even the organizer of a world view), in a way which was or at least tended to be both *coherent* and *uniform*.

Parallelly with the development of early writing and the systematization of knowledge expressed in word lists a coordination of the different elements of mathematical thought can be traced. Metrological notations for entities which were hardly measured in that way in the token-system were developed (*time* can be mentioned), and other metrological systems were extended according to arithmetical principles, among other things with fractional sub-units. Area measures were constructed so as to permit the calculation of the area of a rectangular field from the product of length and breadth (such a system may be less useful to the farmer and even to the taxator than a system based on natural units connected to sowing, ploughing or irrigation, but considered mathematically it is more systematic).

It is only a reconstruction, but on the other hand a reconstruction which makes sociological sense, that this organization of mathematics as a coherent whole (based on arithmetic as the uniting principle) is not solely due to practical 'social needs' for computation; it is a fair guess that it was quite as much a natural product of that same school institution which in other domains acted as a systematizer of knowledge and cunning. Even if practical social needs not only for computation but even for systematization were present (which as far as the systematization is concerned has yet to be proved), it is more than doubtful whether practitioners acting without the background of an institution like the school would be able to elaborate it. So, according to my hypothesis, the creation of *mathematics* in Sumer was specifically a product of that school institution which was able to create knowledge, to create the tools whereby to formulate and to transmit knowledge, and to systematize knowledge.

2 Mathematics in Classical Greece

2.A Historical Summary

Compared with the amount of surviving primary source material for Egyptian and Babylonian mathematics, virtually nothing survives that was physically written by the mathematicians of ancient Greece. Thus, charting this period in the history of mathematics requires a different sort of interpretation and evaluation of what does survive. The most substantial source describing the development of geometry from Thales (c. 600 BC) to Euclid (about 300 BC) is a passage (2.A1) by Proclus (fifth century AD). This is believed to have been largely based on the lost *History of Geometry*, by Eudemus (late fourth century BC), the first historian of mathematics. 2.A2 is a modern commentary illustrating recent textual criticism, and is an ingenious and significant argument that the reference to Pythagoras in Proclus's text is spurious.

2.A1 Proclus's summary

Thales, who had travelled to Egypt, was the first to introduce this science into Greece. He made many discoveries himself and taught the principles for many others to his successors, attacking some problems in a general way and others more empirically. Next after him Mamercus, brother of the poet Stesichorus, is remembered as having applied himself to the study of geometry; and Hippias of Elis records that he acquired a reputation in it. Following upon these men, Pythagoras transformed mathematical philosophy into a scheme of liberal education, surveying its principles from the highest downwards and investigating its theorems in an immaterial and intellectual manner. He it was who discovered the doctrine of proportionals and the structure of the cosmic figures. After him Anaxagoras of Clazomenae applied himself to many questions in geometry, and so did Oenopides of Chios, who was a little younger than Anaxagoras.

Both these men are mentioned by Plato in the *Erastae* as having got a reputation in mathematics. Following them Hippocrates of Chios, who invented the method of squaring lunules, and Theodorus of Cyrene became eminent in geometry. For Hippocrates wrote a book on elements, the first of whom we have any record who did so.

Plato, who appeared after them, greatly advanced mathematics in general and geometry in particular because of his zeal for these studies. It is well known that his writings are thickly sprinkled with mathematical terms and that he everywhere tries to arouse admiration for mathematics among students of philosophy. At this time also lived Leodamas of Thasos, Archytas of Tarentum, and Theaetetus of Athens, by whom the theorems were increased in number and brought into a more scientific arrangement. Younger than Leodamas were Neoclides and his pupil Leon, who added many discoveries to those of their predecessors, so that Leon was able to compile a book of elements more carefully designed to take account of the number of propositions that had been proved and of their utility. He also discovered *diorismi*, whose purpose is to determine when a problem under investigation is capable of solution and when it is not. Eudoxus of Cnidus, a little later than Leon and a member of Plato's group, was the first to increase the number of the so-called general theorems; to the three proportionals already known he added three more and multiplied the number of propositions concerning the 'section' which had their origin in Plato, employing the method of analysis for their solution. Amyclas of Heracleia, one of Plato's followers, Menaechmus, a student of Eudoxus who also was associated with Plato, and his brother Dinostratus made the whole of geometry still more perfect. Theudius of Magnesia had a reputation for excellence in mathematics as in the rest of philosophy, for he produced an admirable arrangement of the elements and made many partial theorems more general. There was also Athenaeus of Cyzicus, who lived about this time and became eminent in other branches of mathematics and most of all in geometry. These men lived together in the Academy, making their inquiries in common. Hermotimus of Colophon pursued further the investigations already begun by Eudoxus and Theaetetus, discovered many propositions in the *Elements*, and wrote some things about locus-theorems. Philippus of Mende, a pupil whom Plato had encouraged to study mathematics, also carried on his investigations according to Plato's instructions and set himself to study all the problems that he thought would contribute to Plato's philosophy.

All those who have written histories bring to this point their account of the development of this science. Not long after these men came Euclid, who brought together the *Elements*, systematizing many of the theorems of Eudoxus, perfecting many of those of Theaetetus, and putting in irrefutable demonstrable form propositions that had been rather loosely established by his predecessors. He lived in the time of Ptolemy the First, for Archimedes, who lived after the time of the first Ptolemy, mentions Euclid. It is also reported that Ptolemy once asked Euclid if there was not a shorter road to geometry than through the *Elements*, and Euclid replied that there was no royal road to geometry. He was therefore later than Plato's group but earlier than Eratosthenes and Archimedes, for these two men were contemporaries, as Eratosthenes somewhere says. Euclid belonged to the persuasion of Plato and was at home in this philosophy; and this is why he thought the goal of the *Elements* as a whole to be the construction of the so-called Platonic figures.

2.A2 W. Burkert on whether Eudemus mentioned Pythagoras

The chief testimony for Pythagoras as a mathematician, always cited in the literature, is in the 'catalogue of geometers' given by Proclus, whose principal source is rightly thought to be Eudemus. 'Pythagoras turned its [geometry's] philosophy into a form of liberal education, seeking its first principles from a higher source and hunting out its laws by a nonmaterialistic and intellectual procedure...' The weight of this pronouncement is enhanced by the prestige of Eudemus as a pupil of Aristotle, as well as by the undeniable fact that the special character of Greek mathematics consists precisely in its theoretical structure, as distinguished from the oriental 'recipes'. To be sure, the passage that follows, ascribing to Pythagoras the discovery of irrationality and of the 'cosmic bodies', is less often accepted; but even in the sentence quoted there are suspicious features. Does the phrase 'nonmaterialistic and intellectual procedure' seem more like the phrase of an early Peripatetic, or like a favourite theme of all Neoplatonists, and especially Proclus? And does Aristotle not say expressly, of the Pythagoreans, 'they apply their propositions to bodies'—bringing out the distinction, in this regard, between them and all genuine Platonists? And does not Eudemus, as far as we know from other fragments, always speak of Pythagoreans, never of Pythagoras, just as Aristotle himself does in philosophical or scientific contexts? A lucky coincidence turns suspicion into certainty: the sentence in question is taken word for word from Iamblichus' *De Communi Mathematica Scientia*—a work that Proclus copies sometimes by the page in his commentary of Euclid. Iamblichus is concerned with 'Pythagorean mathematics'. He mentions that the origins of mathematics lie before Pythagoras, in the work of Egyptians, Assyrians, and Chaldaeans. The distinctive aspect of Pythagoras's work is not only in new discoveries, but above all in the 'purity, subtlety, and exactitude' of his method, and in the way it purifies the soul and leads on to the highest principles and a realm of pure, immaterial Being. Iamblichus admits that Pythagoras and his pupils did not write any of this down, and therefore it is necessary to reconstruct, with considerable effort, 'what they would probably have said if one of them could have taught his doctrine publicly'. But Iamblichus has no doubts about his Neoplatonic theme: 'If we are to pursue mathematics in the Pythagorean manner, we must follow its upward path, full of divinity, which brings purification and perfection'. This is the beginning of a new chapter, in which the significance of mathematics is discussed, both for practical life and in itself. This may be based in part on Aristotle; but, if anything in it is original with Iamblichus, it is the chapter division and the transitional formulae. In the succeeding passage there is nothing about Pythagoras and the Pythagoreans except Iamblichus' introduction and his concluding sentence. 'It was natural, then, that for all these reasons the Pythagoreans honoured the study of mathematics'.

Therefore the often cited sentence about Pythagoras in the 'catalogue of geometers' is not from Eudemus, but is a formulation of Iamblichus, as had been recognized over sixty years ago. Thus its 'authority' is precisely reversed; if, in a context whose significant parts are obviously derived from Eudemus, the passage dealing specifically with Pythagoras has been supplemented with material from Iamblichus, this is an indication that there had been a gap to fill, and that Eudemus did not give enough information about Pythagoras or even none at all. As far as concerns the specific discoveries attributed to Pythagoras, the 'cosmic bodies' are regarded by Proclus as the apex, or the quintessence, of all geometry; and the discovery of irrationality had, long before Proclus, been interpreted in a Platonic sense and bound up with the thrilling story about mathematical secrecy and its betrayal and the ensuing divine punishment. Proclus has merely attributed to Pythagoras the two most famous and, from the Platonic point of view, most significant achievements of geometry, supplying an illustration to accompany the generally phrased sentence from Iamblichus. Nothing is left of the supposed testimony of Eudemus to the achievement of Pythagoras in the foundation of mathematics.

2.B Hippocrates' Quadrature of Lunes

A further extract from Eudemus's *History of Geometry* has been handed down by the commentator Simplicius (sixth century AD), concerning the quadrature of lunes—the area of crescent-like figures—by Hippocrates (late fifth century BC). 'This is one of the most precious sources for the history of Greek geometry before Euclid', in the judgement of Sir Thomas Heath (*History of Greek Mathematics*, Volume I, 1921, p. 182), and is especially interesting as recording in detail so early an attempt on the area of figures with *curved* sides, from which the theorems that must have been known in order to establish this rigorously can be inferred. Simplicius said that he was copying out Eudemus 'word for word, adding only for the sake of clearness a few things taken from Euclid's *Elements*'. In the passage as given here, these additions (as far as they can be determined) have been removed, so what remains is believed to be close to the original Eudemus. How accurate an account Eudemus, writing a century after Hippocrates, gave of this work is hard to say.

[1] The quadratures of lunes, which seemed to belong to an uncommon class of propositions by reason of the close relationship to the circle, were first investigated by Hippocrates, and seemed to be set out in correct form; therefore we shall deal with them at length and go through them. He made his starting-point, and set out as the first of the theorems useful to his purpose, that similar segments of circles have the same ratios as the squares on their bases. And this he proved by showing that the squares on the diameters have the same ratios as the circles.

[2] Having first shown this he described in what way it was possible to square a lune whose outer circumference was a semicircle. He did this by circumscribing about a right-angled isosceles triangle a semicircle and about the base a segment of a circle similar to those cut off by the sides. Since the segment about the base is equal to the sum of those about the sides, it follows that when the part of the triangle above the segment about the base is added to both the lune will be equal to the triangle. Therefore the lune, having been proved equal to the triangle, can be squared. In this way, taking a semicircle as the outer circumference of the lune, Hippocrates readily squared the lune.

[3] Next in order he assumes an outer circumference greater than a semicircle obtained by constructing a trapezium having three sides equal to one another while

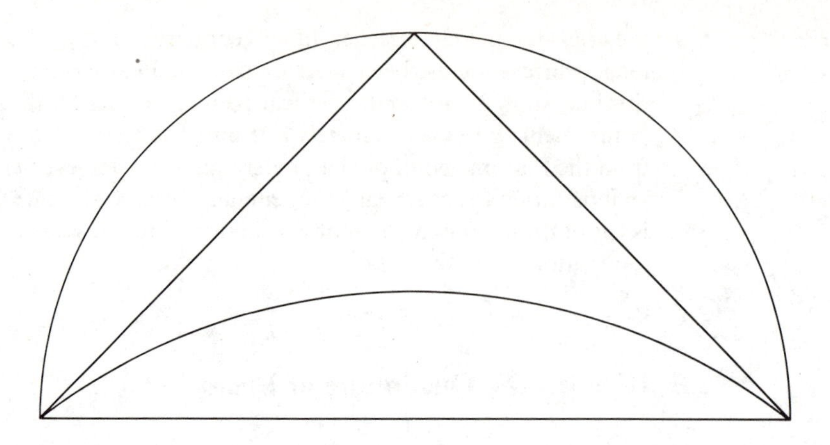

one, the greater of the parallel sides, is such that the square on it is three times the square on each of those sides, and then comprehending the trapezium in a circle and circumscribing about its greatest side a segment similar to those cut off from the circle by the three equal sides. That the said segment is greater than a semicircle is clear if a diagonal is drawn in the trapezium. For this diagonal, subtending two sides of the trapezium, must be such that the square on it is greater than double the square on one of the remaining sides. Therefore the square on *BC* is greater than double the square on either *BA*, *AC*, and therefore also on *CD*. Therefore the square on *BD*, the greatest of the sides of the trapezium, must be less than the sum of the squares on the diagonal and that one of the other sides which is subtended by the said greatest side together with the diagonal. For the squares on *BC*, *CD* are greater than three times, and the square on *BD* is equal to three times, the square on *CD*. Therefore the angle standing on the greatest side of the trapezium is acute. Therefore the segment in which it is is greater than a semicircle. And this segment is the outer circumference of the lune.

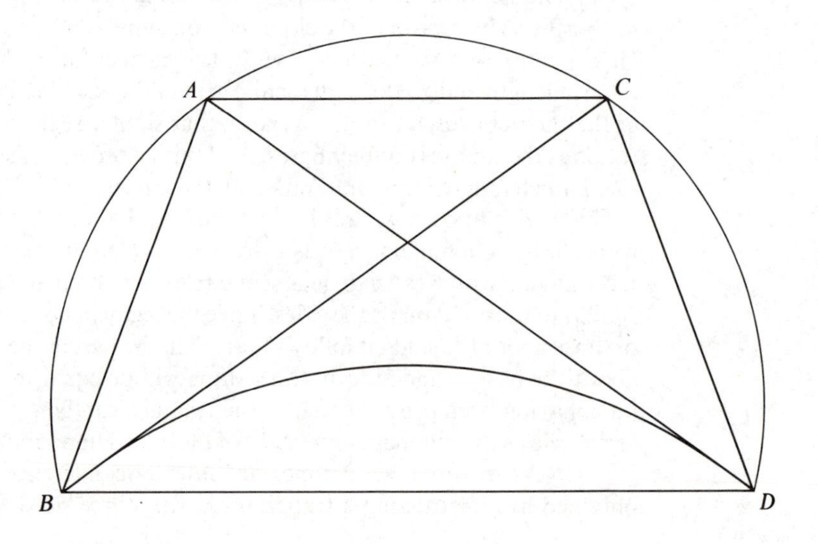

[4] If the outer circumference were less than a semicircle, Hippocrates solved this also, using the following preliminary construction. Let there be a circle with diameter *AB* and centre *K*. Let *CD* bisect *BK* at right angles; and let the straight line *EF* be placed between this and the circumference verging towards *B* so that the square on it is one-and-a-half times the square on one of the radii. Let *EG* be drawn parallel to *AB*, and from *K* let straight lines be drawn joining *E* and *F*. Let the straight line *KF* joined to *F* and produced meet *EG* at *G*, and again let straight lines be drawn from *B* joining *F* and *G*. It is then manifest that *EF* produced will pass through *B*—for by hypothesis *EF* verges towards *B*—and *BG* will be equal to *EK*.

This being so, I say that the trapezium *EKBG* can be comprehended in a circle.

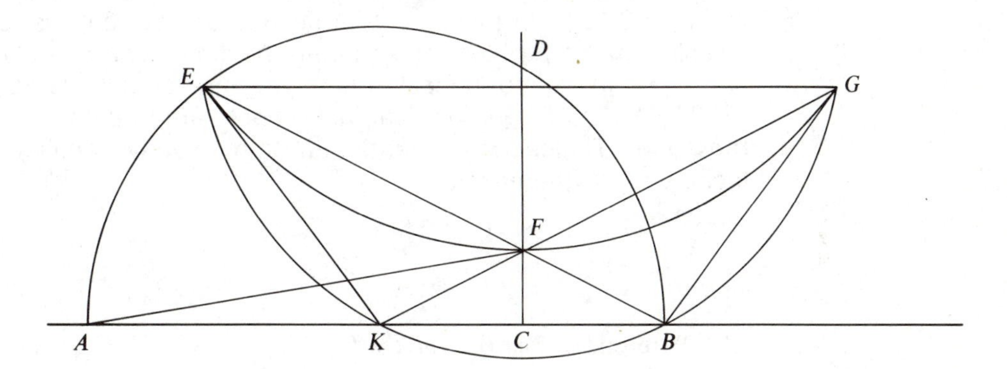

Next let a segment of a circle be circumscribed about the triangle *EFG*; then clearly each of the segments on *EF*, *FG* will be similar to the segments on *EK*, *KB*, *BG*.

This being so, the lune so formed, whose outer circumference is *EKBG*, will be equal to the rectilineal figure composed of the three triangles *BFG*, *BFK*, *EKF*. For the segments cut off from the rectilineal figure, inside the lune, by the straight lines *EF*, *FG* are together equal to the segments outside the rectilineal figure cut off by *EK*, *KB*, *BG*. For each of the inner segments is one-and-a-half times each of the outer, because, by hypothesis, the square on *EF* is one-and-a-half times the square on the radius, that is, the square on *EK* or *KB* or *BG*. Inasmuch then as the lune is made up of the three segments and the rectilineal figure *less* the two segments—the rectilineal figure including the two segments but not the three—while the sum of the two segments is equal to the sum of the three, it follows that the lune is equal to the rectilineal figure. [...]

[5] Thus Hippocrates squared every lune, seeing that he squared not only the lune which has for its outer circumference a semicircle, but also the lune in which the outer circumference is greater, and that in which it is less, than a semicircle.

2.C Two Fifth-century Writers

The two writers represented here are from opposite extremes of tone, style and purpose, but each has a useful bearing on our understanding of mathematical activity in the fifth century BC. The extracts from Parmenides' influential philosophical poem *The Way of Truth* (2.C1), written during the first half of the fifth century and here gleaned from later commentators, are unfortunately rather obscure to us. But they show significant aspects of argument-style and approach utilized by mathematical investigators later: deductive argument from irrefutable foundations, as well as the critical distinction between experience and reason which has characterized mathematical inquiry ever since.

The great Athenian comic dramatist Aristophanes (c. 448–385 BC) occasionally included mathematical activities among the objects of his wide-ranging and uninhibited satire, unwittingly providing unique testimony to late-fifth-century Athenian mathematics. In *The Birds* of 414 BC (2.C2), for instance, a burlesque of the (real) astronomer Meton reveals that the problem of 'squaring the circle' was already familiar. Both here and in Aristophanes' attack on intellectual trends, *The Clouds* (423 BC) (2.C3), geometrical instruments are in evidence, and we note, too, that 'Thales' is evidently an immediately familiar name of the archetypal mathematician.

2.C1 Parmenides' *The Way of Truth*

And the goddess greeted me kindly, and took my right hand in hers, and addressed me with these words: 'Young man, you who come to my house in the company of immortal charioteers with the mares which bear you, greetings. No ill fate has sent you to travel this road—far indeed does it lie from the steps of men—but right and justice. It is proper that you should learn all things, both the unshaken heart of well-rounded truth, and the opinions of mortals, in which there is no true reliance.

Come now, and I will tell you (and you must carry my account away with you when you have heard it) the only ways of enquiry that are to be thought of. The one, that [it] is and that it is impossible for [it] not to be, is the path of Persuasion (for she attends upon Truth); the other, that [it] is not and that it is needful that [it] not be, that I declare to you is an altogether indiscernible track: for you could not know what is not—that cannot be done—nor indicate it.

For never shall this be forcibly maintained, that things that are not are, but you must hold back your thought from this way of enquiry, nor let habit, born of much experience, force you down this way, by making you use an aimless eye or an ear and a tongue full of meaningless sound: judge by reason the strife-encompassed refutation spoken by me.

I shall not allow you to say nor to think from not being: for it is not to be said nor thought that it is not; and what need would have driven it later rather than earlier, beginning from the nothing, to grow? Thus it must either be completely or not at all. Nor will the force of conviction allow anything besides it to come to be ever from not being. Therefore Justice has never loosed her fetters to allow it to come to be or to perish, but holds it fast. And the decision about these things lies in this: it is or it is not.'

2.C2 Aristophanes: Meton squares the circle

(*The* ORACLEMAN *flees, scattering his stock, which* PEISTHETAERUS *flings after him. Meanwhile* METON *enters from the other side.*)

METON: I have come among you—

PEISTHETAERUS: Oh, no, not another! (*Imitating an actor in tragedy.*) How purposed, sir, / Do you thus visit us on buskin'd foot? / What grave intention, what inspir'd design / Counsels your journey? What's the big idea?

METON: I propose to survey the air for you: it will have to be marked out in acres.

PEISTHETAERUS: Good lord, who do you think you are?

METON: Who am I? Why, Meton. *The* Meton. Famous throughout the Hellenic world—you must have heard of my hydraulic clock at Colonus?

PEISTHETAERUS (*eyeing* METON'*s instruments*): And what are those for?

METON: Ah, these are my special rods for measuring the air. You see, the air is shaped, how shall I put it?—like a sort of extinguisher; so all I have to do is to attach this flexible rod at the upper extremity, take the compasses, insert the point here, and—you see what I mean?

PEISTHETAERUS: No.

METON: Well, I now apply the straight rod—so—thus squaring the circle; and there you are. In the centre you have your market place: straight streets leading into it, from here, from here, from here. Very much the same principle, really, as the rays of a star: the star itself is circular, but it sends out straight rays in every direction.

PEISTHETAERUS: Brilliant—the man's a Thales. But—Meton!

METON: Yes?

PEISTHETAERUS: Speaking as a friend (*he lowers his voice*) I think you'd be wise to slip away now.

METON: Why, what's the danger?

PEISTHETAERUS: The people here are like the Lacedaemonians, they don't like strangers. And feeling's running rather high just at the moment.

METON: Party differences?

PEISTHETAERUS: Oh no, far from it: they're quite unanimous.

METON: What's happening, then?

PEISTHETAERUS: There's to be a purge of pretentious humbugs: they're all going to get beaten up. You know what I mean: like this. (*He begins to demonstrate.*)

METON: Perhaps I'd better be going.

PEISTHETAERUS: I'm not sure you're going to get away in time. (*His blows get progressively harder: meanwhile the* CHORUS *advances menacingly.*) Something tells me that someone's going to get beaten up quite soon! (METON *hastily gathers up his instruments and makes for the exit, pursued by the* CHORUS.) I warned you! Go and measure how far it is to somewhere else.

2.C3 Aristophanes: Strepsiades encounters the New Learning

(*There is a two-storey stage-building with large central double-doors. Standing by the doors is a large round jar. All the characters are grotesquely masked.*)

STREPSIADES (*an old man, ruddy-faced, seeking the New Learning*):
I'm down but not out.
I'll offer a prayer to the gods and then go
to enrol in person in the Think-Tank.
Hey! I'm old, absent-minded, slow.
How can I master their logic, their hair-splitting?
On, Strepsiades. Stop trailing your heels.
Knock on their door. (*He does so.*) Porter! PORTER!!

STUDENT (*putting a very pale, cadaverous face round the door*):
Go to hell! Who's that banging on our door?

STREPSIADES:
Strepsiades, Phaedon's son, from Cicynna.

STUDENT:
Clearly illiterate. D'you realise that by banging
at our door when we weren't expecting it you've
caused the miscarriage of a scientific discovery?

STREPSIADES:
I'm sorry. I'm country born and bred.
What was that about a miscarriage?

STUDENT:
Sacred. Hush-hush. Only for registered students.

STREPSIADES:
C'm on. Tell me. I've come
to register as a student in the Think-Tank.

[...]

STUDENT:
Yesterday night there was nothing for dinner.

STREPSIADES:
OK. How did Socrates—um—excogitate some grub?

STUDENT:
First he sprinkled the table with some fine—ash—
took a skewer, bent it to make a pair of compasses—
and hooked someone's cloak from the changing-rooms.

STREPSIADES:
That's real mathematics. Who was Thales anyway?
(*Knocking with increasing vigour*)
Open up! Get a move on! Open up! I want to get inside
the Think-Tank and see Socrates—this very minute.
I *must* have education. Open the bloody door.
(*The double-doors open and a wheeled platform is pushed out. On it is a tableau of fantastic scientific apparatus, including a large map crudely showing the Mediterranean world, and an old bed and a group of students. Their clothes are filthy and ragged, their masks lean and pale. They are staring downwards, some standing rapt and upright, others like ostriches, peering closely at the ground, with their rumps sticking indecorously upwards.*)
Good God! What curious creatures!

STUDENT:
What's up? Do they remind you of anyone?

STREPSIADES:

Spartan troops captured at Pylos.
What on earth are they staring at the ground for?

STUDENT:

They're studying geology.

STREPSIADES:

They're looking for
mushrooms. (*Going up to them*) Here—this isn't work for the Think-Tank.
I'll tell you where to get lovely big ones.
What are those blokes up to with their noses in the ground?

STUDENT:

An advanced course on the Underworld.

STREPSIADES:

What the hell are their arses staring at the sky for?

STUDENT:

They're registered for a special course in astronomy.
(*He hears a creaking as over the top of the stage-building a crane hoists a basket. In it is Socrates, caricatured but clearly recognisable with pop-eyes, snub nose, bald head and beard. The basket can be seen by the audience, not yet by actors*).
Cave or the Beak'll catch you.
(*The tableau jerks into sudden movement*).

STREPSIADES:

Hold on, hold on. (*The tableau freezes.*) Don't let them go.
I've some business to discuss with them.

STUDENT:

Es ist strengstens verboten. It's against the rules
to spend too long in the fresh air.
(*The tableau unfreezes, and the students scamper inside. Strepsiades starts examining the apparatus. He looks at a large globe and some tablets with geometrical diagrams.*)

STREPSIADES:

Good God! What's all this?

STUDENT:

That's astronomy.

STREPSIADES (*with surveying instruments*):

What about this?

STUDENT:

Geometry.

STREPSIADES:

What can you do with it?

STUDENT:

Measure land.

STREPSIADES:

You mean my allotment?

STUDENT:

No, the whole world.

STREPSIADES:

You're joking!
That would be pretty useful in politics.

STUDENT (*with bronze map*):
This is the whole circuit of the earth. D'you see?
Here's Athens (*pointing*).
STREPSIADES (*peering*):
What? I don't believe you.
I can't see a single law-court in session.
STUDENT:
It really is the land of Attica.
STREPSIADES:
Then where's Cicynna? Where are my neighbours?
STUDENT:
Inside. (*Strepsiades is puzzled, shakes the map, looks underneath it.*) Here's Euboea. Look.
Here. Stretching out for ever such a long way.
STREPSIADES:
Oh, yes. It was Pericles helped us put them on the rack.
Well, where the devil's Sparta?
STUDENT:
Where's Sparta? Here.
STREPSIADES:
Much too near. The Think-Tank had better
think again, and move them a lot farther away from us.
STUDENT:
It can't be done.
STREPSIADES:
Then you'll be sorry for it.

2.D The Quadrivium

These extracts illustrate the persistence over some 2000 years of the four-part classification of mathematical sciences that eventually came to constitute a major part of the liberal arts curriculum of mediaeval universities. The classification into arithmetic, music, geometry and astronomy, and indeed the invention or early study of these subjects, came to be seen as due to Pythagoras, though this attribution is to large extent an invention of late antiquity (2.D8). One of the earliest mentions of these subjects together (2.D1) is nonetheless by a Pythagorean, the mathematician Archytas of Taranto (early fourth century BC). The context of a reference by Plato (2.D2) is a discussion of education among the Athenian teachers called Sophists. (For Plato's lengthier discussion of what the quadrivium subjects ought to consist of, see 2.E2.) The word 'quadrivium' seems to have been first used by the Roman scholar-statesman Boethius (c. 480–524 AD), whose influential work (2.D5) is largely a translation of the *Introduction to Arithmetic* (2.D4) by the neo-Pythagorean Nicomachus of Gerasa (c. 100 AD). 2.D6 is an extract from a play by the tenth-century Benedictine nun Hrosvitha of Hildesheim. As is clear from some of the extracts, the modern subject-names for the constituents of the quadrivium are in part misleading—'arithmetic' is rather closer to what we would call number theory, and 'music' to theory of harmony—and the way the terms were interpreted varied from age to age.

2.D1 Archytas

I think that those concerned with the sciences are men of discernment, and it is not strange that they should think correctly about the nature of particular things. For, having judged correctly about the nature of wholes, they were also likely to observe well the several nature of things part by part. And so they have handed down to us clear knowledge of the speed of the heavenly bodies and their risings and settings, of geometry, numbers and, not least, of the science of music. For these sciences seem to be related: they are concerned with the first two kinds of what is, which are related.

2.D2 Plato

'Hippocrates, by becoming a pupil of Protagoras, will, on the very day he joins him, go home a better man, and on each successive day will make similar progress—towards what, Protagoras, and better at what?'

Protagoras heard me out and said: 'You put your questions well, and I enjoy answering good questioners. When he comes to me, Hippocrates will not be put through the same things that another Sophist would inflict on him. The others treat their pupils badly: these young men, who have deliberately turned their backs on specialization, they take and plunge into special studies again, teaching them arithmetic and astronomy and geometry and music'—here he glanced at Hippias—'but from me he will learn only what he has come to learn. What is that subject? The proper care of his personal affairs, so that he may best manage his own household, and also of the State's affairs, so as to become a real power in the city, both as speaker and man of action.'

2.D3 Proclus

The Pythagoreans considered all mathematical science to be divided into four parts: one half they marked off as concerned with quantity, the other half with magnitude; and each of these they posited as twofold. A quantity can be considered in regard to its character by itself or in its relation to another quantity, magnitudes as either stationary or in motion. Arithmetic, then, studies quantity as such, music the relations between quantities, geometry magnitude at rest, spherics magnitude inherently moving.

2.D4 Nicomachus

The ancients, who under the leadership of Pythagoras first made science systematic, defined philosophy as the love of wisdom. Indeed the name itself means this, and before Pythagoras all who had knowledge were called 'wise' indiscriminately—a carpenter, for example, a cobbler, a helmsman, and in a word anyone who was versed in any art or handicraft. Pythagoras, however, restricting the title so as to apply to the knowledge and comprehension of reality, and calling the knowledge of the truth in this the only wisdom, naturally designated the desire and pursuit of this knowledge philosophy, as being desire for wisdom. [...]

Therefore, if we crave for the goal that is worthy and fitting for man, namely, happiness of life—and this is accomplished by philosophy alone and by nothing else, and philosophy, as I said, means for us desire for wisdom, and wisdom the science of the truth in things, and of things some are properly so called, others merely share the name—it is reasonable and most necessary to distinguish and systematize the accidental qualities of things.

Things, then, both those properly so called and those that simply have the name, are some of them unified and continuous, for example, an animal, the universe, a tree, and the like, which are properly and peculiarly called 'magnitudes'; others are discontinuous, in a side-by-side arrangement, and, as it were, in heaps, which are called 'multitudes', a flock, for instance, a people, a heap, a chorus, and the like.

Wisdom, then, must be considered to be the knowledge of these two forms. Since, however, all multitude and magnitude are by their own nature of necessity infinite—for multitude starts from a definite root and never ceases increasing; and magnitude, when division beginning with a limited whole is carried on, cannot bring the dividing process to an end, but proceeds therefore to infinity—and since sciences are always sciences of limited things, and never of infinites, it is accordingly evident that a science dealing either with magnitude, *per se*, or with multitude, *per se*, could never be formulated, for each of them is limitless in itself, multitude in the direction of the more, and magnitude in the direction of the less. A science, however, would arise to deal with something separated from each of them, with quantity, set off from multitude, and size, set off from magnitude.

Again, to start afresh, since of quantity one kind is viewed by itself, having no relation to anything else, as 'even', 'odd', 'perfect', and the like, and the other is relative to something else and is conceived of together with its relationship to another thing, like 'double', 'greater', 'smaller', 'half', 'one and one-half times', 'one and one-third times', and so forth, it is clear that two scientific methods will lay hold of and deal with the whole investigation of quantity; arithmetic, absolute quantity, and music, relative quantity.

And once more, inasmuch as part of 'size' is in a state of rest and stability, and another part in motion and revolution, two other sciences in the same way will accurately treat of 'size', geometry the part that abides and is at rest, astronomy that which moves and revolves.

Without the aid of these, then, it is not possible to deal accurately with the forms of being nor to discover the truth in things, knowledge of which is wisdom, and evidently not even to philosophize properly, for 'just as painting contributes to the menial arts toward correctness of theory, so in truth lines, numbers, harmonic intervals, and the revolutions of circles bear aid to the learning of the doctrines of wisdom', says the Pythagorean Androcydes.

2.D5 Boethius

Among all the men of ancient authority who, following the lead of Pythagoras, have flourished in the purer reasoning of the mind, it is clearly obvious that hardly anyone has been able to reach the highest perfection of the disciplines of philosophy unless the nobility of such wisdom was investigated by him in a certain four-part study, the *quadrivium*, which will hardly be hidden from those properly respectful of expertness. For this is the wisdom of things which are, and the perception of truth gives to these things their unchanging character.

[...] arithmetic considers that multitude which exists of itself as an integral whole; the measures of musical modulation understand that multitude which exists in relation to some other; geometry offers the notion of stable magnitude; the skill of astronomical discipline explains the science of movable magnitude. If a searcher is lacking knowledge of these four sciences, he is not able to find the true; without this kind of thought, nothing of truth is rightly known. This is the knowledge of those things which truly are; it is their full understanding and comprehension. He who spurns these, the paths of wisdom, does not rightly philosophize. [...]

This, therefore, is the *quadrivium* by which we bring a superior mind from knowledge offered by the senses to the more certain things of the intellect. There are various steps and certain dimensions of progressing by which the mind is able to ascend so that by means of the eye of the mind, which (as Plato says) is composed of many corporeal eyes and is of higher dignity than they, truth can be investigated and beheld. This eye, I say, submerged and surrounded by the corporeal senses, is in turn illuminated by the disciplines of the *quadrivium*.

2.D6 Hrosvitha

DISCIPLE: What is music?
PAFNUTIUS: It is a discipline from the quadrivium of philosophy.
DISCIPLE: What is this you call the quadrivium?
PAFNUTIUS: Arithmetic, geometry, music, astronomy.
DISCIPLE: Why a quadrivium?
PAFNUTIUS: Because as from a path or four-fold way, so from this path and from these principles extend the proper progressions of philosophical principles.

2.D7 Roger Bacon

This science [mathematics] of all the parts of philosophy was the earliest discovered. For this was first discovered at the beginning of the human race. Since it was discovered before the flood and then later by the sons of Adam, and by Noah and his sons, as is clear from the prologue to the *Construction of the Astrolabe* according to Ptolemy, and from Albumazar in the larger introduction to astronomy, and from the first book of the *Antiquities*, and this is true as regards all its parts, geometry, arithmetic, music, astronomy. But this would not have been the case except for the fact that this science is earlier than the others and naturally precedes them. Hence it is clear that it should be studied first, that through it we may advance to all the later sciences. [...] But this science is the easiest. This is clearly proved by the fact that mathematics is not beyond the intellectual grasp of any one. For the people at large and those wholly illiterate know how to draw figures and compute and sing, all of which are mathematical operations.

2.D8 W. Burkert on the Pythagorean tradition in education

What is the origin of the firmly rooted conviction that Pythagoreanism was the source of Greek mathematics? This question is easy to answer: it came from the educational tradition. Everyone comes upon the name of Pythagoras for the first time in school mathematics; and this has been true from the earliest stages of the Western cultural tradition. None of the ancient textbooks which formed the basis of the medieval curriculum forgets Pythagoras. He is the companion of Arithmetica in Martianus Capella; and according to Isidore he was the first, among the Greeks, to sketch out the doctrine of number, which was then set forth in detail by Nicomachus. This takes us back to the origin of this tradition; Nicomachus, who is himself called a Pythagorean, begins his *Arithmetic*, which was much used as a schoolbook, with praise of the Master. Boethius' *Arithmetic*, drawn largely from Nicomachus, also names Pythagoras in its first line. Likewise, Gerbert of Aurillac mentions the name of Pythagoras several times in his geometry; and among the patrons of their *ars geometriae* medieval Freemasons include Pythagoras. The *Ars geometriae* bearing the name of Boethius, though obviously not composed before the High Middle Ages, even presents an early version of the Arabic numerals as an invention of the 'Pythagorici', and describes the method of calculating with these *apices* on an abacus, called *mensa Pythagorea*—perhaps the most striking of the anachronisms in which the Pythagorean tradition is so rich. Finally, the early modern period derived the astronomy of Copernicus and Galileo from Pythagoras.

The general belief in the Pythagorean origin of mathematics thus stems from the Neoplatonic and neo-Pythagorean scholastic tradition of late antiquity. In evaluating this it is worth bearing in mind that according to an earnestly meant statement of Iamblichus, even the problem of squaring the circle was solved by Pythagoreans.

2.E Plato

The great renown attached to Plato (c. 427–348 BC) with regard to mathematics is clear from sources such as Eudemus' history (see 2.A1) and Aristoxenus' report of Plato's lecture on the Good (2.E8). Plato's own works not only spell this out further, but also provide some of our best evidence (in lieu of much else) for the mathematical concerns and activities of the early fourth century. The evidence provided is not always easy to interpret unequivocally, though; the passage from Theaetetus (2.E3), for example, has inspired much recent scholarly disputation on what it may tell us of the mathematical developments later found as Book X of Euclid's *Elements* (see 3.E1). The works of Plato are, of course, more than just a primary source for their own time. They are also of almost unparalleled influence on subsequent Western thought, nowhere more so than in his creation myth *Timaeus* (2.E5), whose vision of the creator-god as mathematician has underpinned a stream of cosmological belief down to our own time. And in the *Republic* (2.E2) we find the fullest early account of and rationale for the quadrivium subjects (even though in Plato's conception there are five of them, for reasons he explains in the text).

2.E1 Socrates and the slave boy

MENO: What do you mean when you say that we don't learn anything, but that what we call learning is recollection? Can you teach me that it is so?

SOCRATES: I have just said that you're a rascal, and now you ask me if I can teach you, when I say there is no such thing as teaching, only recollection. Evidently you want to catch me contradicting myself straight away.

MENO: No, honestly, Socrates, I wasn't thinking of that. It was just habit. If you can in any way make clear to me that what you say is true, please do.

SOCRATES: It isn't an easy thing, but still I should like to do what I can since you ask me. I see you have a large number of retainers here. Call one of them, anyone you like, and I will use him to demonstrate it to you.

MENO: Certainly. (*To a slave boy.*) Come here.

SOCRATES: He is a Greek and speaks our language?

MENO: Indeed yes—born and bred in the house.

SOCRATES: Listen carefully then, and see whether it seems to you that he is learning from me or simply being reminded.

MENO: I will.

SOCRATES: Now boy, you know that a square is a figure like this?

(*Socrates begins to draw figures in the sand at his feet. He points to the square ABCD.*)

BOY: Yes.

SOCRATES: It has all these four sides equal?

BOY: Yes.

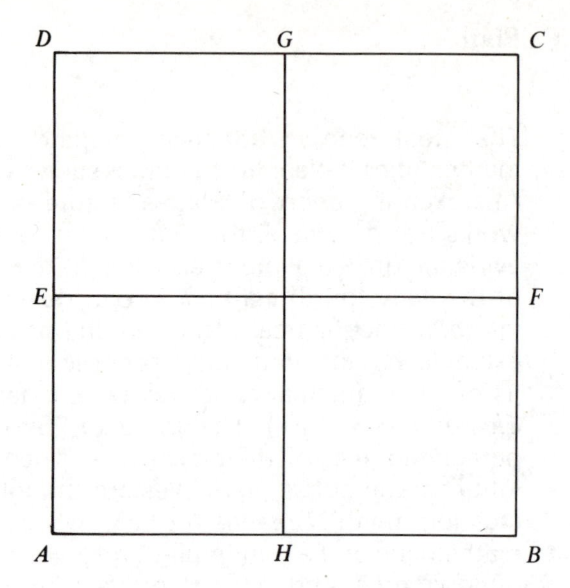

SOCRATES: And these lines which go through the middle of it are also equal? (*The lines EF, GH.*)

BOY: Yes.

SOCRATES: Such a figure could be either larger or smaller, could it not?

BOY: Yes.

SOCRATES: Now if this side is two feet long, and this side the same, how many feet will the whole be? Put it this way. If it were two feet in this direction and only one in that, must not the area be two feet taken once?

BOY: Yes.

SOCRATES: But since it is two feet this way also, does it not become twice two feet?

BOY: Yes.

SOCRATES: And how many feet is twice two? Work it out and tell me.

BOY: Four.

SOCRATES: Now could one draw another figure double the size of this, but similar, that is, with all its sides equal like this one?

BOY: Yes.

SOCRATES: How many feet will its area be?

BOY: Eight.

SOCRATES: Now then, try to tell me how long each of its sides will be. The present figure has a side of two feet. What will be the side of the double-sized one?

BOY: It will be double, Socrates, obviously.

SOCRATES: You see, Meno, that I am not teaching him anything, only asking. Now he thinks he knows the length of the side of the eight-feet square.

MENO: Yes.

SOCRATES: But does he?

MENO: Certainly not.

SOCRATES: He thinks it is twice the length of the other.

MENO: Yes.

SOCRATES: Now watch how he recollects things in order—the proper way to recollect. You say that the side of double length produces the double-sized figure? Like this I mean, not long this way and short that. It must be equal on all sides like the first figure, only twice its size, that is eight feet. Think a moment whether you still expect to get it from doubling the side.

BOY: Yes, I do.

SOCRATES: Well now, shall we have a line double the length of this (*AB*) if we add another the same length at this end (*BJ*)?

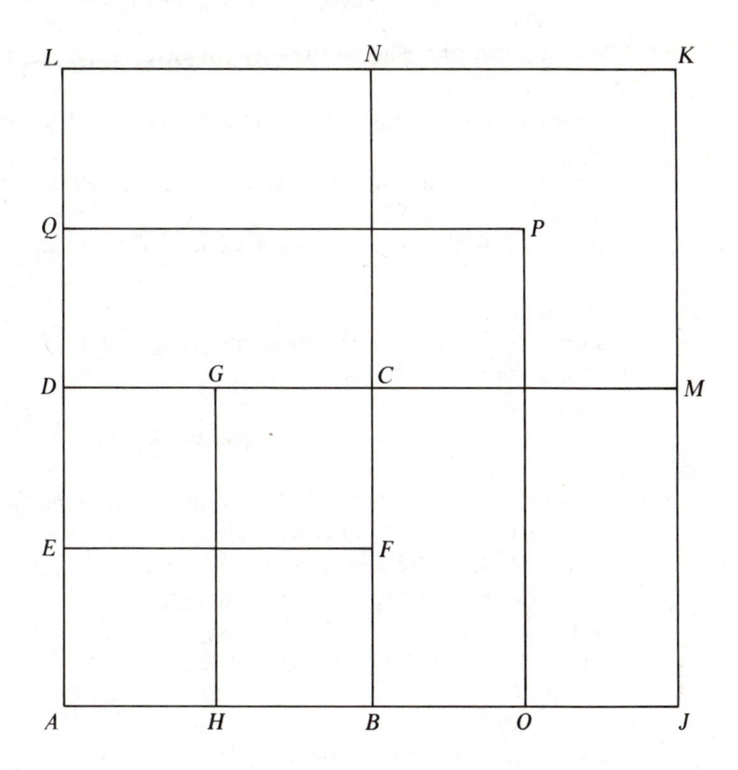

BOY: Yes.

SOCRATES: It is on this line then, according to you, that we shall make the eight-feet square, by taking four of the same length?

BOY: Yes.

SOCRATES: Let us draw in four equal lines (*i.e. counting AJ, and adding JK, KL, and LA made complete by drawing in its second half LD*), using the first as a base. Does this not give us what you call the eight-feet figure?

BOY: Certainly.

SOCRATES: But does it contain these four squares, each equal to the original four-feet one?

(*Socrates has drawn in the lines CM, CN to complete the squares that he wishes to point out.*)

BOY: Yes.

SOCRATES: How big is it then? Won't it be four times as big?

BOY: Of course.

SOCRATES: And is four times the same as twice?
BOY: Of course not.
SOCRATES: So doubling the side has given us not a double but a fourfold figure?
BOY: True.
SOCRATES: And four times four are sixteen, are they not?
BOY: Yes.
SOCRATES: Then how big is the side of the eight-feet figure? This one has given us four times the original area, hasn't it?
BOY: Yes.
SOCRATES: And a side half the length gave us a square of four feet?
BOY: Yes.
SOCRATES: Good. And isn't a square of eight feet double this one and half that?
BOY: Yes.
SOCRATES: Will it not have a side greater than this one but less than that?
BOY: I think it will.
SOCRATES: Right. Always answer what you think. Now tell me: was not this side two feet long, and this one four?
BOY: Yes.
SOCRATES: Then the side of the eight-feet figure must be longer than two feet but shorter than four?
BOY: It must.
SOCRATES: Try to say how long you think it is.
BOY: Three feet.
SOCRATES: If so, shall we add half of this bit (*BO*, *half of BJ*) and make it three feet? Here are two, and this is one, and on this side similarly we have two plus one; and here is the figure you want.
(*Socrates completes the square AOPQ.*)
BOY: Yes.
SOCRATES: If it is three feet this way and three that, will the whole area be three times three feet?
BOY: It looks like it.
SOCRATES: And that is how many?
BOY: Nine.
SOCRATES: Whereas the square double our first square had to be how many?
BOY: Eight.
SOCRATES: But we haven't yet got the square of eight feet even from a three-feet side?
BOY: No.
SOCRATES: Then what length will give it? Try to tell us exactly. If you don't want to count it up, just show us on the diagram.
BOY: It's no use, Socrates, I just don't know.
SOCRATES: Observe, Meno, the stage he has reached on the path of recollection. At the beginning he did not know the side of the square of eight feet. Nor indeed does he know it now, but then he thought he knew it and answered boldly, as was appropriate—he felt no perplexity. Now however he does feel perplexed. Not only does he not know the answer; he doesn't even think he knows.
MENO: Quite true.
SOCRATES: Isn't he in a better position now in relation to what he didn't know?
MENO: I admit that too.

SOCRATES: So in perplexing him and numbing him like the sting-ray, have we done him any harm?

MENO: I think not.

SOCRATES: In fact we have helped him to some extent towards finding out the right answer, for now not only is he ignorant of it but he will be quite glad to look for it. Up to now, he thought he could speak well and fluently, on many occasions and before large audiences, on the subject of a square double the size of a given square, maintaining that it must have a side of double the length.

MENO: No doubt.

SOCRATES: Do you suppose then that he would have attempted to look for, or learn, what he thought he knew (though he did not), before he was thrown into perplexity, became aware of his ignorance, and felt a desire to know?

MENO: No.

SOCRATES: Then the numbing process was good for him?

MENO: I agree.

SOCRATES: Now notice what, starting from this state of perplexity, he will discover by seeking the truth in company with me, though I simply ask him questions without teaching him. Be ready to catch me if I give him any instruction or explanation instead of simply interrogating him on his own opinions.

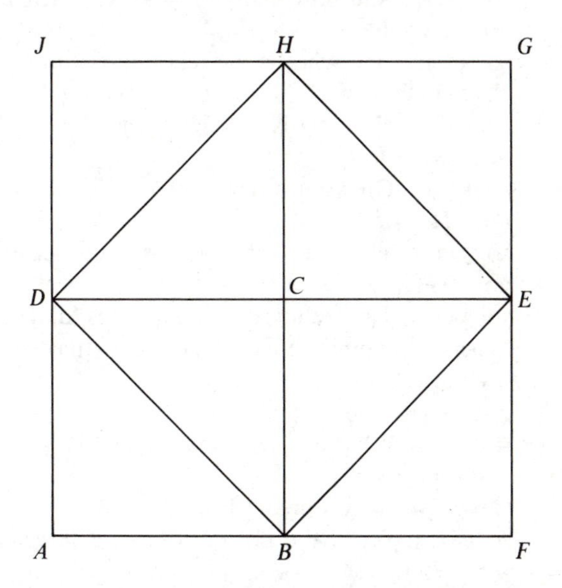

(Socrates here rubs out the previous figures and starts again.)

Tell me, boy, is not this our square of four feet? (*ABCD*.) You understand?

BOY: Yes.

SOCRATES: Now we can add another equal to it like this? (*BCEF*.)

BOY: Yes.

SOCRATES: And a third here, equal to each of the others? (*CEGH*.)

BOY: Yes.

SOCRATES: And then we can fill in this one in the corner? (*DCHJ*.)

BOY: Yes.

SOCRATES: Then here we have four equal squares?
BOY: Yes.
SOCRATES: And how many times the size of the first square is the whole?
BOY: Four times.
SOCRATES: And we want one double the size. You remember?
BOY: Yes.
SOCRATES: Now does this line going from corner to corner cut each of these squares in half?
BOY: Yes.
SOCRATES: And these are four equal lines enclosing this area? (*BEHD*.)
BOY: They are.
SOCRATES: Now think. How big is this area?
BOY: I don't understand.
SOCRATES: Here are four squares. Has not each line cut off the inner half of each of them?
BOY: Yes.
SOCRATES: And how many such halves are there in this figure? (*BEHD*.)
BOY: Four.
SOCRATES: And how many in this one? (*ABCD*.)
BOY: Two.
SOCRATES: And what is the relation of four to two?
BOY: Double.
SOCRATES: How big is this figure then?
BOY: Eight feet.
SOCRATES: On what base?
BOY: This one.
SOCRATES: The line which goes from corner to corner of the square of four feet?
BOY: Yes.
SOCRATES: The technical name for it is 'diagonal'; so if we use that name, it is your personal opinion that the square on the diagonal of the original square is double its area.
BOY: That is so, Socrates.
SOCRATES: What do you think, Meno? Has he answered with any opinions that were not his own?
MENO: No, they were all his.
SOCRATES: Yet he did not know, as we agreed a few minutes ago.
MENO: True.
SOCRATES: But these opinions were somewhere in him, were they not?
MENO: Yes.
SOCRATES: So a man who does not know has in himself true opinions on a subject without having knowledge.
MENO: It would appear so.
SOCRATES: At present these opinions, being newly aroused, have a dream-like quality. But if the same questions are put to him on many occasions and in different ways, you can see that in the end he will have a knowledge on the subject as accurate as anybody's.
MENO: Probably.

SOCRATES: This knowledge will not come from teaching but from questioning. He will recover it for himself.

MENO: Yes.

SOCRATES: And the spontaneous recovery of knowledge that is in him is recollection, isn't it?

MENO: Yes.

SOCRATES: Either then he has at some time acquired the knowledge which he now has, or he has always possessed it. If he always possessed it, he must always have known; if on the other hand he acquired it at some previous time, it cannot have been in this life, unless somebody has taught him geometry. He will behave in the same way with all geometrical knowledge, and every other subject. Has anyone taught him all these? You ought to know, especially as he has been brought up in your household.

MENO: Yes, I know that no one ever taught him.

SOCRATES: And has he these opinions, or hasn't he?

MENO: It seems we can't deny it.

SOCRATES: Then if he did not acquire them in this life, isn't it immediately clear that he possessed and had learned them during some other period?

MENO: It seems so.

SOCRATES: When he was not in human shape?

MENO: Yes.

SOCRATES: If then there are going to exist in him, both while he is and while he is not a man, true opinions which can be aroused by questioning and turned into knowledge, may we say that his soul has been for ever in a state of knowledge? Clearly he always either is or is not a man.

MENO: Clearly.

SOCRATES: And if the truth about reality is always in our soul, the soul must be immortal, and one must take courage and try to discover—that is, to recollect—what one doesn't happen to know, or (more correctly) remember, at the moment.

MENO: Somehow or other I believe you are right.

SOCRATES: I think I am. I shouldn't like to take my oath on the whole story, but one thing I am ready to fight for as long as I can, in word and act: that is, that we shall be better, braver and more active men if we believe it right to look for what we don't know than if we believe there is no point in looking because what we don't know we can never discover.

MENO: There too I am sure you are right.

2.E2 Mathematical studies for the philosopher ruler

'Well, Glaucon', I asked, 'what should men study if their minds are to be drawn from the world of change to reality? Now it occurs to me that we said our rulers must be trained for war when they were young.'

'We did.'

'Then the subject we're looking for must be relevant in war too.'

'How do you mean?'

'It mustn't be useless to soldiers.'

'Not if we can avoid it.'

'Well, we've already arranged for their physical training and their education in literature and music. And of these two, physical training is concerned with the world of change and decay, for the body, which it looks after, grows and declines.'

'Yes, clearly.'

'So it won't be the study we are looking for.'

'No.'

'Then what about the literary education which we described earlier on?'

'That', he reminded me, 'was the complement of their physical education. It gave them a moral training, and used music and rhythm to produce a certain harmony and balance of character rather than knowledge; and its literature, whether fabulous or factual, had a similar ethical content. There was nothing in it to produce the effect you are seeking.'

'Your memory's quite correct,' I said, 'we shan't find what we want there. But where shall we find it, Glaucon? The more practical forms of skill don't seem very elevating—'

'Certainly not. But if we exclude them, as well as physical and literary education, what else is there left?'

'Well, if we can't think of anything outside them, we must find some feature they all share.'

'What do you mean?'

'For example, there is one thing that all occupations, practical, intellectual, or scientific, make use of—one of the first things we must all learn.'

'What?'

'Something quite ordinary—to tell the difference between one, two and three; in a word, to count and calculate. Must not every practical or scientific activity be able to do that?'

'Yes, it must,' he agreed.

'And war as much as any other?'

'Very much so.'

'I wonder if you have noticed what a silly sort of general Agamemnon is made to look on the stage when Palamedes claims to have invented number, and so organized the army at Troy and counted the ships and everything else. It implies that nothing had been counted before and that Agamemnon, apparently, did not know how many feet he had, if he couldn't count. He must have been a funny sort of general!'

'He must indeed', he said, 'if it's really true.'

'So soldiers must learn, as well as other things, how to calculate and count.'

'Yes, of course, if they're to be able to organize an army, indeed if they are to be human at all.'

'I wonder, then,' I asked, 'if you would agree with me that this is probably one of the subjects we are looking for, which naturally stimulates thought, though no one makes proper use of its power to draw men to the truth.' [...]

I Arithmetic

'We can, then, properly lay it down that arithmetic shall be a subject for study by those who are to hold positions of responsibility in our state; and we shall ask them not to be amateurish in their approach to it, but to pursue it till they come to understand, by pure thought, the nature of numbers—they aren't concerned with its usefulness for mere commercial calculation, but for war and for the easier conversion of the soul from the world of becoming to that of reality and truth.'

'Excellent.'

'You know', I said, 'now that we have mentioned arithmetic, it occurs to me what a subtle and useful instrument it is for our purpose, if one studies it for the sake of knowledge and not for commercial ends.'

'How is that?' he asked.

'As we have just said, it draws the mind upwards and forces it to argue about pure numbers, and will not be put off by attempts to confine the argument to collections of visible or tangible objects. You must know how the experts in the subject, if one tries to argue that the unit is divisible, won't have it, but make you look absurd by multiplying it if you try to divide it, to make sure that their unit is never shown to contain a multiplicity of parts.'

'Yes, that's quite true.'

'What do you think they would say, Glaucon, if one were to say to them, "This is very extraordinary—what are these numbers you are arguing about, in which you claim that every unit is exactly equal to every other, and at the same time not divisible into parts?" What do you think their answer would be to that?'

'I suppose they would say that the numbers they mean can be apprehended by thought, but that there is no other way of grasping them.'

'You see therefore,' I pointed out to him, 'that this study looks as if it were really necessary to us, since it so obviously compels the mind to think in order to get at the truth.'

'It certainly does have that effect,' he agreed.

'Another point—have you noticed how those who are naturally good at calculation are nearly always quick at learning anything else, and how the slow-witted, if trained and practised in calculation, always improve in speed even if they get no other benefit? Yet I suppose there's hardly any form of study which comes harder to those who learn or practise it.'

'That is true.'

'For all these reasons, then, we must retain this subject and use it to train our ablest citizens.'

'I agree.'

II Plane geometry

'That's one subject settled, then. Next let us see if the one that follows it is of any use to us.'

'Do you mean geometry?' he asked.

'Exactly.'

'It's obviously useful in war,' he said. 'If a man knows geometry it will make all the difference to him when it comes to pitching camp or taking up a position, or concentrating or deploying an army, or any other military manoeuvre in battle or on the march.'

'For that sort of purpose,' I replied, 'the amount of geometry or calculation needed is small. What we want to find out is whether the subject is on the whole one which, when taken further, has the effect of making it easier to see the Form of the Good. And that, we say, is the tendency of everything which compels the mind to turn to the blessedness of the ultimate reality which it must somehow contrive to see.'

'I agree,' he said.

'So if it compels us to contemplate reality, it will be useful, but otherwise not.'

'That's our view.'

'Well, then, no one with even an elementary knowledge of geometry will dispute that it's a science quite the reverse of what is implied by the terms its practitioners use.'

'Explain.'

'The terms are quite absurd, but they are hard put to it to find others. They talk about "squaring" and "applying" and "adding" and so on, as if they were *doing* something and their reasoning had a practical end, and the subject were not, in fact, pursued for the sake of knowledge.'

'Yes, that's true.'

'And what is more, it must, I think, be admitted that the objects of that knowledge are eternal and not liable to change and decay.'

'Yes, there's no question of that: the objects of geometrical knowledge are eternal.'

'Then it will tend to draw the mind to the truth and direct the philosophers' thought upwards, instead of to our present mundane affairs.'

'It is sure to.'

'Then you must be sure to require the citizens of your ideal state not to neglect geometry. It has considerable incidental advantages too.'

'What are they?' he asked.

'Its usefulness for war, which you have already mentioned,' I replied; 'and there is a certain facility for learning all other subjects in which we know that those who have studied geometry lead the field.'

'They are miles ahead,' he agreed.

'So shall we make this the second subject our young men must study?'

'Yes.'

III Solid geometry

'And the third should be astronomy. Or don't you agree?'

'Yes, I certainly agree. A degree of skill in telling the seasons, months and years is useful not only to the farmer and sailor but also to the soldier.'

'You amuse me,' I said, 'with your obvious fear that the public will disapprove if the subjects you prescribe aren't thought useful. But it is in fact very difficult for people to believe that there is a faculty in the mind of each of us which these studies purify and rekindle after it has been ruined and blinded by other pursuits, though it is more worth preserving than any eye since it is the only organ by which we perceive the truth. Those who agree will think your proposals admirable; but those who have never realized it will probably think you are talking nonsense, as they won't see what other benefit is to be expected from such studies. Make up your mind which party you are going to argue with—or will you ignore both and pursue the argument for your own satisfaction, though without grudging anyone else any benefit he may get from it?'

'That's what I'll do,' he replied; 'I'll go on with the discussion chiefly for my own satisfaction.'

'Then you must go back a bit,' I said, 'as we made a wrong choice of subject to put next to geometry.'

'How was that?'

'We proceeded straight from plane geometry to solid bodies in motion without considering solid bodies first on their own. The right thing is to proceed from second dimension to third, which brings us to cubes and other three-dimensional figures.'

'That's true enough', he agreed, 'but the subject is one which doesn't seem to have been much investigated yet, Socrates.'

'For two reasons,' I replied. 'There is no state which sets any value on it, and so, being difficult, it is not pursued with energy; and research is not likely to progress without a director, who is difficult to find and, even if found, is unlikely to be obeyed in the present intolerant mood of those who study the subject. But under the general direction of a state that set a value on it, their obedience would be assured, and research pressed forward continuously and energetically till the problems were solved. Even now, with all the neglect and inadequate treatment it has suffered from students who do not understand its real uses, the subject is so attractive that it makes progress in spite of all handicaps, and it would not be surprising if its problems were solved.'

'Yes, it has very great attractions,' he said. 'But explain more clearly what you said just now. You said geometry dealt with plane surfaces.'

'Yes.'

'Then you first said astronomy came next, but subsequently went back on what you had said.'

'More haste less speed,' I said. 'In my hurry I overlooked solid geometry, because it's so absurdly undeveloped, and put astronomy, which is concerned with solids in motion, after plane geometry.'

'Yes that's what you did,' he agreed.

'Then let us put astronomy fourth, and assume that the neglect of solid geometry would be made good under state encouragement.'

IV Astronomy

'That is fair enough,' he said. 'And since you have just been attacking me for approving of astronomy for low motives, let me approve of it now on your principles; for it must be obvious to everyone that it, of all subjects, compels the mind to look upwards and leads it from earth to heaven.'

'Perhaps I'm an exception,' I said, 'for I don't agree. I think that, as it's at present handled by those who use it as an introduction to philosophy, it makes us look down, not up.'

'What do you mean?' he asked.

'I think you've a really splendid idea of the study of "higher things",' I replied. 'Perhaps you think that anyone who puts his head back and studies a painted ceiling is using his mind and not his eyes. You may be right, and I may be just simple minded, but I can't believe that the mind is made to look upwards except by studying the ultimate unseen reality. If anyone tries to learn anything about the world of sense whether by gaping upwards or blinking downwards, I don't reckon that the result is *knowledge*—there is no knowledge to be had of such things—nor do I reckon his mind is directed upwards, even if he's lying on his back or floating on the sea.'

'I'm guilty,' he said, 'and deserve to be scolded. But how else do you mean that astronomy ought to be studied if it's to serve our purpose?'

'Like this,' I said. 'The stars in the sky, though we rightly regard them as the finest and most perfect of visible things, are far inferior, just because they are visible, to the true realities; that is, to the movements and bodies in movement whose true relative speeds are to be found in terms of pure numbers and perfect figures, and which are perceptible to reason and thought but not visible to the eye. Do you agree?'

'Yes.'

'Well, then,' I went on, 'we ought to treat the visible splendours of the sky as illustrations to our study of the true realities, just as one might treat a wonderful and carefully drawn design by Daedalus or any other artist or draughtsman. Anyone who knew anything about geometry, and saw such a design, would admire the skill with which it was done, but would think it absurd to study it in the serious hope of learning the truth about proportions such as double or half.'

'It would be absurd to hope for that,' he agreed.

'Isn't the true astronomer in the same position when he watches the movements of the stars?' I asked. 'He will think that the sky and the heavenly bodies have been put together by their maker as well as such things can be; but he will also think it absurd to suppose that there is anything constant or invariable about the relation of day to night, or of day and night to month, or month to year, or, again, of the periods of the other stars to them and to each other. They are all visible and material, and it's absurd to look for exact truth in them.'

'I agree now you put it like that,' he said.

'We shall therefore treat astronomy, like geometry, as setting us problems for solution,' I said, 'and ignore the visible heavens, if we want to make a genuine study of the subject and use it to put the mind's native wit to a useful purpose.'

'You are demanding a lot more work than astronomy at present involves,' he said.

'We shall make other demands like it, I think, if we are to be any use as lawgivers. But,' I asked, 'can you think of any other suitable study?'

'Not at the moment.'

V Harmonics

'All the same, there are several species of motion,' I said. 'I suppose that an expert could enumerate them all; but even I can distinguish two of them.'

'What are they?'

'The one we've been talking about and its counterpart.'

'What's that?'

'I think we may say that, just as our eyes are made for astronomy, so our ears are made for harmony, and that the two are, as the Pythagoreans say, and as we should agree, sister sciences. Isn't that so?'

'Yes.'

'And as the work involved is considerable we will consult them on the subject, and perhaps on others too. But all through we must maintain the principle we laid down when dealing with astronomy, that our pupils must not leave their studies incomplete or stop short of the final objective. They can do this just as much in harmonics as they could in astronomy, by wasting their time on measuring audible concords and notes.'

'Lord, yes, and pretty silly they look,' he said. 'They talk about "intervals" of sound, and listen as carefully as if they were trying to hear a conversation next door. And some say they can distinguish a note between two others, which gives them a minimum unit of measurement, while others maintain that there's no difference between the notes in question. They are all using their ears instead of their minds.'

'You mean those people who torment catgut, and try to wring the truth out of it by twisting it on pegs. I might continue the metaphor and talk about strokes of the bow, and accusations against the strings and their shameless denials—but I'll drop it, because I'm not thinking so much of these people as of the Pythagoreans, who we said would tell us about harmonics. For they do just what the astronomers do; they look for numerical relationships in audible concords, and never get as far as formulating problems and asking which numerical relations are concordant and why.'

'But that would be a fearsome job,' he protested.

'A useful one, none the less,' I said, 'when the object is to discover what is right and good; though not otherwise.'

'That may well be.'

'Yes,' I said, 'for it's only if we can pursue all these studies until we see their kinship and common ground, and can work out their relationship, that they contribute to our purpose and are worth the trouble we spend on them.'

'So I should imagine. But it means a great deal of work.'

2.E3 Theaetetus investigates incommensurability

THEAETETUS: It [to say what knowledge is] looks easy now, Socrates, when you put it like that. There's a point that came up in a discussion I was having recently with your namesake, Socrates here; it rather seems that what you're asking for is something of the same sort.

SOCRATES: What sort of point was it, Theaetetus?

THEAETETUS: Theodorus here was drawing diagrams to show us something about powers—namely that a square of three square feet and one of five square feet aren't commensurable, in respect of length of side, with a square of one square foot; and so on, selecting each case individually, up to seventeen square feet. At that point he somehow got tied up. Well, since the powers seemed to be unlimited in number, it occurred to us to do something on these lines: to try to collect the powers under one term by which we could refer to them all.

SOCRATES: And did you find something like that?

THEAETETUS: I think so; but you must look into it too.

SOCRATES: Tell me about it.

THEAETETUS: We divided all the numbers into two sorts. If a number can be obtained by multiplying some number by itself, we compared it to what's square in shape, and called it square and equal-sided.

SOCRATES: Good.

THEAETETUS: But if a number comes in between—these include three and five, and in fact any number which can't be obtained by multiplying a number by itself, but is obtained by multiplying a larger number by a smaller or a smaller by a larger, so that the sides containing it are always longer and shorter—we compared it to an oblong shape, and called it an oblong number.

SOCRATES: Splendid. But what next?

THEAETETUS: We defined all the lines that square off equal-sided numbers on plane surfaces as lengths, and all the lines that square off oblong numbers as powers, since they aren't commensurable with the first sort in length, but only in respect of the plane figures which they have the power to form. And there's another point like this one in the case of solids.

SOCRATES: That's absolutely excellent, boys. I don't think Theodorus is going to be up on a charge of perjury.

THEAETETUS: Still, Socrates, I wouldn't be able to answer your question about knowledge in the way we managed with lengths and powers. But it seems to me to be something of that sort that you're looking for. So Theodorus does, after all, turn out to have said something false.

SOCRATES: But look here, suppose he'd praised you for running, and said he'd never come across a young man who was so good at it; and then you'd run a race and been beaten by the fastest starter, a man in his prime. Do you think his praise would have been any less true?

THEAETETUS: No.

SOCRATES: And what about knowledge? Do you think it's a small matter to seek it out, as I was saying just now—not one of those tasks which are arduous in every way?

THEAETETUS: Good heavens, no: I think it's really one of the most arduous of tasks.

SOCRATES: Well then, don't lose heart about yourself, and accept that there was something in what Theodorus said. Always do your best in every way; and as for knowledge, do your best to get hold of an account of what, exactly, it really is.

THEAETETUS: If doing my best can make it happen, Socrates, it will come clear.

SOCRATES: Come on, then—because you've just sketched out the way beautifully—try to imitate your answer about the powers. Just as you collected them, many as they are, in one class, try, in the same way, to find one account by which to speak of the many kinds of knowledge.

2.E4 Lower and higher mathematics

SOCRATES: Now we may, I think, divide the knowledge involved in our studies into technical knowledge, and that concerned with education and culture; may we not?

PROTARCHUS: Yes.

SOCRATES: Then taking the technical knowledge employed in handicraft, let us first consider whether one division is more closely concerned with knowledge, and the other less so, so that we are justified in regarding the first kind as the purest, and the second as relatively impure.

PROTARCHUS: Yes, we ought so to regard them.

SOCRATES: Should we then mark off the superior types of knowledge in the several crafts?

PROTARCHUS: How so? Which do you mean?

SOCRATES: If, for instance, from any craft you subtract the element of numbering, measuring, and weighing, the remainder will be almost negligible.

PROTARCHUS: Negligible indeed.

SOCRATES: For after doing so, what you would have left would be guesswork and the exercise of your senses on a basis of experience and rule of thumb, involving the use of that ability to make lucky shots which is commonly accorded the title of art or craft, when it has consolidated its position by dint of industrious practice.

PROTARCHUS: I have not the least doubt you are right.

SOCRATES: Well now, we find plenty of it, to take one instance, in music when it adjusts its concords not by measurement but by lucky shots of a practised finger; in the whole of music, flute-playing and lyre-playing alike, for this latter hunts for the proper length of each string as it gives its note, making a shot for the note, and attaining a most unreliable result with a large element of uncertainty.

PROTARCHUS: Very true.

SOCRATES: Then again we shall find the same sort of thing in medicine and agriculture and navigation and military science.

PROTARCHUS: Quite so.

SOCRATES: Building, however, makes a considerable use of measures and instruments, and the remarkable exactness thus attained makes it more scientific than most sorts of knowledge.

PROTARCHUS: In what respect?

SOCRATES: I am thinking of the building of ships and houses, and various other uses to which timber is put. It employs straight-edge and peg-and-cord, I believe, and compasses and plummet, and an ingenious kind of set-square.

PROTARCHUS: You are perfectly, right, Socrates.

SOCRATES: Let us then divide the arts and crafts so-called into two classes, those akin to music in their activities and those akin to carpentry, the two classes being marked by a lesser and a greater degree of exactness respectively.

PROTARCHUS: So be it.

SOCRATES: And let us take those arts, which just now we spoke of as primary, to be the most exact of all.

PROTARCHUS: I take it you mean the art of numbering, and the others which you mentioned in association with it just now.

SOCRATES: To be sure. But ought we not, Protarchus, to recognize these themselves to be of two kinds? What do you think?

PROTARCHUS: What two kinds do you mean?

SOCRATES: To take first numbering or arithmetic, ought we not to distinguish between that of the ordinary man and that of the philosopher?

PROTARCHUS: On what principle, may I ask, is this discrimination of two arithmetics to be based?

SOCRATES: There is an important mark of difference, Protarchus. The ordinary arithmetician, surely, operates with unequal units: his 'two' may be two armies or two cows or two anythings from the smallest thing in the world to the biggest; while the philosopher will have nothing to do with him, unless he consents to make every single instance of his unit precisely equal to every other of its infinite number of instances.

PROTARCHUS: Certainly you are right in speaking of an important distinction amongst those who concern themselves with number, which justifies the belief that there are two arithmetics.

SOCRATES: Then as between the calculating and measurement employed in building or commerce and the geometry and calculation practised in philosophy—well, should we say there is one sort of each, or should be recognize two sorts?

PROTARCHUS: On the strength of what has been said I should give my vote for there being two.

SOCRATES: Right. Now do you realize our purpose in bringing these matters on to the board?

PROTARCHUS: Possibly, but I should like you to pronounce on the point.

SOCRATES: Well, it seems to me that our discussion, now no less than when we embarked upon it, has propounded a question here analogous to the question about pleasures: it is enquiring whether one kind of knowledge is purer than another, just as one pleasure is purer than another.

PROTARCHUS: Yes, it is quite clear that that has been its reason for attacking this matter.

SOCRATES: Well now, in what preceded had it not discovered that different arts, dealing with different things, possessed different degrees of precision?

PROTARCHUS: Certainly.

SOCRATES: And in what followed did it not first mention a certain art under one single name, making us think it really was one art, and then treat it as two, putting questions about the precision and purity of those two to find out whether the art as practised by the philosopher or by the non-philosopher was the more exact?

PROTARCHUS: I certainly think that is the question which it puts.

SOCRATES: Then, Protarchus, what answer do we give it?

PROTARCHUS: We have got far enough, Socrates, to discern an astonishingly big difference between one kind of knowledge and another in respect of precision.

SOCRATES: Well, will that make it easier for us to answer?

PROTARCHUS: Of course; and let our statement be that the arts which we have had before us are superior to all others, and that those amongst them which involve the effort of the true philosopher are, in their use of measure and number, immensely superior in point of exactness and truth.

SOCRATES: Let it be as you put it.

2.E5 Plato's cosmology

(a) *The four elements*

Now anything that has come to be must be corporeal, visible, and tangible: but nothing can be visible without fire, nor tangible without solidity, and nothing can be solid without earth. So god, when he began to put together the body of the universe, made it of fire and earth. But it is not possible to combine two things properly without a third to act as a bond to hold them together. And the best bond is one that effects the closest unity between itself and the terms it is combining; and this is best done by a continued geometrical proportion. For whenever you have three cube or square numbers with a middle term such that the first term is to it as it is to the third term, and conversely what the third term is to the mean the mean is to the first term, then since the middle becomes first and last and similarly the first and last become middle, it will follow necessarily that all can stand in the same relation to each other, and in so doing achieve unity together. If then the body of the universe were required to be a plane surface with no depth, one middle term would have been enough to connect it with the other terms, but in fact it needs to be solid, and solids always need two connecting middle terms. So god placed water and air between fire and earth, and made them so far as possible proportional to one another, so that air is to water as water is to earth;

and in this way he bound the world into a visible and tangible whole. So by these means and from these four constituents the body of the universe was created to be a unity owing to proportion; in consequence it acquired concord, so that having once come together in unity with itself it is indissoluble by any but its compounder.

(b) *Construction and allocation of the 'Platonic solids' to the elements*

In the first place it is clear to everyone that fire, earth, water, and air are bodies, and all bodies are solids. All solids again are bounded by surfaces, and all rectilinear surfaces are composed of triangles. There are two basic types of triangle, each having one right angle and two acute angles: in one of them these two angles are both half right angles, being subtended by equal sides, in the other they are unequal, being subtended by unequal sides. This we postulate as the origin of fire and the other bodies, our argument combining likelihood and necessity; their more ultimate origins are known to god and to men whom god loves. We must proceed to enquire what are the four most perfect possible bodies which, though unlike one another, are some of them capable of transformation into each other on resolution. If we can find the answer to this question we have the truth about the origin of earth and fire and the two mean terms between them; for we will never admit that there are more perfect visible bodies than these, each in its type. So we must do our best to construct four types of perfect body and maintain that we have grasped their nature sufficiently for our purpose. Of the two basic triangles, then, the isosceles has only one variety, the scalene an infinite number. We must therefore choose, if we are to start according to our own principles, the most perfect of this infinite number. If anyone can tell us of a better choice of triangle for the construction of the four bodies, his criticism will be welcome; but for our part we propose to pass over all the rest and pick on a single type, that of which a pair compose an equilateral triangle. It would be too long a story to give the reason, but if anyone can produce a proof that it is not so we will welcome his achievement. So let us assume that these are the two triangles from which fire and the other bodies are constructed, one isosceles and the other having a greater side whose square is three times that of the lesser. [. . .] We must next describe what geometrical figure each body has and what is the number of its components. We will begin with the construction of the simplest and smallest figure. Its basic unit is the triangle whose hypotenuse is twice the length of its shorter side. If two of these are put together with the hypotenuse as diameter of the resulting figure, and if the process is repeated three times and the diameters and shorter sides of the three figures are made to coincide in the same vertex, the result is a single equilateral triangle composed of six basic units. And if four equilateral triangles are put together, three of their plane angles meet to form a single solid angle, the one which comes next after the most obtuse of plane angles: and when four such angles have been formed the result is the simplest solid figure, which divides the surface of the sphere circumscribing it into equal and similar parts.

The second figure is composed of the same basic triangles put together to form eight equilateral triangles, which yield a single solid angle from four planes. The formation of six such solid angles completes the second figure.

The third figure is put together from one hundred and twenty basic triangles, and has twelve solid angles, each bounded by five equilateral plane triangles, and twenty faces, each of which is an equilateral triangle.

After the production of these three figures the first of our basic units is dispensed with, and the isosceles triangle is used to produce the fourth body. Four such triangles are put together with their right angles meeting at a common vertex to form a square. Six squares fitted together complete eight solid angles, each composed by three plane right angles. The figure of the resulting body is the cube, having six plane square faces.

There still remained a fifth construction, which the god used for arranging the constellations on the whole heaven. [...]

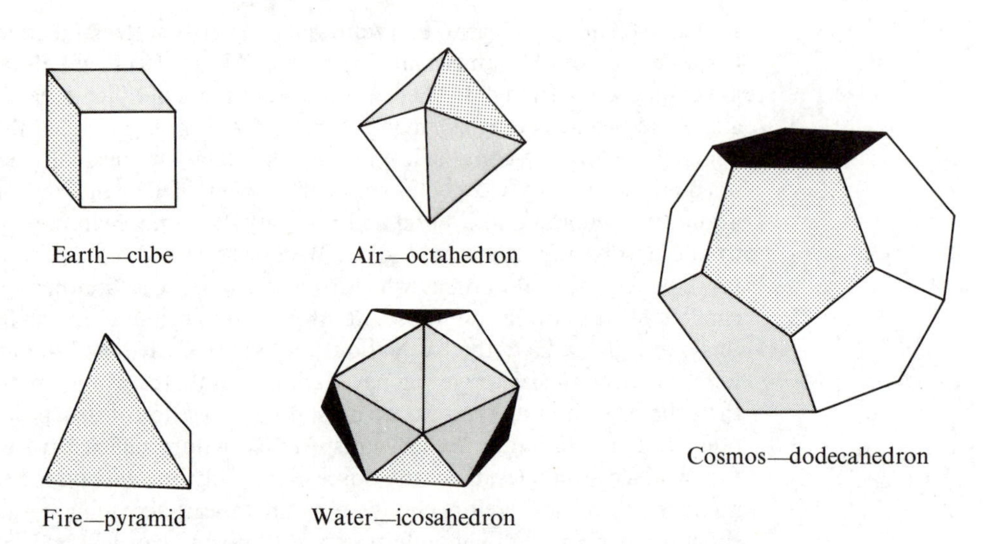

We must proceed to distribute the figures whose origins we have just described between fire, earth, water, and air. Let us assign the cube to earth; for it is the most immobile of the four bodies and the most retentive of shape, and these are characteristics that must belong to the figure with the most stable faces. And of the basic triangles we have assumed, the isosceles has a naturally more stable base than the scalene, and of the equilateral figures composed of them the square is, in whole and in part, a firmer base than the equilateral triangle. So we maintain our principle of likelihood by assigning it to earth, while similarly we assign the least mobile of the other figures to water, the most mobile to fire, and the intermediate to air. And again we assign the smallest figure to fire, the largest to water, the intermediate to air; the sharpest to fire, the next sharpest to air, and the least sharp to water. So to sum up, the figure which has the fewest faces must in the nature of things be the most mobile, as well as the sharpest and most penetrating, and finally, being composed of the smallest number of similar parts, the lightest. Our second figure will be second in all these respects, our third will be third. Logic and likelihood thus both require us to regard the pyramid as the solid figure that is the basic unit or seed of fire; and we may regard the second of the figures we constructed as the basic unit of air, the third of water. We must, of course, think of the individual units of all four bodies as being far too small to be visible, and only becoming visible when massed together in large numbers; and we must assume that the god duly adjusted the proportions between their numbers, their movements, and their other qualities and brought them in every way to the exactest perfection permitted by the willing consent of necessity.

(*c*) *Construction of the world-soul*

God did not of course contrive the soul later than the body, as it has appeared in the narrative we are giving; for when he put them together he would never have allowed the older to be controlled by the younger. Our narrative is bound to reflect much of our own contingent and accidental state. But god created the soul before the body and gave it precedence both in time and value, and made it the dominating and controlling partner. And he composed it in the following way and out of the following constituents. From the indivisible, eternally unchanging Existence and the divisible, changing Existence of the physical world he mixed a third kind of Existence intermediate between them: again with the Same and the Different he made, in the same way, compounds intermediate between their indivisible element and their physical and divisible element: and taking these three components he mixed them into single unity, forcing the Different, which was by nature allergic to mixture, into union with the Same, and mixing both with Existence. Having thus made a single whole of these three, he went on to make appropriate subdivisions, each containing a mixture of Same and Different and Existence. He began the division as follows. He first marked off a section of the whole, and then another twice the size of the first; next a third, half as much again as the second and three times the first, a fourth twice the size of the second, a fifth three times the third, a sixth eight times the first, a seventh twenty-seven times the first. Next he filled in the double and treble intervals by cutting off further sections and inserting them in the gaps, so that there were two mean terms in each interval, one exceeding one extreme and being exceeded by the other by the same fraction of the extremes, the other exceeding and being exceeded by the same numerical amount. These links produced intervals of $\frac{3}{2}$ and $\frac{4}{3}$ and $\frac{9}{8}$ within the previous intervals, and he went on to fill all intervals of $\frac{4}{3}$ with the interval $\frac{9}{8}$; this left, as a remainder in each, an interval whose terms bore the numerical ratio of 256 to 243. And at that stage the mixture from which these sections were being cut was all used up.

He then took the whole fabric and cut it down the middle into two strips, which he placed crosswise at their middle points to form a shape like the letter X; he then bent the ends round in a circle and fastened them to each other opposite the point at which the strips crossed, to make two circles, one inner and one outer. And he endowed them with uniform motion in the same place, and named the movement of the outer circle after the nature of the Same, of the inner after the nature of the Different. The circle of the Same he caused to revolve from left to right, and the circle of the Different from right to left on an axis inclined to it; and made the master revolution that of the Same. For he left the circle of the Same whole and undivided, but slit the inner circle six times to make seven unequal circles, whose intervals were double or triple, three of each; and he made these circles revolve in contrary senses relative to each other, three of them at a similar speed, and four at speeds different from each other and from that of the first three but related proportionately.

And when the whole structure of the soul had been finished to the liking of its framer, he proceeded to fashion the whole corporeal world within it, fitting the two together centre to centre: and the soul was woven right through from the centre to the outermost heaven, which it enveloped from the outside and, revolving on itself, provided a divine source of unending and rational life for all time. The body of the heaven is visible, but the soul invisible and endowed with reason and harmony, being the best creation of the best of intelligible and eternal things.

2.E6 Freeborn studies and Athenian ignorance

ATHENIAN STRANGER: Then there are, of course, still three subjects for the freeborn to study. Calculations and the theory of numbers form one subject; the measurement of length and surface and depth make a second; and the third is the true relation of the movement of the stars one to another. To pursue all these studies thoroughly and with accuracy is a task not for the masses but for a select few—who these should be we shall say later towards the end of our argument, where it would be appropriate—for the multitude it will be proper to learn so much of these studies as is necessary and so much as it can rightly be described a disgrace for the masses not to know, even though it would be hard, or altogether impossible, to pursue with precision all of those studies. [...]

Well, then, the freeborn ought to learn as much of these things as a vast multitude of boys in Egypt learn along with their letters. First there should be calculations of a simple type devised for boys, which they should learn with amusement and pleasure, such as distributions of apples and crowns wherein the same numbers are divided among more or fewer, or distributions of the competitors in boxing and wrestling matches by the method of byes and drawings, or by taking them in consecutive order, or in any of the usual ways. Again, the boys should play with bowls containing gold, bronze, silver and the like mixed together, or the bowls may be distributed as wholes. For, as I was saying, to incorporate in the pupils' play the elementary applications of arithmetic will be of advantage to them later in the disposition of armies, in marches and in campaigns, as well as in household management, and will make them altogether more useful to themselves and more awake. After these things there should be measurements of objects having length, breadth and depth, whereby they would free themselves from that ridiculous and shameful ignorance on all these topics which is the natural condition of all men.

CLEINIAS: And in what, pray, does this ignorance consist?

ATHENIAN STRANGER: My dear Cleinias, when I heard, somewhat belatedly, of our condition in this matter, I also was astonished; such ignorance seemed to me worthy, not of human beings, but of swinish creatures, and I felt ashamed, not for myself alone, but for all the Greeks.

CLEINIAS: Why? Please explain, sir, what you are saying.

ATHENIAN STRANGER: I will indeed do so; or rather I will make it plain to you by asking questions. Pray, answer me one little thing; you know what is meant by *line*?

CLEINIAS: Of course.

ATHENIAN STRANGER: And again by *surface*?

CLEINIAS: Certainly.

ATHENIAN STRANGER: And you know that these are two distinct things, and that *volume* is a third distinct from them?

CLEINIAS: Even so.

ATHENIAN STRANGER: Now does not it appear to you that they are all commensurable one with another?

CLEINIAS: Yes.

ATHENIAN STRANGER: I mean, that line is in its nature measurable by line, and surface by surface, and similarly with volume.

CLEINIAS: Most assuredly.

ATHENIAN STRANGER: But suppose this cannot be said of some of them, neither with more assurance nor with less, but is in some cases true, in others not, and suppose you think it true in all cases; what do you think of your state of mind in this matter?

CLEINIAS: Clearly, that it is unsatisfactory.

ATHENIAN STRANGER: Again, what of the relations of line and surface to volume or of surface and line one to another; do not all we Greeks imagine that they are commensurable in some way or other?

CLEINIAS: We do indeed.

ATHENIAN STRANGER: Then if this is absolutely impossible, though all we Greeks, as I was saying, imagine it possible, are we not bound to blush for them all as we say to them, 'Worthy Greeks, this is one of the things of which we said that ignorance is a disgrace and that to know such necessary matters is no great achievement'?

CLEINIAS: Certainly.

ATHENIAN STRANGER: In addition to these, there are other related points, which often give rise to errors akin to those lately mentioned.

CLEINIAS: What kind of errors do you mean?

ATHENIAN STRANGER: The real nature of commensurables and incommensurables towards one another. A man must be able to distinguish them on examination, or must be a very poor creature. We should continually put such problems to each other—it would be a much more elegant occupation for old people than draughts—and give our love of victory an outlet in pastimes worthy of us.

CLEINIAS: Perhaps so; it would seem that draughts and these studies are not so widely separated.

2.E7 A letter from Plato to Dionysius

Be this my introduction and at the same time a token for you that the letter is from me. Once when you were entertaining the young men of Locri, you occupied a couch a good way from mine. You then rose and came to me with words of greeting that were excellent. I thought so at least and my neighbour at table too, who thereupon—he was one of the cultured circle—put the question: 'I suppose, Dionysius, Plato is a great help to you in your studies?' You replied: 'In much else too, for from the moment that I sent for him, the very fact that I had so sent was at once helpful to me.' Here then is something that we must keep alive. We must see to it that we continue to be more and more helpful to each other. So I am doing my part now to effect this by sending you herewith some Pythagorean treatises and some classifications. I am also sending you a man, as we agreed at the time, who will perhaps be useful to you and Archytas—that is, if Archytas has come to Syracuse. His name is Helicon; he is a native of Cyzicus, a pupil of Eudoxus and well versed in all his teaching. He has also studied under a pupil of Isocrates and under Polyxenus, an associate of Bryson. With all this he has the quality rarely combined with this of possessing social grace and he seems not to be ill-natured. In fact he would impress one rather as being full of fun and good-natured. I say this, however, with misgivings; because I am expressing an opinion about a man, and man, while no mean animal, is a changeable one, with a very few exceptions in a few matters. For even in his case my caution and mistrust led me to make investigations, meeting him personally and inquiring of his fellow-citizens, and no one said anything against the man. But be cautious and test him yourself. By all means, however, if you have the least bit of time for it, take lessons of him, in addition to the rest of your philosophic training. If that is impossible, have someone else thoroughly instructed, so that, when you have time to study, you may do so and not only benefit yourself but add to your reputation; and so I shall go on being constantly helpful to you. So much for these matters.

2.E8 Aristoxenus on Plato's lecture on the Good

It is perhaps well to go through in advance the nature of our inquiry, so that, knowing beforehand the road along which we have to travel, we may have an easier journey, because we will know at what stage we are in, nor shall we harbour to ourselves a false conception of our subject. Such was the condition, as Aristotle often used to tell, of most of the audience who attended Plato's lecture on the Good. Every one went there expecting that he would be put in the way of getting one or other of the things accounted good in human life, such as riches or health or strength or, in fine, any extraordinary gift of fortune. But when they found that Plato's arguments were of mathematics and numbers and geometry and astronomy and that in the end he declared the One to be the Good, they were altogether taken by surprise. The result was that some of them scoffed at the thing, while others found great fault with it.

2.F Doubling the Cube

The problem of constructing a cube double the volume of a given cube seems to have attracted early attention as a problem not readily solvable by the circle-and-line methods of elementary geometry. Two interesting but different accounts are given by later commentators of how the problem originated, each attributing the source of the information to Eratosthenes, the Alexandrian scholar and polymath who lived in the third century BC. The version by Theon (2.F1) is probably more accurate as to what Eratosthenes said, for it has been argued that the letter quoted by Eutocius (2.F3) could not have been written by Eratosthenes. The solution to the problem contained in 2.F3 is almost certainly the work of Eratosthenes, though, and the final bit of the letter (in quotation marks) is believed to quote his actual words. This is one of no fewer than twelve solutions described by Eutocius (sixth century AD) in his *Commentary on Archimedes' Sphere and Cylinder II.1,* a proposition in which Archimedes (see 4.A6) took as given the construction of two mean proportionals. (That is, given two lines a, b, two mean proportionals are lines x, y such that $a{:}x = x{:}y = y{:}b$.) For from an early time the problem of doubling the cube was always solved in its more amenable reduced form of constructing two mean proportionals. It may have been Hippocrates (see 2.B) who first showed that these two problems were related since both Eutocius's account of Eratosthenes (2.F3) and Proclus (2.F2) concur in this. 2.F4 is another of the solutions given by Eutocius, attributed to Menaechmus (mid fourth century BC). This is an especially interesting solution, being apparently the first recorded use of conic sections in Greek mathematics; from which, together with the reference to him by Eratosthenes (towards the end of 2.F3), it is commonly taken that it was Menaechmus who first discovered conic sections and in the context of the present problem. Doubling the cube is one of the 'three classical problems', the other two being trisecting the angle and squaring the circle (see 2.G). Only in the nineteenth century was it finally *proved* that none of these could be solved by circle-and-line constructions.

2.F1 Theon on how the problem (may have) originated

In his work entitled *Platonicus* Eratosthenes says that, when the god announced to the Delians by oracle that to get rid of a plague they must construct an altar double of the existing one, their craftsmen fell into great perplexity in trying to find how a solid could be made double of another solid, and they went to ask Plato about it. He told them that the god had given this oracle, not because he wanted an altar of double the size, but because he wished, in setting this task before them, to reproach the Greeks for their neglect of mathematics and their contempt for geometry.

2.F2 Proclus on its reduction by Hippocrates

'Reduction' is a transition from a problem or a theorem to another which, if known or constructed, will make the original proposition evident. For example, to solve the problem of doubling the cube geometers shifted their inquiry to another on which this depends, namely, the finding of two mean proportionals; and thenceforth they devoted their efforts to discovering how to find two means in continuous proportion between two given straight lines. They say that the first to effect reduction of difficult constructions was Hippocrates of Chios, who also squared the lune and made many other discoveries in geometry, being a man of genius when it came to constructions, if there ever was one.

2.F3 Eutocius's account of its early history, and an instrumental solution

Eratosthenes to King Ptolemy, greetings.

The story goes that one of the ancient tragic poets represented Minos having a tomb built for Glaucus, and that when Minos found that the tomb measured a hundred feet on every side, he said: 'Too small is the tomb you have marked out as the royal resting place. Let it be twice as large. Without spoiling the form quickly double each side of the tomb.'

This was clearly a mistake. For if the sides are doubled the surface is multiplied fourfold and the volume eightfold.

Now geometers, too, sought to find a way to double the given solid without altering its form. This problem came to be known as the duplication of the cube, for, given a cube, they sought to double it. Now when all had sought in vain for a long time, Hippocrates of Chios first discovered that if a way can be found to construct two mean proportionals in continued proportion between two given straight lines, the greater of which is double the lesser, the cube will be doubled. So that his difficulty was resolved into another no less perplexing.

Some time later certain Delians, they say, seeking by order of the oracle to double an altar, fell into the same difficulty. And so they sent representatives to ask the geometers of Plato's school in the Academy to find the solution for them. These geometers zealously tackled the problem of finding two mean proportionals between two given lines. And Archytas of Tarentum is said to have obtained a solution with semicylinders, while Eudoxus used so-called 'curved lines'. All who solved the problem succeeded in finding the deductive proof, but they were not able to demonstrate the construction in a practical and useful way, with the exception of Menaechmus (though he accomplished this only to a very small degree and with difficulty).

Now I have discovered an easy method of finding, by the use of an instrument, not only two but as many mean proportionals as desired between two given lines. With this discovery we shall be able to convert into a cube any given solid whose surfaces are parallelograms, or to change it from one form to another, and, again, to construct a solid of the same form as the given solid but larger, i.e., preserving the similarity. And we shall also be able to apply this in constructing altars and temples. We shall be able, furthermore, to convert our liquid and dry measures, the metretes and the medimnus, into a cube, and from the side of this cube to measure the capacity of other vessels in terms of these measures. My method will also be useful for those who wish to increase the size of catapults and ballistas. For, if the throw is to be increased, all the elements of these engines, the thicknesses, lengths, and the sizes of the openings, wheel casings, and cables must be increased in proportion. But this cannot be done without finding the mean proportionals. I have described below for you the demonstration and the method of construction of my device.

[...] To find the two mean proportionals by an instrument, construct a frame of wood, ivory, or bronze having three equal flat surfaces as thin as possible. Let the middle surface be fixed, and the other two move along grooves; the size and shape of the surfaces may vary as desired. For the proof is not affected.

In order that the required lines may be obtained more accurately, care must be taken that when the surfaces are brought together all parts remain parallel, and fit one another snugly without gaps.

The instrument in bronze is placed on the votive monument beneath the crown of the column and is held fast with lead. Under the instrument the proof is set down concisely with a diagram and after this an epigram. These have been copied below for you, so that you may have them just as they are on the votive column. Of the two figures the second is engraved on the column.

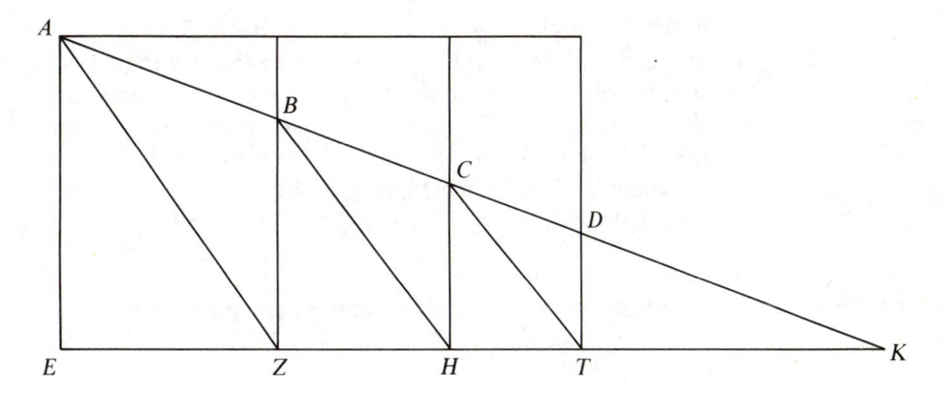

'Given two straight lines, to find two mean proportionals in continued proportion. Let AE and DT be given. I draw together the surfaces in the instrument until points A, B, C, and D are all in a straight line. Consider these points as they are in the diagram: $AK:KB = EK:KZ$ (since AE and BZ are parallel), and $AK:KB = ZK:KH$ (since AZ and BH are parallel).

$\therefore EK:KZ = KZ:KH = AE:BZ = BZ:CH$.

Similarly we shall be able to show that $ZB:CH = CH:DT$.

$\therefore AE, BZ, CH$, and DT are in continued proportion.

That is, two mean proportionals between the two given lines have been found.

Now if the given lines are not equal to AE and DT, by taking AE and DT proportional to the given lines we shall obtain the means between AE and DT and then transfer the results to the given lines. Thus we shall have done what was required. And if it is required to find more mean proportionals, we shall achieve our purpose in each case by constructing one more surface on the instrument than the number of means required. The proof is the same in this case.

If, my friend, you seek to make from a small cube a cube twice as large, and readily convert any solid form into another, here is your instrument. You can, then, measure a fold, or a grain pit, or the broad hollow of a well, if between two rulers you find means the extreme ends of which converge. Do not seek the cumbersome procedure with Archytas' cylinders, or to make the three Menaechmian sections of the cone; seek not the type of curved line described by god-fearing Eudoxus. For with these plates of mine you could readily construct ten thousand means beginning with a small base.

You are a happy father, Ptolemy, because you enjoy youth with your son and have yourself given him all that is precious to the Muses and to kings. May he hereafter, heavenly Zeus, receive the sceptre from your hand: so may it come to pass. And let whoever sees this votive column say: "This is an offering of Eratosthenes of Cyrene".'

2.F4 Eutocius on Menaechmus's use of conic sections

Let the two given straight lines be A, E; it is required to find two mean proportionals between A, E.

Assume it done, and let the means be B, C, and let there be placed in position a straight line DG, with an end point D, and at D let DF be placed equal to C, and let FH

be drawn at right angles and let FH be equal to B. Since the three straight lines A, B, C are in proportion, $A \cdot C = B^2$; therefore the rectangle comprehended by the given straight line A and the straight line C, that is, DF, is equal to the square on B, that is, to the square on FH. Therefore H is on a parabola drawn through D. Let the parallels HK, DK be drawn. Then since the rectangle $B \cdot C$ is given—for it is equal to the rectangle $A \cdot E$—the rectangle $KH \cdot HF$ is given. The point H is therefore on a hyperbola with asymptotes KD, DF. Therefore H is given; and so also is F.

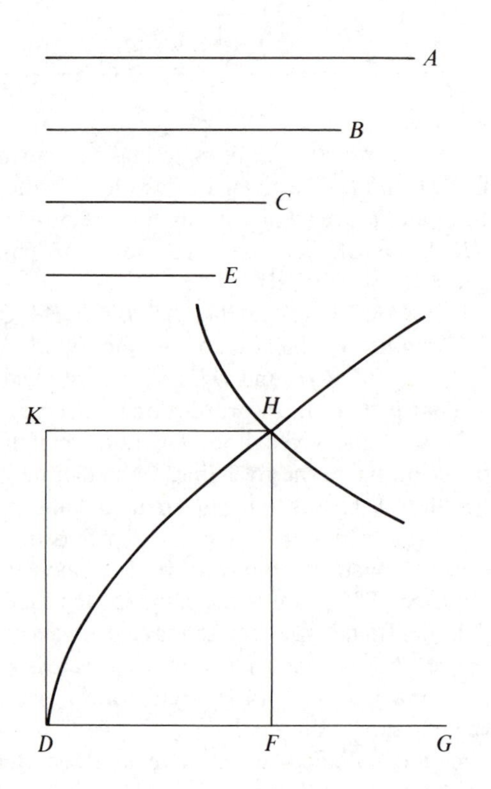

Let the synthesis be made in this manner. Let the given straight lines be A, E, let DG be a straight line given in position with an end point at D, and let there be drawn through D a parabola whose axis is DG, and *latus rectum* A, and let the squares of the ordinates drawn at right angles to DG be equal to the areas applied to A having as their sides the straight lines cut off by them towards D. Let it be drawn, and let it be DH, and let DK be perpendicular to DG, and in the asymptotes KD, DF let there be drawn a hyperbola, such that the straight lines drawn parallel to KD, DF will make an area equal to the rectangle comprehended by A, E. It will then cut the parabola. Let it cut at H, and let HK, HF be drawn perpendicular. Since then $FH^2 = A \cdot DF$, it follows that $A:FH = HF:FD$. Again, since $A \cdot E = HF \cdot FD$, it follows that $A:FH = FD:E$. But $A:FH = FH:FD$. Therefore $A:FH = FH:FD = FD:E$. Let B be placed equal to HF, and C equal to DF. It follows that $A:B = B:C = C:E$. A, B, C, E are therefore in continuous proportion: which was to be found.

2.G Squaring the Circle

The problem of constructing a square equal in area to a circle was sufficiently well known towards the end of the fifth century BC to be satirised in Aristophanes' *The Birds*, of 414 BC (see 2.C2). Two early attempts on the problem of about that period were those of Antiphon (2.G2) and Bryson (2.G3), both known to us only from brief allusions in Aristotle which were expanded on by his commentators. Whatever the details of the arguments may have been, Antiphon and Bryson seem to have tackled the problem directly, hoping to construct a polygon equal to the circle. Later approaches tackled the problem in its reduced form of finding a straight line equal to the circle's circumference. (The equivalence of these two problems, squaring the circle and rectifying its circumference, was shown by Archimedes—see 4.A1.) One of the curves used for this purpose was the quadratrix (2.G4), originally invented by Hippias (c. 400 BC) for solving the third classical problem of trisecting the angle. The use of the quadratrix in squaring the circle is logically problematic, as Pappus points out in 2.G4.

2.G1 Proclus on the origin of the problem

It is my opinion that this problem [Euclid's *Elements*, I.45] is what led the ancients to attempt the squaring of the circle. For if a parallelogram can be found equal to any rectilinear figure, it is worth inquiring whether it is not possible to prove that a rectilinear figure is equal to a circular area.

2.G2 Antiphon's quadrature

(*a*) *Aristotle*

The exponent of any science is not called upon to solve every kind of difficulty that may be raised, but only such as arise through false deductions from the principles of the science: with others than these he need not concern himself. For example, it is for the geometer to expose the quadrature by means of segments, but it is not the business of the geometer to refute the argument of Antiphon.

(*b*) *Themistius's commentary on Aristotle*

For such false arguments as preserve the geometrical hypotheses are to be refuted by geometry, but such as conflict with them are to be left alone. Examples are given by two men who tried to square the circle, Hippocrates of Chios and Antiphon. The attempt of Hippocrates is to be refuted. For, while preserving the principles, he commits a paralogism by squaring only that lune which is described about the side of the square inscribed in the circle, though including every lune that can be squared in the proof. But the geometer could have nothing to say against Antiphon, who inscribed an equilateral triangle in the circle, and on each of the sides set up another triangle, an isosceles triangle with its vertex on the circumference of the circle, and continued this process, thinking that at some time he would make the side of the last triangle, although a straight line, coincide with the circumference.

(*c*) *Simplicius's commentary on Aristotle*

Antiphon drew a circle and inscribed in it a polygonal figure of the sort that can be inscribed [in a circle]. Let us assume that the inscribed figure is a square. Then dividing each of the sides of the square into two halves, he drew perpendiculars from the points of division to the circumference; these perpendiculars obviously divided into halves the corresponding sections of [the circumference of] the circle. Then he drew straight lines from the points [at which the perpendiculars cut the circumference] to the extremities of the sides of the square, so as to get four triangles based on the straight lines [i.e. the sides]; and the whole [resulting] inscribed figure will now be an octagon. And this process he repeated in the same way, dividing each of the sides of the octagon into halves, drawing perpendiculars from the points of division to the circumference and connecting by straight lines the points at which the perpendiculars met the circumference with the extremities of the divided straight lines [i.e. the sides of the octagon]; the inscribed figure he thus obtained was a polygon with sixteen sides. In the same way cutting the sides of the inscribed sixteensided polygon and drawing straight connecting lines he doubled [the number of the sides of] the inscribed polygon; and repeating the process again and again so that in this way, with the progressive exhaustion of the area [of the circle] a polygon would be inscribed in the circle, the sides of which because of their smallness would coincide with the circumference of the circle. Since we can always draw a square equal to any [given] polygon, as we learn in the *Elements*, we shall in fact be constructing a square equal to a circle, seeing that a polygon coinciding with a circle is equal to it.

It is evident that this conclusion infringes geometrical principles; but not, as Alexander said, 'because the geometer assumes as a principle that the circle touches the straight line at one point, and that is the principle infringed by Antiphon'; for the geometer does not *assume* this but *demonstrates* it, in the third book. It is therefore better to say that the principle is this: that it is impossible for a straight line to coincide with a circumference. A line outside [the circle] will touch it at one point only; a line inside at two points and no more; and contact takes place at a point. The progressive division of the area between the straight line and the circumference of the circle will not exhaust it, nor will one ever reach the circumference, if [as is the case] the area is divisible *ad infinitum*. If one *does* reach the circumference that would infringe the geometrical principle which states that magnitudes are divisible *ad infinitum*. And it is this principle that according to Eudemus was infringed by Antiphon.

2.G3 Bryson's quadrature

(*a*) *Aristotle*

The method by which Bryson tried to square the circle, were it ever so much squared thereby, is yet made sophistical by the fact that it has no relation to the matter in hand. [...] The squaring of the circle by means of lunules is not eristic, but the quadrature of Bryson is eristic; the reasoning used in the former cannot be applied to any subject other than geometry alone, whereas Bryson's argument is directed to the mass of people who do not know what is possible and what is impossible in each department, for it will fit any. And the same is true of Antiphon's quadrature.

(*b*) *Alexander's commentary on Aristotle*

But Bryson's quadrature of the circle is a piece of captious sophistry for it does not proceed on principles proper to geometry but on principles of more general application. He circumscribes a square about the circle, inscribes another within the circle, and constructs a third square between the first two. He then says that the circle between the two squares [i.e., the inscribed and the circumscribed] and also the intermediate square are both smaller than the outer square and larger than the inner, and that things larger and smaller than the same things, respectively, are equal. Therefore, he says, the circle has been squared. But to proceed in this way is to proceed on general and false assumptions, general because these assumptions might be generally applicable to numbers, times, spaces, and other fields, but false because both 8 and 9 are smaller and larger than 10 and 7 respectively, but are not equal to each other.

2.G4 Pappus on the quadratrix

For the squaring of the circle a certain line was used by Dinostratus and Nicomedes and certain other more recent geometers, and it takes its name from its special property; for it is called by them the quadratrix, and it is generated in this way.

Let $ABCD$ be a square, and with centre A let the arc BED be described, and let AB be so moved that the point A remains fixed while B is carried along the arc BED; furthermore let BC, while always remaining parallel to AD, follow the point B in its motion along BA, and in equal times let AB, moving uniformly, pass through the angle BAD (that is, the point B pass along the arc BD), and BC pass by the straight line BA (that is, let the point B traverse the length of BA). Plainly then both AB and BC will coincide simultaneously with the straight line AD. While the motion is in progress the straight lines BC, BA will cut one another in their movement at a certain point which continually changes place with them, and by this point there is described in the space between the straight lines BA, AD and the arc BED a concave curve, such as BFG, which appears to be serviceable for the discovery of a square equal to the given circle.

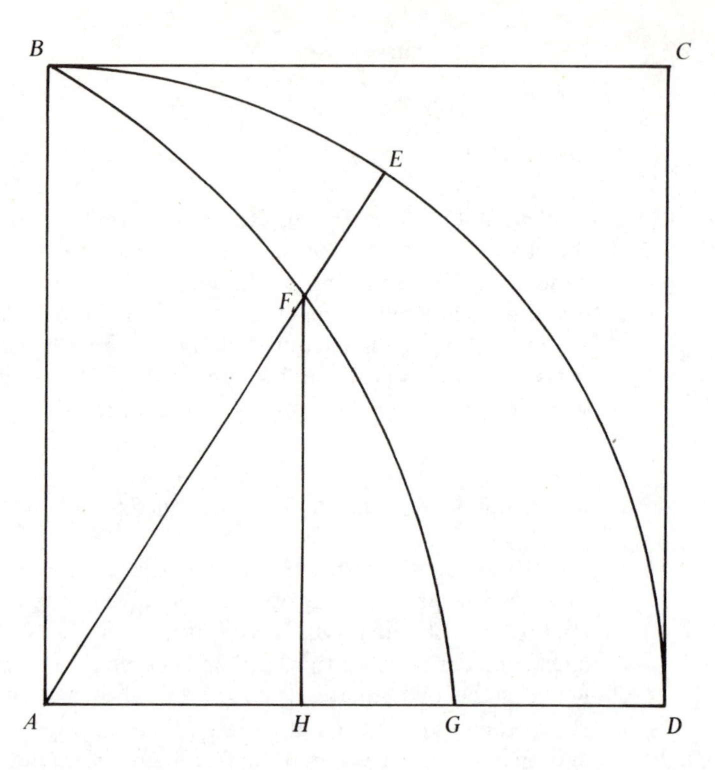

Its principal property is this. If any straight line, such as *AFE*, be drawn to the circumference, the ratio of the whole arc to *ED* will be the same as the ratio of the straight line *BA* to *FH*; for this is clear from the manner in which the line was generated.

With this Sporus is rightly displeased for these reasons. In the first place, the end for which the construction seems to be useful is assumed in the hypothesis. For how is it possible, with two points beginning to move from *B*, to make one of them move along a straight line to *A* and the other along a circumference to *D* in equal time unless first the ratio of the straight line *AB* to the circumference *BED* is known? For it is necessary that the speeds of the moving points should be in this ratio. And how then could one, using unadjusted speeds, make the motions end together, unless this should sometimes happen by chance? But how could this fail to be irrational? Again, the extremity of the curve which they use for the squaring of the circle, that is, the point in which the curve cuts the straight line *AD*, is not found. Let the construction be conceived as aforesaid. When the straight lines *CB*, *BA* move so as to end their motion together, they will coincide with *AD* and will no longer cut each other. In fact, the intersection ceases before the coincidence with *AD*, yet it was this intersection which was the extremity of the curve where it met the straight line *AD*. Unless, indeed, anyone should say the curve is conceived as produced, in the same way that we produce straight lines, as far as *AD*. But this does not follow from the assumptions made; the point *G* can be found only by assuming the ratio of the circumference to the straight line. So unless this ratio is given, we must beware lest, in following the authority of those men who discovered the line, we admit its construction, which is more a matter of mechanics. But first let us deal with that problem which we have said can be proved by means of it.

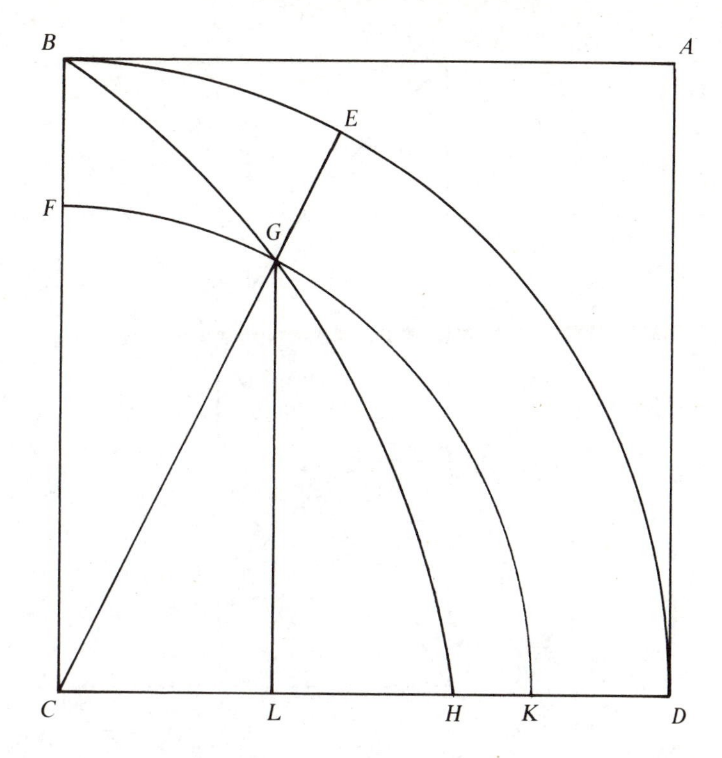

If $ABCD$ is a square and BED the arc of a circle with centre C, while BGH is a quadratrix generated in the aforesaid manner, it is proved that the ratio of the arc DEB towards the straight line BC is the same as that of BC towards the straight line CH. For if it is not, the ratio of the arc DEB towards the straight line BC will be the same as that of BC towards either a straight line greater than CH or a straight line less than CH.

Let it be the former, if possible, towards a greater straight line CK, and with centre C let the arc FGK be drawn cutting the curve at G, and let the perpendicular GL be drawn, and let CG be joined and produced to E. Since therefore the ratio of the arc DEB towards the straight line BC is the same as the ratio of BC, that is CD, towards CK, and the ratio of CD towards CK is the same as that of the arc BED towards the arc FGK (for the arcs of circles are in the same ratio as their diameters), it is clear that the arc FGK is equal to the straight line BC. And since by the property of the curve the ratio of the arc BED towards ED is the same as the ratio of BC towards GL, therefore the ratio of FGK towards the arc GK is the same as the ratio of the straight line BC towards GL. And the arc FGK was proved equal to the straight line BC; therefore the arc GK is also equal to the straight line GL, which is absurd. Therefore the ratio of the arc BED towards the straight line BC is not the same as the ratio of BC towards a straight line greater than CH.

I say that neither is it equal to the ratio of BC towards a straight line less than CH. For, if it is possible, let the ratio be towards KC, and with centre C let the arc FMK be described, and let KG at right angles to CD cut the quadratrix at G, and let CG be joined and produced to E. In similar manner to what has been written above, we shall prove also that the arc FMK is equal to the straight line BC, and that the ratio of the arc BED towards ED, that is, the ratio of FMK towards MK, is the same as that of the

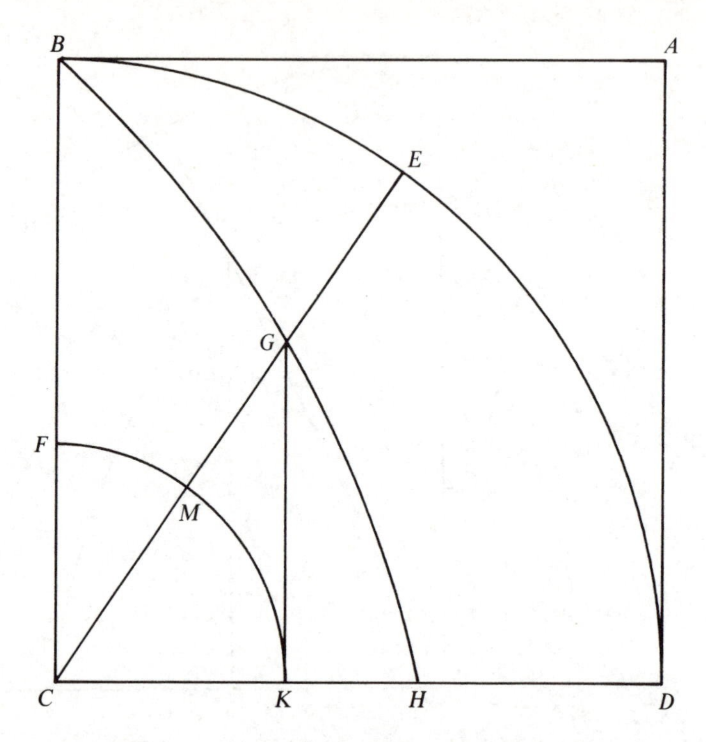

straight line BC towards GK. From this it is clear that the arc MK is equal to the straight line KG, which is absurd. The ratio of the arc BED towards the straight line BC is therefore not the same as the ratio of BC towards a straight line less than CH. Moreover it was proved not the same as the ratio of BC towards a straight line greater than CH; therefore it is the same as the ratio of BC towards CH itself.

This also is clear, that if a straight line is taken as a third proportional to the straight lines HC, CB it will be equal to the arc BED, and four times this straight line will be equal to the circumference of the whole circle. A straight line equal to the circumference of the circle having been found, a square can easily be constructed equal to the circle itself. For the rectangle contained by the perimeter of the circle and the radius is double of the circle, as Archimedes demonstrated.

2.H Aristotle

Aristotle (384–322 BC) was educated from his late teens at Plato's Academy in Athens, and later he set up his own school, the Lyceum. It is believed to be largely the written-up notes of lectures he delivered that form his profound and wide-ranging writings. As with Plato, there is much to be learnt from Aristotle about the mathematics of his day, the period just before Euclid compiled the *Elements*; though—again like Plato—our knowledge has generally to be inferred from remarks made to illustrate arguments about something else. One of Aristotle's favourite examples, for instance, is of the incommensurability of the side and diagonal of a square (2.H6), to which he made many allusions. The fundamental principles of logical thought were among Aristotle's deepest concerns (and is one of the areas in which his influence has been supreme almost ever since). The extracts in 2.H1 reflect this with particular regard to mathematics. The peculiarly Greek concern with what mathematics is and how it differs from other sciences is also discussed by Aristotle. The selection of extracts about this in 2.H3 is interesting for showing that there were already more mathematically related sciences under investigation than we could glean from the quadrivium tradition (see 2.D).

2.H1 Principles of demonstrative reasoning

(a) By first principles in each genus I mean those the truth of which it is not possible to prove. What is *denoted* by the first [terms] and those derived from them is assumed; but, as regards their *existence*, this must be assumed for the principles but proved for the rest. Thus what a unit is, what the straight [line] is, or what a triangle is [must be assumed]; and the existence of the unit and of magnitude must also be assumed, but the rest must be proved. Now of the premisses used in demonstrative sciences some are peculiar to each science and others common [to all], the latter being common by analogy, for of course they are actually useful in so far as they are applied to the subject-matter included under the particular science. Instances of first principles peculiar to a science are the assumptions that a line is of such-and-such a character, and similarly for the straight [line]; whereas it is a common principle, for instance, that, if equals be subtracted from equals, the remainders are equal. But it is enough that each of the common principles is true so far as regards the particular genus [subject-matter]; for [in geometry] the effect will be the same even if the common principle be assumed to be true, not of everything, but only of magnitudes, and, in arithmetic, of numbers.

Now the things peculiar to the science, the existence of which must be assumed, are the things with reference to which the science investigates the essential attributes, e.g. arithmetic with reference to units, and geometry with reference to points and lines. With these things it is assumed that they exist and that they are of such-and-such a nature. But, with regard to their essential properties, what is assumed is only the meaning of each term employed: thus arithmetic assumes the answer to the question what is [meant by] 'odd' or 'even', 'a square' or 'a cube', and geometry to the question what is [meant by] 'the irrational' or 'deflection' or [the so-called] 'verging' [to a point]; but that there are such things is proved by means of the common principles and of what has already been demonstrated. Similarly with astronomy. For every demonstrative science has to do with three things, (1) the things which are assumed to exist, namely the genus [subject-matter] in each case, the essential properties of which the science investigates, (2) the common axioms so-called, which are the primary source of demonstration, and (3) the properties with regard to which all that is assumed is the meaning of the respective terms used.

[...] Now anything that the teacher assumes, though it is matter of proof, without proving it himself, is an hypothesis if the thing assumed is believed by the learner, and it is moreover an hypothesis, not absolutely, but relatively to the particular pupil; but, if the same thing is assumed when the learner either has no opinion on the subject or is of a contrary opinion, it is a postulate. This is the difference between an hypothesis and a postulate; for a postulate is that which is rather contrary than otherwise to the opinion of the learner, or whatever is assumed and used without being proved, although matter for demonstration. Now definitions are not hypotheses, for they do not assert the existence or non-existence of anything, while hypotheses are among propositions. Definitions only require to be understood: a definition is therefore not an hypothesis, unless indeed it be asserted that any audible speech is an hypothesis. An hypothesis is that from the truth of which, if assumed, a conclusion can be established. Nor are the geometer's hypotheses false, as some have said: I mean those who say that 'you should not make use of what is false, and yet the geometer falsely calls the line which he has drawn a foot long when it is not, or straight when it is not straight'. The geometer bases no conclusion on the particular line which he has drawn being that which he has described, but [he refers to] what is *illustrated* by the figures.

(b) Further, with regard to the first principles of demonstration, it is questionable whether they belong to one science or several. By first principles of demonstration I mean the *common opinions* on which all men base their proofs, e.g. that one or other of two contradictories must be true, that it is impossible for the same thing both to be and not to be, and all other propositions of this kind. Does one science deal with these as well as with substance, or do the two things belong to different sciences, and, if so, which of the two must we name as that which we are now looking for?

(c) Some indeed claim to demonstrate this; but this they do through want of education, for not to know of what things we ought to look for demonstration, and what not, simply argues want of education. Now it is impossible that there should be demonstration of absolutely everything, for there would then be an infinite regress, so that even then there would be no proof. But, if there are some things of which we should not look for a demonstration, these persons could not say what principle they regard as more indemonstrable than that above stated. It is in fact possible to prove negatively that even this view is impossible, if the disputant will say anything; and if he will not, it is absurd to attempt to reason with one who has no reason to give for anything .... Such a man, as such, is no better than a vegetable.

2.H2 Geometrical analysis

We deliberate not about ends but about means. For a doctor does not deliberate whether he shall heal, nor an orator whether he shall persuade, nor a statesman whether he shall produce law and order, nor does any one else deliberate about his end. They assume the end and consider how and by what means it is to be attained; and if it seems to be produced by several means they consider by which it is most easily and best produced, while if it is achieved by one only they consider how it will be achieved by this and by what means *this* will be achieved, till they come to the first cause, which in the order of discovery is last. For the person who deliberates seems to investigate and analyse in the way described as though he were analysing a geometrical construction (not all investigation appears to be deliberation—for instance mathematical investigations—but all deliberation is investigation), and what is last in the order of analysis seems to be first in the order of becoming.

2.H3 The distinction between mathematics and other sciences

(a) The mathematician's investigations are about things reached by abstraction; for he investigates things after first eliminating all sensible qualities, such as weight, lightness, hardness and its contrary, also heat and cold and the other sensible contrarieties, leaving only the quantitative and continuous—the latter being continuous one way, two ways, or three ways, as the case may be—and the properties of these things *qua* quantitative and continuous; and he investigates them in relation to nothing else, considering in some cases their relative positions and the facts consequent on these, in other cases their commensurabilities and incommensurabilities, in other cases again their ratios; but nevertheless we lay it down that it is one and the same science which deals with all these things, namely geometry.

(b) Having distinguished in how many senses we speak of nature, we must next consider wherein the mathematician differs from the physicist. For of course physical bodies contain planes, solids, lengths, and points, which are what the mathematician investigates. Again, astronomy is either different from or a part of physics; for it is absurd to suppose that it is the business of the physicist to know what the sun and moon are, but not to know any of their essential attributes, especially when we see that those who discuss nature do, in fact, also discuss the shape of the moon and sun and the question whether the universe and the earth are spherical or not. Now, the mathematician also studies these [figures] but not *qua* limits or boundaries in each case of a natural body. Nor does he investigate their attributes *qua* attributes of such [physical] bodies. In fact, therefore, he treats them as separate; for they are in thought separable from motion and the separation introduces no error. (The partisans of Forms equally do this, but without knowing it; for they postulate separate physical objects which are less susceptible of separation than mathematical objects.) This can be made manifest if we try to frame the definitions of the objects themselves and of their attributes respectively. For the odd, the even, the straight, the curved, as well as number, line, and figure are independent of motion, whereas flesh, bone, man, are not so, the latter terms being analogous to a snub nose, not like the term 'curved'. Other evidence is also furnished by the more physical branches of mathematics such as optics, harmonics, and astronomy. These stand to geometry in a sort of inverse relation. For geometry investigates a physical line but not *qua* physical, whereas optics considers a mathematical line not *qua* mathematical but *qua* physical. And since nature means two things, the form and the matter, we must study it in the way in which we should consider the definition of snubness. That is to say, we must consider natural objects neither without reference to matter nor exclusively with reference to matter.

(c) The Why and the What differ in another way, namely in respect that their investigation belongs to different sciences. This is the case where the two things are so related that one falls under the other; this is the relation in which optics stands to geometry, mechanics to solid geometry, harmonics to arithmetic, and phenomena to astronomy. (Some of these sciences in practice bear the same name, e.g. mathematical and nautical astronomy, mathematical and audible harmonics.) For here it is for the practising observer to know the fact, but for the mathematician to know the cause; it is the mathematicians who are in possession of the demonstrations of the cause, and in many cases they do not even know the fact, just as those who investigate the universal often do not know some of the particular cases coming under it because they have not specially considered them.

(d) Evidence for this is furnished by the fact that the young may be geometers and mathematicians and be wise in such subjects, but they are not thought to be prudent. The reason is that practical wisdom [prudence] is concerned with particular facts as well, which become known as the result of experience, while a young man cannot be experienced [some length of time is required to give experience]. One might indeed inquire further why a boy may be a good mathematician, but cannot be a philosopher or physicist. May we not say it is because the subjects of mathematics are reached by means of abstraction, while the principles of philosophy and physics come from experience; and the young have no conviction of the latter, though they may speak of them, while in mathematics the 'definitions' are plain [free from ambiguity]?

2.H4 The Pythagoreans

In the time of these philosophers [Leucippus and Democritus] and before them the so-called Pythagoreans applied themselves to the study of mathematics and were the first to advance that science; insomuch that, having been brought up in it, they thought that its principles must be the principles of all existing things. Since among these principles numbers are by nature the first, and they thought they found in numbers—more than in fire, earth, and water—many resemblances to things which are and become—thus such-and-such an attribute of numbers is justice, another is soul and mind, another is opportunity, and so on, and again they saw in numbers the attributes and ratios of the musical scales—since, in short, all other things seemed in their whole nature to be assimilated to numbers, while numbers seemed to be the first things in the whole of nature, they supposed the elements of numbers to be the elements of all things, and the whole heaven to be a musical scale and a number.

2.H5 Potential and actual infinities

On the other hand it is clear that, if an infinite does not exist at all, many impossibilities arise: time will have some beginning and end, magnitudes will not be divisible into magnitudes, and number will not be infinite. If therefore, when the case has been set out as above, neither view appears to be admissible, we need an arbitrator; clearly there is a sense in which the infinite exists and another sense in which it does not.

Being means either being potentially, or being actually, and the infinite is possible by way of addition as well as by way of division. Now, as we have explained, magnitude is never actually infinite, but it is infinite by way of division—for it is not difficult to refute the theory of indivisible lines—the alternative that remains, therefore, is that the infinite exists potentially. But in what sense does it exist potentially? You may say, for example, that a given piece of material is potentially a statue, because it will [sometime] be a statue. Not so with something potentially infinite: you must not suppose that it will be actually infinite. Being has many meanings, and we say that the infinite 'is' in the same sense as we say 'it *is* day' or 'the games *are* on', namely in virtue of one thing continually succeeding another. For the distinction between 'potentially' and 'actually' applies to these things too: there *are* Olympian games both in the sense that the contests may take place and that they do take place.

It is clear that the infinite takes different forms, as in time, in generations of men, and in the division of magnitudes. For, generally speaking, the infinite is so in the sense that it is again and again taking on something more, this something being always finite, but different every time. In the case of magnitudes, however, what is taken on during the process stays there; whereas in the case of time and generations of men it passes away, but so that the source of supply never gives out.

The infinite by way of addition is in a manner the same as the infinite by way of division. Within a finite magnitude the infinite by way of addition is realized in an inverse way [to that by way of division]; for, as we see the magnitude being divided *ad infinitum*, so, in the same way, the sum of the successive fractions when added to one another [continually] will be found to tend towards a determinate limit. For if, in a finite magnitude, you take a determinate fraction of it and then add to that fraction in the same ratio, and so on [i.e. so that each part has to be preceding part the same ratio as the part first taken has to the whole], but *not* each time including [in the part taken] one and the same amount of the original whole, you will not traverse [i.e. exhaust] the finite magnitude. But if you increase the ratio so that it always includes one and the same magnitude, whatever it is, you will traverse it, because any finite magnitude can be exhausted by taking away from it continually any definite magnitude however small. In no other sense, then, does the infinite exist; but it does exist in this sense, namely potentially and by way of diminution. In actuality it exists only in the sense in which we say 'it *is* day' or 'the games *are* on', and potentially it exists in the same way as matter, but not independently as the finite does. Thus we may even have a potentially infinite by way of addition of the kind we described, which, as we say, is in a certain way the same as the infinite by way of division; it can always take on someting outside [the total for the time being], but the total will never exceed every determinate magnitude [of the same kind] in the way that, in the direction of division, it passes every determinate magnitude in smallness, and becomes continually smaller and smaller. But in the sense of exceeding every [magnitude] by way of addition, the infinite cannot exist even potentially, unless there exists something actually infinite, but only incidentally so, infinite, that is, in the sense in which natural philosophers declare the body outside the universe to be infinite, whether its substance be air or anything else of the kind. But if it is not possible that there can be a sensible body actually infinite in this sense, it is manifest that neither can there be such a body which is even potentially infinite by way of addition, save in the sense which we have described as the reverse of the infinite by way of division. [...]

Neither does my argument rob mathematicians of their study because it denies that the infinite can exist in such a way as to be actually infinite in the direction of increase, meaning thereby something that cannot be gone through. For, even as things are, mathematicians do not need or make use of it; they only require that the finite straight line shall be as long *as they please*, and that, given a ratio in which the greatest magnitude is cut, another magnitude of any size whatever can be cut in the same ratio. Hence my argument makes no difference to mathematicians for the purpose of their demonstrations; nor does it matter to them whether the infinite exists among existent magnitudes.

2.H6 Incommensurability

(a) Everyone begins, as we said, with wondering that a certain thing should be so, as for example one does in the case of the puppet theatre (if one has not yet found out the explanation), or with reference to the solstices or the incommensurability of the diagonal. For it must seem to everyone matter for wonder that there should exist a thing which is not measurable by the smallest possible measure. The fact is that we have to arrive in the end at the contrary and the better state, as the saying is. This is so in the cases just mentioned when we have learnt about them. A geometer, for instance, would wonder at nothing so much as that the diagonal should prove to be commensurable.

(b) For all who argue *per impossibile* infer by syllogism a false conclusion, and prove the original conclusion hypothetically when something impossible follows from a contradictory assumption, as, for example, that the diagonal [of a square] is incommensurable [with the side] because odd numbers are equal to even if it is assumed to be commensurate. It is inferred by syllogism that odd numbers are equal to even, and proved hypothetically that the diagonal is incommensurate, since a false conclusion follows from the contradictory assumption.

3 Euclid's Elements

The extracts in this chapter are largely from the most successful and persistent of all textbooks, rightly described by its translator Sir Thomas Heath as 'one of the noblest monuments of antiquity'. Of Euclid's life we know virtually nothing, beyond the fact that he lived in Alexandria, probably around 300 BC. While very little of the mathematical content of the *Elements* is believed to be due to Euclid himself, his compilation of strands of Greek mathematical research over the previous two centuries seems to have taken on a definitive status almost immediately.

Of the many commentaries on the *Elements* written in antiquity, the main one to have survived is that on Book I by Proclus, who taught at the Neoplatonic Academy in Athens in the fifth century. The extracts in 3.A and 3.B, chosen to illustrate his blend of historical information, mathematical explication and philosophical rumination, are juxtaposed with the passages from Euclid to which they refer. Proclus lived some seven hundred years after Euclid, but the information he provided, drawn generally from earlier commentators, is an invaluable source for the history of Greek mathematics.

The final section of the chapter, 3.G, is concerned with a recent debate among historians about the interpretation of aspects of Greek geometry—Book II of the *Elements*, in particular.

3.A Introductory Comments by Proclus

(a) It is a difficult task in any science to select and arrange properly the elements out of which all other matters are produced and into which they can be resolved. Of those who have attempted it some have brought together more theorems, some less; some have used rather short demonstrations, others have extended their treatment to great lengths; some have avoided the reduction to impossibility, others proportion; some have devised defenses in advance against attacks upon the starting-points; and in general many ways of constructing elementary expositions have been individually invented. Such a treatise ought to be free of everything superfluous, for that is a hindrance to learning; the selections chosen must all be coherent and conducive to the end proposed, in order to be of the greatest usefulness for knowledge; it must devote great attention both to clarity and to conciseness, for what lacks these qualities confuses our understanding; it ought to aim at the comprehension of its theorems in a general form, for dividing one's subject too minutely and teaching it by bits make knowledge of it difficult to attain. Judged by all these criteria, you will find Euclid's introduction superior to others.

(b) If now anyone should ask what the aim of this treatise is, I should reply by distinguishing between its purpose as judged by the matters investigated and its purpose with reference to the learner. Looking at its subject-matter, we assert that the whole of the geometer's discourse is obviously concerned with the cosmic figures. It starts from the simple figures and ends with the complexities involved in the structure of the cosmic bodies, establishing each of the figures separately but showing for all of them how they are inscribed in the sphere and the ratios that they have with respect to one another. Hence some have thought it proper to interpret with reference to the cosmos the purposes of individual books and have inscribed above each of them the utility it has for a knowledge of the universe. Of the purpose of the work with reference to the student we shall say that it is to lay before him an elementary exposition and a method of perfecting his understanding for the whole of geometry. If we start from the elements, we shall be able to understand the other parts of this science; without the elements we cannot grasp its complexity, and the learning of the rest will be beyond us. The theorems that are simplest and most fundamental and nearest to first principles are assembled here in a suitable order, and the demonstrations of other propositions take them as the most clearly known and proceed from them. In this way also Archimedes in his book on *Sphere and Cylinder* and likewise Apollonius and all other geometers appear to use the theorems demonstrated in this very work as generally accepted starting-points. This, then, is its aim: both to furnish the learner with an introduction to the science as a whole and to present the construction of the several cosmic figures.

3.B Book I

3.B1 Axiomatic foundations

(a) Definitions, postulates and common notions

Definitions

1 A *point* is that which has no part.
2 A *line* is breadthless length.
3 The extremities of a line are points.
4 A *straight line* is a line which lies evenly with the points on itself.

5 A *surface* is that which has length and breadth only.
6 The extremities of a surface are lines.
7 A *plane surface* is a surface which lies evenly with the straight lines on itself.
8 A *plane angle* is the inclination to one another of two lines in a plane which meet one another and do not lie in a straight line.
9 And when the lines containing the angle are straight, the angle is called *rectilineal*.
10 When a straight line set up on a straight line makes the adjacent angles equal to one another, each of the equal angles is *right* and the straight line standing on the other is called a *perpendicular* to that on which it stands.
11 An *obtuse angle* is an angle greater than a right angle.
12 An *acute angle* is an angle less than a right angle.
13 A *boundary* is that which is an extremity of anything.
14 A *figure* is that which is contained by any boundary or boundaries.
15 A *circle* is a plane figure contained by one line such that all the straight lines falling upon it from one point among those lying within the figure are equal to one another;
16 And the point is called the *centre* of the circle.
17 A *diameter* of the circle is any straight line drawn through the centre and terminated in both directions by the circumference of the circle, and such a straight line also bisects the circle.
18 A *semicircle* is the figure contained by the diameter and the circumference cut off by it. And the centre of the semicircle is the same as that of the circle.
19 *Rectilineal figures* are those which are contained by straight lines, *trilateral* figures being those contained by three, *quadrilateral* those contained by four, and *multilateral* those contained by more than four straight lines.
20 Of trilateral figures, an *equilateral triangle* is that which has its three sides equal, an *isosceles triangle* that which has two of its sides alone equal, and a *scalene triangle* that which has its three sides unequal.
21 Further, of trilateral figures, a *right-angled triangle* is that which has a right angle, an *obtuse-angled triangle* that which has an obtuse angle, and an *acute-angled triangle* that which has its three angles acute.
22 Of quadrilateral figures, a *square* is that which is both equilateral and right-angled; an *oblong* that which is right-angled but not equilateral; a *rhombus* that which is equilateral but not right-angled; and a *rhomboid* that which has its opposite sides and angles equal to one another but is neither equilateral nor right-angled. And let quadrilaterals other than these be called *trapezia*.
23 *Parallel* straight lines are straight lines which, being in the same plane and being produced indefinitely in both directions, do not meet one another in either direction.

Postulates

Let the following be postulated:
1 To draw a straight line from any point to any point.
2 To produce a finite straight line continuously in a straight line.
3 To describe a circle with any centre and distance.
4 That all right angles are equal to one another.
5 That, if a straight line falling on two straight lines make the interior angles on the same side less than two right angles, the two straight lines, if produced indefinitely, meet on that side on which are the angles less than the two right angles.

Common notions

1 Things which are equal to the same thing are also equal to one another.
2 If equals be added to equals, the wholes are equal.
3 If equals be subtracted from equals, the remainders are equal.
4 Things which coincide with one another are equal to one another.
5 The whole is greater than the part.

(*b*) *Commentary by Proclus*

On Definition 1

By denying parts to it, then, Euclid signifies to us that the point is the first principle of the entire subject under examination. Negative definitions are appropriate to first principles, as Parmenides teaches us in setting forth the first and ultimate cause by means of negations alone. For every first principle is constituted by a different essence from that of the things dependent on it, and to deny the latter makes evident to us the peculiar property of the principle. For that which is their cause, but not any one of the things of which it is the cause, becomes in a sense knowable through this method of exposition.

On Definition 4

The straight line is a symbol of the inflexible, unvarying, incorruptible, unremitting, and all-powerful providence that is present to all things; and the circle and circular movement symbolize the activity that returns to itself, concentrates on itself, and controls everything in accord with a single intelligible Limit. The demiurgic Nous has therefore set up these two principles in himself, the straight and the circular, and produced out of himself two monads, the one acting in a circular fashion to perfect all intelligible essences, the other moving in a straight line to bring all perceptible things to birth. Since the soul is intermediate between sensibles and intelligibles, she moves in circular fashion insofar as she is allied to intelligible nature but, insofar as she presides over sensibles, exercises her providence in a straight line. So much regarding the similarity of these concepts to the order of being.

On Definition 7

The older philosophers did not think to posit the plane as a species of surface but took the two terms as equivalent for expressing magnitude in two dimensions. Thus the divine Plato said that geometry is the study of planes and contrasted it with stereometry as if he thought surface and plane were the same thing. Likewise also the inspired Aristotle. But Euclid and his successors make the surface the genus and the plane a species of it, as the straight line is a species of line. This is why, by analogy with the straight line, he defines the plane separately from the surface.

On Definition 8

Some of the ancients put the angle in the category of relation, calling it the inclination either of lines or of planes to one another; others place it under quality, saying that, like straight and curved, it is a certain character of a surface or a solid; others refer it to quantity, asserting that it is either a surface or a solid quantity. [...]

But if it is a magnitude and all finite homogeneous magnitudes have a ratio to one another, then all homogeneous angles, at least those in planes, will have a ratio to one another, so that a horned angle will have a ratio to a rectilinear. But all quantities that have a ratio to one another can exceed one another by being multiplied; a horned angle, then, may exceed a rectilinear, which is impossible, for it has been proved [*Elements*, III.16] that a horned angle is less than any rectilinear angle.

And if it is only a quality, like heat or coldness, how can it be divided into equal parts? [...]

As to the third possibility, if the angle is an inclination and in general belongs to the class of relations, it will follow that, when the inclination is one, there is one angle and not more. For if the angle is nothing other than a relation between lines or between planes, how could there be one relation but many angles? [...] Such are the difficulties. [...] Let us follow our 'head' and say that the angle as such is none of the things mentioned but exists as a combination of all these categories, and this is why it presents a difficulty to those who are inclined to make it any one of them.

On Definitions 10–12

The perpendicular thus is also a symbol of directness, purity, undefiled unswerving force, and all such things, a symbol of divine and intelligent measure. For by perpendiculars we measure the altitude of figures, and it is by reference to the right angle that we define the other rectilinear angles, since they have no limiting principle in themselves; they are considered only as exceeding or falling short, each of them being in itself indeterminate. Hence they say that virtue is like rightness, whereas vice is constituted after the fashion of the indeterminate obtuse and acute, possessing both excesses and deficiencies and showing by this more-and-less its own lack of measure. We shall therefore lay it down that the right among rectilinear angles is the image of perfection, undeviating energy, intelligent limit and boundary, and everything similar to them, and that the obtuse and acute angles are likenesses of indefinite change, irrelevant progression, differentiation, partition, and unlimitedness in general. So much for these matters.

To the definitions of the obtuse and acute angles we must add the genus: each of them is rectilinear, one larger and the other smaller than a right angle. Not every angle smaller than a right angle is acute, for the horned angle is smaller than any right angle—indeed smaller than any acute angle—but is not acute; and likewise the semicircular angle is smaller than any right angle but is not acute. The explanation is that these are mixed, not rectilinear angles. Clearly also many angles contained between circular lines appear to be greater than right angles, but they are not for that reason obtuse; for the obtuse angle must be rectilinear. I call attention to this and observe also that in defining a right angle our geometer takes a straight line standing on another straight line and making the adjacent angles equal to one another, whereas he explains the obtuse and the acute angles without assuming a straight line inclined towards one side, referring instead to the right angle; for the right angle is the measure of angles other than right, just as equality is the measure of unequal things.

On Definition 17

The famous Thales is said to have been the first to demonstrate that the circle is bisected by the diameter. The cause of this bisection is the undeviating course of the straight line through the centre; for since it moves through the middle and throughout all parts of its identical movement refrains from swerving to either side, it cuts off equal lengths of the circumference on both sides. If you wish to demonstrate this mathematically, imagine the diameter drawn and one part of the circle fitted upon the other. If it is not equal to the other, it will fall either inside or outside it, and in either case it will follow that a shorter line is equal to a longer. For all the lines from the centre to the circumference are equal, and hence the line that extends beyond will be equal to the line that falls short, which is impossible. The one part, then, fits the other, so that they are equal. Consequently the diameter bisects the circle.

But if from one diameter two semicircles are produced, and if an indefinite number of diameters can be drawn through the centre, it will follow that the number of semicircles is twice infinity. This difficulty is alleged by some persons against the indefinite divisibility of magnitudes. We reply that a magnitude is indefinitely divisible, but not into an infinite number of parts. The latter statement makes an infinite number actual, the former merely potential; the latter assigns existence to the infinite, the other only genesis. With one diameter, then, two semicircles come into being, and the diameters will never be infinite in number, even though they can be taken indefinitely. So the number of semicircles will never be twice infinity; those that are produced at any time will be twice a finite number, for the diameters taken at any time are always finite in number.

On Postulate 5

This ought to be struck from the postulates altogether. For it is a theorem—one that invites many questions, which Ptolemy proposed to resolve in one of his books—and requires for its demonstration a number of definitions as well as theorems. And the converse of it is proved by Euclid himself as a theorem. But perhaps some persons might mistakenly think this proposition deserves to be ranked among the postulates on the ground that the angles' being less than two right angles makes us at once believe in the convergence and intersection of the straight lines. To them Geminus has given the proper answer when he said that we have learned from the very founders of this science not to pay attention to plausible imaginings in determining what propositions are to be accepted in geometry. Aristotle likewise says that to accept probable reasoning from a geometer is like demanding proofs from a rhetorician. And Simmias is made by Plato to say, 'I am aware that those who make proofs out of probabilities are impostors'. So here, although the statement that the straight lines converge when the right angles are diminished is true and necessary, yet the conclusion that because they converge more as they are extended farther they will meet at some time is plausible, but not necessary, in the absence of an argument proving that this is true of straight lines. That there are lines that approach each other indefinitely but never meet seems implausible and paradoxical, yet it is nevertheless true and has been ascertained for other species of lines.

3.B2 The base angles of an isosceles triangle are equal

(*a*) *Proposition 5*

In isosceles triangles the angles at the base are equal to one another, and, if the equal straight lines be produced further, the angles under the base will be equal to one another.

Let ABC be an isosceles triangle having the side AB equal to the side AC; and let the straight lines BD, CE be produced further in a straight line with AB, AC [Post. 2].

I say that the angle ABC is equal to the angle ACB, and the angle CBD to the angle BCE.

Let a point F be taken at random on BD; from AE the greater let AG be cut off equal to AF the less [I. 3]; and let the straight lines FC, GB be joined [Post. 1].

Then, since AF is equal to AG and AB to AC, the two sides FA, AC are equal to the two sides GA, AB, respectively; and they contain a common angle, the angle FAG.

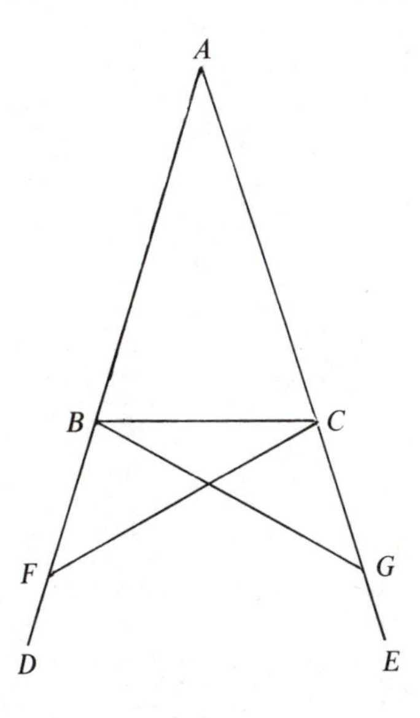

Therefore the base FC is equal to the base GB, and the triangle AFC is equal to the triangle AGB, and the remaining angles will be equal to the remaining angles respectively, namely those which the equal sides subtend, that is, the angle ACF to the angle ABG, and the angle AFC to the angle AGB [I. 4].

And, since the whole AF is equal to the whole AG, and in these AB is equal to AC, the remainder BF is equal to the remainder CG.

But FC was also proved equal to GB; therefore the two sides BF, FC are equal to the two sides CG, GB respectively; and the angle BFC is equal to the angle CGB, while the base BC is common to them; therefore the triangle BFC is also equal to the triangle CGB, and the remaining angles will be equal to the remaining angles respectively, namely those which the equal sides subtend; therefore the angle FBC is equal to the angle GCB, and the angle BCF to the angle CBG.

Accordingly, since the whole angle ABG was proved equal to the angle ACF, and in these the angle CBG is equal to the angle BCF, the remaining angle ABC is equal to the remaining angle ACB; and they are at the base of the triangle ABC.

But the angle FBC was also proved equal to the angle GCB; and they are under the base.

Therefore etc. Q.E.D.

(*b*) *Aristotle*

This is more manifest in geometrical propositions, e.g. the proposition that the base angles of an isosceles triangle are equal. Suppose the straight lines *A*, *B* have been drawn to the centre. Then, if one should assume that the angle *AC* is equal to the angle *BD* without having claimed generally that the angles of semicircles are equal, and again, if one should assume that the angle *C* is equal to the angle *D* without making the additional assumption that every angle of a segment is equal to the other angle of that segment, and if one should then, lastly, assume that, the whole angles being equal and the angles subtracted being equal, the remaining angles *E*, *F* are equal, he will beg the question unless he assumes (generally) that 'if equals be subtracted from equals the remainders are equal'.

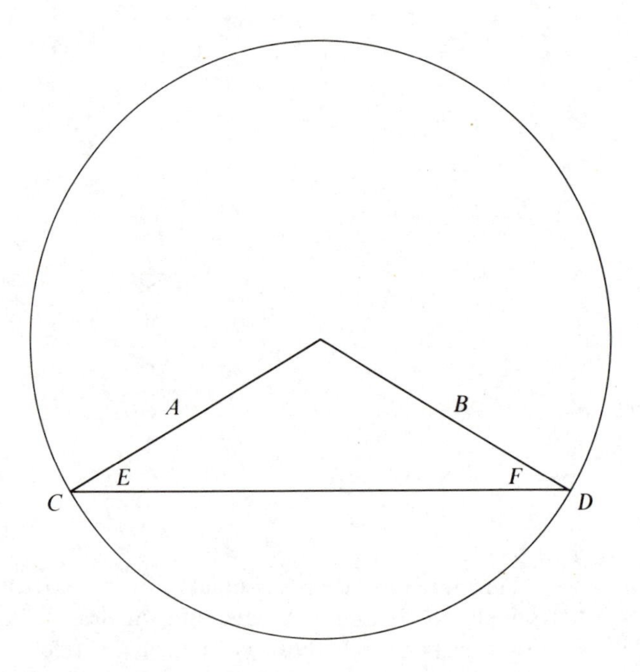

(*c*) *Proclus on Pappus*

Pappus has given a still shorter demonstration that needs no supplementary construction, as follows. Let *ABC* be isosceles with side *AB* equal to side *AC*. Let us think of this triangle as two triangles and reason thus: Since *AB* is equal to *AC* and *AC* is equal to *AB*, the two sides *AB* and *AC* are equal to the two sides *AC* and *AB*, and the angle *BAC* is equal to the angle *CAB* (for they are the same); therefore all the corresponding parts are equal, *BC* to *CB*, the triangle *ABC* to the triangle *ACB*, the angle *ABC* to the angle *ACB*, and angle *ACB* to angle *ABC*. For these are angles subtended by the equal sides *AB* and *AC*. Hence the angles at the base of an isosceles are equal. It looks as if he discovered this method of proof when he noted that in the fourth theorem it was by uniting the two triangles so that they coincide with each other, thus making them one instead of two, that the author of the *Elements* perceived their

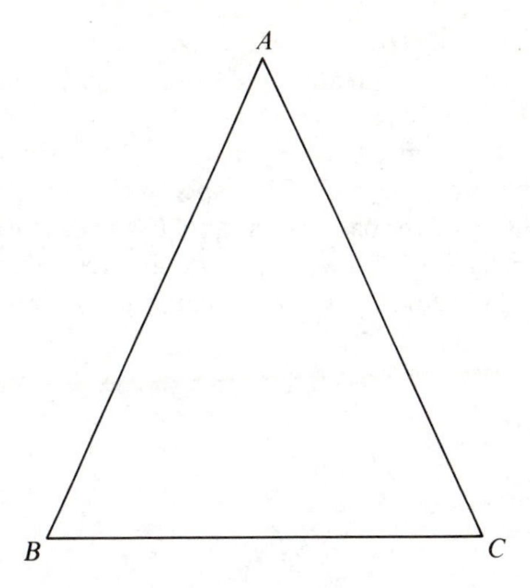

equality in all respects. In the same way, then, it is possible for us, by assumption, to see two triangles in this single one and so prove the equality of the angles at the base.

We are indebted to old Thales for the discovery of this and many other theorems. For he, it is said, was the first to notice and assert that in every isosceles the angles at the base are equal, though in somewhat archaic fashion he called the equal angles similar.

3.B3 Propositions 6, 9, 11 and 20

(*a*) *Proposition 6*

If in a triangle two angles be equal to one another, the sides which subtend the equal angles will also be equal to one another.

Let *ABC* be a triangle having the angle *ABC* equal to the angle *ACB*; I say that the side *AB* is also equal to the side *AC*.

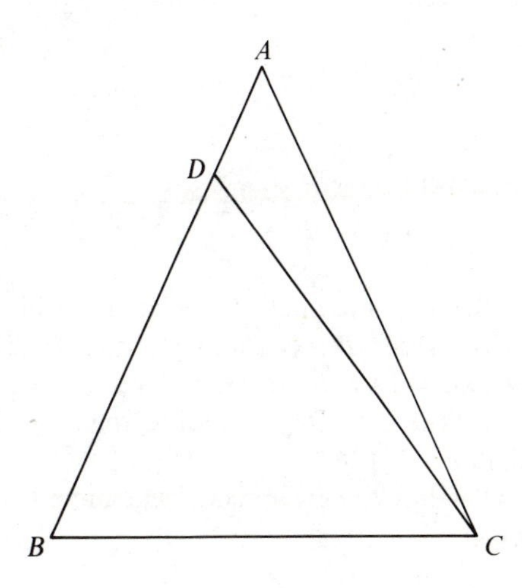

For, if AB is unequal to AC, one of them is greater.

Let AB be greater; and from AB the greater let DB be cut off equal to AC the less; let DC be joined.

Then, since DB is equal to AC, and BC is common, the two sides DB, BC are equal to the two sides AC, CB respectively; and the angle DBC is equal to the angle ACB; therefore the base DC is equal to the base AB, and the triangle DBC will be equal to the triangle ACB, the less to the greater: which is absurd.

Therefore AB is not unequal to AC; it is therefore equal to it.

Therefore etc. Q.E.D.

(*b*) *Proposition 9*

To bisect a given rectilineal angle.

Let the angle BAC be the given rectilineal angle.

Thus it is required to bisect it.

Let a point D be taken at random on AB; let AE be cut off from AC equal to AD [I. 3]; let DE be joined, and on DE let the equilateral triangle DEF be constructed; let AF be joined.

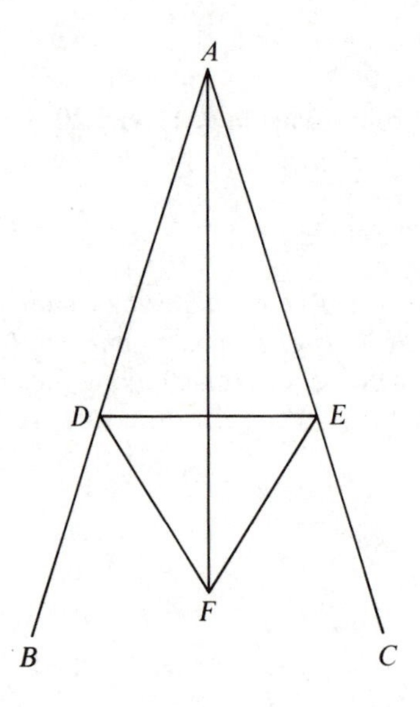

I say that the angle BAC has been bisected by the straight line AF.

For, since AD is equal to AE, and AF is common, the two sides DA, AF are equal to the two sides EA, AF respectively.

And the base DF is equal to the base EF; therefore the angle DAF is equal to the angle EAF [I. 8].

Therefore the given rectilineal angle BAC has been bisected by the straight line AF. Q.E.F.

(*c*) *Proposition 11*

To draw a straight line at right angles to a given straight line from a given point on it.

Let *AB* be the given straight line, and *C* the given point on it.

Thus it is required to draw from the point *C* a straight line at right angles to the straight line *AB*.

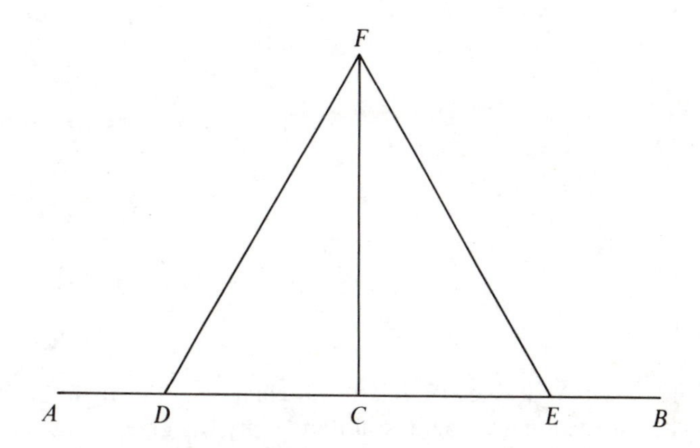

Let a point *D* be taken at random on *AC*; let *CE* be made equal to *CD* [I. 3]; on *DE* let the equilateral triangle *FDE* be constructed [I. 1], and let *FC* be joined; I say that the straight line *FC* has been drawn at right angles to the given straight line *AB* from *C* the given point on it.

For, since *DC* is equal to *CE*, and *CF* is common, the two sides *DC*, *CF* are equal to the two sides *EC*, *CF* respectively; and the base *DF* is equal to the base *FE*; therefore the angle *DCF* is equal to the angle *ECF* [I. 8]; and they are adjacent angles.

But, when a straight line set up on a straight line makes the adjacent angles equal to one another, each of the equal angles is right [Def. 10]; therefore each of the angles *DCF*, *FCE* is right.

Therefore the straight line *CF* has been drawn at right angles to the given straight line *AB* from the given point *C* on it. Q.E.F.

(*d*) *Proposition 20*

In any triangle two sides taken together in any manner are greater than the remaining one.

For let *ABC* be a triangle; I say that in the triangle *ABC* two sides taken together in any manner are greater than the remaining one, namely

BA, *AC* greater than *BC*,
AB, *BC* greater than *AC*,
BC, *CA* greater than *AB*.

For let *BA* be drawn through to the point *D*, let *DA* be made equal to *CA*, and let *DC* be joined.

Then, since *DA* is equal to *AC*, the angle *ADC* is also equal to the angle *ACD* [I. 5]; therefore the angle *BCD* is greater than the angle *ADC* [C.N. 5].

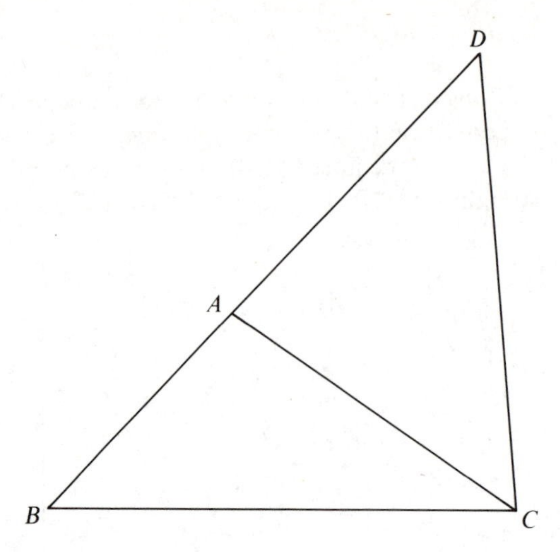

And, since DCB is a triangle having the angle BCD greater than the angle BDC, and the greater angle is subtended by the greater side [I. 19], therefore DB is greater than BC.

But DA is equal to AC; therefore BA, AC are greater than BC.

Similarly we can prove that AB, BC are also greater than CA, and BC, CA than AB.

Therefore etc. Q.E.D.

(*e*) *Proclus on Proposition 20*

The Epicureans are wont to ridicule this theorem, saying it is evident even to an ass and needs no proof; it is as much the mark of an ignorant man, they say, to require persuasion of evident truths as to believe what is obscure without question. Now whoever lumps these things together is clearly unaware of the difference between what is and what is not demonstrated. That the present theorem is known to an ass they make out from the observation that, if straw is placed at one extremity of the sides, an ass in quest of provender will make his way along the one side and not by way of the two others. To this it should be replied that, granting the theorem is evident to sense-perception, it is still not clear for scientific thought. Many things have this character; for example, that fire warms. This is clear to perception, but it is the task of science to find out how it warms, whether by a bodiless power or by physical parts, such as spherical or pyramidal particles. Again it is clear to our senses that we move, but how we move is difficult for reason to explain, whether through a partless medium or from interval to interval, and in this case how we can traverse an infinite number of intervals, for every magnitude is divisible without end. So with respect to a triangle let it be evident to perception that two sides are greater than the third; but how this comes about it is the function of knowledge to say.

This is enough by way of answer to the Epicureans.

3.B4 The angles of a triangle are two right angles

(a) Proposition 32

In any triangle, if one of the sides be produced, the exterior angle is equal to the two interior and opposite angles, and the three interior angles of the triangle are equal to two right angles.

Let *ABC* be a triangle, and let one side of it *BC* be produced to *D*; I say that the exterior angle *ACD* is equal to the two interior and opposite angles *CAB*, *ABC*, and the three interior angles of the triangle *ABC*, *BCA*, *CAB* are equal to two right angles.

For let *CE* be drawn through the point *C* parallel to the straight line *AB* [I. 31].

Then, since *AB* is parallel to *CE*, and *AC* is fallen upon them, the alternate angles *BAC*, *ACE* are equal to one another [I. 29].

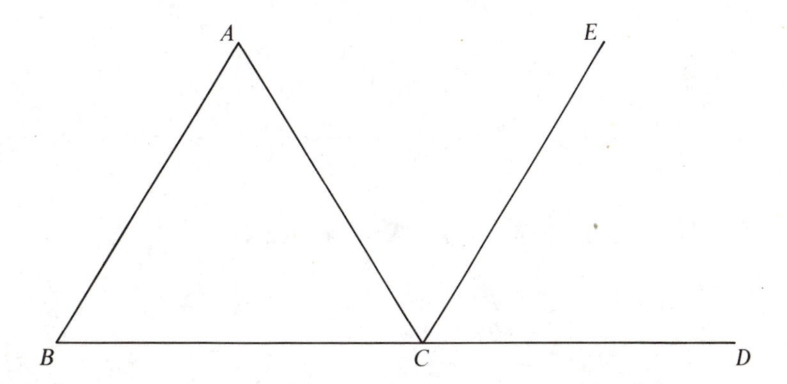

Again, since *AB* is parallel to *CE*, and the straight line *BD* has fallen upon them, the exterior angle *ECD* is equal to the interior and opposite angle *ABC* [I. 29].

But the angle *ACE* was also proved equal to the angle *BAC*; therefore the whole angle *ACD* is equal to the two interior and opposite angles *BAC*, *ABC*.

Let the angle *ACB* be added to each; therefore the angles *ACD*, *ACB* are equal to the three angles *ABC*, *BCA*, *CAB*.

But the angles *ACD*, *ACB* are equal to two right angles [I. 13]; therefore the angles *ABC*, *BCA*, *CAB* are also equal to two right angles.

Therefore etc. Q.E.D.

(b) Aristotle

Propositions too in mathematics are discovered by an activity; for it is by a process of dividing-up that we discover them. [...] Why does the triangle make up two right angles? Because the angles about one point are equal to two right angles. If then the parallel to the side had been drawn up, the reason why would at once have been clear from merely looking at the figure.

(*c*) *Proclus*

Eudemus the Peripatetic attributes to the Pythagoreans the discovery of this theorem, that every triangle has internal angles equal to two right angles, and says they demonstrated it as follows. Let *ABC* be a triangle, and through *A* draw a line *DE* parallel to *BC*. Then since *BC* and *DE* are parallel, the alternate angles are equal, and angle *DAB* is therefore equal to *ABC* and *EAC* to *ACB*. Add the common angle *BAC*. Then angles *DAB*, *BAC*, *CAE*—that is, angles *DAB* and *BAE*, which are two right angles—are equal to the three angles of the triangle *ABC*. Therefore the three angles of a triangle are equal to two right angles. Such is the proof of the Pythagoreans.

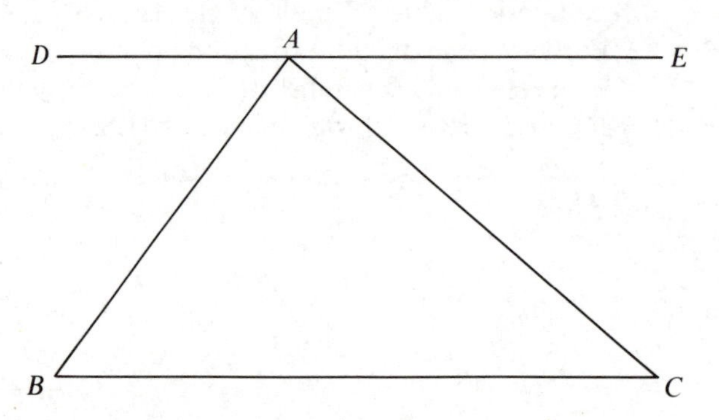

3.B5 Propositions 44, 45, 46 and 47

(*a*) *Proposition 44*

To a given straight line to apply, in a given rectilineal angle, a parallelogram equal to a given triangle.

Let *AB* be the given straight line, *C* the given triangle and *D* the given rectilineal angle; thus it is required to apply to the given straight line *AB*, in an angle equal to the angle *D*, a parallelogram equal to the given triangle *C*.

Let the parallelogram *BEFG* be constructed equal to the triangle *C*, in the angle *EBG* which is equal to *D* [I. 42]; let it be placed so that *BE* is in a straight line with *AB*; let *FG* be drawn through to *H*, and let *AH* be drawn through *A* parallel to either *BG* or *EF* [I. 31].

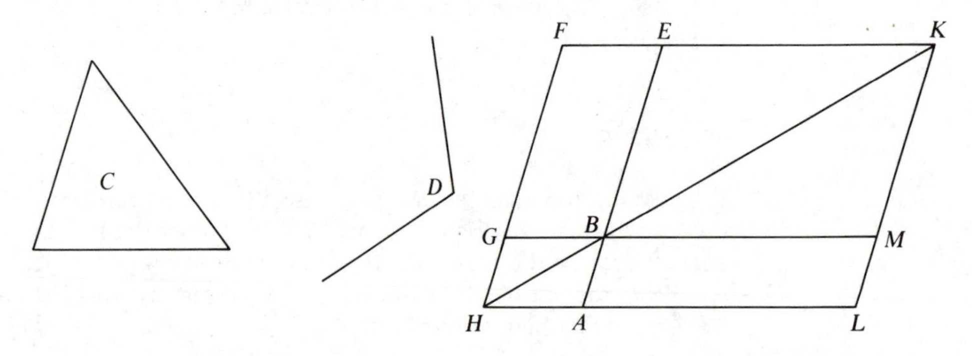

Let *HB* be joined.

Then, since the straight line *HF* falls upon the parallels *AH*, *EF*, the angles *AHF*, *HFE* are equal to two right angles [I. 29].

Therefore the angles *BHG*, *GFE* are less than two right angles; and straight lines produced indefinitely from angles less than two right angles meet [Post. 5]; therefore *HB*, *FE*, when produced, will meet.

Let them be produced and meet at *K*; through the point *K* let *KL* be drawn parallel to either *EA* or *FH* [I. 31], and let *HA*, *GB* be produced to the points *L*, *M*.

Then *HLKF* is a parallelogram, *HK* is its diameter, and *AG*, *ME* are parallelograms, and *LB*, *BF* the so-called complements, about *HK*; therefore *LB* is equal to *BF* [I. 43].

But *BF* is equal to the triangle *C*; therefore *LB* is also equal to *C* [C.N. 1].

And, since the angle *GBE* is equal to the angle *ABM* [I. 15], while the angle *GBE* is equal to *D*, the angle *ABM* is also equal to the angle *D*.

Therefore the parallelogram *LB* equal to the given triangle *C* has been applied to the given straight line *AB*, in the angle *ABM* which is equal to *D*. Q.E.F.

(*b*) *Proclus on Proposition 44*

Eudemus and his school tell us that these things—the application of areas, their exceeding, and their falling short—are ancient discoveries of the Pythagorean muse. It is from these procedures that later geometers took these terms and applied them to the so-called conic lines, calling one of them 'parabola', another 'hyperbola', and the third 'ellipse', although those godlike men of old saw the significance of these terms in the describing of plane areas along a finite straight line. For when, given a straight line, you make the given area extend along the whole of the line, they say you 'apply' the area; when you make the length of the area greater than the straight line itself, then it 'exceeds'; and when less, so that there is a part of the line extending beyond the area described, then it 'falls short'. Euclid too in his sixth book speaks in this sense of 'exceeding' and 'falling short'; but here he needed 'application', since he wished to apply to a given straight line an area equal to a given triangle, in order that we might be able not only to construct a parallelogram equal to a given triangle, but also to apply it to a given finite straight line. For example, when a triangle is given having an area of twelve feet and we posit a straight line whose length is four feet, we apply to the straight line an area equal to the triangle when we take its length as the whole four feet and find how many feet in breadth it must be in order that the parallelogram may be equal to the triangle. Then when we have found, let us say, a breadth of three feet and multiplied the length by the breadth, we shall have the area, that is, if the angle assumed is a right angle. Something like this is the method of 'application' which has come down to us from the Pythagoreans.

(*c*) *Proposition 45*

To construct, in a given rectilineal angle, a parallelogram equal to a given rectilineal figure.

Let *ABCD* be the given rectilineal figure and *E* the given rectilineal angle; thus it is

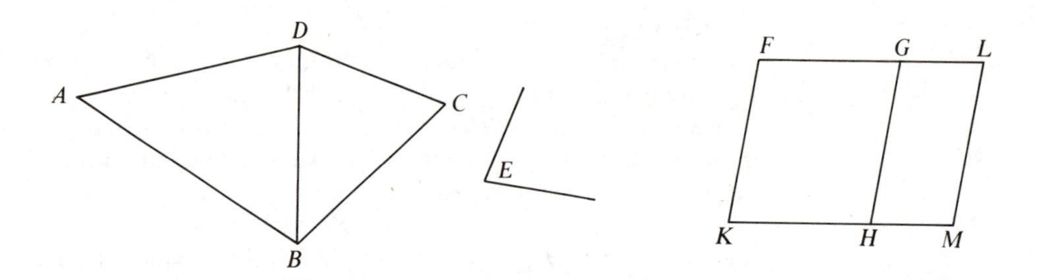

required to construct, in the given angle E, a parallelogram equal to the rectilineal figure $ABCD$.

Let DB be joined, and let the parallelogram FH be constructed equal to the triangle ABD, in the angle HKF which is equal to E [I. 42]; let the parallelogram GM equal to the triangle DBC be applied to the straight line GH, in the angle GHM which is equal to E [I. 44].

Then, since the angle E is equal to each of the angles HKF, GHM, the angle HKF is also equal to the angle GHM [C.N. 1].

Let the angle KHG be added to each; therefore the angles FKH, KHG are equal to the angles KHG, GHM.

But the angles FKH, KHG are equal to two right angles [I. 29]; therefore the angles KHG, GHM are also equal to two right angles.

Thus, with a straight line GH, and at the point H on it, two straight lines KH, HM not lying on the same side make the adjacent angles equal to two right angles; therefore KH is in a straight line with HM [I. 14].

And, since the straight line HG falls upon the parallels KM, FG, the alternate angles MHG, HGF are equal to one another [I. 29].

Let the angle HGL be added to each; therefore the angles MHG, HGL are equal to the angles HGF, HGL [C.N. 2].

But the angles MHG, HGL are equal to two right angles [I. 29]; therefore the angles HGF, HGL are also equal to two right angles [C.N. 1].

Therefore FG is in a straight line with GL [I. 14].

And, since FK is equal and parallel to HG [I. 34], and HG *to* ML also, KF is also equal and parallel to ML [C.N. 1; I. 30]; and the straight lines KM, FL join them (at their extremities); therefore KM, FL are also equal and parallel [I. 33].

Therefore $KFLM$ is a parallelogram.

And, since the triangle ABD is equal to the parallelogram FH, and DBC to GM, the whole rectilineal figure $ABCD$ is equal to the whole parallelogram $KFLM$.

Therefore the parallelogram $KFLM$ has been constructed equal to the given rectilineal figure $ABCD$, in the angle FKM which is equal to the given angle E. Q.E.F.

(*d*) *Proposition 46*

On a given straight line to describe a square.

Let AB be the given straight line; thus it is required to describe a square on the straight line AB.

Let AC be drawn at right angles to the straight line AB from the point A on it [I. 11],

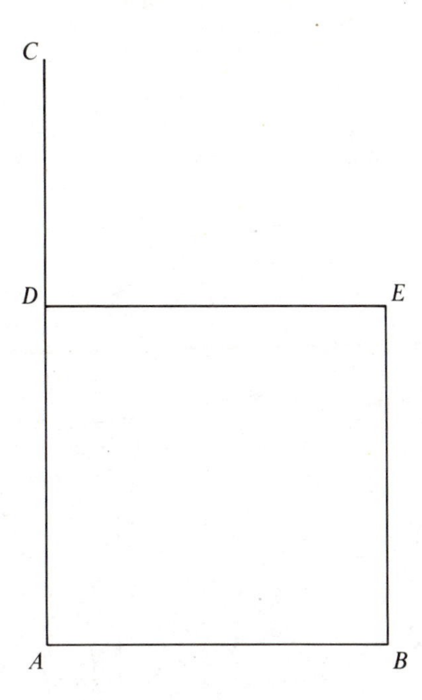

and let *AD* be made equal to *AB*; through the point *D* let *DE* be drawn parallel to *AB*, and through the point *B* let *BE* be drawn parallel to *AD* [I. 31].

Therefore *ADEB* is a parallelogram; therefore *AB* is equal to *DE*, and *AD* to *BE* [I. 34].

But *AB* is equal to *AD*; therefore the four straight lines *BA*, *AD*, *DE*, *EB* are equal to one another; therefore the parallelogram *ADEB* is equilateral.

I say next that it is also right-angled.

For, since the straight line *AD* falls upon the parallels *AB*, *DE*, the angles *BAD*, *ADE* are equal to two right angles [I. 29].

But the angle *BAD* is right; therefore the angle *ADE* is also right.

And in parallelogrammic areas the opposite sides and angles are equal to one another [I. 34]; therefore each of the opposite angles *ABE*, *BED* is also right.

Therefore *ADEB* is right-angled.

And it was also proved equilateral.

Therefore it is a square; and it is described on the straight line *AB*. Q.E.F.

(*e*) *Proposition 47*

In right-angled triangles the square on the side subtending the right angle is equal to the squares on the sides containing the right angle.

Let *ABC* be a right-angled triangle having the angle *BAC* right; I say that the square on *BC* is equal to the squares on *BA*, *AC*.

For let there be described on *BC* the square *BDEC*, and on *BA*, *AC* the squares *GB*, *HC* [I. 46]; through *A* let *AL* be drawn parallel to either *BD* or *CE*, and let *AD*, *FC* be joined.

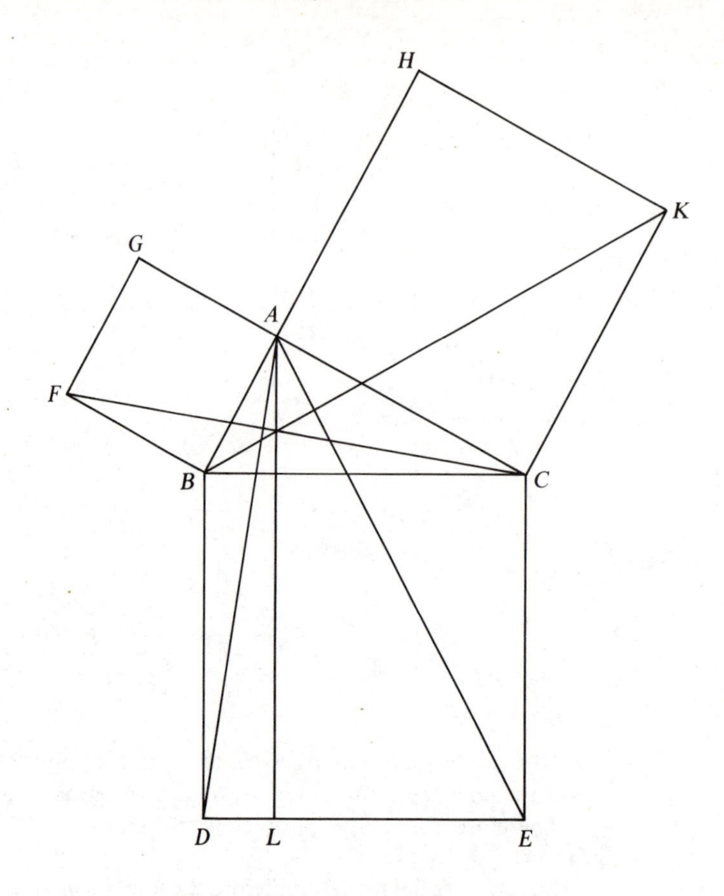

Then, since each of the angles *BAC*, *BAG* is right, it follows that with a straight line *BA*, and at the point *A* on it, the two straight lines *AC*, *AG* not lying on the same side make the adjacent angles equal to two right angles; therefore *CA* is in a straight line with *AG* [I. 14].

For the same reason *BA* is also in a straight line with *AH*.

And, since the angle *DBC* is equal to the angle *FBA*: for each is right: let the angle *ABC* be added to each; therefore the whole angle *DBA* is equal to the whole angle *FBC* [C.N. 2].

And, since *DB* is equal to *BC*, and *FB* to *BA*, the two sides *AB*, *BD* are equal to the two sides *FB*, *BC* respectively; and the angle *ABD* is equal to the angle *FBC*; therefore the base *AD* is equal to the base *FC*, and the triangle *ABD* is equal to the triangle *FBC* [I. 4].

Now the parallelogram *BL* is double of the triangle *ABD*, for they have the same base *BD* and are in the same parallels *BD*, *AL* [I. 41].

And the square *GB* is double of the triangle *FBC*, for they again have the same base *FB* and are in the same parallels *FB*, *GC* [I. 41].

[But the doubles of equals are equal to one another.]

Therefore the parallelogram *BL* is also equal to the square *GB*.

Similarly, if *AE*, *BK* be joined, the parallelogram *CL* can also be proved equal to the square *HC*; therefore the whole square *BDEC* is equal to the two squares *GB*, *HC* [C.N. 2].

And the square *BDEC* is described on *BC*, and the squares *GB*, *HC* on *BA*, *AC*. Therefore the square on the side *BC* is equal to the squares on the sides *BA*, *AC*. Therefore etc. Q.E.D.

(*f*) *Proclus on Proposition 47*

If we listen to those who like to record antiquities, we shall find them attributing this theorem to Pythagoras and saying that he sacrificed an ox on its discovery. For my part, though I marvel at those who first noted the truth of this theorem, I admire more the author of the *Elements*, not only for the very lucid proof by which he made it fast, but also because in the sixth book he laid hold of a theorem even more general than this and secured it by irrefutable scientific arguments. For in that book [VI. 31] he proves generally that in right-angled triangles the figure on the side that subtends the right angle is equal to the similar and similarly drawn figures on the sides that contain the right angle.

3.C Books II–VI

3.C1 Book II: Definitions and Propositions 1, 4, 11 and 14

(*a*) *Definitions*

1 Any rectangular parallelogram is said to be *contained* by the two straight lines containing the right angle.
2 And in any parallelogrammic area let any one whatever of the parallelograms about its diameter with the two complements be called a *gnomon*.

(*b*) *Proposition 1*

If there be two straight lines, and one of them be cut into any number of segments whatever, the rectangle contained by the two straight lines is equal to the rectangles contained by the uncut straight line and each of the segments.

Let *A*, *BC* be two straight lines, and let *BC* be cut at random at the points *D*, *E*; I say that the rectangle contained by *A*, *BC* is equal to the rectangle contained by *A*, *BD*, that contained by *A*, *DE* and that contained by *A*, *EC*.

For let *BF* be drawn from *B* at right angles to *BC* [I. 11]; let *BG* be made equal to *A* [I. 3], through *G* let *GH* be drawn parallel to *BC* [I. 31], and through *D*, *E*, *C* let *DK*, *EL*, *CH* be drawn parallel to *BG*.

Then *BH* is equal to *BK*, *DL*, *EH*.

Now *BH* is the rectangle *A*, *BC*, for it is contained by *GB*, *BC*, and *BG* is equal to *A*; *BK* is the rectangle *A*, *BD*, for it is contained by *GB*, *BD*, and *BG* is equal to *A*; and *DL* is the rectangle *A*, *DE*, for *DK*, that is *BG* is equal to *A* [I. 34].

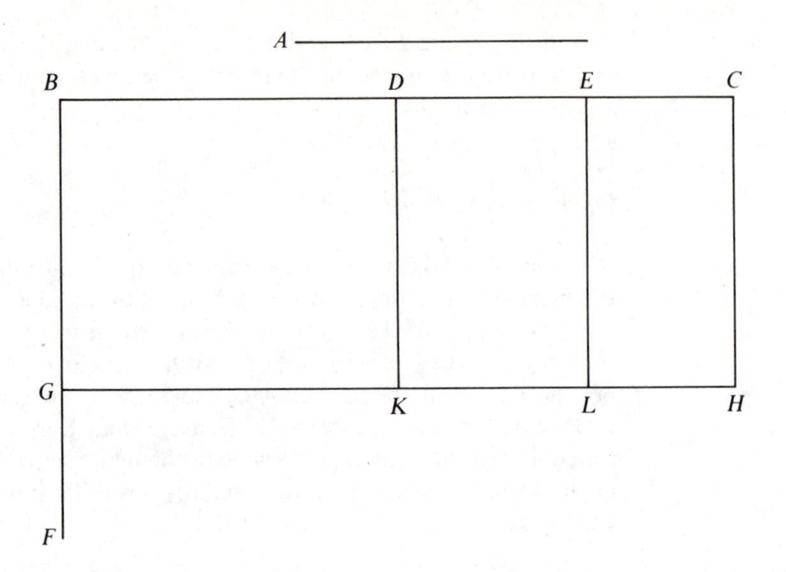

Similarly also EH is the rectangle A, EC.

Therefore the rectangle A, BC is equal to the rectangle A, BD, the rectangle A, DE and the rectangle A, EC.

Therefore etc. Q.E.D.

(c) *Proposition 4*

If a straight line be cut at random, the square on the whole is equal to the squares on the segments and twice the rectangle contained by the segments.

For let the straight line AB be cut at random at C; I say that the square on AB is equal to the squares on AC, CB and twice the rectangle contained by AC, CB.

For let the square $ADEB$ be described on AB [I. 46], let BD be joined; through C let

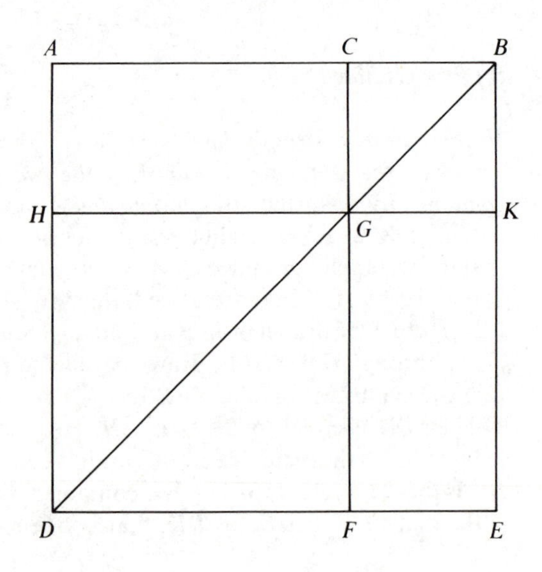

CF be drawn parallel to either *AD* or *EB*, and through *G* let *HK* be drawn parallel to either *AB* or *DE* [I. 31].

Then, since *CF* is parallel to *AD*, and *BD* has fallen on them, the exterior angle *CGB* is equal to the interior and opposite angle *ADB* [I. 29].

But the angle *ADB* is equal to the angle *ABD*, since the side *BA* is also equal to *AD* [I. 5]; therefore the angle *CGB* is also equal to the angle *GBC*, so that the side *BC* is also equal to the side *CG* [I. 6].

But *CB* is equal to *GK*, and *CG* to *KB* [I. 34]; therefore *GK* is also equal to *KB*; therefore *CGKB* is equilateral.

I say next that it is also right-angled.

For, since *CG* is parallel to *BK*, the angles *KBC*, *GCB* are equal to two right angles [I. 29].

But the angle *KBC* is right; therefore the angle *BCG* is also right, so that the opposite angles *CGK*, *GKB* are also right [I. 34].

Therefore *CGKB* is right-angled; and it was also proved equilateral; therefore it is a square; and it is described on *CB*.

For the same reason *HF* is also a square; and it is described on *HG*, that is *AC* [I. 34].

Therefore the squares *HF*, *KC* are the squares on *AC*, *CB*.

Now, since *AG* is equal to *GE*, and *AG* is the rectangle *AC*, *CB*, for *GC* is equal to *CB*, therefore *GE* is also equal to the rectangle *AC*, *CB*.

Therefore *AG*, *GE* are equal to twice the rectangle *AC*, *CB*.

But the squares *HF*, *CK* are also the squares on *AC*, *CB*; therefore the four areas *HF*, *CK*, *AG*, *GE* are equal to the squares on *AC*, *CB* and twice the rectangle contained by *AC*, *CB*.

But *HF*, *CK*, *AG*, *GE* are the whole *ADEB*, which is the square on *AB*.

Therefore the square on *AB* is equal to the squares on *AC*, *CB* and twice the rectangle contained by *AC*, *CB*.

Therefore etc. Q.E.D.

(*d*) *Proposition 11*

To cut a given straight line so that the rectangle contained by the whole and one of the segments is equal to the square on the remaining segment.

Let *AB* be the given straight line; thus it is required to cut *AB* so that the rectangle contained by the whole and one of the segments is equal to the square on the remaining segment.

For let the square *ABDC* be described on *AB* [I. 46]; let *AC* be bisected at the point *E*, and let *BE* be joined; let *CA* be drawn through to *F*, and let *EF* be made equal to *BE*; let the square *FH* be described on *AF*, and let *GH* be drawn through to *K*.

I say that *AB* has been cut at *H* so as to make the rectangle contained by *AB*, *BH* equal to the square on *AH*.

For, since the straight line *AC* has been bisected at *E*, and *FA* is added to it, the rectangle contained by *CF*, *FA* together with the square on *AE* is equal to the square on *EF* [II. 6].

But *EF* is equal to *EB*; therefore the rectangle *CF*, *FA* together with the square on *AE* is equal to the square on *EB*.

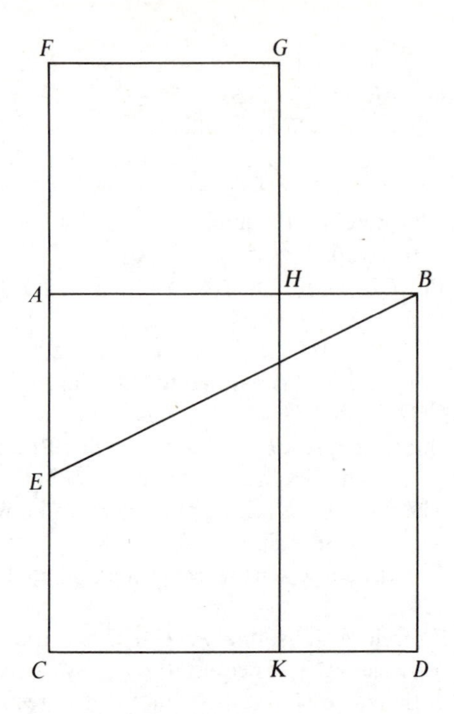

But the squares on BA, AE are equal to the square on EB, for the angle at A is right [I. 47]: therefore the rectangle CF, FA together with the square on AE is equal to the squares on BA, AE.

Let the square on AE be subtracted from each; therefore the rectangle CF, FA which remains is equal to the square on AB.

Now the rectangle CF, FA is FK, for AF is equal to FG; and the square on AB is AD; therefore FK is equal to AD.

Let AK be subtracted from each; therefore FH which remains is equal to HD.

And HD is the rectangle AB, BH, for AB is equal to BD; and FH is the square on AH; therefore the rectangle contained by AB, BH is equal to the square on HA.

Therefore the given straight line AB has been cut at H so as to make the rectangle contained by AB, BH equal to the square on HA. Q.E.F.

(*e*) *Proposition 14*

To construct a square equal to a given rectilineal figure.

Let A be the given rectilineal figure; thus it is required to construct a square equal to the rectilineal figure A.

For let there be constructed the rectangular parallelogram BD equal to the rectilineal figure A [I. 45].

Then, if BE is equal to ED, that which was enjoined will have been done; for a square BD has been constructed equal to the rectilineal figure A.

But, if not, one of the straight lines BE, ED is greater.

Let BE be greater, and let it be produced to F; let EF be made equal to ED, and let BF be bisected at G.

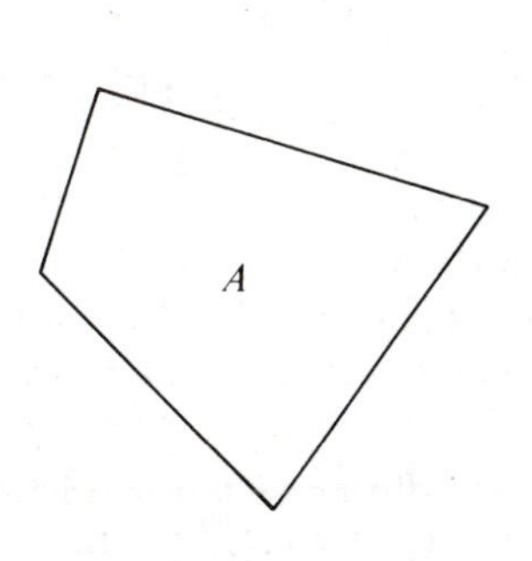

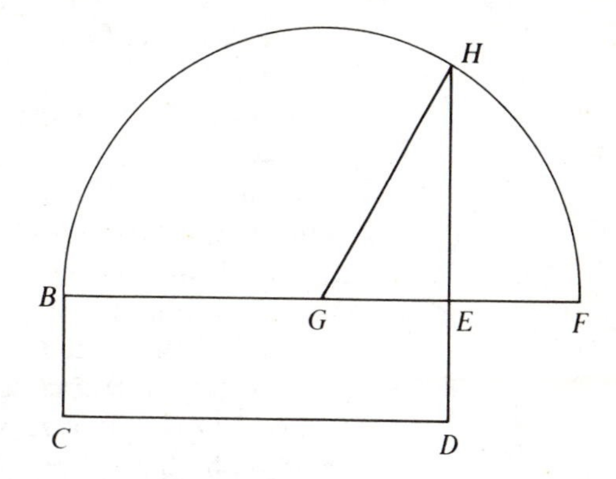

With centre G and distance one of the straight lines GB, GF let the semicircle BHF be described; let DE be produced to H, and let GH be joined.

Then, since the straight line BF has been cut into equal segments at G, and into unequal segments at E, the rectangle contained by BE, EF together with the square on EG is equal to the square on GF [II. 5].

But GF is equal to GH; therefore the rectangle BE, EF together with the square on GE is equal to the square on GH.

But the squares on HE, EG are equal to the square on GH [I. 47]; therefore the rectangle BE, EF together with the square on GE is equal to the squares on HE, EG.

Let the square on GE be subtracted from each; therefore the rectangle contained by BE, EF which remains is equal to the square on EH.

But the rectangle BE, EF is BD, for EF is equal to ED; therefore the parallelogram BD is equal to the square on HE.

And BD is equal to the rectilineal figure A.

Therefore the rectilineal figure A is also equal to the square which can be described on EH.

Therefore a square, namely that which can be described on EH, has been constructed equal to the given rectilineal figure A. Q.E.F.

3.C2 Book III: Definitions and Proposition 16

(*a*) *Definitions*

1 *Equal circles* are those the diameters of which are equal, or the radii of which are equal.

2 A straight line is said to *touch a circle* which, meeting the circle and being produced, does not cut the circle.

3 *Circles* are said to *touch one another* which, meeting one another, do not cut one another.

4 In a circle straight lines are said *to be equally distant from the centre* when the perpendiculars drawn to them from the centre are equal.

5 And that straight line is said to be *at a greater distance* on which the greater perpendicular falls.

6 A *segment of a circle* is the figure contained by a straight line and a circumference of a circle.
7 An *angle of a segment* is that contained by a straight line and a circumference of a circle.
8 An *angle in a segment* is the angle which, when a point is taken on the circumference of the segment and straight lines are joined from it to the extremities of the straight line which is the *base of the segment*, is contained by the straight lines so joined.
9 And, when the straight lines containing the angle cut off a circumference, the angle is said to *stand upon* that circumference.
10 A *sector of a circle* is the figure which, when an angle is constructed at the centre of the circle, is contained by the straight lines containing the angle and the circumference cut off by them.
11 *Similar segments of circles* are those which admit equal angles, or in which the angles are equal to one another.

(*b*) *Proposition 16*

The straight line drawn at right angles to the diameter of a circle from its extremity will fall outside the circle, and into the space between the straight line and the circumference another straight line cannot be interposed; further the angle of the semicircle is greater, and the remaining angle less, than any acute rectilineal angle.

Let *ABC* be a circle about *D* as centre and *AB* as diameter; I say that the straight line drawn from *A* at right angles to *AB* from its extremity will fall outside the circle.

For suppose it does not, but, if possible, let it fall within as *CA*, and let *DC* be joined.

Since *DA* is equal to *DC*, the angle *DAC* is also equal to the angle *ACD* [I. 5].

But the angle *DAC* is right; therefore the angle *ACD* is also right: thus, in the triangle *ACD*, the two angles *DAC*, *ACD* are equal to two right angles: which is impossible [I. 17].

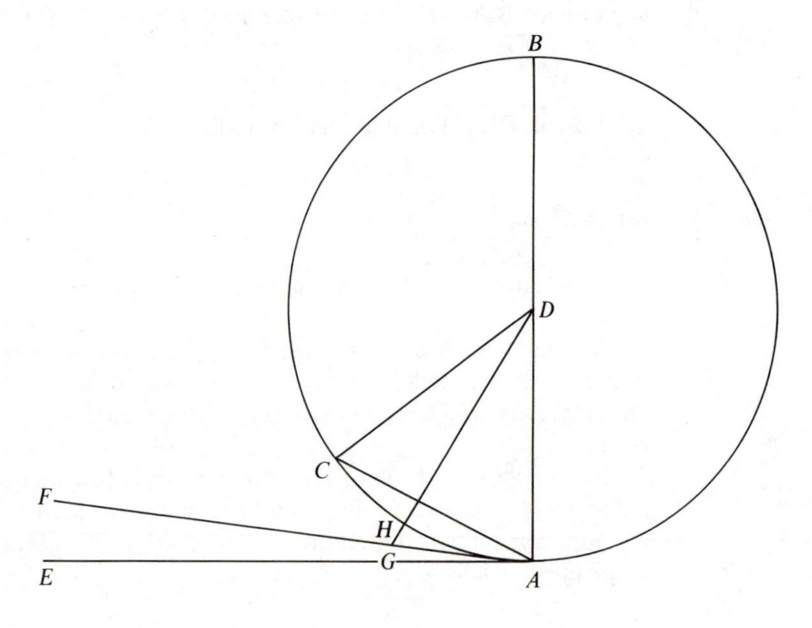

Therefore the straight line drawn from the point *A* at right angles to *BA* will not fall within the circle.

Similarly we can prove that neither will it fall on the circumference; therefore it will fall outside.

Let it fall as *AE*; I say next that into the space between the straight line *AE* and the circumference *CHA* another straight line cannot be interposed.

For, if possible, let another straight line be so interposed, as *FA*, and let *DG* be drawn from the point *D* perpendicular to *FA*.

Then, since the angle *AGD* is right, and the angle *DAG* is less than a right angle, *AD* is greater than *DG* [I. 19].

But *DA* is equal to *DH*; therefore *DH* is greater than *DG*, the less than the greater: which is impossible.

Therefore another straight line cannot be interposed into the space between the straight line and the circumference.

I say further that the angle of the semicircle contained by the straight line *BA* and the circumference *CHA* is greater than any acute rectilineal angle, and the remaining angle contained by the circumference *CHA* and the straight line *AE* is less than any acute rectilineal angle.

For, if there is any rectilineal angle greater than the angle contained by the straight line *BA* and the circumference *CHA*, and any rectilineal angle less than the angle contained by the circumference *CHA* and the straight line *AE*, then into the space between the circumference and the straight line *AE* a straight line will be interposed such as will make an angle contained by straight lines which is greater than the angle contained by the straight line *BA* and the circumference *CHA*, and another angle contained by straight lines which is less than the angle contained by the circumference *CHA* and the straight line *AE*.

But such a straight line cannot be interposed; therefore there will not be any acute angle contained by straight lines which is greater than the angle contained by the straight line *BA* and the circumference *CHA*, nor yet any acute angle contained by straight lines which is less than the angle contained by the circumference *CHA* and the straight line *AE*.—

Porism From this it is manifest that the straight line drawn at right angles to the diameter of a circle from its extremity touches the circle. Q.E.D.

3.C3 Book V: definitions

1 A magnitude is a *part* of a magnitude, the less of the greater, when it measures the greater.
2 The greater is a *multiple* of the less when it is measured by the less.
3 A *ratio* is a sort of relation in respect of size between two magnitudes of the same kind.
4 Magnitudes are said to *have a ratio* to one another which are capable, when multiplied, of exceeding one another.
5 Magnitudes are said to *be in the same ratio*, the first to the second and the third to the

fourth, when, if any equimultiples whatever be taken of the first and third, and any equimultiples whatever of the second and fourth, the former equimultiples alike exceed, are alike equal to, or alike fall short of, the latter equimultiples respectively taken in corresponding order.
6 Let magnitudes which have the same ratio be called *proportional.*

3.C4 Book VI: Definitions and Propositions 13, 30 and 31

(a) Definitions

1 *Similar rectilineal figures* are such as have their angles severally equal and the sides about the equal angles proportional.
2 A straight line is said to have been *cut in extreme and mean ratio* when, as the whole line is to the greater segment, so is the greater to the less.

(b) Proposition 13

To two given straight lines to find a mean proportional.

Let *AB*, *BC* be the two given straight lines; thus it is required to find a mean proportional to *AB*, *BC*.

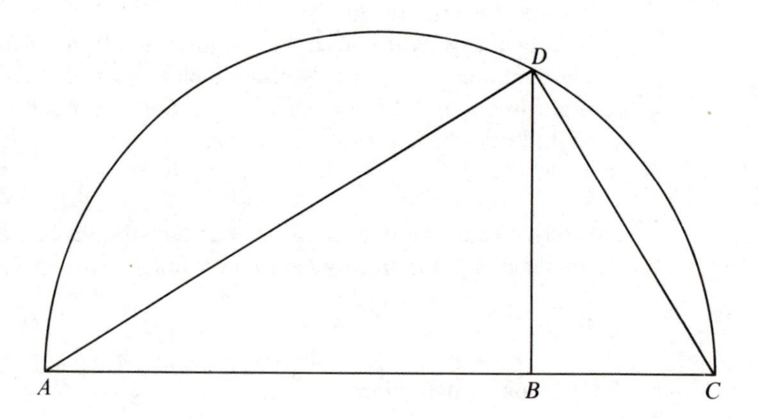

Let them be placed in a straight line, and let the semicircle *ADC* be described on *AC*; let *BD* be drawn from the point *B* at right angles to the straight line *AC*, and let *AD*, *DC* be joined.

Since the angle *ADC* is an angle in a semicircle, it is right [III. 31].

And, since, in the right-angled triangle *ADC*, *DB* has been drawn from the right angle perpendicular to the base, therefore *DB* is a mean proportional between the segments of the base *AB*, *BC* [VI. 8, Por.].

Therefore to the two given straight lines *AB*, *BC* a mean proportional *DB* has been found. Q.E.F.

(*c*) *Proposition 30*

To cut a given finite straight line in extreme and mean ratio.

Let *AB* be the given finite straight line; thus it is required to cut *AB* in extreme and mean ratio.

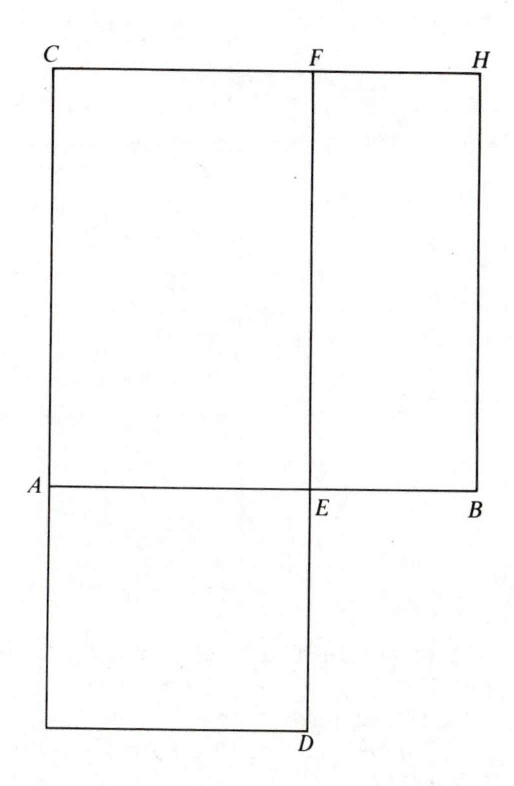

On *AB* let the square *BC* be described; and let there be applied to *AC* the parallelogram *CD* equal to *BC* and exceeding by the figure *AD* similar to *BC* [VI. 29].

Now *BC* is a square; therefore *AD* is also a square.

And, since *BC* is equal to *CD*, let *CE* be subtracted from each; therefore the remainder *BF* is equal to the remainder *AD*.

But it is also equiangular with it; therefore in *BF*, *AD* the sides about the equal angles are reciprocally proportional [VI. 14]; therefore, as *FE* is to *ED*, so is *AE* to *EB*.

But *FE* is equal to *AB*, and *ED* to *AE*.

Therefore, as *BA* is to *AE*, so is *AE* to *EB*.

And *AB* is greater than *AE*; therefore *AE* is also greater than *EB*.

Therefore the straight line *AB* has been cut in extreme and mean ratio at *E*, and the greater segment of it is *AE*. Q.E.F.

(*d*) *Proposition 31*

In right-angled triangles the figure on the side subtending the right angle is equal to the similar and similarly described figures on the sides containing the right angle.

Let *ABC* be a right-angled triangle having the angle *BAC* right; I say that the figure on *BC* is equal to the similar and similarly described figures on *BA*, *AC*.

Let *AD* be drawn perpendicular.

Then since, in the right-angled triangle *ABC*, *AD* has been drawn from the right angle at *A* perpendicular to the base *BC*, the triangles *ABD*, *ADC* adjoining the perpendicular are similar both to the whole *ABC* and to one another [VI. 8].

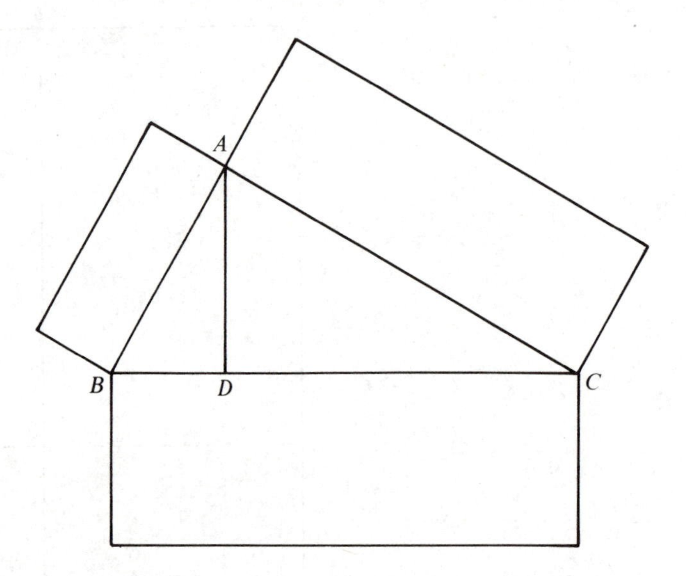

And, since *ABC* is similar to *ABD*, therefore, as *CB* is to *BA*, so is *AB* to *BD* [VI. Def. 1].

And, since three straight lines are proportional, as the first is to the third, so is the figure on the first to the similar and similarly described figure on the second [VI. 19, Por.]

Therefore, as *CB* is to *BD*, so is the figure on *CB* to the similar and similarly described figure on *BA*.

For the same reason also, as *BC* is to *CD*, so is the figure on *BC* to that on *CA*; so that, in addition, as *BC* is to *BD*, *DC*, so is the figure on *BC* to the similar and similarly described figures on *BA*, *AC*.

But *BC* is equal to *BD*, *DC*; therefore the figure on *BC* is also equal to the similar and similarly described figures on *BA*, *AC*.

Therefore etc. Q.E.D.

3.D The Number Theory Books

3.D1 Book VII: definitions

1 An *unit* is that by virtue of which each of the things that exist is called one.

2 A *number* is a multitude composed of units.
3 A number is a *part* of a number, the less of the greater, when it measures the greater;
4 but *parts* when it does not measure it.
5 The greater number is a *multiple* of the less when it is measured by the less.
6 An *even number* is that which is divisible into two equal parts.
7 An *odd number* is that which is not divisible into two equal parts, or that which differs by an unit from an even number.
8 An *even-times even number* is that which is measured by an even number according to an even number.
9 An *even-times odd number* is that which is measured by an even number according to an odd number.
10 An *odd-times odd number* is that which is measured by an odd number according to an odd number.
11 A *prime number* is that which is measured by an unit alone.
12 Numbers *prime to one another* are those which are measured by an unit alone as a common measure.
13 A *composite number* is that which is measured by some number.
14 Numbers *composite to one another* are those which are measured by some number as a common measure.
15 A number is said to *multiply* a number when that which is multiplied is added to itself as many times as there are units in the other, and thus some number is produced.
16 And, when two numbers having multiplied one another make some number, the number so produced is called *plane*, and its *sides* are the numbers which have multiplied one another.
17 And, when three numbers having multiplied one another make some number, the number so produced is *solid*, and its *sides* are the numbers which have multiplied one another.
18 A *square number* is equal multiplied by equal, or a number which is contained by two equal numbers.
19 And a *cube* is equal multiplied by equal and again by equal, or a number which is contained by three equal numbers.
20 Numbers are *proportional* when the first is the same multiple, or the same part, or the same parts, of the second that the third is of the fourth.
21 *Similar plane* and *solid* numbers are those which have their sides proportional.
22 A *perfect number* is that which is equal to its own parts.

3.D2 Book IX: Propositions 20, 21 and 22

(*a*) *Proposition 20*

Prime numbers are more than any assigned multitude of prime numbers.

Let A, B, C be the assigned prime numbers; I say that there are more prime numbers than A, B, C.

For let the least number measured by A, B, C be taken, and let it be DE; let the unit DF be added to DE.

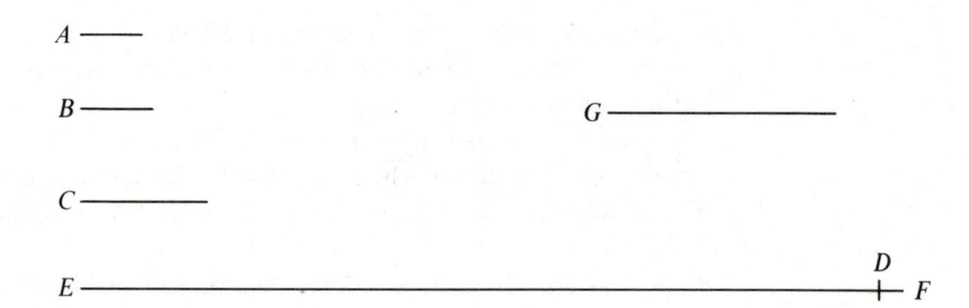

Then EF is either prime or not.

First, let it be prime; then the prime numbers A, B, C, EF have been found which are more than A, B, C.

Next, let EF not be prime; therefore it is measured by some prime number [VII. 31].

Let it be measured by the prime number G.

I say that G is not the same with any of the numbers A, B, C.

For, if possible, let it be so.

Now A, B, C measure DE; therefore G also will measure DE.

But it also measures EF.

Therefore G, being a number, will measure the remainder, the unit DF: which is absurd.

Therefore G is not the same with any one of the numbers A, B, C.

And by hypothesis it is prime.

Therefore the prime numbers A, B, C, G have been found which are more than the assigned multitude of A, B, C. Q.E.D.

(*b*) *Proposition 21*

If as many even numbers as we please be added together, the whole is even.

For let as many even numbers as we please, AB, BC, CD, DE, be added together; I say that the whole AE is even.

For, since each of the numbers AB, BC, CD, DE is even, it has a half part [VII. Def. 6]; so that the whole AE also has a half part.

But an even number is that which is divisible into two equal parts [*id.*]; therefore AE is even. Q.E.D.

(*c*) *Proposition 22*

If as many odd numbers as we please be added together, and their multitude be even, the whole will be even.

For let as many odd numbers as we please, AB, BC, CD, DE, even in multitude, be added together; I say that the whole AE is even.

For, since each of the numbers AB, BC, CD, DE is odd, if an unit be subtracted from

each, each of the remainders will be even [VII. Def. 7]; so that the sum of them will be even [IX. 21].

But the multitude of the units is also even.

Therefore the whole *AE* is also even [IX. 21]. Q.E.D.

3.D3 Book IX: Proposition 36

(*a*) *Proposition 36*

If as many numbers as we please beginning from an unit be set out continuously in double proportion, until the sum of all becomes prime, and if the sum multiplied into the last make some number, the product will be perfect.

For let as many numbers as we please, *A*, *B*, *C*, *D*, beginning from an unit be set out in double proportion, until the sum of all becomes prime, let *E* be equal to the sum, and let *E* by multiplying *D* make *FG*; I say that *FG* is perfect.

For, however many *A*, *B*, *C*, *D* are in multitude, let so many *E*, *HK*, *L*, *M* be taken in double proportion beginning from *E*; therefore, *ex aequali*, as *A* is to *D*, so is *E* to *M* [VII. 14].

Therefore the product of *E*, *D* is equal to the product of *A*, *M* [VII. 19].

And the product of *E*, *D* is *FG*; therefore the product of *A*, *M* is also *FG*.

Therefore *A* by multiplying *M* has made *FG*; therefore *M* measures *FG* according to the units in *A*.

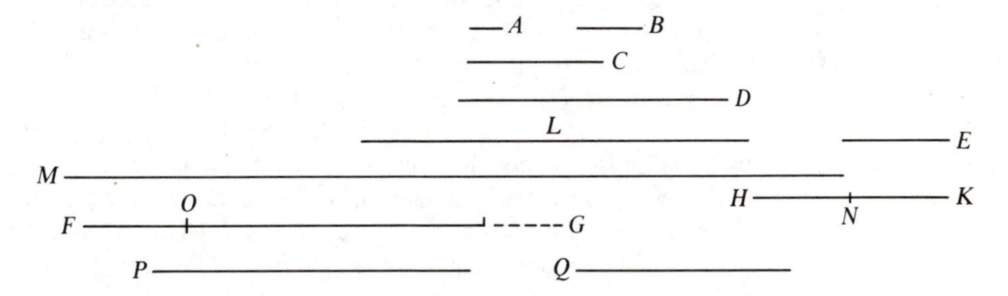

And *A* is a dyad; therefore *FG* is double of *M*.

But *M*, *L*, *HK*, *E* are continuously double of each other; therefore *E*, *HK*, *L*, *M*, *FG* are continuously proportional in double proportion.

Now let there be subtracted from the second *HK* and the last *FG* the numbers *HN*, *FO*, each equal to the first *E*; therefore, as the excess of the second is to the first, so is the excess of the last to all those before it [IX. 35].

Therefore, as *NK* is to *E*, so is *OG* to *M*, *L*, *KH*, *E*.

And *NK* is equal to *E*; therefore *OG* is also equal to *M*, *L*, *HK*, *E*.

But *FO* is also equal to *E*, and *E* is equal to *A*, *B*, *C*, *D* and the unit.

Therefore the whole *FG* is equal to *E*, *HK*, *L*, *M* and *A*, *B*, *C*, *D* and the unit; and it is measured by them.

I say also that FG will not be measured by any other number except A, B, C, D, E, HK, L, M and the unit.

For, if possible, let some number P measure FG, and let P not be the same with any of the numbers A, B, C, D, E, HK, L, M.

And, as many times as P measures FG, so many units let there be in Q; therefore Q by multiplying P has made FG.

But, further, E has also by multiplying D made FG; therefore, as E is to Q, so is P to D [VII. 19].

And, since A, B, C, D are continuously proportional beginning from an unit, therefore D will not be measured by any other number except A, B, C [IX. 13].

And, by hypothesis, P is not the same with any of the numbers A, B, C; therefore P will not measure D.

But, as P is to D, so is E to Q; therefore neither does E measure Q [VII. Def. 20].

And E is prime; and any prime number is prime to any number which it does not measure [VII. 29].

Therefore E, Q are prime to one another.

But primes are also least [VII. 21], and the least numbers measure those which have the same ratio the same number of times, the antecedent the antecedent and the consequent the consequent [VII. 20]; and, as E is to Q, so is P to D; therefore E measures P the same number of times that Q measures D.

But D is not measured by any other number except A, B, C; therefore Q is the same with one of the numbers A, B, C.

Let it be the same with B.

And, however many B, C, D are in multitude, let so many E, HK, L be taken beginning from E.

Now E, HK, L are in the same ratio with B, C, D; therefore, *ex aequali*, as B is to D, so is E to L [VII. 14].

Therefore the product of B, L is equal to the product of D, E [VII. 19].

But the product of D, E is equal to the product of Q, P; therefore the product of Q, P is also equal to the product of B, L.

Therefore, as Q is to B, so is L to P [VII. 19].

And Q is the same with B; therefore L is also the same with P: which is impossible, for by hypothesis P is not the same with any of the numbers set out.

Therefore no number will measure FG except A, B, C, D, E, HK, L, M and the unit.

And FG was proved equal to A, B, C, D, E, HK, L, M and the unit; and a perfect number is that which is equal to its own parts [VII. Def. 22]; therefore FG is perfect. Q.E.D.

(*b*) *Nicomachus on perfect numbers*

Now when a number, comparing with itself the sum and combination of all the factors whose presence it will admit, neither exceeds them in multitude nor is exceeded by them, then such a number is properly said to be perfect, as one which is equal to its own parts. Such numbers are 6 and 28; for 6 has the factors half, third, and sixth, 3, 2, and 1, respectively, and these added together make 6 and are equal to the original number, and neither more nor less. Twenty-eight has the factors half, fourth, seventh, fourteenth, and twenty-eighth, which are 14, 7, 4, 2 and 1; these added together make

28, and so neither are the parts greater than the whole nor the whole greater than the parts, but their comparison is in equality, which is the peculiar quality of the perfect number.

It comes about that even as fair and excellent things are few and easily enumerated, while ugly and evil ones are widespread, so also the superabundant and deficient numbers are found in great multitude and irregularly placed—for the method of their discovery is irregular—but the perfect numbers are easily enumerated and arranged with suitable order; for only one is found among the units, 6, only one other among the tens, 28, and a third in the rank of the hundreds, 496 alone, and a fourth within the limits of the thousands, that is, below ten thousand, 8,128. And it is their accompanying characteristic to end alternately in 6 or 8, and always to be even.

There is a method of producing them, neat and unfailing, which neither passes by any of the perfect numbers nor fails to differentiate any of those that are not such, which is carried out in the following way.

You must set forth the even-times even numbers from unity, advancing in order in one line, as far as you please: 1, 2, 4, 8, 16, 32, 64, 128, 256, 512, 1,024, 2,048, 4,096 .... Then you must add them together, one at a time, and each time you make a summation observe the result to see what it is. If you find that it is a prime, incomposite number, multiply it by the quantity of the last number added, and the result will always be a perfect number. If, however, the result is secondary and composite, do not multiply, but add the next and observe again what the resulting number is; if it is secondary and composite, again pass it by and do not multiply; but add the next; but if it is prime and incomposite, multiply it by the last term added, and the result will be a perfect number; and so on to infinity. In similar fashion you will produce all the perfect numbers in succession, overlooking none.

For example, to 1 I add 2, and observe the sum, and find that it is 3, a prime and incomposite number in accordance with our previous demonstrations; for it has no factor with denominator different from the number itself, but only that with denominator agreeing. Therefore I multiply it by the last number to be taken into the sum, that is, 2; I get 6, and this I declare to be the first perfect number in actuality, and to have those parts which are beheld in the numbers of which it is composed. For it will have unity as the factor with denominator the same as itself, that is, its sixth part; and 3 as the half, which is seen in 2, and conversely 2 as its third part.

Twenty-eight likewise is produced by the same method when another number, 4, is added to the previous ones. For the sum of the three, 1, 2, and 4, is 7, and is found to be prime and incomposite, for it admits only the factor with denominator like itself, the seventh part. Therefore I multiply it by the quantity of the term last taken into the summation, and my result is 28, equal to its own parts, and having its factors derived from the numbers already adduced, a half corresponding to 2; a fourth, to 7; a seventh, to 4; a fourteenth to offset the half; and a twenty-eighth, in accordance with its own nomenclature, which is 1 in all numbers.

When these have been discovered, 6 among the units and 28 in the tens, you must do the same to fashion the next. Again add the next number, 8, and the sum is 15. Observing this, I find that we no longer have a prime and incomposite number, but in addition to the factor with denominator like the number itself, it has also a fifth and a third, with unlike denominators. Hence I do not multiply it by 8, but add the next number, 16, and 31 results. As this is a prime, incomposite number, of necessity it will be multiplied, in accordance with the general rule of the process, by the last number added, 16, and the result is 496, in the hundreds; and then comes 8,128 in the thousands, and so on, as far as it is convenient for one to follow.

3.E Books X–XIII

3.E1 Book X: Definitions, Propositions 1 and 9, and Lemma 1

(*a*) *Definitions*

1 Those magnitudes are said to be *commensurable* which are measured by the same measure, and those *incommensurable* which cannot have any common measure.
2 Straight lines are *commensurable in square* when the squares on them are measured by the same area, and *incommensurable in square* when the squares on them cannot possibly have any area as a common measure.
3 With these hypotheses, it is proved that there exist straight lines infinite in multitude which are commensurable and incommensurable respectively, some in length only, and others in square also, with an assigned straight line. Let then the assigned straight line be called *rational*, and those straight lines which are commensurable with it, whether in length and in square or in square only, *rational*, but those which are incommensurable with it *irrational*.
4 And let the square on the assigned straight line be called *rational* and those areas which are commensurable with it *rational*, but those which are incommensurable with it *irrational*, and the straight lines which produce them *irrational*, that is, in case the areas are squares, the sides themselves, but in case they are any other rectilineal figures, the straight lines on which are described squares equal to them.

(*b*) *Proposition 1*

Two unequal magnitudes being set out, if from the greater there be subtracted a magnitude greater than its half, and from that which is left a magnitude greater than its half, and if this process be repeated continually, there will be left some magnitude which will be less than the lesser magnitude set out.

Let AB, C be two unequal magnitudes of which AB is the greater: I say that, if from AB there be subtracted a magnitude greater than its half, and from that which is left a magnitude greater than its half, and if this process be repeated continually, there will be left some magnitude which will be less than the magnitude C.

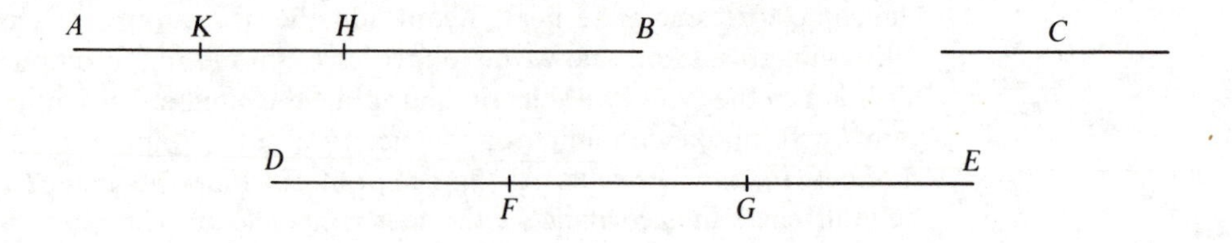

For C if multiplied will sometime be greater than AB [cf. V. Def. 4].

Let it be multiplied, and let DE be a multiple of C, and greater than AB; let DE be divided into the parts DF, FG, GE equal to C, from AB let there be subtracted BH greater than its half, and, from AH, HK greater than its half, and let this process be repeated continually until the divisions in AB are equal in multitude with the divisions in DE.

Let, then, AK, KH, HB be divisions which are equal in multitude with DF, FG, GE.

Now, since DE is greater than AB, and from DE there has been subtracted EG less than its half, and, from AB, BH greater than its half, therefore the remainder GD is greater than the remainder HA. And, since GD is greater than HA, and there has been subtracted, from GD, the half GF, and, from HA, HK greater than its half, therefore the remainder DF is greater than the remainder AK.

But DF is equal to C; therefore C is also greater than AK.

Therefore AK is less than C.

Therefore there is left of the magnitude AB the magnitude AK which is less than the lesser magnitude set out, namely C. Q.E.D.

And the theorem can be similarly proved even if the parts subtracted be halves.

(c) *Proposition 9*

The squares on straight lines commensurable in length have to one another the ratio which a square number has to a square number; and squares which have to one another the ratio which a square number has to a square number will also have their sides commensurable in length. But the squares on straight lines incommensurable in length have not to one another the ratio which a square number has to a square number; and squares which have not to one another the ratio which a square number has to a square number will not have their sides commensurable in length either.

For let A, B be commensurable in length; I say that the square on A has to the square on B the ratio which a square number has to a square number.

A B

C

D

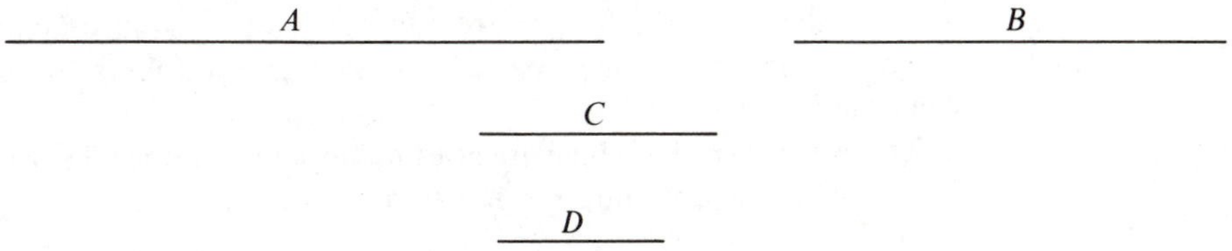

For, since A is commensurable in length with B, therefore A has to B the ratio which a number has to a number [X. 5].

Let it have to it the ratio which C has to D.

Since then, as A is to B, so is C to D, while the ratio of the square on A to the square on B is duplicate of the ratio of A to B, for similar figures are in the duplicate ratio of their corresponding sides [VI. 20, Por.]; and the ratio of the square on C to the square on D is duplicate of the ratio of C to D, for between two square numbers there is one mean proportional number, and the square number has to the square number the ratio duplicate of that which the side has to the side [VIII. 11]; therefore also, as the square on A is to the square on B, so is the square on C to the square on D.

Next, as the square on A is to the square on B, so let the square on C be to the square on D; I say that A is commensurable in length with B.

For since, as the square on A is to the square on B, so is the square on C to the square on D, while the ratio of the square on A to the square on B is duplicate of the ratio of A to B, and the ratio of the square on C to the square on D is duplicate of the ratio of C to D, therefore also, as A is to B, so is C to D.

Therefore A has to B the ratio which the number C has to the number D; therefore A is commensurable in length with B [X. 6].

Next, let A be incommensurable in length with B; I say that the square on A has not to the square on B the ratio which a square number has to a square number.

For, if the square on A has to the square on B the ratio which a square number has to a square number, A will be commensurable with B.

But it is not; therefore the square on A has not to the square on B the ratio which a square number has to a square number.

Again, let the square on A not have to the square on B the ratio which a square number has to a square number; I say that A is incommensurable in length with B.

For, if A is commensurable with B, the square on A will have to the square on B the ratio which a square number has to a square number.

But it has not; therefore A is not commensurable in length with B.

Therefore etc.

Porism And it is manifest from what has been proved that straight lines commensurable in length are always commensurable in square also, but those commensurable in square are not always commensurable in length also.

Lemma It has been proved in the arithmetical books that similar plane numbers have to one another the ratio which a square number has to a square number [VIII. 26], and that, if two numbers have to one another the ratio which a square number has to a square number, they are similar plane numbers [Converse of VIII. 26].

And it is manifest from these propositions that numbers which are not similar plane numbers, that is, those which have not their sides proportional, have not to one another the ratio which a square number has to a square number.

For, if they have, they will be similar plane numbers: which is contrary to the hypothesis.

Therefore numbers which are not similar plane numbers have not to one another the ratio which a square number has to a square number.

(*d*) *Lemma 1*

To find two square numbers such that their sum is also square.

Let two numbers AB, BC be set out, and let them be either both even or both odd.

Then since, whether an even number is subtracted from an even number, or an odd number from an odd number, the remainder is even [IX. 24, 26], therefore the remainder AC is even.

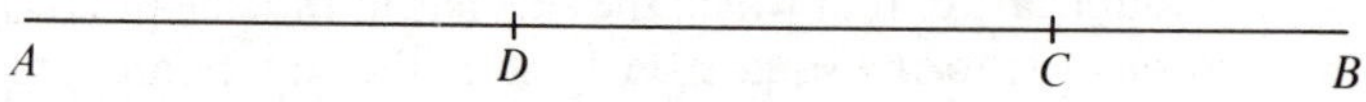

A D C B

Let AC be bisected at D.

Let AB, BC also be either similar plane numbers, or square numbers, which are themselves also similar plane numbers.

Now the product of AB, BC together with the square on CD is equal to the square on BD [II. 6].

And the product of AB, BC is square, inasmuch as it was proved that, if two similar plane numbers by multiplying one another make some number, the product is square [IX. 1].

Therefore two square numbers, the product of AB, BC, and the square on CD, have been found which, when added together, make the square on BD.

And it is manifest that two square numbers, the square on BD and the square on CD, have again been found such that their difference, the product of AB, BC, is a square, whenever AB, BC are similar plane numbers.

But when they are not similar plane numbers, two square numbers, the square on BD and the square on DC, have been found such that their difference, the product of AB, BC, is not square. Q.E.D.

3.E2 Book XI: Definitions

1 A *solid* is that which has length, breadth, and depth.

[...]

14 When, the diameter of a semicircle remaining fixed, the semicircle is carried round and restored again to the same position from which it began to be moved, the figure so comprehended is a *sphere*.

15 The *axis of the sphere* is the straight line which remains fixed and about which the semicircle is turned.

16 The *centre of the sphere* is the same as that of the semicircle.

17 A *diameter of the sphere* is any straight line drawn through the centre and terminated in both directions by the surface of the sphere.

18 When, one side of those about the right angle in a right-angled triangle remaining fixed, the triangle is carried round and restored again to the same position from which it began to be moved, the figure so comprehended is a *cone*.

And, if the straight line which remains fixed be equal to the remaining side about the right angle which is carried round, the cone will be *right-angled*; if less, *obtuse-angled*; and if greater, *acute-angled*.

19 The *axis of the cone* is the straight line which remains fixed and about which the triangle is turned.

20 And the *base* is the circle described by the straight line which is carried round.

21 When, one side of those about the right angle in a rectangular parallelogram remaining fixed, the parallelogram is carried round and restored again to the same position from which it began to be moved, the figure so comprehended is a *cylinder*.

22 The *axis of the cylinder* is the straight line which remains fixed and about which the parallelogram is turned.

23 And the *bases* are the circles described by the two sides opposite to one another which are carried round.

24 *Similar cones and cylinders* are those in which the axes and the diameters of the bases are proportional.
25 A *cube* is a solid figure contained by six equal squares.
26 An *octahedron* is a solid figure contained by eight equal and equilateral triangles.
27 An *icosahedron* is a solid figure contained by twenty equal and equilateral triangles.
28 A *dodecahedron* is a solid figure contained by twelve equal, equilateral, and equiangular pentagons.

3.E3 Book XII: Proposition 2

Circles are to one another as the squares on the diameters.

Let *ABCD*, *EFGH* be circles, and *BD*, *FH* their diameters; I say that, as the circle *ABCD* is to the circle *EFGH*, so is the square on *BD* to the square on *FH*.

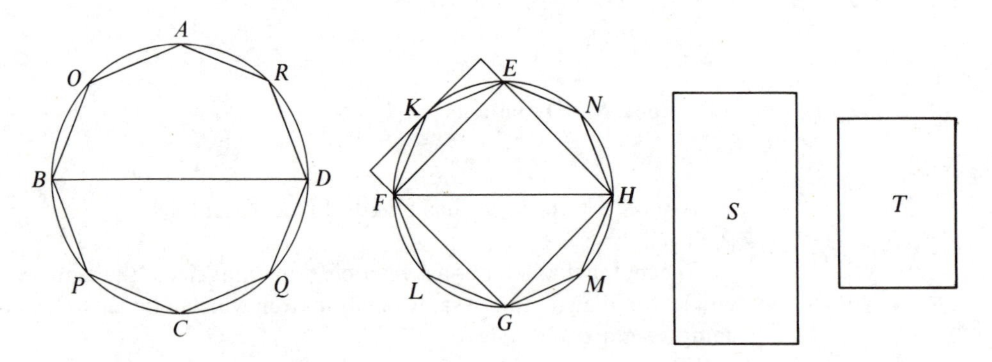

For, if the square on *BD* is not to the square on *FH* as the circle *ABCD* is to the circle *EFGH*, then, as the square on *BD* is to the square on *FH*, so will the circle *ABCD* be either to some less area than the circle *EFGH*, or to a greater.

First, let it be in that ratio to a less area *S*.

Let the square *EFGH* be inscribed in the circle *EFGH*; then the inscribed square is greater than the half of the circle *EFGH*, inasmuch as, if through the points *E*, *F*, *G*, *H* we draw tangents to the circle, the square *EFGH* is half the square circumscribed about the circle, and the circle is less than the circumscribed square; hence the inscribed square *EFGH* is greater than the half of the circle *EFGH*.

Let the circumference *EF*, *FG*, *GH*, *HE* be bisected at the points *K*, *L*, *M*, *N*, and let *EK*, *KF*, *FL*, *LG*, *GM*, *MH*, *HN*, *NE* be joined; therefore each of the triangles *EKF*, *FLG*, *GMH*, *HNE* is also greater than the half of the segment of the circle about it, inasmuch as, if through the points *K*, *L*, *M*, *N* we draw tangents to the circle and complete the parallelograms on the straight lines *EF*, *FG*, *GH*, *HE*, each of the triangles *EKF*, *FLG*, *GMH*, *HNE* will be half of the parallelogram about it, while the segment about it is less than the parallelogram; hence each of the triangles *EKF*, *FLG*, *GMH*, *HNE* is greater than the half of the segment of the circle about it.

Thus, by bisecting the remaining circumferences and joining straight lines, and by doing this continually, we shall leave some segments of the circle which will be less than the excess by which the circle *EFGH* exceeds the area *S*.

For it was proved in the first theorem of the tenth book that, if two unequal magnitudes be set out, and if from the greater there be subtracted a magnitude greater than the half, and from that which is left a greater than the half, and if this be done continually, there will be left some magnitude which will be less than the lesser magnitude set out.

Let segments be left such as described, and let the segments of the circle *EFGH* on *EK*, *KF*, *FL*, *LG*, *GM*, *MH*, *HN*, *NE* be less than the excess by which the circle *EFGH* exceeds the area *S*.

Therefore the remainder, the polygon *EKFLGMHN*, is greater than the area *S*.

Let there be inscribed, also, in the circle *ABCD* the polygon *AOBPCQDR* similar to the polygon *EKFLGMHN*; therefore, as the square on *BD* is to the square on *FH*, so is the polygon *AOBPCQDR* to the polygon *EKFLGMHN* [XII. 1].

But, as the square on *BD* is to the square on *FH*, so also is the circle *ABCD* to the area *S*; therefore also, as the circle *ABCD* is to the area *S*, so is the polygon *AOBPCQDR* to the polygon *EKFLGMHN* [V. 11]; therefore, alternately, as the circle *ABCD* is to the polygon inscribed in it, so is the area *S* to the polygon *EKFLGMHN* [V. 16].

But the circle *ABCD* is greater than the polygon inscribed in it; therefore the area *S* is also greater than the polygon *EKFLGMHN*.

But it is also less: which is impossible.

Therefore, as the square on *BD* is to the square on *FH*, so is not the circle *ABCD* to any area less than the circle *EFGH*.

Similarly we can prove that neither is the circle *EFGH* to any area less than the circle *ABCD* as the square on *FH* is to the square on *BD*.

I say next that neither is the circle *ABCD* to any area greater than the circle *EFGH* as the square on *BD* is to the square on *FH*.

For, if possible, let it be in that ratio to a greater area *S*.

Therefore, inversely, as the square on *FH* is to the square on *DB*, so is the area *S* to the circle *ABCD*.

But, as the area *S* is to the circle *ABCD*, so is the circle *EFGH* to some area less than the circle *ABCD*; therefore also, as the square on *FH* is to the square on *BD*, so is the circle *EFGH* to some area less than the circle *ABCD* [V. 11]: which was proved impossible.

Therefore, as the square on *BD* is to the square on *FH*, so is not the circle *ABCD* to any area greater than the circle *EFGH*.

And it was proved that neither is it in that ratio to any area less than the circle *EFGH*; therefore, as the square on *BD* is to the square on *FH*, so is the circle *ABCD* to the circle *EFGH*.

Therefore etc. Q.E.D.

3.E4 Book XIII: statements of Propositions 13–18 and a final result

(a) Statements of Propositions 13–18

13 To construct a pyramid, to comprehend it in a given sphere, and to prove that the square on the diameter of the sphere is one and a half times the square on the side of the pyramid.

14 To construct an octahedron and comprehend it in a sphere, as in the preceding case; and to prove that the square on the diameter of the sphere is double of the square on the side of the octahedron.

15 To construct a cube and comprehend it in a sphere, like the pyramid; and to prove that the square on the diameter of the sphere is triple of the square on the side of the cube.

16 To construct an icosahedron and comprehend it in a sphere; like the aforesaid figures; and to prove that the side of the icosahedron is the irrational straight line called minor.

17 To construct a dodecahedron and comprehend it in a sphere, like the aforesaid figures, and to prove that the side of the dodecahedron is the irrational straight line called apotome.

18 To set out the sides of the five figures and to compare them with one another.

(*b*) *A final result*

I say next that *no other figure, besides the said five figures, can be constructed which is contained by equilateral and equiangular figures equal to one another.*

For a solid angle cannot be constructed with two triangles, or indeed planes.

With three triangles the angle of the pyramid is constructed, with four the angle of the octahedron, and with five the angle of the icosahedron; but a solid angle cannot be formed by six equilateral and equiangular triangles placed together at one point, for, the angle of the equilateral triangle being two-thirds of a right angle, the six will be equal to four right angles; which is impossible, for any solid angle is contained by angles less than four right angles [XI. 21].

For the same reason, neither can a solid angle be constructed by more than six plane angles.

By three squares the angle of the cube is contained, but by four it is impossible for a solid angle to be contained, for they will again be four right angles.

By three equilateral and equiangular pentagons the angle of the dodecahedron is contained; but by four such it is impossible for any solid angle to be contained, for, the angle of the equilateral pentagon being a right angle and a fifth, the four angles will be greater than four right angles: which is impossible.

Neither again will a solid angle be contained by other polygonal figures by reason of the same absurdity.

Therefore etc. Q.E.D.

3.F Scholarly and Personal Discovery of Euclid's Text

3.F1 A. Aaboe on the textual basis

The problems confronting us when we wish to establish a firm textual basis for the study of Greek mathematics are entirely different from the ones we met in Babylonian mathematics. There our texts—the clay tablets—might be broken or damaged, and the terminology might be obscure and understandable only from the context. But one thing was beyond doubt, and that was the authenticity of the texts, for these were the very tablets the Babylonians themselves had written.

Let us now take Euclid's *Elements* as an example illustrating how different the situation is when we deal with Greek mathematical texts. It was written about 300 BC, but the earliest manuscripts containing the Greek text date from the tenth-century AD, i.e. they are much closer in time to us than to Euclid.

Thus even our oldest texts are copies of copies of copies many times removed, and from these we must try to establish what Euclid himself wrote. This is a detective problem of no mean proportions, and classical scholars have developed refined techniques for solving it. The procedure is, in crude outline, as follows:

We compare manuscripts X and Y. If Y has all the errors and peculiarities of X and in addition some of its own, it is a fair assumption that Y is a copy, or a copy of a copy of X. If X and Y have a number of errors in common and each some of its own, they are probably both derived from a common archetype Z, which may be lost but reconstructible. In this fashion the extant manuscripts can be arranged in families, each family represented by an archetype. From the archetypes the original text is then reconstructed.

J. L. Heiberg, the Danish classical scholar who with unbelievable industry gave us the definitive editions of most of the Greek mathematical texts, found that the extant Euclid manuscripts fall into two families. All but one are descendants of an edition by Theon of Alexandria, a busy editor and commentator of the fourth century AD. One manuscript seems, however, to be derived mainly from a version free of Theon's recensions, but based on a later copy of Euclid than the one Theon had used. Taking these and other facts into account, Heiberg succeeded in establishing as trustworthy a Greek text of Euclid's *Elements* as possible, and it was published between 1883 and 1888. This edition formed the basis of all later investigations and translations of Euclid, e.g. T. L. Heath's English version.

But Euclid's *Elements* was, of course, known in the Western World long before Heiberg's edition. Already in the reign of the Calif Harun ar-Rashid (786–809), whose fame has been sustained by the Tales of the Arabian Nights, Euclid was rendered in Arabic by al-Hajjaj, and several Arabic versions followed, some of them drastically abbreviated and others taking great liberties with the Greek original. In the twelfth century some of the Arabic versions were introduced in Europe in Latin translations (Adelard, Gerard of Cremona), and many other Latin translations appeared in the thirteenth and fourteenth centuries. In 1482 the first printed edition of Euclid (the Campanus translation) was published, and the first Latin translation from the Greek, by Zamberti, appeared in 1505. A Greek text was printed in 1533.

We see here a characteristic pattern: first, translations from the Greek into Arabic in the ninth century, then the Latinization of Arabic versions in the twelfth (the time of the Crusades, during which the contacts between Christians and Muslims were not always bloody), then the printing of the Latin versions towards the end of the fifteenth century, followed closely by Latin translations from the Greek and the Greek text itself (we are now in the Renaissance), and finally, the scholarly definitive edition of the Greek text during the latter half of the nineteenth century.

This could be the history of almost any Greek mathematical, or indeed, technical text, and it illustrates well the tastes and interests of the various periods. There may be variations of different sorts—some of Archimedes' and Apollonius' works, for example, are preserved only in their Arabic versions—and the dates may shift a bit, but in the main the pattern holds.

3.F2 Later personal impacts

(*a*) *Thomas Hobbes* (*1628*)

He was 40 years old before he looked on Geometry; which happened accidentally. Being in a Gentleman's Library, Euclid's *Elements* lay open, and 'twas the *47 El. libri I*. He read the Proposition. *By G—*, sayd he (he would now and then sweare an emphaticall Oath by way of emphasis) *this is impossible*! So he reads the Demonstration of it, which referred him back to such a Proposition; which proposition he read. That referred him back to another, which he also read. *Et sic deinceps* [and so on] that at last he was demonstratively convinced of that trueth. This made him in love with Geometry.

(*b*) *Bertrand Russell* (*1883*)

At the age of eleven, I began Euclid, with my brother as my tutor. This was one of the great events of my life, as dazzling as first love. I had not imagined that there was anything so delicious in the world. After I had learned the fifth proposition, my brother told me that it was generally considered difficult, but I had found no difficulty whatever. This was the first time it had dawned upon me that I might have some intelligence. From that moment until Whitehead and I finished *Principia Mathematica*, when I was thirty-eight, mathematics was my chief interest, and my chief source of happiness. Like all happiness, however, it was not unalloyed. I had been told that Euclid proved things, and was much disappointed that he started with axioms. At first I refused to accept them unless my brother could offer me some reason for doing so, but he said: 'If you don't accept them we cannot go on', and as I wished to go on, I reluctantly admitted them *pro tem.* The doubt as to the premisses of mathematics which I felt at that moment remained with me, and determined the course of my subsequent work.

The beginnings of Algebra I found far more difficult, perhaps as a result of bad teaching. I was made to learn by heart: 'The square of the sum of two numbers is equal to the sum of their squares increased by twice their product'. I had not the vaguest idea what this meant, and when I could not remember the words, my tutor threw the book at my head, which did not stimulate my intellect in any way.

3.G Historians Debate Geometrical Algebra

3.G1 B. L. van der Waerden

When one opens Book II of the *Elements*, one finds a sequence of propositions, which are nothing but geometric formulations of algebraic rules. So, e.g. II.1:

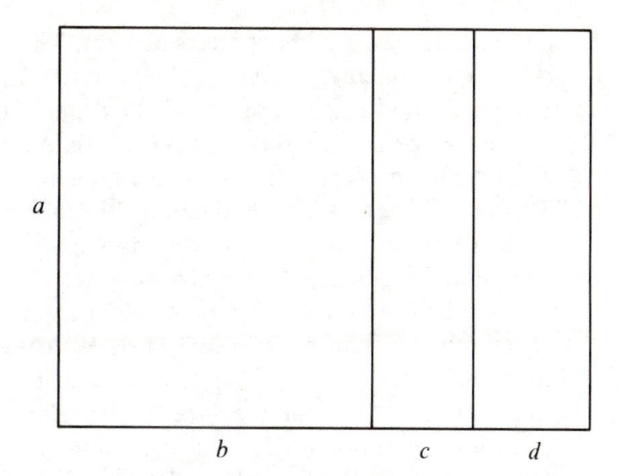

Figure 1

If there be two straight lines, and one of them be cut into any number of segments whatever, the rectangle contained by the two straight lines is equal to the rectangles contained by the uncut straight line and each of the segments,

corresponds to the formula [see Figure 1]:

$$a(b + c + \ldots) = ab + ac + \ldots$$

II.2 and II.3 are special cases of this proposition. II.4 corresponds to the formula

$$(a + b)^2 = a^2 + b^2 + 2ab.$$

The proof can be read off immediately from Figure 2. In II.7, one recognizes the analogous formula for $(a - b)^2$. We have here, so to speak, the start of an algebra

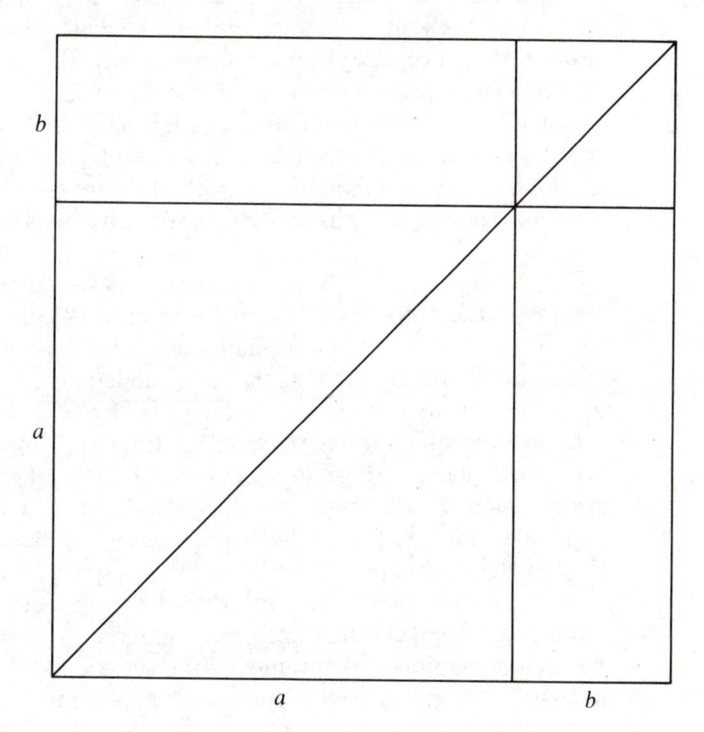

Figure 2

textbook, dressed up in geometrical form. The magnitudes under consideration are always line segments; instead of 'the product *ab*', one speaks of 'the rectangle formed by *a* and *b*', and in place of a^2, of 'the square on *a*'.

Quite properly, Zeuthen speaks in this connection of a 'geometric algebra'. Throughout Greek mathematics, one finds numerous applications of this 'algebra'. The line of thought is always algebraic, the formulation geometric. The greater part of the theory of polygons and polyhedra is based on this method; the entire theory of conic sections depends on it. Theaetetus in the fourth century, Archimedes and Apollonius in the third are perfect virtuosos on this instrument.

Presently we shall make clear that this geometric algebra is the continuation of Babylonian algebra. The Babylonians also used the terms 'rectangle' for xy and 'square' for x^2, but besides these and alternating with them, such arithmetic expressions as multiplication, root extraction, etc. occur as well. The Greeks, on the other hand, consistently avoid such expressions, except in operations on integers and on simple fractions; everything is translated into geometric terminology. But since it is indeed a translation which occurs here and the line of thought is algebraic, there is no danger of misrepresentation, if we reconvert the derivations into algebraic language and use modern notations. From now on we shall therefore quite coolly replace expressions such as 'the square on *a*', 'the rectangle formed by *a* and *b*' by the modern symbols a^2 and ab, whenever they simplify the presentation.

3.G2 Sabetai Unguru

One of the central concepts for the understanding of ancient Greek mathematics has customarily been, at least since the time of Paul Tannery and Hieronymus Georg Zeuthen, the concept of 'geometric algebra'. What it amounts to is the view that Greek mathematics, especially after the discovery of the 'irrational' by the Pythagorean school, is *algebra* dressed up, primarily for the sake of rigor, in geometrical garb. The reasoning of Greek mathematics, the line of attack of its various problems, the solutions provided to those problems, etc. all are essentially *algebraic*, though, to be sure, for reasons that have never been fully elaborated, attired in geometrical accouterments.

[...] I believe such a view is offensive, naive, and historically untenable. It is certainly indefensible on the basis of the historical record, i.e., on the basis of a study of the documents of Greek mathematics, undertaken *not* from the point of view of the achievements, results, and methods of modern mathematics, which, it should be unequivocally understood, are completely irrelevant in attempting to understand Greek mathematics for its own sake, but from the standpoint adopted by the ancient Greek mathematicians themselves, inasmuch as this standpoint could be grasped by a modern mind. To read ancient mathematical texts with modern mathematics in mind is the safest method for misunderstanding the character of ancient mathematics, in which philosophical presuppositions and metaphysical commitments played a much more fundamental and decisive role than they play in modern mathematics. To assume that one can apply automatically and indiscriminately to any mathematical content the modern manipulative techniques of algebraic symbols is the surest way to fail to understand the inherent differences built into the mathematics of different eras.

3.G3 B. L. van der Waerden

We (Zeuthen and his followers) feel that the Greeks started with algebraic problems and translated them into geometric language. Unguru thinks that we argued like this: We found that the theorems of Euclid II can be translated into modern algebraic formalism, and that they are easier to understand if thus translated, and this we took as '*the* proof that this is what the ancient mathematician had in mind'. Of course, this is nonsense. We are not so weak in logical thinking! The fact that a theorem can be translated into another notation does not prove a thing about what the author of the theorem had in mind.

No, our line of thought was quite different. We studied the wording of the theorems and tried to reconstruct the original ideas of the author. We found it *evident* that these theorems did not arise out of geometrical problems. We were not able to find any interesting geometrical problem that would give rise to theorems like II.1–4. On the other hand, we found that the explanation of these theorems as arising from algebra worked well. Therefore we adopted the latter explanation.

Now it turns out, to my great surprise, that what we, working mathematicians, found evident, is not evident to Unguru. [...]

Thābit ibn Qurra, a contemporary of Al-Khwārizmī, was an excellent geometer and astronomer, fully conversant with the work of Euclid. In a little known treatise, Thābit pointed out that the solution of the three types of quadratic equations according to 'the Algebra people' is equivalent to the 'application of areas with excess or defect' as presented by Euclid.

The example of Thābit shows that Unguru is completely wrong in thinking that mathematicians like Zeuthen came to their opinions about Greek geometric algebra only because they translated Euclid's propositions into modern algebraic symbolism. It is true that Zeuthen was able to use modern symbolism, but Thābit was not, and yet he arrived at the same conclusion as Zeuthen, namely that Al-Khwārizmī's solution of quadratic equations is equivalent to Euclid's procedure.

Unguru, like many non-mathematicians, grossly overestimates the importance of symbolism in mathematics. These people see our papers full of formulae, and they think that these formulae are an essential part of mathematical thinking. We, working mathematicians, know that in many cases the formulae are not at all essential, only convenient. The treatise of Thābit offers a good illustration of this thesis.

3.G4 Sabetai Unguru

Those who perceive an algebraic substructure bolstering Greek mathematics claim that the Greeks started with algebraic problems but, then, translated them into a geometric format. They have reached this conclusion, according to van der Waerden, by studying 'the wording of the theorems' and by trying 'to reconstruct the original ideas of the author. We found it *evident* that these theorems did not arise out of geometrical problems [!]. We were not able to find any interesting geometrical problem that would give rise to theorems like II.1–4. On the other hand, we found that the explanation of these theorems as arising from algebra worked well. Therefore we adopted the latter explanation.'

But what evidence does van der Waerden present to demonstrate that 'these theorems did not arise out of geometrical problems'? The answer, he tells us, is that no 'interesting geometrical problem' leads to them. How does he know? Answer: he could not find any. But the conclusion is unwarranted, since even if it is true that no interesting geometrical problem led to them, it does not follow that noninteresting geometrical problems did not lead to them either. Furthermore, what *is* an interesting geometrical problem? Van der Waerden does not say, but the answer is implicit in what follows: 'we found that the explanation of these theorems as arising from algebra worked well. Therefore....' An interesting geometrical problem, then, seems to be a problem the assumption of which 'works well' in explaining the origin of the theorems under discussion. And van der Waerden has decided arbitrarily (since he could not check all possible geometrical theorems and problems) that there are no 'interesting geometrical problems' working well under the circumstances. On the other hand, what works well is the assumption of an underlying algebraic foundation to Greek geometry. What does 'working well' mean, then? Again, no answer is provided, but it would clearly seem to mean something removing difficulties and enabling one to cut through to the root and thus come up with 'simple', 'convincing', straightforward explanations. Ultimately, then, in the paragraph under discussion, van der Waerden does say that Greek geometry (at least some important parts of it) is, taken by itself, unfathomable, puzzling, weird, and that one can get rid of these unsavoury features by assuming a hidden algebraic basis to it. Therefore, 'the Greeks started with algebraic problems and translated them into geometric language'. Q.E.D.

Leaving aside the circularity of the entire argument, and the conflation of logic and history that it involves, van der Waerden's assertions represent an unconscious but nevertheless clear-cut vindication of the argument that the real roots of the methodological position embodied in the concept 'geometric algebra' lie in the modern mathematician's ability to read geometric texts algebraically without any historical qualms.

3.G5 Ian Mueller

The scope of this book precludes discussion of all the issues raised by the Zeuthen–Neugebauer interpretation of geometric algebra. However, it may be possible to get clear on some issues. A precondition for doing so is an understanding of what is claimed in calling a piece of Greek mathematics algebraic. The development of modern algebra can be viewed as the culmination of the structural conception of mathematics. But that conception is very explicit in the vocabulary of algebra, in which terms like 'group' and 'field' designating arbitrary structures satisfying given formal conditions are prominent. Normally too, at least at the more elementary level, the algebraist will emphasize the diversity of possible mathematical embodiments of a group or a field or other algebraic domain. It is quite clearly out of the question that one could establish an algebraic interpretation of much of Greek mathematics on the basis of vocabulary or explicit procedures. Rather, such an interpretation depends upon reading texts like

Book II as saying something other than what they appear to be saying. I have already argued that the structural conception of mathematics, which is the core of algebra, is essentially foreign to Euclid. To invoke this argument in the present discussion would perhaps be to beg the question, but the argument surely does provide *prima facie* grounds for doubting the viability of the algebraic interpretation.

However, it should be pointed out that the notion of algebra which seems to be relevant to the issues raised here is not so much the abstract theories of modern algebra as their forerunner, the application of standard algebraic formulas and procedures to the solution of quantitative problems. Zeuthen has no difficulty in showing how important Greek mathematical results can be construed as geometric embodiments of such algebraic solutions. However, he knows perfectly well that there is no such thing as a Greek algebraic equation in our sense. There are only geometric theorems and problems in the manner of II.1. Nor are there any such equations in Babylonian mathematics, most of which consists of step by step solutions to particular numerical problems on the basis of a general procedure. It is presumably for this reason that Neugebauer speaks of a translation of Babylonian *methods* into the language of geometric algebra.

[...] I would like to sum up and perhaps clarify what seem to me the fundamental issues in the interpretation of geometric algebra. There can be no question of the consistency and coherence of the algebraic interpretation of geometric algebra. Nor can there be any question that geometric algebra plays a role in the *Elements*, the *Conica*, and elsewhere very like the role of algebra in analytic geometry. Sometimes proponents of the algebraic interpretation appear to be proclaiming little more than this fact. But this minimal, correct claim would be compatible with a geometric interpretation, since there is nothing to prevent geometric results from playing the role in descriptive geometry which algebraic ones play in analytic geometry. The claims which seem to me clearly to distinguish the algebraic interpretation from the geometric one are:

1 The lines and areas of geometric algebra represent arbitrary quantities;
2 Geometric algebra is a translation of Babylonian algebraic methods;
3 The 'line of thought' in much of Greek mathematics is 'at bottom purely algebraic'.

The truth of any one of these claims would seem to me sufficient to establish the algebraic interpretation to the exclusion of the geometric one. Ultimately the first of these claims reduces to the view that geometric algebra is intended to be applicable to numbers. Most of the direct evidence for this view is relatively late, the earliest being perhaps the scholia which interpret propositions in Book II numerically. These scholia and other evidence make it certain that the possibility of applying geometric algebra to arithmetic problems was an established fact by the first century AD, but it is difficult to know how much light the later situation throws on the *Elements* themselves. Zeuthen's view that there was a pre-Euclidean arithmetic algebra is in a sense confirmed by our present understanding of Babylonian mathematics; but, in terms of Greek evidence, Zeuthen's view is essentially historical conjecture. Within the *Elements* themselves there are three apparent arithmetic applications of Book II. Here I wish only to suggest the necessity of distinguishing between the recognition and use of arithmetic-geometric analogies and algebraic thought. It seems relatively clear that geometric ideas played a substantial role in early Greek arithmetic thinking which may well have been based entirely on the representation of numbers as plane arrays of units. This way of dealing with numbers would obviously facilitate the recognition of analogies between geometric results and arithmetic ones and would also suggest the possibility of exploiting geometric procedures in arithmetic. Such analogical thinking is to be distinguished from the algebraic approach of combining the treatment of distinct disciplines by abstracting the common features of the objects they deal with. We shall see that Euclid does something like this in Book V where he treats magnitudes in general. However, I shall argue that even there he is concerned only with the geometric, which he separates from the arithmetic in keeping with the prevalent Greek opinion that the infinitely divisible or continuous and the discrete are radically different kinds of things.

Taken literally, claim 2 gives one the picture of Greek mathematicians methodically rewriting a given body of mathematics. However, this literal reading involves the difficulty that the Greeks place great emphasis on proof whereas, as far as one can tell, the Babylonians never felt it necessary to justify their procedures. Thus the real notion of Greek translation of Babylonian algebra has to involve both the representation of methods as theorems and problems and the supplying of proofs. This notion of translation is considerably more flexible than the ordinary one invoked in saying, for example, that arithmetic can be translated into set theory. This flexibility becomes particularly problematic in light of the fact that we will presumably never know in what form the Greeks might have come to know Babylonian mathematics, if indeed they did know it in the fifth and fourth centuries BC. One is left with the choice between a precise hypothesis of methodical translation which the evidence would not seem to justify, and a looser hypothesis which would not seem to be clearly preferable to other alternatives, e.g., the assumption that geometric algebra is the Greek embodiment of a generally shared knowledge or that the Babylonians and the Greeks reached equivalent results independently.

Claim 3 rather obviously needs to be made more precise. In the *Elements* we find transformations of areas into other equal ones by means of the addition and subtraction of lengths and areas and the geometric analogues of multiplication, division, and the extraction of square roots. The fact that similar operations are fundamental in algebra does not seem to me sufficient to settle the question whether the operations in the *Elements* are aptly described as algebraic. The paradigms of algebraic reasoning are abstract structural argument, on the one hand, and the manipulation of equations, on the other. Neither of these paradigms appears to be basically Euclidean. Indeed, however one wishes to describe the results proved in Book II, the proofs themselves show no sense of the connection between the propositions involved. This fact suggests strongly that Euclid is approaching his subject by looking at the geometric properties of particular spatial configurations and not by considering abstract relations between quantities or formal relations between expressions. The study of other parts of the *Elements* may reverse this impression, but for the moment there is no good reason to assume that Euclid is reasoning algebraically.

3.G6 John L. Berggren

My own view is that to establish geometrized algebra as a historical fact still requires that considerable research be done on the time and method of transmission of

Babylonian mathematical knowledge to the Greek world. Some Babylonian ideas—for example, the gnomon—seem to have been transmitted at an early date, whereas other notions—that of degree measurement of angles and the sexagesimal system, even in the modified form in which the Greeks used it, for example—seem to have arrived after Euclid wrote. When, in this interval, one is to date the importation and geometrization of Babylonian algebra is a historical question to be settled not by conjecture but by research. If the event cannot be located historically one must recognize the possibility that it may not have occurred.

4 Archimedes and Apollonius

4.A Archimedes

Though some of his works are lost, we have more of the writings of Archimedes than of any other great mathematician of antiquity. This is a tribute to the high regard in which his work was held, as well as to his productivity. It seems that his works, sent mostly from Syracuse, where he lived, to the mathematical community in Alexandria, continued to be studied sufficiently to ensure their preservation. (For details of their transmission down to us, see 4.B4.) This selection of extracts is representative of his range of mathematical interests—he left few writings about his well-known mechanical inventions—besides containing valuable historical evidence in the prefatory letters. We have followed the ordering of works suggested by Wilbur Knorr (*Archive for History of Exact Sciences*, 19 (1978) pp. 211–290), an ordering that enables the development of Archimedes' mathematical thought and approach to be studied. Archimedes was killed by a Roman soldier in 212 BC.

4.A1 Measurement of a circle

Proposition 1

The area of any circle is equal to a right-angled triangle in which one of the sides about the right angle is equal to the radius, and the other to the circumference, of the circle.

Let *ABCD* be the given circle, *K* the triangle described.

Then, if the circle is not equal to *K*, it must be either greater or less.

I. If possible, let the circle be greater than *K*.

Inscribe a square *ABCD*, bisect the arcs *AB*, *BC*, *CD*, *DA*, then bisect (if necessary) the halves, and so on, until the sides of the inscribed polygon whose angular points are the points of division subtend segments whose sum is less than the excess of the area of the circle over *K*.

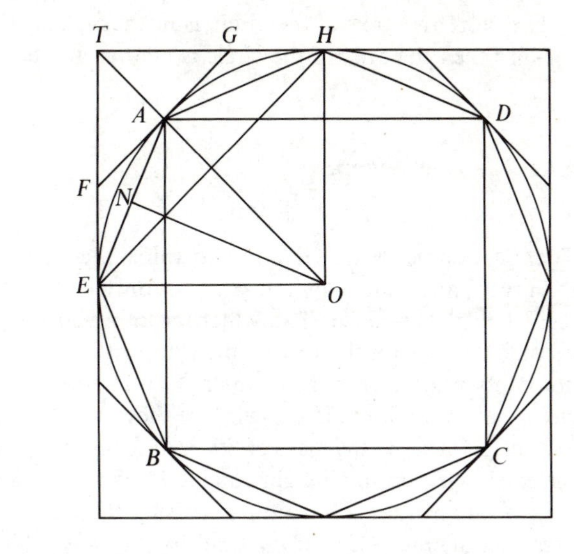

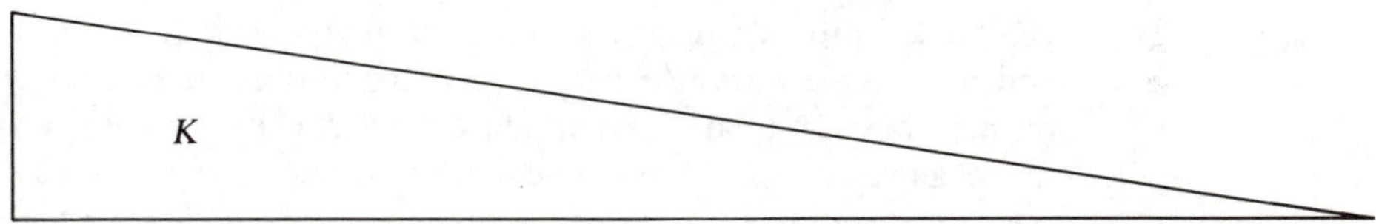

Thus the area of the polygon is greater than K.

Let AE be any side of it, and ON the perpendicular on AE from the centre O.

Then ON is less than the radius of the circle and therefore less than one of the sides about the right angle in K. Also the perimeter of the polygon is less than the circumference of the circle, i.e. less than the other side about the right angle in K.

Therefore the area of the polygon is less than K; which is inconsistent with the hypothesis.

Thus the area of the circle is not greater than K.

II. If possible, let the circle be less than K.

Circumscribe a square, and let two adjacent sides, touching the circle in E, H, meet in T. Bisect the arcs between adjacent points of contact and draw the tangents at the points of bisection. Let A be the middle point of the arc EH, and FAG the tangent at A.

Then the angle TAG is a right angle.

Therefore $TG > GA > GH$.

It follows that the triangle FTG is greater than half the area $TEAH$.

Similarly, if the arc AH be bisected and the tangent at the point of bisection be drawn, it will cut off from the area GAH more than one-half.

Thus, by continuing the process, we shall ultimately arrive at a circumscribed polygon such that the spaces intercepted between it and the circle are together less than the excess of K over the area of the circle.

Thus the area of the polygon will be less than K.

Now, since the perpendicular from O on any side of the polygon is equal to the radius of the circle, while the perimeter of the polygon is greater than the circumference of the circle, it follows that the area of the polygon is greater than the triangle K; which is impossible.

Therefore the area of the circle is not less than K.

Since then the area of the circle is neither greater nor less than K, it is equal to it.

4.A2 The sand-reckoner

There are some, King Gelon, who think that the number of the sand is infinite in multitude; and I mean by the sand not only that which exists about Syracuse and the rest of Sicily but also that which is found in every region whether inhabited or uninhabited. Again there are some who, without regarding it as infinite, yet think that no number has been named which is great enough to exceed its multitude. And it is clear that they who hold this view, if they imagined a mass made up of sand in other respects as large as the mass of the earth, including in it all the seas and the hollows of the earth filled up to a height equal to that of the highest of the mountains, would be many times further still from recognising that any number could be expressed which exceeded the multitude of the sand so taken. But I will try to show you by means of geometrical proofs, which you will be able to follow, that, of the numbers named by me and given in the work which I sent to Zeuxippus, some exceed not only the number of the mass of sand equal in magnitude to the earth filled up in the way described, but also that of a mass equal in magnitude to the universe. Now you are aware that 'universe' is the name given by most astronomers to the sphere whose centre is the centre of the earth and whose radius is equal to the straight line between the centre of the sun and the centre of the earth. This is the common account, as you have heard from astronomers. But Aristarchus of Samos brought out a book consisting of some hypotheses, in which the premisses lead to the result that the universe is many times greater than that now so called. His hypotheses are that the fixed stars and the sun remain unmoved, that the earth revolves about the sun in the circumference of a circle, the sun lying in the middle of the orbit, and that the sphere of the fixed stars, situated about the same centre as the sun, is so great that the circle in which he supposes the earth to revolve bears such a proportion to the distance of the fixed stars as the centre of the sphere bears to its surface. Now it is easy to see that this is impossible; for, since the centre of the sphere has no magnitude, we cannot conceive it to bear any ratio whatever to the surface of the sphere. We must however take Aristarchus to mean this: since we conceive the earth to be, as it were, the centre of the universe, the ratio which the earth bears to what we describe as the 'universe' is the same as the ratio which the sphere containing the circle in which he supposes the earth to revolve bears to the sphere of the fixed stars. For he adapts the proofs of his results to a hypothesis of this kind, and in particular he appears to suppose the magnitude of the sphere in which he represents the earth as moving to be equal to what we call the 'universe'.

I say then that, even if a sphere were made up of the sand, as great as Aristarchus supposes the sphere of the fixed stars to be, I shall still prove that, of the numbers named in the *Principles*, some exceed in multitude the number of the sand which is equal in magnitude to the sphere referred to, provided that the following assumptions be made.

1 *The perimeter of the earth is about 3,000,000 stadia and not greater*

It is true that some have tried, as you are of course aware, to prove that the said perimeter is about 300,000 stadia. But I go further and, putting the magnitude of the earth at ten times the size that my predecessors thought it, I suppose its perimeter to be about 3,000,000 stadia and not greater.

2 *The diameter of the earth is greater than the diameter of the moon, and the diameter of the sun is greater than the diameter of the earth.*

In this assumption I follow most of the earlier astronomers.

3 *The diameter of the sun is about 30 times the diameter of the moon and not greater.*

It is true that, of the earlier astronomers, Eudoxus declared it to be about nine times as great, and Pheidias my father twelve times, while Aristarchus tried to prove that the diameter of the sun is greater than 18 times but less than 20 times the diameter of the moon. But I go even further than Aristarchus, in order that the truth of my proposition may be established beyond dispute, and I suppose the diameter of the sun to be about 30 times that of the moon and not greater.

4 *The diameter of the sun is greater than the side of the chiliagon inscribed in the greatest circle in the [sphere of the] universe.*

I make this assumption because Aristarchus discovered that the sun appeared to be about $\frac{1}{720}$th part of the circle of the zodiac, and I myself tried, by a method which I will now describe, to find experimentally the angle subtended by the sun and having its vertex at the eye. [...]

Orders and periods of numbers

I. We have traditional names for numbers up to a myriad (10,000); we can therefore express numbers up to a myriad myriads (100,000,000). Let these numbers be called numbers of the *first order*.

Suppose the 100,000,000 to be the unit of the *second order*, and let the *second order* consist of the numbers from that unit up to $(100{,}000{,}000)^2$.

Let this again be the unit of the *third order* of numbers ending with $(100{,}000{,}000)^3$; and so on, until we reach the 100,000,000*th order* of numbers ending with $(100{,}000{,}000)^{100{,}000{,}000}$, which we will call P.

II. Suppose the numbers from 1 to P just described to form the *first period*.

Let P be the unit of the *first order of the second period*, and let this consist of the numbers from P up to $100{,}000{,}000P$.

Let the last number be the unit of the *second order of the second period*, and let this end with $(100{,}000{,}000)^2P$.

We can go on in this way till we reach the *100,000,000th order of the second period* ending with $(100{,}000{,}000)^{100{,}000{,}000}P$, or P^2.

III. Taking P^2 as the unit of the *first order of the third period*, we proceed in the same way till we reach the *100,000,000th order of the third period* ending with P^3.

IV. Taking P^3 as the unit of the *first order of the fourth period*, we continue the same process until we arrive at the *100,000,000th order of the 100,000,000th period* ending with $P^{100{,}000{,}000}$. [This last number is expressed by Archimedes as 'a myriad-myriad units of the myriad-myriad-th order of the myriad-myriad-th period', which is easily seen to be 100,000,000 times the product of $(100{,}000{,}000)^{99{,}999{,}999}$ and $P^{99{,}999{,}999}$, i.e. $P^{100{,}000{,}000}$.]

Octads

Consider the series of terms in continued proportion of which the first is 1 and the second 10 [i.e. the geometrical progression 1, 10^1, 10^2, 10^3, . . .]. The *first octad* of these terms [i.e. 1, 10^1, 10^2,... 10^7] fall accordingly under the *first order of the first period* above described, the *second octad* [i.e. 10^8, 10^9,... 10^{15}] under the *second order of the first period*, the first term of the octad being the unit of the corresponding order in each case. Similarly for the *third octad*, and so on. We can, in the same way, place any number of octads.

Theorem

If there be any number of terms of a series in continued proportion, say A_1,A_2, A_3, ...A_m, ...A_n, ...A_{m+n-1},... *of which* $A_1 = 1$, $A_2 = 10$ [so that the series forms the geometrical progression 1, 10^1, 10^2, ... 10^{m-1}, ... 10^{n-1}, ... 10^{m+n-2},...], *and if any two terms as* A_m, A_n *be taken and multiplied, the product* $A_m \cdot A_n$ *will be a term in the same series and will be as many terms distant from* A_n *as* A_m *is distant from* A_1*; also it will be distant from* A_1 *by a number of terms less by one than the sum of the numbers of terms by which* A_m *and* A_n *respectively are distant from* A_1.

Take the term which is distant from A_n by the same number of terms as A_m is distant from A_1. This number of terms is m (the first and last being both counted). Thus the term to be taken is m terms distant from A_n, and is therefore the term A_{m+n-1}.

We have therefore to prove that $A_m \cdot A_n = A_{m+n-1}$.

Now terms equally distant from other terms in the continued proportion are proportional.

Thus $\dfrac{A_m}{A_1} = \dfrac{A_{m+n-1}}{A_n}$.

But $A_m = A_m \cdot A_1$, since $A_1 = 1$.

Therefore $A_{m+n-1} = A_m \cdot A_n$.

The second result is now obvious, since A_m is m terms distant from A_1, A_n is n terms distant from A_1, and A_{m+n-1} is $(m+n-1)$ terms distant from A_1. [...] It follows that the number of grains of sand which would be contained in a sphere equal to the sphere of the fixed stars $< (10{,}000)^3 \times 1{,}000$ units of *seventh order* $<$ (13th term of series) $\times$ (52nd term of series) $<$ 64th term of series [i.e. 10^{63}] $<$ [10^7 or] 10,000,000 units of *eighth order* of numbers.

Conclusion

I conceive that these things, King Gelon, will appear incredible to the great majority of people who have not studied mathematics, but that to those who are conversant therewith and have given thought to the question of the distances and sizes of the earth, the sun and moon and the whole universe, the proof will carry conviction. And it was for this reason that I thought the subject would be not inappropriate for your consideration.

4.A3 Quadrature of the parabola

(a) *Letter to Dositheus*

When I heard that Conon, who was my friend in his lifetime, was dead, but that you were acquainted with Conon and withal versed in geometry, while I grieved for the loss not only of a friend but of an admirable mathematician, I set myself the task of communicating to you, as I had intended to send to Conon, a certain geometrical theorem which had not been investigated before but has now been investigated by me, and which I first discovered by means of mechanics and then exhibited by means of geometry. Now some of the earlier geometers tried to prove it possible to find a rectilineal area equal to a given circle and a given segment of a circle; and after that they endeavoured to square the area bounded by the section of the whole cone and a straight line, assuming lemmas not easily conceded, so that it was recognised by most people that the problem was not solved. But I am not aware that any one of my predecessors has attempted to square the segment bounded by a straight line and a section of a right-angled cone [a parabola], of which problem I have now discovered the solution. For it is here shown that every segment bounded by a straight line and a section of a right-angled cone [a parabola] is four-thirds of the triangle which has the same base and equal height with the segment, and for the demonstration of this property the following lemma is assumed: that the excess by which the greater of [two] unequal areas exceeds the less can, by being added to itself, be made to exceed any given finite area. The earlier geometers have also used this lemma; for it is by the use of this same lemma that they have shown that circles are to one another in the duplicate ratio of their diameters, and that spheres are to one another in the triplicate ratio of their diameters, and further that every pyramid is one third part of the prism which has the same base with the pyramid and equal height; also, that every cone is one third part of the cylinder having the same base as the cone and equal height they proved by assuming a certain lemma similar to that aforesaid. And, in the result, each of the aforesaid theorems has been accepted no less than those proved without the lemma. As therefore my work now published has satisfied the same test as the propositions referred to, I have written out the proof and send it to you, first as investigated by means of mechanics, and afterwards too as demonstrated by geometry. Prefixed are, also, the elementary propositions in conics which are of service in the proof.

(b) *Proposition 24*

Every segment bounded by a parabola and a chord Qq is equal to four-thirds of the triangle which has the same base as the segment and equal height.

Suppose $K = \frac{4}{3}\Delta PQq$, where P is the vertex of the segment; and we have then to prove that the area of the segment is equal to K.

For, if the segment be not equal to K, it must either be greater or less.

I. Suppose the area of the segment greater than K.

If then we inscribe in the segments cut off by PQ, Pq triangles which have the same base and equal height, i.e. triangles with the same vertices R, r as those of the segments, and if in the remaining segments we inscribe triangles in the same manner, and so on,

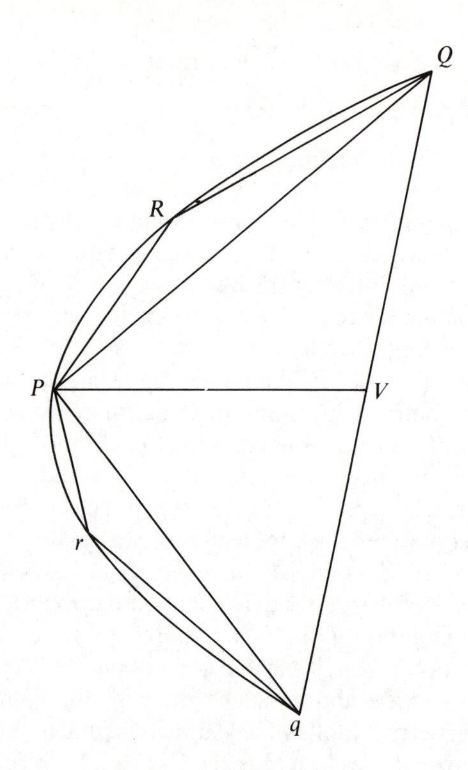

we shall finally have segments remaining whose sum is less than the area by which the segment PQq exceeds K.

Therefore the polygon so formed must be greater than the area K; which is impossible, since [Prop. 23] $A + B + C + \ldots + Z < \frac{4}{3}A$, where $A = \Delta PQq$.

Thus the area of the segment cannot be greater than K.

II. Suppose, if possible, that the area of the segment is less than K.

If then $\Delta PQq = A, B = \frac{1}{4}A, C = \frac{1}{4}B$, and so on, until we arrive at an area X such that X is less than the difference between K and the segment, we have $A + B + C + \ldots + X + \frac{1}{3}X = \frac{4}{3}A$ [Prop. 23] $= K$.

Now, since K exceeds $A + B + C + \ldots + X$ by an area less than X, and the area of the segment by an area greater than X, it follows that $A + B + C + \ldots + X >$ (the segment); which is impossible, by Prop. 22 above.

Hence the segment is not less than K.

Thus, since the segment is neither greater nor less than K, (area of segment PQq) $= K = \frac{4}{3}\Delta PQq$.

4.A4 On the equilibrium of planes: Book I

Propositions 6 and 7

Two magnitudes, whether commensurable [Proposition 6] *or incommensurable* [Proposition 7] *balance at distances reciprocally proportional to the magnitudes.*

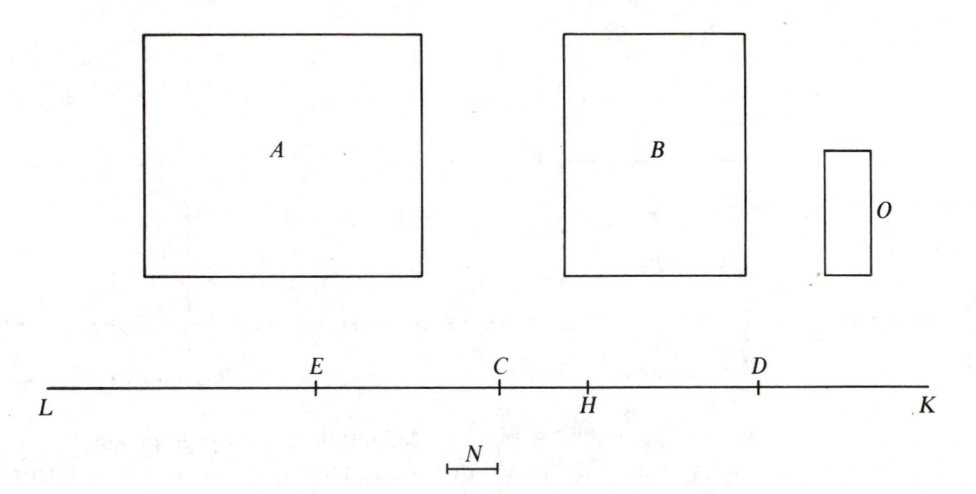

I. Suppose the magnitudes A, B to be commensurable, and the points A, B to be their centres of gravity. Let DE be a straight line so divided at C that $A:B = DC:CE$.

We have then to prove that, if A be placed at E and B at D, C is the centre of gravity of the two taken together.

Since A, B are commensurable, so are DC, CE. Let N be a common measure of DC, CE. Make DH, DK each equal to CE, and EL (on CE produced) equal to CD. Then $EH = CD$, since $DH = CE$. Therefore LH is bisected at E, as HK is bisected at D.

Thus LH, HK must each contain N an even number of times.

Take a magnitude O such that O is contained as many times in A as N is contained in LH, whence $A:O = LH:N$.

But $B:A = CE:DC = HK:LH$.

Hence, *ex aequali*, $B:O = HK:N$, or O is contained in B as many times as N is contained in HK.

Thus O is a common measure of A, B.

Divide LH, HK into parts each equal to N, and A, B into parts each equal to O. The parts of A will therefore be equal in number to those of LH, and the parts of B equal in number to those of HK. Place one of the parts of A at the middle point of each of the parts N of LH, and one of the parts of B at the middle point of each of the parts N of HK.

Then the centre of gravity of the parts of A placed at equal distances on LH will be at E, the middle point of LH [Prop. 5, Cor. 2], and the centre of gravity of the parts of B placed at equal distances along HK will be at D, the middle point of HK.

Thus we may suppose A itself applied at E, and B itself applied at D.

But the system formed by the parts O of A and B together is a system of equal magnitudes even in number and placed at equal distances along LK. And, since $LE = CD$, and $EC = DK$, $LC = CK$, so that C is the middle point of LK. Therefore C is the centre of gravity of the system ranged along LK.

Therefore A acting at E and B acting at D balance about the point C.

II. Suppose the magnitudes to be incommensurable, and let them be $(A + a)$ and B respectively. Let DE be a line divided at C so that $(A + a):B = DC:CE$.

Then, if $(A + a)$ placed at E and B placed at D do not balance about C, $(A + a)$ is either too great to balance B, or not great enough.

Suppose, if possible, that $(A + a)$ is too great to balance B. Take from $(A + a)$ a

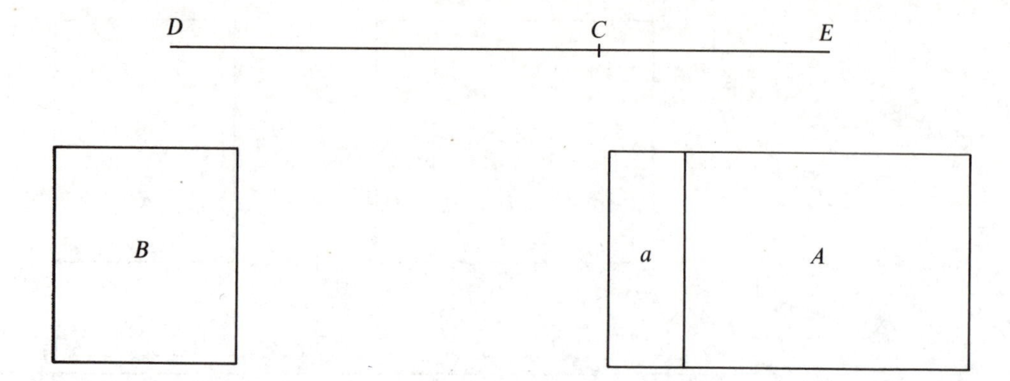

magnitude a smaller than the deduction which would make the remainder balance B, but such that the remainder A and the magnitude B are commensurable.

Then, since A, B are commensurable, and $A:B < DC:CE$, A and B will not balance [Prop. 6], but D will be depressed.

But this is impossible, since the deduction a was an insufficient deduction from $(A + a)$ to produce equilibrium, so that E was still depressed.

Therefore $(A + a)$ is not too great to balance B; and similarly it may be proved that B is not too great to balance $(A + a)$.

Hence $(A + a)$, B taken together have their centre of gravity at C.

4.A5 On the sphere and cylinder: Book I

(a) *Letter to Dositheus*

On a former occasion I sent you the investigations which I had up to that time completed, including the proofs, showing that any segment bounded by a straight line and a section of a right-angled cone [a parabola] is four-thirds of the triangle which has the same base with the segment and equal height. Since then certain theorems not hitherto demonstrated have occurred to me, and I have worked out the proofs of them. They are these: first, that the surface of any sphere is four times its greatest circle; next, that the surface of any segment of a sphere is equal to a circle whose radius is equal to the straight line drawn from the vertex of the segment to the circumference of the circle which is the base of the segment; and, further, that any cylinder having its base equal to the greatest circle of those in the sphere, and height equal to the diameter of the sphere, is itself [i.e. in content] half as large again as the sphere, and its surface also [including its bases] is half as large again as the surface of the sphere. Now these properties were all along naturally inherent in the figures referred to, but remained unknown to those who were before my time engaged in the study of geometry. Having, however, now discovered that the properties are true of these figures, I cannot feel any hesitation in setting them side by side both with my former investigations and with those of the theorems of Eudoxus on solids which are held to be most irrefragably established, namely, that any pyramid is one third part of the prism which has the same base with the pyramid and equal height, and that any cone is one third part of the cylinder which has the same base with the cone and equal height. For, though these properties also were naturally inherent in the figures all along, yet they were in fact unknown to all the many able geometers who lived before Eudoxus, and had not been observed by any one. Now, however, it will be open to those who possess the requisite ability to examine these discoveries of mine. They ought to have been published while Conon was still alive, for I should conceive that he would best have been able to grasp them and to pronounce upon them the appropriate verdict; but, as I judge it well to communicate them to those who are conversant with mathematics, I send them to you with the proofs written out, which it will be open to mathematicians to examine.

(*b*) *Proposition 34*

Any sphere is equal to four times the cone which has its base equal to the greatest circle in the sphere and its height equal to the radius of the sphere.

Let the sphere be that of which $ama'm'$ is a great circle.

If now the sphere is not equal to four times the cone described, it is either greater or less.

I. If possible, let the sphere be greater than four times the cone.

Suppose V to be a cone whose base is equal to four times the great circle and whose height is equal to the radius of the sphere.

Then, by hypothesis, the sphere is greater than V; and two lines β, γ can be found (of which β is the greater) such that $\beta:\gamma <$ (volume of sphere)$:V$.

Between β and γ place two arithmetic means δ, ε.

As before, let similar regular polygons with sides $4n$ in number be circumscribed about and inscribed in the great circle, such that their sides are in a ratio less than $\beta:\delta$.

Imagine the diameter aa' of the circle to be in the same straight line with a diameter of both polygons, and imagine the latter to revolve with the circle about aa', describing the surfaces of two solids of revolution. The volumes of these solids are therefore in the triplicate ratio of their sides [Prop. 32].

Thus (vol. of outer solid):(vol. of inscribed solid) $< \beta^3:\delta^3$ (by hypothesis) $< \beta:\gamma$ (*a fortiori* since $\beta:\gamma > \beta^3:\delta^3$) $<$ (volume of sphere): V (*a fortiori*).

But this is impossible, since the volume of the circumscribed solid is greater than that of the sphere [Prop. 28], while the volume of the inscribed solid is less than V [Prop. 27].

Hence the sphere is not greater than V, or four times the cone described in the enunciation.

II. If possible, let the sphere be less than V.

In this case we take β, γ (β being the greater) such that $\beta:\gamma < V$:(volume of sphere).

The rest of the construction and proof proceeding as before, we have finally (volume of outer solid):(volume of inscribed solid) $< V$:(volume of sphere).

But this is impossible, because the volume of the outer solid is greater than V [Prop. 31, Cor.], and the volume of the inscribed solid is less than the volume of the sphere.

Hence the sphere is not less than V.

Since then the sphere is neither less nor greater than V, it is equal to V, or to four times the cone described in the enunciation.

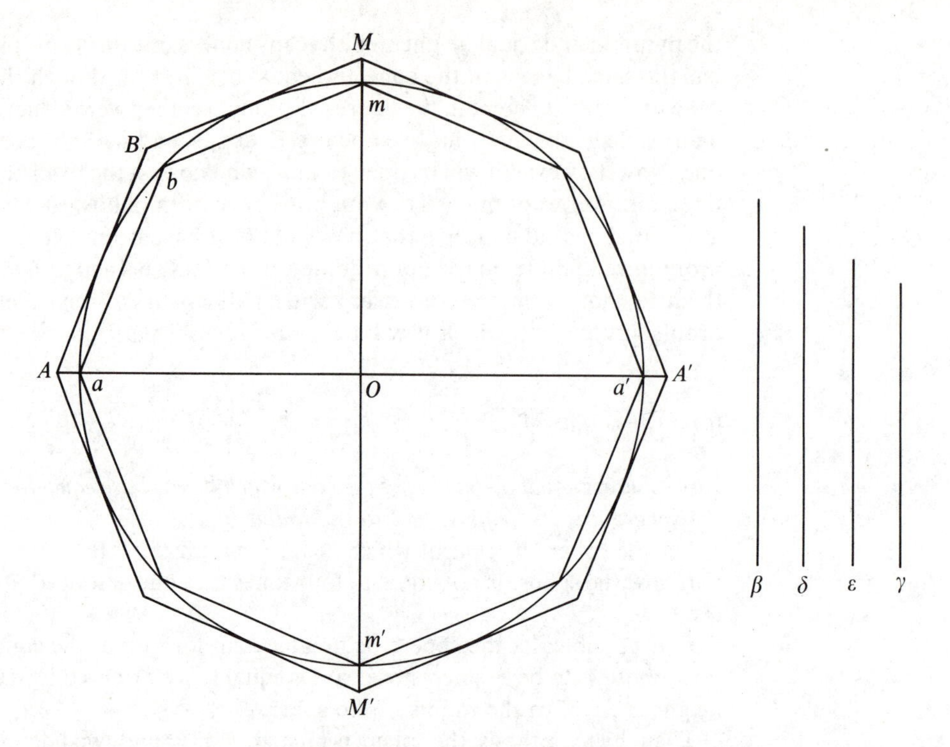

Corollary From what has been proved it follows that *every cylinder whose base is the greatest circle in a sphere and whose height is equal to the diameter of the sphere is $\frac{3}{2}$ of the sphere, and its surface together with its bases is $\frac{3}{2}$ of the surface of the sphere.*

For the cylinder is three times the cone with the same base and height [Euclid's *Elements*, XII. 10], i.e. six times the cone with the same base and with height equal to the radius of the sphere.

But the sphere is four times the latter cone [Prop. 34]. Therefore the cylinder is $\frac{3}{2}$ of the sphere.

Again, the surface of a cylinder [excluding the bases] is equal to a circle whose radius is a mean proportional between the height of the cylinder and the diameter of its base [Prop. 13].

In this case the height is equal to the diameter of the base and therefore the circle is that whose radius is the diameter of the sphere, or a circle equal to four times the great circle of the sphere.

Therefore the surface of the cylinder with the bases is equal to six times the great circle.

And the surface of the sphere is four times the great circle [Prop. 33]; whence (surface of cylinder with bases) = $\frac{3}{2}$ (surface of sphere).

4.A6 On the sphere and cylinder: Book II

Proposition 1

Given a cone or a cylinder, to find a sphere equal to the cone or to the cylinder.

If V be the given cone or cylinder, we can make a cylinder equal to $\frac{3}{2}V$. Let this cylinder be the cylinder whose base is the circle on AB as diameter and whose height is OD.

Now, if we could make another cylinder, equal to the cylinder (OD) but such that its height is equal to the diameter of its base, the problem would be solved, because this latter cylinder would be equal to $\frac{3}{2}V$, and the sphere whose diameter is equal to the height (or to the diameter of the base) of the same cylinder would then be the sphere required [I. 34, Cor.].

Suppose the problem solved, and let the cylinder (CG) be equal to the cylinder (OD), while EF, the diameter of the base, is equal to the height CG.

Then, since in equal cylinders the heights and bases are reciprocally proportional,

$$AB^2 : EF^2 = CG : OD = EF : OD. \qquad (1)$$

Suppose MN to be such a line that

$$EF^2 = AB \cdot MN. \qquad (2)$$

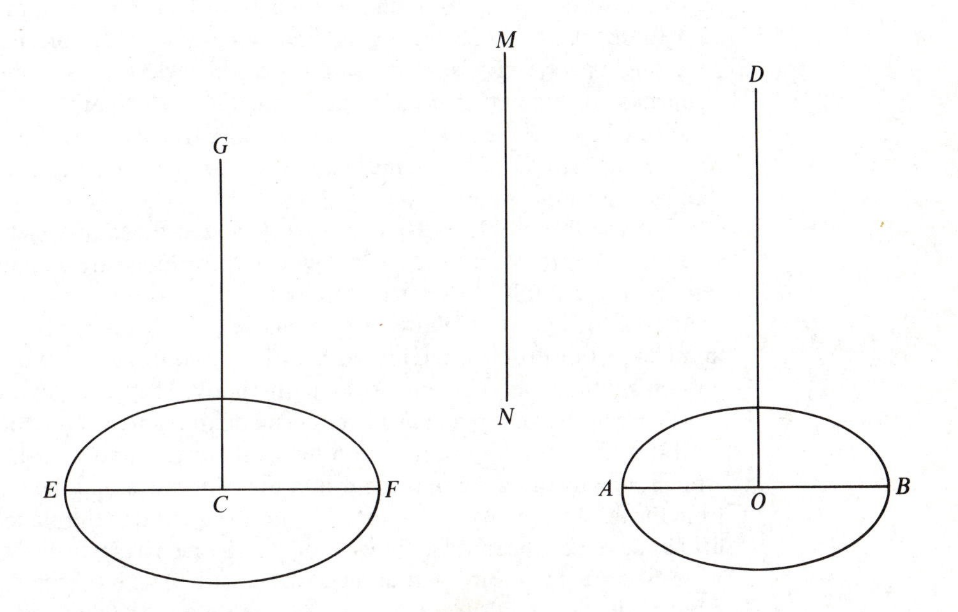

Hence $AB : EF = EF : MN$, and, combining (1) and (2), we have $AB : MN = EF : OD$, or $AB : EF = MN : OD$.

Therefore $AB : EF = EF : MN = MN : OD$, and EF, MN *are two mean proportionals between* AB, OD.

The synthesis of the problem is therefore as follows. Take two mean proportionals EF, MN between AB and OD, and describe a cylinder whose base is a circle on EF as diameter and whose height CG is equal to EF.

Then, since $AB : EF = EF : MN = MN : OD$, $EF^2 = AB \cdot MN$, and therefore $AB^2 : EF^2 = AB : MN = EF : OD = CG : OD$; whence the bases of the two cylinders (OD), (CG) are reciprocally proportional to their heights.

Therefore the cylinders are equal, and it follows that cylinder $(CG) = \frac{3}{2}V$.

The sphere on EF as diameter is therefore the sphere required, being equal to V.

4.A7 On spirals

(*a*) *Letter to Dositheus*

Of most of the theorems which I sent to Conon, and of which you ask me from time to time to send you the proofs, the demonstrations are already before you in the books brought to you by Heracleides; and some more are also contained in that which I now send you. Do not be surprised at my taking a considerable time before publishing these proofs. This has been owing to my desire to communicate them first to persons engaged in mathematical studies and anxious to investigate them. In fact, how many theorems in geometry which have seemed at first impracticable are in time successfully worked out! Now Conon died before he had sufficient time to investigate the theorems referred to; otherwise he would have discovered and made manifest all these things, and would have enriched geometry by many other discoveries besides. For I know well that it was no common ability that he brought to bear on mathematics, and that his industry was extraordinary. But, though many years have elapsed since Conon's death, I do not find that any one of the problems has been stirred by a single person. I wish now to put them in review one by one, particularly as it happens that there are two included among them which are impossible of realisation [and which may serve as a warning] how those who claim to discover everything but produce no proofs of the same may be confuted as having actually pretended to discover the impossible.

What are the problems I mean, and what are those of which you have already received the proofs, and those of which the proofs are contained in this book respectively, I think it proper to specify. [...]

After these came the following propositions about the *spiral*, which are as it were another sort of problem having nothing in common with the foregoing; and I have written out the proofs of them for you in this book. They are as follows. If a straight line of which one extremity remains fixed be made to revolve at a uniform rate in a plane until it returns to the position from which it started, and if, at the same time as the straight line revolves, a point move at a uniform rate along the straight line, starting from the fixed extremity, the point will describe a spiral in the plane. I say then that the area bounded by the spiral and the straight line which has returned to the position from which it started is a third part of the circle described with the fixed point as centre and with radius the length traversed by the point along the straight line during the one revolution. And, if a straight line touch the spiral at the extreme end of the spiral, and another straight line be drawn at right angles to the line which has revolved and resumed its position from the fixed extremity of it, so as to meet the tangent, I say that the straight line so drawn to meet it is equal to the circumference of the circle. Again, if the revolving line and the point moving along it make several revolutions and return to the position from which the straight line started, I say that the area added by the spiral in the third revolution will be double of that added in the second, that in the fourth three times, that in the fifth four times, and generally the areas added in the later revolutions will be multiples of that added in the second revolution according to the successive numbers, while the area bounded by the spiral in the first revolution is a sixth part of that added in the second revolution. Also, if on the spiral described in one revolution two points be taken and straight lines be drawn joining them to the fixed extremity of the revolving line, and if two circles be drawn with the fixed point as centre and radii the lines drawn to the fixed extremity of the straight line, and the shorter of the two lines be produced, I say that (1) the area bounded by the circumference of the greater circle in the direction of [the part of] the spiral included between the straight lines, the spiral [itself] and the produced straight line will bear to (2) the area bounded by the circumference of the lesser circle, the same [part of the] spiral and the straight line joining their extremities the ratio which (3) the radius of the lesser circle together with two thirds of the excess of the radius of the greater circle over the radius of the lesser bears to (4) the radius of the lesser circle together with one third of the said excess.

The proofs then of these theorems and others relating to the spiral are given in the present book. Prefixed to them, after the manner usual in other geometrical works, are the propositions necessary to the proofs of them. And here too, as in the books previously published, I assume the following lemma, that, if there be [two] unequal lines or [two] unequal areas, the excess by which the greater exceeds the less can, by being [continually] added to itself, be made to exceed any given magnitude among those which are comparable with [it and with] one another.

(*b*) *Proposition 1*

If a point move at a uniform rate along any line, and two lengths be taken on it, they will be proportional to the times of describing them.

Two unequal lengths are taken on a straight line, and two lengths on another straight line representing the times; and they are proved to be proportional by taking equimultiples of each length and the corresponding time after the manner of Euclid's *Elements*, V, Def. 5.

(*c*) *Proposition 2*

If each of two points on different lines respectively move along them each at a uniform rate, and if lengths be taken, one on each line, forming pairs, such that each pair are described in equal times, the lengths will be proportional.

This is proved at once by equating the ratio of the lengths taken on one line to that of the times of description, which must also be equal to the ratio of the lengths taken on the other line.

(*d*) *Proposition 3*

Given any number of circles, it is possible to find a straight line greater than the sum of all their circumferences.

For we have only to describe polygons about each and then take a straight line equal to the sum of the perimeters of the polygons.

(e) Proposition 4

Given two unequal lines, viz. a straight line and the circumference of a circle, it is possible to find a straight line less than the greater of the two lines and greater than the less.

For, by the Lemma, the excess can, by being added a sufficient number of times to itself, be made to exceed the lesser line.

Thus e.g., if $c > l$ (where c is the circumference of the circle and l the length of the straight line), we can find a number n such that $n(c - l) > l$.

Therefore $c - l > l/n$, and $c > l + l/n > l$.

Hence we have only to divide l into n equal parts and add one of them to l. The resulting line will satisfy the condition.

(f) Definitions

1 If a straight line drawn in a plane revolve at a uniform rate about one extremity which remains fixed and return to the position from which it started, and if, at the same time as the line revolves, a point move at a uniform rate along the straight line beginning from the extremity which remains fixed, the point will describe a *spiral* in the plane.

2 Let the extremity of the straight line which remains fixed while the straight line revolves be called the *origin* of the spiral.

3 And let the position of the line from which the straight line began to revolve be called the *initial line* in the revolution.

4 Let the length which the point that moves along the straight line describes in one revolution be called the *first distance*, that which the same point describes in the second revolution the *second distance*, and similarly let the distances described in further revolutions be called after the number of the particular revolution.

5 Let the area bounded by the spiral described in the first revolution and the *first distance* be called the *first area*, that bounded by the spiral described in the second revolution and the *second distance* the *second area*, and similarly for the rest in order.

6 If from the origin of the spiral any straight line be drawn, let that side of it which is in the same direction as that of the revolution be called *forward*, and that which is in the other direction *backward*.

7 Let the circle drawn with the *origin* as centre and the *first distance* as radius be called the *first circle*, that drawn with the same centre and twice the radius the *second circle*, and similarly for the succeeding circles.

(g) Proposition 18

If OA be the initial line, A the end of the first turn of the spiral, and if the tangent to the spiral at A be drawn, the straight line OB drawn from O perpendicular to OA will meet the said tangent in some point B, and OB will be equal to the circumference of the 'first circle'.

Let AKC be the 'first circle'. Then, since the 'backward' angle between OA and the tangent at A is acute [Prop. 16], the tangent will meet the 'first circle' in a second point C. And the angles CAO, BOA are together less than two right angles; therefore OB will meet AC produced in some point B.

Then, if c be the circumference of the first circle, we have to prove that $OB = c$.

If not, OB must be either greater or less than c.

(1) If possible, suppose $OB > c$.

Measure along OB a length OD less than OB but greater than c.

We have then a circle AKC, a chord AC in it less than the diameter, and a ratio $AO:OD$ which is greater than the ratio $AO:OB$ or (what is, by similar triangles, equal to it) the ratio of $\frac{1}{2}AC$ to the perpendicular from O on AC. Therefore [Prop. 7] we can draw a straight line OPF, meeting the circle in P and CA produced in F, such that $FP:PA = AO:OD$.

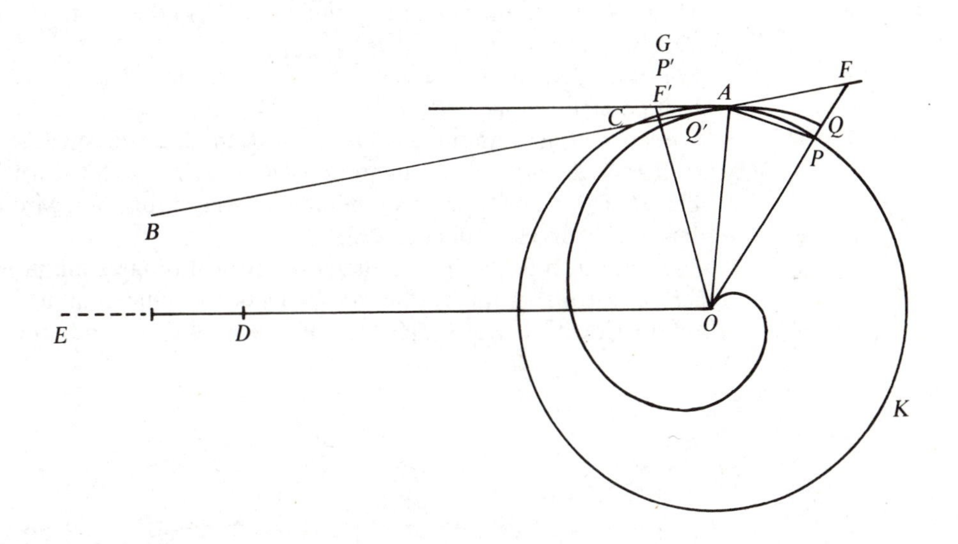

Thus, alternately, since $AO = PO$, $FP:PO = PA:OD < (\text{arc}\,PA):c$, since $(\text{arc}\,PA) > PA$, and $OD > c$.

Componendo, $FO:PO < (c + \text{arc}\,PA):c < OQ:OA$, where OF meets the spiral in Q [Prop. 15].

Therefore, since $OA = OP$, $FO < OQ$; which is impossible.

Hence $OB \not> c$.

(2) If possible, suppose $OB < c$.

Measure OE along OB so that OE is greater than OB but less than c.

In this case, since the ratio $AO:OE$ is less than the ratio $AO:OB$ (or the ratio of $\frac{1}{2}AC$ to the perpendicular from O on AC), we can [Prop. 8] draw a line $OF'P'G$, meeting AC in F', the circle in P', and the tangent at A to the circle in G, such that $F'P':AG = AO:OE$.

Let $OP'G$ cut the spiral in Q'.

Then we have, alternately, $F'P':P'O = AG:OE > (\text{arc}\ AP'):c$,
because $AG > (\text{arc}\ AP')$, and $OE < c$.

Therefore $F'O:P'O < (\text{arc}\,AKP'):c < OQ':OA$ [Prop. 14].

But this is impossible, since $OA = OP'$, and $OQ' < OF'$.

Hence $OB \not< c$.

Since therefore OB is neither greater nor less than c, $OB = c$.

(*h*) *Proposition 24*

The area bounded by the first turn of the spiral and the initial line is equal to one-third of the 'first circle' [$= \frac{1}{3}\pi(2\pi a)^2$, where the spiral is $r = a\theta$].

[The same proof shows equally that, *if OP be any radius vector in the first turn of the spiral, the area of the portion of the spiral bounded thereby is equal to one-third of that sector of the circle drawn with radius OP which is bounded by the initial line and OP, measured in the 'forward' direction from the initial line.*]

Let O be the origin, OA the initial line, A the extremity of the first turn.

Draw the 'first circle', i.e. the circle with O as centre and OA as radius.

Then, if C_1 be the area of the first circle, R_1 that of the first turn of the spiral bounded by OA, we have to prove that $R_1 = \frac{1}{3}C_1$.

For, if not, R_1 must be either greater or less than $\frac{1}{3}C_1$.

I. If possible, suppose $R_1 < \frac{1}{3}C_1$.

We can then circumscribe a figure about R_1 made up of similar sectors of circles such that, if F be the area of this figure, $F - R_1 < \frac{1}{3}C_1 - R_1$, whence $F < \frac{1}{3}C_1$.

Let OP, OQ, ... be the radii of the circular sectors, beginning from the smallest. The radius of the largest is of course OA.

The radii then form an ascending arithmetical progression in which the common difference is equal to the least term OP. If n be the number of the sectors, we have [by Prop. 10, Cor. 1] $n \cdot OA^2 < 3(OP^2 + OQ^2 + \cdots + OA^2)$; and, since the similar sectors

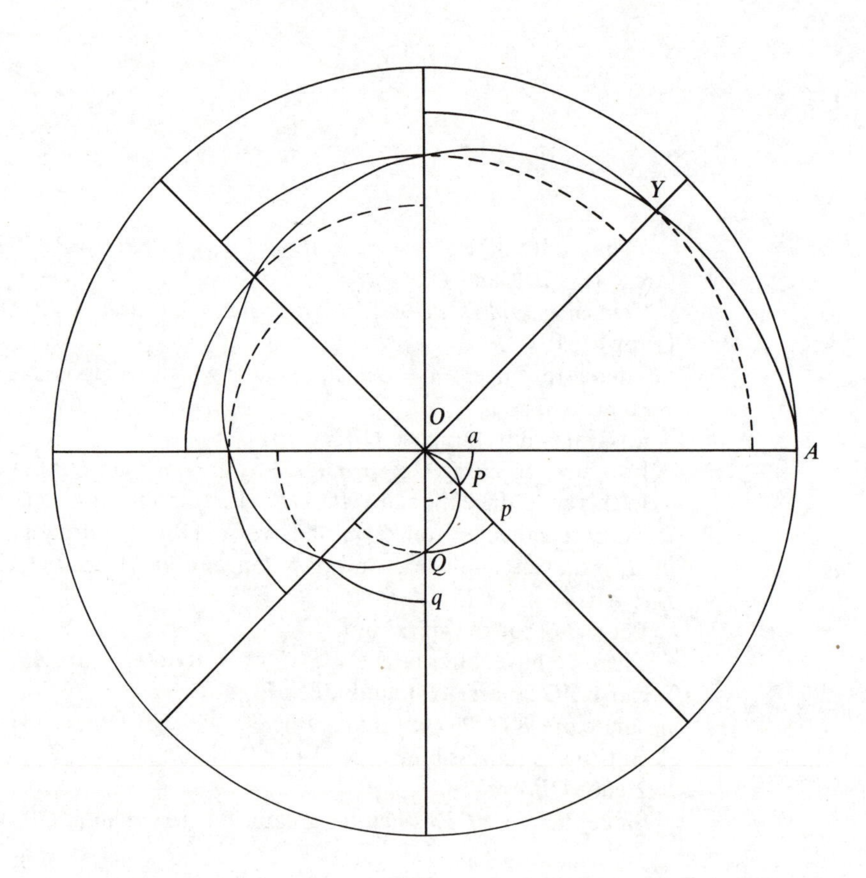

are proportional to the squares on their radii, it follows that $C_1 < 3F$, or $F > \frac{1}{3}C_1$. But this is impossible, since F was less than $\frac{1}{3}C_1$.

Therefore $R_1 \not< \frac{1}{3}C_1$.

II. If possible, suppose $R_1 > \frac{1}{3}C_1$.

We can then *inscribe* a figure made up of similar sectors of circles such that, if f be its area, $R_1 - f < R_1 - \frac{1}{3}C_1$, whence $f > \frac{1}{3}C_1$.

If there are $(n - 1)$ sectors, their radii, as $OP, OQ, \ldots$, form an ascending arithmetical progression in which the least term is equal to the common difference, and the greatest term, as OY, is equal to $(n - 1)OP$.

Thus [Prop. 10, Cor. 1] $n \cdot OA^2 > 3(OP^2 + OQ^2 + \cdots + OY^2)$, whence $C_1 > 3f$, or $f < \frac{1}{3}C_1$; which is impossible, since $f > \frac{1}{3}C_1$.

Therefore $R_1 \not> \frac{1}{3}C_1$.

Since then R_1 is neither greater nor less than $\frac{1}{3}C_1$, $R_1 = \frac{1}{3}C_1$.

4.A8 On conoids and spheroids

(*a*) *Letter to Dositheus*

In this book I have set forth and send you the proofs of the remaining theorems not included in what I sent you before, and also of some others discovered later which, though I had often tried to investigate them previously, I had failed to arrive at because I found their discovery attended with some difficulty. And this is why even the propositions themselves were not published with the rest. But afterwards, when I had studied them with greater care, I discovered what I had failed in before.

Now the remainder of the earlier theorems were propositions concerning the right-angled conoid [paraboloid of revolution]; but the discoveries which I have now added relate to an obtuse-angled conoid [hyperboloid of revolution] and to spheroidal figures, some of which I call *oblong* and others *flat*.

I. Concerning the *right-angled conoid* it was laid down that, if a section of a right-angled cone [a parabola] be made to revolve about the diameter [axis] which remains fixed and return to the position from which it started, the figure comprehended by the section of the right-angled cone is called a *right-angled conoid*, and the diameter which has remained fixed is called its *axis*, while its *vertex* is the point in which the axis meets the surface of the conoid. And if a plane touch the right-angled conoid, and another plane drawn parallel to the tangent plane cut off a segment of the conoid, the *base* of the segment cut off is defined as the portion intercepted by the section of the conoid on the cutting plane, the *vertex* [of the segment] as the point in which the first plane touches the conoid, and the *axis* [of the segment] as the portion cut off within the segment from the line drawn through the vertex of the segment parallel to the axis of the conoid.

The questions propounded for consideration were

(1) why, if a segment of the right-angled conoid be cut off by a plane at right angles to the axis, will the segment so cut off be half as large again as the cone which has the same base as the segment and the same axis, and

(2) why, if two segments be cut off from the right-angled conoid by planes drawn in any manner, will the segments so cut off have to one another the duplicate ratio of their axes.

[...] After prefixing therefore the theorems and directions which are necessary for the proof of them, I will then proceed to expound the propositions themselves to you.

(b) Propositions 21, 22

Any segment of a paraboloid of revolution is half as large again as the cone or segment of a cone which has the same base and the same axis.

Let the base of the segment be perpendicular to the plane of the paper, and let the plane of the paper be the plane through the axis of the paraboloid which cuts the base of the segment at right angles in BC and makes the parabolic section BAC.

Let EF be that tangent to the parabola which is parallel to BC, and let A be the point of contact.

Then: (1) if the plane of the base of the segment is perpendicular to the axis of the paraboloid, that axis is the line AD bisecting BC at right angles in D; (2) if the plane of the base is not perpendicular to the axis of the paraboloid, draw AD parallel to the axis of the paraboloid. AD will then bisect BC, but not at right angles.

Draw through EF a plane parallel to the base of the segment. This will touch the paraboloid at A, and A will be the vertex of the segment, AD its axis.

The base of the segment will be a circle with diameter BC or an ellipse with BC as major axis.

Accordingly a cylinder or a frustum of a cylinder can be found passing through the circle or ellipse and having AD for its axis [Prop. 9]; and likewise a cone or a segment of a cone can be drawn passing through the circle or ellipse and having A for vertex and AD for axis [Prop. 8].

Suppose X to be a cone equal to $\frac{3}{2}$ (cone or segment of cone ABC). The cone X is therefore equal to half the cylinder or frustum of a cylinder EC [cf. Prop. 10].

We shall prove that the volume of the segment of the paraboloid is equal to X.

If not, the segment must be either greater or less than X.

I. If possible, let the segment be greater than X.

We can then inscribe and circumscribe, as in the last proposition, figures made up of cylinders or frusta of cylinders with equal height and such that

$$(\text{circumscribed figure}) - (\text{inscribed figure}) < (\text{segment}) - X.$$

Let the greatest of the cylinders or frusta forming the circumscribed figure be that whose base is the circle or ellipse about BC and whose axis is OD, and let the smallest of them be that whose base is the circle or ellipse about PP' and whose axis is AL.

Let the greatest of the cylinders forming the inscribed figure be that whose base is the circle or ellipse about RR' and whose axis is OD, and let the smallest be that whose base is the circle or ellipse about PP' and whose axis is LM.

Produce all the plane bases of the cylinders or frusta to meet the surface of the complete cylinder or frustum EC.

Now, since (circumscribed figure) – (inscribed figure) < (segment) – X, it follows that

$$(\text{inscribed figure}) > X. \qquad (\alpha)$$

Next, comparing successively the cylinders or frusta with heights equal to OD and respectively forming parts of the complete cylinder or frustum EC and of the inscribed

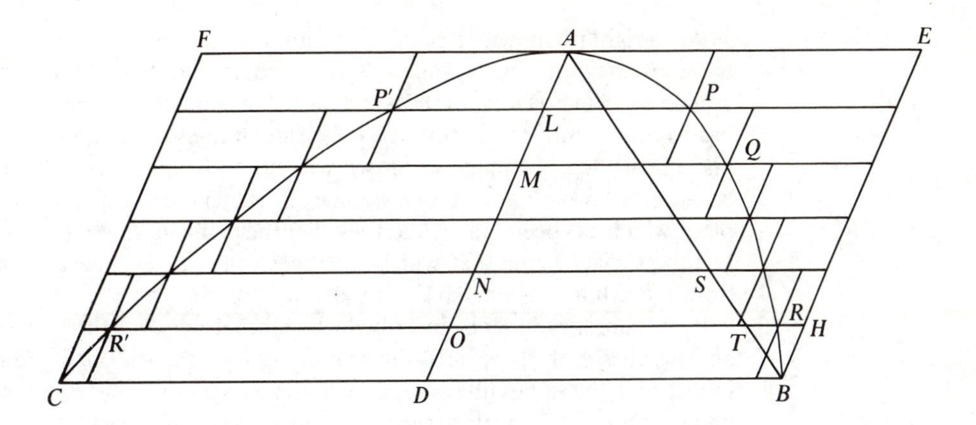

figure, we have (first cylinder or frustum in EC) : (first in inscribed figure) = $BD^2:RO^2 = AD:AO = BD:TO$, where AB meets OR in T.

And (second cylinder or frustrum in EC) : (second in inscribed figure) = $HO:SN$, in like manner, and so on.

Hence [Prop. 1] (cylinder or frustum EC) : (inscribed figure) = $(BD + HO + \cdots):(TO + SN + \cdots)$, where BD, HO,... are all equal, and BD, TO, SN,... diminish in arithmetical progression.

But [Lemma preceding Prop. 1] $BD + HO + \cdots > 2(TO + SN + \cdots)$.

Therefore (cylinder or frustum EC) > 2 (inscribed figure), or X > (inscribed figure); which is impossible, by (α) above.

II. If possible, let the segment be less than X.

In this case we inscribe and circumscribe figures as before, but such that (circumscribed figure) − (inscribed figure) < X − (segment), whence it follows that

$$\text{(circumscribed figure)} < X. \qquad (\beta)$$

And, comparing the cylinders or frusta making up the complete cylinder or frustum CE and the *circumscribed* figure respectively, we have (first cylinder or frustum in CE):(first in circumscribed figure) = $BD^2:BD^2 = BD:BD$.

(second in CE) : (second in circumscribed figure) = $HO^2:RO^2 = AD:AO = HO:TO$, and so on.

Hence [Prop. 1] (cylinder or frustum CE) : (circumscribed figure) = $(BD + HO + \cdots):(BD + TO + \cdots) < 2:1$ [Lemma preceding Prop. 1], and it follows that X < (circumscribed figure); which is impossible, by (β).

Thus the segment, being neither greater nor less than X, is equal to it, and therefore to $\frac{3}{2}$ (cone or segment of cone ABC).

4.A9 The method treating of mechanical problems

(a) Letter to Eratosthenes

I sent you on a former occasion some of the theorems discovered by me, merely writing out the enunciations and inviting you to discover the proofs, which at the moment I did not give. The enunciations of the theorems which I sent were as follows:

1 If in a right prism with a parallelogrammic base a cylinder be inscribed which has its bases in the opposite parallelograms, and its sides [i.e. four generators] on the remaining planes [faces] of the prism, and if through the centre of the circle which is the base of the cylinder and [through] one side of the square in the plane opposite to it a plane be drawn, the plane so drawn will cut off from the cylinder a segment which is bounded by two planes and the surface of the cylinder, one of the two planes being the plane which has been drawn and the other the plane in which the base of the cylinder is, and the surface being that which is between the said planes; and the segment cut off from the cylinder is one sixth part of the whole prism.
2 If in a cube a cylinder be inscribed which has its bases in the opposite parallelograms and touches with its surface the remaining four planes [faces], and if there also be inscribed in the same cube another cylinder which has its bases in other parallelograms and touches with its surface the remaining four planes [faces], then the figure bounded by the surfaces of the cylinders, which is within both cylinders, is two-thirds of the whole cube.

Now these theorems differ in character from those communicated before; for we compared the figures then in question, conoids and spheroids and segments of them, in respect to size, with figures of cones and cylinders: but none of those figures have yet been found to be equal to a solid figure bounded by planes; whereas each of the present figures bounded by two planes and surfaces of cylinders is found to be equal to one of the solid figures which are bounded by planes. The proofs then of these theorems I have written in this book and now send to you. Seeing moreover in you, as I say, an earnest student, a man of considerable eminence in philosophy, and an admirer [of mathematical inquiry], I thought fit to write out for you and explain in detail in the same book the peculiarity of a certain method, by which it will be possible for you to get a start to enable you to investigate some of the problems in mathematics by means of mechanics. This procedure is, I am persuaded, no less useful even for the proof of the theorems themselves; for certain things first became clear to me by a mechanical method, although they had to be demonstrated by geometry afterwards because their investigation by the said method did not furnish an actual demonstration. But it is of course easier, when we have previously acquired, by the method, some knowledge of the questions, to supply the proof than it is to find it without any previous knowledge. This is a reason why, in the case of the theorems the proof of which Eudoxus was the first to discover, namely that the cone is a third part of the cylinder, and the pyramid of the prism, having the same base and equal height, we should give no small share of the credit to Democritus who was the first to make the assertion with regard to the said figure though he did not prove it. I am myself in the position of having first made the discovery of the theorem now to be published [by the method indicated], and I deem it necessary to expound the method partly because I have already spoken of it and I do not want to be thought to have uttered vain words, but equally because I am persuaded that it will be of no little service to mathematics; for I apprehend that some, either of my contemporaries or of my successors, will, by means of the method when once established, be able to discover other theorems in addition, which have not yet occurred to me.

First then I will set out the very first theorem which became known to me by means of mechanics, namely that

Any segment of a section of a right-angled cone [i.e. a parabola] *is four-thirds of the triangle which has the same base and equal height*, and after this I will give each of the other theorems investigated by the same method. Then, at the end of the book, I will give the geometrical [proofs of the propositions].

[I premise the following propositions which I shall use in the course of the work.]

1 If from [one magnitude another magnitude be subtracted which has not the same centre of gravity, the centre of gravity of the remainder is found by] producing [the straight line joining the centres of gravity of the whole magnitude and of the subtracted part in the direction of the centre of gravity of the whole] and cutting off from it a length which has to the distance between the said centres of gravity the ratio which the weight of the subtracted magnitude has to the weight of the remainder [*On the Equilibrium of Planes*, I. 8].

2 If the centres of gravity of any number of magnitudes whatever be on the same straight line, the centre of gravity of the magnitude made up of all of them will be on the same straight line [*Ibid*, I. 5].

3 The centre of gravity of any straight line is the point of bisection of the straight line [*Ibid*, I. 4].

4 The centre of gravity of any triangle is the point in which the straight lines drawn from the angular points of the triangle to the middle points of the (opposite) sides cut one another [*Ibid*, I. 13, 14].

5 The centre of gravity of any parallelogram is the point in which the diagonals meet [*Ibid*, I. 10].

6 The centre of gravity of a circle is the point which is also the centre [of the circle].

7 The centre of gravity of any cylinder is the point of bisection of the axis.

8 The centre of gravity of any cone is [the point which divides its axis so that] the portion [adjacent to the vertex is] triple [of the portion adjacent to the base].

[All these propositions have already been] proved. [Besides these I require also the following proposition, which is easily proved:

If in two series of magnitudes those of the first series are, in order, proportional to those of the second series and further], the magnitudes [of the first series], either all or some of them, are in any ratio whatever [to those of a third series], and if the magnitudes of the second series are in the same ratio to the corresponding magnitudes [of a fourth series], then the sum of the magnitudes of the first series has to the sum of the selected magnitudes of the third series the same ratio which the sum of the magnitudes of the second series has to the sum of the [correspondingly] selected magnitudes of the fourth series [*On Conoids and Spheroids*, Prop. 1].

(b) *Proposition 1*

Let *ABC* be a segment of a parabola bounded by the straight line *AC* and the parabola *ABC*, and let *D* be the middle point of *AC*. Draw the straight line *DBE* parallel to the axis of the parabola and join *AB*, *BC*.

Then shall the segment *ABC* be $\frac{4}{3}$ of the triangle *ABC*.

From *A* draw *AKF* parallel to *DE*, and let the tangent to the parabola at *C* meet *DBE* in *E* and *AKF* in *F*. Produce *CB* to meet *AF* in *K*, and again produce *CK* to *H*, making *KH* equal to *CK*.

Consider *CH* as the bar of a balance, *K* being its middle point.

Let *MO* be any straight line parallel to *ED*, and let it meet *CF*, *CK*, *AC* in *M*, *N*, *O* and the curve in *P*.

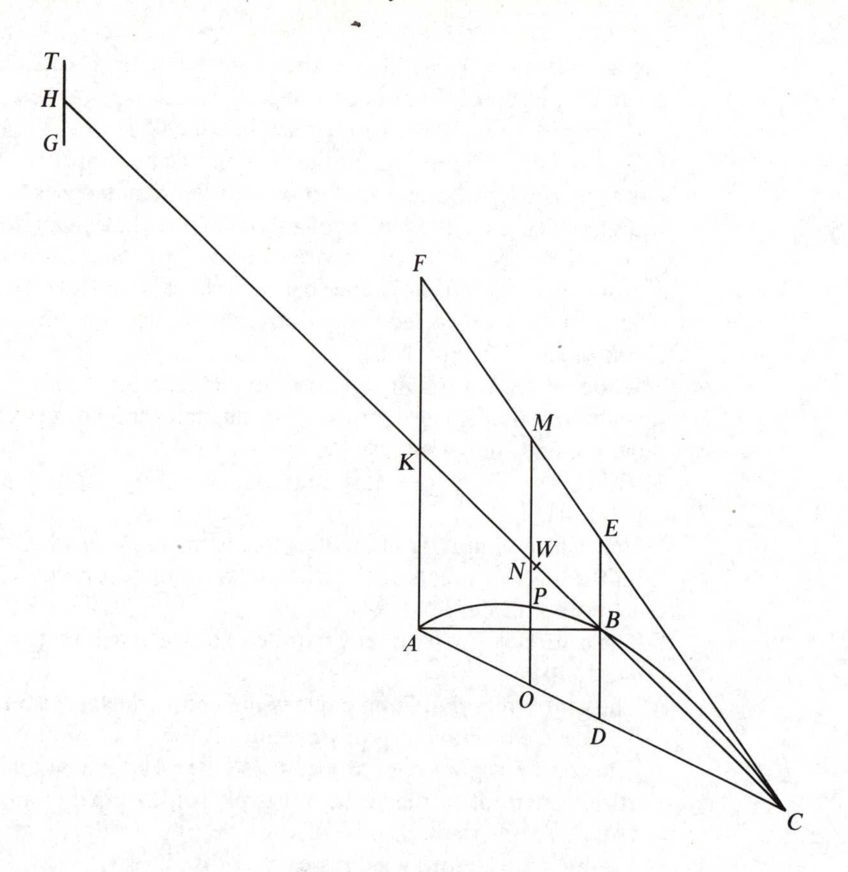

Now, since CE is a tangent to the parabola and CD the semi-ordinate, $EB = BD$; for this is proved in the Elements [of Conics].

Since FA, MO are parallel to ED, it follows that $FK = KA$, $MN = NO$.

Now, by the property of the parabola, proved in a lemma, $MO : OP = CA : AO$ [*Quadrature of Parabola*, Prop. 5] $= CK : KN$ [Euclid's *Elements*, VI. 2] $= HK : KN$.

Take a straight line TG equal to OP, and place it with its centre of gravity at H, so that $TH = HG$; then, since N is the centre of gravity of the straight line MO, and $MO : TG = HK : KN$, it follows that TG at H and MO at N will be in equilibrium about K [*On the Equilibrium of Planes*, I. 6, 7].

Similarly, for all other straight lines parallel to DE and meeting the arc of the parabola, (1) the portion intercepted between FC, AC with its middle point on KC and (2) a length equal to the intercept between the curve and AC placed with its centre of gravity at H will be in equilibrium about K.

Therefore K is the centre of gravity of the whole system consisting (1) of all the straight lines as MO intercepted between FC, AC and placed as they actually are in the figure and (2) of all the straight lines placed at H equal to the straight lines as PO intercepted between the curve and AC.

And, since the triangle CFA is made up of all the parallel lines like MO, and the segment CBA is made up of all the straight lines like PO within the curve, it follows that the triangle, placed where it is in the figure, is in equilibrium about K with the segment CBA placed with its centre of gravity at H.

Divide KC at W so that $CK = 3KW$; then W is the centre of gravity of the triangle

ACF; for this is proved in the books on equilibrium [*On the Equilibrium of Planes*, I. 15].

Therefore $\triangle ACF$:(segment ABC) = $HK:KW = 3:1$.

Therefore segment $ABC = \frac{1}{3}\triangle ACF$.

But $\triangle ACF = 4\triangle ABC$.

Therefore segment $ABC = \frac{4}{3}\triangle ABC$.

Now the fact here stated is not actually demonstrated by the argument used; but that argument has given a sort of indication that the conclusion is true. Seeing then that the theorem is not demonstrated, but at the same time suspecting that the conclusion is true, we shall have recourse to the geometrical demonstration which I myself discovered and have already published.

(c) *Proposition 4*

Any segment of a right-angled conoid [i.e a paraboloid of revolution] *cut off by a plane at right angles to the axis is* $1\frac{1}{2}$ *times the cone which has the same base and the same axis as the segment.*

This can be investigated by our method, as follows.

Let a paraboloid of revolution be cut by a plane through the axis in the parabola BAC; and let it also be cut by another plane at right angles to the axis and intersecting the former plane in BC. Produce DA, the axis of the segment, to H, making HA equal to AD.

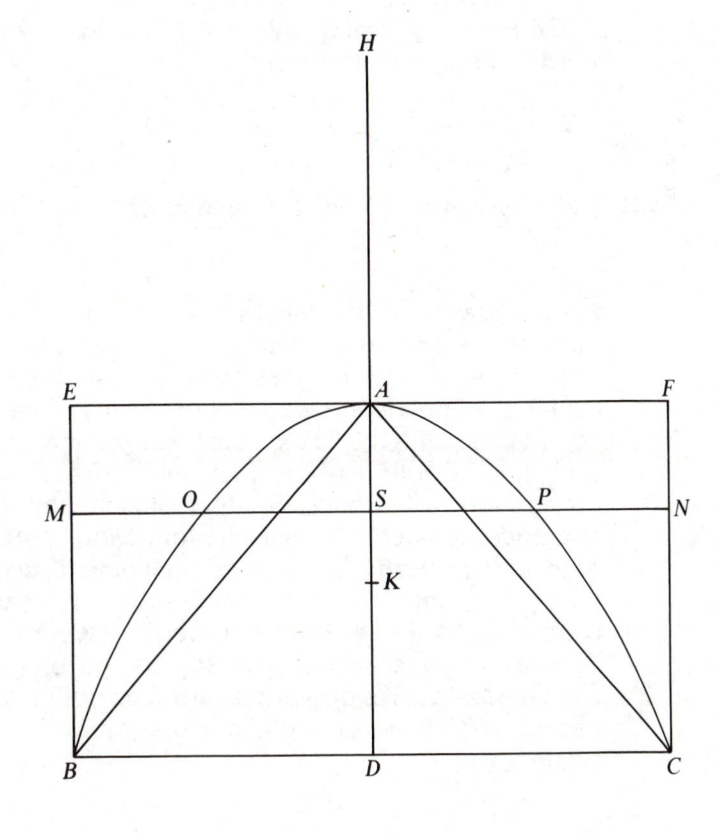

Imagine that HD is the bar of a balance, A being its middle point.

The base of the segment being the circle on BC as diameter and in a plane perpendicular to AD, imagine (1) a cone drawn with the latter circle as base and A as vertex, and (2) a cylinder with the same circle as base and AD as axis.

In the parallelogram EC let any straight line MN be drawn parallel to BC, and through MN let a plane be drawn at right angles to AD; this plane will cut the cylinder in a circle with diameter MN and the paraboloid in a circle with diameter OP.

Now, BAC being a parabola and BD, OS ordinates, $DA:AS = BD^2:OS^2$, or $HA:AS = MS^2:SO^2$.

Therefore $HA:AS =$ (circle, radius MS):(circle, radius OS) = (circle in cylinder):(circle in paraboloid).

Therefore the circle in the cylinder, in the place where it is, will be in equilibrium about A with the circle in the paraboloid, if the latter is placed with its centre of gravity at H.

Similarly for the two corresponding circular sections made by a plane perpendicular to AD and passing through any other straight line in the parallelogram which is parallel to BC.

Therefore, as usual, if we take all the circles making up the whole cylinder and the whole segment and treat them in the same way, we find that the cylinder, in the place where it is, is in equilibrium about A with the segment placed with its centre of gravity at H.

If K is the middle point of AD, K is the centre of gravity of the cylinder; therefore $HA:AK =$ (cylinder):(segment).

Therefore cylinder = 2(segment).

And cylinder = 3(cone ABC) [Euclid's *Elements*, XII. 10];
therefore segment = $\frac{3}{2}$(cone ABC).

4.B Later Accounts of the Life and Works of Archimedes

Partly because of the dramatic quality of his death, partly because of his interest in mechanical inventions, the life and personality of Archimedes has left a lasting impression on subsequent generations. The well-evidenced knowledge we have of details of his life is somewhat slight in comparison with the liveliness of later legends, but the latter are interesting both for the folk-memory they may enshrine and as testimony to later conceptions of what a mathematician should be like. Of the many references to Archimedes in the following centuries, we have chosen two of the most important: Vitruvius's celebrated account (4.B2) of Archimedes taking a bath and Plutarch's account (4.B1) of aspects of Archimedes' life, death, works and opinions, based on we know not what sources. Sir Thomas Heath's account (4.B4) of the progress of the Greek text of Archimedes' works down the centuries outlines the involved and tenuous path by which we come to our present knowledge of what Archimedes wrote.

4.B1 Plutarch

Plutarch is describing the siege of Syracuse, which started in 214 BC, led by the Roman general Marcellus.

Marcellus directed a fleet of sixty quinquiremes, which were equipped with many different kinds of weapons and missiles. In addition he had built a siege-engine which was mounted on a huge platform supported by eight galleys lashed together, and with this he sailed up to the city walls, confident that the size and the imposing spectacle of his armament together with his personal prestige would combine to overawe the Syracusans. But he had reckoned without Archimedes, and the Roman machines turned out to be insignificant not only in the philosopher's estimation, but also by comparison with those which he had constructed himself. Archimedes did not regard his military inventions as an achievement of any importance, but merely as a by-product, which he occasionally pursued for his own amusement, of his serious work, namely the study of geometry. He had done this in the past because Hiero, the former ruler of Syracuse, had often pressed and finally persuaded him to divert his studies from the pursuit of abstract principles to the solution of practical problems, and to make his theories more intelligible to the majority of mankind by applying them through the medium of the senses to the needs of everyday life.

It was Eudoxus and Archytas who were the originators of the now celebrated and highly prized art of mechanics. They used it with great ingenuity to illustrate geometrical theorems, and to support by means of mechanical demonstrations easily grasped by the senses propositions which are too intricate for proof by word or diagram. For example, to solve the problem of finding two mean proportional lines, which are necessary for the construction of many other geometrical figures, both mathematicians resorted to mechanical means, and adapted to their purposes certain instruments named mesolabes taken from conic sections. Plato was indignant at these developments, and attacked both men for having corrupted and destroyed the ideal purity of geometry. He complained that they had caused her to forsake the realm of disembodied and abstract thought for that of material objects, and to employ instruments which required much base and manual labour. For this reason mechanics came to be separated from geometry, and as the subject was for a long time disregarded by philosophers, it took its place among the military arts.

However this may be, Archimedes in writing to Hiero, who was both a relative and a friend of his, asserted that with any given force it was possible to move any given weight, and then, carried away with enthusiasm at the power of his demonstration, so we are told, went on to enlarge his claim, and declared that if he were given another world to stand on, he could move the earth. Hiero was amazed, and invited him to put his theorem into practice and show him some great weight moved by a tiny force. Archimedes chose for his demonstration a three-masted merchantman of the royal fleet, which had been hauled ashore with immense labour by a large gang of men, and he proceeded to have the ship loaded with her usual freight and embarked a large number of passengers. He then seated himself at some distance away and without using any noticeable force, but merely exerting traction with his hand through a complex system of pulleys, he drew the vessel towards him with as smooth and even a motion as if she were gliding through the water. The king was deeply impressed, and recognising the potentialities of his skill, he persuaded Archimedes to construct for him a number of engines designed both for attack and defence, which could be employed in any kind of siege warfare. Hiero himself never had occasion to use these, since most of his life was spent at peace amid festivals and public ceremonies, but when the present war broke out, the apparatus was ready for the Syracusans to use and its inventor was at hand to direct its employment.

When the Romans first attacked by sea and land, the Syracusans were struck dumb with terror and believed that nothing could resist the onslaught of such powerful forces. But presently Archimedes brought his engines to bear and launched a tremendous barrage against the Roman army. This consisted of a variety of missiles, including a great volley of stones which descended upon their target with an incredible noise and velocity. There was no protection against this artillery, and the soldiers were knocked down in swathes and their ranks thrown into confusion. At the same time huge beams were run out from the walls so as to project over the Roman ships: some of them were then sunk by great weights dropped from above, while others were seized at the bows by iron claws or by beaks like those of cranes, hauled into the air by means of counterweights until they stood upright upon their sterns, and then allowed to plunge to the bottom, or else they were spun round by means of windlasses situated inside the city and dashed against the steep cliffs and rocks which jutted out under the walls, with great loss of life to the crews. Often there would be seen the terrifying spectacle of a ship being lifted clean out of the water into the air and whirled about as it hung there, until every man had been shaken out of the hull and thrown in different directions, after which it would be dashed down empty upon the walls. As for the enormous siege-engine which Marcellus brought up, mounted on eight galleys as I have described, and known as a *sambuca* because of its resemblance to the musical instrument of that name, a stone weighing a hundred pounds was discharged while it was still approaching the city wall, immediately followed by a second and a third. These descended on their target with a thunderous crash and a great surge of water, shattered the platform on which the machine was mounted, loosened the bolts which held it together, and dislodged the whole framework from the hulks which supported it. Marcellus, finding his plan of attack thus brought to a standstill, drew off his ships as quickly as possible and ordered his land forces to retire.

After this he held a council of war and formed a new plan to move up as closely as possible to the walls under cover of darkness. The Romans calculated that the cables which Archimedes used for his siege-engines imparted such a tremendous velocity to the missiles they discharged that these would go flying over their heads, but that at close quarters, where a low trajectory was required, they would be ineffective. However, Archimedes, it seems, had long ago foreseen such a possibility and had designed engines which were suitable for any distance and missiles to match them. He had had a large number of loopholes made in the walls, and in these he placed short-range weapons known as scorpions, which were invisible to the attacker, but could be discharged as soon as he arrived at close quarters.

So when the Romans crept up close to the walls expecting to surprise the enemy, they were again greeted by a hail of missiles. Huge stones were dropped on them almost perpendicularly, and it seemed as if they were faced by a curtain of darts along the whole length of the wall, so that the attackers soon fell back. But here too, even while they were hurrying, as they hoped, out of danger, they came under fire from the medium-range catapults which caused heavy losses among them: at the same time many of their ships were dashed against one another, and all this while they were helpless to retaliate. Archimedes had mounted most of his weapons under the cover of the city walls, and the Romans began to believe that they were fighting against a supernatural enemy, as they found themselves constantly struck down by opponents whom they could never see.

Marcellus, however, escaped unhurt from this assault and afterwards made fun of his own siege-experts and engineers. 'We may as well give up fighting this geometrical Briareus,' he said, 'who uses our ships like cups to ladle water out of the sea, who has whipped our *sambuca* and driven it off in disgrace, and who can outdo the hundred-handed giants of mythology in hurling so many different missiles at us at once.' For the truth was that all the rest of the Syracusans merely provided the manpower to operate Archimedes' inventions, and it was his mind which directed and controlled every manoeuvre. All other weapons were discarded, and it was upon his alone that the city relied both for attack and defence. At last the Romans were reduced to such a state of alarm that if they saw so much as a length of rope or a piece of timber appear over the top of the wall, it was enough to make them cry out, 'Look, Archimedes is aiming one of his machines at us!' and they would turn their backs and run. When Marcellus saw this, he abandoned all attempts to capture the city by assault, and settled down to reduce it by blockade.

As for Archimedes, he was a man who possessed such exalted ideals, such profound spiritual vision, and such a wealth of scientific knowledge that, although his inventions had earned him a reputation for almost superhuman intellectual power, he would not deign to leave behind him any writings on his mechanical discoveries. He regarded the business of engineering, and indeed of every art which ministers to the material needs of life, as an ignoble and sordid activity, and he concentrated his ambition exclusively upon those speculations whose beauty and subtlety are untainted by the claims of necessity. These studies, he believed, are incomparably superior to any others, since here the grandeur and beauty of the subject matter vie for our admiration with the cogency and precision of the methods of proof. Certainly in the whole science of geometry it is impossible to find more difficult and intricate problems handled in simpler and purer terms than in his works. Some writers attribute this to his natural genius. Others maintain that a phenomenal industry lay behind the apparently effortless ease with which he obtained his results. The fact is that no amount of mental effort of his own would enable a man to hit upon the proof of one of Archimedes' theorems, and yet as soon as it is explained to him, he feels that he might have discovered it himself, so smooth and rapid is the path by which he leads us to the required conclusion. So it is not at all difficult to credit some of the stories which have been told about him; of how, for example, he often seemed so bewitched by the song of some inner and familiar Siren that he would forget to eat his food or take care of his person; or how when he was carried by force, as he often was to the bath for his body to be washed and anointed, he would trace geometrical figures in the ashes and draw diagrams with his finger in the oil which had been rubbed over his skin. Such was the rapture which his work inspired in him, so as to make him truly the captive of the Muses. And although he was responsible for many discoveries of great value, he is said to have asked his friends and relatives to place on his tomb after his death nothing more than the shape of a cylinder enclosing a sphere, with an inscription explaining the ratio by which the containing solid exceeds the contained.

Such was Archimedes' character, and in so far as it rested with him, he kept himself and his city unconquered.

[But eventually the city was captured.]

But what distressed Marcellus most of all was the death of Archimedes. As fate would have it the philosopher was by himself, engrossed in working out some calculation by means of a diagram, and his eyes and his thoughts were so intent upon the problem that he was completely unaware that the Romans had broken through the defences, or that the city had been captured. Suddenly a soldier came upon him and ordered him to accompany him to Marcellus. Archimedes refused to move until he had worked out his problem and established his demonstration, whereupon the soldier flew into a rage, drew his sword, and killed him. According to another account, the Roman came up with a drawn sword and threatened to kill him there and then: when Archimedes saw him, he begged him to stay his hand for a moment, so that he should not leave his theorem imperfect and without its demonstration, but the soldier paid no attention and dispatched him at once. There is yet a third story to the effect that Archimedes was on his way to Marcellus bringing some of his instruments, such as sundials and spheres and quadrants, with the help of which the dimensions of the sun could be measured by the naked eye, when some soldiers met him, and believing that he was carrying gold in the box promptly killed him. At any rate it is generally agreed that Marcellus was deeply affected by his death, that he abhorred the man who had killed him as if he had committed an act of sacrilege, and that he sought out Archimedes' relatives and treated them with honour.

4.B2 Vitruvius

In the case of Archimedes, although he made many wonderful discoveries of diverse kinds, yet of them all, the following, which I shall relate, seems to have been the result of a boundless ingenuity. Hiero, after gaining the royal power in Syracuse, resolved, as a consequence of his successful exploits, to place in a certain temple a golden crown which he had vowed to the immortal gods. He contracted for its making at a fixed price, and weighed out a precise amount of gold to the contractor. At the appointed time the latter delivered to the king's satisfaction an exquisitely finished piece of handiwork, and it appeared that in weight the crown corresponded precisely to what the gold had weighed.

But afterwards a charge was made that gold had been abstracted and an equivalent weight of silver had been added in the manufacture of the crown. Hiero, thinking it an outrage that he had been tricked, and yet not knowing how to detect the theft, requested Archimedes to consider the matter. The latter, while the case was still on his mind, happened to go to the bath, and on getting into a tub observed that the more his body sank into it the more water ran out over the tub. As this pointed out the way to explain the case in question, without a moment's delay, and transported with joy, he jumped out of the tub and rushed home naked, crying with a loud voice that he had found what he was seeking; for as he ran he shouted repeatedly in Greek, 'Εὕρηκα, εὕρηκα'.

Taking this as the beginning of his discovery, it is said that he made two masses of the same weight as the crown, one of gold and the other of silver. After making them, he filled a large vessel with water to the very brim, and dropped the mass of silver into it. As much water ran out as was equal in bulk to that of the silver sunk in the vessel. Then, taking out the mass, he poured back the lost quantity of water, using a pint measure, until it was level with the brim as it had been before. Thus he found the weight of silver corresponding to a definite quantity of water.

After this experiment, he likewise dropped the mass of gold into the full vessel and, on taking it out and measuring as before, found that not so much water was lost, but a smaller quantity: namely, as much less as a mass of gold lacks in bulk compared to a mass of silver of the same weight. Finally, filling the vessel again and dropping the crown itself into the same quantity of water, he found that more water ran over for the crown than for the mass of gold of the same weight. Hence, reasoning from the fact that more water was lost in the case of the crown than in that of the mass, he detected the mixing of silver with the gold, and made the theft of the contractor perfectly clear.

4.B3 John Wallis (1685)

It is to me a thing unquestionable, That the Ancients had somewhat of like nature with our Algebra; from whence many of their prolix and intricate Demonstrations were derived. And I find other modern Writers of the same opinion with me therein. [...]

But this their Art of Invention, they seem very studiously to have concealed: contenting themselves to demonstrate by Apagogical Demonstrations, (or reducing to Absurdity, if denied,) without shewing us the method, by which they first found out those Propositions, which they thus demonstrate by other ways.

Of which, Nuñes or Nonius in his *Algebra* (in Spanish) fol.114.b. speaks thus: 'O how well had it been if those Authors, who have written in Mathematics, had delivered to us their Inventions, in the same way, and with the same Discourse, as they were found out! And not as Aristotle says of Artificers in Mechanics, who shew us the Engines they have made, but conceal the Artifice, to make them the more admired! The method of Invention, in divers Arts, is very different from that of Tradition, wherein they are delivered. Nor are we to think, that all these Propositions in Euclid and Archimedes were in the same way found out, as they are now delivered to us.'

4.B4 Sir Thomas Heath

Heron, Pappus and Theon all cite works of Archimedes which no longer survive, a fact which shows that such works were still extant at Alexandria as late as the third and fourth centuries AD. But it is evident that attention came to be concentrated on two works only, the *Measurement of a Circle* and *On the Sphere and Cylinder*. Eutocius (c. 500 AD) only wrote commentaries on these works and on the *Plane Equilibriums*, and he does not seem even to have been acquainted with the *Quadrature of the Parabola* or the work *On Spirals*, although these have survived. Isidorus of Miletus revised the commentaries of Eutocius on the *Measurement of a Circle* and the two books *On the Sphere and Cylinder*, and it would seem to have been in the school of Isidorus that these treatises were turned from their original Doric into the ordinary language, with alterations designed to make them more intelligible to elementary pupils. But neither in Isidorus' time nor earlier was there any collected edition of Archimedes' works, so that those which were less read tended to disappear.

In the ninth century Leon, who restored the University of Constantinople, collected together all the works that he could find at Constantinople, and had the manuscript written (the archetype, Heiberg's A) which, through its derivatives, was up to the discovery of the Constantinople manuscript (C) containing *The Method*, the only source for the Greek text. Leon's manuscript came, in the twelfth century, to the Norman Court at Palermo, and thence passed to the House of Hohenstaufen. Then, with all the library of Manfred, it was given to the Pope by Charles of Anjou after the battle of Benevento in 1266. It was in the Papal Library in the years 1269 and 1311, but, some time after 1368, passed into private hands. In 1491 it belonged to Georgius Valla, who translated from it the portions published in his posthumous work *De Expetendis et Fugiendis Rebus* (1501), and intended to publish the whole of Archimedes with Eutocius' commentaries. On Valla's death in 1500 it was bought by Albertus Pius, Prince of Carpi, passing in 1530 to his nephew, Rodolphus Pius, in whose possession it remained till 1544. At some time between 1544 and 1564 it disappeared, leaving no trace.

The greater part of A was translated into Latin in 1269 by William of Moerbeke at the Papal Court at Viterbo. This translation, in William's own hand, exists at Rome (Cod. Ottobon. lat. 1850, Heiberg's B), and is one of our prime sources, for, although the translation was hastily done and the translator sometimes misunderstood the Greek, he followed its wording so closely that his version is, for purposes of collation, as good as a Greek manuscript. William used also, for his translation, another manuscript from the same library which contained works not included in A. This manuscript was a collection of works on mechanics and optics; William translated from it the two books *On Floating Bodies*, and it also contained the *Plane Equilibriums* and the *Quadrature of the Parabola*, for which books William used both manuscripts.

The four most important extant Greek manuscripts (except C, the Constantinople manuscript discovered in 1906) were copied from A. The earliest is E, the Venice manuscript (Marcianus 305), which was written between the years 1449 and 1472. The next is D, the Florence manuscript (Laurent. XXVIII. 4), which was copied in 1491 for Angelo Poliziano, permission having been obtained with some difficulty in consequence of the jealousy with which Valla guarded his treasure. The other two are G (Paris. 2360) copied from A after it had passed to Albertus Pius, and H (Paris. 2361) copies in 1544 by Christopherus Auverus for Georges d'Armagnac, Bishop of Rodez. These four manuscripts, with the translation of William of Moerbeke (B), enable the readings of A to be inferred.

A Latin translation was made at the instance of Pope Nicholas V about the year 1450 by Jacobus Cremonensis. It was made from A, which was therefore accessible to Pope Nicholas though it does not seem to have belonged to him. Regiomontanus made a copy of this translation about 1468 and revised it with the help of E (the Venice manuscript of the Greek text) and a copy of the same translation belonging to Cardinal Bessarion, as well as another 'old copy' which seems to have been B.

The *editio princeps* was published at Basel (*apud Hervagium*) by Thomas Gechauff Venatorius in 1544. The Greek text was based on a Nürnberg MS (Norimberg. Cent. V, app. 12) which was copied in the sixteenth century from A but with interpolations derived from B; the Latin translation was Regiomontanus's revision of Jacobus Cremonensis (Norimb. Cent. V, 15).

A translation by F. Commandinus published at Venice in 1558 contained the *Measurement of a Circle*, *On Spirals*, the *Quadrature of the Parabola*, *On Conoids and Spheroids*, and the *Sand-reckoner*. This translation was based on the Basel edition, but Commandinus also consulted E and other Greek manuscripts.

Torelli's edition (Oxford, 1792) also followed the *editio princeps* in the main, but Torelli also collated E. The book was brought out after Torelli's death by Abram Robertson, who also collated five more manuscripts, including D, G and H. The collation, however, was not well done, and the edition was not properly corrected when in the press.

The second edition of Heberg's text of all the works of Archimedes with Eutocius's commentaries, Latin translation, apparatus criticus, &c., is now available (1910–15) and, of course, supersedes the first edition (1880–1) and all others. It naturally includes *The Method*, the fragment of the *Stomachion*, and so much of the Greek text of the two books *On Floating Bodies* as could be restored from the newly discovered Constantinople manuscript.

4.C Diocles

Diocles flourished at the beginning of the second century BC, so was a contemporary of Apollonius. His work *On Burning Mirrors* was cited by the sixth-century commentator Eutocius, but otherwise was thought lost until its recent rediscovery in a fifteenth-century Arabic manuscript. The extracts here are from the translation by G. J. Toomer. 4.C1, the start of the work (omitting the Arabic salutation), affords an interesting insight into conditions of mathematical activity in the early Hellenistic period, supplementing the inferences we can make from the earlier letters of Archimedes. Proposition 1 of the work (4.C2) is the first proof of the focal property of the parabola: that parallel rays (as from the sun) are all reflected from a parabolic mirror to the same point, what we now (since Kepler) call the focus. Proposition 12 (4.C3) is part of a discussion of solutions to the classical problem of doubling the cube, in its reduced form of finding two mean proportionals, and is where Diocles introduced the curve later known by the name *cissoid.*

4.C1 Introduction to *On Burning Mirrors*

He said: Pythion the Thasian geometer wrote a letter to Conon in which he asked him how to find a mirror surface such that when it is placed facing the sun the rays reflected from it meet the circumference of a circle. And when Zenodorus the astronomer came down to Arcadia and was introduced to us, he asked us how to find a mirror surface such that when it is placed facing the sun the rays reflected from it meet a point and thus cause burning. So we want to explain the answer to the problem posed by Pythion and to that posed by Zenodorus; in the course of this we shall make use of the premisses established by our predecessors. One of those two problems, namely the one requiring the construction of a mirror which makes all the rays meet in one point, is the one which was solved practically by Dositheus. The other problem, since it was only theoretical, and there was no argument worthy to serve as proof in its case, was not solved practically. We have set out a compilation of the proofs of both these problems and elucidated them.

The burning-mirror surface submitted to you is the surface bounding the figure produced by a section of a right-angled cone [i.e. parabola] being revolved about the line bisecting it [i.e. its axis]. It is a property of that surface that all the rays are reflected to a single point, namely the point [on the axis] whose distance from the surface is equal to a quarter of the line which is the parameter of the squares on the perpendiculars drawn to the axis [i.e. the ordinates]. Whenever one increases that surface by a given amount, there will be a [corresponding] increase in the above-mentioned conic section. So the rays reflected from that additional [surface] will also be reflected to exactly the same point, and thus they will increase the intensity of the heat around that point. The intensity of the burning in this case is greater than that generated from a spherical surface, for from a spherical surface the rays are reflected to a straight line, not to a point, although people used to guess that they are reflected to the centre; the rays which meet at one place in that [i.e. a spherical] surface are reflected from the surface [consisting] of a spherical segment less than half the sphere, and [even] if the mirror consists of half the sphere or more than half, only those rays reflected from less than half the sphere are reflected to that place.

4.C2 Diocles proves the focal property of the parabola

Let there be a parabola KBM, with axis AZ, and let half the parameter of the squares on the ordinates be line BH. Let BE on the axis be equal to BH, and let BE be bisected at point D. Let us draw a line tangent to the section at an arbitrary point, namely line θA, and draw line θG as ordinate to AZ. Then we know that $AB = BG$ and that the line drawn from θ perpendicular to θA meets AZ beyond E. So let us draw $Z\theta$ perpendicular to θA, and join θD.

Then $GZ = BH$ and $HB = BE$, so $GZ = BE$.

We subtract GE, common (to GZ and BE), then the remainder $GB = EZ$.

But $GB = BA$, so $AB = EZ$.

And $BD = DE$, because BE is bisected at D, so the sum $AD = DZ$.

And because triangle $A\theta Z$ is right-angled and its base AZ is bisected at D, $AD = D\theta = DZ$.

So $\angle O = \angle X$ and $\angle A = \angle PQ$.

So let a line parallel to AZ pass through θ, namely line θS.

Then $\angle O = \angle R$, which is alternate to it, and $\angle O = \angle X$, so $\angle X = \angle R$ also.

And $\angle PQX = \angle RT$, right angles, so $\angle T = \angle PQ$, remainders.

So when line $S\theta$ meets line $A\theta$ it is reflected to point D, forming equal angles, PQ and T, between itself and the tangent $A\theta$. Hence it has been shown that if one draws from any point on KBM a line tangent to the section, and draws the line connecting the point of tangency with point D, e.g. line θD, and draws line $S\theta$ parallel to AZ, then in

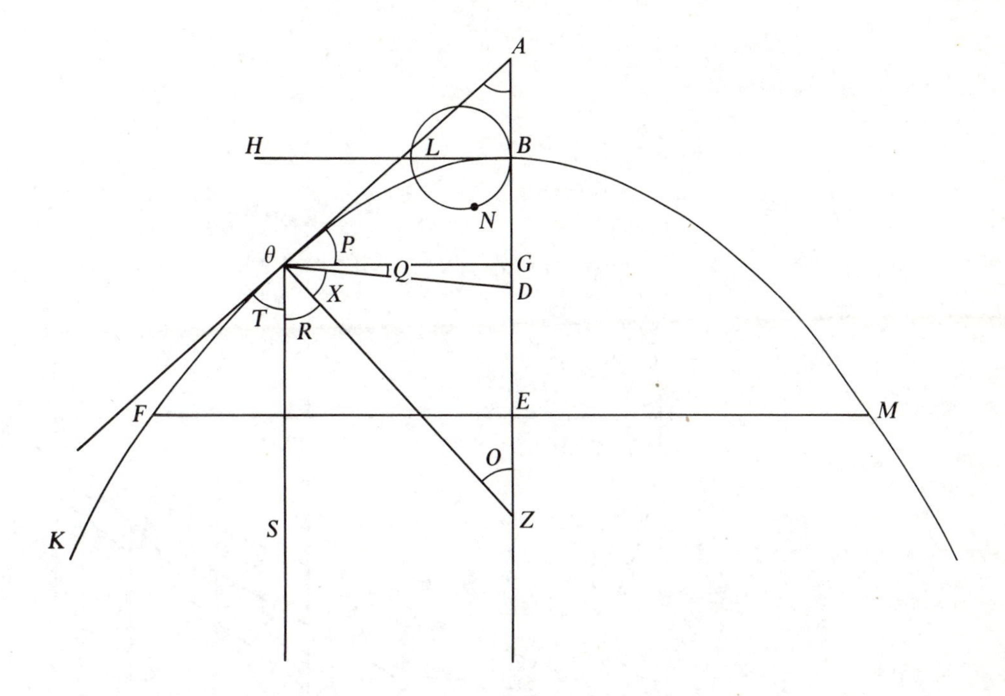

that case line $S\theta$ is reflected to point D, i.e. the line passing through point θ is reflected at equal angles from the tangent to the section. And all parallel lines from all points on KBM have the same property, so, since they make equal angles with the tangents, they go to point D.

Hence, if AZ is kept stationary, and KBM revolved [about it] until it returns to its original position, and a concave surface of brass is constructed on the surface described by KBM, and placed facing the sun, so that the sun's rays meet the concave surface, they will be reflected to point D, since they are parallel to each other. And the more the [reflecting] surface is increased, the greater will be the number of rays reflected to point D.

4.C3 Diocles introduces the cissoid

Let there again be a circle $ABGD$, with two diameters AB, GD, cutting one another at right angles. Let us cut off from the circle successive equal arcs DZ, ZH, $H\theta$, and draw perpendiculars ZK, HL, θM to line AB. Cut off from the other quadrant of the circle [i.e. AD], beginning from point D, arcs equal in size and also in number to arcs DZ, ZH, θH, namely arcs DN, NS, SO. Let the line joining B to N cut ZK at P, and the line joining B to S cut HL at Q, and the line joining B to O cut θM at R. Then it has been shown in the preceding proposition that ZK and KB are continuous proportionals between AK and KP. Similarly HL and LB are continuous proportionals between AL and LQ, and θM and MB are continuous proportionals between AM and MR. So if we construct the perpendiculars closer than those we mentioned, and mark points on them as we marked P, Q and R, and draw through all these points by means of the

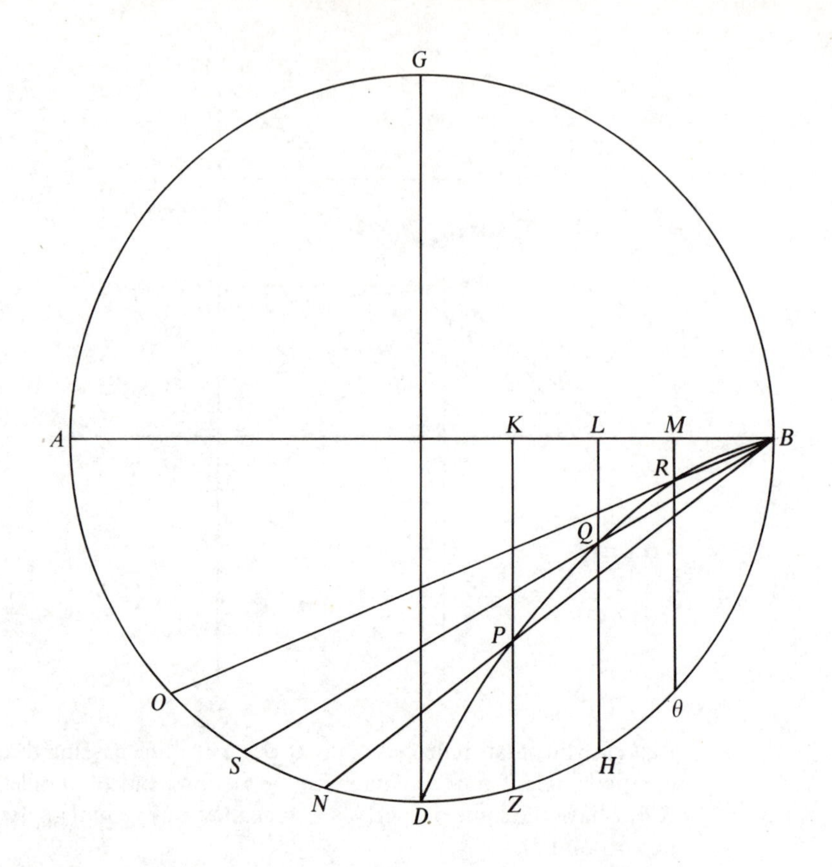

curved ruler line $BRQPD$, then it is obvious that if we mark on it [line $BRQPD$] a point [e.g.] P, and draw perpendicular PK from it [P] to AB, the result is that ZK and KB are continuous proportionals between AK and KP.

4.D Apollonius

Revered in antiquity and since as 'The Great Geometer', Apollonius contributed to most aspects of the geometry of his day (late third to early second century BC), as well as to mathematical astronomy. But his only major work to have substantially survived is *Conics*, which put the subject—the family of curves most deeply investigated in Greek times—on a newly coherent, unified foundation (4.D3–4.D4). Most of our knowledge of Apollonius's life is only what can be deduced from his letters prefacing the various books of the *Conics* (4.D1–4.D2). The other extracts (4.D5–4.D7) illustrate just a few of the results and methods found in this massive and influential work.

4.D1 General preface to *Conics*: to Eudemus

If you are restored in body, and other things go with you to your mind, well and good; and we too fare pretty well. At the time I was with you in Pergamum, I observed you were quite eager to be kept informed of the work I was doing in conics. And so I have sent you this first book revised, and we shall dispatch the others when we are satisfied with them. For I don't believe you have forgotten hearing from me how I worked out the plan for these conics at the request of Naucrates, the geometer, at the time he was with us in Alexandria lecturing, and how on arranging them in eight books we immediately communicated them in great haste because of his near departure, not revising them but putting down whatever came to us with the intention of a final going over. And so finding now the occasion of correcting them, one book after another, we publish them. And since it happened that some others among those frequenting us got acquainted with the first and second books before the revision, don't be surprised if you come upon them in a different form.

Of the eight books the first four belong to a course in the elements. The first book contains the generation of the three sections and of the opposite branches, and the principal properties in them worked out more fully and universally than in the writings of others. The second book contains the properties having to do with the diameters and axes and also the asymptotes, and other things of a general and necessary use for limits of possibility. And what I call diameters and what I call axes you will know from this book. The third book contains many incredible theorems of use for the construction of solid loci and for limits of possibility of which the greatest part and the most beautiful are new. And when we had grasped these, we knew that the three-line and four-line locus had not been constructed by Euclid, but only a chance part of it and that not very happily. For it was not possible for this construction to be completed without the additional things found by us. The fourth book shows in how many ways the sections of a cone intersect with each other and with the circumference of a circle, and contains other things in addition none of which has been written up by our predecessors, that is in how many points the section of a cone or the circumference of a circle and the opposite branches meet the opposite branches. The rest of the books are fuller in treatment. For there is one dealing more fully with maxima and minima, and one with equal and similar sections of a cone, and one with limiting theorems, and one with determinate conic problems. And so indeed, with all of them published, those happening upon them can judge them as they see fit.

4.D2 Prefaces to Books II, IV and V

(a) Book II: to Eudemus

I have sent you my son Apollonius bringing you the second book of the conics as arranged by us. Go through it then carefully and acquaint those with it worthy of sharing in such things. And Philonides, the geometer, I introduced to you in Ephesus, if ever he happen about Pergamum, acquaint him with it too. And take care of yourself, to be well.

(*b*) *Book IV: to Attalus*

Some time ago I expounded and sent to Eudemus of Pergamum the first three books of my conics which I have compiled in eight books, but, as he has passed away, I have resolved to dedicate the remaining books to you because of your earnest desire to possess my works. I am sending you on this occasion the fourth book. It contains a discussion of the question, in how many points at most it is possible for sections of cones to meet one another and the circumference of a circle, on the assumption that they do not coincide throughout, and further in how many points at most a section of a cone or the circumference of a circle can meet the hyperbola with two branches, [or two double-branch hyperbolas can meet one another]; and, besides these questions, the book considers a number of others of a similar kind. Now the first question Conon expounded to Thrasydaeus, without, however, showing proper mastery of the proofs, and on this ground Nicoteles of Cyrene, not without reason, fell foul of him. The second matter has merely been mentioned by Nicoteles, in connexion with his controversy with Conon, as one capable of demonstration; but I have not found it demonstrated either by Nicoteles himself or by any one else. The third question and the others akin to it I have not found so much as noticed by any one. All the matters referred to, which I have not found anywhere, required for their solution many and various novel theorems, most of which I have, as a matter of fact, set out in the first three books, while the rest are contained in the present book. These theorems are of considerable use both for the syntheses of problems and for *diorismi*. Nicoteles indeed, on account of his controversy with Conon, will not have it that any use can be made of the discoveries of Conon for the purpose of *diorismi*; he is, however, mistaken in this opinion, for, even if it is possible, without using them at all, to arrive at results in regard to limits of possibility, yet they at all events afford a readier means of observing some things, e.g. that several or so many solutions are possible, or again that no solution is possible; and such foreknowledge secures a satisfactory basis for investigations, while the theorems in question are again useful for the analyses of *diorismi*. And, even apart from such usefulness, they will be found worthy of acceptance for the sake of the demonstrations themselves, just as we accept many other things in mathematics for this reason and for no other.

(*c*) *Book V: to Attalus*

In this fifth book I have laid down propositions relating to *maximum* and *minimum* straight lines. You must know that my predecessors and contemporaries have only superficially touched upon the investigation of the shortest lines, and have only proved what straight lines touch the sections and, conversely, what properties they have in virtue of which they are tangents. For my part, I have proved these properties in the first book (without however making any use, in the proofs, of the doctrine of the shortest lines), inasmuch as I wished to place them in close connexion with that part of the subject in which I treat of the production of the three conic sections, in order to show at the same time that in each of the three sections countless properties and necessary results appear, as they do with reference to the original [transverse] diameter. The propositions in which I discuss the shortest lines I have separated into classes, and I have dealt with each individual case by careful demonstration; I have also connected the investigation of them with the investigation of the greatest lines above mentioned, because I considered that those who cultivate this science need them for obtaining a knowledge of the analysis, and determination of limits of possibility, of problems as well as for their synthesis: in addition to which, the subject is one of those which seem worthy of study for their own sake.

4.D3 Book I: first definitions

1 If from a point a straight line is joined to the circumference of a circle which is not in the same plane with the point, and the line is produced in both directions, and if, with the point remaining fixed, the straight line being rotated about the circumference of the circle returns to the same place from which it began, then the generated surface composed of the two surfaces lying vertically opposite one another, each of which increases indefinitely as the generating straight line is produced indefinitely, I call a conic surface, and I call the fixed point the vertex, and the straight line drawn from the vertex to the centre of the circle the axis.
2 And the figure contained by the circle and by the conic surface between the vertex and the circumference of the circle I call a cone, and the point which is also the vertex of the surface I call the vertex of the cone, and the straight line drawn from the vertex to the centre of the circle the axis, and the circle the base of the cone.
3 I call right cones those having axes perpendicular to their bases, and oblique those not having axes perpendicular to their bases.
4 Of any curved line which is in one plane I call that straight line the diameter which, drawn from the curved line, bisects all straight lines drawn to this curved line parallel to some straight line; and I call the end of that straight line [the diameter] situated on the curved line the vertex of the curved line, and I say that each of these parallels is drawn ordinatewise to the diameter.
5 Likewise of any two curved lines lying in one plane I call that straight line the transverse diameter which cuts the two curved lines and bisects all the straight lines drawn to either of the curved lines parallel to some straight line; and I call the ends of the diameter situated on the curved lines the vertices of the curved lines; and I call that straight line the upright diameter which, lying between the two curved lines, bisects all the straight lines intercepted between the curved lines and drawn parallel to some straight line; and I say that each of the parallels is drawn ordinatewise to the diameter.
6 The two straight lines each of which being a diameter bisects the straight lines parallel to the other I call the conjugate diameters of a curved line and of two curved lines.
7 And I call that straight line the axis of a curved line and of two curved lines which being a diameter of the curved line or lines cuts the parallel straight lines at right angles.
8 And I call those straight lines the conjugate axes of a curved line and of two curved lines which being conjugate diameters cut the straight lines parallel to each other at right angles.

4.D4 Apollonius introduces the parabola, hyperbola and ellipse

(a) Book I: Proposition 11

If a cone is cut by a plane through its axis, and also cut by another plane cutting the base of the cone in a straight line perpendicular to the base of the axial triangle, and if further the diameter of the section is parallel to one side of the axial triangle, then any straight line which is drawn from the section of the cone to its diameter parallel to the common section of the cutting plane and of the cone's base, will equal in square the rectangle contained by the straight line cut off by it on the diameter beginning from the section's vertex and by another straight line which has the ratio to the straight line between the angle of the cone and the vertex of the section that the square on the base of the axial triangle has to the rectangle contained by the remaining two sides of the triangle. And let such a section be called a parabola.

Let there be a cone whose vertex is the point *A*, and whose base is the circle *BC*, and let it be cut by a plane through its axis, and let it make as a section the triangle *ABC* [I. 3]. And let it also be cut by another plane cutting the base of the cone in the straight line *DE* perpendicular to the straight line *BC*, and let it make as a section on the surface of the cone the line *DFE*, and let the diameter of the section *FG* [I. 7 and Def. 4] be

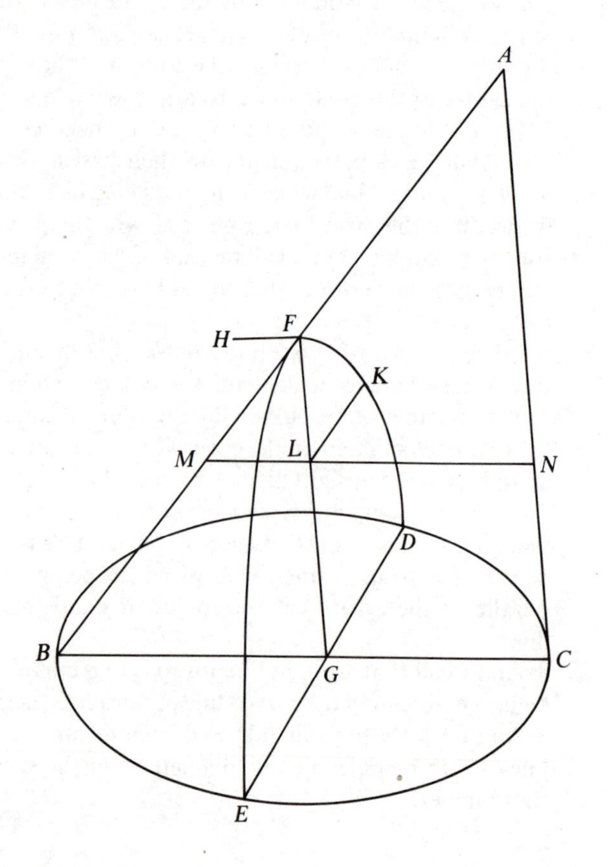

parallel to one side AC of the axial triangle. And let the straight line FH be drawn from the point F perpendicular to the straight line FG, and let it be contrived that sq. BC:rect. BA, AC::FH:FA.

And let some point K be taken at random on the section, and through K let the straight line KL be drawn parallel to the straight line DE.

I say that sq. KL = rect. HF, FL.

For let the straight line MN be drawn through L parallel to the straight line BC. And the straight line DE is also parallel to the straight line KL. Therefore the plane through KL and MN is parallel to the plane through BC and DE [Euclid's *Elements*, XI. 15], that is to the base of the cone. Therefore the plane through KL and MN is a circle whose diameter is MN (I. 4). And KL is perpendicular to MN since DE is also perpendicular to BC [Euclid, XI. 10]. Therefore rect. ML, LN = sq. KL [Euclid, III. 31; VI. 8, Porism].

And since sq. BC:rect. BA, AC::HF:FA,
and sq. BC:rect. BA, AC comp. BC:CA, CA:BA [Euclid, VI. 23],
therefore HF:FA comp. BC:CA, BC:BA.

But BC:CA::MN:NA::ML:LF [Euclid, VI. 4],
and BC:BA::MN:MA::LM:MF::NL:FA [Euclid, VI. 2].

Therefore HF:FA comp. ML:LF, NL:FA.

But rect. ML, LN:rect. LF, FA comp. ML:LF, LN:FA [Euclid, VI. 23].

Therefore HF:FA::rect. ML, LN:rect. LF, FA.

But, with the straight line FL taken as common height,
HF:FA::rect. HF, FL:rect. LF, FA [Euclid, VI. 1],
therefore rect. ML, LN:rect. LF, FA::rect. HF, FL:rect. LF, FA [Euclid, V. 11].

Therefore rect. ML, LN = rect. HF, FL [Euclid, V. 9].

But rect. ML, LN = sq. KL, therefore also sq. KL = rect. HF, FL.

And let such a section be called a parabola, and let HF be called the straight line to which the straight lines drawn ordinatewise to the diameter FG are applied in square, and let it also be called the upright side.

(*b*) *Book I: Proposition 12 (statement)*

If a cone is cut by a plane through its axis, and also by another plane cutting the base of the cone in a straight line perpendicular to the base of the axial triangle, and if the diameter of the section produced meets one side of the axial triangle beyond the vertex of the cone, then any straight line which is drawn from the section to its diameter parallel to the common section of the cutting plane and of the cone's base, will equal in square some area applied to a straight line to which the straight line added along the diameter of the section and subtending the exterior angle of the triangle has the ratio that the square on the straight line drawn from the cone's vertex to the triangle's base parallel to the section's diameter has to the rectangle contained by the sections of the base which this straight line makes when drawn, this area having as breadth the straight line cut off on the diameter beginning from the section's vertex by this straight line from the section to the diameter and exceeding by a figure similar and similarly situated to the rectangle contained by the straight line subtending the exterior angle of the triangle and by the parameter. And let such a section be called an hyperbola.

(c) *Book I: Proposition 13 (statement)*

If a cone is cut by a plane through its axis, and is also cut by another plane on the one hand meeting both sides of the axial triangle, and on the other extended neither parallel to the base nor subcontrariwise, and if the plane the base of the cone is in, and the cutting plane meet in a straight line perpendicular either to the base of the axial triangle or to it produced, then any straight line which is drawn from the section of the cone to the diameter of the section parallel to the common section of the planes, will equal in square some area applied to a straight line to which the diameter of the section has the ratio that the square on the straight line drawn from the cone's vertex to the triangle's base parallel to the section's diameter has to the rectangle contained by the intercepts of this straight line [*on the base*] *from the sides of the triangle, an area having as breadth the straight line cut off on the diameter beginning from the section's vertex by this straight line from the section to the diameter, and deficient by a figure similar and similarly situated to the rectangle contained by the diameter and parameter. And let such a section be called an ellipse.*

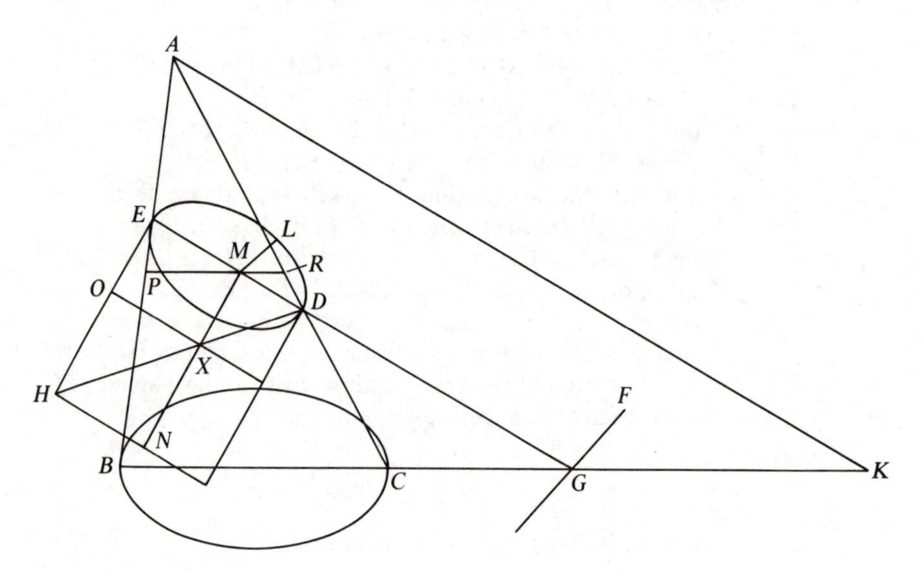

4.D5 Some results on tangents and diameters

(a) *Book I: Proposition 46*

If a straight line touching a parabola meets the diameter, the straight line drawn through the point of contact parallel to the diameter in the direction of the section bisects the straight lines drawn in the section parallel to the tangent.

Let there be a parabola whose diameter is the straight line *ABD*, and let the straight line *AC* touch the section [I. 24], and through *C* let the straight line *HCM* be drawn parallel to the straight line *AD* [I. 26], and let some point *L* be taken at random on the section, and let the straight line *LNFE* [I. 18, 22] be drawn parallel to *AC*.

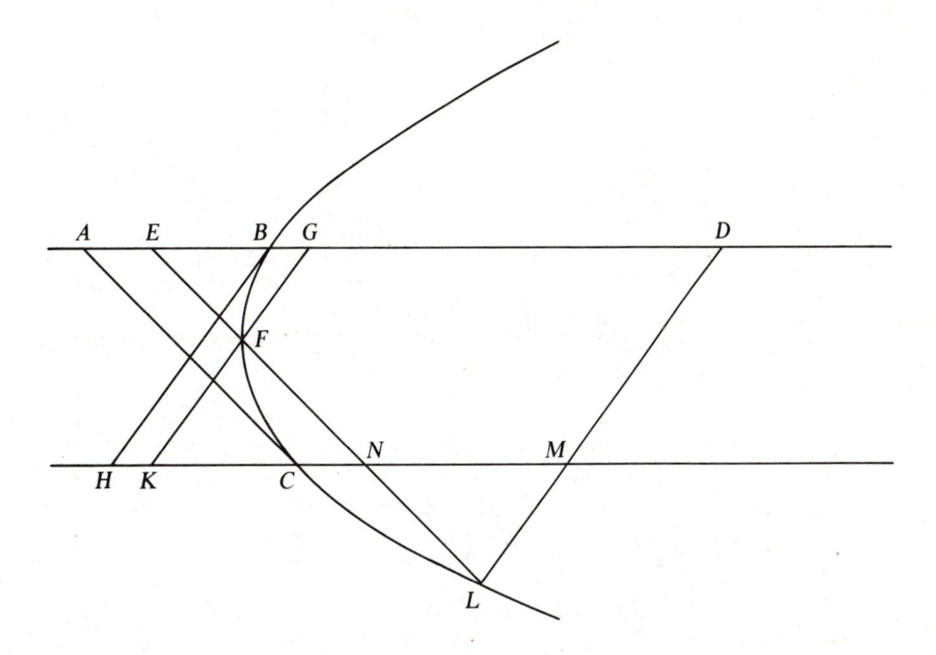

I say that $LN = NF$.

Let the straight lines BH, KFG, and LMD be drawn ordinatewise. Since then by the things already shown in the forty-second theorem [I. 42] trgl. ELD = pllg. BM, and trgl. EFG = pllg. BK, therefore the remainders pllg. GM = quadr. $LFGD$.

Let the common pentagon $MDGFN$ be subtracted; therefore the remainders trgl. KFN = trgl. LMN.

And KF is parallel to LM; therefore $FN = LN$ [Euclid's *Elements*, VI. 22, Lemma].

(*b*) *Book I: Proposition 47*

If a straight line touching an hyperbola or ellipse or circumference of a circle meets the diameter, and through the point of contact and the centre a straight line is drawn in the direction of the section, it bisects the straight lines drawn in the section parallel to the tangent. [Figures overleaf.]

Let there be an hyperbola or ellipse or circumference of a circle whose diameter is the straight line AB and centre C, and let the straight line DE be drawn tangent to the section, and let the straight line CE be joined and produced, and let a point N be taken at random on the section, and through N let the straight line $HNOG$ be drawn parallel.

I say that $NO = OG$.

For let the straight lines XNF, BL, and GMK be dropped ordinatewise.

Therefore by things already shown in the forty-third theorem [I. 43] trgl. HNF = quadr. $LBFX$, and trgl. GHK = quadr. $LBKM$.

Therefore the remainders quadr. $NGKF$ = quadr. $MKFX$.

Let the common pentagon $ONFKM$ be subtracted; therefore the remainders trgl. OMG = trgl. NXO.

And the straight line MG is parallel to the straight line NX; therefore $NO = OG$ [Euclid's *Elements*, VI. 22, Lemma].

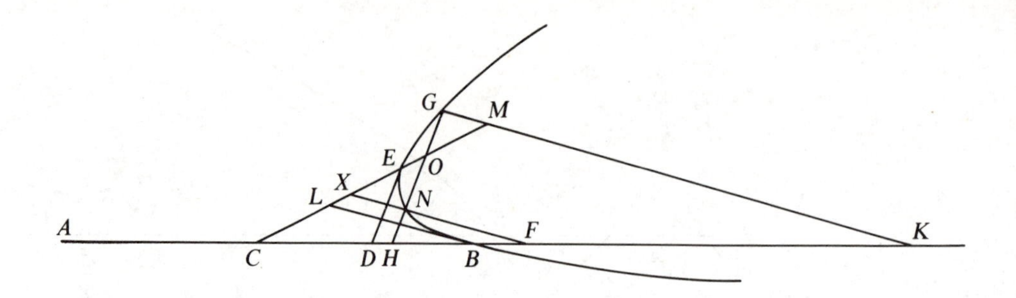

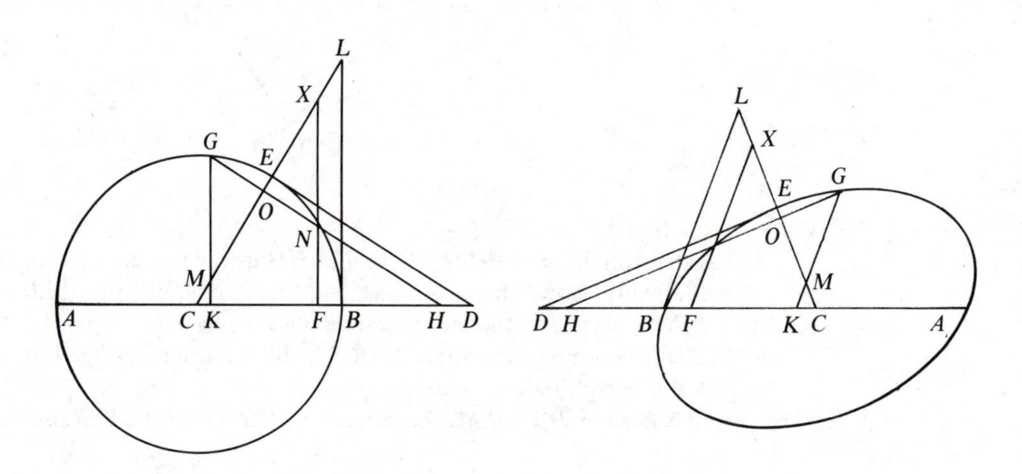

(c) *Book II: Proposition 29*

If in a section of a cone or circumference of a circle two tangents meet, the straight line drawn from their point of meeting to the midpoint of the straight line joining the points of contact is a diameter of the section.

Let there be a section of a cone or circumference of a circle to which let the straight lines *AB* and *AC*, meeting at *A*, be drawn tangent, and let *BC* be joined and bisected at *D*, and let *AD* be joined.

I say that it is a diameter of the section.

For if possible, let *DE* be a diameter, and let *EC* be joined; then it will cut the section [I. 35, 36]. Let it cut it at *F*, and through *F* let *FKG* be drawn parallel to *CDB*. Since then $CD = DB$ also $FH = HG$. And since the tangent at *L* is parallel to *BC* [II. 5, 6], and *FG* is also parallel to *BC*, therefore also *FG* is parallel to the tangent at *L*.

Therefore $FH = HK$ [I. 46, 47]; and this is impossible. Therefore *DE* is not a diameter. Then likewise we could show that there is no other except *AD*.

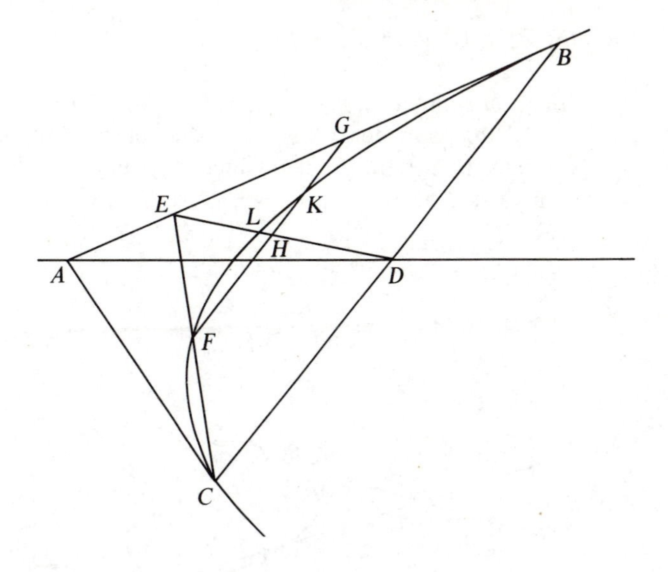

4.D6 How to find diameters, centres and tangents

(*a*) *Book II: Proposition 44*

Given a section of a cone, to find a diameter.

Let there be the given conic section on which are the points A, B, C, D and E. Then it is required to find a diameter.

Let it have been done, and let it be CH. Then with DF and EH drawn ordinatewise and produced, $DF = FB$, and $EH = HA$ [First Def. I. 4]. If then we fix the straight lines BD and EA in position to be parallel, the points H and F will be given. And so HFC will be given in position.

Then it will be constructed thus: let there be the given conic section on which are the points A, B, C, D and E, and let the straight lines BD and AE be drawn parallel, and be bisected at F and H. And the straight line FH joined will be a diameter of the section [First Def. I. 4]. And in the same way we could also find an indefinite number of diameters.

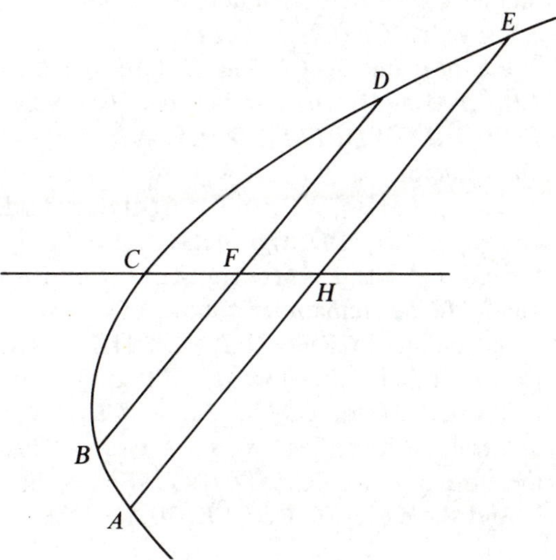

(b) Book II: Proposition 45

Given an ellipse or hyperbola, to find the centre.

And this is evident; for if two diameters of the section, *AB* and *CD*, are drawn through [II. 44], the point at which they cut each other will be the centre of the section, as indicated below.

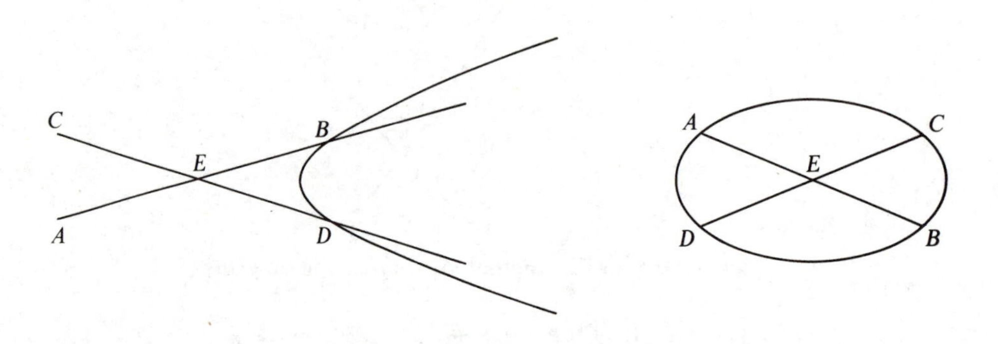

(c) Book III: Proposition 37

If two straight lines touching a section of a cone or circumference of a circle or opposite sections meet, and a straight line is joined to their points of contact, and from the point of meeting of the tangents some straight line is drawn across cutting the line [of the section] at two points, then as the whole straight line is to the straight cut off outside, so will the segments produced by the straight line joining the points of contact be to each other.

Let there be the section of a cone *AB* and tangents *AC* and *CB* and let *AB* be joined, and let *CDEF* be drawn across.

I say that *CF*:*CD*::*FE*:*ED*.

Let the diameters *CH* and *AK* be drawn through *C* and *A*, and through *F* and *D*, *DP*, *FR*, *LFM* and *NDO* parallel to *AH* and *LC*. Since then *LFM* is parallel to *XDO*,
FC:*CD*::*LF*:*XD*::*FM*:*DO*::*LM*:*XO*;
and therefore sq. *LM*:sq. *XO*::sq. *FM*:sq. *DO*.

But sq. *LM*:sq. *XO*::trgl. *XCO* [Euclid's *Elements*, VI. 19],
and sq. *FM*:sq. *DO*::trgl. *FRM*:trgl. *DPO*;
therefore also trgl. *LMC*:trgl.*XCO*::trgl. *FRM*:trgl. *DPO*::remainder quadr. *LCRF*:remainder quadr. *XCP*.

But quadr. *LCRF* = trgl. *ALK* [III. 2; III. 11], and quadr. *XCPD* = trgl. *ANX* [III. 2; III. 11]; therefore sq. *LM*:sq. *XO*::trgl. *ALK*:trgl. *ANX*.

But sq. *LM*:sq. *XO*::sq. *FC*:sq. *CD*,
and trgl. *ALK*:trgl. *ANX*::sq. *LA*:sq. *AX*::sq. *FE*:sq. *ED*;
therefore also sq. *FC*:sq. *CD*::sq. *FE*:sq. *ED*.

And therefore *FC*:*CD*::*FE*:*ED*.

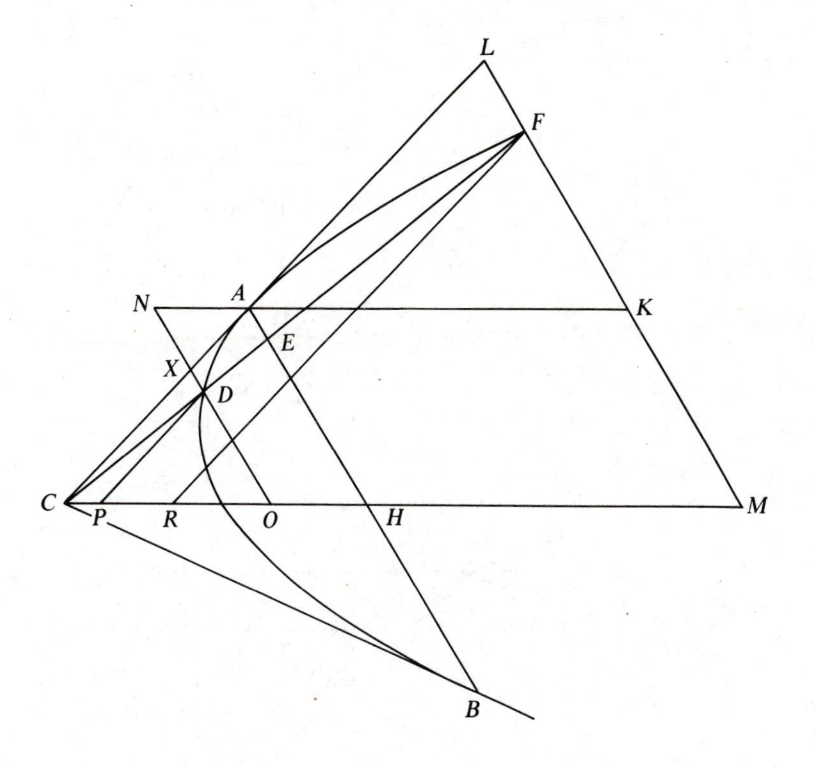

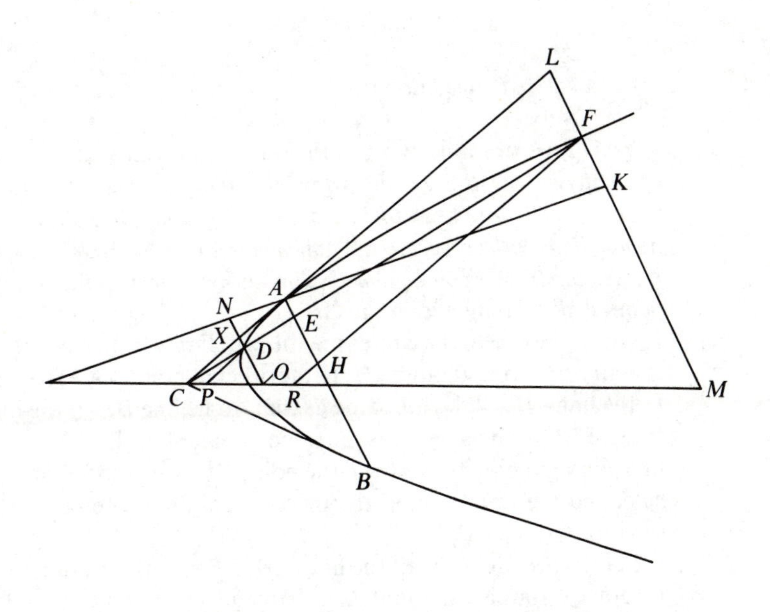

[More figures overleaf.]

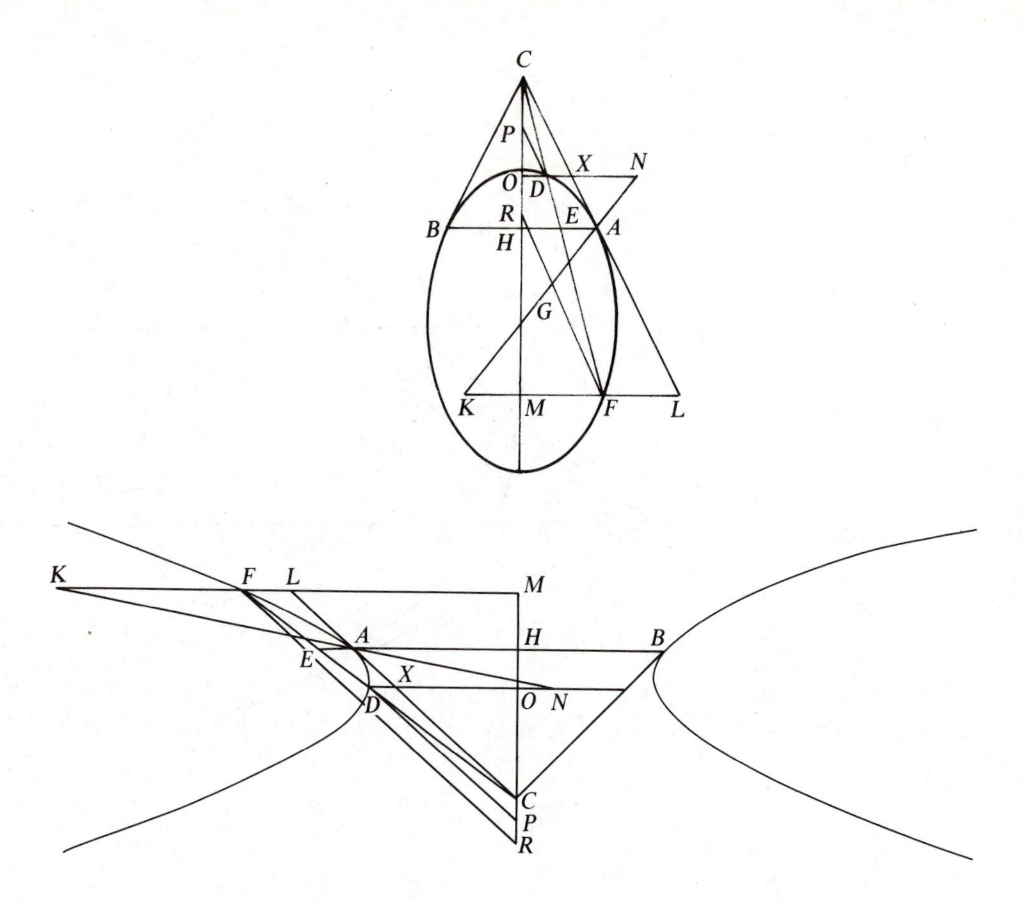

(*d*) *Book IV: Proposition 9*

If one draws two lines through the same point which both cut a section of a cone or the circumference of a circle in two points, and if one divides their interior segments in the ratio of the whole lines to the exterior segments, in such a way that the segments are homologous with respect to the same point, the line through the points of division will meet the section in two points, and the lines drawn through the meeting points to the exterior point shall be tangents to the curve.

Let AB be one of the curves mentioned. Draw from a point D lines DE, DF which cut the curve, the one at points H, E, the other at points G, F, so that the ratio of the line EL to the line LH will be the same as that of the line DE to the line HD, while the ratio of the line FK to the line KG is the same as that of the line DF to the line DG. I say that the line joining them, drawn from the point L to the point K, will meet the section on each side, and the lines joining the meeting points to the point D will be tangents to the section.

In fact, because each of the lines ED, DF cuts the section in two points, it is possible to draw through the point D a diameter of the section and, consequently, to draw tangents on each side [II. 49]. Draw the tangents DB, DA and let the line joining them, BA, if possible not pass through the points L, K but through one of them or neither of them.

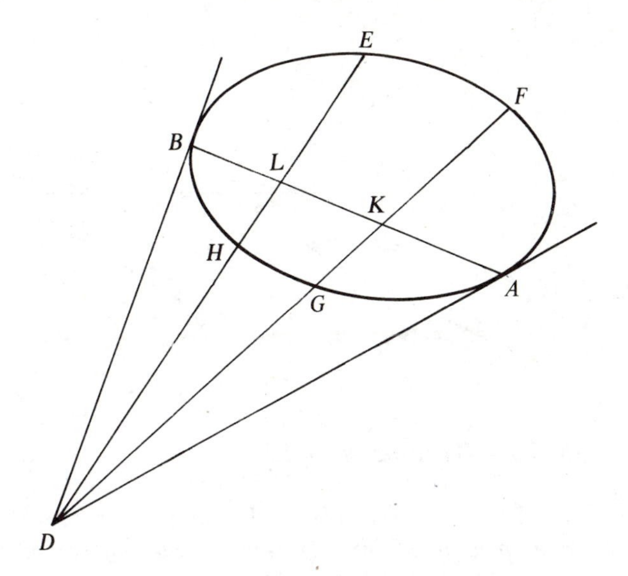

First of all, suppose this line passes only through the point L and cuts the line FG at a point M; from this the line FM is to the line MG as the line FD is to the line DG [III. 37]. Which cannot happen, for it has been supposed that the line FK is to the line KG as the line FD is to the line DG. On the other hand, if the line passes through neither of the points L, K, it will display what cannot happen on either of the lines DE, DF.

4.D7 Focal properties of hyperbolas and ellipses

(a) Book III: Proposition 51

If a rectangle equal to the fourth part of the figure is applied from both sides to the axis of an hyperbola or opposite sections and exceeding by a square figure, and straight lines are deflected from the resulting points of application to either one of the sections, then the greater of the two straight lines exceeds the less by exactly as much as the axis.

For let there be an hyperbola or opposite sections whose axis is AB and centre C, and let each of the rectangles AD, DB and AE, EB be equal to the fourth part of the figure, and from points E and D let the straight lines EF and FD be deflected to the line of the section.

I say that $EF = FD + AB$.

Let FKH be drawn tangent through F, and GCH through C parallel to FD; therefore $\angle KHG = \angle KFD$; for they are alternate. And $\angle KFD = \angle GFH$ [III. 48]; therefore $GF = GH$.

But $GF = GE$, since also $AE = BD$ and $AC = CB$ and $EC = CD$; and therefore $GH = EG$.

And so $FE = 2GH$.

And since it has been shown [III. 50] $CH = CB$, therefore $FE = 2(GC + CB)$.

But $FD = 2GC$, and $AB = 2CB$; therefore $FE = FD + AB$.

And so EF is greater than FD by AB.

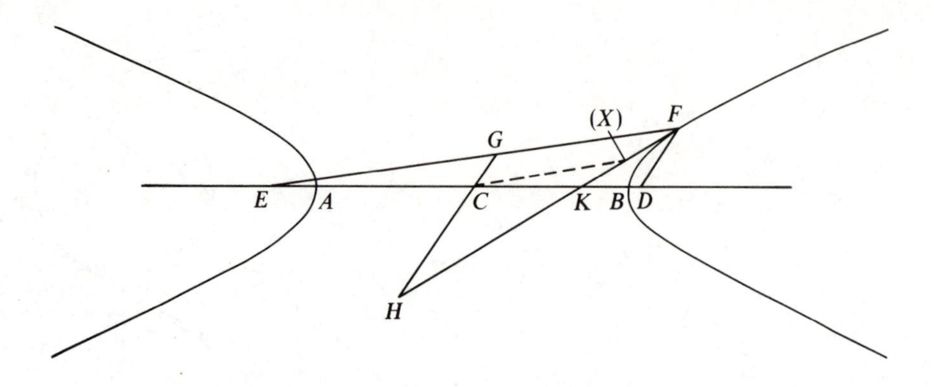

(b) *Book III: Proposition 52*

If in an ellipse a rectangle equal to the fourth part of the figure is applied from both sides to the major axis and deficient by a square figure, and from the points resulting from the application straight lines are deflected to the line of the section, then they will be equal to the axis.

Let there be an ellipse, whose major axis is AB, and let each of the rectangles AC, CB and AD, DB be equal to the fourth part of the figure, and from C and D let the straight lines CE and ED have been deflected to the line of the section.

I say that $CE + ED = AB$.

Let FEH be drawn tangent, and G be centre and through it let GKH be drawn parallel to CE. Since then $\angle CEF = \angle HEK$ [III. 48], and $\angle CEF = \angle EHK$, therefore also $\angle EHK = \angle HEK$.

Therefore also $HK = KE$.

And since $AG = GB$, and $AC = DB$, therefore also $CG = GD$; and so also $EK = KD$.

And for this reason $ED = 2HK$, and $EC = 2KG$, and $ED + EC = 2GH$.

But also $AB = 2GH$ [III. 50]; therefore $AB = ED + EC$.

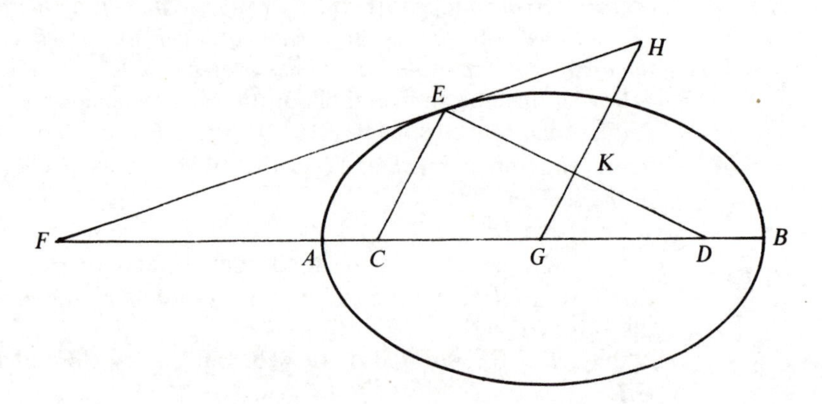

5 Mathematical Traditions in the Hellenistic Age

Despite the overwhelmingly geometric appearance of the previous three chapters, other mathematical activities were taking place, of various levels and kinds, in classical Greece and thereafter in the world culturally dominated by Alexandria. These readings give some indication of the variety. The range of mathematical sciences is well described by Proclus (5.A1), following the writer Geminus (first century BC) and apparently reflecting actual practice more realistically than did the quadrivium classification (see 2.D). Several of these sciences are exemplified in the rest of 5.A (though we have left out, for reasons of space, some very important ones such as astronomy and mathematical geography). 5.B further illustrates the practice of the commentators of late antiquity, to whom in either their editing or their commentating capacity we owe almost all our knowledge of Greek mathematics. 5.C and 5.D deal with more numerical matters, picking up on the tradition of problems with numbers as answers. These problems are of a kind readily handled nowadays by generally simple algebraic techniques, so trying to see them in their own terms is an interesting challenge for the modern reader. This is especially so in the case of Diophantus (third century AD), a mathematician of great subsequent influence.

5.A The Mathematical Sciences

5.A1 Proclus on the divisions of mathematical science

[After describing 'the doctrine of the Pythagoreans and their fourfold division of the mathematical sciences', Proclus continued:]

But others, like Geminus, think that mathematics should be divided differently; they think of one part as concerned with intelligibles only and of another as working with perceptibles and in contact with them. By intelligibles, of course, they mean those objects that the soul arouses by herself and contemplates in separation from embodied forms. Of the mathematics that deals with intelligibles they posit arithmetic and geometry as the two primary and most authentic parts, while the mathematics that attends to sensibles contains six sciences: mechanics, astronomy, optics, geodesy, canonics, and calculation. Tactics they do not think it proper to call a part of mathematics, as others do, though they admit that it sometimes uses calculation, as in the enumeration of military forces, and sometimes geodesy, as in the division and measurement of encampments. Much less do they think of history and medicine as parts of mathematics, even though writers of history often bring in mathematical theorems in describing the lie of certain regions or in calculating the size, breadth, or perimeters of cities, and physicians often clarify their own doctrines by such methods, for the utility of astronomy to medicine is made clear by Hippocrates and all who speak of seasons and places. So also the master of tactics will use the theorems of mathematics, even though he is not a mathematician, if he should ever want to lay out a circular camp to make his army appear as small as possible, or a square or pentagonal or some other form of camp to make it appear very large.

These, then, are the species of general mathematics. Geometry in its turn is divided into plane geometry and stereometry. There is no special branch of study devoted to points and lines, inasmuch as no figure can be constructed from them without planes or solids; and it is always the function of geometry, whether plane or solid, either to construct figures or to compound or divide figures already constructed. In the same way arithmetic is divided into the study of linear numbers, plane numbers, and solid numbers; for it examines number as such and its various kinds as they proceed from the number one, investigating the generation of plane numbers, both similar and dissimilar, and progressions to the third dimension. Geodesy and calculation are analogous to these sciences, since they discourse not about intelligible but about sensible numbers and figures. For it is not the function of geodesy to measure cylinders or cones, but heaps of earth considered as cones and wells considered as cylinders; and it does not use intelligible straight lines, but sensible ones, sometimes more precise ones, such as rays of sunlight, sometimes coarser ones, such as a rope or a carpenter's rule. Nor does the student of calculation consider the properties of number as such, but of numbers as present in sensible objects; and hence he gives them names from the things being numbered, calling them sheep numbers or cup numbers. He does not assert, as does the arithmetician, that something is least; nevertheless with respect to any given class he assumes a least, for when he is counting a group of men, one man is his unit. Again optics and canonics are offshoots of geometry and arithmetic. The former science uses visual lines and the angles made by them; it is divided into a part specifically called optics, which explains the illusory appearances presented by objects seen at a distance, such as the converging of parallel lines or the rounded appearance of square towers, and general catoptrics, which is concerned with the various ways in which light is reflected. The latter is closely bound up with the art of representation and studies what is called 'scene-painting', showing how objects can be represented by images that will not seem disproportionate or shapeless when seen at a distance or on an elevation. The science of canonics deals with the perceptible ratios between notes of the musical scales and discovers the divisions of the monochord, everywhere relying on sense-perception and, as Plato says, 'putting the ear ahead of the mind'.

In addition to these there is the science called mechanics, a part of the study of perceptible and embodied forms. Under it comes the art of making useful engines of war, like the machines that Archimedes is credited with devising for defence against the besiegers of Syracuse, and also the art of wonder-working, which invents figures moved sometimes by wind, like those written about by Ctesibius and Heron, sometimes by weights, whose imbalance and balance respectively are responsible for movement and rest, as the *Timaeus* shows, and sometimes by cords and ropes in imitation of the tendons and movements of living beings. Under mechanics also falls the science of equilibrium in general and the study of the so-called centre of gravity, as well as the art of making spheres imitating the revolutions of the heavens, such as was cultivated by Archimedes, and in general every art concerned with the moving of material things. There remains astronomy, which inquires into the cosmic motions, the sizes and shapes of the heavenly bodies, their illuminations and distances from the earth, and all such matters. This art draws heavily on sense-perception and coincides in large measure with the science of nature. The parts of astronomy are gnomonics, which occupies itself with marking the divisions of time by the placing of sun-dials; meteorology, which determines the different risings of the heavenly bodies and their distances from one another and teaches many and varied details of astronomical theory; and dioptrics, which fixes the positions of the sun, moon, and stars by means of special instruments. Such are the traditions we have received from the writings of the ancients regarding the divisions of mathematical science.

5.A2 Pappus on mechanics

The science of mechanics, my dear Hermodorus, has many important uses in practical life, and is held by philosophers to be worthy of the highest esteem, and is zealously studied by mathematicians, because it takes almost first place in dealing with the nature of the material elements of the universe. For it deals generally with the stability and movement of bodies [about their centres of gravity], and their motions in space, inquiring not only into the causes of those that move in virtue of their nature, but forcibly transferring [others] from their own places in a motion contrary to their nature; and it contrives to do this by using theorems appropriate to the subject matter. The mechanicians of Heron's school say that mechanics can be divided into a *theoretical* and a *manual* part; the theoretical part is composed of geometry, arithmetic, astronomy and physics, the manual of work in metals, architecture, carpentering and painting and anything involving skill with the hands. The man who had been trained from his youth in the aforesaid sciences as well as practised in the aforesaid arts, and in addition has a versatile mind, would be, they say, the best architect and inventor of mechanical devices. But as it is impossible for the same person to familiarize himself with such mathematical studies and at the same time to learn the above-mentioned arts, they instruct a person wishing to undertake practical tasks in mechanics to use the resources given to him by actual experience in his special art.

Of all the [mechanical] arts the most necessary for the purposes of practical life are: (1) that of the *makers of mechanical powers*, they themselves being called mechanicians by the ancients—for they lift great weights by mechanical means to a height contrary to nature, moving them by a lesser force; (2) that of the *makers of engines of war*, they also being called mechanicians—for they hurl to a great distance weapons made of stone and iron and such-like objects, by means of the instruments, known as catapults, constructed by them; (3) in addition, that of the men who are properly called *makers of engines*—for by means of instruments for drawing water which they construct water is more easily raised from a great depth; (4) the ancients also describe as mechanicians the *wonder-workers*, of whom some work by means of pneumatics, as Heron in his *Pneumatica*, some by using strings and ropes, thinking to imitate the movements of living things, as Heron in his *Automata* and *Balancings*, some by means of floating bodies, as Archimedes in his book *On Floating Bodies*, or by using water to tell the time, as Heron in his *Hydria*, which appears to have affinities with the science of sun-dials; (5) they also describe as mechanicians the *makers of spheres*, who know how to make models of the heavens, using the uniform circular motion of water.

Archimedes of Syracuse is acknowledged by some to have understood the cause and reason of all these arts; for he alone applied his versatile mind and inventive genius to all the purposes of ordinary life, as Geminus the mathematician says in his book *On the Classification of Mathematics*. Carpus of Antioch says somewhere that Archimedes of Syracuse wrote only one book on mechanics, that on the construction of spheres, not regarding any other matters of this sort as worth describing. Yet that remarkable man is universally honoured and held in esteem, so that his praises are still loudly sung by all men, but he himself on purpose took care to write as briefly as seemed possible on the most advanced parts of geometry and subjects connected with arithmetic; and he obviously had so much affection for these sciences that he allowed nothing extraneous to mingle with them. Carpus himself and certain others also applied geometry to some arts, and with reason; for geometry is in no way injured, but is capable of giving content to many arts by being associated with them, and, so far from being injured, it is obviously, while itself advancing those arts, appropriately honoured and adorned by them.

5.A3 Optics

(a) Vitruvius on scene-painting

In the first place Agatharcus, in Athens, when Aeschylus was bringing out a tragedy, painted a scene, and left a commentary about it. This led Democritus and Anaxagoras to write on the same subject, showing how, given a centre in a definite place, the lines should naturally correspond with due regard to the point of sight and the divergence of the visual rays, so that by this deception a faithful representation of the appearance of buildings might be given in painted scenery, and so that, though all is drawn on a vertical flat façade, some parts may seem to be withdrawing into the background, and others to be standing out in front.

(b) Euclid on optics

Definitions
Let it be assumed:
1 That the rectilinear rays proceeding from the eye diverge indefinitely;

2 That the figure contained by a set of visual rays is a cone of which the vertex is at the eye and the base at the surface of the objects seen;
3 That those things are seen upon which visual rays fall and those things are not seen upon which visual rays do not fall;
4 That things seen under a larger angle appear larger, those under a smaller angle appear smaller, and those under equal angles appear equal;
5 That things seen by higher visual rays appear higher, and things seen by lower visual rays appear lower;
6 That, similarly, things seen by rays further to the right appear further to the right, and things seen by rays further to the left appear further to the left;
7 That things seen under more angles are seen more clearly.

Proposition VI

Parallel lines when seen from a distance appear to be an unequal distance apart.

Let *AB* and *GD* be two parallels, and *E* be the eye. I hold that *AB* and *GD* seem to be an unequal distance apart, and that the interval between them at a point nearer the eye seems greater than at a point more remote from the eye.

Let *EB*, *EZ*, *ET*, *ED*, *EH*, and *EK* be visual rays. Draw *BD*, *ZH*, and *TK*.

Now since $\angle BED > \angle ZEH$, *BD* appears greater than *ZH*.

Again, since $\angle ZEH > \angle TEK$, *ZH* appears greater than *TK*.

That is, $BD > ZH > TK$ *in appearance.*

The intervals, then, between parallels will not appear equal but unequal.

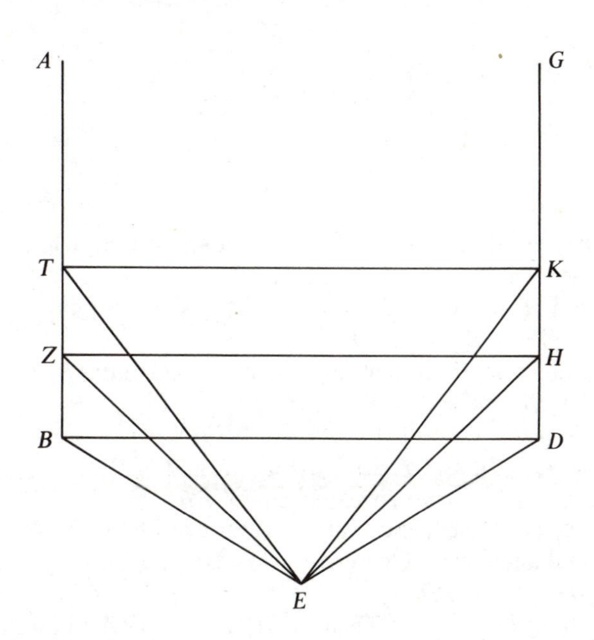

(c) *Heron on catoptrics (the theory of mirrors)*

Here the fundamental law of reflection is proved by Heron (first century AD). It was known several centuries earlier. An earlier use of it by Diocles can be found in 4.C2.

That rays incident upon polished bodies are reflected has, then, in our opinion, been adequately proved. Now by the same reasoning, that is, by a consideration of the speed of the incidence and the reflection, we shall prove that these rays are reflected at equal angles in the case of plane and spherical mirrors. For our proof must again make use of minimum lines. I say, therefore, that of all incident rays [from a given point] reflected to a given point by plane and spherical mirrors the shortest are those that are reflected at equal angles; and if this is the case the reflection at equal angles is in conformity with reason.

Consider AB a plane mirror, G the eye, and D the object of vision. Let a ray GA be incident upon this mirror. Draw AD, and let $\angle EAG = \angle BAD$. Let another ray GB also be incident upon the mirror. Draw BD. I say that $GA + AD < GB + BD$.

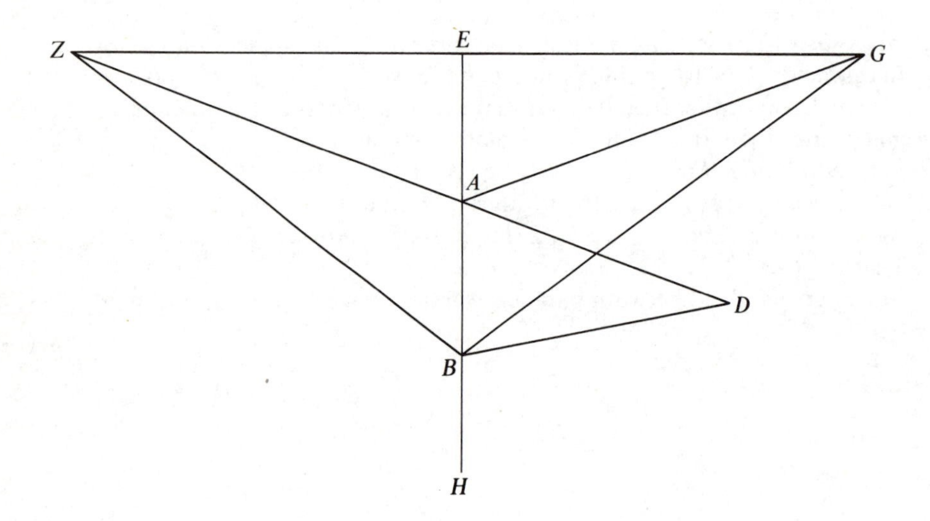

Draw GE from G perpendicular to AB, and prolong GE and AD until they meet, say at Z. Draw ZB.

Now $\angle BAD = \angle EAG$, and $\angle ZAE = \angle BAD$ (as vertical angles).

Therefore $\angle ZAE = \angle EAG$.

And since the angles at E are right angles, $ZA = AG$ and $ZB = BG$.

But $ZD < ZB + BD$ and $ZA = AG$, $ZB = BG$.

Therefore $GA + AD < GB + BD$.

Now $\angle EAG = \angle BAD$, and $\angle EBG < \angle EAG$, and $\angle HBD > \angle BAD$.

Therefore $\angle HBD$ is, *a fortiori*, greater than $\angle EBG$.

Let AB be the surface of a spherical mirror, G the eye, and D the object seen. Let GA and AD make equal angles with the mirror, while GB and BD make unequal angles. I say that $GA + AD < GB + BD$.

Draw EAZ tangent at A. Then $\angle HAE = \angle BAZ$, and the remainder $\angle EAG = \angle ZAD$.

If ZD be drawn, $GA + AD < GZ + ZD$, as was proved above.

But $GZ + ZD < GB + BD$.

Therefore $GA + AD < GB + BD$.

In general, then, in the case of mirrors [both plane and spherical], one must consider

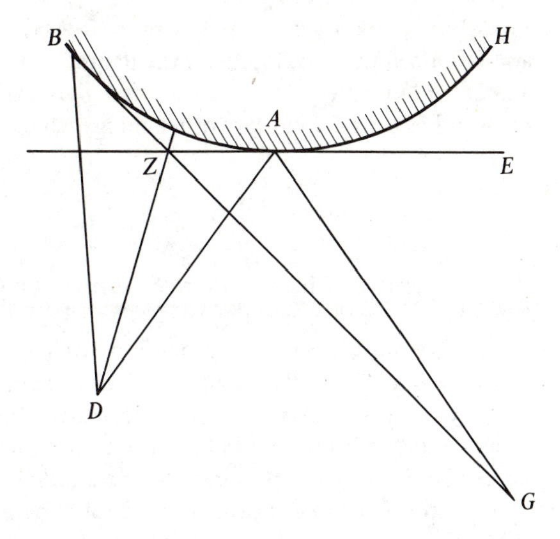

whether there is or is not a point from which incident rays may be reflected at equal angles in such a way that the ray incident from the organ of vision and the ray reflected to the object of vision, when added together, make a sum less than that of all other pairs of rays similarly incident and reflected.

5.A4 Music

(*a*) *Archytas*

There are three proportions in music, the arithmetic, the geometric, and the subcontrary or so-called harmonic. We have an arithmetic proportion when three terms are related with respect to excess, as follows: the first exceeds the second by as much as the second exceeds the third. In this proportion the ratio of the two larger terms is smaller, while the ratio of the two smaller terms is larger. We have a geometric proportion when the first term is to the second as the second is to the third. The ratio of the two larger terms is equal to that of the two smaller. The subcontrary proportion, which we call harmonic, is that in which the terms are such that if the first exceeds the second by a certain part of the first, the second will exceed the third by the same part of the third. In this proportion the ratio of the two larger terms is larger, while the ratio of the two smaller terms is smaller.

(*b*) *Nicomachus*

The first three proportions, then, which are acknowledged by all the ancients, Pythagoras, Plato, and Aristotle, are the arithmetic, geometric, and harmonic; and there are three others subcontrary to them, which do not have names of their own, but are called in more general terms the fourth, fifth, and sixth forms of mean; after which the moderns discover four others as well, making up the number ten, which, according to the Pythagorean view, is the most perfect possible. It was in accordance with this number indeed that not long ago the ten relations were observed to take their proper number, the so-called ten categories, the divisions and forms of the extremities of our hands and feet, and countless other things which we shall notice in the proper place.

(c) *Aristotle*

Why are a double fifth and a double fourth not concordant, whereas a double octave is? Is it because a fifth is in the ratio of 3 to 2, and a fourth in that of 4 to 3? Now in a series of three numbers in the ratio of 3 to 2 or 4 to 3, the two extreme numbers will have no ratio to one another; for neither will they be in a superparticular ratio nor will one be a multiple of the other. But, since the octave is in a ratio of 2 to 1, if it be doubled the extreme numbers would be in a fourfold ratio. So, since a concord is a compound of sounds which are in a proper ratio to one another, and sounds which are at an interval of two octaves from one another are in a ratio to one another (while double fourths and double fifths are not), the sounds constituting the double octave would give a concord (while the others would not) for the reasons given above.

(d) *Theon of Smyrna*

Pythagoras is reputed to have discovered the numerical ratio of sounds that are consonant with one another, namely, the fourth, in the ratio of 4:3, the fifth, in the ratio of 3:2, the octave, in the ratio of 2:1, the octave plus fourth, in the ratio of 8:3 (which is multisuperparticular, for it is $2 + \frac{2}{3}$), the octave plus fifth, in the ratio of 3:1, the double octave, having the ratio of 4:1; and of the other intervals, those that encompass a tone, in the ratio of 9:8, and those that encompass what is now called a semitone, formerly a diesis, in the ratio of 256 to 243.

He investigated these ratios on the basis of the length and thickness of strings, and also on the basis of the tension obtained by turning the pegs or by the more familiar method of suspending weights from the strings. And in the case of wind instruments the basis was the diameter of the bore, or the greater or lesser intensity of the breath. Also the bulk and weights of discs and vessels were examined. Now whichever of these criteria is chosen in connection with any one of the aforesaid ratios, other conditions being equal, the consonance which corresponds to the ratio selected will be produced.

For the present let it suffice for us to illustrate by means of the lengths of strings, using the so-called monochord. For if the single string in the monochord is divided into four equal parts, the sound produced by the whole length of the string forms with the sound produced by three quarters of the string (the ratio being 4:3) the consonance of a fourth. Again, the sound produced by the whole string forms with that produced by half the string (the ratio being 2:1) the consonance of an octave. And the sound produced by the whole string forms with that produced by one quarter of the string (the ratio being 4:1) the consonance of a double octave.

Again, the sound produced by three quarters of the string forms with that produced by half the string (the ratio being 3:2) the consonance of a fifth. The sound produced by three quarters of the string forms with that produced by one quarter (the ratio being 3:1) the consonance of octave plus fifth. If the string is divided into 9 equal parts, the sound produced by the whole string and that produced by 8 parts (the ratio being 9:8) encompass the interval of one tone.

All the consonances are contained in the tetractys consisting of 1, 2, 3, and 4. For in these numbers are the consonances of the fourth, the fifth, the octave [the octave plus fifth, and the double octave], that is, the ratios of 4:3, 3:2, 2:1, 3:1, and 4:1.

5.A5 Heron on geometric mensuration

There is a general method for finding, without drawing a perpendicular, the area of any triangle whose three sides are given. For example, let the sides of the triangle be 7, 8 and 9. Add together 7, 8 and 9; the result is 24. Take half of this, which gives 12. Take away 7; the remainder is 5. Again, from 12 take away 8; the remainder is 4. And again 9; the remainder is 3. Multiply 12 by 5; the result is 60. Multiply this by 4; the result is 240. Multiply this by 3; the result is 720. Take the square root of this and it will be the area of the triangle. Since 720 has not a rational square root, we shall make a close approximation to the root in this manner. Since the square nearest to 720 is 729, having a root 27, divide 27 into 720; the result is $26\frac{2}{3}$; add 27; the result is $53\frac{2}{3}$. Take half of this; the result is $26\frac{1}{2} + \frac{1}{3}(= 26\frac{5}{6})$. Therefore the square root of 720 will be very nearly $26\frac{5}{6}$. For $26\frac{5}{6}$ multiplied by itself gives $720\frac{1}{36}$; so that the difference is $\frac{1}{36}$. If we wish to make the difference less than $\frac{1}{36}$, instead of 729 we shall take the number now found, $720\frac{1}{36}$, and by the same method we shall find an approximation differing by much less than $\frac{1}{36}$.

The geometrical proof of this is as follows: *In a triangle whose sides are given to find the area*. Now it is possible to find the area of the triangle by drawing one perpendicular and calculating its magnitude, but let it be required to calculate the area without the perpendicular.

Let ABC be the given triangle, and let each of AB, BC, CA be given; to find the area. Let the circle DEF be inscribed in the triangle with centre G [Euclid's *Elements* IV. 9], and let AG, BG, CG, DG, EG, FG be joined. Then [Euclid I. 41] $BC \cdot EG = 2 \cdot \text{triangle } BGC$, $CA \cdot FG = 2 \cdot \text{triangle } AGC$, $AB \cdot DG = 2 \cdot \text{triangle } ABG$. Therefore the rectangle contained by the perimeter of the triangle ABC and EG, that is the radius of the circle DEF, is double of the triangle ABC. Let CB be produced and let BH be placed equal to AD; then CBH is half of the perimeter of the triangle ABC because $AD = AF$, $DB = BE$, $FC = CE$ [by Euclid III. 17]. Therefore $CH \cdot EG = \text{triangle } ABC$. But $CH \cdot EG = \sqrt{CH^2 \cdot EG^2}$; therefore $(\text{triangle } ABC)^2 = HC^2 \cdot EG^2$.

Let GL be drawn perpendicular to CG and BL perpendicular to CB, and let CL be joined. Then since each of the angles CGL, CBL is right, a circle can be described about the quadrilateral $CGBL$ [Euclid III. 31]; therefore the angles CGB, CLB are together equal to two right angles [Euclid III. 22]. But the angles CGB, AGD are together equal to two right angles because the angles at G are bisected by AG, BG, CG and the angles CGB, AGD together with AGC, DGB are equal to four right angles; therefore, the angle AGD is equal to the angle CLB. But the right angle ADG is equal to the right angle CBL; therefore the triangle AGD is similar to the triangle CBL.

Therefore $BC:BL = AD:DG = BH:EG$, and so [Euclid V. 16] $BC:BH = BL:EG = BK:KE$, because BL is parallel to GE; hence [Euclid V. 18]

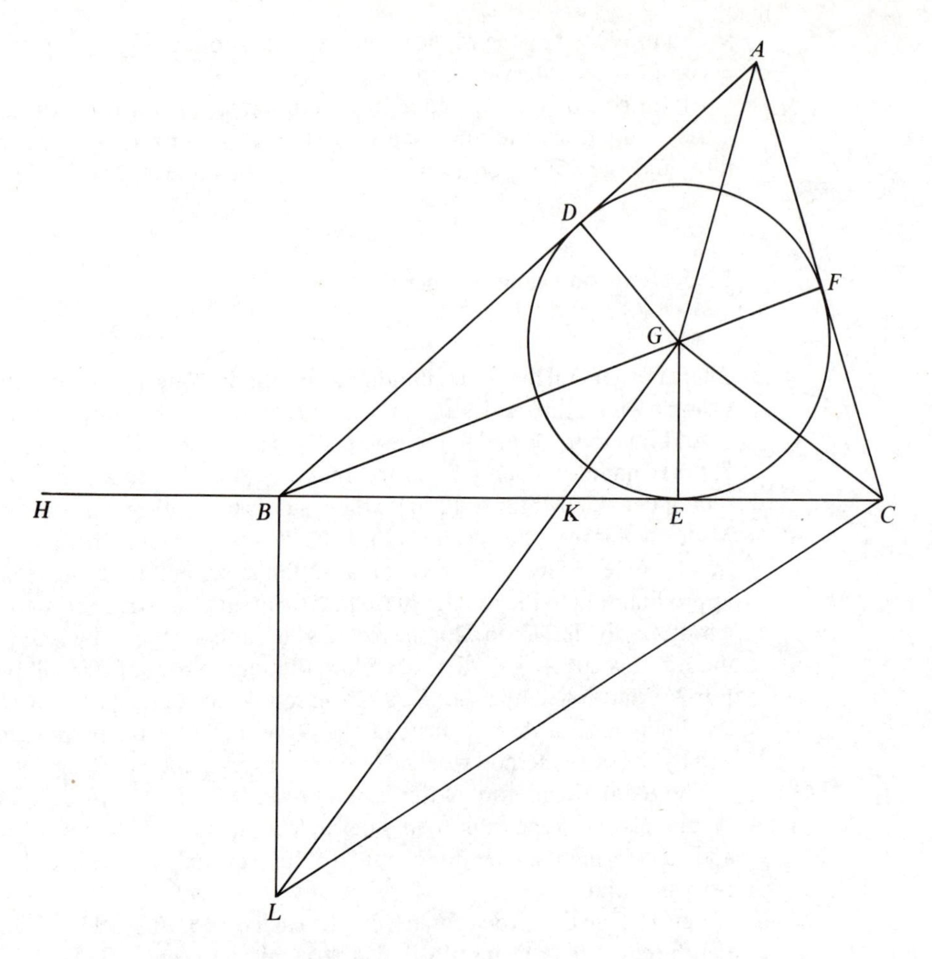

$CH:BH = BE:EK$. Therefore $CH^2:CH \cdot HB = BE \cdot EC:CE \cdot EK = BE \cdot EC:EG^2$, for in a right-angled triangle EG has been drawn from the right angle perpendicular to the base; therefore $CH^2 \cdot EG^2$, whose square root is the area of the triangle ABC, is equal to $(CH \cdot HB)(CE \cdot EB)$. And each of CH, HB, BE, CE is given; for CH is half of the perimeter of the triangle ABC, while BH is the excess of half the perimeter over CB, BE is the excess of half the perimeter over AC, and CE is the excess of half the perimeter over AB, inasmuch as $EC = CF$, $BH = AD = AF$. Therefore the area of the triangle ABC is given.

5.A6 Geodesy: Vitruvius on two useful theorems of the ancients

First of all, among the many very useful theorems of Plato, I will cite one as demonstrated by him. Suppose there is a place or a field in the form of a square and we are required to double it. This has to be effected by means of lines correctly drawn, for it will take a kind of calculation not to be made by means of mere multiplication. The following is the demonstration. A square place ten feet long and ten feet wide gives an area of one hundred feet. Now if it is required to double the square, and to make one of two hundred feet, we must ask how long will be the side of that square so as to get from this the two hundred feet corresponding to the doubling of the area. Nobody can find this by means of arithmetic. For if we take fourteen, multiplication will give one hundred and ninety-six feet; if fifteen, two hundred and twenty-five feet.

Therefore, since this is inexplicable by arithmetic, let a diagonal line be drawn from angle to angle of that square of ten feet in length and width, dividing it into two triangles of equal size, each fifty feet in area. Taking this diagonal line as the length, describe another square. Thus we shall have in the larger square four triangles of the same size and the same number of feet as the two of fifty feet each which were formed by the diagonal line in the smaller square. In this way Plato demonstrated the doubling by means of lines.

Then again, Pythagoras showed that a right angle can be formed without the contrivances of the artisan. Thus, the result which carpenters reach very laboriously, but scarcely to exactness, with their squares, can be demonstrated to perfection from the reasoning and methods of his teaching. If we take three rules, one three feet, the second four feet, and the third five feet in length, and join these rules together with their tips touching each other so as to make a triangular figure, they will form a right angle. Now if a square be described on the length of each one of these rules, the square on the side of three feet in length will have an area of nine feet; of four feet, sixteen; of five, twenty-five.

Thus the area in number of feet made up of the two squares on the sides three and four feet in length is equalled by that of the one square described on the side of five. When Pythagoras discovered this fact, he had no doubt that the Muses had guided him in the discovery, and it is said that he very gratefully offered sacrifice to them.

This theorem affords a useful means of measuring many things, and it is particularly serviceable in the building of staircases in buildings, so that the steps may be at the proper levels.

Suppose the height of the story, from the flooring above to the ground below, to be divided into three parts. Five of these will give the right length for the stringers of the stairway. Let four parts, each equal to one of the three composing the height between the upper story and the ground, be set off from the perpendicular, and there fix the lower ends of the stringers. In this manner the steps and the stairway itself will be properly placed.

5.B The Commentating Tradition

5.B1 Theon on the purpose of his treatise

Everyone would agree that he could not understand the mathematical arguments used by Plato unless he were practised in this science; and that the study of these matters is neither unintelligent nor unprofitable in other respects Plato himself would seem to make plain in many ways. One who had become skilled in all geometry and all music and astronomy would be reckoned most happy on making acquaintance with the writings of Plato, but this cannot be come by easily or readily, for it calls for a very great deal of application from youth upwards. In order that those who have failed to become practised in these studies, but aim at a knowledge of his writings, should not wholly fail in their desires, I shall make a summary and concise sketch of the mathematical theorems which are specially necessary for readers of Plato, covering not only arithmetic and music and geometry, but also their application to stereometry and astronomy, for without these studies, as he says, it is not possible to attain the best life, and in many ways he makes clear that mathematics should not be ignored.

5.B2 Proclus on critics of geometry

Now that we have summed up these matters, it remains for us to examine the propositions that come after the principles. Up to this point we have been dealing with the principles, and it is against them that most critics of geometry have raised objections, endeavouring to show that these parts are not firmly established. Of those in this group whose arguments have become notorious some, such as the Sceptics, would do away with all knowledge, like enemy troops destroying the crops of a foreign country, in this case a country that has produced philosophy, whereas others, like the Epicureans, propose only to discredit the principles of geometry. Another group of critics, however, admit the principles but deny that the propositions coming after the principles can be demonstrated unless they grant something that is not contained in the principles. This method of controversy was followed by Zeno of Sidon, who belonged to the school of Epicurus and against whom Posidonius has written a whole book and shown that his views are thoroughly unsound.

The disputes about the principles have been fairly well disposed of in our preceding exposition, and Zeno's attack will concern us a little later. For the present let us briefly review the definitions of theorem and problem, the distinction between them, the parts of each and the kinds into which they can be divided, and then turn to the exposition of the matters demonstrated by the author of the *Elements*. We shall select the more elegant of the comments made on them by the ancient writers, though we shall cut short their endless loquacity and present only what is most competent and relevant to scientific procedures, giving greater attention to the working out of fundamentals than to the variety of cases and lemmas which, we observe, usually attract the attention of the younger students of the subject.

5.B3 Pappus on analysis and synthesis

The so-called *Treasury of Analysis*, my dear Hermodorus, is, in short, a special body of doctrine furnished for the use of those who, after going through the usual elements, wish to obtain power to solve problems set to them involving curves, and for this purpose only is it useful. It is the work of three men, Euclid the writer of the *Elements*, Apollonius of Perga and Aristaeus the elder, and proceeds by the method of analysis and synthesis.

Now *analysis* is a method of taking that which is sought as though it were admitted and passing from it through its consequences in order to something which is admitted as a result of synthesis; for in analysis we suppose that which is sought to be already done, and we inquire what it is from which this comes about, and again what is the antecedent cause of the latter, and so on until, by retracing our steps, we light upon something already known or ranking as a first principle; and such a method we call analysis, as being a reverse solution.

But in *synthesis*, proceeding in the opposite way, we suppose to be already done that which was last reached in the analysis, and arranging in their natural order as consequents what were formerly antecedents and linking them one with another, we finally arrive at the construction of what was sought; and this we call synthesis.

Now analysis is of two kinds, one, whose object is to seek the truth, being called *theoretical*, and the other, whose object is to find something set for finding, being called *problematical*. In the theoretical kind we suppose the subject of the inquiry to exist and to be true, and then we pass through its consequences in order, as though they also were true and established by our hypothesis, to something which is admitted; then, if that which is admitted be true, that which is sought will also be true, and the proof will be the reverse of the analysis, but if we come upon something admitted to be false, that which is sought will also be false. In the problematical kind we suppose that which is set as already known, and then we pass through its consequences in order, as though they were true, up to something admitted; then, if what is admitted be possible and can be done, that is, if it be what the mathematicians call *given*, what was originally set will also be possible, and the proof will again be the reverse of the analysis, but if we come upon something admitted to be impossible, the problem will also be impossible.

So much for analysis and synthesis.

This is the order of the books in the aforesaid *Treasury of Analysis*. Euclid's *Data*, one book, Apollonius' *Cutting-off of a Ratio*, two books, *Cutting-off of an Area*, two books, *Determinate Section*, two books, *Contacts*, two books, Euclid's *Porisms*, three books, Apollonius' *Vergings*, two books, his *Plane Loci*, two books, *Conics*, eight books, Aristaeus' *Solid Loci*, five books, Euclid's *Surface Loci*, two books, Eratosthenes' *On Means*, two books. In all there are thirty-three books, whose contents as far as Apollonius' *Conics* I have set out for your examination, including not only the number of the propositions, the conditions of possibility and the cases dealt with in each book, but also the lemmas which are required; indeed, I believe that I have not omitted any inquiry arising in the study of these books.

5.B4 Pappus on three types of geometrical problem

When ancient geometers desired to divide a given rectilinear angle into three equal parts, they were baffled for the following reason. There are, we say, three types of problem in geometry, the so-called 'plane', 'solid', and 'linear' problems. Those that can be solved with straight line and circle are properly called 'plane' problems, for the lines by which such problems are solved have their origin in a plane. Those problems that are solved by the use of one or more sections of the cone are called 'solid' problems. For it is necessary in the construction to use the surfaces of solid figures, that is to say, of cones. There remains the third type, the so-called 'linear' problem. For the construction in these cases curves other than those already mentioned are required, curves having a more varied and forced origin and arising from more irregular surfaces and from complex motions. Of this character are the curves discovered in the so-called 'surface loci' and numerous others even more involved discovered by Demetrius of Alexandria in his *Treatise on Curves* and by Philo of Tyana from the interweaving of plectoids and of other surfaces of every kind. These curves have many wonderful properties. More recent writers have indeed considered some of them worthy of more extended treatment, and one of the curves is called 'the paradoxical curve' by Menelaus. Other curves of the same type are spirals, quadratrices, cochloids, and cissoids.

Now it is considered a serious type of error for geometers to seek a solution to a plane problem by conics or linear curves and, in general, to seek a solution by a curve of the wrong type. Examples of this are to be found in the problem of the parabola in the fifth book of Apollonius' *Conics*, and in the use of a solid *neusis* with respect to a circle in Archimedes' work on the spiral. For in the latter case it is possible without the use of anything solid to prove Archimedes' theorem, viz., that the circumference of the circle traced at the first turn is equal to the straight line drawn at right angles to the initial line and meeting the tangent to the spiral.

In view of the existence of these different classes of problem, the geometers of the past who sought by planes to solve the aforesaid problem of the trisection of an angle, which is by its nature a solid problem, were unable to succeed. For they were as yet unfamiliar with the conic sections and were baffled for that reason. But later with the help of the conics they trisected the angle using the following *neusis* for the solution.

[Then Pappus shows that a *neusis* (the line ED in what follows) can be constructed by means of the intersection of a circle and a hyperbola.]

Now that this has been proved, the given rectilinear angle may be trisected as follows.

Let ABC be an acute angle, and from any point [of AB] draw the perpendicular AC. Complete parallelogram FC and produce FA. Now, since FC is a rectangular parallelogram, let line ED be inserted between AE and AC, verging toward B and equal to twice AB. It has been shown above that this is possible.

Now I say that $\angle EBC = \frac{1}{3}\angle ABC$.

For bisect DE at G and draw AG.

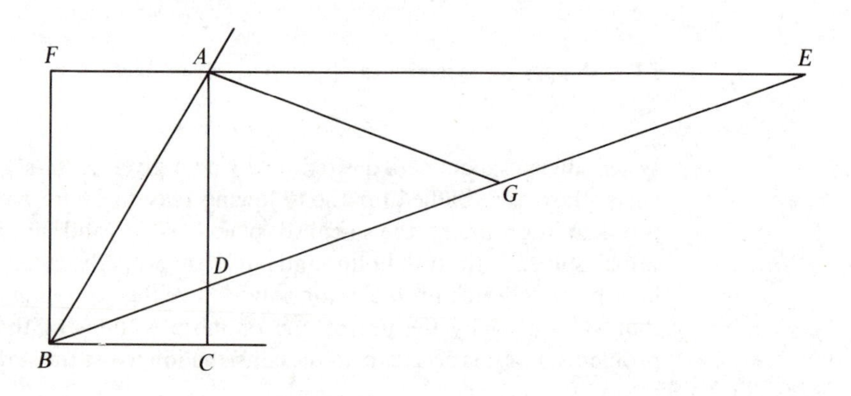

$DG = GA = GE.$
$\therefore DE = 2AG = 2AB.$
$\therefore AB = AG.$
$\therefore \angle ABD = \angle AGD.$
But $\angle AGD = 2\angle AED = 2\angle DBC.$
$\therefore \angle ABD = 2\angle DBC.$
And if we bisect $\angle ABD$, we shall have trisected $\angle ABC$.

5.B5 Pappus on the sagacity of bees

Though God has given to men, most excellent Megethion, the best and most perfect understanding of wisdom and mathematics, He has allotted a partial share to some of the unreasoning creatures as well. To men, as being endowed with reason, He granted that they should do everything in the light of reason and demonstration, but to the other unreasoning creatures He gave only this gift, that each of them should, in accordance with a certain natural forethought, obtain so much as is needful for supporting life. This instinct may be observed to exist in many other species of creatures, but it is specially marked among bees. Their good order and their obedience to the queens who rule in their commonwealths are truly admirable, but much more admirable still is their emulation, their cleanliness in the gathering of honey, and the forethought and domestic care they give to its protection. Believing themselves, no doubt, to be entrusted with the task of bringing from the gods to the more cultured part of mankind a share of ambrosia in this form, they do not think it proper to pour it carelessly into earth or wood or any other unseemly and irregular material, but, collecting the fairest parts of the sweetest flowers growing on the earth, from them they prepare for the reception of the honey the vessels called honeycombs, [with cells] all equal, similar and adjacent, and hexagonal in form.

That they have contrived this in accordance with a certain geometrical forethought we may thus infer. They would necessarily think that the figures must all be adjacent one to another and have their sides common, in order that nothing else might fall into the interstices and so defile their work. Now there are only three rectilineal figures which would satisfy the condition, I mean regular figures which are equilateral and equiangular, inasmuch as irregular figures would be displeasing to the bees. For equilateral triangles and squares and hexagons can lie adjacent to one another and have their sides in common without irregular interstices. For the space about the same point can be filled by six equilateral triangles and six angles, of which each is $\frac{2}{3}\cdot$right angle, or by four squares and four right angles, or by three hexagons and three angles of a hexagon, of which each is $1\frac{1}{3}\cdot$right angle. But three pentagons would not suffice to fill the space about the same point, and four would be more than sufficient; for three angles of the pentagon are less than four right angles (inasmuch as each angle is $1\frac{1}{5}\cdot$right angle), and four angles are greater than four right angles. Nor can three heptagons be placed about the same point so as to have their sides adjacent to each other; for three angles of a heptagon are greater than four right angles (inasmuch as each is $1\frac{3}{7}\cdot$right angle). And the same argument can be applied even more to polygons with a greater number of angles. There being, then, three figures capable by themselves of filling up the space around the same point, the triangle, the square and the hexagon, the bees in their wisdom chose for their work that which has the most angles, perceiving that it would hold more honey than either of the two others.

Bees, then, know just this fact which is useful to them, that the hexagon is greater than the square and the triangle and will hold more honey for the same expenditure of material in constructing each. But we, claiming a greater share in wisdom than the bees, will investigate a somewhat wider problem, namely that, *of all equilateral and equiangular plane figures having an equal perimeter, that which has the greater number of angles is always greater, and the greatest of them all is the circle having its perimeter equal to them.*

5.B6 Proclus on other commentators

Proclus ended his *Commentary on the First Book of Euclid's Elements* with these words.

As for us, if we are able to go through the remaining books in the same fashion, we shall have much to thank the gods for; but if other concerns draw us aside, we ask those who are admirers of this science to expound the remaining books by the same method, aiming always at what is important and can be clearly divided, since the commentaries now in circulation contain great and manifold confusion and contribute nothing to the exposition of causes, to dialectical judgement, or to philosophical understanding.

5C Problems whose Answers are Numbers

5.C1 Proclus on Pythagorean triples

There are two sorts of right-angled triangles, isosceles and scalene. In isosceles triangles you cannot find numbers that fit the sides; for there is no square number that is the double of a square number, if you ignore approximations, such as the square of seven which lacks one of being double the square of five. But in scalene triangles it is possible to find such numbers, and it has been clearly shown that the square on the side subtending the right angle may be equal to the squares on the sides containing it. Such is the triangle in the *Republic*, in which sides of three and four contain the right angle and five subtends it, so that the square on five is equal to the squares on those sides. For this is twenty-five, and of those the square of three is nine and that of four sixteen. The statement, then, is clear for numbers.

Certain methods have been handed down for finding such triangles, one of them attributed to Plato, the other to Pythagoras. The method of Pythagoras begins with odd numbers, positing a given odd number as being the lesser of the two sides containing the angle, taking its square, subtracting one from it, and positing half of the remainder as the greater of the sides about the right angle; then adding one to this, it gets the remaining side, the one subtending the angle. For example, it takes three, squares it, subtracts one from nine, takes the half of eight, namely, four, then adds one to this and gets five; and thus is found the right-angled triangle with sides of three, four, and five. The Platonic method proceeds from even numbers. It takes a given even number as one of the sides about the right angle, divides it into two and squares the half, then by adding one to the square gets the subtending side, and by subtracting one from the square gets the other side about the right angle. For example, it takes four, halves it and squares the half, namely, two, getting four; then subtracting one it gets three and adding one gets five, and thus it has constructed the same triangle that was reached by the other method. For the square of this number is equal to the square of three and the square of four taken together.

5.C2 Problems in Heron's *Geometrica*

(*a*) *Given the sum of the diameter, perimeter and area of a circle, to find each of them separately*. It is done thus: Let the given sum be 212. Multiply this by 154; the result is 32 648. To this add 841, making 33 489, whose square root is 183. From this take away 29, leaving 154, whose eleventh part is 14; this will be the diameter of the circle. If you wish to find the circumference, take 29 from 183, leaving 154; double this, making 308, and take the seventh part, which is 44; this will be the perimeter. *To find the area*. It is done thus: Multiply the diameter, 14, by the perimeter, 44, making 616; take the fourth part of this, which is 154; this will be the area of the circle. The sum of the three numbers is 212.

(*b*) *To find two rectangles such that the area of the first is three times the area of the second*. I proceed thus: Take the cube of 3, making 27; double this, making 54. Now take away 1, leaving 53. Then let one side be 53 feet and the other 54 feet. As for the other rectangle, [I proceed] thus: Add together 53 and 54, making 107 feet: multiply this by 3, [making 321; take away 3], leaving 318. Then let one side be 318 feet and the other 3 feet. The area of the one will be 954 feet and of the other 2862 feet.

(*c*) In a right-angled triangle the sum of the area and the perimeter is 280 feet; *to separate the sides and find the area*. I proceed thus: Always look for the factors; now 280 can be factorized into 2.140, 4.70, 5.56, 7.40, 8.35, 10.28, 14.20. By inspection, we find 8 and 35 fulfil the requirements. For take one-eighth of 280, getting 35 feet. Take 2 from 8, leaving 6 feet. Then 35 and 6 together make 41 feet. Multiply this by itself, making 1681 feet. Now multiply 35 by 6, getting 210 feet. Multiply this by 8, getting 1680 feet. Take this away from the 1681, leaving 1, whose square root is 1. Now take the 41 and subtract 1, leaving 40, of which the half is 20; this is the perpendicular, 20 feet. And again take 41 and add 1, getting 42 feet, of which the half is 21; and let this be the base, 21 feet. And take 35 and subtract 6, leaving 29 feet. Now multiply the perpendicular and the base together, [getting 420], of which the half is 210 feet; and the three sides comprising the perimeter amount to 70 feet; add them to the area, getting 280 feet.

5.C3 The cattle problem

A problem that Archimedes solved and, in a letter to Eratosthenes of Cyrene, sent to those who were working on such problems in Alexandria.

Compute the number of cattle of the Sun, O stranger, and if you are wise apply your wisdom and tell me how many once grazed on the plains of the island of Sicilian Thrinacia, divided into four herds by differences in the colour of their skin—one milk-white, the second sleek and dark-skinned, the third tawny-coloured, and the fourth dappled.

In each herd there was a great multitude of bulls, and there were these ratios. The number of white bulls, mark well, O stranger, was equal to one-half plus one-third the number of dark-skinned, in addition to all the tawny-coloured; the dark-skinned bulls were equal to one-fourth plus one-fifth the number of dappled, in addition to all the tawny-coloured. The number of dappled bulls, observe, was equal to one-sixth plus one-seventh the white, in addition to all the tawny-coloured.

Now for the cows there were these conditions: the number of white cows was exactly equal to one-third plus one-fourth of the whole dark-skinned herd; the number of dark-skinned cows, again, was equal to one-fourth plus one-fifth of the whole dappled herd, bulls included; the number of dappled cows was exactly equal to one-fifth plus one-sixth of the whole tawny-coloured herd as it went to pasture; and the number of tawny-coloured cows was equal to one-sixth plus one-seventh of the whole white herd.

Now if you could tell me, O stranger, exactly how many were the cattle of the Sun, not only the number of well-fed bulls, but the number of cows as well, of each colour, you would be known as one neither ignorant nor unskilled in numbers, but still you would not be reckoned among the wise. But come now, consider these other facts, too, about the cattle of the Sun.

When the white bulls were mingled with the dark-skinned, their measure in length and depth was equal as they stood unmoved, and the broad plains of Thrinacia were all covered with their number. And, again, when the tawny-coloured bulls were joined with the dappled ones they stood in perfect triangular form beginning with one and widening out, without the addition or need of any of the bulls of other colours.

Now if you really comprehend this problem and solve it giving the number in all the herds, go forth a proud victor, O stranger, adjudged, mark you, all-powerful in this field of wisdom.

5.C4 Problems from *The Greek Anthology*

Problem 3

Cypris thus addressed Love, who was looking down-cast: 'How, my child, hath sorrow fallen on thee?' And he answered: 'The Muses stole and divided among themselves, in different proportions, the apples I was bringing from Helicon, snatching them from my bosom. Clio got the fifth part, and Euterpe the twelfth, but divine Thalia the eighth. Melpomene carried off the twentieth part, and Terpsichore the fourth, and Erato the seventh; Polyhymnia robbed me of thirty apples, and Urania of a hundred and twenty, and Calliope went off with a load of three hundred apples. So I come to thee with lighter hands, bringing these fifty apples that the goddesses left me.'

Problem 49

Make me a crown weighing sixty minae, mixing gold and brass, and with them tin and much-wrought iron. Let the gold and bronze together form two-thirds, the gold and tin together three-fourths, and the gold and iron three-fifths. Tell me how much gold you must put in, how much brass, how much tin, and how much iron, so as to make the whole crown weigh sixty minae.

Problem 50

Throw me in, silversmith, besides the bowl itself, the third of its weight, and the fourth, and the twelfth; and casting them into the furnace stir them, and mixing them all up take out, please, the mass, and let it weigh one mina.

Problem 126

This tomb holds Diophantus. Ah, how great a marvel! The tomb tells scientifically the measure of his life. God granted him to be a boy for the sixth part of his life, and adding a twelfth part to this, He clothed his cheeks with down; He lit him the light of wedlock after a seventh part, and five years after his marriage He granted him a son. Alas! late-born wretched child; after attaining the measure of half his father's life, chill Fate took him. After consoling his grief by this science of numbers for four years he ended his life.

Problem 132

This is Polyphemus the brazen Cyclops, and as if on him someone made an eye, a mouth, and a hand, connecting them with pipes. He looks quite as if he were dripping water and seems also to be spouting it from his mouth. None of the spouts are irregular; that from his hand when running will fill the cistern in three days only, that from his eye in one day, and his mouth in two-fifths of a day. Who will tell me the time it takes when all three are running?

Problem 136

Brick-makers, I am in a great hurry to erect this house. Today is cloudless, and I do not require many more bricks, but I have all I want but three hundred. Thou alone in one day couldst make as many, but thy son left off working when he had finished two hundred, and thy son-in-law when he had made two hundred and fifty. Working all together, in how many hours can you make these?

Problem 137

Let fall a tear as you pass by; for we are those guests of Antiochus whom his house slew when it fell, and God gave us in equal shares this place for a banquet and a tomb. Four of us from Tegea lie here, twelve from Messene, five from Argos, and half of the banqueters were from Sparta, and Antiochus himself. A fifth of the fifth part of those who perished were from Athens, and do thou, Corinth, weep for Hylas alone.

Answers

3 Love had 3360 apples.
49 Gold $30\frac{1}{2}$, brass $9\frac{1}{2}$, tin $14\frac{1}{2}$, iron $5\frac{1}{2}$ minae.
50 The bowl weighed $\frac{3}{5}$ of a mina.
126 Diophantus was 84 when he died.
132 $\frac{6}{23}$ of a day.
136 $\frac{2}{5}$ of a day.
137 50 guests.

5.C5 An earlier and a later problem

(*a*) *Problem 79 from the Rhind papyrus*

A house inventory			
		houses	7
		cats	49
1	2 801	mice	343
2	5 602	spelt	2 401
4	11 204	hekat	16 807
Total	19 607	Total	19 607

(*b*) *Harley MS 7316* (*c. 1730*)

As I was going to St Ives,
I met a man with seven wives,
Each wife had seven sacks,
Each sack had seven cats,
Each cat had seven kits:
Kits, cats, sacks, and wives,
How many were there going to St Ives?

5.D Diophantus

Almost nothing is known of Diophantus's life, but around 250 AD is our best estimate of when he flourished, and it may be that some facts about his life are given in Problem 126 of *The Greek Anthology* (see 5.C4). Until recently it was believed that only six books of his *Arithmetica* had survived, but a further four have now been found, in Arabic translation, whose location in Diophantus's original work seems to have been between Books I–III and what have come down to us (in Greek) as Books IV–VI. In these extracts the latter are referred to using the code (G) and the newly discovered Arabic books by (A). Two translations of I.7 are included, to enable a comparison to be made between a literal translation and the more succinct summary of the mathematical content appearing in Sir Thomas Heath's edition.

5.D1 Book I.7

(*a*) *A literal translation by I. E. Drabkin*

From the same number to subtract two given numbers so that the remainders will have a given ratio to one another.

Let the numbers to be subtracted from the same number be 100 and 20, and let the larger remainder be three times that of the smaller.

Let the required number be $1x$. If I subtract from it 100, the remainder is $1x - 100$ units; if I subtract from it 20, the remainder is $1x - 20$ units. Now the larger remainder will have to be three times the smaller. Therefore three times the smaller will be equal to the larger. Now three times the smaller is $3x - 300$ units; and this is equal to $1x - 20$ units.

Let the deficiency be added in both cases. $3x$ equal $1x + 280$ units. If we subtract equals from equals, $2x$ equals 280 units, and x is 140 units.

Now as to our problem. I have set the required number as $1x$; it will therefore be 140 units. If I subtract from it 100, the remainder is 40; and if I subtract from it 20, the remainder is 120. And the larger remainder is three times the smaller.

(*b*) *A summary translation by Sir Thomas Heath*

From the same [required] number to subtract two given numbers, so as to make the remainders have to one another a given ratio.

Given numbers 100, 20, given ratio 3:1.

Required number x. Therefore $x - 20 = 3(x - 100)$, and $x = 140$.

5.D2 Book I.27

To find two numbers such that their sum and product are given numbers.

Necessary condition. The square of half the sum must exceed the product by a square number.

Given sum 20, given product 96.

$2x$ the difference of the required numbers.

Therefore the numbers are $10 + x$, $10 - x$.

Hence $100 - x^2 = 96$.

Therefore $x = 2$, and the required numbers are 12, 8.

5.D3 Book II.8

To divide a given square number into two squares.

Let it be required to divide 16 into two squares.

And let the first square $= x^2$; then the other will be $16 - x^2$; it shall be required therefore to make $16 - x^2 =$ a square.

I take a square of the form $(mx - 4)^2$, m being any integer and 4 the root of 16; for example, let the side be $2x - 4$, and the square itself $4x^2 + 16 - 16x$. Then $4x^2 + 16 - 16x = 16 - x^2$. Add to both sides the negative terms and take like from like. Then $5x^2 = 16x$, and $x = \frac{16}{5}$.

One number will therefore be $\frac{256}{25}$, the other $\frac{144}{25}$, and their sum is $\frac{400}{25}$ or 16, and each is a square.

Pierre de Fermat (1601–1665) wrote the following against II.8 in his copy of Diophantus:

> On the other hand it is impossible to separate a cube into two cubes, or a biquadrate into two biquadrates, or generally *any power except a square into two powers with the same exponent*. I have discovered a truly marvellous proof of this, which however the margin is not large enough to contain.

5.D4 Book III.10

To find three numbers such that the product of any pair of them added to a given number gives a square.

Let the given number be 12. Take a square (say 25) and subtract 12. Take the difference (13) for the product of the first and second numbers, and let these numbers be $13x$, $1/x$ respectively.

Again subtract 12 from another square, say 16, and let the difference (4) be the product of the second and third numbers.

Therefore the third number $= 4x$.

The third condition gives $52x^2 + 12 =$ a square; now $52 = 4 \cdot 13$, and 13 is not a square; but, if it were a square, the equation could easily be solved.

Thus we must find two numbers to replace 13 and 4 such that their product is a square, while either $+12$ is also a square.

Now the product is a square if both are squares; hence we must find two squares such that either $+12 =$ a square.

This is easy and, as we said, it makes the equation easy to solve.

The squares 4, $\frac{1}{4}$ satisfy the condition.

Retracing our steps, we now put $4x$, $1/x$ and $x/4$ for the numbers, and we have to solve the equation $x^2 + 12 = \text{square} = (x + 3)^2$, say.

Therefore $x = \frac{1}{2}$, and $(2, 2, \frac{1}{8})$ is a solution.

5.D5 Book IV(A).3

We wish to find two square numbers the sum of which is a cubic number.

We put x^2 as the smaller square and $4x^2$ as the greater square. The sum of the two squares is $5x^2$, and this must be equal to a cubic number. Let us make its side any number of x's we please, say x again, so that the cube is x^3. Therefore, $5x^2$ is equal to x^3. As the side which contains the x^2's is the lesser in degree, we divide the whole by x^2; hence x is equal to 5. Then, since we assumed the smaller square to be x^2, and since x^2 arises from the multiplication of x—which we found to be 5—by itself, x^2 is 25. And, since we put for the greater square $4x^2$, it is 100. The sum of the two squares is 125, which is a cubic number with 5 as its side.

Therefore, we have found two square numbers the sum of which is a cubic number, namely 125. This is what we intended to find.

5.D6 Book IV(A).9

We wish to find two cubic numbers which comprise a square.

We set $4x$ as the side of the greater cube and x as the side of the smaller cube. Then the greater cube is $64x^3$, the smaller x^3, and the number they comprise is $64x^6$; this must be equal to a square number. We put as its side x^2's, the coefficient of which is equal to the side of the square arising from the multiplication of the 64 by the 4, namely 256, having as its side 16. Therefore, we put as the side of the square $16x^2$, so that the square is $256x^4$. Then $64x^6$ equals $256x^4$. So we divide the whole by x^4, since the x^4's are the lower in degree of the two sides; the division of the $64x^6$ by x^4 gives $64x^2$, while we obtain 256 from the division of the $256x^4$ by x^4. Therefore, $64x^2$ equals 256, hence x^2 equals 4; x^2 being a square, as well as 4, their sides are thus equal; the side of x^2 being x, and that of 4 being 2, x is 2. Then, since we set x as the side of the smaller cube, the smaller cube is 8, and since we set $4x$, i.e., 8, as the side of the larger cube, the larger cube is 512. When we multiply it by the smaller cube, the result is the number they comprise, namely 4096, which is a square having 64 as its side.

Therefore, we have found two cubic numbers which comprise a square number, namely 8 and 512. This is what we intended to find.

Suppose now we intend to find a cubic number such that we obtain, after dividing it by a cube, a square number; we shall look for a square number such that, after multiplying it by another cubic number—which we also seek—a cubic number results from the multiplication. This being found, the result of the multiplication of the one by the other will be the desired cubic number.

Likewise if we intend to find a square number such that the division of it by a square results in a cube: we shall treat it inversely to what precedes.

And similarly for anything we seek involving a division which is of the preceding kind: for these two [cases] are [in reality] one, since division is merely the inverse of multiplication.

5.D7 Book VI(A).11

We wish to find a cubic number such that if we add it to its square, the result is a square number.

We put x as the side of the cubic number, so that the cubic number is x^3. Adding x^3 to its square, that it, [to] x^6, we obtain $x^6 + x^3$, which must be a square. Let us put for its side a number of x^3's such that, when we subtract from their square x^6, the remainder is a cube; such is $3x^3$: when we subtract x^6 from the square of $3x^3$, we obtain $8x^6$, which is a cubic number. Hence, if we equate $8x^6$ with a cubic number, the problem will be soluble and the treatment will not be impossible. Let us multiply the $3x^3$ by themselves, so we obtain $9x^6$, which then equals $x^6 + x^3$. We remove the x^6 which is common, so $8x^6$ equals x^3. The division of the two sides by x^3 gives $8x^3$ equal to 1; hence x^3 is $\frac{1}{8}$, or one part of 8. If we increase this by its square, that is, [by] one part of 64 parts of the unit, the result is 9 parts of 64 parts of the unit, which is a square number with 3 parts of 8 as its side.

Therefore, we have found a number fulfilling the condition imposed upon us, and this is one part of 8 parts of the unit. This is what we intended to find.

5.D8 Book V(G).9

To divide unity into two parts such that, if the same given number be added to either part, the result will be a square.

Necessary condition. The given number must not be odd and the double of it + 1 must not be divisible by any prime number which, when increased by 1, is divisible by 4 [i.e. any prime number of the form $4n - 1$].

Given number 6. Therefore 13 must be divided into two squares each of which > 6. If then we divide 13 into two squares the difference of which < 1, we solve the problem.

Take half of 13 or $6\frac{1}{2}$, and we have to add to $6\frac{1}{2}$ a small fraction which will make it a square, or, multiplying by 4, we have to make $\frac{1}{x^2} + 26$ a square, i.e. $26x^2 + 1 =$ a square $= (5x + 1)^2$, say, whence $x = 10$.

That is, in order to make 26 a square, we must add $\frac{1}{100}$, or, to make $6\frac{1}{2}$ a square, we must add $\frac{1}{400}$, and $\frac{1}{400} + 6\frac{1}{2} = (\frac{51}{20})^2$.

Therefore we must divide 13 into two squares such that their sides may be as nearly as possible equal to $\frac{51}{20}$.

Now $13 = 2^2 + 3^2$. Therefore we seek two numbers such that 3 *minus* the first $= \frac{51}{20}$, so that the first $= \frac{9}{20}$, and 2 *plus* the second $= \frac{51}{20}$, so that the second $= \frac{11}{20}$.

We write accordingly $(11x + 2)^2$, $(3 - 9x)^2$ for the required squares [substituting x for $\frac{1}{20}$].

The sum $= 202x^2 - 10x + 13 = 13$.

Therefore $x = \frac{5}{101}$, and the sides are $\frac{257}{101}, \frac{258}{101}$.

Subtracting 6 from the squares of each, we have, as the parts of unity, $\frac{4843}{10201}, \frac{5358}{10201}$.

5.D9 Book VI(G).19

To find a right-angled triangle such that its area added to one of the perpendiculars gives a square, while the perimeter is a cube.

Make a right-angled triangle from some indeterminate odd number, say $2x + 1$; then the altitude $= 2x + 1$, the base $= 2x^2 + 2x$, and the hypotenuse $= 2x^2 + 2x + 1$.

Since the perimeter = a cube, $4x^2 + 6x + 2 = (4x + 2)(x + 1) =$ a cube; and, if we divide all the sides by $x + 1$, we have to make $4x + 2$ a cube.

Again, the area + one perpendicular = a square.

Therefore $\dfrac{2x^3 + 3x^2 + x}{(x + 1)^2} + \dfrac{2x + 1}{x + 1} =$ a square;

that is $\dfrac{2x^3 + 5x^2 + 4x + 1}{x^2 + 2x + 1} = 2x + 1 =$ a square.

But $4x + 2 =$ a cube; therefore we must find a cube which is double of a square; this is of course 8.

Therefore $4x + 2 = 8$, and $x = 1\frac{1}{2}$.

The required triangle is $(\frac{8}{5}, \frac{15}{5}, \frac{17}{5})$.

5.D10 Book VI(G).21

To find a right-angled triangle such that its perimeter is a square, while its perimeter added to its area gives a cube.

Form a right-angled triangle from x, 1.

The perpendiculars are then $2x$, $x^2 - 1$, and the hypotenuse $x^2 + 1$.

Hence $2x^2 + 2x$ should be a square, and $x^3 + 2x^2 + x$ a cube.

It is easy to make $2x^2 + 2x$ a square; let $2x^2 + 2x = m^2x^2$; therefore $x = 2/(m^2 - 2)$.

By the second condition, $\dfrac{8}{(m^2-2)^3} + \dfrac{8}{(m^2-2)^2} + \dfrac{2}{m^2-2}$ must be a cube, i.e.

$\dfrac{2m^4}{(m^2-2)^3} =$ a cube.

Therefore $2m^4 =$ a cube, or $2m =$ a cube $= 8$, say.

Thus $m = 4$, $x = \frac{2}{14} = \frac{1}{7}$, $x^2 = \frac{1}{49}$.

But one of the perpendiculars of the triangle is $x^2 - 1$, and we cannot subtract 1 from $\frac{1}{49}$.

Therefore we must find another value for x greater than 1; hence $2 < m^2 < 4$.

And we have therefore to find a cube such that $\frac{1}{4}$ of the square of it is greater than 2, but less than 4.

If z^3 be this cube, $2 < \frac{1}{4}z^6 < 4$, or $8 < z^6 < 16$.

This is satisfied by $z^6 = \frac{729}{64}$, or $z^3 = \frac{27}{8}$.

Therefore $m = \frac{27}{16}$, $m^2 = \frac{729}{256}$, and $x = \frac{512}{217}$, the square of which is > 1.

Thus the triangle is known $\left[\frac{1024}{217}, \frac{215055}{47089}, \frac{309233}{47089}\right]$.

6 Islamic Mathematics

Islamic mathematicians not only made many original contributions, but by their scholarship they kept much Greek science alive and in due course transmitted it to an awakening West. Our selection reflects them in both of these roles. The Banu Musa, or sons of Moses, were a family of wealthy patrons of learning in the ninth century—a glimpse of their philosophy is given in 6.A1. Learning was important, as the expanding empire required sophisticated handling, and the Arabs set to it diligently; but, as Omar Khayyam was later to remark, mathematicians would at times have to compete for the favours of superstitious princes with charlatans and astrologers—the lover of truth could be rebuffed with hurtful sarcasms (quoted in Youshkevitch: see 6.C3). *Plus ça change*, one might observe. Naturally, one of the first things the Arab mathematicians did was to turn to the Greek body of learning; some of their reflections on this are given in 6.A2 and 6.A3. But they also developed their own original style, as can be seen from the boldly ungeometric approach to problem solving proposed during the ninth century by al-Khwarizmi and abu Kamil, who are among the first two algebraists (see 6.B1 and 6.B2)—although it is interesting to see them have recourse to geometry to prove the validity of their methods. Omar Khayyam (eleventh to twelfth century) also displays the same duality in his treatment of cubic equations (6.B3). As later in the West, once Arab mathematicians mastered Greek geometry they began to claim that it contains flaws, notably the parallel postulate. These matters are surveyed in 6.C3, and we take a closer look at the work of two notable investigators, al-Haytham (tenth to eleventh century) and Omar Khayyam, in 6.C1 and 6.C2.

6.A Commentators and Translators

6.A1 The Banu Musa

Because we have seen that there is fitting need for the knowledge of the measure of surface figures and of the volume of bodies, and we have seen that there are some things, a knowledge of which is necessary for this field of learning but which—as it appears to us—no one up to our time understands, and [that] there are some things we have pursued because certain of the ancients who lived in the past had sought understanding of them and yet knowledge has not come down to us, nor does any one of those we have examined understand, and [that] there are some things which some of the early savants understood and wrote about in their books but knowledge of which, although coming down to us, is not common in our time—for all these reasons it has seemed to us that we ought to compose a book in which we demonstrate the necessary part of this knowledge that has become evident to us.

And if we consider some of those things which the ancients posed and the knowledge of which has become public among men of our time but which we need for the proof of something we pose in our book, we shall merely call it to mind and it will not be necessary for us in our book to describe it [in detail], since knowledge of it is common; for this reason we seek only a brief statement. On the other hand, if we consider something which the ancients posed and which is not well remembered nor excellently known but the explanation of which we need in our book, then we shall put it in our book, relating it to its author. It will be evident from what we shall recount concerning the composition of our book that one who wishes to read and understand it must be well instructed in the books of geometry in common usage among men of our time.

[...] And everything which we have put in our book is our own teaching except the knowledge of finding the measure of the circumference of a circle from its diameter, for that is the work of Archimedes, and except the knowledge of placing two quantities between two quantities so that they are [all] in continued proportion. For although we have posed in our book in regard to the matter [of the two mean proportionals] the method that Menelaus fashioned, we put forth in addition our own method concerning it. And further we posed how to trisect an angle.

And indeed the understanding of all these things we have recounted in our book is of great moment for all those who seek a knowledge of geometry and computation, and the use [of these things] is vital and they are necessary for those who seek this knowledge. For the knowledge of the surface and volume of a sphere which is one of the things [presented here] is properly a part of those things which no one of our time, as far as I have seen, knows how to compute by a method according with the truth [which is] in one who claims to know the demonstration of his method. This book has been completed with the help of God.

6.A2 Al-Sijzi

I am astonished that anybody who pursues and occupies himself with the art of geometry, even though he acquires it from the excellent Ancients, thinks that there are weakness and shortcomings in them; and especially when he is a beginner and a student, with so little knowledge of it that he imagines that he can achieve with very little effort things, which he believes to be easy to handle and easily understood, although that was far beyond the understanding of those who are trained in this art and skilled in it. [...]

What makes it necessary to believe in the weakness of the excellent Archimedes, with his superiority in geometry over the rest of the geometers? He reached such a high level in geometry that the Greeks called him 'the geometer Archimedes'. None among the Ancients nor any of the later geometers were called by his name because of his excellence in geometry. He took great pains to find out useful things. By his power he completed the tools, the instruments and the mechanical procedures. He established the lemmata for the heptagon and followed a path leading to success. By his power we have understood the heptagon just as Heron understood the machines by his [Archimedes'] power and his hard work in mathematical matters.

This being so, and in spite of his excellence, superiority and high level in the art of geometry, this evil erring man [Abu'l-Jud] finds fault with him. He refers to the first group of his [Abu'l-Jud's] corrupt, false lemmata, which are far from the path leading to success, and by means of which one cannot arrive at the construction of the heptagon, and [he refers] to the false argument with which he misled himself. He thought that he could mislead somebody, but, by God, [he] only [misled] those who have no mastery of geometry, or of the introduction to it. Then, in addition to this, he accused Archimedes of things which are ignominious even for those who have a minimum of intelligence, not to speak of the geometers. He maintains that the lemma set forth by Archimedes is more difficult than what is sought, he says that his [Archimedes'] method is ugly, and accuses him of improperly assuming [something].

It is a wonderful achievement, the proof which Archimedes discovered for the lemmata of the heptagon. But he did not write it down in his book to prevent anyone, such as this outcast, who is not worthy of it, from making use of it. When I had acquired instruction from the knowledge of Archimedes, the lemmata of Apollonius, and in particular from my contemporaries such as al-Ala' ibn Sahl, I was also eager [to know] of this noble, abstruse proposition, and the division of the rectilineal angle into three equal parts, which I achieved with very little effort by means of the first treatise of the Book of Apollonius on *Conics*.

Now I shall describe the affair. I shall quote the words of this person who misleads himself, so that it may serve as an education for beginners. I shall describe the wickedness in his own words and the mistake in what he constructed. Then I shall follow it with the lemmata of the heptagon. I shall follow that with the construction of the heptagon. I shall finish the book with the division of the rectilineal angle into three equal parts. To God belongs success.

6.A3 Omar Khayyam

Before we talk about the manifest of the desired by conic sections, we bring in an idea to persuade the readers of this paper to acquisition of knowledge, proof of facts which we describe for them, and the gratitude for blessings that the great God has bestowed upon his slaves, because mentioning the blessings is a great thanksgiving to the benefactor. As is in the Koran, 'But to thy Lord's blessing thou relate'. The reader of this paper should not think that we brought the discussion to this point for ostentation since this is a habit of poor boasters and selfish people. Egotism is suitable for sordid persons because their minds have no capacity but for a slight part of sciences. When they learn that small part, then they think that all sciences end with that and only that.

We seek refuge with God from what our concupiscence may manifest, something which may seduce us and may prevent us from reaching the goal and deliverance.

I say what is called *square square* (x^4) by algebraists in continuous magnitude is a theoretical fact. It does not exist in reality in any way. The use of the words *square square* (x^4), *square cube* (x^5), *cube cube* (x^6), and so forth in continuous magnitudes are for the number of these magnitudes because numbers and values are all of the same kind as magnitudes. Students of philosophy are responsible for the explanation of this fact.

Whatever algebraists use that exist in reality as continuous magnitude are four: number, *object* (x), *square* (x^2), and *cube* (x^3). A number is the one which is a state of mind and independent of all magnitudes. This does not exist in reality. The number is something in mind and general. It only comes into existence when it is denoted by a material cause.

The *object* (x) in continuous values is denoted by a straight line. The *square* (x^2) is denoted by a quadrilateral of equal sides with right angles whose side is equal to a straight line called *object* (x). The *cube* (x^3) is a solid which is bounded by six equal surfaces of four sides whose sides are equal, angles are right angles, each side is the same straight line called *object* (x), and each surface is a square which is called *square* (x^2). Thus the cube is found by multiplying *object* (x) by itself, and then multiplying this product by *object* (x). The explanation for the construction and its proof is due to Euclid in Proposition 27 of Book XI of the *Elements*.

Square square (x^4) which is known by algebraists as the product of *square* (x^2) by *square* (x^2) has no meaning in continuous values. How is it possible to multiply a square which is an area by itself? A surface is of two dimensions and the product of two dimensions by two dimensions would be four dimensions, and an object with more than three dimensions is impossible.

Thus, whatever is obtained by algebra is obtained by these four things. Whoever thinks algebra is a trick in obtaining unknowns has thought it in vain. No attention should be paid to the fact that algebra and geometry are different in appearance. Algebras (*jabbre* and *maghabeleh*) are geometric facts which are proved by Propositions 5 and 6 of Book II of the *Elements*.

Whoever says that *square square* and three *squares* is equal to twenty-eight, then he gets half of *squares* and multiplies it by itself; then he adds the number to it, he takes the square root and it becomes five and half, then he subtracts half of the number of *squares* from it and the remainder is four; he says that is *square* and thus *square square* is sixteen, and he is convinced that *square square* is obtained by algebra. His imagination is quite weak because he has not obtained the *square square*, but he has obtained *square*. That is as if that *square* and three *objects* be equal to twenty-eight. Then the *object* is obtained according to Book II. He insists that the *object* which he has obtained is the square root of *square square*. This is a secret with whose aids one will become aware of other secrets.

Now we get back to the subject that we were discussing. We say equations of first class in three places, that is, numbers, *objects* (x), and *squares* (x^2) have six forms; three of them are singletons and the other three are polynomials. Their unknowns can be obtained by Book II of the *Elements*. This has been explained in books of algebraists. But whenever *cubes* (x^3) come in, and among them and other places there is an equation, we need solid geometry, and especially conics and conic sections because a *cube* (x^3) is a solid. The singletons are three:

(1) $x^3 = ax^2$, this is the same as $x = a$.
(2) $x^3 = ax$, this is the same as $x^2 = a$.
(3) $x^3 = a$,

where there is no solution for this problem in this case except the numerical solution which is extracting the cube root, or a geometrical solution which is constructing a parallelepiped equal to a given parallelepiped. In this case we need conic sections. For people who do not know conics, certain instruments are used.

The polynomials of it have two sorts which are of three or four terms. Polynomials of three terms are:

$x^3 + ax^2 = b$, which cannot be solved but with conic sections.
$x^3 + ax^2 = bx$, which is the same as $x^2 + ax = b$.
$x^3 + a = bx$, which is not solved but with conic sections.
$x^3 + a = bx^2$, which cannot be solved except with conic sections.
$x^3 + ax = b$, which is not solved but with conic sections.
$x^3 + ax = bx^2$, which is the same as $x^2 + a = bx$.
$ax^2 + bx = x^3$, which is the same as $ax + b = x^2$.
$ax^2 + b = x^3$, which is not solved except with conic sections.
$ax + b = x^3$, which cannot be solved but with conic sections.

What is mentioned here has nine sorts of three terms. Three of them are solved by Book II of the *Elements* and the other six cannot be solved except by conic sections.

The equations of four terms are:

$$x^3 = ax^2 + bx + c,$$
$$x^3 + ax = b + cx^2,$$
$$x^3 + ax^2 + b = cx,$$
$$x^3 + ax^2 + bx = c,$$
$$x^3 + ax^2 = bx + c,$$
$$x^3 + ax = bx^2 + c,$$
$$x^3 + a = bx^2 + cx.$$

These are seven sorts of four term equations. None can be solved except with conic sections.

Thus of polynomial equations (having more than two terms) there are thirteen kinds which cannot be solved except with conic sections. There is only one kind of singleton which is not solved except with conic sections. That is:

$$x^3 = a.$$

Ancient mathematicians of other languages have not discovered these ideas and nothing has reached us through translation to our language. Among recent mathematicians of our language, the first person who ran into an equation of three terms of these fourteen cases was Mahani, the geometer, who tried to solve what Archimedes has supposed obvious in Proposition 4 of the book *Sphere and Cylinder*. That is, Archimedes has said that two lines AB and BC are known in values and are connected to each other in one direction. The ratio of BC to BE is known. Then,

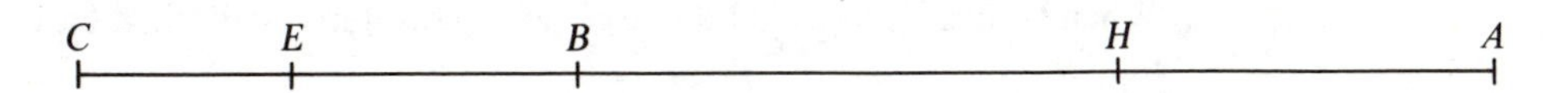

according to what has been said in the book of *Constructions*, *CE* is known. Then he has said that we choose the ratio of *HC* to *CE* as the ratio of the square of *AB* to the square of *AH*. But he has not said how this idea is done. This problem requires conic sections. In the book *Sphere and Cylinder* aside from this, there is nothing requiring conic sections. For this reason Archimedes has considered it obvious. Proposition four is the division of a sphere into a given ratio by a plane.

Mahani used the notations of the algebraist. Since the analysis led to an equation containing numbers, *squares*, and *cubes*, he could not solve it using conic sections. Then he considered the problem impossible. In spite of his superior knowledge, he could not solve the problem.

Then Aboo Jaffar Khazen came, found a solution for the problem, and wrote a paper on it.

Aboo Nassre ibn Aragh, the master and commander of the faithful, from Khawarazm, tried the problem of Archimedes about the construction of the side of a heptagon in a circle (that is, a square with the mentioned property). He also used notations of algebraists. Consequently the analysis led to *cubes* and *squares* equal to numbers. He solved this equation with conic sections. There is no doubt that this man was a great mathematician.

The problem that Aboo Sohl Koohi, Aboo-al-Wafa Bozejani, Aboo Hamed Saghani, and their friends in Bagdad at Azd-ed-Doleh court were not able to solve is: We want to divide ten into two parts so that the sum of the squares of them plus the ratio of the larger to the smaller will be equal to seventy. The analysis of this problem leads to *squares* equal to *cubes*, *objects*, and numbers. These learned men were perplexed about this problem for a long time until Aboo-al-Wojood solved it and it was treasured in the library of the Kings of Samani.

Thus these equations are three forms of polynomials but two of them are of three terms and one of four terms. There is only one singleton which is the equation of *cube* and numbers. These equations were solved by learned men before us, but we have not received any of the other forms with details of their work. If the opportunity arises and I can succeed, I shall bring all of these fourteen forms with all their branches and cases, and how to distinguish whatever is possible or impossible so that a paper, containing elements which are greatly useful in this art will be prepared. For this I shall hope for God's help and rely on him because, in all events, he is the helper and power belongs to him. His grandeur is glorious.

Now, after the introduction of these ideas, let us go back to our problem. That is, to find a cube that with two hundred times its side is equal to twenty squares of its side and two thousand.

6.B Algebra

6.B1 Al-Khwarizmi on the algebraic method

(a) Calculating by completion and reduction

When I considered what people generally want in calculating, I found that it always is a number.

I also observed that every number is composed of units, and that any number may be divided into units.

I observed that the numbers which are required in calculating by Completion and Reduction are of three kinds, namely, roots, squares and simple numbers relative to neither root nor square.

A root is any quantity which is to be multiplied by itself, consisting of units, or numbers ascending, or fractions descending.

A square is the whole amount of the root multiplied by itself.

A simple number is any number which may be pronounced without reference to root or square.

A number belonging to one of these three classes may be equal to a number of another class; you may say, for instance, 'squares are equal to roots', or 'squares are equal to numbers', or 'roots are equal to numbers'.

[Al-Khwarizmi then dealt with examples of these cases before continuing as follows.]

I found that these three kinds: namely, roots, squares and numbers, may be combined together, and thus three compound species arise; that is, 'squares and roots equal to numbers'; 'squares and numbers equal to roots'; 'roots and numbers equal to squares'.

Roots and squares are equal to numbers: for instance, 'one square, and ten roots of the same, amount to thirty-nine dirhems'; that is to say, what must be the square which, when increased by ten of its own roots, amounts to thirty-nine? The solution is this: you halve the number of the roots, which in the present instance yields five. This you multiply by itself; the product is twenty-five. Add this to thirty-nine; the sum is sixty-four. Now take the root of this, which is eight, and subtract from it half the number of the roots, which is five; the remainder is three. This is the root of the square which you sought for; the square itself is nine. [...]

Squares and numbers are equal to roots: for instance, 'a square and twenty-one in numbers are equal to ten roots of the same square'. That is to say, what must be the amount of a square, which, when twenty-one dirhems are added to it, becomes equal to the equivalent of ten roots of that square? Solution: halve the number of the roots; the moiety is five. Multiply this by itself; the product is twenty-five. Subtract from this the twenty-one which are connected with the square; the remainder is four. Extract its root; it is two. Subtract this from the moiety of the roots, which is five; the remainder is three. This is the root of the square which you required, and the square is nine. Or you may add the root to the moiety of the roots; the sum is seven; this is the root of the square which you sought for, and the square itself is forty-nine.

When you meet with an instance which refers you to this case, try its solution by addition, and if that do not serve, then subtraction certainly will. For in this case both addition and subtraction may be employed, which will not answer in any other of the three cases in which the number of the roots must be halved. And know that, when in a question belonging to this case you have halved the number of the roots and multiplied the moiety by itself, if the product be less than the number of dirhems connected with the square, then the instance is impossible; but if the product be equal to the dirhems by themselves, then the root of the square is equal to the moiety of the roots alone, without either addition or subtraction.

(b) Geometrical demonstrations

We have said enough so far as numbers are concerned, about the six types of equations. Now, however, it is necessary that we should demonstrate geometrically the truth of the same problems which we have explained in numbers. Therefore our first proposition is this, that a square and 10 roots equal 39 units.

The proof is that we construct a square of unknown sides, and let this square figure represent the square (second power of the unknown) which together with its root you wish to find. Let the square, then, be *ab*, of which any side represents one root. When we multiply any side of this by a number (or numbers) it is evident that that which results from the multiplication will be a number of roots equal to the root of the same number (of the square). Since then ten roots were proposed with the square, we take a fourth part of the number ten and apply to each side of the square an area of equidistant sides, of which the length should be the same as the length of the square first described and the breadth $2\frac{1}{2}$, which is a fourth part of 10. Therefore four areas of equidistant sides are applied to the first square, *ab*. Of each of these the length is the length of one root of the square *ab* and also the breadth of each is $2\frac{1}{2}$, as we have just said. These now are the areas *c*, *d*, *e*, *f*. Therefore it follows from what we have said that

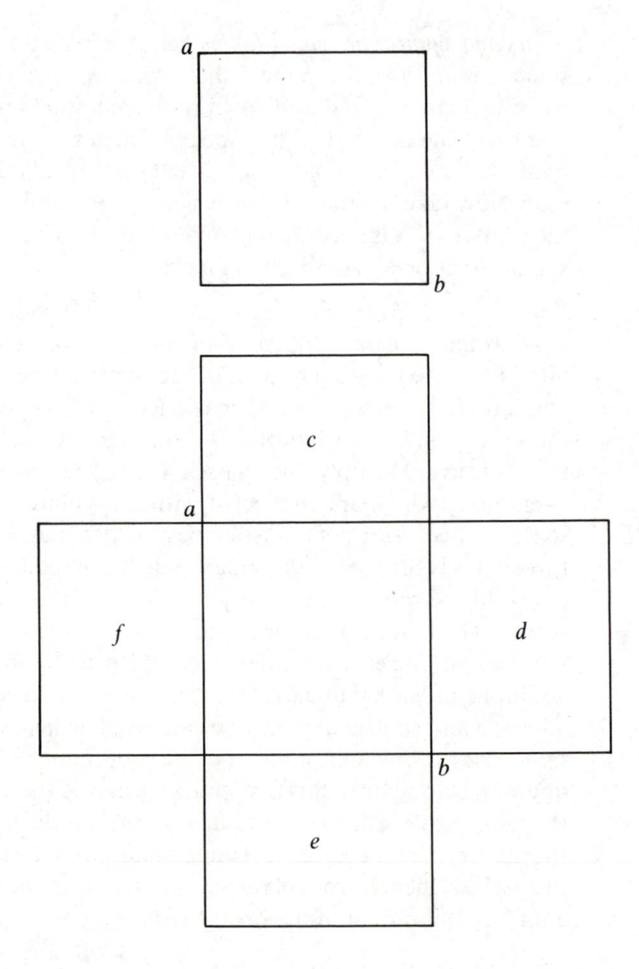

there will be four areas having sides of unequal length, which also are regarded as unknown. The size of the areas in each of the four corners, which is found by multiplying $2\frac{1}{2}$ by $2\frac{1}{2}$, completes that which is lacking in the larger or whole area. Whence it is we complete the drawing of the larger area by the addition of the four products, each $2\frac{1}{2}$ by $2\frac{1}{2}$; the whole of this multiplication gives 25.

And now it is evident that the first square figure, which represents the square of the unknown [x^2], and the four surrounding areas [$10x$] make 39. When we add 25 to this, that is, the four smaller squares which indeed are placed at the four angles of the square *ab*, the drawing of the larger square, called *GH*, is completed. Whence also the

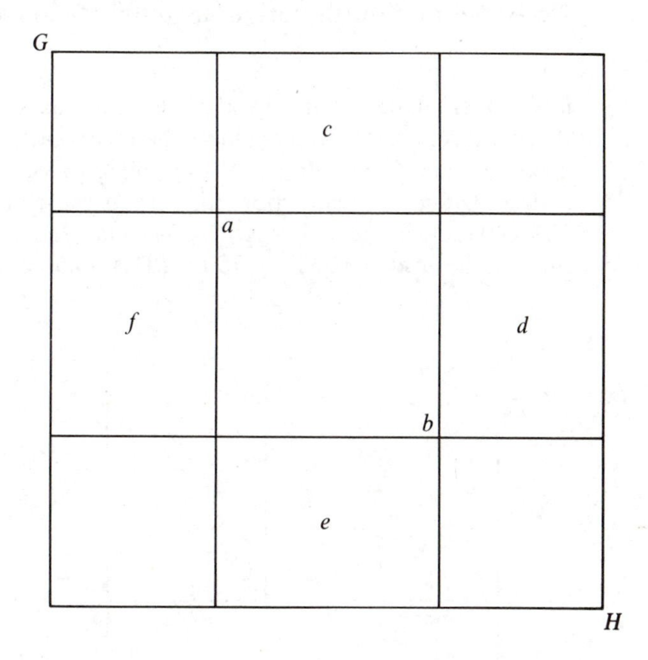

sum total of this is 64, of which 8 is the root, and by this is designated one side of the completed figure. Therefore when we subtract from 8 twice the fourth part of 10, which is placed at the extremities of the larger square *GH*, there will remain but 3. Five being subtracted from 8, 3 necessarily remains, which is equal to one side of the first square *ab*.

This 3 then expresses one root of the square figure, that is, one root of the proposed square of the unknown, and 9 the square itself. Hence we take half of 10 and multiply this by itself. We then add the whole product of the multiplication to 39, that the drawing of the larger square *GH* may be completed; for the lack of the four corners rendered incomplete the drawing of the whole of this square. Now it is evident that the fourth part of any number multiplied by itself and then multiplied by four gives the same number as half of the number multiplied by itself. Therefore if half of the root is multiplied by itself, the sum total of this multiplication will wipe out, equal, or cancel the multiplication of the fourth part by itself and then by 4.

6.B2 Abu-Kamil on the algebraic method

Abu-Kamil wrote his *Algebra* around 900 AD, and based it on al-Khwarizmi's. We give his geometric explanation of the solution of $x^2 + 21 = 10x$ and an indication of his treatment of the algebraic identities underlying his manipulations, selecting only two of his examples:

$$(1)\ (a \pm x)(b \pm x) = ab \pm (a + b)x + x^2.$$
$$(2)\ (a + x)(b - x) = ab + (b - a)x - x^2.$$

They, and others like them, were also proved by geometrical arguments. Note that by 'fourth part' abu-Kamil means x^2.

(a) All this will be explained and considered. Take the number which is together with the square, or 21, and which is more than the square. Construct the square as a square quadrilateral *ABGD*. Add the 21 to it; it is surface *ABHL*. This surface is larger than surface *ABGD* by construction. Line *BL*, because of this, is greater than *BD*. Surface *HD* is 10 roots of the surface *ABGD*. Then line *LD* is 10 and surface *HB* equals 21. It is equal to the product of *LB* by *BD* for *BD* is equal to *BA*. Divide *LD* in half by point *C*.

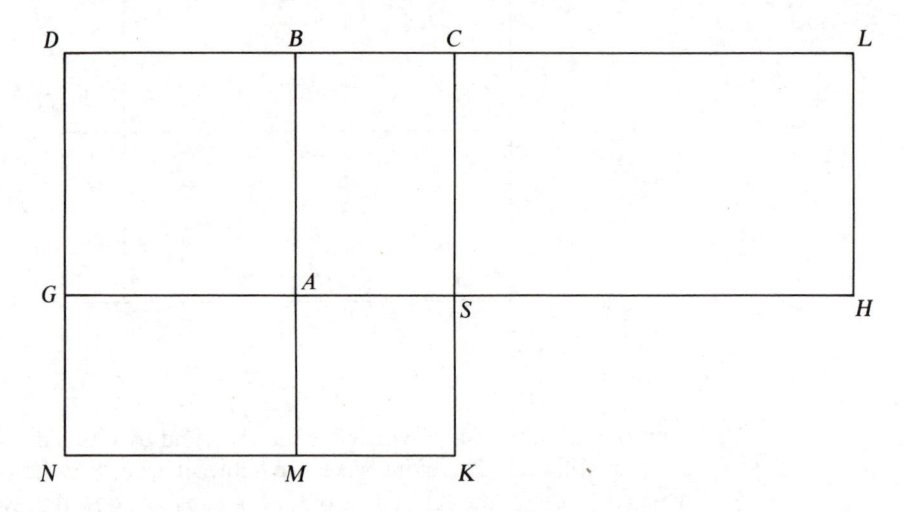

Already, it is divided into two unequal parts by point *B*. Thus, the product of *LB* by *BD* added to the square on *CB* is equal to the square on *CD* as Euclid stated in the second part of his book. But the product of line *CD* by itself is 25 since its length is 5. Line *LB* times *BD* is 21, as I have shown. The square on line *CB* is 4; its side is 2. But line *CD* is 5. Line *BD*, then, is 3. This is the root of the square; the square is 9. If one wishes that I demonstrate conclusively what I have said, I construct the square quadrilateral *KD* on line *CD*. Surface *KD* is 25 since the line *CD* is 5. Surface *CG* is equal to surface *CH* since line *LC* equals line *CD* and surface *AC* equals surface *AN*. Also, the 3 surfaces *AC*, *AD*, and *AN* are equal to the surface *HB* which is the product of *LB* by *BD*, or 21. Thus, surface *KA* remains as 4; it is a square quadrilateral since line *KN* equals line *KC* and line *CS* equals line *NM*. Line *KS* remains equal to *KM*. Line *MK* is 2; it is equal to line

CB. Line *CB* is then 2. Line *BD* remains, then, as 3; it is the root of the square. The square is 9. This is what it was desired to show.

(b) These are the six cases which are related in this book. Their solutions are explained and also their results. Three of them are simple: a square is equal to roots, a square is equal to numbers, and roots are equal to numbers. The three of them which are compound are: a square and roots are equal to numbers; a square and numbers are equal to roots; and roots and numbers are equal to a square. Many arithmeticians and algebraists cannot help teaching some of them. For every case of the six, there are problems which are taught by algebraists.

I shall begin first with the multiplication of things, one by one; things and numbers by themselves; and with things alone or numbers alone except that which we have described and in such manner that whoever wishes to read this book will not remain without knowing about it.

I shall explain in which fashion things, which are the roots, are multiplied one by one when they are simple or when they are together with numbers, [whether] they are subtracted from numbers or the numbers are subtracted from them, and on whichever side they are added one to another or subtracted one from another.

When the roots are added to numbers or subtracted from them, on whichever side they are arranged, then the fourth part is added. The fourth part, then, is the product of the roots, one by the other, or the product of the numbers, one by the other, since two sets of two numbers are multiplied four times. For every one of the first two terms will be multiplied by every one of the latter two terms, or four times. When the roots are subtracted from numbers, the fourth part is added; it is the product of the roots, one by the other. When one is added and the other subtracted, then the fourth part is subtracted; it is the product of the roots, one by the other.

When numbers are subtracted from roots, the fourth part is added; the fourth part then is the product of one of the numbers by the other. When one of the two numbers is added to the roots and the other subtracted from the roots, then the fourth part is subtracted; it is the product of one of the two numbers by the other. When roots are added to numbers and one number is subtracted from roots, then the fourth part is subtracted; it is the product of the added roots by the subtracted number. Thus, we have related, in regard to the fourth part, its meaning as I have seen it according to mathematicians beginning with it and in multiplication.

Perhaps, the fourth part, apart from what has been written, is only a practical rule in that the product, when the two [terms] are [subtracted], is added; the subtracted times the added is subtracted; the added times the added is added. Know that things times things equal squares; things times numbers equal things. Things are roots and roots are things; they are two names with but one meaning.

6.B3 Omar Khayyam on the solution of cubic equations

A solid cube plus squares plus edges equal to a number.

We draw *BH* to represent the side of a square equal to the given sum of the edges, and construct a solid whose base is the square of *BH*, and which equals the given

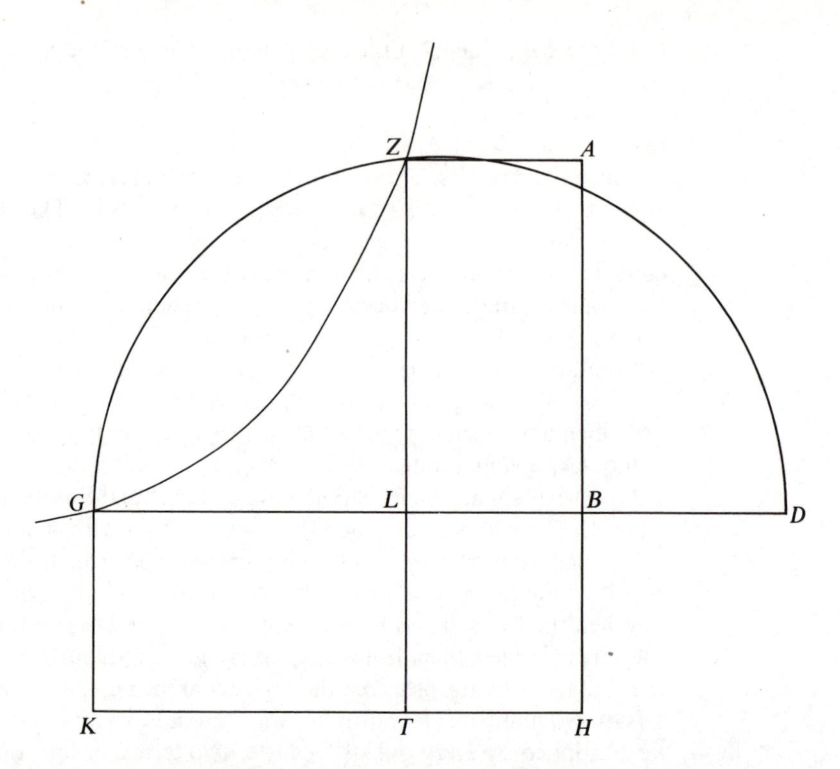

number. Let its height *BG* be perpendicular to *BH*. We draw *BD* equal to the given sum of the squares and along *BG* produced, and draw on *DG* as diameter a semicircle *DZG*, and complete the area *BK*, and draw through the point *G* a hyperbola with the lines *BH* and *HK* as asymptotes. It will intersect the circle at the point *G* because it intersects the line tangential to it [the circle], i.e., *GK*. It must therefore intersect it [the circle] at another point. Let it intersect it [the circle] at *Z* whose position would then be known, because the positions of the circle and the conic are known. From *Z* we draw perpendiculars *ZT* and *ZA* to *HK* and *HA*. Therefore the area *ZH* equals the area *BK*. Now make *HL* common. There remains [after subtraction of *HL*] the area *ZB* equal to the area *LK*. Thus the proportion of *ZL* to *LG* equals the proportion of *HB* to *BL*, because *HB* equals *TL*; and their squares are also proportional. But the proportion of the square of *ZL* to the square of *LG* is equal to the proportion of *DL* to *LG*, because of the circle. Therefore the proportion of the square of *HB* to the square of *BL* would be equal to the proportion of *DL* to *LG*. Therefore the solid whose base is the square of *HB* and whose height is *LG* would equal the solid whose base is the square of *BL* and whose height is *DL*. But this latter solid is equal to the cube of *BL* plus the solid whose base is the square of *BL* and whose height is *BD*, which is equal to the given sum of the squares. Now we make common [we add] the solid whose base is the square of *HB* and whose height is *BL*, which is equal to the sum of the roots. Therefore the solid whose base is the square of *HB* and whose height is *BG*, which we drew equal to the given number, is equal to the solid cube of *BL* plus [a sum] equal to the given sum of its edges plus [a sum] equal to the given sum of its squares; and that is what we wished to demonstrate.

Thus this class has no variations, and none of its problems is impossible, and it has been solved by the properties of the hyperbola together with the properties of the circle.

6.C The Foundations of Geometry

6.C1 Al-Haytham on the parallel postulate

Al-Haytham proposed to prove the postulate (Postulate 5 of Euclid's *Elements*, see 3.B1) using only the first twenty-eight propositions of Euclid's *Elements*, which, as he observed, are independent of it.

Let us start with a premise for that, and that is: 'When two straight lines are produced from the two extremities of a finite straight line, containing two right angles with the first line, then every perpendicular dropped from one of these two lines on the other is equal to the first line, which contained two right angles with these two lines.' Thus, every perpendicular dropped from one of the afore-mentioned lines on the other contains a right angle with the line from which it was dropped. An example of this is as follows: there is extended from the two extremities *A*,*B* of line *AB* two lines *AG*, *BD*, and the two angles *GAB*, *DBA* are each right. Then point *G* is assumed on line *AG* and from it perpendicular *GD* is dropped on line *BD*. I say, then, that line *GD* is equal to line *AB*. The proof of that is that nothing else is possible.

If it were possible, then let it not be equal. If *GD* is not equal to *AB*, it is either greater than it or less. Let it first be greater than it. Line *GA* is produced in a straight line in the direction of *A*, and let it be *AE*. *BD* is also extended rectilinearly in the direction of *B*, and let it be *BT*. We cut off *EA* equal to *AG*. From point *E* a perpendicular is dropped on line *BT*, and let it be *ET*. Let us connect the two lines *GB*, *BE*. Since line *GA* is equal to line *EA* and line *AB* is common, the two lines *GA*, *AB* are equal to the two lines *EA*, *AB* and angles *GAB*, *EAB* are equal because they are right. Therefore, base *GB* is equal to base *EB*, and triangle *GAB* is equal to triangle *EAB*, and the rest of the angles are equal to the rest of the angles. Therefore angle *GBA* is equal to angle *EBA*; the whole angle *ABD* is equal to the whole angle *ABT*. It remains that angle *GBD* is equal to angle *EBT*, and angle *GDB* is equal to angle *ETB*, since they are right. Therefore, triangle *GDB* is equal to triangle *ETB* since two angles of one of them are equal to two angles of the other, and the two sides *GB*, *BE* are equal. Therefore, line *GD* is equal to line *ET*. But *GD* had been greater than *AB*, so *ET* is greater than *AB*.

Let us imagine line *ET* moving along line *TB*, while, during its motion, it is perpendicular to it, so that angle *ETB*, throughout the motion of *ET*, is always right. When point *T*, by the movement of line *ET* ends up at point *B*, line *ET* will coincide with line *BA*, since the two angles *ETB*, *ABD* are equal (because each of them is a right angle). When line *ET* coincides with line *BA*, point *E* will be outside line *AB* and higher than point *A*, since line *ET* had been shown as being greater than line *AB*. Therefore, let line *ET*, while it coincides with line *BA* be line *BH*. Line *BH* can be imagined also after this position moving in the direction of *GD*, and it is in the equivalent of its first position. Then, when point *B*, by the motion of line *BH* ends up at point *D*, line *BH* coincides with line *DG* since the two angles *HBT*, *GDB* are equal because they are right. When line *BH* coincides with line *GD*, point *H* coincides with point *G*, since line *HB* is line *ET* and line *ET* is equal to line *GD*. When line *BH* (which is line *ET*) arrives at line

GD and coincides with it, line *ET* will have moved over line *TD*, and point *T* will have ended up at point *D*. Point *E* will have ended up at point *G*. It was shown above in defining parallel lines that the higher end of every line moving in this way traces a straight line; therefore point *E* traces a straight line during the movement of line *ET* over line *TB*. Let the line point *E* traces be line *EHG*; thus, line *EHG* is a straight line, but line *EAG* is a straight line by assumption. Point *H* has been shown to be higher than point *A*, so line *EHG* is other than line *EAG*. But the two points *E*,*G* are common to these two lines and are straight, therefore two straight lines would contain a space; but this is impossible. The impossibility necessarily follows from our assumption that line *GD* is greater than line *AB*. Therefore, line *GD* cannot be greater than line *AB*.

[Similarly he then showed that it could not be less.]

6.C2 Omar Khayyam's critique of al-Haytham

A part of wisdom, the easiest one, is called mathematics. Few of the matters are quite obvious, but sometimes in geometry a simple matter hides even from a sound and keen mind and an excellent intuition.

This part of wisdom, mathematics, is based on a book of wisdom called logic. It discusses things based on common sense such as that the whole is larger than a part of it. But for axioms there is no proof from common sense.

For a long time I had a strong desire in studying and research in sciences to distinguish some from others, particularly, the book [Euclid's] *Elements of Geometry* which is the origin of all mathematics, and discusses point, line, surface, angle, etc. There are many postulates which should be accepted without proof, such as through two given points there passes one and only one straight line. But there are doubtful matters, among them, the greatest one which has never been proved, i.e., 'Two straight lines intersect if they meet a given line in two distinct points such that the sum of the angles on the one side of the given line between the two points is less than two right angles', has been taken to be true.

I have seen many books which have objected to this idea, among the earlier ones Heron and Autolycus, and the later ones al-Khazen, al-Sheni, al-Neyrizi, etc. None has given a proof. Then I have seen the book of Ibn Haytham, God bless his soul, called the solution of doubt. This postulate among other things was accepted without proof. There are many other things which are foreign to this field such as: If a straight line segment moves so that it remains perpendicular to a given line, and one end of it remains on the given line, then the other end of it draws a parallel.

There are many things wrong here. How could a line move remaining normal to a given line? How could a proof be based on this idea? How could geometry and motion be connected? Motion is only allowed for a single element. A line is generated by a motion of a point and a surface is generated by a motion of a line. Euclid says that a sphere is generated by rotation of a half circle. This solid is bounded by a surface whose points are equidistant to a fixed point inside of it. Also, Euclid uses a straight line segment with a fixed end. He rotates this line segment around the fixed end of it, in a flat surface, to get a circle. But none of these is comparable with Ibn Haytham's idea.

6.C3 Youshkevitch on the history of the parallel postulate

The distinguished Russian historian of mathematics A. P. Youshkevitch published his *History of Mathematics in the Middle Ages* in 1961, since when it has become the standard one-volume work and has been translated, in whole or in part, into many languages, though, sadly, not into English. To give some impression of the riches it describes, we here give some extracts from his account of Arab mathematicians' studies into the foundations of geometry.

(a) The studies of the fifth postulate of Euclid, which relates to parallel lines, are of great importance. The Greeks had already tried to prove this postulate for a period of more than four centuries.

The mathematicians of Islam undertook the study of the theory of parallels and that of proportion a little after the Arabic translation of Euclid's *Elements*. The first works which we know elaborating a theory of parallels are those of an author already mentioned, a contemporary and collaborator of al-Khwarizmi: al-Gauhari, an astronomer and an original mathematician from Farab (now Otrar in the Soviet Republic of Kazakstan). Thanks to an exposition of Nasir ad-Din at-Tusi we know of some of his work on *The Rectification of a Book of the Elements*. Al-Gauhari there proposes a demonstration of the fifth postulate in which he implicitly assumes that if the alternating angles determined by an arbitrary line cutting two others are equal, then every other line which cuts those two will likewise make equal angles. Al-Gauhari uses this postulate implicitly in his first proposition according to which two such lines do not cut and are in fact equidistant. Following this argument, al-Gauhari deduces [two propositions, one of which is that] one can always draw through any point in the interior of an angle a line which meets both sides of the angle (from this he then deduces the fifth postulate). This last proposition is particularly remarkable, for the famous 'proof' of Euclid's fifth postulate proposed in 1800 by A. M. Legendre rests on its implicit admission as a hypothesis.

(b) In the first of [his] two works Tabit ibn Qurra stays very close to al-Gauhari. [But] in his second work Qurra starts the question from a very different point of view. He first of all establishes in a very detailed manner the necessity of using the concept of motion in geometrical proofs (he was perhaps led to this idea by his thorough knowledge of the works of Archimedes and because he was equally interested in mechanics). [...] The conceptual link between the kinematic proof of Qurra and that of al-Haytham, who was not only a mathematician but also a physicist, is obvious, even though he does not mention his predecessor. [Youshkevitch goes on to describe al-Haytham's work, and then comments as follows.] One of the important consequences of the work of al-Haytham was to clarify the reciprocal relationship that exists between the parallel postulate and the angle sum of a quadrilateral. In Euclid's *Elements* this relationship was not studied completely: it was only shown that it follows from the parallel postulate that the angle sum of a quadrilateral is equal to four right-angles. It is also necessary to note that al-Haytham stated another proposition which plays an important role in the theory of parallels. This is the proposition that two intersecting lines cannot be parallel to the same line. In his second commentary on the *Elements* al-Haytham shows that this proposition reduces to a 'proposition proved by Euclid' but which is 'clearer, to our mind'. He includes it in the proof of Proposition 29 of Book I of *Elements*.

The theory of parallels was then developed by Omar Khayyam. In his *Commentary on the Problematic Postulates of a Book of Euclid* he is first of all in disagreement with al-Haytham: following the example of Aristotle and Euclid he opposes the use of motion in geometry. [...] Omar Khayyam's theory of parallels contains some weaknesses, longeurs, and inexactitudes. For a start, the principle which he proposes is contained in two propositions, each of which is equivalent to Euclid's fifth postulate, in such a way that you can omit whichever one you choose. We do not intend to make a critique of Omar Khayyam's reflections, but to underline certain important points which have a long history.

First of all Omar Khayyam states the axiom of Archimedes and the axiom cited above [i.e. two lines which get closer to one another must cut and it is impossible that they should eventually move away from one another in the direction in which they appear to run together]. He deduces from this axiom, just as did al-Haytham, that two perpendiculars to a line are equidistant. Then he establishes eight propositions which can replace Proposition 29 of Book I of *Elements*. Omar Khayyam here describes 'Saccheri's quadrilateral' formed from a straight line AB, equal perpendiculars AC and BD erected at the extremities of AB, and the line CD. Omar Khayyam proves in Proposition 1 that the two upper angles of this quadrilateral are equal. In Proposition 2 he proves that the perpendicular EG erected at the middle of the lower horizontal is also perpendicular to the upper horizontal and divides it in two equal parts. Proposition 3 of Omar Khayyam is of capital importance. Three hypotheses are examined in it, i.e.: (1) the upper angles of the quadrilateral are acute; (2) they are obtuse; (3) they are right angles. The first two hypotheses separately lead to a contradiction, thanks to the use of the new postulate of parallels [which Youshkevitch then goes on to describe before continuing as follows] so only the hypothesis of a right angle can be retained.

(c) The next step in the development in the theory of parallels was taken by Nasir ad-Din at-Tusi, of whom we know three works that treat this problem. [...] In his *Discussion which Removes Doubts about Parallel Lines* at-Tusi explains in detail and rather literally—although he omits some essential passages—the theories of parallels of al-Gauhari, al-Haytham, and Omar Khayyam. Each theory is submitted to a critical analysis and, by way of a conclusion, at-Tusi develops his own theory of parallels. It is based largely on that of al-Haytham and partially on those of al-Gauhari. [...]

In the second version of his *Exposition of Euclid* at-Tusi takes another path. Instead of the postulate of the first version he raises two hypotheses:

(1) Suppose there are two lines AB and CD such that the perpendiculars EF, GH, KL dropped on AB from points on CD always make unequal adjacent angles, always acute in the direction of B and obtuse in the direction of A. Then the two lines AB and CD approach one another until they cut on the side of the acute angles, and they move away from one another on the side of the obtuse angles; i.e. the length of the

perpendiculars diminishes on the side of the points *B* and *D* and increases on the sides of the points *A* and *C*.

(2) If, conversely, the length of the perpendiculars diminishes in the direction of the points *B* and *D* and increases in the direction of the points *A* and *C*, so that the lines *AB* and *CD* approach one another in the direction of the points *B* and *D*, then they move away in the opposite direction, and each perpendicular makes two angles with the line *AB* one of which is acute and the other is obtuse, the acute angles being situated on the side of the points *B* and *D* and the obtuse angles on the other side.

By means of these two hypotheses at-Tusi tries to show that in the quadrilateral mentioned above [i.e. the one also described by Omar Khayyam] all the angles are right angles. [...] He proves this proposition by a *reductio ad absurdum*. Strictly speaking at-Tusi inserts, without noticing, a proposition at this point which is equivalent to the fifth postulate.

(d) The ideas of Omar Khayyam and Nasir ad-Din at-Tusi occupy a particularly important place in the theory of parallels and in the prehistory of non-Euclidean geometry. However, they were far from thinking that it was possible to create a geometry that differed from Euclidean geometry. They were purely and simply trying to resolve the problem of parallels by basing themselves on propositions which seemed to them to be more evident. That said, they made important discoveries. We have already spoken of one of them, i.e. the reciprocal dependence of the postulate of parallels and the angle sum of a quadrilateral and consequently of a triangle. Another important point it is useful to recall is their attempt to refute the hypotheses of acute and obtuse angles by reducing them to a contradiction.

The work of Omar Khayyam was unknown for a long time: it was published for the first time in 1936, in Tehran, in an Arabic edition. The second version of Nasir ad-Din at-Tusi's *Exposition of Euclid* appeared in Rome, first in Arabic in 1594, then in an (incomplete) Latin edition in 1657. At-Tusi's demonstration given above was known to Wallis, who gave it in his own work on the fifth postulate. It was likewise known to Saccheri.

7 Mathematics in Mediaeval Europe

7.A The Thirteenth and Fourteenth Centuries

Latin translations from Arabic works in the twelfth century, and the influence of trade with the Arabic world, were among the factors that led to a revival of mathematics in the Latin West in the thirteenth and fourteenth centuries. Fibonacci, or Leonardo of Pisa (c. 1170–1240), is perhaps the best known of the mathematicians of his day, if only for what has become known as the Fibonacci series. The selections from his work given here illustrate his role in transmitting mathematics to the Latin West and his considerable abilities as a mathematician. In the *Liber Abaci* (1202, 1228) he introduced the Hindu-Arabic numeral system and much else from Arabic mathematics as well. 7.A1 (a), on double false position, reflects his awareness of the Arabic origins of algebra. 7.A1 (b) is typical of what Fibonacci called 'tree problems'—simple algebraic problems of the same form as Egyptian problems, which can be solved by linear equations. The birds problem (7.A1 (c)) illustrates Fibonacci's work in indeterminate analysis; the lion, the leopard and the bear (7.A1 (d)), as its name suggests, is a more gory version of the kind of problem found in *The Greek Anthology* (5.C4).

The algebra of Jordanus de Nemore (c. 1220; see 7.A2) has only recently been published; Jordanus used a rudimentary algebraic symbolism long before other European mathematicians. 7.A3, from M. Biagio (c. 1340), illustrates the Italian manuscript tradition of the thirteenth and fourteenth centuries, in which algebra was very much part of practical arithmetic. It has been argued that this mathematical tradition, by diffusing mathematical knowledge and competence more widely, was a seed-bed for the later development of mathematics in its own right.

7.A1 Leonardo Fibonacci

(a) *Double false position*

Elchataieym in Arabic is rendered into Latin as 'duarum falsarum posicionum regula' [in English, 'the rule of double false position'], by means of which the solutions of almost all questions can be found. [...] Now, the two false positions are taken at will. This means that sometimes they are both smaller than is correct, sometimes both larger, sometimes one larger and the other smaller: and the truth of the solutions is found from the proportion of the difference of one position from the other, that is what happens in the rule of fourth proportion, where three numbers are involved; from which a fourth (unknown) [number], that is the truth of the solution, is to be found; the first number of these is the difference of one number of false position from the other. The second is the approximation, which becomes truth by means of this difference. The third is the remainder, which is for approximating to the truth. We wish to show how they work in the rule of weighing, so that by the demonstration of how these differences work subtly in weighing you may be able to understand the subtle solution of other questions by elchataieym.

(b) *Tree problem*

There is a tree, $\frac{1}{4}$ and $\frac{1}{3}$ of which lie below ground; and are 21 *palmi*: we are asked for the length of the tree: because $\frac{1}{4}$ and $\frac{1}{3}$ are found in 12, suppose the tree to be divided into 12 equal parts; of which a third, and a quarter, that is 7 parts, are [i.e. make] 21 *palmi*: so that as is the proportion of 7 to 21, so will be [the proportion of] 12 parts to the length of the tree. And because, when four numbers are proportional, the first multiplied by the fourth is equal to the second multiplied by the third: so if you multiply the second of the numbers mentioned, 21, by the third 12, and divide by the first number mentioned, that is by 7, they give 36 for the fourth (unknown) number, that is, for the length of the tree: or because 21 is three times 7, take three times 12, and you will similarly have 36.

There is another method we use, namely that for the unknown thing you put any number, chosen at will, which can be divided exactly into the fractions that are proposed in the question: and according to how the question is posed, with this proposed number you try to find the proportion that occurs in the solution of the question. For example: the number we are asked to find in this question is the length of the tree: therefore suppose it to be 12, since this can be divided exactly by 3, and by 4, which are given as divisors: and because it is said $\frac{1}{4}$ and $\frac{1}{3}$ of the tree are 21, take $\frac{1}{4}$ and $\frac{1}{3}$ of the 12 you supposed, they will be [i.e. will add up to] 7; and if this [sum] had chanced to be 21 we should have arrived at the required answer, namely that the tree would be 21 *palmi*. But because 7 is not 21; it happens that as 7 is in proportion to 21, so the supposed tree will be to the one we seek, that is as 12 to 36: therefore we might say: for 12, which I suppose, we obtain 7; what should I suppose so that we obtain 21? And when it is expressed this way, [we see that] we should multiply together the numbers at the end, that is 12 by 21; and the sum [*sic*, although he means product] should be divided by the remaining number.

(c) Birds problem

Someone bought sparrows 3 for a penny (*denarius*), and turtle-doves 2 for a penny, and a dove 1 for 2 pence, and he had 30 birds of these kinds for 30 pence. We want to know how many birds he bought of each kind: I first supposed 30 sparrows for 10 pence, and kept back 20 pence, which make up the difference between 10 pence and 30; and I changed one of the sparrows into a turtle-dove, and the increase [in the money spent] brought about by that change was $\frac{1}{6}$ of a penny; because the sparrow was worth $\frac{1}{3}$ of a penny, and the turtle-dove was worth $\frac{1}{2}$ of a penny, that is $\frac{1}{6}$ of a penny more than the price of a sparrow: and, again, I changed one of the sparrows into a dove, and by that change improved my position by $1\frac{2}{3}$, that is by the difference there is between $\frac{1}{3}$ of a penny and 2 pence; and I made six of these $1\frac{2}{3}$s, and by making six of them obtained 10: and according to this I should change sparrows into turtle-doves and doves, until this change yielded the 20 pence which I kept back earlier: so I made six of them, and by so doing obtained 120; which I divided into two parts, one of which could be divided exactly by 10 and the other by 1; and the total of [the results of] the two divisions was not to be so large as 30; and the first part was 110, and the other 10: and I divided the first part, that is 110, by 10, and the second by 1, and I had 11 doves and 10 turtle-doves: taking these from the 30 kinds, there remained 9 for the number of the sparrows; which sparrows are worth 3 pence, and the 10 turtle-doves are worth 5 pence, and the 11 doves are worth 22 pence; and so from these three kinds of birds we shall have 30 for 30 pence, as was required.

And if we wish to have 29 birds for 29 pence, we may operate in the same way, that is we take the price of 29 sparrows, the cheapest birds, from the 29 pence, and the remaining [money] is taken six times, and thus gives 116; which we again divide into two parts, one of which is to be exactly divisible by 10, and the other by 1; and the sum of [the results of] the two divisions is not as large as 29; which parts can be constructed in two ways: firstly so that the first part is 110, and the second 6; and when 110 is divided by 10 we obtain 11 doves; and when 6 is divided by 1 we obtain 6 turtle-doves; taking these [i.e. their sum] from 29, there remains 12 for the number of the sparrows: or [i.e. secondly] we shall divide 116 into 100 and 16; and we shall divide 100 by 10, and 16 by 1, and we shall have 10 doves and 16 turtle-doves; and the remainder, to make the number up to 29, that is 3, will be sparrows; and thus we have solved this question in two ways.

And if we wish to have 15 birds for 15 pence I shall show this is not possible without a fractional number of birds. For example, if I were to subtract the price of 15 sparrows from 15 pence; and I take the remaining pence six times, which gives 60, this cannot be divided into two parts, one of which is to be divisible by 10 and the other by 1, so that from these two divisions [i.e. from the sum of the quotients] we obtain a number less than 15: for instance: if I divide 60 into 50 and 10; and divide 50 by 10, and 10 by 1, the results of the two divisions are 5 and 10; which together add up to 15, that is to the sum of all the birds; and thus there will be no sparrow in this purchase; because 5 doves are worth 10 pence, and 10 turtle-doves are worth 5 pence; and thus from these two kinds of birds alone we have 15 birds for 15 pence: and nor is there any other number less than 60 and more than 50 which can be divided exactly by 10; and a smaller number has no place here; for if we were to put 40 for one part there would remain 20 for the other part: from which [we have] that if 40 is divided by 10, and 20 by 1, there results from the sum of the quotients of the two divisions 24 birds: which has no place [here], since there must be 15 [birds]. But if we wished to consider fractions of birds, we would divide the abovementioned 60 into 55 and 5, and divide 55 by 10, giving us $5\frac{1}{2}$ doves: and divide 5 by 1, giving us 5 turtle-doves. So subtracting $5\frac{1}{2}$ doves as 5 turtle-doves from the 15 birds, there will remain $4\frac{1}{2}$ sparrows, whose price is 1 penny and a half; and the price of 5 turtle-doves is $2\frac{1}{2}$ pence; and the price of $5\frac{1}{2}$ doves is 11 pence; and thus from these three kinds of birds we have 15 birds for 15 pence.

(*d*) *The lion, the leopard and the bear*

A lion would eat one sheep in four hours; and a leopard [would eat it] in 5 hours; and a bear [would eat it] in 6: we are asked, if a single sheep were to be thrown to them, how many hours would they take to devour it? You will do this: for 4 hours, in which the lion eats a sheep, put $\frac{1}{4}$; and for the 5 hours the leopard takes put $\frac{1}{5}$; and for the 6 hours the bear takes, put $\frac{1}{6}$: and because $\frac{1}{6}$, $\frac{1}{5}$ and $\frac{1}{4}$ are found [exactly] in 60, suppose that in 60 hours they will devour the sheep. Then consider how many sheep a lion would eat in the 60 hours: since in four hours it devours one sheep, it is obvious that it would eat 15 sheep in the 60 hours; and the leopard would eat 12 as a fifth of 60 is 12. Similarly the bear would eat 10; since 10 is $\frac{1}{6}$ of 60. Therefore in 60 hours they [i.e. all three animals together] would eat 15 plus 12 plus 10 sheep, that is 37. So you will say: for the 60 hours, which I suppose, they will eat 37 sheep. What [time] should I suppose so that they will eat only one sheep? So multiply one by 60, and divide by 37, which gives $1\frac{23}{37}$. And in that number [of hours] they will have eaten up the sheep.

7.A2 Jordanus de Nemore on problems involving numbers

(*a*) *If a given number is separated into two parts whose difference is known, then each of the parts can be found.*

Since the lesser part and the difference equal the larger, the lesser with another equal to itself together with the difference make the given number. Subtracting therefore the difference from the total, what remains is twice the lesser. Halving this yields the smaller and, consequently, the greater part.

For example, separate 10 into two parts whose difference is 2. If that is subtracted from 10, 8 remains, whose half is 4. This is the smaller number and the other is 6.

(*b*) *If a given number is separated into two parts such that the product of the parts is known, then each of the parts can be found.*

Let the given number a be separated into x and y so that the product of x and y is given as b. Moreover, let the square of $x + y$ be e, and the quadruple of b be f. Subtract this from e to get g, which will then be the square of the difference of x and y. Take the square root of g and call it h. h is also the difference of x and y. Since h is known, then x and y can be found.

The mechanics of this is easily done thus. For example, separate 10 into two numbers whose product is 21. The quadruple of this is 84, which subtracted from the square of 10, namely from 100, yields 16. 4 is the root of this and also the difference of the two parts. Subtracting this from 10 to get 6, which halved yields 3, the lesser part; and the greater is 7.

(c) *If only one of two parts of a number is known, provided the sum of the product of the parts and the square of the unknown part is given, then the number can be found.*

Let the parts of the number be x and b, with b given. Also given is a, the sum of the product of the parts, and the square of x. Add z, equal to x, to $x + b$ so that the entire $x + b + z$ can be separated into $x + b$ and z. Now since $x + b$ times z equals the given a, and the difference of $x + b$ and z is the given b, then $x + b$ and z are found as are x and $x + b$.

For example, let 6 be one of the parts and 40 the sum of the product and the square. Double 40 and redouble to get 160. Add to this 36 to obtain 196 whose root is 14. From this subtract 6 and halve the remainder to yield 4. This is the unknown part that with 6 makes the desired number 10.

(d) *If the sum of the squares of the two parts of a given number together with the square of their difference is known, then both parts can be found.*

If the sum is subtracted from the square of the given number, what remains is twice the product of the two parts less the square of their difference. This becomes the sum of the squares of the parts less twice the square of their difference, and finally it is the given sum less thrice the square of the difference. When, therefore, the remainder is subtracted from the given sum, take one third of what is left. The root of this is the difference that was sought. Hence, all can be found.

For example, square the two parts of 10, and adding them to the square of their difference yields 56. Subtract this from 100 to get 44, which in turn is subtracted from 56. The remainder is 12, whose third is 4. The root of this is 2, the difference of the parts. Therefore the larger number is 6 and the smaller is 4.

7.A3 M. Biagio: a quadratic equation masquerading as a quartic

Find two quantities, such that one is the square root of the other, and when each quantity is multiplied by itself and these multiplications added together, they make 110. Do it by position: let the first quantity be one thing. It follows that the other will be one square. Now multiply one thing by itself. It makes one square. And then you will multiply one square by itself, to make one square of a square. Add to one square: they make one square of a square and one square, which is equal to 110. This rule has not yet been written; it will be demonstrated in the first chapter of the following book. You must first reduce to one the [coefficient of the] square of the square, which is itself. And then halve the [coefficient of the] squares, which will give us $\frac{1}{2}$ and multiply it by itself, which will give us $\frac{1}{4}$, and add it to the number, which will give us $110\frac{1}{4}$. Take the root of this: $10\frac{1}{2}$. Subtract from it the half of the squares. The remainder is 10, and this is the value of the square. So the value of the thing is the root of 10. And because the first number was set as one thing, you will say that it is the root of 10, and the other was 10.

7.B The Fifteenth Century

The three fifteenth-century mathematicians represented here give, in a small way, some idea of the 'state of the art' at the time. Johannes Regiomontanus (1436–1476) was notable both for his work in trigonometry and for his programme for publishing mathematical works, cut short by his death. The selections from *On Triangles* (7.B1) are from his work on trigonometry. Book I, Theorems 20 and 28 make explicit use of the sine function; in Book II, Theorem 1, the law of sines is stated.

Little is known of the life of Nicolas Chuquet (c. 1440–c. 1488). He taught in Lyons, and his mathematical manuscripts appear to have been an offshoot of his teaching, with both elementary and advanced parts; his work illustrates how the practical tradition could contribute to the development of more advanced mathematics. 7.B2 gives some of his algebraic notation; he indicated powers of an unknown through exponents. However, Chuquet's mathematical work had almost no influence on either his contemporaries or subsequent mathematicians, unlike the *Summa* of Luca Pacioli (c. 1445–1517). This is a lengthy and comprehensive work covering a wide variety of mathematical and quasi-mathematical topics, including arithmetic, algebra, a summary of Euclid's geometry, mercantile matters such as tables of moneys, weights and measures, and a treatise on double-entry book-keeping. 7.B3(a) is from Pacioli's summary of the first part of the *Summa*, which indicates its contents. In 7.B3(b) Pacioli introduces algebra and summarizes the topics he considers in this connection. 7.B3(c) gives Pacioli's discussion of the six standard forms of quadratic equations. The influence of Arab mathematical works is obvious.

7.B1 Johannes Regiomontanus on triangles

(*a*) *Book I, Theorem 20. In every right triangle, if we describe a circle with centre a vertex of an acute angle and radius the length of the longest side, then the side subtending this acute angle is the right sine of the arc adjacent to that side and opposite the given angle; the third side is equal to the sine of the complement of the arc.*

If a right triangle ABC is given with C the right angle and A an acute angle, around the vertex of which a circle BED is described with the longest side—that is, the side opposite the largest angle—as radius, and if side AC is extended sufficiently to meet the circumference of the circle at point E, then side BC opposite angle BAC is the sine of arc BE subtending the given angle, and furthermore the third side AC is equal to the right sine of the complement of arc BE.

[Then, extending BC to CD, just as by definition the entire line BD is the chord of arc BD, so also its half, namely line BC, is the sine of the half-arc BE opposite angle BAE or BAC.]

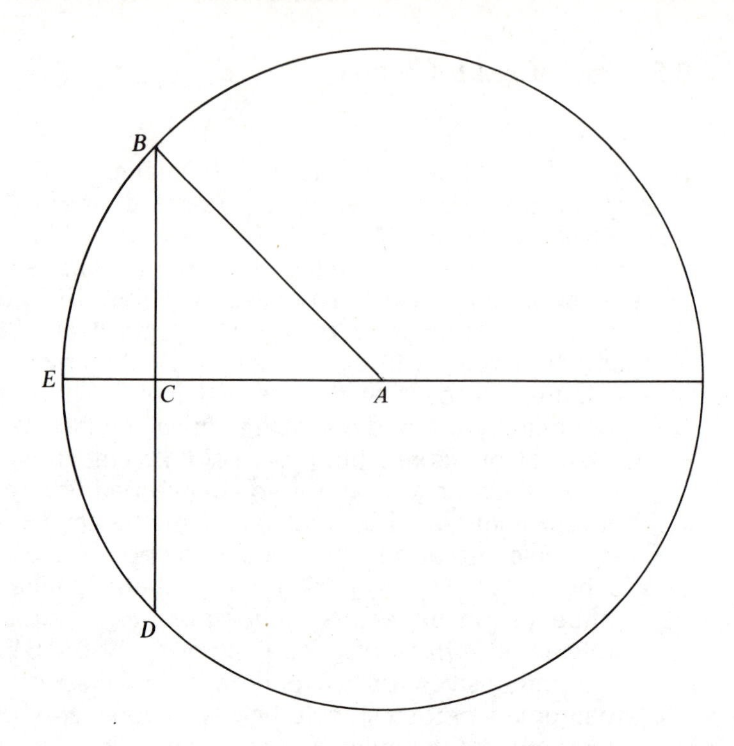

(*b*) *Book I, Theorem 28. When the ratio of two sides of a right triangle is given, its angles can be ascertained.*

One of the two sides is opposite the right angle or else none is. First, if side *AB*, whose ratio to side *AC* is known, is opposite right angle *ACB*, then the angles of this triangle become known.

[For instance, if in triangle *ABC*, $AB:BC = 9:7$, then multiply 7 into $R = 60\,000$ (the whole right sine) and divide by 9. The quotient, 46 667, corresponds to about 51°3′, the value of angle *ABC*.]

(*c*) *Book II, Theorem 1. In every rectilinear triangle the ratio of one side to another side is as that of the right sine of the angle opposite one of the sides to the right sine of the angle opposite the other side.*

As we said elsewhere, the sine of an angle is the sine of the arc subtending that angle. Moreover, these sines must be related through one and the same radius of the circle or through several equal radii. Thus, if triangle *ABG* is a rectilinear triangle, then the ratio of side *AB* to side *AG* is as that of the sine of angle *AGB* to the sine of angle *ABG*; similarly, that of side *AB* to *BG* is as that of the sine of angle *AGB* to the sine of angle *BAG*.

If triangle *ABG* is a right triangle, we will provide the proof directly from Theorem I.28 above. However, if it is not a right triangle yet the two sides *AB* and *AG* are equal, the two angles opposite the sides will also be equal and hence their sines will be equal. Thus from the two sides themselves it is established that our proposition is verified. But if one of the two sides is longer than the other—for example, if *AG* is longer—then *BA* is

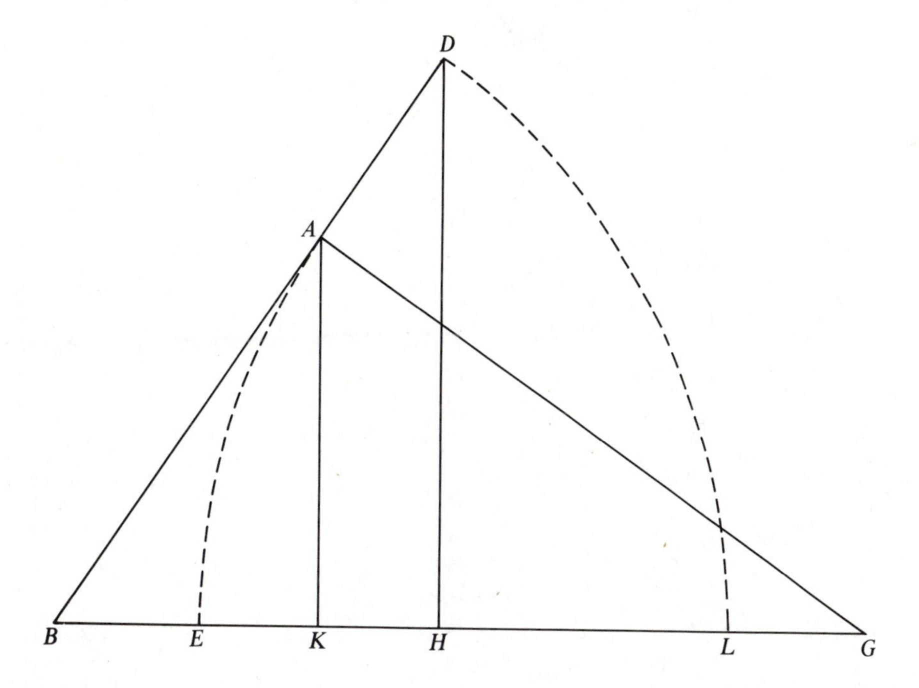

drawn all the way to D, until the whole line BD is equal to side AG. Then around the two points B and G as centres, two equal circles are understood to be drawn with the lengths of lines BD and GA as radii respectively. The circumferences of these circles intersect the base of the triangle at points L and E, so that arc DL subtends angle DBL, or ABG, and arc AE subtends angle AGE, or AGB. Finally two perpendiculars AK and DH from the two points A and D, fall upon the base. Now it is evident that DH is the right sine of angle ABG and AK is the right sine of angle AGB. Moreover, by VI.4 of Euclid, the ratio of AB to BD, and therefore to AG, is as that of AK to DH. Hence what the proposition asserts is certain.

7.B2 Nicolas Chuquet on exponents

As Boethius says in his first book and in the first chapter, the science of numbers is very great, and among the sciences of the quadrivium it is the one in the pursuit of which every man ought to be diligent. And elsewhere he says that the science of numbers ought to be preferred as an acquisition before all others, because of its necessity and because of the great secrets and other mysteries which there are in the properties of numbers. All sciences partake of it, and it has need of none. [...]

To understand the reason why denomination of number is added to denomination, and to have knowledge of the order of numbers which was mentioned in the first chapter, it is necessary to set down several proportional numbers beginning with 1 and arranged in a continuous sequence, like 1, 2, 4, 8, 16, 32, etc. or 3, 9, 27, etc.

Numbers	Denomination
1	0
2	1
4	2
8	3
16	4
32	5
64	6
128	7
256	8
512	9
1 024	10
2 048	11
4 096	12
8 192	13
16 384	14
32 768	15
65 536	16
131 072	17
262 144	18
524 288	19
1 048 576	20

Now it is necessary to know that 1 represents and is in the place of numbers, whose denomination is 0. 2 represents [...] the first terms, whose denomination is 1. 4 holds the place of the second terms, whose denomination is 2. And 8 is in the place of the third terms, 16 holds the place of the fourth terms, 32 represents the fifth terms, and so for the others. Now whoever multiplies 1 by 1, it comes to 1, and because 1 multiplied by 1 does not change at all, neither does any other number when it is multiplied by 1 increase or diminish, and for this consideration, whoever multiplies a number by a number, it comes to a number, whose denomination is 0. And whoever adds 0 to 0 makes 0. Afterwards, whoever multiplies 2, which is the first number, by 1, which is a number, the multiplication comes to 2; then afterwards, whoever adds their denominations, which are 0 and 1, it makes 1; thus the multiplication comes to 2^1. And from this it comes that when one multiplies numbers by first terms or vice versa, it comes to first terms. Also whoever multiplies 2^1 by 2^1, it comes to 4 which is a second number. Thus the multiplication amounts to 4^2. For 2 multiplied by 2 makes 4 and adding the denominations, that is, 1 with 1, makes 2. And from this it comes that whoever multiplies first terms by first terms, it comes to second terms. Likewise whoever multiplies 2^1 by 4^2, it comes to 8^3. For 2 multiplied by 4 and 1 added with 2 makes 8^3. And thus whoever multiplies first terms by second terms, it comes to third terms. Also, whoever multiplies 4^2 by 4^2, it comes to 16 which is a fourth number, and for this reason whoever multiplies second terms by second terms, it comes to fourth terms. Likewise whoever multiplies 4 which is a second number by 8 which is a third number makes 32 which is a fifth number. And thus whoever multiplies second terms by third terms or vice versa, it comes to fifth terms. And third terms by fourth terms comes to 7th terms, and fourth terms by fourth terms, it comes to 8th terms, and so for the others. In this discussion there is manifest a secret which is in the proportional numbers. It is that whoever multiplies a proportional number by itself, it comes to the number of the double of its denomination, as, whoever multiplies 8 which is a third number by itself, it comes to 64 which is a sixth. And 16 which is a fourth number multiplied by itself should come to 256, which is an eighth. And whoever multiplies 128 which is the 7th proportional by 512 which is the 9th, it should come to 65 536 which is the 16th.

7.B3 Luca Pacioli

(*a*) *On the content of his* Summa

The whole of this book is divided into five principal parts. In the first, numbers are discussed, in every way you would expect in simple and speculative practice. That is, writing and reading the characters, division, multiplication, addition, subtraction, and all sorts of progressions with very worthy rules newly induced, and very subtle cases; and the extraction of roots with numbers and with instruments and by geometrical methods, with their approximations. The philosophical algorisms discuss these things, from which, through this, there will always be knowledge of whole numbers, fractions, roots, binomials, their conjugates, and roots of roots, and every method of solving every proposed problem by algebra. And proportions and proportionalities and division, multiplication, addition, subtraction, which are necessary to perspective, music, astrology, cosmography, architecture, law, and medicine. With every substantiation from the fifth book of Euclid. And the rules of false position with their explanations. And of irrational lines, with which the whole of the tenth book of Euclid deals, with their practical methods of operation with clear demonstrations, always worked in such a way that everyone can learn them with great ease. And all these things with what follows will be according to the ancient and also modern mathematicians. Mostly from the very perspicacious philosopher of Megara, Euclid, and Severinus Boethius, and from our modern mathematicians, Leonardo Pisano [Fibonacci], Giordano, Biagio of Parma, Johannes de Sacrobosco, and Prodocimo of Padua, from whom I take the major part of this volume. You will have the table of this part and all the others, one by one below, set down in order, according to their distinctions, treatises, articles, and pages, according to how they differ, part by part. And in this first part are also contained all commercial occurrences of problems and rules, that is by hundredweights, thousands, pounds, ounces, investments, sales, profits, losses, journeys or transportations of goods, weights, measures, and money from place to place. And calculation of prices, with limitations of profit, loss, tares, gifts, uses, import and export duties in different places, taxes on sales made through brokers, carriage, fares, stabling, and whatever other exactions there may be, such as hiring, rents, household salaries, agents' fees, and workmen's wages. Appreciation, depreciation, gold, silver, copper, lightness and heaviness of all weights, superfluity and scarcity of all measures; lengths, widths, heights, and thicknesses, according to the commercial custom.

(*b*) *On algebra*

I do not think that I need now to defer any longer the greatest part necessary to the practice of arithmetic and also of geometry, called commonly 'the greater art', or 'the rule of the thing', or 'algebra and almucabala', called by us 'speculative practice'. Because higher things are contained in it than in the lesser art or business practice, as will be shown as we proceed. Such as roots and their squares, both simple and compound ways that occur; and in binomials and their conjugates; and in plus and minus abstractly upheld in its operation. These things cannot happen ordinarily in questions or problems of trade, unless so many new impositions were made, raising the subject to a higher level of operation. Wanting to deal with this subject in such a way as to proceed in an orderly manner, we will divide this section into seven principal parts. In the first of these we will speak of the two terms found to be convenient to this kind of operation, one called 'plus' and the other 'minus', and how between them they are used in their workings, and why they were invented. They are the most necessary of all to this practice. Then in the second we will deal with roots in every way that they are used and worked, with their definitions and divisions or disparities. In the third we will tell of binomials and also conjugates or residues, and of their disparities and operations. And of the notion of the fifteen lines, with which all the tenth book of Euclid principally deals. In the fourth we shall show certain ways in multiplication, and consequently in division, of ordering the *cosa*, and squares and cubes and higher powers, set down in the way of common tables, so that in the multiplication of these powers it can be found more easily what one of these quantities times another will generate. And the same with dividing one by another, what should come of such a division, which will be obvious from the information on multiplication. In the fifth will be set down the basic equations of algebra and almucabala with their distinctions between simple and compound, and together with these, the proof of each of them, and where their strength comes from, with clear and open diagrams. In the sixth we shall demonstrate the method of constructing the basic equation appropriate to whatever equation the performer may come across, so that he can with great ease make reply to such an equation, which will conform to the solution of that basic equation, for there are six fundamental and principal forms (as will be described) and infinite then are those which have their origin in these. And they proceed according to the proportion of these six forms. In the seventh and last principal part we shall explain why it is not possible to solve every question by algebra, and in consequence we shall give information of each impossibility, according to the way of its equation. And thus, by looking at any equation one will know its possibility. And after all these things, several very useful questions will be set.

(*c*) *On quadratic equations*

We have seen what these terms mean and represent in the practice of arithmetic. Now we must see in how many ways they can be made equal, one to the other, and the other to the one, and two of them to one of them, and one to two of them. On this I say that they can be made equal to each other in six ways. First, the square to the things. Second, the square to the numbers. Third, thing or things to numbers. Fourth, the square and the thing can be put equal to the number. Fifth, the thing and the number can be equal to the square. Sixth, the square and the number can be equal to the thing. Other than in these six ways described it is not possible to have any equation in them. And in regard to these six equations, there have been formed six rules, which are commonly called the six basic equations. And of these six, three give a standard to the first three of the equations, that is, of one to the other and the other to the one, which are called simple. And the other three are standard forms of the other three equations, that is, of two equal to one and conversely. And these three are called compound, that is, that two together are always equal to the third, which happens (as we have said) in three ways, whence the rules or cases of one and the other. These are the standard three of the simple equations. [...]

The three compound rules of algebraic equations

1 When the squares and the things are equal to a number, first you must reduce all the equation to one square, that is if there is less than one square you must equally restore and make good. And if there is more than one square you must reduce to one square, and reducing is done by dividing the whole of the equation by the amount of the squares. And when you have done this, halve the things, and multiply one half by itself. The number is added to this product, and the root of this sum minus the half of the things is the value of the thing required.

2 When the thing and the number are equal to squares. First (the same thing is done as above in the preceding, that is) reduce the whole equation to one square, removing anything greater than one square equally and geometrically, and making up anything less on both sides likewise geometrically, by doing this: (as I said above) it suffices to divide the whole of the equation by the quantity of squares, and then it will be reduced to one square. Having done this you will halve the things, and multiply one half by itself, and to the result add the number. And the root of this sum will always be the value of the thing, when the half of the things is added to it.

3 When the square and the number are equal to the things. In that case (do as we said above, that is) reduce the whole equation to one square, that is, divide the whole of it by the quantity of the squares, and then halve the things and multiply one half by itself. And from the result always subtract the number which is found in the equation, and the root of the remainder added to the half of the things, or indeed subtracted from the half of the things, will be the value of the thing.

Geometrical demonstration of the first compound basic equation

[...] the other three compound equations certainly need a cautious declaration and demonstration, so that their truthfulness will be matched more openly, that is that one must observe in their setting up what is contained in their statements above. And here in the following I intend to demonstrate them one by one in an orderly way. And first we shall demonstrate the truth of the one where the squares and the things are equal to a number. For example, 1 square and 10 things are equal to 39, which are straightforward numbers. [...] Let $abcd$ be a tetragon which has each side greater than the number 5. And on the side ab is marked the point e in such a way that be is exactly the number 5 and the remainder ea is an unknown quantity. And in the same way, on the side ad is marked the point f in such a way that fd is again the number 5, and af is again an unknown excess over 5. And in the same way, on the side bc is marked the point g, so that again cg is 5, and the remainder bg is similarly unknown. And on the side *cd*, the point *h* in the same way, that is so that *ch* is 5, and *hd* unknown. Thus of all these 4 straight lines, each will be known to be 5.

[Pacioli goes on to deduce very carefully that the side *ab* is 8, and so the root, *ea*, is 3.]

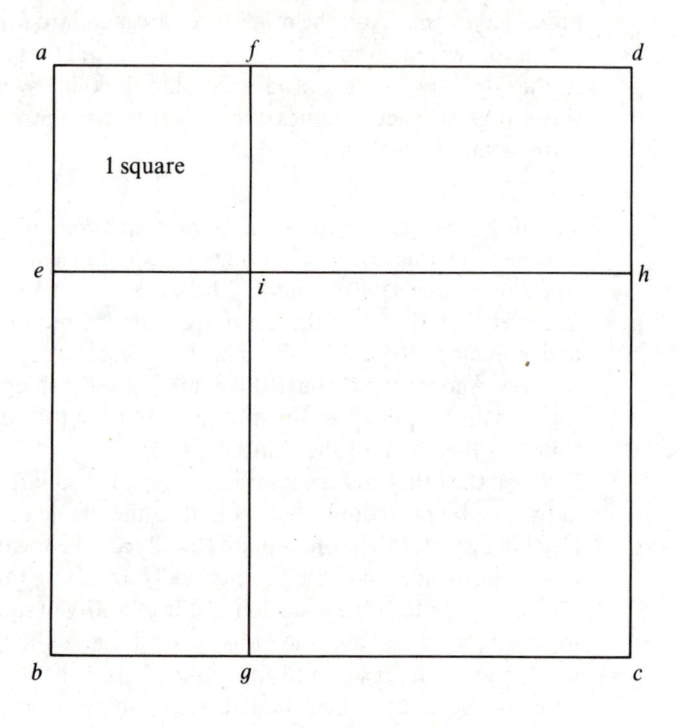

8 Sixteenth-century European Mathematics

8.A The Development of Algebra in Italy

The first person to solve cubic equations algebraically was Scipione del Ferro (1465?–1525), who was professor of mathematics at Bologna. At some stage he entrusted the solution method to his pupil Antonio Maria Fior, who proceeded to live off it by challenging others to contests at mathematical problem-solving. As can be seen from 8.A1, Fior thought it worth while to put all his mathematical eggs into this cubical basket, which rather suggests that he was not a very good mathematician but very confident of his secret. Indeed, because all his problems are of the form 'cube and things equal to numbers' ($x^3 + px = q$), it seems very likely that this is the only kind of cubic del Ferro had taught him to solve. But in 1535 he was unlucky enough to challenge Niccolò Tartaglia (1506?–1559) who, on the night of 12–13 February, worked out the solution for himself, and so won the contest. Apparently he declined the thirty dinners at Fior's expense he had thereby won. The news of his success soon spread, and Gerolamo Cardano (1501–1576) heard of it. After much patient lobbying, they met in 1539 in Milan, and Tartaglia divulged the method. Tartaglia's later claims as to what was said on this occasion form 8.A2. (The use of verses to remember complicated items was not unusual at the time, and was proof against theft.) But when Cardano and his pupil Ludovico Ferrari (1522–1565) learned that the solution method had been known to del Ferro, and found that they had new things to say, both about other types of cubic and about quartic equations (which Ferrari had discovered how to solve), they decided to publish (see 8.A4). Tartaglia was furious, and a prolonged battle was waged between them, some of which can be glimpsed in 8.A3. Some of Ferrari's problems, in particular, show that he had a broad and philosophically rich attitude to mathematics.

Even by Rafael Bombelli's time, a mere generation later, things had begun to change. Bombelli's attitude to algebra was heavily influenced by the rediscovery of ancient texts then well under way (see Section 8.B) and, in particular, his own deep study of Diophantus. He acknowledges this in his Preface (8.A5(a)). We also give a glimpse of his profound ideas about complex numbers (8.A5(b)), to which he was led by the awkward fact that when a cubic equation has only one real root, the algebraic method appears to break down. Bombelli saw that the way out was to recognize that the algebra yields the answer as a sum of two complex conjugate numbers and so as a real number after all.

8.A1 Antonio Maria Fior's challenge to Niccolò Tartaglia (1535)

These are the thirty problems proposed by me Antonio Maria Fior to you Master Niccolò Tartaglia.

1 Find me a number such that when its cube root is added to it, the result is six, that is 6. [This is equivalent to the equation $x^3 + x = 6$.]
2 Find me two numbers in double proportion such that when the square of the larger number is multiplied by the smaller, and this product is added to the two original numbers, the result is forty, that is 40. [Equivalent to the equation $(2x)^2 \cdot x + x + 2x = 40$, i.e. $4x^3 + 3x = 40$.]
3 Find me a number such that when it is cubed, and the said number is added to this cube, the result is five. [Equivalent to $x^3 + x = 5$.]
[...]
15 A man sells a sapphire for 500 ducats, making a profit of the cube root of his capital. How much is this profit? [$x^3 + x = 500$.]
[...]
17 There is a tree, 12 *braccia* high, which was broken into two parts at such a point that the height of the part which was left standing was the cube root of the length of the part that was cut away. What was the height of the part that was left standing? [$x^3 + x = 12$.]
[...]
30 There are two bodies of 20 triangular faces [icosahedra] whose corporeal areas added together make 700 *braccia*, and the area of the smaller is the cube root of the larger. What is the smaller area? [$x^3 + x = 700$.]

8.A2 Tartaglia's account of his meeting with Gerolamo Cardano (1539)

CARDANO: I hold it very dear that you have come now, when his Excellency the Signor Marchese has ridden as far as Vigevano, because we will have the opportunity to talk, and to discuss our affairs together until he returns. Certainly you have, alas, been unkind in not wishing to give me the rule that you discovered, on the case of the thing and the cube equal to a number, even after my greatest entreaties for it.

TARTAGLIA: I tell you, I am not so unforthcoming merely on account of the solution, nor of the things discovered through it, but on account of those things which it is possible to discover through the knowledge of it, for it is a key which opens the way to the ability to investigate boundless other cases. And if it were not that at present I am busy with the translation of Euclid into Italian (and at the moment I have translated as far as his thirteenth book), I would already have found a general rule for many other cases. But as soon as I have completed this work on Euclid that I have already begun, I intend to compose a book on the practice [of arithmetic], and together with it a new algebra, in which I have resolved not only to publish to every man all my discoveries of new cases already mentioned, but many others which I hope to find; and, more, I want to demonstrate the rule that enables one to investigate boundless other cases, which I hope will be a useful and beautiful thing. And this is the reason which makes me refuse them to everyone, because at present I am not working on them (being, as I said, busy with Euclid), and if I teach them to any speculative person (as is your Excellency), he could easily with such clear information find other solutions (it being easy to combine it with the things already discovered), and publish it, as inventor. And to do that would spoil all my plans. Thus this is the principal reason that has made me so unkind to your Excellency, so much more as you are at present having your book printed on a similar subject, and even though you wrote to me that you want to give out these discoveries of mine under my name, acknowledging me as the inventor. Which in effect does not please me on any account, because I want to publish these discoveries of mine in my books, and not in another person's books.

CARDANO: And I also wrote to you that if you did not consent to my publishing them, I would keep them secret.

TARTAGLIA: It is enough that I did not choose to believe that.

CARDANO: I swear to you, by God's holy Gospels, and as a true man of honour, not only never to publish your discoveries, if you teach me them, but I also promise you, and I pledge my faith as a true Christian, to note them down in code, so that after my death no one will be able to understand them. If you want to believe me now, then believe me, if not, leave it be.

TARTAGLIA: If I did not give credit to all your oaths, I would certainly deserve to be judged a faithless man, but since I have decided to ride to Vigevano to call upon his Excellency the Signor Marchese, because it is now three days that I have been here, and I am sorry to have waited for him so long, when I have returned I promise to demonstrate everything to you.

CARDANO: Since you have decided anyway to ride as far as Vigevano after the Signor Marchese, I want to give you a letter to give to his Excellency, so that he should know who you are. But before you go, I want you to show me the rule for these solutions of yours, as you have promised me.

TARTAGLIA: I am satisfied. But I want you to know, that, to enable me to remember the method in any unforeseen circumstance, I have arranged it as a verse in rhyme, because if I had not taken this precaution, I would frequently have forgotten it, and although my telling it in rhyme is not very concise, it has not bothered me, because it is enough that it serves to bring the rule to mind every time that I recite it. And I want to write down this verse for you in my own hand, so that you can be sure that I am giving you the invention accurately and well.

When the cube and the things together
Are equal to some discrete number,
[To solve $x^3 + cx = d$,]

Find two other numbers differing in this one.
Then you will keep this as a habit
That their product should always be equal
Exactly to the cube of a third of the things.
[Find u, v such that $u - v = d$ and $uv = (c/3)^3$.]
The remainder then as a general rule
Of their cube roots subtracted
Will be equal to your principal thing.
[Then $x = \sqrt[3]{u} - \sqrt[3]{v}$.]
In the second of these acts,
When the cube remains alone,
[In the second case, to solve $x^3 = cx + d$,]
You will observe these other agreements:
You will at once divide the number into two parts
So that the one times the other produces clearly
The cube of a third of the things exactly.
[Find u, v such that $u + v = d$ and $uv = (c/3)^3$.]
Then of these two parts, as a habitual rule,
You will take the cube roots added together,
And this sum will be your thought.
[Then $x = \sqrt[3]{u} + \sqrt[3]{v}$.]
The third of these calculations of ours
Is solved with the second if you take good care,
As in their nature they are almost matched.
[The third case, to solve $x^3 + d = cx$, is similar to the second.]
These things I found, and not with sluggish steps,
In the year one thousand five hundred, four and thirty.
With foundations strong and sturdy
In the city girdled by the sea.

This verse speaks so clearly that, without any other example, I believe that your Excellency will understand everything.

CARDANO: How well I will understand it, and I have almost understood it at the present. Go if you wish, and when you have returned, I will show you then if I have understood it.

TARTAGLIA: Now, remember your Excellency, and just do not forget your faithful promise, because if by unhappy chance it is broken, that is if you publish these solutions, whether in this book that you are having printed at the moment; or even, in another one different from this one, you publish them giving my name and acknowledging me as the real inventor, I promise you and I swear to publish immediately another book which will not be very agreeable for you.

CARDANO: Do not doubt that I will keep my promise. Go, and be sure that you give this my letter to the Signor Marchese on my behalf.

TARTAGLIA: Now please do not forget.

CARDANO: Go, straight away.

TARTAGLIA: By my faith, but for that, I do not want to go to Vigevano. I would much rather turn in the direction of Venice, let the matter go as it will.

8.A3 Tartaglia versus Ludovico Ferrari (1547)

Ferrari to Tartaglia

[...]
15 Find me two numbers such that when they are added together, they make as much as the cube of the lesser added to the product of its triple with the square of the greater; and the cube of the greater added to its triple times the square of the lesser makes 64 more than the sum of these two numbers. [Find a, b such that $a + b = b^3 + 3ba^2$ and $a^3 + 3ab^2 = 64 + a + b$.]
[...]
17 Divide eight into two parts such that their product multiplied by their difference comes to as much as possible, proving everything.
[...]
21 Find me six quantities in continuous proportion starting with one, such that the double of the second with the triple of the third is equal to the root of the sixth. [Find the sequence $1, r, r^2 \ldots r^5$ such that $2r + 3r^2 = \sqrt{r^5}$.]
22 As far as appertains to mathematics, I require the exposition of that place in Plato's *Timaeus*, which, in Latin, begins, '*Fuit autem talis illa partitio*' [He began the division as follows] as far as these words, '*Postquam igitur secundum creatoris*' [Therefore afterwards according to the creator]. [The passage in question is 2.E5(c).]
23 There is a cube such that its sides and its surfaces added together are equal to the proportional quantity between the said cube and one of its faces. What is the size of the cube? [Find x, y such that $12x + 6x^2 = y$, where $\frac{x^3}{y} = \frac{y}{x^2}$ (i.e. $y^2 = x^5$).]
[...]
27 There is a right-angled triangle, such that when the perpendicular is drawn, one of the sides with the opposite part of the base makes 30, and the other side with the other part makes 28. What is the length of one of the sides? [Find AB, AC or BC such that $AB + DC = 30$ and $AC + BD = 28$.]
[...]
30 Is unity a number or not?

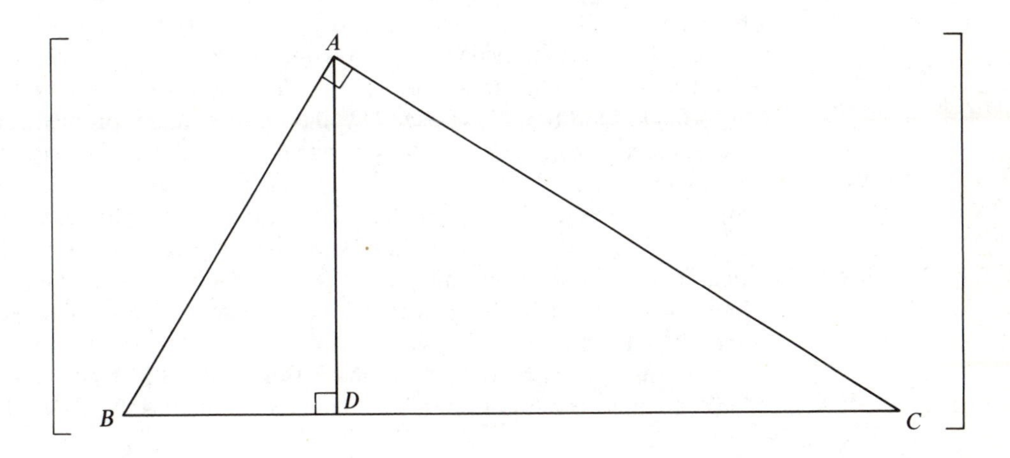

Tartaglia to Ferrari

In your fifteenth problem, you required me to find you two numbers such that, when added together they should make as much as the lesser added to the multiplication of its triple with the square of the greater, and that the cube of the greater together with the multiplication of its triple with the square of the lesser should come to 64 more than the sum of the said two numbers. I reply to you by saying that the greater number, or quantity, will be the cube root of the whole of 4 plus $\sqrt{15\frac{215}{216}}$, plus the cube root of the whole of 4 minus $\sqrt{15\frac{215}{216}}$, plus 2 [i.e. $\sqrt[3]{(4+\sqrt{15\frac{215}{216}})}+\sqrt[3]{(4-\sqrt{15\frac{215}{216}})}+2$]. And the lesser will be the same, minus 2, that is, it will be the cube root of the whole of 4 plus $\sqrt{15\frac{215}{216}}$, plus the cube root of the whole of 4 minus $\sqrt{15\frac{215}{216}}$, minus 2 [$\sqrt[3]{(4+\sqrt{15\frac{215}{216}})}+\sqrt[3]{(4-\sqrt{15\frac{215}{216}})}-2$]. [...]

In your seventeenth problem, you asked me to divide eight into two parts such that their product multiplied by their difference should come to as much as possible. I reply to you that the greater part was 4 plus $\sqrt{5\frac{1}{3}}$ and the lesser was 4 minus $\sqrt{5\frac{1}{3}}$, the product is $10\frac{2}{3}$, which multiplied by the difference, which is $\sqrt{21\frac{1}{3}}$, makes $\sqrt{2423\frac{7}{27}}$, and this is fruit of our tree with which you thought to make war on me, but you failed in that intention. [...]

In your twenty-first problem you required me to find you six quantities in continuous proportion starting with one, and such that the double of the second one with the triple of the third should be equal to the root of the sixth. I reply to you, that the first was (as you required) one, the second of them will be the cube root of the whole of 47 plus $\sqrt{12}$, plus the cube root of the whole of 47 minus $\sqrt{12}$, plus 3 [$\sqrt[3]{(47+\sqrt{12})}+\sqrt[3]{(47-\sqrt{12})}+3$]. The others can be found in the usual way, but since, I accept, there is no skill in that, I omit that labour. [...]

In your twenty-third problem you told me that there is a cube of which the sides and surfaces added together are equal to the quantity in mean proportion between the said cube and one of its faces, and you required of me the size of the cube. I reply to you, that the side of the said cube will be the cube root of the whole of 2664 plus $\sqrt{19008}$, plus the cube root of the whole of 2664 minus $\sqrt{19008}$, plus 12 [$\sqrt[3]{(2664+\sqrt{19008})}+\sqrt[3]{(2664-\sqrt{19008})}+12$]. [...]

To your twenty-seventh problem, where you say, 'There is a right-angled triangle, in which, when the perpendicular is drawn, one of the sides with the opposite part of the base makes 30, and the other [side] with the other [part] makes 28', and you ask me the length of one of the sides. I reply to you, that you have proposed this to me in order that I should elucidate to you what you do not understand, and that it is true, in your *Ars Magna*, Question 14, on page 71 you propose a similar problem, namely, you want one of the sides with the opposite part to make 29, and the other with the other to make 31. In which triangle the sides fixed by you are rational, that is, one would be 20, the other 15, the base 25, the perpendicular 12, the greater part of the base 16, and the lesser 9. And in the end you did not know how to solve such a problem by a general rule. It is a very shameful thing, to put forward such a question in public, and not to know how to solve it by a general rule. I have the same opinion of your Problems 26 and 19, but I reserved to myself my reply to you on them, in front of the referees.

I inform you, moreover, that in my first solutions I interposed one for you which looks credible, but for all that is not true nor properly solved. And I did this for two reasons. One, to ascertain whether such a question was understood by you or not; the other, to make clear to everyone that in this dispute of ours there is no need for any other referees but ourselves, as I asserted in my first reply. Because, if you had not been ignorant of the solution of that question of yours, you would immediately have noticed its falsity, and you would have made me aware of, or demonstrated, the error in my solution. And I would have assented, to show that in these mathematical disputes, one cannot oppose nor contradict the truth. And then I chose to send it to you to see if you knew the correct solution, but as far as I can see you did not notice that falsely completed solution, which leads me to believe, and to hold for certain, that you are ignorant of the solution to that question.

Ferrari to Tartaglia

My seventeenth problem, if you look at it closely, says, divide eight into two parts such that their product multiplied by their difference comes to as much as possible, proving everything. But you, just like a forger, not only in obscure things but also in open and public things, omit, when reporting my words, the part that matters, and which you did not dare to ignore, namely these two words, 'proving everything', and then having already misused the thing in your way, you simply divide eight into two parts, which I do not wish to tell you, whether they are the correct ones or not. But I will tell you, that, in not proving that which my problem requires, you are still leaving it unsolved, and so you show yourself to be that man of integrity and honour, that I meant in my last *Cartello*. [...]

As for the twenty-second. You at first say that it is not a question for a mathematician. To which I reply, that, if by a mathematician you mean someone like you, that is, someone who spends the whole time on roots, fifth powers, cubes, and other trifles, then you are quite right. And I promise you that if it were up to me to reward you, taking example from the custom of Alexander, I would load you up so much with roots and radishes, that you would never eat anything else in your life. But if by a mathematician you mean a man expert in arithmetic, geometry, astrology, and music, and all the other arts that depend on these ones, as were all the ancient mathematicians; and nowadays there are a few who not only possess the aforementioned arts, but also know how to use them in every other science, as they are required. I tell you that the problem is a mathematical one, and one of the finest that could be posed. Because, whoever understands well the reasoning behind all these numbers, and the reasoning behind these crossings and rotations of lines, and then (which is more important) what follows from these things, will understand the finest passage in the whole of mathematics and philosophy together. I will tell you candidly that this is not a subject for you; and so, interpreting only very clumsily, and some lines on purpose into Italian, you have passed along, leaving the problem wholly unsolved.

8.A4 Gerolamo Cardano

(*a*) *Cardano's aims in writing the* Ars Magna

This art originated with Mahomet the son of Moses the Arab. Leonardo of Pisa is a trustworthy source for this statement. There remain, moreover, four propositions of his with their demonstrations, which we will ascribe to him in their proper places. After a long time, three derivative propositions were added to these. They are of uncertain authorship, though they were placed with the principal ones by Luca Paccioli. I have also seen another three, likewise derived from the first, which were discovered by some unknown person. Notwithstanding the latter are much less well known than the others, they are really more useful, since they deal with the solution of [equations containing] a cube, a constant, and the cube of a square.

In our own days Scipione del Ferro of Bologna has solved the case of the cube and first power equal to a constant, a very elegant and admirable accomplishment. Since this art surpasses all human subtlety and the perspicuity of mortal talent and is a truly celestial gift and a very clear test of the capacity of men's minds, whoever applies himself to it will believe that there is nothing that he cannot understand. In emulation of him, my friend Niccolò Tartaglia of Brescia, wanting not to be outdone, solved the same case when he got into a contest with his [Scipione's] pupil, Antonio Maria Fior, and, moved by my many entreaties, gave it to me. For I had been deceived by the words of Luca Paccioli, who denied that any more general rule could be discovered than his own. Notwithstanding the many things which I had already discovered, as is well known, I had despaired and had not attempted to look any further. Then, however, having received Tartaglia's solution and seeking for the proof of it, I came to understand that there were a great many other things that could also be had. Pursuing this thought and with increased confidence, I discovered these others, partly by myself and partly through Ludovico Ferrari, formerly my pupil. Hereinafter those things which have been discovered by others have their names attached to them; those to which no name is attached are mine. The demonstrations, except for the three by Mahomet and the two by Ludovico, are all mine. Each is individually set out under a proper heading and, following the rule, an illustration is added.

Although a long series of rules might be added and a long discourse given about them, we conclude our detailed consideration with the cubic, others being merely mentioned, even if generally, in passing. For as *positio* [the first power] refers to a line, *quadratum* [the square] to a surface, and *cubum* [the cube] to a solid body, it would be very foolish for us to go beyond this point. Nature does not permit it. Thus, it will be seen, all those matters up to and including the cubic are fully demonstrated, but the others which we will add, either by necessity or out of curiosity, we do not go beyond barely setting out. In everything, however, the worth of the preceding books [of this work], especially the third and fourth books, should be kept in mind, lest I be thought trifling when I repeat or obscure when I skip over something.

(*b*) *On the solution to a cubic equation*

Scipio del Ferro of Bologna about thirty years ago invented [the method set forth in] this chapter, [and] communicated it to Antonio Maria Florido of Venice, who when he once engaged in a contest with Nicolo Tartalea of Brescia announced that Nicolo also invented it; and he [Nicolo] communicated it to us when we asked for it, but suppressed the demonstration. With this aid we sought the demonstration, and found it, though with great difficulty, in the manner which we set out in the following.

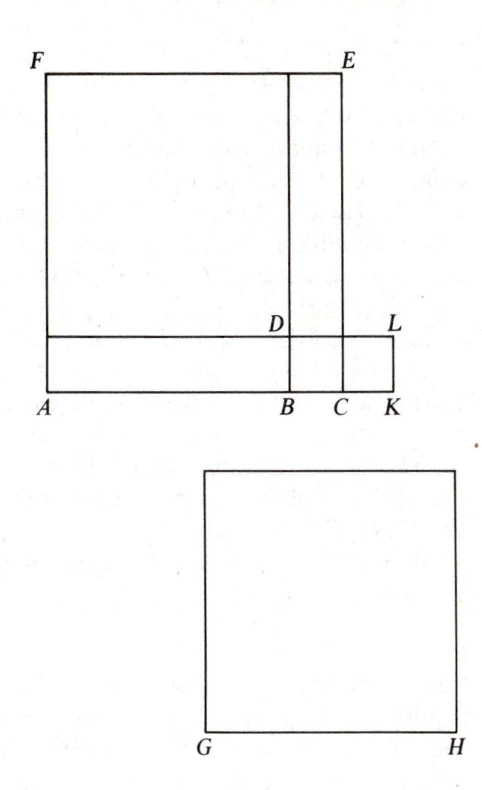

Demonstration

For example, let the cube of *GH* and six times the side *GH* be equal to 20. I take two cubes *AE* and *CL* whose difference shall be 20, so that the product of the side *AC* by the side *CK* shall be 2,—i.e., a third of the number of 'things'; and I lay off *CB* equal to *CK*, then I say that if it is done thus, the remaining line *AB* is equal to *GH* and therefore to the value of the 'thing,' for it was supposed of *GH* that it was so [i.e. equal to x], therefore I complete, after the manner of the first theorem of the 6th chapter of this book, the solids *DA*, *DC*, *DE*, *DF*, so that we understand by *DC* the cube of *BC*, by *DF* the cube of *AB*, by *DA* three times *CB* times the square of *AB*, by *DE* three times *AB* times the square of *BC*. Since therefore from *AC* times *CK* the result is 2, from 3 times *AC* times *CK* will result 6, the number of 'things': and therefore from *AB* times 3 *AC* times *CK* there results 6 'things' *AB*, or 6 times *AB*, so that 3 times the product of *AB*, *BC*, and *AC* is 6 times *AB*. But the difference of the cube *AC* from the cube *CK*, and likewise from the cube *BC*, equal to it by hypothesis, is 20; and from the first theorem of the 6th chapter, this is the sum of the solids *DA*, *DE*, and *DF*, so that these three solids make 20. But taking *BC minus*, the cube of *AB* is equal to the cube of *AC* and 3 times *AC* into the square of *CB* and minus the cube of *BC* and minus 3 times *BC* into the square of *AC*. By the demonstration, the difference between 3 times *CB* times the square of *AC*, and 3 times *AC* times the square of *BC*, is [3 times] the product of *AB*, *BC*, and *AC*. Therefore since this, as has been shown, is equal to 6 times *AB*, adding 6 times *AB* to that which results from *AC* into 3 times the square of *BC* there results 3 times *BC* times the square of *AC*, since *BC* is minus. Now it has been shown that the product of *CB* into 3 times the square of *AC* is minus; and the remainder which is equal to that is plus, hence 3 times *CB* into the square of *AC* and 3 times *AC* into the square of *CB* and 6 times *AB* make nothing. Accordingly, by common sense, the difference between the cubes *AC* and *BC* is as much as the totality of the cube of *AC*, and 3 times *AC* into the square of *CB*, and 3 times *CB* into the square of *AC* (minus), and the cube of *BC* (minus), and 6 times *AB*. This therefore is 20, since the difference of the cubes *AC* and *CB* was 20. Moreover, by the second theorem of the 6th chapter, putting *BC* minus, the cube of *AB* will be equal to the cube of *AC* and 3 times *AC* into the square of *BC* minus the cube of *BC* and minus 3 times *BC* into the square of *AC*. Therefore the cube of *AB*, with 6 times *AB*, by common sense, since it is equal to the cube of *AC* and 3 times *AC* into the square of *CB*, and minus 3 times *CB* into the square of *AC*, and minus the cube of *CB* and 6 times *AB*, which is now equal to 20, as has been shown, will also be equal to 20. Since therefore the cube of *AB* and 6 times *AB* will equal 20, and the cube of *GH*, together with 6 times *GH*, will equal 20, by common sense and from what has been said in the 35th [of Book I] and 31st of the 11th Book of [Euclid's] *Elements*, *GH* will be equal to *AB*, therefore *GH* is the difference of *AC* and *CB*. But *AC* and *CB*, or *AC* and *CK*, are numbers or lines containing an area equal to a third part of the number of 'things' whose cubes differ by the number in the equation, wherefore we have the

Rule

Cube the third part of the number of 'things', to which you add the square of half the number of the equation, and take the root of the whole, that is, the square root, which you will use, in the one case adding the half of the number which you just multiplied by itself, in the other case subtracting the same half, and you will have a 'binomial' and 'apotome' respectively; then subtract the cube root of the apotome from the cube root of the binomial, and the remainder from this is the value of the 'thing'. In the example, the cube and 6 'things' equals 20; raise 2, the 3rd part of 6, to the cube, that makes 8; multiply 10, half the number, by itself, that makes 100; add 100 and 8, that makes 108; take the root, which is $\sqrt{108}$, and use this, in the first place adding 10, half the number, and in the second place subtracting the same amount, and you will have the binomial $\sqrt{108} + 10$, and the apotome $\sqrt{108} - 10$; take the cube root of these and subtract that of the apotome from that of the binomial, and you will have the value of the 'thing', $\sqrt[3]{(\sqrt{108} + 10)} - \sqrt[3]{(\sqrt{108} - 10)}$.

(c) *From Cardano's autobiography*

My father, in my earliest childhood, taught me the rudiments of arithmetic, and about that time made me acquainted with the arcana; whence he had come by this learning I know not. This was about my ninth year. Shortly after, he instructed me in the elements of the astrology of Arabia, meanwhile trying to instill in me some system of theory for memorizing, for I had been but poorly endowed with ability to remember. After I was twelve years old he taught me the first six books of Euclid, but in such a manner that he expended no effort on such parts as I was able to understand by myself.

This is the knowledge I was able to acquire and learn without any elementary schooling, and without a knowledge of the Latin tongue. [...]

Among the extraordinary, though quite natural circumstances of my life, the first and most unusual is that I was born in this century in which the whole world became known; whereas the ancients were familiar with but little more than a third part of it. [...] The conviction grows that, as a result of these discoveries, the fine arts will be neglected and but lightly esteemed, and certainties will be exchanged for uncertainties. These things may be true sometime or other, but meanwhile we shall rejoice as in a flower-filled meadow. For what is more amazing than pyrotechnics? Or than the fiery bolts man has invented, so much more destructive than the lightning of the gods?

Nor of thee, O Great Compass, will I be silent, for thou dost guide us over boundless seas, through gloomy nights, through the wild storms seafarers dread, and through the pathless wilderness.

The fourth marvel is the invention of the typographic art, a work of man's hands, and the discovery of his wit—a rival, forsooth, of the wonders wrought by divine intelligence. What lack we yet unless it be the taking of Heaven by storm! Oh, the madness of men to give heed to vanity rather than the fundamental things of life! Oh, what arrogant poverty of intellectual humility not to be moved to wonder! [...]

After I had been prompted by a recurring dream, I wrote my books *On Subtlety* which I enlarged after the first printing, and then, having included even more new material I had them sent to press for the third edition. From this I turned to the *Book of the Great Art* which I composed while Giovanni da Colla was contending with me, and also Tartaglia, from whom I had received the first chapter. He preferred to find in me a rival, and a better man at that, than an associate bound to him by gratitude, and of all men most devotedly his friend.

While I was traveling down the Loire River and had nothing to do I wrote my *Commentaries on Ptolemy* in the year 1552. I added *On Proportions* and the *Regula Aliza* to the *Ars Magna* in 1568 and published the work. Following this I also corrected and had rewritten the *Treatise on Arithmetic*, two volumes entitled *New Geometry*, and a work on *Music*, but this last I corrected and revised six years later, that is in 1574. The several volumes *On a Variety of Matters* I published in 1558; they consisted of the remaining material of the work *On Subtlety*, which I was not able to put in order nor whip into shape on account of the host of cares which beset me. For my sons were little inclined to obedience, or conformity; my financial returns were next to nothing; the demands of reading gave me no respite; what with domestic arrangements, the practice of my profession throughout the city, prescriptions, letters and so many other distractions, there was not time to breathe, much less any opportunity for revision of my writings.

8.A5 Rafael Bombelli

(*a*) *From the Preface to* Algebra

I have decided first to consider the majority of the authors who up to now have written about [algebra], so that I can fill in what they have missed out. They are very many, and among them certainly Mohammed ibn Musa, an Arab, is believed to be the first, and there is a little book of his, but of very small value. I believe that the word 'algebra' came from him, because some years ago, Brother Luca [Pacioli] of Borgo San Sepolcro of the Minorite order, having set himself the task of writing on this science, as much in Latin as in Italian, said that the word 'algebra' was Arabic, and means in our language 'position', and that the science came from the Arabs. Many who have written after him have believed and said likewise, but in recent years, a Greek work on this discipline has been discovered in the Library of our Lord in the Vatican, composed by a certain Diophantus of Alexandria, a Greek author, who lived at the time of Antoninus Pius. When it had been shown to me by Master Antonio Maria Pazzi, from Reggio, public lecturer in mathematics in Rome, and we had judged him an author very intelligent in numbers (although he did not treat irrational numbers, but in him alone is evident a perfect order of working), he and I, in order to enrich the world with such a complete work, set ourselves to translate it, and we have translated five books (of the seven). The remainder we have not been able to finish because of the troubles that have occurred to us, the one and the other. In this work we have found that he cites the Indian authors many times, and thus I have been made aware that this discipline belonged to the Indians before the Arabs. After him (after a long interval of time) Leonardo Pisano wrote, in Latin, but after him there was no one who said anything useful until the above-mentioned Brother Luca, who indeed (even though he was a careless writer and for that reason made some mistakes) was nonetheless the first to shed light on this science, even though there are some who set themselves up and take to themselves all the credit, wickedly finding fault with the few errors of the Brother and saying nothing of his good work. Then in our times both Italians and foreigners have written, such as Oroncio, Scribelio, and Boglione, all French; young Stifel, a German, and a certain Spagnard, who wrote copiously in his own language, but truly there has been no one who has opened the secret of algebra other than Cardano of Milan in his *Ars Magna*, where he spoke a lot about this science, but in the telling was obscure. He treated it likewise in certain of his broadsheets, which with Ludovico Ferrari, from our Bologna, he wrote against Niccolò Tartaglia of Brescia. In these there are extremely fine and ingenious problems in this science, but with such little modesty from Tartaglia (who of his nature was so accustomed to speaking evil, that when he thought he had made an honoured sage of himself, he had actually been talking slander) that he offended almost all honourable minds, seeing clearly how he talked nonsense about both Cardano and Ferrari, geniuses of these our times more divine than human. There are others still who have written on the subject, but if I wanted to name them all I would have enough to do. But because their books have been of little benefit I will say nothing of them, and only (as before) say that having thus seen how much of the subject has been dealt with by the above-mentioned authors, I too then in my turn have put together the present work for the common good, dividing it into three books.

(*b*) *On imaginary numbers*

I have found another kind of cube root of a compound expression very different from the other kinds, which results from the case of the cube equal to so many [things] and a number, when the cube of the third of the things is greater than the square of the half of the number, as will happen in this case. This kind of square root has different arithmetical operations from the others and a different denomination, because when the cube of the third of the things is greater than the square of the half of the number, the excess can be called neither plus nor minus. But I shall call it 'plus of minus' when it is to be added, and when it is to be subtracted I shall call it 'minus of minus', and this operation is most necessary, more than the other cube roots of compound expressions, in regard to the cases of powers of powers [fourth powers], accompanied by cubes, or things, or both together, which are the equations where this sort of root appears much more than the other. This will seem to many to be more artificial than real, and I held the same opinion myself, until I found the geometrical demonstration (as will be shown in the proof of the above case on a plane surface). And first I shall deal with multiplication, setting down the rule of plus and minus.

Plus times plus of minus, makes plus of minus.	$[(+1)(i) = +i]$
Minus times plus of minus, makes minus of minus.	$[(-1)(i) = -i]$
Plus times minus of minus, makes minus of minus.	$[(+1)(-i) = -i]$
Minus times minus of minus, makes plus of minus.	$[(-1)(-i) = +i]$
Plus of minus times plus of minus, makes minus.	$[(+i)(+i) = -1]$
Plus of minus times minus of minus, makes plus.	$[(+i)(-i) = +1]$
Minus of minus times minus of minus, makes minus.	$[(-i)(-i) = -1]$

8.B Renaissance Editors

The two most outstanding Renaissance editors of ancient texts were Federigo Commandino (1509–1575) (see 8.B1–8.B2) and Francesco Maurolico (1494–1575) (see 8.B3). A large number of ancient texts were tracked down, patiently read and put before the growing public in handsome editions by these two, who commanded both the classical and the mathematical skills necessary for such work. The letter (8.B1) from their English colleague, the sage John Dee (for more on whom see 9.B), gives a good impression of the high importance they attached to their task, and the thoughtfulness with which they worked. The identity of the 'most famous' of all Arabs, Machomet Bagdedine, to whom John Dee refers, is uncertain (and has become more so with time—an interesting comment on the results of research in the history of mathematics). Perhaps the best statement of Maurolico's views was the one he addressed to his patron Juan de Vega in 1556. It lends itself not to extract but to summary, such as Paul Rose provides in 8.B3.

8.B1 John Dee to Federigo Commandino

Having now for many years set my selfe chiefly (my most learned Frederick) how to preserve from utter ruine, the most famous moniments of our Ancestors (such as I could) in all the more curious or elegant kinds of Philosophy; least that either so worthy men should be robbed of their due renown, or we longer want the most abundant benefit of such books. I say that I so bestowing my pains among other most ancient writings of Philosophers, did at length happen upon this small Book, written indeed, in a very blinde ilfavoured character, and also by reason of its age, hard to read. But that I might the better see, I used helps to my sight: and by often study and exercise therein, I got the knack of reading it. Whereupon being hereby better persuaded of the excellencie and worth of the Book, I earnestly wished that the same might forthwith be communicated to the Society of Philosophers. But while I was pondering this in my minde, I found none more worthy than your self (my Commandine) in this our age, to enjoy these our Labours; who have also your own selfe revived certain most excellent Works of Archimedes and Ptolemee almost lost, and to have brought them forth into publick view in a most magnificent dresse. Therefore this little Book, as a perpetual pledge, even of the affection wherewith I ever embrace you, I commit to you and your trust; and do earnestly beseech you, that you will not suffer this our common labour to go forth into the World destitute of that adornment wherewith you are wont to send abroad others. Yea, I surely hope (if I well know you and your endeavours) that you will some time or other so enrich this subject, as that you will neither permit it to rest in a Pentagonal form; nor suffer the Solids themselves long to want the like sections by Plains. Verily, if you would but a little put forward these things, they will by themselves go on to the other kinds of Superficies. But that they may be applyed to Solids they will require your sound knowledge, and more than ordinary pains in the Mathematicks. As for the Authors name, I would have you understand, that to the very old copy from whence I writ it, the name Machomet Bagdedine was put [. . .] rather [. . .] this may be deemed a book of our Euclide, all whose Books were long since turned out of the Greeke into the Syriack and Arabic Tongues. Whereupon, It being found some time or other to want its Title with the Arabians or Syrians, was easily attributed by the transcribers to that most famous Mathematician among them, Machomet: Which I am able to prove by many ancient testimonies, to be often done in Moniments of the Ancients; and certain friends of mine doe know (that I may bring one of many) that we have by this means restored one small Book, incomparable in occult and mysticall Philosophie, of that most ancient and excellent Phylosopher Anaxagoras, everywhere now through many ages enobled with the name of Aristotle to Anaxagoras himselfe, and that by the most sure proofs. Yea further, we could not yet perceive so great acutenesse of any Machomet in the Mathematicks, from their moniments which we enjoy, as everywhere appears in these Problems. Moreover, that Euclid himselfe wrote one Book [. . .] that is to say, of Divisions, as may be evidenced from Proclus's Commentaries upon his first of *Elements*: and we know none other extant under this title, nor can we finde any, which for the excellence of its treatment, may more rightfully or worthily be ascribed to Euclid. Finally, I remember that in a certain very ancient piece of Geometry, I have read a place cited out of this little Book in expresse words, even as from a most certain work of Euclid. Therefore we have thus briefly declared our opinions for the present, which we desire may carry with them so much weight as they have truth in them. [. . .]

But whatsoever that Book of Euclid was concerning Divisions, certainly this is such an one as may be both very profitable for the studies of many, and also bring much honour and renown to every most noble ancient Mathematician [. . .] I will now direct my discourse to you, who are herein to be greatly intreated by me, that you would with all possible diligence advance your most weighty and usefull labours, which you did yesterday most courteously shew to me in your Study. For so you will make the fairest way to perpetuate your fame, who have in so few years, put forth many Books, so happily, so neatly, and so many of your own: who alone in our time dost adorn the most excellent Princes of Mathematicians, Archimedes, Apollonius and Ptolomeus, with their due lustre. So you will restore a new and wonderfull livelynesse to

Mathematicall Learning much decayed [...] Now the purpose of Travell calls me away, lest I should be put to undergo a greater trouble of this scorching Season round about us, before I can shelter my selfe in the Roman Shades. Farewell, therefore, the honour and renown of Mathematicians; farewel my most courteous Commandine.

8.B2 Bernardino Baldi on Commandino

Having suffered such blows, and assured, by experience, of the uncertainty of fortune, he [Commandino] returned to his native land with a mind to find calm and the expectation of quiet and virtuous leisure: which he thought he could do, since he had already seen both his daughters married and had put his family affairs in order. So he expected to complete many works he had already begun; but then Francesco Maria [della Rovere], the son of our Duke Guido Ubaldo, a young man of noble spirit, knowing how well such sciences are suited to one who plays the part of a ruler and turns his hand to the arts of war, would not permit Federico to live cloistered within the walls of his family home; but, making him the most honourable proposals, wished, as his father had done before, to call him [Commandino] to his service: when he had entered [the ruler's household], in reading Euclid's *Elements* to the prince he gave him much satisfaction by his interpretation [of the text]. Hence, the prince, thinking it unjust to deprive the world of the things he had heard in private, persuaded Federico to translate the work and write a commentary on it. So Commandino, desirous of acting for the common good, and in part obeying his lord's commands, putting aside the translations of Pappus, of Theodosius, of Heron, of Autolycus and of Aristarchus, gave himself wholeheartedly to translating and writing a commentary on Euclid: nor did he labour in vain; for in a short time they [the translation and commentary] were printed at Pesaro and it became clear how great a benefit it was to the world that he had turned his hand to them. Of which, among many others, Christopher Clavius bears witness, affirming that, of all those who up until our time had devoted themselves to studies of this author's [Euclid's] *Elements*, Commandino alone had restored them to their original clarity, in accordance with the meaning and the tradition of ancient commentators; and had not fallen into those errors he had discovered and noted in [the works of] many others. Federico adorned this book with most perceptive glosses and commentaries, partly original to himself and partly taken from the best books by these [earlier] scholars. Similarly, he added some prefaces so eloquent that anyone who reads them must recognize how outstanding he [Commandino] was in more highly-regarded arts, particularly in the other branches of philosophy. So they were printed; and since the works were written at the instance of Francesco Maria, and on account of his persuasion, so Commandino dedicated and consecrated them to his name.

There was then at the prince's court Alderano Cibo, the son of the Marchese of Massa; a young man of most lively spirit and enamoured of the beauty of these studies [i.e. mathematics]. Federico, observing his inclination, by way of encouraging his desire dedicated to him [his translation of] the work of Aristarchus of Samos, a most ancient and renowned Greek writer, in which [work] the author demonstrates the sizes and distances of the Sun and the Moon, and together with [it in the same volume] the scholia of Pappus, both parts having commentaries by Commandino himself.

At about the same time, an English nobleman from London, called John Dee, a most learned man, a student of the antique and a lover of these studies [i.e. mathematics], being on his way to Rome, attracted by Federico's reputation came to Urbino, with the sole purpose of making his acquaintance and paying him a visit; when he was received there by him with great courtesy he found the reality much greater than what he had learned by repute. The said John brought with him a little book that had not been printed, inscribed with the name Macometto Bagdedino, which deals with the division of areas; and which he [Dee] had rescued, with much patience, from the shadows of antiquity and the barbarities of the Arabs. Hence, wishing it to see the light, he judged it the best opportunity of carrying out his intention to leave the book in the hands of Commandino: which he did, accompanying it with a most elaborate [i.e. complimentary] letter; in which, among many other things, he wrote these words: 'I found none more worthy than your self (my Commandino) in this our age, to enjoy these our Labours; who have also your own self revived certain most excellent Works of Archimedes and Ptolemeus almost lost, and have brought them forth into publick view'. This short work extended only so far as the division of the pentagon; so Federico, not content, as he himself says, that this author's treatise should close with no more than the division of that [polygon], reduced to two very short problems all that the author had gathered into many, and demonstrated the manner in which one might divide all other areas *ad infinitum*: which done, judging the book worthy of a prince, he printed it, and dedicated it to the name of Francesco Maria, in the year 1570. Later this little book was translated into the vernacular [i.e. Italian], and published, by Fulvio Vianni de' Malatesti da Montefiore a young man of very noble spirit. [. . .] The heirs saw to it that the printing of Heron's work was completed, and it was dedicated to the cardinal of Urbino, since that had been Commandino's intention while he lived. The works which death had forced him to leave unfinished, or which he had not been able to publish, were these: the six books of Pappus's Collections; all the other works of Euclid; two books of Theodosius, one on habitations and the other on days and nights; two books by Autolycus, on rising and setting and another on the moving sphere; the work of Leonardo Pisano and that of Fra Luca [Pacioli], which he intended to correct and modernize.

8.B3 Paul Rose on Francesco Maurolico

Regarding the writings of the Greek mathematicians as the necessary foundation of mathematics, Maurolico has taken special care to acquire their most accurate texts. However, he will retain whatever has been well rendered or added by translators. Moreover, if he himself finds a more correct or direct method of proving something, he will not hesitate to apply it to the texts, with the permission of other scholars. (As we have already seen in the cases of Apollonius and Archimedes, Maurolico is not a purist for the classical text, but rather favours any reconstruction of the text which will promote the advancement of mathematics.)

After remarking that mathematics reaches its conclusions without controversy, and that, if it is neglected, then all certitude vanishes, Maurolico goes on to classify the various mathematical sciences in a similar pattern to that adopted in the earlier *Prologi*. He then begins a survey of the main authors in each division.

Rejecting the common notion that Euclid the mathematician is to be identified with the philosopher, Euclid of Megara, Maurolico praises the author of the *Elements* for having been the first to place in order the beginnings of mathematics. But the versions of Campanus and Zamberti are both unsatisfactory. Campanus changed the definitions and added much for his own purposes. Of the four conclusions at the end of Book IV, only two are acceptable. Nonetheless, Zamberti was unjust in his attacks on Campanus, for he himself was ensnared by the faults of the Greek text, being ignorant of the mathematics involved. But if such learned men as Campanus have been misled, it is little wonder that so many defective Greek and Latin manuscripts can be found. Whether the fault of scribe or translator, errors abound and especially in regard to the numbers and diagrams in the text. A little sleepiness and old errors are propagated, new ones introduced. Wherefore Maurolico has taken care to collate all the manuscripts of the *Elements* which he can lay his hands on. Nor has he hesitated to introduce shorter proofs of his own to the text. At the same time, he has felt free to omit proofs in other writers, such as those in Jordanus's *Arithmetica* which redound with *fastidium* rather than *speculatio*. (Again, texts are not sacred, but rather mathematics.) [...]

The next writer on geometry to be surveyed is Apollonius whose writings are so essential to an understanding of Archimedes and of the theory of sundials. Memmo had translated Apollonius, but working from a frail manuscript, had omitted much that he could not read. Proceeding word by word, he had lost the sense frequently. The same was to be said of Giorgio Valla's translations of Greek words of geometry, so rare are those who understand well this part of philosophy. Memmo's edition, Maurolico writes, was also marred by numerous printing errors, particularly in the diagrams, so that not even Apollonius himself could have corrected with ease this work. Maurolico has, however, devoted much effort to restoring the *Conics*. He leaves his readers to judge how well he has succeeded in adding more direct proofs and necessary propositions and figures.

The same has been done for Archimedes. Curiously, however, Maurolico states that he has omitted the *Arenarius*, since he believes it to be deserving of neglect. He also mentions some printed copies of Archimedes which are extant. Although he does not specify the editors, he seems to be referring to the 1544 Basle edition since he appears to be unaware of *On Floating Bodies* which had appeared in the 1543 Tartaglia edition.

Coming to optics, wherein are combined geometrical and physical arguments, Maurolico draws attention to the works of Euclid and Ptolemy, and cites also with approval Pecham's *Perspectiva* and Roger Bacon. Witelo's huge work, on the other hand, contains more of *fastidium* than of *speculatio*.

In astronomy Maurolico holds the *Almagest* to be the most illustrious work. But, naturally, he approves of the excellent ordering of the proofs done by Peurbach and Regiomontanus, as well as some improvements made by Geber. Other useful adjuncts to the *Almagest* are Alfragan, Thabit, Albategnius, Campanus, Proclus, Euclid's *Phenomena*, Autolycus, Theodosius, and Peurbach's *Theoricae Novae*. (Hence, there is no anti-Arabic or antimedieval bias *per se* in Maurolico's mind.) But Sacrobosco and Gerard of Cremona (Sabbioneta) Maurolico finds completely unsuitable for learning astronomy, as Regiomontanus has already forcibly shown. Regiomontanus has also indicated that the *Tabulae Alfonsinae*, as excellent as they may be, stand greatly in need of correction.

Having surveyed the basic authorities in mathematics, Maurolico examines the contributions of the moderns to the literature. He starts by saying that most modern books are to be passed over in silence, inasmuch as they are full of obscure curiosities, rather than of agreeable utility. Thus, he deems Luca Pacioli's *Summa* to be prolix and poor as Cardano has already pointed out. Cardano is one mathematician that Maurolico exempts from his condemnation of the moderns, even though his writings sometimes give more labour than delight. In general, says Maurolico, the moderns would do much better if, instead of pouring out immaturely their own works, they were to examine carefully the books of the ancients.

8.C Algebra at the Turn of the Century

The themes we have been following in this chapter come together interestingly in the work of both Simon Stevin (1548–1620) (8.C1) and François Viète (1540–1603) (8.C2). In these brief extracts they can be seen struggling to reconcile the new algebraic methods with their recently refurbished inheritance from the Greeks. Both men seem to have wanted to claim a deep root in the classics while at the same time making new advances, usefully and most obviously in the field of notation. Both were fascinated by the work of Diophantus because it seemed to suggest that, after all, algebra was Greek, and also because they found the *Arithmetica* so interesting. With such as them, and Thomas Harriot (9.D) in England, the Renaissance in mathematics concluded its journey from Italy to the emerging powers of northern Europe.

8.C1 Simon Stevin

(*a*) *On a notation for powers*

When the ancients had perceived the merit of the progression of numbers such as 2, 4, 8, 16, 32 etc. or 3, 9, 27, 81, 243 etc., the first multiplied by itself yields as a product the second of the order, then the second once multiplied by the first yields the third of the order, and the third multiplied by the first yields the fourth of the order, and so forth; for 2 by itself makes 4, the same by 2 makes 8, and this same by 2 makes 16 etc. Similarly 3 by itself makes 9, the same by 3 makes 27, and this same one by 3 makes 81 etc. They saw that it was necessary to give proper names to these numbers, by which they could be clearly indicated, in calling the first in the order *Prime*, which we shall indicate by ①, and the second in the order they named *Second*, which we shall denote by ②, and so forth, by example:

① 2. ② 4. ③ 8. ④ 16. ⑤ 32. ⑥ 64 etc.

Further:

① 3. ② 9. ③ 27. ④ 81. ⑤ 243. ⑥ 729 etc.

Then seeing that this first number was like the side of a square, and the second its square, and the third the cube of the first etc., and that this similarity of numbers and of magnitudes revealed several secrets of numbers, they also attributed to them the names of magnitudes, calling the first *Side*, the second *Square*, the third *Cube*, and consequently all these numbers in general *Geometric Numbers*.

(*b*) *On the term 'equation'*

Since appropriate names are of great importance in the sciences, and indeed most especially difficult, it is not at all wrong that we should choose them rather than unsuitable ones: what will here be the invention of the fourth proportional of quantities, is commonly called equation; we call it thus because it is more fitting to scholarship. For since there are always three terms given, to which one seeks a fourth proportional (as will appear in its place) why should not this indeed be called invention of fourth proportional, as in all others? As for what will be said to me, that it is also an equation, indeed I concede it and not only with respect to algebraic quantities, but in all others. For example, 6 aunes cost 4 pounds, how much are 3 aunes? One finds its fourth proportional 2 pounds, which is also an equation, because one equates to the value of 3 aunes the value of 2 pounds; however it is not at all in usage to call it an equation; but one calls it (and with good reason, since it is more apt) invention of the fourth proportional: and for the same reason we call it thus here, so that the great mystery of the proportion of quantities, with all the causes of things, may be simpler and clearer than ever formerly. For this word equation evokes apprentices, which is some singular matter, that however is common to everyday arithmetic, for we seek to give a fourth proportional to three terms. But that which they call equation by no means consists of an equality of absolute quantities, but of an equality of their values. Thus this proportion consists in the value of quantities, as the similar is everyday, with common bodily objects. For example one ox is worth two sheep with eight pounds, therefore one sheep is worth four pounds, which are four proportional terms, not according to their quantity, with respect to which the product of the extremes is by no means equal to the product of the means, but according to their value, for since a 16 pound value of ox, to a 16 pound value of 2 sheep with 8 pounds, thus a 4 pound value of 1 sheep to a 4 pound value of the fourth term, are proportional terms, which we shall put in order to make it clear, in this way:

1 ox	2 sheep + 8 pounds	1 sheep	4 pounds
16 pounds	16 pounds	4 pounds	4 pounds

The same is meant also by quantities: for when we say, 1 ② is equal, or worth 2 ① + 8, therefore 1 ① is worth 4, these are four proportional terms; but with respect do their values, of which the product of the extremes is only equal to the product of the means. Their presentation in the same way as the preceding is thus:

1 ②	2 ① + 8	1 ①	4
16	16	4	4

(c) *On the history of cubic equations*

The inventors of these rules of three about quantities have been:

Mohammed ben Musa	① equal to ⓪. Their derivatives. ② equal to ①⓪.
Some unknown author	Their derivatives.
Some other unknown author	③ equal to ①⓪. ③ equal to ②⓪.
Lodovico Ferrari	④ equal to ③②①⓪.

As for Diophantus, it appears that in his time the inventions of Mohammed had only become known, as can be gathered from his first six books. It is true that he solves marvellous questions, as we shall state in its place, but he usually carries out his operations with an admirable subtlety, thus that the first and second term become ① equal to ⓪, or their derivations, and never, but rarely, to ② equal to ①⓪.

The derivatives of ② equal to ①⓪, invented by the aforesaid first unknown author, are described by Luca Pacioli.

As for the inventions of the second unknown author, Cardano declares in writing that he found them; but they were not in one iota divulged. Also Scipione del Ferro of Bologna found the first kind, which is of ③ equal to ①⓪. Niccolò Tartaglia of Brescia pursued this, but on the occasion of some dispute he had in this matter, with Antonio Maria Fior, a Venetian, and a disciple of the aforesaid Scipione, in which he discovered something by which Niccolò conjectured and found it. Which after many pleas from Cardano, he told him of it, and it was a ground-work for Cardano, by means of which he came, at the end of several geometrical demonstrations to ③ equal to ②①, ⓪, and their derivatives, which he set down in a book entitled *Ars Magna*.

But the invention of Ludovico Ferrari has recently been published in Italian by Rafael Bombelli, great arithmetician of our time.

(d) *On decimal fractions*

The argument

The dime has two parts, that is Definitions & Operations. By the first definition is declared what *dime* is, by the second, third, and fourth what *commencement*, *prime*, *second*, etc. and *dime numbers* are. The operation is declared by four propositions: the addition, subtraction, multiplication, and division of dime numbers. The order whereof may be successively represented by this Table.

The dime has two parts	Definitions, as what is	dime, commencement, prime, second, etc. dime number.
	Operations or practice of the	addition, subtraction, multiplication, division.

And to the end the premises may the better be explained, [. . .] adjoined [is material on] the use of the dime in many things by certain examples, and also definitions and operations, to teach such as do not already know the use and practice of numeration, and the four principles of common arithmetic in whole numbers, namely addition, subtraction, multiplication, & division, together with the Golden Rule, sufficient to instruct the most ignorant in the usual practice of this art of dime or decimal arithmetic.

The first definition

Dime is a kind of arithmetic, invented by the tenth progression, consisting in characters of ciphers, whereby a certain number is described and by which also all accounts which happen in human affairs are dispatched by whole numbers, without fractions or broken numbers.

Let the certain number be one thousand one hundred and eleven, described by the characters of ciphers thus 1111, in which it appears that each 1 is the 10th part of his precedent character 1; likewise in 2378 each unity of 8 is the tenth of each unity of 7, and so of all the others. But because it it convenient that the things whereof we would speak have names, and that this manner of computation is found by the consideration of such tenth or dime progression, that is that it consists therein entirely, as shall hereafter appear, we call this treatise fitly by the name of *dime*, whereby all accounts happening in the affairs of man may be wrought and effected without fractions or broken numbers, as hereafter appears.

The second definition

Every number propounded is called commencement, *whose sign is thus* ⓪.

By example, a certain number is propounded of three hundred sixty-four: we call them the 364 *commencements*, described thus 364 ⓪, and so of all other like.

The third definition

And each tenth part of the unity of the commencement *we call the* prime, *whose sign is thus* ①, *and each tenth part of the unity of the prime we call the* second, *whose sign is* ②, *and so of the other: each tenth part of the unity of the precedent sign, always in order one further.*

As 3 ① 7 ② 5 ③ 9 ④, that is to say: 3*primes*, 7 *seconds*, 5 *thirds*, 9 *fourths*, and so proceeding infinitely, but to speak of their value, you may note that according to this definition the said numbers are $\frac{3}{10}$, $\frac{7}{100}$, $\frac{5}{1000}$, $\frac{9}{10000}$, together $\frac{3759}{10000}$, and likewise 8 ⓪ 9 ① 3 ② 7 ③ are worth 8, $\frac{9}{10}$, $\frac{3}{100}$, $\frac{7}{1000}$, together $8\frac{937}{1000}$, and so of other like. Also you may understand that in this *dime* we use no fractions, and that the multitude of signs, except ⓪, never exceed 9, as for example not 7 ① 12 ②, but in their place 8 ① 2 ②, for they value as much.

The fourth definition

The numbers of the second and third definitions beforegoing are generally called dime numbers.

8.C2 François Viète

(a) Introduction to the analytical art

I owe my life, or if there is anything dearer to me than life, entirely to you; and now, O divine Mélusine, I owe to you especially the whole study of Mathematics, to which I have been spurred on both by your love for it and by the very great skill you have in that art, nay more, the comprehensive knowledge in all sciences which can never be sufficiently admired in one of your sex who is of so royal and noble a race. O princess most to be revered, those things which are new are wont in the beginning to be set forth rudely and formlessly and must then be polished and perfected in succeeding centuries. Behold, the art which I present is new, but in truth so old, so spoiled and defiled by the barbarians, that I considered it necessary, in order to introduce an entirely new form into it, to think out and publish a new vocabulary, having gotten rid of all its pseudo-technical terms lest it should retain its filth and continue to stink in the old way, but since till now ears have been little accustomed to them, it will be hardly avoidable that many will be offended and frightened away at the very threshold. And yet underneath the Algebra or Almucabala which they lauded and called 'the great art', all Mathematicians recognized that incomparable gold lay hidden, though they used to find very little. There were those who vowed hecatombs and made sacrifices to the Muses and Apollo if any one would solve some one problem or other of the order of such problems as we solve freely by the score, since our art is the surest finder of all things mathematical. Now that the thing has come to pass, *will they be bound by their vows*? However, it would be right for me not now to commend my own wares, but in all moderation yours and those which have been acquired and renewed through your beneficence, to bear witness to my desire that whatever glory is due on account of the felicity of your rule should not be snatched away. For it is not the same in mathematics as in other studies, that everyone's opinion is free and free his judgement. Here things are done by rule and effort, and neither the persuasions of rhetoricians nor the pleadings of lawyers are of use. The metal which I bring forth yields the kind of gold which they wanted for so long a time.

(b) On the meaning and components of analysis and on matters useful to zetetics

There is a certain way of searching for the truth in mathematics that Plato is said first to have discovered. Theon called it analysis, which he defined as assuming that which is sought as if it were admitted [and working] through the consequences [of that assumption] to what is admittedly true, as opposed to synthesis, which is assuming what is [already] admitted [and working] through the consequences [of that assumption] to arrive at and to understand that which is sought.

Although the ancients propounded only [two kinds of] analysis, zetetics and poristics, to which the definition of Theon best applies, I have added a third, which may be called rhetics or exegetics. It is properly zetetics by which one sets up an equation or proportion between a term that is to be found and the given terms, poristics by which the truth of a stated theorem is tested by means of an equation or proportion, and exegetics by which the value of the unknown term in a given equation or proportion is determined. Therefore the whole analytic art, assuming this three-fold function for itself, may be called the science of correct discovery in mathematics.

Now whatever pertains to zetetics begins, in accordance with the art of logic, with syllogisms and enthymemes the premises of which are those fundamental rules with which equations and proportions are established. These are derived from axioms and from theorems created by analysis itself. Zetetics, however, has its own method of proceeding. It no longer limits its reasoning to numbers, a shortcoming of the old analysts, but works with a newly discovered symbolic logistic which is far more fruitful and powerful than numerical logistic for comparing magnitudes with one another. It rests on the law of homogeneous terms first and then sets up, as it were, a formal series or scale of terms ascending or descending proportionally from class to class in keeping with their nature and, [by this series,] designates and distinguishes the grades and natures of terms used in comparisons.

(c) *On the function of rhetics*

An equation having been set up with a magnitude that is to be found, rhetics or exegetics, which is the remaining part of analysis and pertains most especially to the general ordering of the art (as it logically should since the other two [are more concerned with] patterns than with rules), performs its function. It does so both with numbers, if the problem to be solved concerns a term that is to be extracted numerically, and with lengths, surfaces or bodies, if it is a matter of exhibiting a magnitude itself. In the latter case the analyst turns geometer by executing a true construction after having worked out a solution that is analogous to the true. In the former he becomes an arithmetician, solving numerically whatever powers, either pure or affected, are exhibited. He brings forth examples of his art, either arithmetic or geometric, in accordance with the terms of the equation that he has found or of the proportion properly derived from it.

It is true that not every geometric construction is elegant, for each particular problem has its own refinements. It is also true that [that construction] is preferred to any other that makes clear not the structure of a work from an equation but the equation from the structure; thus the structure demonstrates itself. So a skillful geometer, although thoroughly versed in analysis, conceals the fact and, while thinking about the accomplishment of his work, sheds light on and explains his problem synthetically. Then, as an aid to the arithmeticians, he sets out and demonstrates his theorem with the equation or proportion he sees in it.

9 Mathematical Sciences in Tudor and Stuart England

9.A Robert Record

Robert Record (c. 1510–1558), author of the first sequence of mathematical textbooks in English, was one of the best authors of such texts. Record's arithmetic text *The Ground of Artes* (1543) remained in print, in successive editions, for over a century and a half, and few textbook writers have had so engaging a style. All but his geometry text *The Pathway to Knowledge* (1551) are written in dialogue form, with periodic excursions into rhythmic, rhymed prose (e.g. the end of 9.A1 (b)) and throughout show great pedagogical thoughtfulness. It is perhaps not surprising that a writer who emphasized reason over authority (9.A3) and chose examples to bring out political comment (9.A1 (c)) should have ended his days in prison; his arrest forms the poignant dramatic conclusion (see 9.A4 (c)) to his last book, the algebra text *The Whetstone of Witte* (1557).

9.A1 *The Ground of Artes*

(a) From The Preface unto the Kynges Majestie (1552 edition)

Seeing Arithmetik is so many waies needful unto the fyrst plantyng of a common welthe, it must needes be as much required to the preservation of it also. [...] What shall wee saye for the statutes of this Realme, whyche bee the onlye stay of good ordre in maner nowe? As touchyng the measuring of ground by lengthe and bredthe there is a good and an ancient statut made by arte of Arithmetik, and now it shalbe to lyttle use, if by the same arte it be not practised and tried: for the assize of bred and drinke, the two most common and most necessarye thinges for sustentation of man, there was a goodly ordinaunce in the Lawe made, whiche by Ignoraunce hath so growen out of knowledge and use, that fewe men doo understande it: and therefore the statute bookes wonderfully corrupted, and the commons cruelly oppressed, not withstandinge some menne have written, that it is to doubtful a matter to execute those assises by those statutes, by reason they depend of the standerd of the coyne, whiche is muche chaunged frome the state of that tyme, whanne those statutes were made. [...]

Agayn there is an auncient order for assise of fyre wood, and coales, which was renewed not many yeres paste, and nowe howe Avarice and Ignorance doth canvas that statute, it is to pitifull to talke of, and more myserable to feele.

Farthermore for the statute of coynage, and the standerde therof, if the people understoode rightely the statute, they shuld not nor wolde not (as they often do) gather an excuse for their foly therby. But as I sayde, these statutes by wysedome and good knowledge of Arithmetike were made, and by the same must they be continued. And let Ignorance no more meddle with the use of them, then it dyd wyth the making of them. Oh in how miserable case is that realme, where the ministers and interpreters of the lawes, are destitute of all good sciences, whiche be the keyes of the lawes? Howe can they eyther make good lawes, or mayntayne them, that lacke that true knowledg, whereby to judge them?

(*b*) *The profit of Arithmetick*

MASTER: [...] Wherefore as without numbring a man can do almost nothing, so with the help of it, you may attain to all things.

SCHOLAR: Yes, sir, why then it were best to learn the Art of numbring, first of all other learning, and then a man need learn no more, if all other come with it.

MASTER: Nay not so: but if it be first learned, then shall a man be able (I mean) to learn, perceive, and attain to other Sciences; which without it he could never get.

SCHOLAR: I perceive by your former words, that Astronomy and Geometry depend much on the help of numbring: but that other Sciences, as Musick, Physick, Law, Grammer, and such like, have any help of Arithmetick, I perceive not.

MASTER: I may perceive your great Clerk-linesse by the ordering of your Sciences: but I will let that passe now, because it toucheth not the matter that I intend, and I will shew you how Arithmetick doth profit in all these somewhat grosly, according to your small understanding, omitting other reasons more substantiall.

First (as you reckon them) Musick hath not onely great help of Arithmetick, but is made, and hath his perfectnesse of it: for all Musick standeth by number and proportion: And in Physick, beside the calculation of criticall dayes, with other things, which I omit, how can any man judge the pulse rightly, that is ignorant of the proportion of numbers?

And so for the Law, it is plain, that the man that is ignorant of Arithmetick, is neither meet to be a Judge, neither an Advocate, nor yet a Proctor. For how can hee well understand another mans cause, appertaining to distribution of goods, or other debts, or of summes of money, if he be ignorant of Arithmetick? This oftentimes causeth right to bee hindered, when the Judge either delighteth not to hear of a matter that hee perceiveth not, or cannot judge for lack of understanding: this commeth by ignorance of Arithmetick.

Now, as for Grammer, me thinketh you would not doubt in what it needeth number, sith you have learned that Nouns of all sorts, Pronouns, Verbs, and Participles are distinct diversly by numbers: besides the variety of Nouns of Numbers, and Adverbs. And if you take away number from Grammer, then is all the quantity of Syllables lost. And many other ways doth number help Grammer. Whereby were all kindes of Meeters found and made? was it not by number?

But how needfull Arithmetick is to all parts of Philosophy, they may soon see, that do read either Aristotle, Plato, or any other Philosophers writings. For all their examples almost, and their probations, depend of Arithmetick. It is the saying of Aristotle, that hee that is ignorant of Arithmetick, is meet for no Science. And Plato his Master wrote a little sentence over his Schoolhouse door, Let none enter in hither (quoth he) that is ignorant of Geometry. Seeing hee would have all his Scholars expert in Geometry, much rather he would the same in Arithmetick, without which Geometry cannot stand.

And how needfull Arithmetick is to Divinity, it appeareth, seeing so many Doctors gather so great mysteries out of number, and so much do write of it. And if I should go about to write all the commodities of Arithmetick in civill acts, as in governance of Common-weales in time of peace, and in due provision & order of Armies, in time of war, for numbering of the Host, summing of their wages, provision of victuals, viewing of Artillery, with other Armour; beside the cunningest point of all, for casting of ground, for encamping of men, with such other like: And how many wayes also Arithmetick is conducible for all private Weales, of Lords and all Possessioners, of Merchants, and all other occupiers, and generally for all estates of men, besides Auditors, Treasurers, Receivers, Stewards, Bailiffes, and such like, whose Offices without Arithmetick are nothing: If I should (I say) particularly repeat all such commodities of the noble Science of Arithmetick, it were enough to make a very great book.

SCHOLAR: No, no sir, you shall not need: For I doubt not, but this, that you have said, were enough to perswade any man to think this Art to be right excellent and good, and so necessary for man, that (as I think now) so much as man lacketh of it, so much hee lacketh of his sense and wit.

MASTER: What, are you so farre changed since, by hearing these few commodities in generall: by likelihood you would be farre changed if you knew all the particular Commodities.

SCHOLAR: I beseech you Sir, reserve those Commodities that rest yet behinde unto their place more convenient: and if yee will bee so good as to utter at this time this excellent treasure, so that I may be somewhat inriched thereby, if ever I shall be able, I will requite your pain.

MASTER: I am very glad of your request, and will do it speedily, sith that to learn it you bee so ready.

SCHOLAR: And I to your authority my wit do subdue; whatsoever you say, I take it for true.

MASTER: That is too much; and meet for no man to bee beleeved in all things, without shewing of reason. Though I might of my Scholar some credence require, yet except I shew reason, I do it not desire.

(c) *A discussion of sheep*

SCHOLAR: There is supposed a lawe made that (for furtheryng of tyllage) every man that doth kepe shepe, shall for every 10 shepe eare and sowe one acre of grounde, and for his allowance in sheepe pasture there is appointed for every 4 shepe one acre of pasture. Nowe is there a ryche shepemaister whyche hath 7000 akers of grounde, and woulde gladlye kepe as manye sheepe as he myght by that statute. I demaunde howe many shepe shall he kepe?

MASTER: Answere to the question your selfe.

SCHOLAR: Fyrste I suppose he maye kepe 500 shepe, and for them he shall have in pasture, after the rate of 4 shepe to an acre, 125 akers, and in earable grounde 50 acres that is 175 in all, but this errour is to litell by 6825. Therefore I gesse agayn that he maye kepe 1000 shepe, that is in pasture 250 akers, and in tyllage 100 akers, which maketh 350, that is to lytle by 6650. These bothe erroures with theyr positions I sette downe as you see, and multiply in crosse 6825 by 1000, and it maketh 6825000. Then

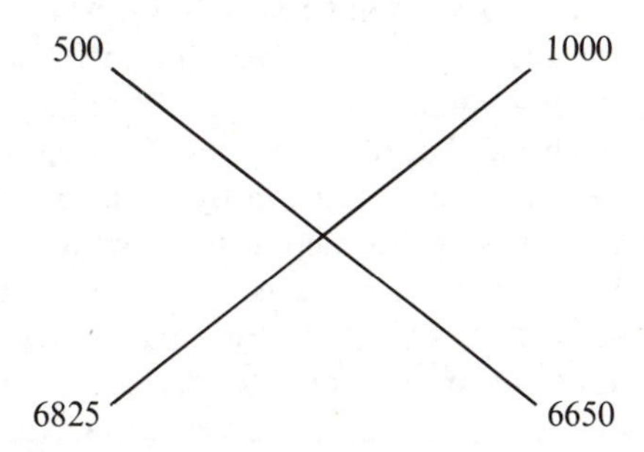

I multiply 6650 by 500, and it doth amount to 3325000, which sum I do subtract out of the fyrst, & there remaineth 3500000, as the dividende. Also I doo subtract the lesser errour out of the greater, and so remayneth 175, by which I divide the said dividende, and the quotient will be 20000, so that I see that by this rate he that hathe 7000 acres of ground may keepe 20000 shepe: & therby I conjecture that many menne may kepe so many shepe: for many men (as the common talke is) have so many acres of grounde.

MASTER: That talk is not likely, for so muche grounde is in compas above $48\frac{3}{4}$ miles. But leave this talke & returne to your questions, least your pointing be scarse wel taken.

SCHOLAR: In dede I doo remembre that the Egyptians did grudg so much against shepards, till at length thei smarted for it, & yet they were but smal shepemaisters to some men that be now, and the shepe are waxen so fierce nowe and so myghtye, that none can withstande them but the lyon.

MASTER: I perceave you talke as you hear some other, but to the work of your question [and the Master goes on to explain a simpler way of doing the calculation].

9.A2 *The Pathway to Knowledge*

(*a*) *The commodities of Geometry*

Sith Merchauntes by shippes great riches do winne,
  I may with good righte at their seate beginne.
The Shippes on the sea with Saile and with Ore,
  Were first founde, and styll made, by Geometries lore.

Their Compas, their Carde, their Pulleis, their Ankers,
  Were founde by the skill of witty Geometers.
To sette forth the Capstocke, and eche other part,
  Wold make a great showe of Geometries arte.
Carpenters, Carvers, Joiners and Masons,
  Painters and Limners with suche occupations,
Broderers, Goldesmithes, if they be cunning,
  Must yelde to Geometrye thankes for their learning.
The Carte and the Plowe, who doth them well marke,
  Are made by good Geometrye. And so in the warke
Of Tailers and Shoomakers, in all shapes and fashion,
  The woorke is not praised, if it wante proportion.
So weavers by Geometrye hade their foundacion,
  Their Loome is a frame of straunge imaginacion.
The wheele that doth spinne, the stone that doth grind,
  The Myll that is driven by water or winde,
Are workes of Geometrye straunge in their trade,
  Fewe could them devise, if they were unmade.
And all that is wrought by waight or by measure,
  Without proofe of Geometry can never be sure.
Clockes that be made the times to devide,
  The wittiest invencion that ever was spied,
Now that they are common they are not regarded,
  The artes man contemned, the woorke unrewarded.
But if they were scarse, and one for a shewe,
  Made by Geometrye, then should men know,
That never was arte so wonderfull witty,
  So needfull to man, as is good Geometry.

(*b*) *Bisecting an angle*

To divide an angle of right lines into ii equal partes.

First open your compasse as largely as you can, so that it do not excede the length of the shortest line that incloseth the angle. Then set one foote of the compasse in the verye point of the angle and with the other fote draw a compassed arch from the one lyne of the angle to the other, that arch shall you devide in halfe, and then draw a line from the angle to the middle of the arch, and so the angle is divided into ii equall partes. *Example.* Let the triangle be A, B, C, then set I one foot of the compasse in B, and with the other I draw the arch D, E, which I part into ii equall parts in F, and then draw a line from B, to F, & so I have mine intent. [See figure opposite.]

9.A3 *The Castle of Knowledge*

No man can worthely praise Ptolemye, his travell being so great, his diligence so exacte in observations, and conference with all nations, and all ages, and his reasonable examination of all opinions, with demonstrable confirmation of his owne assertion, yet

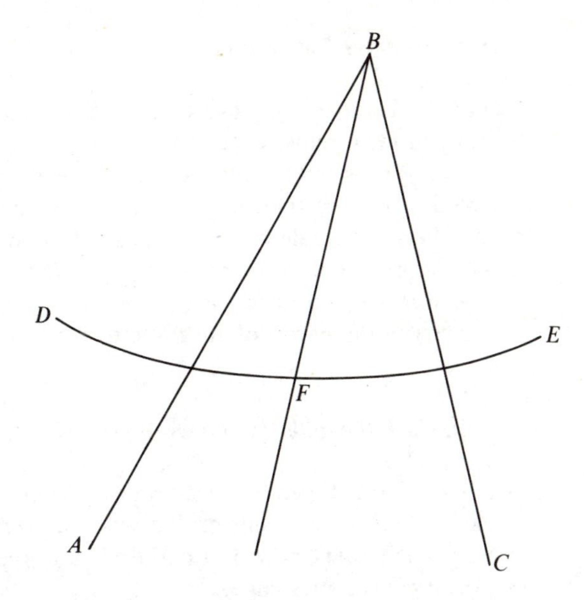

muste you and all men take heed, that both in him and in al mennes workes, you be not abused by their autoritye, but evermore attend to their reasons, and examine them well, ever regarding more what is saide, and how it is proved, then who saieth it: for autoritie often times deceaveth many menne.

9.A4 *The Whetstone of Witte*

(*a*) *Preface*

I maie truely saie, that if any imperfection bee in nomber, it is bicause that nomber, can scarsely nomber, the commodities of it self. For the moare that any experte man, doeth weigh in his mynde the benifites of it, the more of them shall he see to remain behinde. And so shall he well perceiue, that as nomber is infinite, so are the commodities of it as infinite. And if any thyng doe or maie exceade the whole worlde, it is nomber, whiche so farre surmounteth the measure of the worlde, that if there were infinite worldes, it would at the full cōprehend them all. This nomber also hath other prerogatiues, aboue all naturalle thynges, for neither is there certaintie in any thyng without it, nother good agremente where it wanteth. Whereof no man can doubte, that hath been accustomed in the Bookes of Plato, Aristotell, and other aunciente Philosophers, where he shall see, how thei searche all secrete knowledge and hid misteries, by the aide of nomber. For not onely the constitution of the whole worlde, dooe thei referre to nomber, but also the composition of manne, yea and the verie substaunce of the soule. Of whiche thei professe to knowe no moare, then thei cā by the benifite of nomber attaine. Furthermore, for knowledge and certaintie in any other thynge, that mannes witte can reche vnto, there is noe possibilitie without nomber. It is confessed emongeste all men, that knowe what learnyng meaneth, that besides the Mathematicalle artes, there is noe vnfallible knoweledge, excepte it bee borowed of them.

(*b*) *On numbers and fractions*

SCHOLAR: Yet one thyng more I must demaunde of you, Why Euclide, and the other learned men, refuse to accompte fractions amongest nombers.

MASTER: Bicause all nombers doe consiste of a multitude of unities: and every proper fraction is lesse than a unitie, and therefore can not fractions exactly be called nombers: but maie bee called rather fractions of nombers.

SCHOLAR: In deede now that I doe waie the mater more exactly, it appereth that a fraction is not properly a nomber, but a connexion and conference of nombers, declaryng the partes of an unitie.

(*c*) *In which the author is taken to prison*

SCHOLAR: Now I perceive that in Addition, and Subtraction of Surdes, the last nombers that did result of that woorke, were *universalle rootes.*

MASTER: You saie truthe. But harke what meaneth that hastie knockyng at the doore?

SCHOLAR: It is a messenger.

MASTER: What is the message? tel me in mine eare Yea sir is that the mater? Then is there noe remedie, but that I must neglect all studies, and teaching, for to withstande those daungers. My fortune is not so good, to have quiete tyme to teache.

SCHOLAR: But my fortune and my fellowes, is moche worse, that your unquietness, so hindereth our knowledge. I praie God amende it.

MASTER: I am inforced to make an eande of this mater: But yet will I promise you, that whiche you shall chalenge of me, when you see me at better laiser: That I will teache you the whole arte of *universalle rootes.* And the extraction of rootes in all *Square Surdes*: with the demonstration of theim, and all the former workes.

If I mighte have been quietly permitted, to reste but a litle while longer, I had determined not to have ceased, till I had ended all these thinges at large. But now farewell. And applie your studie diligently in this that you have learned. And if I maie gette any quietnesse reasonable, I will not forget to performe my promise with an augmentation.

SCHOLAR: My harte is so oppressed by this sodaine unquietnesse, that I can not expresse my grief. But I will praie, with all theim that love honeste knowledge, that God of his mercie, will sone ende your troubles, and graunte you soche reste, as your travell doeth merite. And al that love learnyng: saie thereto. Amen.

MASTER: Amen, and Amen.

9.B John Dee

The life of John Dee (1527–1608) followed the not untypical Elizabethan trajectory of attaining high influence and acclaim, yet dying in poverty and disgrace. His activities were so varied that historians still adopt widely differing perspectives, as the extracts in 9.B3 indicate. His most influential work, and one of the most significant Renaissance accounts of the value and scope of the mathematical sciences, was his 'very fruitfull Praeface' (9.B1) to Henry Billingsley's translation of Euclid's *Elements*. The latter was itself a very scholarly edition, the first such to appear in English. It is interesting to compare a proposition from this work (9.B2) with the equivalent in Record's earlier and more craftsman-oriented geometry text (9.A2(b)) (and with Sir Thomas Heath's modern translation in 3.B3(b)).

9.B1 Mathematicall Praeface to Henry Billingsley's Euclid

(a) The Elements of Geometrie of the most auncient Philosopher Euclide of Megara. Faithfully (now first) translated into the Englishe toung, by H. Billingsley, Citizen of London. Whereunto are annexed certaine Scholies, Annotations, and Inventions, of the best Mathematiciens, both of time past, and in this our age.

With a very fruitfull Praeface made by M. J. Dee, specifying the chiefe Mathematicall Sciences, what they are, and wherunto commodious: where, also, are disclosed certaine new Secrets Mathematicall and Mechanicall, untill these our daies, greatly missed.

(b) To the unfained lovers of truthe, and constant Studentes of Noble Sciences, John Dee of London, hartily wisheth grace from heaven, and most prosperous successe in all their honest attemptes and exercises.

[...] Of *Mathematicall* thinges, are two principall kindes: namely, *Number*, and *Magnitude*. [...] Neither *Number*, nor *Magnitude*, have any Materialitie. First, we will consider of *Number*, and of the Science *Mathematicall*, to it appropriate, called *Arithmetike*: and afterward of *Magnitude*, and his Science, called *Geometrie*. But that name contenteth me not: whereof a word or two hereafter shall be sayd. How Immateriall and free from all matter, *Number* is, who doth not perceave? yea, who doth not wonderfully wonder at it? [...] And therefore the great & godly Philosopher *Anitius Boetius*, sayd: [...] All thinges (*which from the very first originall being of thinges, have bene framed and made*) *do appeare to be Formed by the reason of Numbers. For this was the principall example or patterne in the minde of the Creator.*

(c) But farder understand, that vulgar Practisers, have Numbers, otherwise, in sundry Considerations: and extend their name farder, than to Numbers, whose least part is an *Unit*. For the common Logist, Reckenmaster, or Arithmeticien, in hys using of Numbers: of an Unit, imagineth lesse partes: and calleth them *Fractions*. As of an *Unit*, he maketh an halfe, and thus noteth it, $\frac{1}{2}$. And so of other, (infinitely diverse) partes of an *Unit*. Yea and farder, hath, *Fractions of Fractions. &c.* And, forasmuch, as, *Addition*, *Substraction*, *Multiplication*, *Division* and *Extraction of Rotes*, are the chief, and sufficient partes of *Arithmetike*: which is, the *Science that demonstrateth the properties, of Numbers, and all operations, in numbers to be performed*: How often, therfore, these five operations, do, for the most part, of their execution, differre from the five operations of like generall property and name, in our Whole numbers practisable, So often, (for a more distinct doctrine) we, vulgarly account and name it, an other kynde of *Arithmetike*. [...] Practise hath led *Numbers* farder, and hath framed them, to take upon them, the shew of *Magnitudes* propertie: Which is

Incommensurabilitie and *Irrationalitie*. (For in pure *Arithmetike*, an *Unit*, is the common measure of all Numbers.) And, here, Numbers are become, as Lynes, Playnes and Solides: some tymes *Rationall*, some tymes *Irrationall*. And have propre and peculier characters, (as $\sqrt{ʒ}$. $\sqrt{ℭ}$. and so of other. Which is to signifie *Rote Square, Rote Cubik: and so forth:*) & propre and peculier fashions in the five principall partes: Wherfore the practiser, estemeth this, a diverse *Arithmetike* from the other. [. . .] And yet (beside all this) Consider: the infinite desire of knowledge, and incredible power of mans Search and Capacitye: how, they, joyntly have waded farder (by mixtyng of speculation and practise) and have found out, and atteyned to the very chief perfection (almost) of *Numbers* Practicall use. Which thing, is well to be perceived in that great Arithmeticall Arte of *Æquation*: commonly called the *Rule of Coss*. or *Algebra*. The Latines termed it, *Regulam Rei & Census*, that is, the *Rule of the thyng and his value*. With an apt name: comprehendyng the first and last pointes of the worke.

(d) This science of *Magnitude*, his properties, conditions, and appertenances: commonly, now is, and from the beginnyng, hath of all Philosophers, ben called *Geometrie*. But, veryly, with a name to base and scant, for a Science of such dignitie and amplenes. And, perchaunce, that name, by common and secret consent, of all wisemen, hitherto hath ben suffred to remayne: that it might carry with it a perpetuall memorye, of the first and notablest benefite, by that Science, to common people shewed: Which was, when Boundes and meres of land and ground were lost, and confounded (as in *Egypt*, yearely, with the overflowyng of *Nilus*, the greatest and longest river in the world) or, that ground bequeathed, were to be assigned: or, ground sold, were to be layd out: or (when disorder prevailed) that Commons were distributed into severalties. For, where, upon these & such like occasions, Some by ignorance, some by negligence, Some by fraude, and some by violence, did wrongfully limite, measure, encroach, or challenge (by pretence of just content, and measure) those landes and groundes: great losse, disquietnes, murder, and warre did (full oft) ensue: Till, by Gods mercy, and mans Industrie, The perfect Science of Lines, Plaines, and Solides (like a divine Justicier) gave unto every man, his owne. [. . .] But, well you may perceive by *Euclides Elementes*, that more ample is our Science, then to measure Plaines: and nothyng lesse therin is tought (of purpose) then how to measure Land. An other name, therfore, must nedes be had, for our Mathematicall Science of Magnitudes: which regardeth neither clod, nor turff: neither hill, nor dale: neither earth nor heaven: but is absolute *Megethologia*: not creping on ground, and dasseling the eye, with pole perche, rod or lyne: but liftyng the hart above the heavens, by invisible lines, and immortall beames meteth with the reflexions, of the light incomprehensible: and so procureth Joye, and perfection unspeakable. Of which true use of our *Megethica*, or *Megethologia*, *Divine Plato* seemed to have good taste, and judgement: and (by the name of *Geometrie*) so noted it: and warned his Scholers therof: as, in hys seventh *Dialog*, of the Common wealth, may evidently be sene.

(e) No man, therfore, can doute, but toward the atteyning of knowledge incomparable, and Heavenly Wisedom: Mathematicall Speculations, both of Numbers and Magnitudes: are meanes, aydes, and guides: ready, certaine, and necessary. From henceforth, in this my Praeface, will I frame my talke, to *Plato* his fugitive Scholers: or, rather, to such, who well can, (and also wil,) use their utward senses, to the glory of God, the benefite of their Countrey, and their owne secret contentation, or honest preferment, on this earthly Scaffold. To them, I will orderly recite, describe & declare a great Number of Artes, from our two Mathematicall fountaines, derived into the fieldes of *Nature*. Wherby, such Sedes, and Rotes, as lye depe hyd in the ground of *Nature*, are refreshed, quickened, and provoked to grow, shote up, floure, and give frute, infinite, and incredible.

(f) The Arte of Navigation, demonstrateth how, by the shortest good way, by the aptest Direction, & in the shortest time, a sufficient Ship, betwene any two places (in passage Navigable) assigned: may be conducted: and in all stormes, & naturall disturbances chauncying, how, to use the best possible meanes, whereby to recover the place first assigned. What nede, the *Master Pilote*, hath of other Artes, here before recited, it is easie to know: as, of *Hydrographie*, *Astronomie*, *Astrologie*, and *Horometrie*. Presupposing continually, the common Base, and foundacion of all: namely *Arithmetike* and *Geometrie*. So that, he be hable to understand, and Judge his own necessary Instrumentes, and furniture Necessary: Whether they be perfectly made or no: and also can, (if nede be) make them, hym selfe. As Quadrantes, The Astronomers Ryng, The Astronomers Staffe, The Astrolabe Universall. An Hydrographicall Globe. Charts Hydrographicall, true, (not with parallell Meridians). The Common Sea Compas: The Compas of Variacion: The Proportionall, and Paradoxall Compasses (of me Invented, for our two Moscovy Master Pilotes, at the request of the Company) Clockes with spryng: houre, halfe houre, and three houre Sandglasses: & sundry other Instrumentes: And also, be hable, on Globe, or Playne to describe the Paradoxall Compasse: and duely to use the same, to all maner of purposes, whereto it was invented. And also, be hable to Calculate the Planetes places for all Tymes.

Moreover, with Sonne Mone or Sterre (or without) be hable to define the Longitude & Latitude of the place, which he is in: So that, the Longitude & Latitude of the place, from which he sayled, be given: or by him, be knowne. whereto, appertayneth expert meanes, to be certified ever, of the Ships way. &c. [...] Sufficiently, for my present purpose, it doth appeare, by the premisses, how *Mathematicall*, the *Arte* of *Navigation*, is: and how it nedeth and also useth other *Mathematicall Artes*: And now, if I would go about to speake of the manifold Commodities, commyng to this Land, and others, by Shypps and *Navigation*, you might thinke, that I catch at occasions, to use many wordes, where no nede is.

Yet, this one thyng may I, (justly) say. In *Navigation*, none ought to have greater care, to be skillfull, then our English Pylotes. And perchaunce, Some, would more attempt: And other Some, more willingly would be aydyng, if they wist certainely, What Priviledge, God had endued this Iland with, by reason of Situation, most commodious for Navigation, to Places most Famous & Riche.

(g) And where I was willed, somewhat to alledge, why, in our vulgare Speche, this part of the Principall Science of *Geometrie*, called *Euclides Geometricall Elementes*, is published, to your handlyng: being unlatined people, and not Universitie Scholers: Verily, I thinke it nedelesse.

For, the Honour, and Estimation of the Universities, and Graduates, is, hereby, nothing diminished. Seing, from, and by their Nurse Children, you receave all this Benefite: how great soever it be.

Neither are their Studies, hereby, any whit hindred. No more, then the Italian *Universities*, as *Academia Bononiensis*, *Ferrariensis*, *Florentina*, *Mediolanensis*, *Patavina*, *Papiensis*, *Perusina*, *Pisana*, *Romana*, *Senensis*, or any one of them, finde them selves, any deale, disgraced, or their Studies any thing hindred, by *Frater Lucas de Burgo* [Pacioli], or by *Nicolaus Tartalea*, who in vulgar Italian language, have published, not onely *Euclides Geometrie*, but of *Archimedes* somewhat: and in Arithmetike and Practicall Geometrie, very large volumes, all in their vulgar speche. [...]

Besides this, how many a Common Artificer, is there, in these Realmes of England and Ireland, that dealeth with Numbers, Rule, & Cumpasse: Who, with their owne Skill and experience, already had, will be hable (by these good helpes and informations) to finde out, and devise, new workes, straunge Engines, and Instrumentes: for sundry purposes in the Common Wealth? or for private pleasure? and for the better maintayning of their own estate? I will not (therefore) fight against myne owne shadowe. For, no man (I am sure) will open his mouth against this Enterprise. [...] To none (therefore) will I make any *Apologie*, for a vertuous acte doing: and for commending, or setting forth, Profitable Artes to English men, in the English toung. [...]

And that you may the easier perceive, and better remember, the principall pointes, whereof my Preface treateth, I will give you the Groundplatt of my whole discourse, in a Table annexed: from the first to the last, somewhat Methodically contrived.

If hast, hath caused my poore pen, any where, to stumble: You will, (I am sure) in part of recompence, (for my earnest and sincere good will to pleasure you), Consider the rockish huge mountaines, and the perilous unbeaten wayes, which (both night and day, for the while) it hath toyled and labored through, to bryng you this good Newes, and Comfortable profe, of Vertues frute.

So, I Commit you unto God's Mercyfull direction, for the rest: hartely besechyng hym, to prosper your Studyes, and honest Intentes: to his Glory, & the Commodity of our Countrey. *Amen*.

9.B2 Bisecting an angle, from Henry Billingsley's Euclid

To devide a rectiline angle geven, into two equall partes.

Suppose that the rectiline angle geven be *BAC*. It is required to devide the angle *BAC* into two equal partes. In the line *AB* take a point at all adventures, & let the same be *D*. And (by the third proposition) from the lyne *AC* cutte of the line *AE* equall to *AD*. And (by the first peticion) draw a right line from the point *D* to the point *E*. And (by the first proposition) upon the line *DE* describe an equilater triangle and let the same be *DFE*, and (by the first peticion) drawe a right line from the poynte *A* to the point *F*. Then I say that the angle *BAC* is by the line *AF* devided into two equal partes. For, forasmuch as *AD* is equall to *AE*, and *AF* is common to them both: therfore these two *DA* and *AF*, are equall to these two *EA* and *AF*, the one to the other. But (by the first proposition) the base *DF* is equall to the base *EF*: wherefore (by the 8. proposition) the angle *DAF* is equal to the angle *FAE*. Wherefore the rectiline angle geven, namely, *BAC* is devided into two equal partes by the right line *AF*: Which was required to be done.

In this proposition is not taught to devide a right lined angle into mo partes then two: albeit to devide an angle, so it be a right angle, into three partes, it is not hard. And it is taught of Vitellio in his first boke of *Perspective*, the 28. proposition. For to devide an acute angle into three equal partes, is (as saith Proclus) impossible: unless it be by

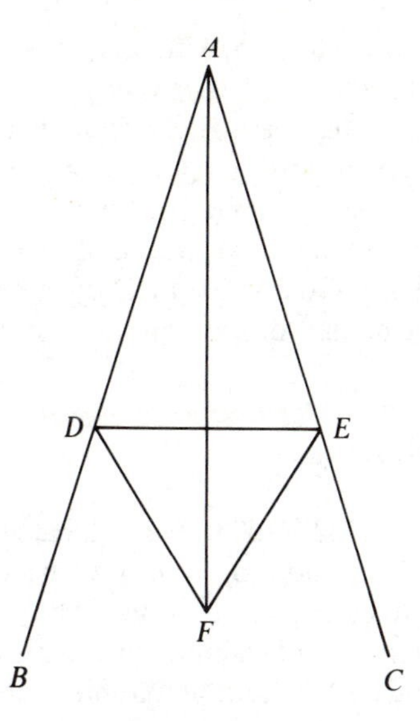

the helpe of other lines which are of a mixt nature. Which thing Nicomedes did by such lines which are called *Concoides linea*, who first serched out the invention, nature, & properties of such lines. And others did it by other meanes, as by the helpe of quadrant lines invented by Hippias & Nicomedes. Others by *Helices* or Spiral lines invented of Archimedes. But these are things of much difficulty and hardnes, and not here to be intreated of.

9.B3 Views of John Dee

(*a*) *John Aubrey*

Hee had a very faire cleare rosie complexion; a long beard as white as milke; he was tall and slender; a very handsome man. His Picture in a wooden cutt is at the end of Billingsley's *Euclid*. He wore a Gowne like an Artist's gowne, with hanging sleeves, and a slitt; a mighty good man he was.

Old Goodwife Faldo (a Natif of Mortlak in Surrey) did know Dr Dee, and told me that he did entertain the Polonian Ambassador at his howse in Mortlak, and dyed not long after; and that he shewed the Eclipse with a darke Roome to the said Ambassador. She beleeves that he was eightie years old when he dyed. She sayd, he kept a great many Stilles goeing. That he layd the storme. That the Children dreaded him because he was accounted a Conjurer. He recovered the Basket of Cloathes stollen, when she and his daughter (both Girles) were negligent: she knew this.

He used to distill Egge-shells, and 'twas from hence that Ben. Johnson had his hint of the *Alkimist*, whom he meant.

He was sent Ambassador for Queen Elizabeth (Goody Faldo thinkes) into Poland. The Emperour of Muscovia, upon report of the great learning of the Mathematician, invited him to Mosco, with offer of two thousand pound a yeare, and from Prince Boris one thousand markes; to have his Provision from the Emperor's Table, to be honourably received, and accounted as one of the chief men in the Land. All of which Dee accepted not.

Arthur Dee, his sonne, a Physitian at Norwych and intimate friend of Sir Thomas Browne, M.D., told Dr Bathurst that (being but a Boy) he used to play at Quoits with the Plates of Gold made by Projection in the Garret of Dr Dee's Lodgings in Prague and that he had more than once seen the Philosopher's Stone.

(*b*) *E. G. R. Taylor*

Dee, John (1527–1608), was educated at St John's College, Cambridge, and made a foundation Fellow of Trinity College (1546), which suggests that his *mathematical genius* early attracted attention. The practical applications of astronomy and geometry were foremost in his mind, whether for casting nativities or advancing navigation, for reforming the calendar or mapping subterranean mines. His visits to Louvain, Brussels and Paris (1547–50) made him acquainted with the foremost Continental mathematicians, among whom the designing, description and use of instruments in the service of geodesy, cartography, dialling, gunnery, etc., was taken for granted as part of their work. On Dee's return to London he was introduced by John Cheke to Sir William Cecil and so to Court circles, where preparations for maritime expansion were being set on foot. For the next 30 years he gave advice and instruction to pilots and navigators and collected a great mathematical and scientific library, besides a variety of mathematical instruments. The impact of his *Preface* to the English Euclid (translated by Henry Billingsley) upon young men of the middle class, sons of tradesmen and craftsmen, was very great, setting out as it did the ways in which geometry could advance technique and foster inventions. Peter Ramus, whom he met in Paris, named him publicly as one who would ornament a chair in Mathematics, should the Queen found such chairs at the two universities, as Peter eloquently but fruitlessly advised her to do in his *Scholae Mathematicae*. Writing in 1574 to the Earl of Leicester, Dr Richard Forster, his astrologer-physician, described Dee as a very Atlas bearing upon his shoulders the sole weight of the revival in England of the mathematical arts, and in 1590 Tycho Brahe sent greetings to him and to Thomas Digges as 'most noble, excellent and learned' mathematicians.

(*c*) *Frances Yates*

Dee belonged emphatically to the Renaissance Hermetic tradition, brought up to date with new developments, and which he further expanded in original and important directions. Dee was, in his own right, a brilliant mathematician, and he related his study of number to the three worlds of the Cabalists. In the lower elemental world he studied number as technology and applied science and his *Preface* to Euclid provided a brilliant survey of the mathematical arts in general. In the celestial world, his study of number was related to astrology and alchemy, and in his *Monas Hieroglyphica* he believed that he had discovered a formula for a combined cabalist, alchemical, and mathematical science which would enable its possessor to move up and down the scale of being from the lowest to the highest spheres. And in the supercelestial sphere, Dee believed that he had found the secret of conjuring angels by numerical computations in the cabalist tradition. Dee as 'Rosicrucian' is thus a figure typical of the late Renaissance magus who combined 'Magia, Cabala, and Alchymia' to achieve a world-view in which advancing science was strangely mingled with angelology.

9.C The Value of Mathematical Sciences

9.C1 Roger Ascham (1570)

Some wits, moderate enough by nature, be many times marred by overmuch study and use of some sciences, namely, Music, Arithmetic, and Geometry. These sciences, as they sharpen men's wits overmuch, so they change men's manners oversore, if they be not moderately mingled, and wisely applied to some good use of life. Mark all mathematical heads, which be only and wholly bent to those sciences, how solitary they be themselves, how unfit to live with others, and how unapt to serve in the world.

9.C2 William Kempe (1592)

Take away arithmetic, ye take away the merchant's eye, whereby he seeth his direction in buying and selling; ye take away the goldsmith's discretion, whereby he mixeth his metals in due quantities; ye take away the captain's dexteritie, whereby he embattaileth his army in convenient order; finally ye take from all sorts of men, the faculty of executing their functions aright. Arithmetic then teacheth unto us matters in divinity, judgeth civil causes uprightly, cureth diseases, searcheth out the nature of things created, singeth sweetly, buyeth, selleth, maketh accompts, weigheth metals and worketh them, skirmisheth with the enemy, goeth on warfare, and setteth her hand almost to every good work, so profitable is she to mankind.

9.C3 Gabriel Harvey (1593)

He that remembreth Humfrey Cole, a Mathematicall Mechanician, Matthew Baker a Ship Wright, John Shute an Architect, Robert Norman a Navigatour, William Bourne a Gunner, John Hester a Chimist, or any like cunning, and subtile Empirique, (Cole, Baker, Shute, Norman, Bourne, Hester will be remembered, when greater Clarkes shalbe forgotten) is a prowd man, if he contemne expert artisans, or any sensible industrious practitioner, howsoever unlectured in Schooles, or Unlettered in bookes. [. . .] and what profounde Mathematician, like Digges, Hariot, or Dee, esteemeth not the pregnant Mechanician? Let every man in his degree enjoy his due: and let the brave engineer, fine Daedalist, skilfull Neptunist, marvelous Vulcanist, and every Mercuriall occupationer, that is, every Master of his craft, and every Doctour of his mystery, be respected according to the uttermost extent of his publique service, or private industry.

9.C4 Thomas Hylles (1600)

No State, no age, no man, nor child, but here may wisdome win
For numbers teach the parts of speech, wher children first begin.
And number bears so great a sway even from the most to least
That who in numbring hath no skill, is numbred for a beast:
For what more beastly can be thoght? nay what more blockish than
Then man to want the onely art, which proper is to man,
For many creatures farre excell mankind in many things,
But never none could number yet, save man in whom it springs.
If numbring then be (almost) al, betweene a man and beast,
Come learne o men to number then, which arte is here profest
If martial man thou minde to be, or office do expect.
In court or country where thou dwelst, or if thou do elect,
In Phisicke and Philosophie, or law to spend thy dayes,
Assure thyself without this arte, thou never canst have praise.
I overpasse Astronomie, and Geometrie also,
Cosmographie, Geographie, and many others mo,
And musick with her dulcet tunes, all which without this arte,
Thou never canst attayne unto, nor scarce to any part,
Ne canst thou be an auditor, or make a true survey,
Nor make a common reckoning, if numbers be away.
But if thou willte a merchant be, then make this booke thy Muse,
Wher thou shalt find rules fit for thee, as thou canst wish or chuse.
And having onely handie craft, yet herein mayst thou finde,
Such things as may oft serve thy turne, and much enrich thy minde.
Nay if thou but a shepherd be, it wil thee sore accumber
To do thy duty as thou shouldst without the help of number.
To number all the benefits, that number brings to man,
Would be too long here to rehearse, and more than well I can,
Wherefore to speake one woorde for all, and let the rest alone,
Without this art man is no man, but like a block or stone.

9.C5 Francis Bacon (1603)

The mathematics are either pure or mixed. To the pure mathematics are those sciences belonging which handle quantity determinate, merely severed from any axioms of natural philosophy; and these are two, geometry and arithmetic; the one handling quantity continued, and the other dissevered. Mixed hath for subject some axioms or parts of natural philosophy, and considereth quantity determined, as it is auxiliary and incident unto them. For many parts of nature can neither by invented with sufficient subtilty, nor demonstrated with sufficient perspicuity, nor accommodated unto use with sufficient dexterity, without the aid and intervening of the mathematics; of which sort are perspective, music, astronomy, cosmography, architecture, enginery, and divers others. In the mathematics I can report no deficience, except it be that men do not sufficiently understand the excellent use of the pure mathematics, in that they do remedy and cure many defects in the wit and faculties intellectual. For if the wit be too dull, they sharpen it; if too wandering, they fix it; if too inherent in the sense, they abstract it. So that as tennis is a game of no use in itself, but of great use in respect it maketh a quick eye and a body ready to put itself into all postures; so in the mathematics, that use which is collateral and intervenient is no less worthy than that which is principal and intended. And as for the mixed mathematics, I may only make this prediction, that there cannot fail to be more kinds of them, as nature grows further disclosed.

9.D Thomas Harriot

No mathematical work by Thomas Harriot (1560–1621) was published in his lifetime, and his reputation subsequently has oscillated as his papers were studied or forgotten. It is still difficult to present a fair overview of the work of arguably Britain's 'greatest mathematical scientist before Newton' (9.D7(b)), as the long-urged edition of his papers has not yet materialized. The readings here illustrate the respect in which he was held by contemporaries (9.D1, and also 9.C3), as well as a friend's warning that he should publish his work (9.D4). There was a lengthy controversy in the late seventeenth century, with heated nationalist overtones, stimulated by John Wallis's claim (9.D6) that René Descartes had, in effect, plagiarized Harriot's algebra. Only recently have historians again begun to make serious study of Harriot's work (9.D7).

9.D1 Dedicatory poem by George Chapman

To my admired and soule-loved friend Mayster of all essential and true knowledge, M. Harriots

To you whose depth of soule measures the height,
And all dimensions of all workes of weight,
Reason being ground, structure and ornament,
To all inventions, grave and permanent,
And your cleare eyes the Spheres where *Reason* moves;
[...]
O had your perfect eye Organs to pierce
Into that Chaos whence this stiffled verse
By violence breaks: where Gloweworme like doth shine
In nights of sorrow, this hid soul of mine:
And how her genuine formes struggle for birth,
Under the claws of this fowle Panther earth.
Then under all those formes you should discerne
My love to you, in my desire to learn
Skill and the love of skill do ever kisse.
No band of love so stronge as knowledge is:

Which who is he that may not learne of you,
Whom learning doth with his light's throne endow?

9.D2 A sonnet by Harriot

This somewhat mysterious poem Harriot took great pains over, and there is material relating to it in at least three places in his manuscripts. It seems to concern the sign rule for multiplication.

If more by more must needs make more
Then lesse by more makes lesse of more
And lesse by lesse makes lesse of lesse
If more be more and lesse be lesse

Yet lesse of lesse makes lesse or more
Use which is best keep both in store
If lesse of lesse you will make lesse
Then bate the same from that is lesse.

But if the same you will make more
Then add to it the signe of more.
The rule of more is best to use
Yet for some cause the other choose

So both are one, for both are true
Of this inough and so adeu.

9.D3 Examples of Harriot's algebra

(*a*) *The roots of a cubic equation*

From the original

$$\begin{array}{c|l} a-b & \amalg\ aaa - baa - bca \\ a-c & \qquad\quad - caa - bda \\ a-d & \qquad\quad - daa - cda - bcd. \\ \hline \end{array}$$

It is deduced that I put b or c or d equal to a itself.

This symbolism on the left means $(a-b)(a-c)(a-d)$ in modern notation. Harriot showed that this has no roots apart from b, c and d.

(*b*) *A quartic equation with imaginary roots*

$$\left.\begin{array}{r} b - a \\ c - a \\ df + aa \\ \hline \end{array}\right| \sqsubset\!\sqsupset 0000 \qquad \begin{array}{l} a \sqsubset\!\sqsupset b,\, c \\ a \sqsubset\!\sqsupset \sqrt{-df} \end{array}$$

(*c*) *Another quartic equation with imaginary roots*

$$12 \sqsubset\!\sqsupset + 8a - 13aa + 8aaa - aaaa$$

$$a \sqsubset\!\sqsupset 2,\, 6 \qquad \left.\begin{array}{l} a \sqsubset\!\sqsupset - \sqrt{-1} \\ a \sqsubset\!\sqsupset + \sqrt{-1} \end{array}\right\} \text{noetic roots}$$

This equation has no hypostatic roots other than 2 and 6. [...] But it has two other noetic ones.

(*d*) *A quadratic equation with complex roots*

$$25 \sqsubset\!\sqsupset 6a - aa$$

$$a \sqsubset\!\sqsupset 3 + \sqrt{-16}$$

$$a \sqsubset\!\sqsupset 3 - \sqrt{-16}$$

9.D4 Letter to Harriot from William Lower (1610)

Doe you not here startle, to see every day some of your inventions taken from you; for I remember longe since you told me as much, that the motions of the planets were not perfect circles. So you taught me the curious way to observe weight in Water, and within a while after Ghetaldi comes out with it, in print. a little before Vieta prevented you of the Gharland for the great Invention of Algebra. al these were your deues and manie others that I could mention; and yet too great reservednesse hath robd you of these glories. but although the inventions be greate, the first and last I meane, yet when I survei your storehouse, I see they are the smallest things, and such as in Comparison of manie others are of smal or no value. Onlie let this remember you, that it is possible by too much procrastination to be prevented in the honor of some of your rarest inventions and speculations. Let your Countrie and friends injoye the comforts they would have in the true and great honor you would purchase your selfe by publishing some of your choise works.

9.D5 John Aubrey's brief life of Harriot

Mr Hariot went with Sir Walter Raleigh into Virginia, and haz writt the *Description of Virginia*, which is printed. Dr Pell tells me that he finds amongst his papers, an Alphabet that he had contrived for the American Language, like Devills.

When the Earle of Northumberland and Sir Walter Raleigh were both Prisoners in the Tower, they grew acquainted, and Sir Walter Raleigh recommended Mr Hariot to him, and the Earle setled an Annuity of two hundred pounds a yeare on him for his life, which he enjoyed. But to Hues (who wrote *de Usu Globorum*) and to Mr Warner he gave an Annuity but of sixty pounds per annum. These three were usually called the Earle of Northumberland's three Magi. They had a Table at the Earle's chardge, and the Earle himselfe had them to converse with him, singly or together.

Sir Francis Stuart had heard Mr Hariot say that he had seen nine Cometes, and had predicted Seaven of them, but did not tell them how. 'Tis very strange: *excogitent Astronomi*.

He did not like (or valued not) the old storie of the Creation of the World. He could not beleeve the old position; he would say *ex nihilo nihil fit* [nothing comes of nothing]. But a *nihilum* killed him at last: for in the top of his Nose came a little red speck (exceeding small) which grew bigger and bigger, and at last killed him. I suppose it was that which the Chirurgians call a *noli me tangere* [touch me not].

He made a Philosophical Theologie, wherin he cast-off the Old Testament, and then the New-one would (consequently) have no Foundation. He was a Deist. His Doctrine he taught to Sir Walter Raleigh, Henry Earle of Northumberland, and some others. The Divines of those times look't on his manner of death as a Judgement upon him for nullifying the Scripture.

9.D6 John Wallis on Harriot and Descartes

(*a*) *From Wallis's* Treatise of Algebra (*1685*)

Mr. Harriot was contemporary with Mr. Oughtred, (but elder than he, and died before him,) and left many good things behind him in writing. Of which there is nothing hitherto made publick, but only his *Algebra* or *Analitice*, which was published by Mr. Warner, soon after that of Mr. Oughtred, in the same Year 1631.

He alters the way of Notation, used by Vieta and Oughtred, for another more convenient.

And he hath also made a strange improvement of *Algebra*, by discovering the true construction of *Compound Equations*, and how they be raised by a Multiplication of *Simple Equations*, and may therefore be resolved into such.

By this means he shews the number of Roots (real or imaginary) in every Equation, and the Ingredients of all the Coefficients, in each degree of Affection.

He shews also how to increase or diminish the *Roots*, (yet unknown) by any Excess, or in any Proportion assigned; to destroy some of the intermediate Terms, to turn Negative Roots into Affirmative, or these into those; with many other things very advantagious in the practice of *Algebra*.

And amongst other things, teacheth (thereby) to resolve, not only *Quadraticks*, but all *Cubick* Equations; even those whose roots have, by others, been thought *Inexplicable*, and but *Imaginary*.

In sum, he hath taught (in a manner) all that hath since passed for the *Cartesian* method of Algebra; there being scarce anything of (pure) *Algebra* in Des Cartes which was not before in Harriot; from whom Des Cartes seems to have taken what he hath, (that is purely *Algebra*), but without naming him.

(b) Letter from Samuel Morland to Wallis, January 1688

Some time ago I read in the elegant and truly precious book that you have written on *Algebra*, about Descartes, this philosopher so extolled above all for having arrived at a very perfect system by his own powers, without the aid of others, this Descartes, I say, who has received in geometry very great light from our Oughtred and our Harriot, and has followed their track though he carefully suppressed their names. I stated this in a conversation with a professor in Utrecht (where I reside at present). He requested me to indicate to him the page-numbers in the two authors which justified this accusation. I admitted that I could not do so. The *Géométrie* of Descartes is not sufficiently familiar to me, although with Oughtred I am fairly familiar. I pray you therefore that you will assume this burden. Give me at least those references to passages of the two authors from the comparison of which the plagiarism by Descartes is the most striking.

(c) Wallis's reply, March 1688

I nowhere give him the name of a plagiarist; I would not appear so impolite. However this I say, the major part of his algebra (if not all) is found before him in other authors (notably in our Harriot) whom he does not designate by name. That algebra may be applied to geometry, and that it is in fact so applied, is nothing new. Passing the ancients in silence, we state that this has been done by Vieta, Ghetaldi, Oughtred and others, before Descartes. They have resolved by algebra and specious arithmetic [literal arithmetic] many geometrical problems. ... But the question is not as to application of algebra to geometry (a thing quite old), but of the Cartesian algebra considered by itself.

9.D7 Recent historical accounts

(a) Jon V. Pepper

To what extent Harriot's later work on meridional parts, completed in 1614, is *equivalent* to the integration

$$\int_0^\lambda \sec\theta\,\mathrm{d}\theta = k \ln\tan\left(\tfrac{1}{4}\pi + \tfrac{1}{2}\lambda\right),$$

is a matter of careful interpretation. It certainly gets the correct numerical results, not done before, nor repeated for a century or more, and amounts to a sound evaluation of the logarithmic tangent function; tangents are deliberately and consciously used, but as the work predates Napier's *Descriptio* (1614), it is hardly fair to ask if logarithms were known to be involved as such. Whether Harriot realized that his tables could be used for calculations involving tangents, as Napier's could for sines, is not yet clear. Certainly the exponential nature of the function is clearly visible and recognized. If we grant, as there is not then much difficulty in so doing, the direct relation between the logarithmic tangent and Harriot's obtaining of it *as* a logarithmic tangent, there is still left the question of its origin as the limit of a sum of secants. Curiously, Harriot's original approximate table, derived directly as such a sum, is much closer to this than his later work, which is based on a very profound means of *avoidance* of a limiting sum of secants as such. This is far from saying that limits or even limits of sums are avoided, as the bases of Harriot's methods include several such results. The simplest view to take is that his method was a special device for a particular problem, which led both to an anticipation of some rather specialized logarithm tables, and to the creation of much highly original mathematics.

(*b*) *D. T. Whiteside*

Harriot in fact possessed a depth and variety of technical expertise which gives him good title to have been England's—Britain's—greatest mathematical scientist before Newton. In mathematics itself he was master equally of the classical synthetic methods of the Greek geometers Euclid, Apollonius, Archimedes and Pappus, and of the recent algebraic analysis of Cardano, Bombelli, Stevin and Viète. In optics he departed from Alhazen, Witelo and Della Porta to make first discovery of the sine-law of refraction at an interface, deriving an exact, quantitative theory of the rainbow, and also came to found his physical explanation of such phenomena upon a sophisticated atomic substratum. In mechanics he went some way to developing a viable notion of rectilinear impact, and adapted the measure of uniform deceleration elaborated by such medieval 'calculators' as Heytesbury and Alvarus Thomas correctly to deduce that the ballistic path of a projectile travelling under gravity and a unidirectional resistance effectively proportional to speed is a tilted parabola—this years before Galileo had begun to examine the simple dynamics of unresisted free fall. In astronomy he was as accurate, resourceful and assiduous an observer through his telescopic 'trunks'—even anticipating Galileo in pointing them to the Moon—as he was knowledgeable in conventional Copernican theory and wise to the nuances of Kepler's more radical hypotheses of celestial motion in focal elliptical orbits. He further applied his technical expertise to improving the theory and practice of maritime navigation; determined the specific gravities and optical dispersions of a wide variety of liquids and some solids; and otherwise busied himself with such more conventional occupations of the Renaissance *savant* as making alchemical experiment and creating an improved system of 'secret' writing.

9.E Logarithms

9.E1 John Napier's Preface to *A Description of the Admirable Table of Logarithms*

Seeing there is nothing (right well beloved students in the Mathematics) that is so troublesome to Mathematicall practise, nor that doth more molest and hinder Calculators, than the Multiplications, Divisions, square and cubical Extractions of great numbers, which besides the tedious expence of time, are for the most part subject to many slippery errors. I began therefore to consider in my minde, by what certaine and ready Art I might remove those hindrances. And having thought upon many things to this purpose, I found at length some excellent briefe rules to be treated of (perhaps) hereafter. But amongst all, none more profitable than this, which together with the hard and tedious Multiplications, Divisions, and Extractions of rootes, doth also cast away from the worke it selfe, even the very numbers themselves that are to be multiplied, divided and resolved into rootes, and putteth other numbers in their place, which performe as much as they can do, onely by Addition and Subtraction, Division by two or Division by three; which secret invention, being (as all other good things are) so much the better as it shall be the more common; I thought good heretofore to set forth in Latine for the publique use of Mathematicians. But now some of our Countrymen in this Island well affected to these studies, and the more publique good, procured a most learned Mathematician to translate the same into our vulgar English tongue, who after he had finished it sent the coppy of it to me, to be seene and considered on by myself. I having most willingly and gladly done the same, finde it to be most exact and precisely conformable to my minde and the originall. Therefore it may please you who are inclined to these studies, to receive it from me and the Translator, with as much good will as we recommend it unto you. Fare yee well.

9.E2 Henry Briggs on the early development of logarithms

Wonder not, that these logarithms are different from those which the excellent baron of Marchiston published in his Admirable Canon. For when I explained the doctrine of them to my auditors at Gresham college in London, I remarked that it would be much more convenient, the logarithm of the sine total or radius being 0 (as in the *Canon Mirificus*), if the logarithm of the tenth part of the said radius, namely, of 5° 44′ 21″, were 100000 &c; and concerning this I presently wrote to the author; also, as soon as the season of the year and my public teaching would permit, I went to Edinburgh, where, being kindly received by him, I staid a whole month. But when we began to converse about the alteration of them, he said that he had formerly thought of it, and wished it; but that he chose to publish those that were already done, till such time as his leisure and health would permit him to make others more convenient. And as to the nature of the change, he thought it more expedient that 0 should be made the logarithm of 1: and 100000 &c. the logarithm of radius; which I could not but acknowledge was much better. Therefore, rejecting those which I had before prepared, I proceeded, at his exhortation, to show him the principle of them; and should have been glad to do the same the third summer, if it had pleased God to spare him so long.

9.E3 William Lilly on the meeting of Napier and Briggs

I will acquaint you with one memorable Story related unto me by Mr John Marr, an excellent Mathematician and Geometrician, whom I conceive you remember; he was Servant to King James and Charles the First.

At first, when the Lord Napier, or Marchiston, made publick his *Logarithms*, Mr Briggs, then Reader of the Astronomy lecture at Gresham College in London, was so surprized with Admiration of them, that he could have no Quietness in himself, untill he had seen that noble Person the Lord Marchiston, whose only Invention they were;

he acquaints John Marr herewith, who went into Scotland before Mr Briggs, purposely to be there when these Two so learned Persons should meet: Mr Briggs appoints a certain Day when to meet at Edinborough, but failing thereof; the Lord Napier was doubtful he would not come: It happened one Day as John Marr and the Lord Napier were speaking of Mr Briggs; 'Ah, John (saith Marchiston), Mr Briggs will not now come'; at the very one knocks at the gate; John Marr hasted down, and it proved Mr Briggs, to his great Contentment; he brings Mr Briggs up into my Lord's chamber, where almost one quarter of an hour was spent, each beholding other almost with Admiration before one word was spoke, at last Mr Briggs began.

'My lord, I have undertaken this long Journey purposely to see your Person, and to know by what Engine of Wit or Ingenuity you came first to think of this most excellent Help unto Astronomy, *viz*., the *Logarithms*; but, my Lord, being by you found out, I wonder nobody else found it out before, when now known it is so easy.' He was nobly entertained by the Lord Napier, and every Summer after that, during the Lord's being alive, this venerable Man, Mr Briggs, went purposely into Scotland to visit him.

9.E4 Charles Hutton on Johannes Kepler's construction of logarithms

Kepler here, first of any, treats of logarithms in the true and genuine way of the measures of ratios, or proportions, as he calls them, and that in a very full and scientific manner: and this method of his was afterwards followed and abridged by Mercator, Halley, Cotes, and others, as we shall see in the proper places. Kepler first erects a regular and purely mathematical system of proportions, and the measures of proportions, treated at considerable length in a number of propositions, which are fully and chastely demonstrated by genuine mathematical reasoning, and illustrated by examples in numbers. This part contains and demonstrates both the nature and the principles of the structure of logarithms. And in the second part he applies those principles in the actual construction of his table, which contains only 1000 numbers, and their logarithms, in the form as we before described: and in this part he indicates the various contrivances employed in deducing the logarithms of proportions one from another, after a few of the leading ones had been first formed, by the general and more remote principles. He uses the name *logarithms*, given them by the inventor, being the most proper, as expressing the very nature and essence of those artificial numbers, and containing as it were a definition in the very name of them; but without taking any notice of the inventor, or of the origin of those useful numbers.

As this tract [of 1625] is very curious and important in itself, and is besides very rare and little known, instead of a particular description only, I shall here give a brief translation of both the parts, omitting only the demonstrations of the propositions, and some rather long illustrations of them. The book is dedicated to Philip, landgrave of Hesse, but is without either preface or introduction, and commences immediately with the subject of the first part, which is entitled *The Demonstration of the Structure of Logarithms*; and the contents of it are as follow:

Postulate 1 That all proportions equal among themselves, by whatever variety of couplets of terms they may be denoted, are measured or expressed by the same quantity.

Axiom 1 If there be any number of quantities of the same kind, the proportion of the extremes is understood to be composed of all the proportions of every adjacent couplet of terms, from the first to the last.

Proposition 1 The mean proportional between two terms, divides the proportion of those terms into two equal proportions.

Axiom 2 Of any number of quantities regularly increasing, the means divide the proportion of the extremes into one proportion more than the number of the means.

Postulate 2 That the proportion between any two terms is divisible into any number of parts, until those parts become less than any proposed quantity.

An example of this section is then inserted in a small table, in dividing the proportion which is between 10 and 7 into 1073741824 equal parts, by as many mean proportionals wanting one, namely, by taking the mean proportional between 10 and 7, then the mean between 10 and this mean, and the mean between 10 and the last, and so on for 30 means, or 30 extractions of the square root, the last or 30th of which roots is 99999999966782056900; and the 30th power of 2, which is 1073741824, shows into how many parts the proportion between 60 and 7, or between 1000&c, and 700&c, is divided by 1073741824 means, each of which parts is equal to the proportion between 1000&c, and the 30th mean 999&c, that is, the proportion between 1000&c, and 999&c, is the 1073741824th part of the proportion between 10 and 7. Then by assuming the small difference 00000000033217943100, for the measure of the very small element of the proportion of 10 to 7, or for the measure of the proportion of 1000&c, to 999&c, or for the logarithm of this last term, and multiplying it by 1073741824, the number of parts, the product gives 35667.49481.37222.14400, for the logarithm of the less term 7 or 700&c.

Postulate 3 That the extremely small quantity or element of a proportion may be measured or denoted by any quantity whatever; as, for instance, by the difference of the terms of that element.

Proposition 2 Of three continued proportionals, the difference of the two first has to the difference of the latter two, the same proportion which the first term has to the 2d, or the 2d to the 3d.

[...]

Proposition 20 When four numbers are proportional, the first to the second as the third to the fourth, and the proportions of 1000 to each of the three former are known, there will also be known the proportion of 1000 to the fourth number.

Corollary 1 By this means other chiliads are added to the former.

Corollary 2 Hence arises the method of performing the Rule-of-Three, when 1000 is not one of the terms. Namely, from the sum of the measures of the proportions of 1000 to the second and third, take that of 1000 to the first, and the remainder is the measure of the proportion of 1000 to the fourth term.

Definition The measure of the proportion between 1000 and any less number as before described, and expressed by a number, is set opposite to that less number in the chiliad, and is called its *logarithm*, that is, the number (*arithmos*) indicating the proportion (*logos*) which 1000 bears to that number, to which the logarithm is annexed.

9.E5 John Keil on the use of logarithms

The Mathematicks formerly received considerable Advantages; first, by the Introduction of the Indian Characters, and afterwards by the Invention of Decimal Fractions; yet has it since reaped at least as much from the Invention of Logarithms, as from both the other two. The Use of these, every one knows, is of the greatest Extent, and runs through all Parts of Mathematicks. By their Means it is that Numbers almost infinite, and such as are otherwise impracticable, are managed with Ease and Expedition. By their assistance the Mariner steers his Vessel, the Geometrician investigates the Nature of the higher Curves, the Astronomer determines the Places of the Stars, the Philosopher accounts for other Phenomena of Nature; and lastly, the Usurer computes the Interest of his Money.

9.E6 Edmund Stone on definitions of logarithms

Dr Wallis, in his *History of Algebra*, calls Logarithms the Indexes of the Ratio's of Numbers to one another.—Dr Halley, in the *Philosophical Transactions*, No 216, says, they are the Exponents of the Ratio's of Unity to Numbers.—So also Mr Cotes, in his *Harmonia Mensurarum*, says, they are the Numerical Measures of Ratio's; but all these convey but a very confused Notion of Logarithms. Nay, if what the great Dr Barrow says, in one of his *Mathematical Lectures*, be admitted for Truth, (where he treats of the Nature of a Ratio, and denies it to be any manner of Quantity.) those Gentlemen's Definitions must be either Nonsense, or very near it. [...]

Mr Mercator's *Logarithmotechnia*, set forth An. 1668, was the first public Treatise of the Construction of Logarithms by the Hyperbola, that is, by help of infinite Series', nearly expressing the Asymptotical Hyperbolic Spaces in Number. And after him Dr Gregory, and others did the same thing; but no one has shown how to perform the Business so perspicuous and elegant as Sir Isaac, as will easily appear upon comparing his Way above mentioned with any other extant—Dr Halley too (in *Trans Philos* No 216) has given their Nature and Construction (after a sort) without any mention of the Hyperbola; tho' it is evident, that all the while he had the Hyperbola and the Mensuration of the asymptotical Spaces under consideration; but rather than expressly mention them, because he will not use Geometrical Figures in an Affair purely Arithmetical (as Mr Jones, in his Synopsis, says) he perplexes and strains his Reader's Imagination with several almost unintelligible Ways of Expression; such as an infinite number n of equal Ratio's or Ratiunculae, in a continued Scale of Proportions between the two terms of any ratio, as 1 and $1 + x$ or $\overline{1 + x}^n$. [...] These and several other are the unintelligible, or at least obscure Expressions of the Doctor in his Logarithmical Doctrine; all which are entirely avoided, and the whole seems clear to any Arithmetician and Geometrician of the least Capacity from the Consideration of the Hyperbola, as above mentioned.

Mr Cotes too, at the beginning of his *Harmon. Mensur.* has done this business in imitation of Dr Halley, altho' more short, yet with the same Obscurity: for I appeal to any one, even of his greatest Admirers, if they know what he would be at in his first Problem, viz to find the Measure of a Ratio from the terms of a problem itself, (which should always be done) without having first known something of the matter from other Principles, as the Hyperbola, &c.

9.F William Oughtred

Though for most of his life rector of a country parish near Guildford, William Oughtred (1575–1660) was one of the most influential mathematical teachers of the seventeenth century. Many who were later distinguished mathematicians came to him for lessons, as Aubrey describes (9.F4), and his *Clavis Mathematicae* was a powerful demonstration of the economical expression possible in the new symbolical algebra.

9.F1 Oughtred's *Clavis Mathematicae*

(*a*) *Preface to the new edition*

Many yeares since, I being imployed by the late illustrious Earle of Arundel, to instruct one of his sons in the Mathematicks, penned for his use in Latine, a method of precepts for the more ready attaining, not a superficial notion, but a well-grounded understanding of those mysterious sciences, and of the ancient Writers thereof. This afterward at the request of divers learned and judicious men, I published under the name of *Clavis Mathematicae*. Which Treatise being not written in the usuall syntheticall manner, nor with verbous expressions, but in the inventive way of Analitice, and with symboles or notes of things instead of words, seemed unto many very hard; though indeed it was but their owne diffidence, being scared by the newnesse of the delivery; and not any difficulty in the thing it selfe. For this specious and symbolicall manner, neither racketh the memory with multiplicity of words, nor chargeth the phantasie with comparing and laying things together; but plainly presenteth to the eye the whole course and processe of every operation and argumentation.

Now my scope and intent in the first Edition of that my *Key* was, and in this New Filing, or rather forging of it, is, to reach out to the ingenious lovers of these Sciences, as it were Ariadnes thread, to guide them through the intricate Labyrinth of these studies, and to direct them for the more easie and full understanding of the best and antientest Authors: such as are Euclides, Archimedes, Apollonius Pergaeus that Great Geometer, Diophantus, Ptolomaeus, and the rest: That they may not only learn their propositions, which is the highest point of Art that most students aime at; but also may perceive with what solertiousnesse, by what engines of Aequations, Interpretations, Comparations, Reductions, and Disquisitions, those antient Worthies have beautified, enlarged, and first found out this most excellent Science.

Truly when I was conversant in reading their bookes, and with wonder observed their most witty demonstrations, so skilfully framed out of principles, as one would little expect or thinke, but laid together with divine Artifice: I was even amazed, whence possibly any power of imagination should be able to sustaine so immense a pile of consequences, and cause that so many things, so far asunder distant, could be at once present to the minde, and as with one consent joyne and lay themselves together for the structure of one argument. Wherefore that I might more cleerly behold the things themselves, I uncasing the Propositions and Demonstrations out of their covert of words, designed them in notes and Species appearing to the very eye. After that by comparing the divers affections of Theoremes in equality, proportion, affinity and dependence, I tryed to educe new out of them. Lastly, by framing like questions problematically, and in way of Analysis, as if they were already done, resolving them into their principles, I sought out reasons and means whereby they might be effected. And by this course of practice, not without long time, and much industry, I found out this way for the helpe and facilitation of Art.

(*b*) *On the benefit of specious arithmetic*

This *Specious Arithmetic* is more appliable to the Analyticall art, (in which by taking the thing sought as knowne, we finde out that we seeke) than that Numerous. For in the Numerous, the numbers with which we worke, are so, as it were, swallowed up into that new which is brought forth, that they quite vanish, not leaving any print or footstep of themselves behinde them. But in the Specious, the species remaine without any change, shewing the processe of the whole worke: and so doe not onely resolve the question in hand; but also teach a generall Theoreme for the solution of like questions in other magnitudes given.

9.F2 John Wallis on Oughtred's *Clavis*

Mr William Oughtred (our Country-man) in his *Clavis Mathematicae*, (or Key of Mathematicks,) first published in the Year 1631, follows Vieta (as he did Diophantus) in the use of the Cossick Denominations; omitting (as he had done) the names of *Sursolids*, and contenting himself with those of *Square* and *Cube*, and the Compounds of these.

But he doth abridge Vieta's Characters or Species, using only the letters q, c, &c. which in Vieta are expressed (at length) by *Quadrate*, *Cube*, &c. For though when Vieta first introduced this way of Specious Arithmetick, it was more necessary (the thing being new,) to express it in words at length: Yet when the thing was once received in practise, Mr Oughtred (who affected brevity, and to deliver what he taught as briefly as might be, and reduce all to a short view,) contented himself with single Letters instead of those words.

Thus what Vieta would have written

$$\frac{A\ \textit{Quadrate},\ \text{into}\ B\ \textit{Cube},}{CDE\ \textit{Solid},}\ \textit{Equal to FG Plane},$$

would with him be thus expressed

$$\frac{A_q B_c}{CDE} = FG.$$

And the better to distinguish upon the first view, what quantities were Known, and what Unknown, he doth (usually) denote the Known to *Consonants*, and the Unknown by *Vowels*; as Vieta (for the same reason) had done before him.

I know there are who find fault with his *Clavis*, as too obscure, because so short, but without cause; for his words be always full, but not Redundant, and need only a little attention in the Reader to weigh the force of every word, and the Syntax of it; [...] And this, when once apprehended, is much more easily retained, than if it were expressed with the prolixity of some other Writers; where a Reader must first be at the pains to weed out a great deal of superfluous Language, that he may have a short prospect of what is material; which is here contracted for him in a short Synopsis. [...]

Mr Oughtred in his *Clavis*, contents himself (for the most part) with the solution of Quadratick Equations, without proceeding (or very sparingly) to Cubick Equations, and those of Higher Powers; having designed that Work for an *Introduction* into *Algebra* so far, leaving the Discussion of Superior Equations for another work. . . . He contents himself likewise in Resolving Equations, to take notice of the *Affirmative* or *Positive Roots*; omitting the *Negative* or *Ablative Roots*, and such as are called *Imaginary* or *Impossible Roots*. And of those which, he calls *Ambiguous* Equations, (as having more Affirmative Roots than one,) he doth not (that I remember) any where take notice of more than *Two* Affirmative Roots; (Because in Quadratick Equations, which are those he handleth, there are indeed no more.) Whereas yet in *Cubick* Equations, there may be *Three*, and in those of Higher Powers, yet more. Which Vieta was well aware of, and mentioneth in some of his Writings; and of which Mr Oughtred could not be ignorant.

9.F3 Letters on the value of Oughtred's *Clavis*

(*a*) *Wallis to John Collins, 5 February 1666*

But for the goodness of the book [*Clavis*] in itself, it is that (I confess) which I look upon as a very good book, and which doth in as little room deliver as much of the fundamental and useful part of geometry (as well as of arithmetic and algebra) as any book I know; and why it should not be now acceptable I do not see. It is true, that as in other things so in mathematics, fashions will daily alter, and that which Mr Oughtred designed by great letters may be now by others be designed by small; but a mathematician will, with the same ease and advantage, understand *Ac*, and a^3 or *aaa*. Nor will Euclid or Archimedes cease to be classic authors and in request, though some of their considerable propositions be, by Mr Oughtred and others, delivered now in a more advantageous way, according to men's present apprehensions. And the like I judge of Mr Oughtred's *Clavis*, which I look upon (as those pieces of Vieta who first went in that way) as lasting books and classic authors in this kind.

(*b*) *Collins to Wallis*

As for Mr Oughtred's method of symbols, this I say to it; it may be proper for you as a commentator to follow it, but divers I know, men of inferior rank that have good skill in algebra, that neither use nor approve it. One Anderson, a weaver, [...] Mr Dary, the tobacco cutter, a knowing man in algebra, [...] I might name Wadley, a lighterman, and may acquiesce in these men's judgments, or at least in Dr Pell's, who hath said it is unworthy the present age to continue it, as rendering easy matters obscure. Is not A^5 sooner wrote than *Aqc*? Let *A* be 2. the cube of 2 is 8, which squared is 64: one of the questions between Maghet Grisio and Gloriosus is whether 64 = *Acc* or *Aqc*. The Cartesian method tells you it is A^6, and decides the matter.

9.F4 John Aubrey on Oughtred

His Father taught to write at Eaton, and was a Scrivener; and understood common Arithmetique, and 'twas no small helpe and furtherance to his son to be instructed in it when a schoole-boy.

He was a little man, had black haire, and blacke eies (with a great deal of spirit). His head was always working. He would drawe lines and diagrams on the Dust.

He was more famous abroad for his learning, and more esteemed, then at home. Severall great Mathematicians came over into England on purpose to converse with him. His countrey neighbours (though they understood not his worth) knew that there must be extraordinary worth in him, that he was so visited by Foreigners.

When Mr Seth Ward, M.A. and Mr Charles Scarborough, D.M. came as in Pilgrimage, to see him and admire him, Mr Oughtred had against their comeing prepared a good dinner, and also he had dressed himselfe, thus: an old red russet cloath-cassock that had been black in dayes of yore, girt with a old leather girdle, an old fashion russet hatt, that had been a Bever, *tempore Reginae Elizabethae*. When learned Foraigners came and sawe how privately he lived, they did admire and blesse themselves, that a person of so much worth and learning should not be better provided for.

Seth Ward, M.A., a Fellow of Sydney Colledge in Cambridge (now Bishop of Sarum) came to him, and lived with him halfe a yeare (and he would not take a farthing for his diet) and learned all his Mathematiques of him. Sir Jonas More was with him a good while, and learn't; he was but an ordinary Logist before. Sir Charles Scarborough was his Scholar; so Dr John Wallis was his Scholar; so was Christopher Wren his scholar; so was Mr Smethwyck, R.S.S. But he did not so much like any as those that tugged and tooke paines to worke out Questions. He taught all free.

One Mr Austin (a most ingeniose man) was his scholar, and studyed so much that he became mad, fell a laughing, and so dyed, to the great griefe of the old Gentleman. Mr Stokes, another scholar, fell mad, and dream't that the good old Gentleman came to him, and gave him good advice, and so he recovered, and is still well.

He was an Astrologer, and very lucky in giving his Judgements on Nativities; he confessed that he was not satisfied how it came about that one might foretell by the Starres, but so it was that it fell out true as he did often by his experience find; he did beleeve that some genius or spirit did help.

The Countrey people did beleeve that he could conjure, and 'tis like enough that he might be well enough contented to have them thinke so.

He was a great lover of Chymistry, which he studyed before his son Ben can remember, and continued it; and told John Evelyn, of Detford, Esq, R.S.S., not above a yeare before he dyed, that if he were but five yeares (or three yeares) younger, he doubted not to find out the Philosopher's stone. It was made of the harshest cleare water that he could gett, which he lett stand to putrify, and evaporated by cimmering.

He was a good Latinist and Graecian, as appears in a little Treatise of his against one Delamaine, a Joyner, who was so sawcy to write against him (I thinke about his *circles of Proportion*).

Nicolaus Mercator went to see him a few yeares before he dyed. 'Twas about Midsommer, and the weather was very hott, and the old gentleman had a good fire, and used Mr Mercator with much humanity (being exceedingly taken with his excellent Mathematicall Witt) and one piece of his courtesie was, to be mighty importunate with him to sett on his upper hand next the fire; he being cold (with age) thought he had been so too.

Before he dyed he burned a world of Papers, and sayd that the world was not worthy of them; he was so superb [proud]. He burned also severall printed bookes, and would not stirre, till they were consumed. His son Ben was confident he understood Magique.

He dyed the 13th day of June, 1660, in the yeare of his age eighty-eight plus odde dayes. Ralph Greatrex, his great friend, the Mathematicall Instrument-maker, sayed he conceived he dyed with joy for the comeing-in of the King, which was the 29th of May before. *And are yee sure he is restored? Then give me a glasse of Sack to drinke his Sacred Majestie's health.* His spirits were then quite upon the wing to fly away.

9.G Brief Lives

Among our most perceptive and lively insights into the richness of seventeenth-century intellectual life are the notes made by the scholar John Aubrey (1626–1697) on the lives of his contemporaries and previous generations. His information is of varying reliability, but of value in other ways for showing the beliefs and concerns of seventeenth-century antiquarian scholarship; for instance, Aubrey's attentiveness to what became of mathematicians' books and papers is notable.

9.G1 Thomas Allen (1542–1632)

The great Dudley, Earle of Leicester, made use of him for casting of Nativities, for he was the best Astrologer of his time. Queen Elizabeth sent for him to have his advice about the new Star that appeared in the Swan or Cassiopeia (but I think the Swan) to which he gave his Judgement very learnedly.

In those darke times, Astrologer, Mathematician, and Conjurer were accounted the same things; and the vulgar did verily beleeve him to be a Conjurer. He had a great many Mathematicall Instruments and Glasses in his Chamber, which did also confirme the ignorant in their opinion.

9.G2 Sir Henry Savile (1549–1622)

He was very munificent, as appeares by the two Lectures he has given to Astronomy and Geometry. Bishop Seth Ward, of Sarum, has told me that he first sent for Mr Gunter, from London (being of Oxford University) to have been his Professor of Geometrie: so he came and brought with him his Sector and Quadrant, and fell to resolving of Triangles and doeing a great many fine things. Said the grave Knight, *Doe you call this reading of Geometrie? This is shewing of tricks, man!* and so dismisst him with scorne, and sent for Henry Briggs, from Cambridge.

I have heard Dr Wallis say, that Sir H. Savill has sufficiently confuted Joseph Scaliger *de Quadratura Circuli*, in the very margent of the booke: and that sometimes when J. Scaliger sayes $AB = CD$ ex constructione, Sir H. Savill writes sometimes in the margent, *Et Dominatio vestra est Asinus ex constructione* [and your rule is an ass by construction]. One sayes of Jos Scaliger, that where he erres, he erres so ingeniosely, that one had rather erre with him then hit the mark with Clavius.

He had travelled very well, and had a generall acquaintance with the Learned men abroad; by which meanes he obtained from beyond sea, out of their Libraries, severall rare Greeke MSS, which he had copied by an excellent Amanuensis for the Greeke character. He gave his Collection of Mathematicall Bookes to a peculiar little Library belonging to the Savilian Professors.

9.G3 Walter Warner (1550–1640)

This Walter Warner was both Mathematician and Philosopher, and 'twas he that putt-out Thomas Hariot's *Algebra*, though he mentions it not.

Walter had but one hand (borne so): Dr John Pell thinks a right hand; his mother was frighted, which caused this deformity, so that instead of a left hand, he had only a Stump with five warts upon it, instead of a hand and fingers. He wore a cuffe on it like a pockett. The Doctor never sawe his stump, but Mr Warner's man told him so.

Mr Walter Warner made an Inverted Logarithmicall Table, i.e. whereas Briggs' table fills his Margin with Numbers encreasing by Unites, and over-against them setts their Logarithms, which because of incommensurability must needs be either abundant or deficient; Mr Warner (like a Dictionary of the Latine before the English) fills the Margin with Logarithmes encreasing by Unites, and setts to every one of them so many continuall meane proportionalls between one and 10, and they for the same reason must also have the last figure incompleat. These, which, before Mr John Pell grew acquainted with Mr Warner, were ten thousand, and at Mr Warner's request were by Mr Pell's hands, or direction, made a hundred-thousand.

Quaere Dr Pell, what is the use of those Inverted Logarithmes? for W. Warner would not doe such a thing in vaine.

9.G4 Edmund Gunter (1581–1626)

Captain Ralph Gretorex, Mathematical-Instrument Maker in London, sayd that he was the first that brought Mathematicall Instruments to perfection. His *Booke of the Quadrant, Sector, and Crosse-staffe* did open men's understandings and made young men in love with that Studie. Before, the Mathematicall Sciences were lock't-up in the Greeke and Latin tongues; and so lay untoucht, kept safe in some Libraries. After Mr Gunter published his Booke, these Sciences sprang up amain, more and more to that height it is at now (1690).

When he was a Student at Christchurch, it fell to his lott to preach the Passion Sermon, which some old divines that I knew did heare, but 'twas sayd of him then in the University that our Saviour never suffered so much since his Passion as in that sermon, it was such a lamentable one—*Non omnia possumus omnes* [all things are not possible to all men]. The world is much beholding to him for what he hath donne well.

9.G5 Thomas Hobbes (1588–1679)

I have heard Mr Hobbes say that he was wont to draw lines on his thigh and on the sheetes, abed, and also multiply and divide.

He would often complain that Algebra (though of great use) was too much admired, and so followed after, that it made men not contemplate and consider so much the nature and power of Lines, which was a great hinderance to the Groweth of Geometrie; for that though algebra did rarely well and quickly, and easily in right lines, yet 'twould not *bite* in *solid* (I thinke) Geometry. Quod N.B.

'Twas pitty that Mr Hobbs had not began the study of the Mathematics sooner, els he would not have layn so open. But one may say of him, as one sayes of Jos. Scaliger, that where he erres, he erres so ingeniosely, that one had rather erre with him then hitt the marke with Clavius.

After he began to reflect on the Interest of the King of England as touching his affaires between him and parliament, for ten yeares together his thoughts were much, or almost altogether, unhinged from the Mathematiques; which was a great putt-back to his Mathematicall improvement; for in ten yeares (or better) discontinuance of that Study (especially) one's Mathematiques will become very rusty.

When he was at Florence, he contracted a friendship with the famous Galileo Galileo, whom he extremely venerated and magnified; and not only as he was a prodigious Witt, but for his sweetnes of nature and manners. They pretty well resembled one another as to their countenances, as by their Pictures doeth appear; were both cheerfull and melancholique-sanguine; and had both a consimilitie of Fate, to be hated and persecuted by the Ecclesiastiques.

He had a high esteeme for the Royall Societie, having sayd that Naturall Philosophy was removed from the Universities to Gresham Colledge, meaning the Royall Societie that meetes there; and the Royall Societie (generally) had the like for him: and he would long since have been ascribed a Member there, but for the sake of one or two persons, whom he tooke to be his enemies: viz. Dr Wallis (surely their Mercuries are in opposition) and Mr Boyle. I might add Sir Paul Neile, who disobliges everybody.

9.G6 Sir Charles Cavendish (1591–1654)

Sir Charles Cavendish was the younger Brother to William, Duke of Newcastle. He was a little, weake, crooked man, and nature having not adapted him for the Court nor Campe, he betooke himself to the Study of the Mathematiques, wherin he became a great Master. His father left him a good Estate, the revenue wherof he expended on bookes and on learned men.

He had collected in Italie, France, &c, with no small chardge, as many Manuscript Mathematicall bookes as filled a Hoggeshead, which he intended to have printed; which if he had lived to have donne, the growth of Mathematicall Learning had been 30 yeares or more forwarder then 'tis. But he died of the Scurvey, contracted by hard study, about 1652, and left an Attorney of Clifford's Inne, his Executor, who shortly after died, and left his Wife Executrix, who sold this incomparable Collection aforesaid, by weight to the past-board makers for Wast-paper. A good Caution for those that have good MSS to take care to see them printed in their lifetimes.

He writt several things in Mathematiques for his owne pleasure.

9.G7 René Descartes (1596–1650)

He was so eminently learned that all learned men made visits to him, and many of them would desire him to shew them his Instruments (in those dayes mathematicall learning lay much in the knowledge of Instruments, and, as Sir Henry Savile sayd, in doeing of tricks) he would drawe out a little Drawer under his Table, and shew them a paire of Compasses with one of the Legges broken; and then, for his Ruler, he used a sheet of paper folded double.

Mr Hobbes was wont to say that had Des Cartes kept himselfe wholy to Geometrie that he had been the best Geometer in the world but that his head did not lye for Philosophy.

9.G8 Edward Davenant

He was to his dyeing day of great diligence in study, well versed in all kinds of Learning, but his Genius did most strongly encline him to the Mathematiques, wherin he has written (in a hand as legible as print) MSS in 4to a foot high at least. I have often heard him say (jestingly) that he would have a man knockt in the head that should write anything in Mathematiques that had been written of before. I have heard Sir Christopher Wren say that he does beleeve he was the best Mathematician in the world about 30 or 35+ yeares agoe. But being a Divine he was unwilling to print, because the world should not know how he had spent the greatest part of his time.

I have writt to his Executor, that we may have the honour and favour to conserve his MSS in the Library of the Royal Societie, and to print what is fitt. I hope I shall obtaine my desire. He had a noble Library, which was the aggregate of his Father's, the Bishop's, and his owne.

He was very ready to teach and instruct. He did me the favour to informe me first in Algebra. His daughters were Algebrists.

9.G9 Seth Ward (1617–1689)

Seth Ward, Lord Bishop of Sarum, was borne at Huntingford, a small market-towne in Hartfordshire, anno Domini 1618 (when the great blazing Starre appeared). His Father was an Attorney there, and of very honest repute. (Dr Guydos, physician of Bath, says that anciently there was but One Attorney in Somerset, and he was so poor, that he went a foot to London; and now, 1689, they swarme there like Locusts: they go to Market and breed Contention.) His father taught him common arithmetique, and his Genius lay much to the Mathematiques, which being naturall to him, he quickly and easily attained.

At sixteen yeares old he went to Sydney colledge in Cambridge.

Sir Charles Scarborough, M.D. (then an ingeniose young student, and Fellowe of Caius Colledge in Cambridge) was his great acquaintance; both students in mathematiques; which the better to perfect, they went to Mr William Oughtred, at Albury in Surrey, to be enformed by him in his *Clavis Mathematica*, which was then a booke of *Aenigmata*. Mr Oughtred treated them with exceeding humanity, being pleased at his heart when an ingeniose young man came to him that would ply his Algebra hard. When they returned to Cambridge, they read the *Clavis Mathematica* to their Pupills, which was the first time that that booke was ever read in a University. Mr Laurence Rooke, a good mathematician and algebrist (and I thinke had also been Mr Oughtred's disciple) was his great acquaintance.

As he is the pattern of humility and courtesie, so he knowes when to be severe and austere; and he is not one to be trampled or worked upon. He is a Batchelour, and of a most magnificent and munificent mind.

He hath been a Benefactor to the Royall Societie (of which he was one of the first Members and Institutors: the beginning of Philosophical Experiments was at Oxon, 1649, by Dr Wilkins, Seth Ward, Ralph Bathurst, &c). He also gave a noble pendulum Clock to the Royall Societie (which goes a weeke) to perpetuate the memory of his deare and learned friend, Mr Laurence Rooke, who tooke his sicknesse of which he dyed by setting up so often for Astronomicall Observations.

I searcht all Seth, Episcopus Sarum's, papers that were at his house at Knightsbridge neer London where he dyed. The custome is, when the Bishop of Sarum dies, that the Deane and Chapter lock-up his Studie and put a Seale on it. His scatterd papers I rescued from being used by the Cooke since his death; which was destinated with other good papers and letters to be put under pies.

9.G10 Sir Jonas Moore (1617–1679)

Sir Jonas Moore was borne at Whitelee in Lancashire, towards the Bishoprick of Durham. He was inclined to Mathematiques when a boy, which some kind friends of his putt him upon, and instructed him in it, and afterwards Mr Oughtred more fully enformed him; and then he taught Gentlemen in London, which was his livelyhood.

He was one of the most accomplisht Gentlemen of his time; a good Mathematician, and a good Fellowe.

Sciatica: he cured it by boyling his Buttock.

The Duke of Yorke said that: Mathematicians and Physicians had no Religion: which being told to Sir Jonas More, he presented his duty to the D.Y. and wished, with all his heart that his Highnesse *were a Mathematician too*: this was since he was supposed to be a Roman Catholic.

9.H Advancement of Mathematics

9.H1 John Pell's *Idea of Mathematics*

(*a*) *John Pell to Samuel Hartlib*

The summe of what I have heretofore written or spoken to you, concerning *the advancement of the Mathematickes*, is this: As long as men want *will*, *wit*, *meanes* or *leisure* to attend those studies, it is no marvaile if they make no great progresse in them. To remedy which I conceive *these meanes* not to be amisse.

1. To write a *Consiliarius Mathematicus*, (so I call it) answering to these 3 questions:
   Q.1. What *fruit* or *profit* ariseth from the study of *Mathematics*?
   Q.2. What *helpes* are there for the attaining this profitable knowledge?
   Q.3. What *order* is to be observed in using these helpes?

   To this purpose it should containe
   1. A plaine and popular *discourse* of the *extent* of the Mathematics, with the *profit* that redounds, first to the *Student* himself, and then to the *Country* wherein there are many such grounded Artists.
   2. A *Catalog* of Mathematicians and their workes in this order:
      1. A Synopsis of all the severall *kindes* of Mathematicall writings, either *extant* in print, or *accessible* Manuscripts in public Libraries, with severall numbers set to every kinde.
      2. A Chronicall *Catalog* of all *Mathematicians names* that ever were of note, according to the order of the yeares *when they lived*, with the yeare when any of the workes were *first printed*.
      3. A *Catalog* of the *writings themselves*, in the order of yeares in which they were *printed* in any language: And this I would contrive thus: First, the yeere of our *Lord*, and then the names of all the Mathematicall bookes printed that yeare in any country or language, after the usuall manner of Catalogs; but
         1. *adding* the volume, that is, not onely what fold [4°, 8° &c.] but also the number of leaves, that we may estimate the bulke of the booke.

2. *præfixing* before the title, the yeare to which you must looke backe to know either when it was written, or when it was last before printed, in that or any other language.
3. *setting* in the margent, after the title,
    1. the yeare wherein it was the *next time* printed:
    2. the *number of reference* to the *Synopses* in the first page; By which numbers one may presently runne over all the bookes of one sort, of this or that particular subject.

3. A *Counsell* directing a student to the *best* bookes in every kinde; In what *order*, and *how* to reade them, What to observe, what to beware of in some Mathematicasters, how to *proceed* and keepe all. [...]

To which end it would be good

2. To erect a *Public Library*, containing all those bookes, and one instrument of every sort that hath beene invented, with sufficient revenue
    1. to buy one copy of all those that shall be printed yearly in other countries, and
    2. to maintaine a Library keeper of great judgement, to whom it may belong

   To peruse all books of such subjects, to be printed within that country, and
    1. suppresse whatsoever is not according to Art, that Learners be not abused, and
    2. admonish the writers, if they bring nothing but stale stuffe.

   [...]

   And this is the best course that I can thinke on for the making use of *such helpes as we have already*. If men desire *better* helpes, let them employ fit Artists
3. To write and publish these *three new Treatises*
    1. *Pandectæ Mathematicæ*, Comprehending as cleerly, orderly, thriftily, and ingenuously as may be, whatsoever may be gathered out of all those Mathematicall bookes and inventions that were before us, or that may be inferred as Consectaries thereon; *citing*, at the end of every period or proposition, the ancientest Author in which it is found, and *branding* all later writers if they be taken stealing, or borrowing without acknowledgement, or [which is worse] expresly arrogating to themselves any other mans inventions. This would bring that *great Library* into farre lesse roome, to the saving of more *labour*, *time* and *cost*, to all after-students, than men can yet well-imagine. But because this also would be too great and cumbersome to carry about us, Let there be composed
    2. *Comes Mathematicus*, Comprehending in a pocket-booke, [and therefore as briefly as may be] the usefullest Tables and the Precepts for their use, in solving all Problemes, whether purely-Mathematicall, or applied to such practices as mens various occasions may require.

   And lastly, that in this kinde of Learning also, we be no longer tyed to bookes, Let there be composed
    3. *Mathematicus autarchus*, or An instruction, shewing how any Mathematician that will take the paines, may prepare himselfe, so, as that he may, though he be utterly destitute of bookes or instruments, resolve any Mathematicall Probleme as exactly as if he had a complete *Library* by him.

And this is the *Idéa*, which I have long framed to my selfe, according to my fashion, with whom this passeth for an undoubted truth, that the surest way to come to all possible excellency *in any thing*, is to propose to our selves the perfectest *Idéa's* that we can imagine, then to seeke the meanes tending thereto, as rationally as may be, and to prosecute it with indefatigable dilligence; yet, if the *Idéa* prove too high for us, to rest our selves content with *approximation*.

As for this present *Idéa*, I am so farre from counting it meerely-impossible, that I see not why it might not be performed by one man, without any assistants, provided that he were neither *distracted* with cares for his maintenance, nor *diverted* by other employments.

The excellency of this worke, makes me wish mine owne nation the honour of *first* undertaking and perfecting this designe, And I conceive I have some reason to hope that it will be so. For, though I know few or none that are both able and willing to promote designes of this nature, yet can I not therefore be persuaded that this Kingdome is so destitute of learned Nobility and Gentry, that there can be found none to countenance and advance *this* part of Learning, even in *this* way, if they could see it possible and likely to be effected.

As for the *Library* and *Catalog*, there can be no doubt but they may easily be had, if money be not wanting. Nor is it unlikely that divers of this nation (if they be set apart for it) are able to compose the other 3 *new Treatises*; For though I know no such, yet I persuade my selfe there may be found amongst us men able to encounter all the difficulties, and to endure all the labour, that they must needs meet with in the raising of so great a *Fabricke*. And I the rather beleeve that there are many such, because for mine *owne part*, notwithstanding the want of Counsell and helpes in that study, and the innumerable diversions and distractions that I have had, I am neverthelesse come to such a confidence of my understanding the depth of that study, that, were I to pen those *Pandects*, I should lay heavier lawes upon my selfe, than I have already mentioned; namely, *First* to lay downe such an exact *Method* or description of the processe of Mans reason in inventions, that *afterward* it should be imputed meerly to my negligence and disobedience to my owne lawes, [and not to their insufficiency] if, from my first grounds, seeds, or principles, I did not, in an orderly way, according to that praescribed Method, deduce, not onely all that ever is to be found in our Antecessors writings, and whatsoever they may seeme to have thought on, but also all the Mathematicall inventions, Theoremes, Problemes and Precepts, that it is possible for the working wits of our successors to light upon, and that in one certaine, unchanged order, from the first seeds of Mathematics, to their highest and noblest applications, as well as to the meanest and most ordinary. Not setting them down at random as they come in my head, as those before us have done, so that they seeme to have light upon their Problemes and the solutions of them by chance, not to have found them by one perpetuall, constant, invariable processe of Art. Yet such an Art may men invent, if they accustome themselves, as I have long done, to consider, not onely the *usefulnesse* of mens workes, and the *meaning* and *truth* of their writings, but also *how it came to passe* that they fell upon such thoughts, and that they proposed to themselves such ends, or found out such meanes for them.

Were these *Pandects* thus made and finished, I suppose it is manifest, that by their orderly, rationall and uniforme compleatnesse, above all that hath beene hitherto written, they would spare after-students much *labour and time* that is now spent in *seeking* out of bookes, and *disorderly reading* them, and *struggling* with their cloudy expressions, unapt representations, different Methods, confusions, tautologies, impertinencies, falshoods by paralogismes and pseudographemes, uncertainties because of insufficient demonstrations, &c. besides much *cost* also, now throwne away upon the multitude of bookes, the greater part whereof they had perhaps beene better never to have seene.

And it may be some would like the Method of that worke so well, as to extend it farther, and apply it to other studies; *in speculation* imitating this my warinesse, that *no falshood be admitted, and no truth omitted*; and *for practice* enuring themselves, *Any subject* being propounded, to determine the number of all the Problemes that can be conceived concerning it, and *Any Probleme* being propounded, demonstratively to shew *either all the meanes* of its solution, *or the impossibility* of it: and if *so*, then whether it be *not yet*, or *not at all* possible.

(*b*) *Response from Marin Mersenne*

Instead of all that apparatus which the author of the *Idea of Mathematics* proposes in order to collect the various mathematical writings, it seems to me more advisable to select only the best and most worthy, which would scarcely exceed twelve in number. Beginning first with ancient authors whose works are still extant, such as Euclid, Apollonius, Archimedes, Theodosius, Pappus, Ptolemy, with their other fragments and manuscripts which have not yet seen the light, some of which are with Golius at Leyden and some preserved at Rome. Then to these might be added more recent works like Vieta's for algebra, Clavius's five volumes and our Herigon's recently edited five volumes, thus having the best and neglecting the rest. Similarly of the opticians might be selected Vitellis, Kepler, Aquilon and de Villes. Of the arithmeticians, after Diophantus the best are Cardan, Tartaglea and your Napier. Of spherical triangles and logarithmic calculations, there are Briggs, Gordan, Pitiscus, Snell and our Morinus. Of Astronomy, after Ptolemy and some Arabs, should be obtained table-makers like Alphonsus, John Regiomontanus, Kepler and our Duret. As few as eight or ten authors should be selected for fortification and music. Similarly for mechanics (forces of motion, machines and water-works) ten or twelve authors might be abridged for those interested. If twelve wise men cooperating together were to undertake this task each could compose a suitable volume on one science, supplementing with material lacking in the others and deleting anything redundant, together we would doubtless have set out briefly in twelve volumes all that could be desired. In my opinion all pure and mixed mathematics could be included in these twelve volumes. Similarly we could give the nicer parts of philosophy in three volumes and all the liberal and mechanical arts in three more so that learning might be obtained cheaply. It would not help to have all mathematical instruments but better to have four or five of the best, such as Astrolabes, Quadrants, Goniometers, &c.

(*c*) *Response from René Descartes*

I inspected the *Mathematical Idea* only incidentally and now only recollect that there was nothing with which I greatly disagreed; I much approved both the mathematical apparatus listed and the self-sufficient mathematician or *autarchus* there described. In almost the same sense I distinguish between the History and the Science of Mathematics. By the History I mean all that already discovered and contained in books; by Science, the skill of resolving all questions and thence of investigating on ones own whatever may be discovered in that Science by human ingenuity. Whoever possesses this faculty has little need of other assistance and can properly be called self-sufficient. . . . Now it is very desirable that this mathematical history, now scattered in many volumes and still incomplete, should be collected in one book, and so eliminate the need to seek or purchase many books. Since authors transcribe many things from one author, anything extant could be found in any moderately equipped library. Diligence in collecting all things is not so important as the judgment to reject the superfluous and the ability to fill in the gaps in our knowledge. [. . .] If such a book were available it would be easy for anybody to learn all Mathematical history, and much of the science also.

9.H2 Letters between John Pell and Sir Charles Cavendish

(a) Pell to Cavendish, August 1644

Des Cartes himself is gone into France. Monsieur Hardy tells us, in a letter lately written, that Des Cartes met him in Paris, and blamed him for offering so much mony to our Arabicke professor at Utrecht, for his Arabicke manuscript of Apollonius. Which Mr Hardy interprets as a signe of envy in Des Cartes, as being unwilling that we should esteeme the ancients, or admire any man but himselfe for the doctrine of lignes courbes.

But I think France alone will afford me argument for a large letter, and therefore I leave it till the next time.

Come we therefore to England. And first for Mr Warner's Analogickes, of which you desire to know whether they be printed. You remember that his papers were given to his kinsman, a merchant in London, who sent his partner to bury the old man: himselfe being hindred by a politicke gout, which made him keepe out of their sight that urged him to contribute to the parliament's assistance, from which he was exceedingly averse. So he was looked upon as one that absented himselfe out of malignancy, and his partner managed the whole trade. Since my comming over, the English merchants heere tell me that both he and his partner are broken, and now they both keepe out of sight, not as malignants, but as bankrupts. But this you may better inquire among our Hamburg merchants. In the meane time I am not a little afraid that all Mr Warner's papers, and no small share of my labours therein, are seazed upon, and most unmathematically divided between the sequestrators and creditors, who (being not able to ballance the account where there appeare so many numbers, and much troubled at the sight of so many crosses and circles in the superstitious Algebra and that blacke art of Geometry) will, no doubt, determine once in their lives to become figure-casters, and so vote them all to be throwen into the fire, if some good body doe not reprieve them for pye-bottoms, for which purposes you know analogicall numbers are incomparably apt, if they be accurately calculated.

(b) Cavendish to Pell, November 1644

Manye thankes for your letter, wherein you write that you have Apollonius's 3 bookes of conicks in Arabick, more than wee had in the Greeck, and 36 authors more. I hope some of those are of the mathemathicks. Howsoever I dout not but they are worthie the press. I like extreamelie both the proposition and demonstration of Apollonius in your letter; and to my aprehension the expression of the same proposition in Mersennus his book is perplexed and no demonstration translated. I wonder Goleas hath not published it all this whyle; yet being nowe in your handes, I am not sorie he did not; for I assure myself wee shall nowe have it with more advantage than the loss of so much time. Though I doute not but your explication of Diophantus will put us in to a more sure waye of analiticks than formerlie, yet I suppose there is so much to be added and explained concerning analiticks that it will require a large volume, and I hope you continue your intention of publishing such a worke, which I beseech you thinke seriouslie of to publish with all convenient speeds; for it is a worcke worthie of you.

(*c*) *Pell to Cavendish, March 1646*

Last Thursday Des Cartes came into our Auditory and hearde me reade, though when I had done he excused it, saying that if his guide had knowen my chamber so well as my publike houre and place, he would rather have come thither to me: he went with me to my lodging, where we had long discourse of Mathematicall matters, though I sought not so much to speake myselfe as to give him occasion to speake. He thinks it needlesse to say any thing more of my refutation than he hath done and is contented that any body turne it into Latine and give it to the printer. I perceive he demonstrates not willingly. He sayes he hath penned very few demonstrations in his life (understand after the style of the old Grecians which he affects not) THAT he never had an Euclide of his owne but in 4 dayes, 30 yeares agoe. (He is now 50 yeares old, wanting 3 months.) He posted over it and so restored it to him that had lent it him. He then thought all the theorems very manifest, but cares not to remember any more of them than 47.I *de triangulo rectangulo* and 4.VI *de triangulis similibus*. He hath a highe opinion of Euclid and Apollonius for writing so largely that which he conceives may be put into so little roome. He suspects Diophantus might be excellent in the books which are loste, but in most of his questions which he hath wrought, he finds they might have been solved with lesse adoe than Diophantus makes. Of all the Ancients he magnifies none but Archimedes, who he says, in his bookes *de Sphaera & Cylindro* and a piece or two more, shows himself *fuisse bonum Algebraicum & habuisse vere-magnum ingenium*. I will not trouble you of what he said of Vieta, Fermat and Roberval and Golius: Of Mr Hobbes I durst make no mention to him. I showed him the titles of P. Vincentius his bookes. He told me that there was another Jesuit (as I remember in Antwerp) that threatened to write against him as soon as he came forth affirming that he had seene some theorems false in the beginning of his seconde booke. Whereupon a third man sent them to Descartes desiring his judgement. He replyed that they were all true, but that vulgar wits could not so fully apprehend them as to acknowledge their truth. He says he had no other instructor for Algebra than the reading of Clavy Algebra above 30 yeares agoe. That not longe after coming into Denmarke, he visited Longomontanus and proffered to demonstrate to him the ground of his error. They spent one whole day together, shut up in a chamber alone. In the evening when they should parte, he perceived that Longomontanus understood none of his reasons. So he thought it not worth while to goe to him any more. He praises my way of dealing with him in rationall numbers, utterly excluding all mention or thought of Surds, and thinkes that if

Longomontanus cannot understand that paper he can understand nothing. And therefore wondered to heare that he had written twice against me. Of Mathematicall things he is resolved to write no more. He sayes the ingenious will finde them out alone, and for the rest it is but lost labour to write. To this purpose he said so much, that when he was gone I found my will to write something abated; but me thought I had a greater opinion of my skill than I had before, of which alteration you can perhaps hardly guesse the cause.

9.H3 Hobbes and Wallis (1656)

(*a*) *Hobbes*

Truly, were it not that I must defend my reputation, I should not have showed the world how little there is of sound doctrine in any of your books. For when I think how dejected you will be for the future, and how the grief of so much time irrecoverably lost, together with the conscience of taking so great a stipend, for mis-teaching the young men of the University, and the consideration of how much your friends will be ashamed of you, will accompany you for the rest of your life, I have more compassion for you than you have deserved. Your treatise of the Angle of Contact, I have before confuted in a very few leaves. And for that of your Conic Sections, it is so covered over with the scab of symbols, that I had not the patience to examine whether it be well or ill demonstrated.

(*b*) *Wallis*

As for my Treatise of Conick Sections, you say, it is so covered over with the Scab of Symbols, that you had not the patience to examine whether it be well or ill demonstrated. A very fine way of confutation; and with much ease. You have not the patience to examine it, (that is, in plain English, you do not understand it,) Ergo I have performed nothing in any of my Books. [...] But, Sir, must I be bound to tell you a tale, and find you ears too? Is it not lawfull for me to write Symbols, till you can understand them? Sir, they were not written for you to read, but for them that can.

9.H4 The mathematical education of John Wallis

Wallis was a schoolboy of fifteen at the start of the period to which this extract from his autobiography refers.

At Christmass 1631 (a season of the year when Boys use to have a vacancy from School) I was, for about a fortnight, at home with my Mother at *Ashford*. I there found that a younger Brother of mine (in Order to a Trade) had for about 3 months, been learning (as they call'd it) to *Write and Cipher*, or *Cast account* (and he was a good proficient for that time.) When I had been there a few days; I was inquisitive to know what it was, they so called. And (to satisfie my curiosity) my Brother did (during the Remainder of my stay there before I return'd to School) shew me what he had been Learning in those 3 months. Which was (beside the writing a fair hand) the *Practical* part of *Common Arithmetick* in *Numeration*, *Addition*, *Substraction*, *Multiplication*, *Division*, *The Rule of Three* (*Direct and Inverse*), *the Rule of Fellowship* (*with and without*, *Time*), *the Rule of False-Position*, *Rules of Practise and Reduction of Coins* and some other little things. Which when he had shewed me by steps, in the same method that he had learned them: and I had wrought over all the *Examples* which he before had done in his book; I found no difficulty to understand it and I was very well pleased with it: and thought it ten days or a fortnight well spent. This was my first insight into *Mathematicks*; and all the *Teaching* I had.

This suiting my humor so well; I did thenceforth prosecute it, (at School and in the University) not as a formal Study, but as a pleasing Diversion, at spare hours; as books of *Arithmetick*, or others *Mathematical* fel occasionally in my way. For I had none to direct me, what books to read, or what to seek, or in what method to proceed. For Mathematicks, (at that time, with us) were scarce looked upon as *Accademical* studies, but rather *Mechanical*; as the business of *Traders*, *Merchants*, *Seamen*, *Carpenters*, *Surveyors of Lands*, or the like; and perhaps some *Almanak-makers in London*. And amongst more than Two hundred Students (at that time) in our College, I do not know of any Two (perhaps not any) who had more of *Mathematicks* than I, (if so much) which was then but little; And but very few, in that whole University. For the Study of *Mathematicks* was at that time more cultivated in *London* than in the Universities.

9.H5 Samuel Pepys learns arithmetic

Pepys was twenty-nine when he wrote these entries in his diary, having graduated from the university of Cambridge nearly ten years earlier.

4th July, 1662
By and by comes Mr Cooper, Mate of the *Royall Charles*, of whom I entend to learn Mathematiques. After an hour's being with him at Arithmetique, my first attempt being to learn the multiplication table, then we parted till tomorrow.

5th July
At my office all afternoon and then my maths [. . .] at night with Mr Cooper; and so to supper and bed.

8th July
Cooper being there, ready to attend me; so he and I to work till it was dark.

9th July
Up by 4-aclock and at my multiplication table hard, which is all the trouble I meet withal in my arithmetique.

10 July
Up by 4-aclock and before I went to the office, I practised my arithmetique.

11th July
Up by 4-aclock and hard at my multiplication table which I am now almost master of.

12th July
At night with Cooper at Arithmetique [...]

13th July
Having by some mischance hurt my cods [... I] keep my bed all this morning.

14th July
Up by 4-aclock and to my Arithmetique [...]

18th July
[...] and then came Cooper for my Mathematiques; but in good earnest my head is so full of business that I cannot understand it.

10 Mathematics and the Scientific Revolution

There are many possible views of the scientific revolution, and as many possibilities for its architects, but most would give prominence to Kepler and Galileo. From our standpoint, which seeks to bring out the mathematical significance of their work, they appear as a fascinatingly contrasting pair. Johannes Kepler (1571–1630) was a gifted mathematician, well versed in the recently published editions of Euclid, Apollonius, and Archimedes. For him, explanations in science should have two aspects: a thorough grounding in geometry, and a plausible physical cause. Ideally they should contribute to a picture of the universe which, being harmonious, had thereby a better chance of expressing God's design. This belief in harmony led him first (in 1598) to propose his famous model of the solar system based on the idea of nesting the five regular solids inside one another. The model explains why there are six planets, and it has the sun at the centre because Kepler felt on both physical and mystical grounds that that was the right place for it. Later his struggle with Mars led him to his famous discovery of elliptical orbits (1609); we reprint his summary of this 'war', as he called it, in 10.A1. But elliptical orbits did not destroy his allegiance to the shell model afforded by the regular solids. As you can see in 10.A2 and 10.A3, he was willing only to modify that idea, in favour of other, more musical but still equally geometrical ideas about harmony.

There is, of course, much to Kepler that we cannot present. Only pictures of some of his stellated regular solids are here to remind us of his remarkable geometric eye (see 10.A3(c)) and none of his ideas about music. Nor have we presented any of his thoughts about the computation of volumes which belong to the prehistory of the calculus (and for which the reader may consult the *Source Book* by Dirk Struik—see Introduction). But we have tried to emphasize his neo-Platonism (10.A4), for the centrality of mathematics in his work has both interesting specific features and general lessons from which later mathematicians and scientists could learn.

The situation with Galileo Galilei (1564–1642) is quite different. His philosophy of science was much more pragmatic and his use for mathematics more mundane than Kepler's. To be sure, he argued that you read nature as a book 'written in the language of mathematics' (see 10.B1); but what Galileo chiefly read were measured numbers, even if he may only have read them in his mind's eye. For him the role of mathematics was to organize observations securely into a deductive hierarchy, and this was increasingly his aim as he grew older. Our selections from the *Two New Sciences* (1638) make this aspect of his work plain, at the cost, perhaps, of playing down his inventiveness as a theoretical physicist.

It is not too much to claim that one of the greatest achievements of both Kepler and Galileo was to identify a path from patient observation of natural phenomena to the discovery of natural laws. The confidence that one could repose in these laws was in large part due to the way they were argued for mathematically; the new sciences developed in such a way as to amplify this involvement with mathematics.

10.A Johannes Kepler

10.A1 Planetary motion

[1] It is extremely difficult nowadays to write mathematical books, especially astronomical ones. For unless use is made of an exact precision in the propositions, explanations, demonstrations, and conclusions the book will not be mathematical; if such use is made, then the reading of it becomes very difficult, especially in Latin, which lacks articles and that charm characteristic of Greek. [...] This is why there are nowadays few suitable readers; most generally despise and reject such works. How many mathematicians are there who will undertake the labour of reading through the *Conics* of Apollonius of Perga? And yet his material is of a kind which lends itself to explanation by figures and lines much better than Astronomy does.

[2] I myself, who pass for a mathematician, tire my brain in reading my own work, when I try to evoke from the figures the demonstrations which I myself formerly expressed in the figures and the text. And if I should mitigate the obscurity of the material by introducing roundabout expressions, then I should seem, mathematically speaking, to fall into the opposite fault of prolixity.

[3] And prolixity of expression lends itself to obscurity no less than concise brevity does. This latter escapes the eyes of the mind, while the former distracts them; the latter lacks light, while the former suffers from a superabundance of it; the latter does not affect the sight, the former clearly overwhelms it.

[Kepler next describes the table of contents, which he has designed as a careful guide to the whole book, and then continues as follows.]

[4] Let the reader understand that there are two sects of astronomers: the one distinguished by having Ptolemy as their leader and by the allegiance of many of the ancients; the other assigned to the moderns, though the more ancient. The former sect treats each of the planets separately and assigns the causes of the motion of each to its

individual sphere, while the latter compares the planets together, deducing from one and the same common cause whatever is found to be common to their motions. This latter sect is further subdivided; for Copernicus like Aristarchus of old finds in the motion of the earth, our dwelling place, the cause of the planets' apparent stations and retrogradations, and with this I agree; but Tycho Brahe finds it in the sun, in whose vicinity is the connection (as he says) or focus (which is not to be understood as physical, but as quantitative) of the eccentric circles of all the five planets, and this focus or centre revolves about the motionless earth with the body of the sun. [...]

[5] Now my design in this work is chiefly to correct astronomical theory (especially in respect of the motion of Mars) in each of its three forms, so that the results of computations from tables may be found to agree with the celestial appearances, which hitherto could not be done with sufficient certainty. For example, in August, 1608 Mars was a little less than four degrees beyond the place assigned to it in the *Prutenic Tables*. In August and September, 1593 this error amounted to a little less than five degrees, which now by my computations is quite removed.

[6] Now while setting this out [in the book] and continuing successfully, I digress into Aristotle's metaphysics, or rather celestial physics, and I inquire into the causes of the natural motions [of the planets]. From the discussion of these things certain pretty plain conclusions emerge in time, by which Copernicus' theory of the world is, with a few changes, upheld, while the other two theories are condemned as false. [...]

[7] The first step toward the investigation of the physical causes of the motions [of the planets] was that I should demonstrate that the focus of their eccentrics lies in no other place than the very centre of the body of the sun, contrary to the belief of both Copernicus and of Tycho. [...]

[8] When these matters have been demonstrated by an infallible method, the first step toward ascertaining the physical causes has been confirmed and a new one toward them prepared, very clearly if the views of Copernicus and Tycho are adopted, obscurely but at least plausibly in the Ptolemaic system.

[9] For either the earth is moved, or the sun; and it is certainly demonstrated that the body which moves is moved nonuniformly; that is to say, slowly when it is remote from the body at rest, quickly when it approaches the latter closely.

[10] Accordingly, there now immediately appears a means of distinguishing in physics between the three opinions—by conjecture, it is true, but with no less certainty than is found in the conjecture of physicians concerning physiology, or those in any other part of physics.

[11] First of all the Ptolemaic system is exploded. For who will believe that there are as many theories for the sun (exactly similar to one another, true, and equal) as there are planets? It seems that a single theory of the sun is enough to enable Tycho to perform the same functions, and it is a very generally accepted maxim in physics that nature uses as few means as possible.

[12] That Copernicus' system is superior to Tycho's as regards celestial physics is proved on many grounds. [...] For Tycho no less than for Ptolemy, each planet is moved not only by its own proper motion but also by the true motion of the sun, the two being combined together in a mixed motion, forming helices; whence it follows that there are no solid spheres, as Tycho has very solidly demonstrated. Copernicus, on the other hand, has quite freed the five planets from this irrelevant motion, as one caused by a deception arising from the circumstances in which it is observed. Hence, in

Tycho's system as in Ptolemy's, motions are multiplied to no purpose.
[13] Secondly, if there are no solid spheres, the lot of the intelligences and moving spirits becomes exceedingly hard when they are ordered to look after so many things as they must if they are to drive the planets with double, mixed motions. For at the very least they must endeavour at one and the same time to look after the principles, centres, and periods of both motions. But if the earth moves, many things can be accomplished, as I show, by powers which are not animate, but physical, and especially magnetical. [...]
[14] It is further demonstrated that the law governing the speed of the earth's motion (assuming it moves) is quantitatively related to its approach to and recession from the sun. And it is the same with the remainder of the planets; their motion is increased or decreased by their approach to or recession from the sun. The demonstration of these things is to this extent geometrical.
[15] From this very reliable demonstration may be drawn a physical hypothesis—that the source of the motions of the five planets is in the sun itself. It is further very probable that the source of the Earth's motion is in the same place as the source of the motion of the remaining five planets: that is, in the sun also. That the earth moves is, accordingly, probable, since there is a similar apparent cause of its motion.
[16] On the contrary, for this and many other reasons, it is highly probable that the sun is fixed in its place in the centre of the world, for in it is the source of the motion of at least five planets. [...] And the source of all motion is more likely to remain at rest than to move.
[Some pages later Kepler goes on to discuss how this motion might be caused, as follows.]
[17] The true theory of heaviness (or gravity) rests upon the following axioms:
All corporeal substance, insofar as it is corporeal, is capable of being at rest in every place where it is alone, outside the orbit of the power of related bodies.
[18] Heaviness is a mutual corporeal disposition between related bodies toward union or conjunction (which, in the order of things, the magnetic faculty is also), so that it is much rather the case that the earth attracts a stone than that the stone seeks the earth. [...]
[19] If the earth should cease to attract its oceans, the water in all its seas would fly up and flow round the body of the moon.
[20] The sphere of attractive power in the moon extends as far as the earth and, imperceptibly in land-locked seas, but measurably in the wide oceans [...] draws up the waters under the torrid zone. [...]
[21] And it follows, if the attractive power of the moon extends as far as the earth, the attractive power of the earth must be even more likely to extend to the moon and far beyond, so that nothing in any way composed of terrestrial matter can, by being borne aloft, ever escape from the mighty embrace of this attractive power. [...]
[22] Since no projectile ascends one-hundredth-thousandth part of the diameter of the earth above its surface, nor do the clouds or smoke containing a very slight amount of earthy matter rise as much as one-two-thousandth part, it follows that the resistance of these clouds, smoke, or projectiles, and their natural inclination to repose, can do nothing to obstruct this embrace [of the earth], in proportion to which any resistance belonging to these things is negligible. And so what is thrown upward returns to its [former] place without experiencing any interference from the motion of the earth, for

this cannot be neutralized, but draws along with itself whatever flies through the air, constraining them by its magnetic power no less than if those bodies were in actual contact with it. [...]

[23] I began to say that in this work I treat the whole of astronomy according to physical causes, not according to fictitious hypotheses. I have striven toward this end in two steps [of argument]: by discovering, first, that the eccentrics of the planets focus upon the sun, and second, that the theory of the earth required an equant circle and the bisection of its eccentricity.

[24] Here now is the third step: for it was demonstrated [...] that the eccentricity of the equant of Mars is also to be precisely bisected; long ago Copernicus and Tycho were in doubt about this.

[25] Whence by induction from all the planets [...] it is demonstrated that, since there are no solid spheres (as Tycho showed from the paths of comets), the body of the sun is the source of the power which drives all the planets. I have also reasoned in such a way that it becomes clear that the sun, remaining in its place but rotating on its axis, emits through the breadth of the universe an immaterial *species* of its body analogous to the immaterial *species* of its light. This *species* of the rotation of the body of the sun is itself also rotated like a very rapid vortex throughout the breadth of the world, and it bears along with itself in its revolution the bodies of the planets with a more or less intense violence, according to the law of the greater or less density of its emission.

[26] Having set forth this common power by which the planets—each in its circle—are borne around the sun, it follows logically in my reasoning that individual moving agents should be attributed to each planet and located in their globes, since I have already rejected the solid spheres on the authority of Tycho.

[27] It is incredible how much labour the moving powers constituted by this way of argument caused me in the fourth part [of the book], giving false distances of the planets from the sun, irreconcilable with the observations, when I tried to work out the equations of the eccentric. This was not because these moving powers were wrongly invoked, but because I had forced them to tramp around in circles like donkeys in a mill, being bewitched by common opinion. Restrained by these fetters, they could not do their work.

[28] Nor was my burdensome task finished before I had taken the fourth step in the building of my physical hypotheses; after most laborious demonstrations and analyses of many observations, I learned that the path of a planet in the sky is not a circle, but is rather an oval course, perfectly elliptical.

[29] Geometry then came to my aid and taught me that such a path would be traced if we assign to each of the moving agents of the planets the task of librating its body in a straight line toward the sun. Not only this, but equations of the eccentric that are accurate and in conformity with the observations are effected by such a libration.

[30] Thus at length the coping stone was placed on the building, and it was demonstrated geometrically that a libration of this sort could be performed by a corporeal magnetic faculty. So these very moving agents of the planets are shown to be most probably nothing but attributes of the planetary bodies themselves, like the attributes of a magnet such as seeking for the Pole and attracting iron; thus the whole theory of the celestial motions may be based upon merely corporeal, that is magnetic, faculties, except for the gyration of the body of the sun remaining in its space, for which a vital faculty seems to be needed. [...]

10.A2 Celestial harmony

In this extract Kepler compares Ptolemy's *Harmonica*, a classical work on music theory, with his own achievement in *Harmonice Mundi*.

There, beyond my expectations and with the greatest wonder, I found approximately the whole third book given over to the same consideration of celestial harmony, fifteen hundred years ago. But indeed astronomy was far from being of age as yet; and Ptolemy, in an unfortunate attempt, could make others subject to despair, as being one who, like Scipio in Cicero, seemed to have recited a pleasant Pythagorean dream rather than to have aided philosophy. But both the crudeness of the ancient philosophy and this exact agreement in our meditations, down to the last hair, over an interval of fifteen centuries, greatly strengthened me in getting on with the job. For what need is there of many men? The very nature of things, in order to reveal herself to mankind, was at work in the different interpreters of different ages, and was the finger of God—to use the Hebrew expression; and here, in the minds of two men, who had wholly given themselves up to the contemplation of nature, there was the same conception as to the configuration of the world, although neither had been the other's guide in taking this route. But now since the first light eight months ago, since broad day three months ago, and since the sun of my wonderful speculation has shone fully a very few days ago: nothing holds me back. I am free to give myself up to the sacred madness, I am free to taunt mortals with the frank confession that I am stealing the golden vessels of the Egyptians, in order to build of them a temple for my God, far from the territory of Egypt. If you pardon me, I shall rejoice; if you are enraged, I shall bear up. The die is cast, and I am writing the book—whether to be read by my contemporaries or by posterity matters not. Let it await its reader for a hundred years, if God Himself has been ready for His contemplator for six thousand years.

10.A3 The regular solids

(*a*) *A celestial model*

Let the reader recall from my *Mysterium Cosmographicum*, which I published twenty-two years ago, that the number of the planets, or orbits about the Sun, was derived by the most wise Creator from the five solid figures, about which Euclid so many centuries ago wrote the book which, since it is made up of a series of propositions, is called *Elements*. That there cannot be more regular bodies, that regular plane figures, that is, cannot unite into a solid in more than five ways, was made clear in the second book of the present work.

As regards the relations of the planetary orbits, the relation between two neighbouring orbits is always such that, as will easily be seen, each one of the orbits approximates one of the terms of the ratio which exists between the orbits of one of the five solid bodies; the ratio, that is, of the orbit circumscribed about the figure to the orbit inscribed. For when, following the observations of Brahe, I had completed the demonstration of the distances, I discovered this fact: if the angles of the cube are applied to the innermost circle of Saturn, the centres of the planes nearly touch the middle circle of Jupiter, and if the angles of the tetrahedron rest on the innermost circle of Jupiter, the centres of the planes of the tetrahedron nearly touch the outermost circle of Mars; also, if the angles of the octahedron rise from any one of the circles of Venus (for all three are reduced to a very narrow space), the centres of the planes of the octahedron enter and descend below the outermost circle of Mercury; finally, coming to the ratios which exist between the orbits of the dodecahedron and the orbits of the icosahedron, which ratios are equal to each other, we find that the nearest of all to these are the ratios or distances between the circles of Mars and the Earth and between those of the Earth and Venus, and these ratios also, if we reckon from the innermost circle of Mars to the middle circle of the Earth and from the middle circle of the Earth to the middle circle of Venus, are similarly equal to each other; for the middle distance of the Earth is the mean proportional between the smallest distance of Mars and the middle distance of Venus; but these two ratios between the circles of the planets are still larger than are the ratios of those two sets of orbits in the figures, so that the centres of the planes of the dodecahedron do not touch the outermost circle of the Earth, nor do the centres of the planes of the icosahedron touch the outermost circle of Venus; and this hiatus is not filled up by the semidiameter of the orbit of the Moon, added to the greatest distance of the Earth and taken away from the smallest distance. But there is a certain other relation connected with a figures that I notice: if an enlarged dodecahedron to which I have given the name *echinus* (hedgehog) as being formed of twelve five-cornered stars and thereby being very near to the five regular bodies, if, I say, this dodecahedron should place its twelve points on the innermost circle of Mars, then the sides of the pentagons, which are, respectively, the bases of the different radii or points, touch the middle circle of Venus.

Briefly: the cube and the octahedron enter somewhat their conjugate planetary orbits, the dodecahedron and the icosahedron do not quite reach their conjugate orbits, the tetrahedron just touches both orbits; in the first case there is a deficiency, in the second case an excess, in the last case an equality, in the distances of the planets.

From these considerations it is apparent that the exact relations of the planetary distances were not derived from the regular figures alone; for the Creator, the very fountainhead of geometry, who, as Plato says, practises geometry eternally, does not deviate from his archetype. And indeed this fact might be gathered from the consideration that all the planets change their distances through definite periods of time; so that each one has two notable distances from the Sun, the maximum and minimum; and there may be made between every two planets a fourfold comparison of their distances from the Sun, comparisons of their maximum and of the minimum distances, and comparisons of their mutually opposed distances, those that are farthest apart and those that are nearest together; thus, of all the combinations of two neighbouring planets, the comparisons are twenty in number, while on the other hand the solid figures are but five. It is reasonable to believe, however, that the Creator, if he paid attention to the relation of the orbits in their general aspect, paid attention also to the relation of the varying distances of the individual orbits in detail, and that these acts of attention were the same in both cases and were connected with each other. When we duly consider this fact, we shall certainly arrive at the conclusion that for establishing the diameters and the eccentricities of the orbits there are required several principles in combination, besides the principle of the five regular bodies.

(*b*) *A planetary model*

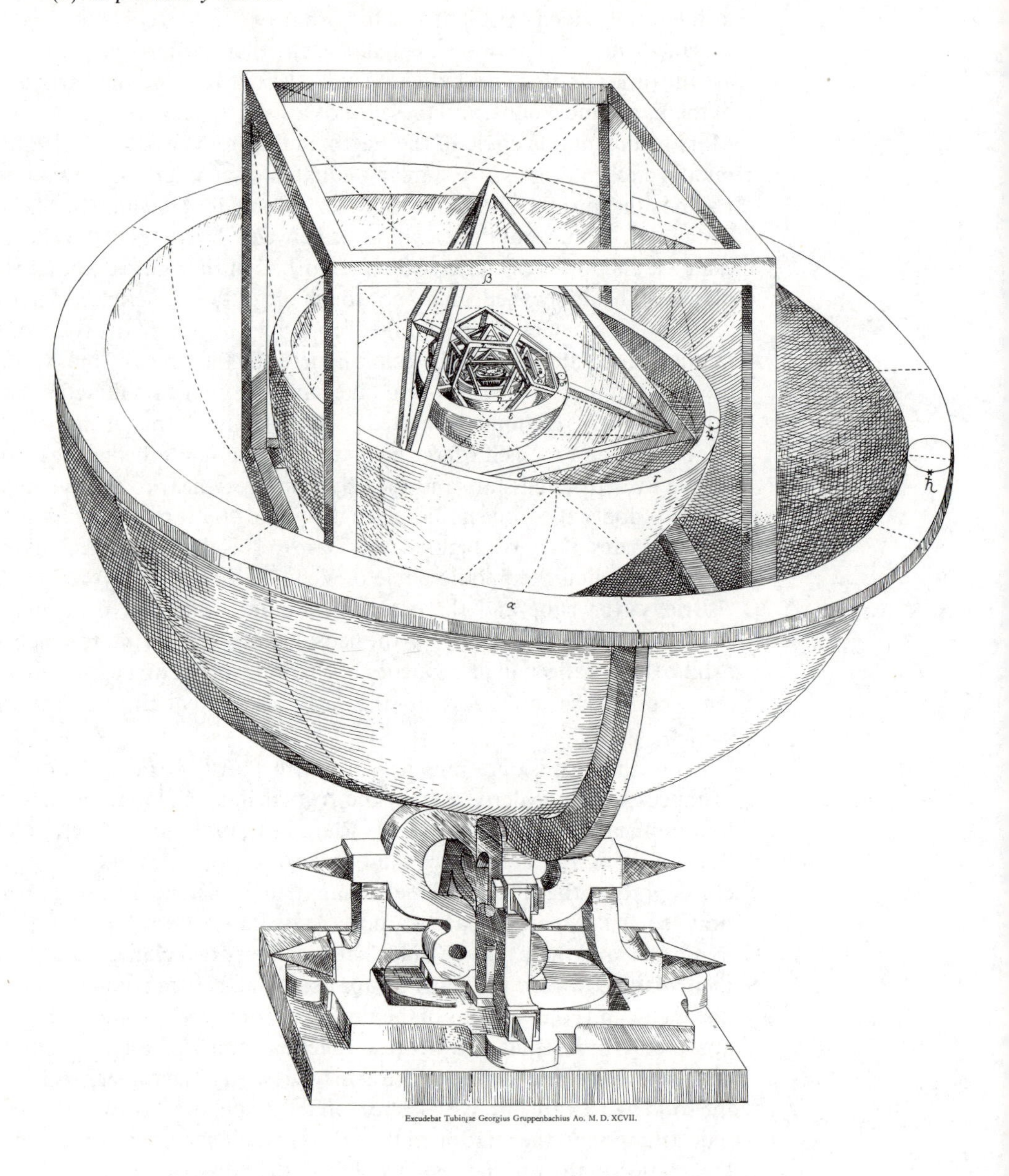

(*c*) *The stellated solids*

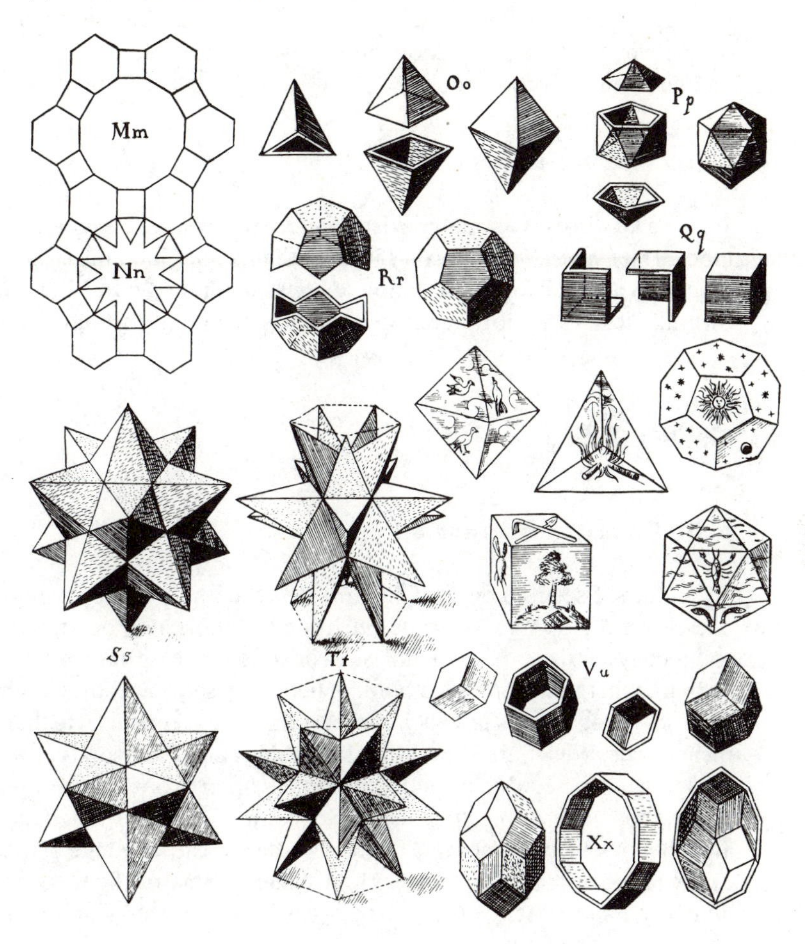

10.A4 The importance of geometry

(*a*) *The idea of geometric quantities* (*1596*)

These figures pleased me because they are quantities, that is, something which existed before the skies. For quantities were created before the beginning, together with substance; but the sky was only created on the second day. [. . .] The ideas of quantities have been and are in God for all eternity, they are God himself; they are therefore also present as archetypes in all minds created in God's likeness. On this point both the pagan philosophers and the teachers of the church agree.

(*b*) *The aims of his scientific work* (*1605*)

My aim is to show that the heavenly machine is not a kind of divine, live being, but a kind of clockwork (and he who believes that a clock has a soul, attributes the maker's glory to the works), insofar as nearly all the manifold motions are caused by a most simple, magnetic, and natural force, just as all the motions of the clock are caused by a simple weight. And I also show how these physical causes are to be given numerical and geometrical expression.

(c) *On what geometry is (1619)*

Why waste words? Geometry existed before the creation, is co-eternal with the mind of God, *is God himself* (what exists in God that is not God himself?); geometry provided God with a model for the Creation and was implanted into Man, together with God's own likeness—and not merely conveyed to his mind through the eyes.

10.B Galileo Galilei

10.B1 On mathematics and the world

(a) In Sarsi I seem to discern the firm belief that in philosophising one must support oneself on the opinion of some celebrated author, as if our minds ought to remain completely sterile and barren unless wedded to the reasoning of someone else. Possibly he thinks that philosophy is a book of fiction by some author, like the *Iliad* or *Orlando Furioso*—productions in which the least important thing is whether what is written in them is true. Well, Sarsi, that is not how things are. Philosophy is written in this grand book the universe, which stands continually open to our gaze. But the book cannot be understood unless one first learns to comprehend the language and to read the alphabet in which it is composed. It is written in the language of mathematics, and its characters are triangles, circles, and other geometric figures, without which it is humanly impossible to understand a single word of it; without these, one wanders about in a dark labyrinth.

(b) I say that the human intellect does understand some propositions perfectly, and thus in these it has as much absolute certainty as has Nature herself. Those are of the mathematical sciences alone; that is, geometry and arithmetic, in which the Divine intellect indeed knows infinitely more propositions than we do, since it knows all. Yet with regard to those few which the human intellect does understand, I believe that its knowledge equals the Divine in objective certainty—for here it succeeds in understanding necessity, than which there can be no greater certainty.

(c) When you apply a material sphere to a material plane in the concrete, you apply a sphere which is not perfect to a plane which is not perfect, and you say these do not touch at a single point alone. But I tell you that even in the abstract, an immaterial sphere that is not a perfect sphere can touch an immaterial plane that is not perfectly flat in not one point, but over part of its surface—so what happens here in the concrete happens in the same way in the abstract.

It would indeed be news to me if bookkeeping in abstract numbers did not correspond to concrete coins of gold and silver or to merchandise. Just as an accountant who wants his calculations to deal with sugar, silk and wool must discount boxes, bales, and packings, so the philosopher-geometer, when he wants to recognise in the concrete those effects which he has proved in the abstract, must deduct the material hindrances; and if he is able to do that, I assure you that material things are in no less agreement than arithmetical computations. The errors, then, reside not in abstractness or concreteness, but in a bookkeeper who does not understand how to balance his books.

10.B2 The regular motion of the pendulum

SALVIATI: We come now to the other questions, relating to pendulums, a subject which may appear to many exceedingly arid, especially to those philosophers who are continually occupied with the more profound questions of nature. Nevertheless, the problem is one which I do not scorn. I am encouraged by the example of Aristotle whom I admire especially because he did not fail to discuss every subject which he thought in any degree worthy of consideration.

Impelled by your queries I may give you some of my ideas concerning certain problems in music, a splendid subject, upon which so many eminent men have written: among these is Aristotle himself who has discussed numerous interesting acoustical questions. Accordingly, if on the basis of some easy and tangible experiments, I shall explain some striking phenomena in the domain of sound, I trust my explanations will meet your approval.

SAGREDO: I shall receive them not only gratefully but eagerly. For, although I take pleasure in every kind of musical instrument and have paid considerable attention to harmony, I have never been able to fully understand why some combinations of tones are more pleasing than others, or why certain combinations not only fail to please but are even highly offensive. Then there is the old problem of two stretched strings in unison; when one of them is sounded, the other begins to vibrate and to emit its note; nor do I understand the different ratios of harmony and some other details.

SALVIATI: Let us see whether we cannot derive from the pendulum a satisfactory solution of all these difficulties. And first, as to the question whether one and the same pendulum really performs its vibrations, large, medium, and small, all in exactly the same time, I shall rely upon what I have already heard from our Academician. He has clearly shown that the time of descent is the same along all chords, whatever the arcs which subtend them, as well along an arc of 180° (i.e., the whole diameter) as along one of 100°, 60°, 10°, 2°, $\frac{1}{2}$°, or 4′. It is understood, of course, that these arcs all terminate at the lowest point of the circle, where it touches the horizontal plane.

If now we consider descent along arcs instead of their chords then, provided these do not exceed 90°, experiment shows that they are all traversed in equal times; but these times are greater for the chord than for the arc, an effect which is all the more remarkable because at first glance one would think just the opposite to be true. For since the terminal points of the two motions are the same and since the straight line included between these two points is the shortest distance between them, it would seem reasonable that motion along this line should be executed in the shortest time;

but this is not the case, for the shortest time—and therefore the most rapid motion—is that employed along the arc of which this straight line is the chord.

As to the times of vibration of bodies suspended by threads of different lengths, they bear to each other the same proportion as the square roots of the lengths of the thread; or one might say the lengths are to each other as the squares of the times; so that if one wishes to make the vibration-time of one pendulum twice that of another, he must make its suspension four times as long. In like manner, if one pendulum has a suspension nine times as long as another, this second pendulum will execute three vibrations during each one of the first; from which it follows that the lengths of the suspending cords bear to each other the [inverse] ratio of the squares of the number of vibrations performed in the same time.

SAGREDO: You give me frequent occasion to admire the wealth and profusion of nature when, from such common and even trivial phenomena, you derive facts which are not only striking and new but which are often far removed from what we would have imagined. Thousands of times I have observed vibrations especially in churches where lamps, suspended by long cords, had been inadvertently set into motion; but the most which I could infer from these observations was that the view of those who think that such vibrations are maintained by the medium is highly improbable: for, in that case, the air must needs have considerable judgment and little else to do but kill time by pushing to and fro a pendent weight with perfect regularity. But I never dreamed of learning that one and the same body, when suspended from a string a hundred cubits long and pulled aside through an arc of 90° or even 1° or $\frac{1}{2}°$, would employ the same time in passing through the least as through the largest of these arcs; and, indeed, it still strikes me as somewhat unlikely. Now I am waiting to hear how these same simple phenomena can furnish solutions for those acoustical problems—solutions which will be at least partly satisfactory.

SALVIATI: First of all one must observe that each pendulum has its own time of vibration so definite and determinate that it is not possible to make it move with any other period than that which nature has given it. For let any one take in his hand the cord to which the weight is attached and try, as much as he pleases, to increase or diminish the frequency of its vibrations; it will be time wasted.

10.B3 Naturally accelerated motion

For I think no one believes that swimming or flying can be accomplished in a manner simpler or easier than that instinctively employed by fishes and birds.

When, therefore, I observe a stone initially at rest falling from an elevated position and continually acquiring new increments of speed, why should I not believe that such increases take place in a manner which is exceedingly simple and rather obvious to everybody? If now we examine the matter carefully we find no addition or increment more simple than that which repeats itself always in the same manner. This we readily understand when we consider the intimate relationship between time and motion; for just as uniformity of motion is defined by and conceived through equal times and equal spaces (thus we call a motion uniform when equal distances are traversed during equal time-intervals), so also we may, in a similar manner, through equal time-intervals, conceive additions of speed as taking place without complication; thus we may picture to our mind a motion as uniformly and continuously accelerated when, during any equal intervals of time whatever, equal increments of speed are given to it. Thus if any equal intervals of time whatever have elapsed, counting from the time at which the moving body left its position of rest and began to descend, the amount of speed acquired during the first two time-intervals will be double that acquired during the first time-interval alone; so the amount added during three of these time-intervals will be treble; and that in four, quadruple that of the first time-interval. To put the matter more clearly, if a body were to continue its motion with the same speed which it had acquired during the first time-interval and were to retain this same uniform speed, then its motion would be twice as slow as that which it would have if its velocity had been acquired during *two* time-intervals.

And thus, it seems, we shall not be far wrong if we put the increment of speed as proportional to the increment of time; hence the definition of motion which we are about to discuss may be stated as follows: A motion is said to be uniformly accelerated, when starting from rest, it acquires, during equal time-intervals, equal increments of speed.

10.B4 The time and distance laws for a falling body

Theorem I, Proposition I

The time in which any space is traversed by a body starting from rest and uniformly accelerated is equal to the time in which that same space would be traversed by the same body moving at a uniform speed whose value is the mean of the highest speed and the speed just before acceleration began.

Let us represent by the line *AB* the time in which the space *CD* is traversed by a body which starts from rest at *C* and is uniformly accelerated; let the final and highest value of the speed gained during the interval *AB* be represented by the line *EB* drawn at right angles to *AB*; draw the line *AE*, then all lines drawn from equidistant points on *AB* and parallel to *BE* will represent the increasing values of the speed, beginning with the instant *A*. Let the point *F* bisect the line *EB*; draw *FG* parallel to *BA*, and *GA* parallel to *FB*, thus forming a parallelogram *AGFB* which will be equal in area to the triangle *AEB*, since the side *GF* bisects the side *AE* at the point *I*; for if the parallel lines in the triangle *AEB* are extended to *GI*, then the sum of all the parallels contained in the quadrilateral is equal to the sum of those contained in the triangle *AEB*; for those in the triangle *IEF* are equal to those contained in the triangle *GIA*, while those included in the trapezium *AIFB* are common. Since each and every instant of time in the time-interval *AB* has its corresponding point on the line *AB*, from which points parallels drawn in and limited by the triangle *AEB* represent the increasing values of the growing velocity, and since parallels contained within the rectangle represent the values of a speed which is not increasing, but constant, it appears, in like manner, that the momenta assumed by the moving body may also be represented, in the case of the accelerated motion, by the increasing parallels of the triangle *AEB*, and, in the case of the uniform motion, by the parallels of the rectangle *GB*. For, what the momenta may lack in the first part of the accelerated motion (the deficiency of the momenta being represented by the parallels of the triangle *AGI*) is made up by the momenta represented by the parallels of the triangle *IEF*.

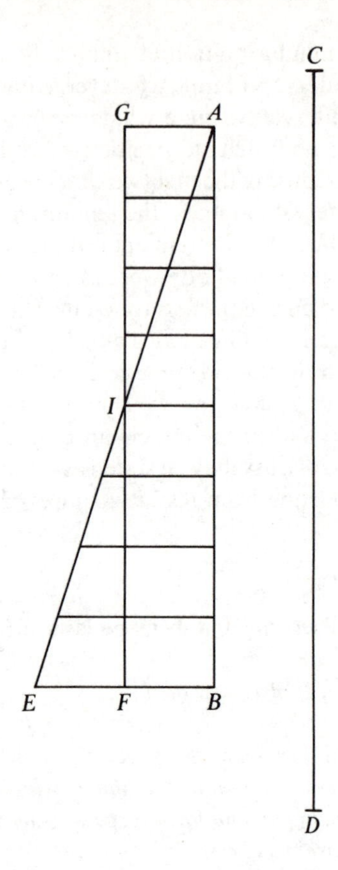

Hence it is clear that equal spaces will be traversed in equal times by two bodies, one of which, starting from rest, moves with a uniform acceleration, while the momentum of the other, moving with uniform speed, is one-half its maximum momentum under accelerated motion. Q.E.D.

Theorem II, Proposition II

The spaces described by a body falling from rest with a uniformly accelerated motion are to each other as the squares of the time-intervals employed in traversing these distances.

Let the time beginning with any instant *A* be represented by the straight line *AB* in which are taken any two time-intervals *AD* and *AE*. Let *HI* represent the distance through which the body, starting from rest at *H*, falls with uniform acceleration. If *HL* represents the space traversed during the time-interval *AD*, and *HM* that covered during the interval *AE*, then the space *MH* stands to the space *LH* in a ratio which is the square of the ratio of the time *AE* to the time *AD*; or we may say simply that the distances *HM* and *HL* are related as the squares of *AE* and *AD*.

Draw the line *AC* making any angle whatever with the line *AB*; and from the points *D* and *E*, draw the parallel lines *DO* and *EP*; of these two lines, *DO* represents the greatest velocity attained during the interval *AD*, while *EP* represents the maximum velocity acquired during the interval *AE*. But it has just been proved that so far as

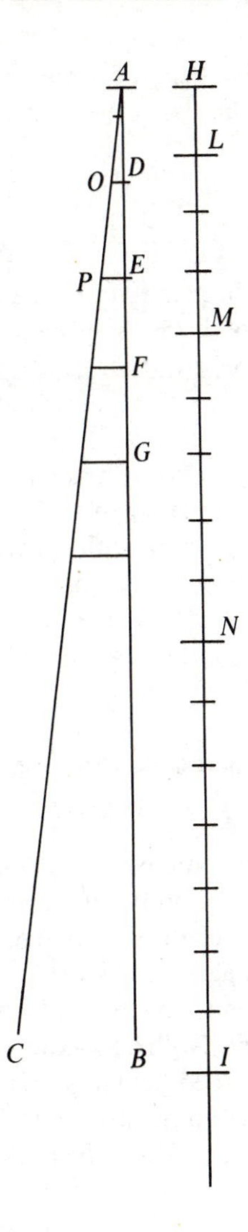

distances traversed are concerned it is precisely the same whether a body falls from rest with a uniform acceleration or whether it falls during an equal time-interval with a constant speed which is one-half the maximum speed attained during the accelerated motion. It follows therefore that the distances *HM* and *HL* are the same as would be traversed, during the time-intervals *AE* and *AD*, by uniform velocities equal to one-half those represented by *DO* and *EP* respectively. If, therefore, one can show that the distances *HM* and *HL* are in the same ratio as the squares of the time-intervals *AE* and *AD*, our proposition will be proven.

But in the fourth proposition of the first book it has been shown that the spaces traversed by two particles in uniform motion bear to one another a ratio which is equal to the product of the ratio of the velocities by the ratio of the times. But in this case the ratio of the velocities is the same as the ratio of the time-intervals (for the ratio of *AE* to *AD* is the same as that of $\frac{1}{2}EP$ to $\frac{1}{2}DO$ or of *EP* to *DO*). Hence the ratio of the spaces traversed is the same as the squared ratio of the time-intervals. Q.E.D.

Evidently then the ratio of the distances is the square of the ratio of the final velocities, that is, of the lines *EP* and *DO*, since these are to each other as *AE* to *AD*.

Corollary I

Hence it is clear that if we take any equal intervals of time whatever, counting from the beginning of the motion, such as *AD*, *DE*, *EF*, *FG*, in which the spaces *HL*, *LM*, *MN*, *NI* are traversed, these spaces will bear to one another the same ratio as the series of odd numbers, 1, 3, 5, 7; for this is the ratio of the differences of the squares of the lines [which represent time], differences which exceed one another by equal amounts, this excess being equal to the smallest line [viz. the one representing a single time-interval]: or we may say [that this is the ratio] of the differences of the squares of the natural numbers beginning with unity.

While, therefore, during equal intervals of time the velocities increase as the natural numbers, the increments in the distances traversed during these equal time-intervals are to one another as the odd numbers beginning with unity.

10.B5 The parabolic path of a projectile

We can now resume the text and see how he demonstrates his first proposition in which he shows that a body falling with a motion compounded of a uniform horizontal and a naturally accelerated one describes a semi-parabola.

Let us imagine an elevated horizontal line or plane *ab* along which a body moves with uniform speed from *a* to *b*. Suppose this plane to end abruptly at *b*; then at this point the body will, on account of its weight, acquire also a natural motion downwards along the perpendicular *bn*. Draw the line *be* along the plane *ba* to represent the flow, or measure, of time; divide this line into a number of segments, *bc*, *cd*, *de*, representing equal intervals of time; from the points *b*, *c*, *d*, *e*, let fall lines which are parallel to the

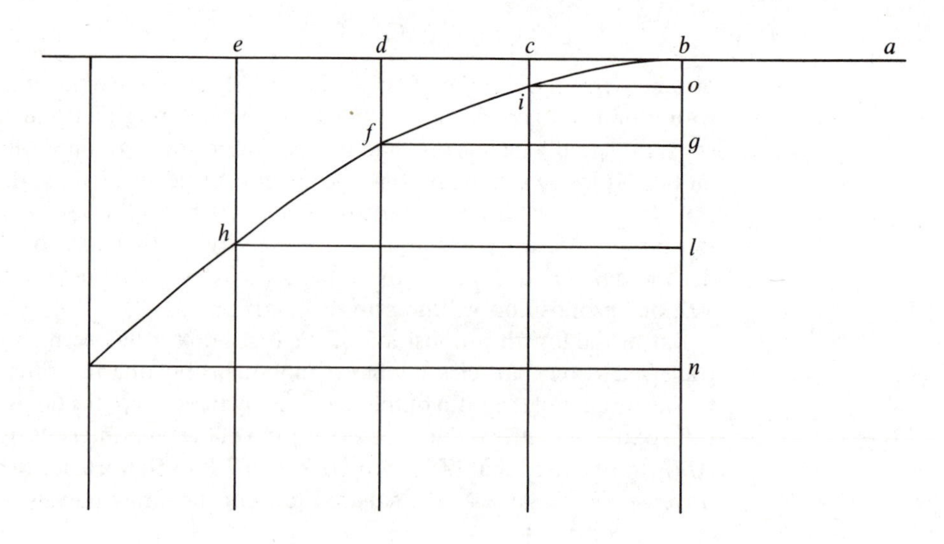

perpendicular *bn*. On the first of these lay off any distance *ci*, on the second a distance four times as long, *df*; on the third, one nine times as long, *eh*; and so on, in proportion to the squares of *cb*, *db*, *eb*, or, we may say, in the squared ratio of these same lines. Accordingly we see that while the body moves from *b* to *c* with uniform speed, it also falls perpendicularly through the distance *ci*, and at the end of the time-interval *bc* finds itself at the point *i*. In like manner at the end of the time-interval *bd*, which is the double of *bc*, the vertical fall will be four times the first distance *ci*; for it has been shown in a previous discussion that the distance traversed by a freely falling body varies as the square of the time; in like manner the space *eh* traversed during the time *be* will be nine times *ci*; thus it is evident that the distances *eh*, *df*, *ci* will be to one another as the squares of the lines *be*, *bd*, *bc*. Now from the points *i*, *f*, *h* draw the straight lines *io*, *fg*, *hl* parallel to *be*; these lines *hl*, *fg*, *io* are equal to *eb*, *db* and *cb*, respectively; so also are the lines *bo*, *bg*, *bl* respectively equal to *ci*, *df*, and *eh*. The square of *hl* is to that of *fg* as the line *lb* is to *bg*; and the square of *fg* is to that of *io* as *gb* is to *bo*; therefore the points *i*, *f*, *h*, lie on one and the same parabola. In like manner it may be shown that, if we take equal time-intervals of any size whatever, and if we imagine the particle to be carried by a similar compound motion, the positions of this particle, at the ends of these time-intervals, will lie on one and the same parabola. Q.E.D.

11 Descartes, Fermat and Their Contemporaries

11.A René Descartes

For René Descartes (1596–1650), the study of mathematics was only part of a larger task, that of creating a new and reliable body of knowledge, and even in mathematics he saw himself as merely refining existing methods (or so he tells us in 11.A1). It is not even easy to capture his originality precisely. But no single mathematical work in the first half of the seventeenth century generated quite so much excitement or taught its readers so much as the appendix on geometry Descartes added to his *Discourse on Method* (1637). Like his predecessor François Viète, Descartes advocated using letters and formal manipulations with symbols to analyse geometrical problems, but his algebraic analysis was both literally and conceptually easier to use. The elimination of dimensional considerations (see 11.A2) turns out to have lifted a weight off everyone's shoulders, simple though it may seem. It enabled Descartes to propose a general way of approaching problems in the theory of equations which was particularly fruitful (11.A3). As H. J. M. Bos has argued (passages from his interesting paper of 1981 are quoted in 11.A10), Descartes took his problems from geometry and therefore sought to answer them there, by giving a geometrical construction for the unknown. So his algebra had eventually to return one to geometry, which it did by providing certain arguments about curves. In this way Descartes reworked geometry, for by passing from curves to their equations he necessarily did a great deal to show how curves could be studied in their own right.

Indeed one particular problem seems to have been crucial to his development: the Pappus problem of the locus to three or four lines (see 11.A4). It seems that it was when he realized in 1632 that his methods made this problem easy that he realized how far he had come, as his letter to Marin Mersenne (11.A5) attests. When he discussed the problem in *Geometry,* it provided the occasion for demonstrating how well algebraic analysis deals with problems whose solutions are curves, and in this way

Descartes suggested how many new curves can be created and described (11.A6–11.A8). As Bos discusses, Descartes' method, being limited to algebraic curves, was hard for him to pin down, for Descartes needed an *a priori* definition of geometric which was to prove elusive. Later workers were not so devoted to a philosophical position and could gratefully accept a technique so well adapted to such a large class of curves; we have included in the next section two prefaces by Philippe de la Hire (11.B2–11.B3) to show how the method was argued for, when it was used, and why it was otherwise avoided.

Perhaps the most important problems in geometry at this time concerned the findings of tangents to curves and the areas enclosed by curves, together with the related problems of volumes and centres of gravity. Algebraic analysis played a role here too, as is abundantly clear in the work of Pierre de Fermat (1601–1665) (see 11.C1–11.C3), but all the work rested ultimately on some geometric intuition, even Descartes's strikingly algebraic approach to tangency (11.A9). The contribution of algebra was not only to facilitate the work, but also to aid in the formation of generalizations which were to culminate in the formal algorithms of the calculus as invented by Isaac Newton and Wilhelm Leibniz. It is in this sense that Descartes's *Geometry* made possible the other great mathematical breakthrough of the century.

11.A1 Descartes's method

[1] Among the branches of philosophy, I had, when younger, studied logic, and among those of mathematics, geometrical analysis and algebra; three arts or sciences which should have been able to contribute something to my design. But in examining them I noticed that as far as logic was concerned, its syllogisms and most of its other methods serve rather to explain to another what one already knows, or even, as in the art of Lully, to speak freely and without judgement of what one does not know, than to learn new things. Although it does contain many true and good precepts, they are interspersed among so many others that are harmful or superfluous that it is almost as difficult to separate them as to bring forth a Diana or a Minerva from a block of virgin marble. Then, as far as the analysis of the Greeks and the algebra of the moderns is concerned, besides the fact that they deal with abstractions and speculations which appear to have no utility, the first is always so limited to the consideration of figures that it cannot exercise the understanding without greatly fatiguing the imagination, and the last is so limited to certain rules and certain numbers that it has become a confused and obscure art which perplexes the mind instead of a science which educates it. In consequence I thought that some other method must be found to combine the advantages of these three and to escape their faults. Finally, just as the multitude of laws frequently furnishes an excuse for vice, and a state is much better governed with a few laws which are strictly adhered to, so I thought that instead of the great number of precepts of which logic is composed, I would have enough with the four following ones, provided that I made a firm and unalterable resolution not to violate them even in a single instance.

[2] The first rule was never to accept anything as true unless I recognized it to be certainly and evidently such: that is, carefully to avoid all precipitation and prejudgement, and to include nothing in my conclusions unless it presented itself so clearly and distinctly to my mind that there was no reason or occasion to doubt it.

The second was to divide each of the difficulties which I encountered into as many parts as possible, and as might be required for an easier solution.

The third was to think in an orderly fashion when concerned with the search for truth, beginning with the things which were simplest and easiest to understand, and gradually and by degrees reaching toward more complex knowledge, even treating, as though ordered, materials which were not necessarily so.

The last was, both in the process of searching and in reviewing when in difficulty, always to make enumerations so complete, and reviews so general, that I would be certain that nothing was omitted.

[3] Those long chains of reasoning, so simple and easy, which enabled the geometricians to reach the most difficult demonstrations, had made me wonder whether all things knowable to men might not fall into a similar logical sequence. If so, we need only refrain from accepting as true that which is not true, and carefully follow the order necessary to deduce each one from the others, and there cannot be any propositions so abstruse that we cannot prove them, or so recondite that we cannot discover them. It was not very difficult, either, to decide where we should look for a beginning, for I knew already that one begins with the simplest and easiest to know. Considering that among all those who have previously sought truth in the sciences, mathematicians alone have been able to find some demonstrations, some certain and evident reasons, I had no doubt that I should begin where they did, although I expected no advantage except to accustom my mind to work with truths and not to be satisfied with bad reasoning. I do not mean that I intended to learn all the particular branches of mathematics; for I saw that although the objects they discuss are different, all these branches are in agreement in limiting their consideration to the relationships or proportions between their various objects. I judged therefore that it would be better to examine these proportions in general, and use particular objects as illustrations only in order to make their principles easier to comprehend, and to be able the more easily to apply them afterwards, without any forcing, to anything for which they would be suitable. I realized that in order to understand the principles of relationships I would sometimes have to consider them singly, and sometimes comprehend and remember them in groups. I thought I could consider them better singly as relationships between lines, because I could find nothing more simple or more easily pictured to my imagination and my senses. But in order to remember and understand them better when taken in groups, I had to express them in numbers, and in the smallest numbers possible. Thus I took the best traits of geometrical analysis and algebra, and corrected the faults of one by the other.

11.A2 The elementary arithmetical operations

And I shall not hesitate to introduce these arithmetical terms into geometry, for the sake of greater clearness.

For example, let AB be taken as unity, and let it be required to multiply BD by BC. I have only to join the points A and C, and draw DE parallel to CA; then BE is the product of BD and BC.

If it be required to divide BE by BD, I join E and D, and draw AC parallel to DE; then BC is the result of the division.

If the square root of GH is desired, I add, along the same straight line, FG equal to unity; then, bisecting FH at K, I describe the circle FIH about K as a centre, and draw from G a perpendicular and extend it to I, and GI is the required root. I do not speak here of cube root, or other roots, since I shall speak more conveniently of them later.

11.A3 The general method for solving any problem

[1] If, then, we wish to solve any problem, we first suppose the solution already effected, and give names to all the lines that seem needful for its construction—to those that are unknown as well as to those that are known. Then, making no distinction between known and unknown lines, we must unravel the difficulty in any way that shows most naturally the relations between these lines, until we find it possible to express a single quantity in two ways. This will constitute an equation, since the terms of one of these two expressions are together equal to the terms of the other.
[2] We must find as many such equations as there are supposed to be unknown lines; but if, after considering everything involved, so many cannot be found, it is evident that the question is not entirely determined. In such a case we may choose arbitrarily lines of known length for each unknown line to which there corresponds no equation.
[3] If there are several equations, we must use each in order, either considering it alone or comparing it with the others, so as to obtain a value for each of the unknown lines; and so we must combine them until there remains a single unknown line which is equal to some known line, or whose square, cube, fourth power, fifth power, sixth power, etc., is equal to the sum or difference of two or more quantities, one of which is known, while the others consist of mean proportionals between unity and this square, or cube, or fourth power, etc., multiplied by other known lines. I may express this as follows:

$$z = b,$$

$$\text{or} \quad z^2 = -az + b^2,$$

$$\text{or} \quad z^3 = az^2 + b^2z - c^3,$$

$$\text{or} \quad z^4 = az^3 - c^3z + d^4, \quad \text{etc.}$$

That is, z, which I take for the unknown quantity, is equal to b; or, the square of z is equal to the square of b diminished by a multiplied by z; or, the cube of z is equal to a multiplied by the square of z, plus the square of b multiplied by z, diminished by the cube of c; and similarly for the others.
[4] Thus, all the unknown quantities can be expressed in terms of a single quantity, whenever the problem can be constructed by means of circles and straight lines, or by conic sections, or even by some other curve of degree not greater than the third or fourth.
[5] But I shall not stop to explain this in more detail, because I should deprive you of the pleasure of mastering it yourself, as well as of the advantage of training your mind

by working over it, which is in my opinion the principal benefit to be derived from this science. Because, I find nothing here so difficult that it cannot be worked out by any one at all familiar with ordinary geometry and with algebra, who will consider carefully all that is set forth in this treatise.

[6] I shall therefore content myself with the statement that if the student, in solving these equations, does not fail to make use of division wherever possible, he will surely reach the simplest terms to which the problem can be reduced.

[7] And if it can be solved by ordinary geometry, that is, by the use of straight lines and circles traced on a plane surface, when the last equation shall have been entirely solved there will remain at most only the square of an unknown quantity, equal to the product of its root by some known quantity, increased or diminished by some other quantity also known. Then this root or unknown line can easily be found. For example, if I have $z^2 = az + b^2$, I construct a right triangle NLM with one side LM, equal to b, the square root of the known quantity b^2, and the other side, LN, equal to $\frac{1}{2}a$, that is, to half the other known quantity which was multiplied by z, which I supposed to be the unknown line. Then prolonging MN, the hypotenuse of this triangle, to O, so that NO is equal to NL, the whole line OM is the required line z. This is expressed in the following way:

$$z = \tfrac{1}{2}a + \sqrt{\tfrac{1}{4}a^2 + b^2}\,.$$

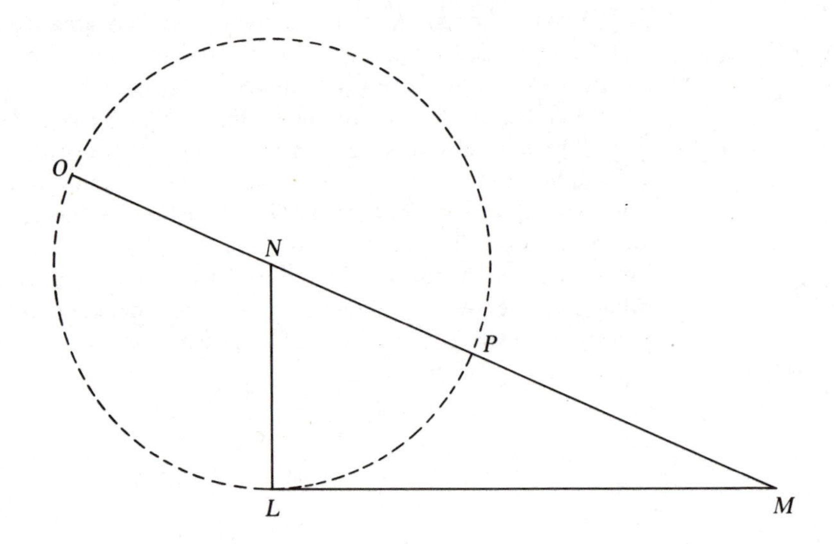

11.A4 Pappus on the locus to three, four or several lines

Apollonius, who completed the four books of Euclid's *Conics* and added another four, gave us eight books of *Conics*. Aristaeus, who wrote the still extant five books of *Solid Loci* supplementary to the *Conics*, called the three conics sections of an acute-angled, right-angled and obtuse-angled cone respectively. [...] Apollonius says in his third book that the 'locus with respect to three or four lines' had not been fully worked out by Euclid, and in fact neither Apollonius himself nor anyone else could have added anything to what Euclid wrote, using only those properties of conics which had been proved up to Euclid's time; as Apollonius himself bears witness when he says that the locus could not be fully investigated without the propositions that he had been compelled to work out for himself. Now Euclid regarded Aristaeus as deserving credit for his contributions to conics, and did not try to anticipate him or to overthrow his system; for he showed scrupulous fairness and exemplary kindness towards all who were able in any degree to advance mathematics, and was never offensive, but aimed at accuracy, and did not boast like the other. Accordingly he wrote so much about the locus as was possible by means of the *Conics* of Aristaeus, but did not claim finality for his proofs. If he had done so, we should have been obliged to censure him, but as things are he is in no wise to blame, seeing that Apollonius himself is not called to account, though he left the most part of his *Conics* incomplete. Moreover Apollonius was able to add the lacking portion of the theory of the locus through having become familiar beforehand with what had been written about it by Euclid, and through having spent much time with Euclid's pupils at Alexandria, whence he derived his scientific habit of mind.

Now this 'locus with respect to three and four lines', the theory of which he is so proud of having expanded—though he ought rather to acknowledge his debt to the original author—is of this kind. If three straight lines be given in position, and from one and the same point straight lines be drawn to meet the three straight lines at given angles, and if the ratio of the rectangle contained by two of the straight lines towards the square on the remaining straight line be given, then the point will lie on a solid locus given in position, that is on one of the three conic sections. And if straight lines be drawn to meet at given angles four straight lines given in position, and the ratio of the rectangle contained by two of the straight lines so drawn towards the rectangle contained by the remaining two be given, then in the same way the point will lie on a conic section given in position.

If from any point straight lines be drawn to meet at given angles five straight lines given in position, and the ratio be given between the volume of the rectangular parallelepiped contained by three of them to the volume of the rectangular parallelepiped contained by the remaining two and a given straight line, the point will lie on a curve given in position. If there be six straight lines, and the ratio be given between the volume of the aforesaid solid formed by three of them to the volume of the solid formed by the remaining three, the point will again lie on a curve given in position. If there be more than six straight lines, it is no longer permissible to say 'if the ratio be given between some figure contained by four of them to some figure contained by the remainder', since no figure can be contained in more than three dimensions. It is true that some recent writers have agreed among themselves to use such expressions, but they have no clear meaning when they multiply the rectangle contained by these straight lines with the square on that or the rectangle contained by those. They might, however, have expressed such matters by means of the composition of ratios, and have given a general proof both for the aforesaid propositions and for further propositions after this manner: *If from any point straight lines be drawn to meet at given angles straight lines given in position, and there be given the ratio compounded of that which one straight line so drawn bears to another, that which a second bears to a second, that which a third bears to a third, and that which the fourth bears to a given straight line—if there be seven, or, if there be eight, that which the fourth bears to the fourth—the point will lie on a curve given in position;* and similarly, however many the straight lines be, and whether odd or even. Though, as I said, these propositions follow the locus on four lines, [geometers] have by no means solved them to the extent that the curve can be recognized.

11.A5 Descartes to Marin Mersenne (1632)

In my last letter I did not thank you for the demonstration of the two geometrical means which you sent me: but I had not yet received your letters, and I must tell you that Mr. Mydorge also found the demonstration for them, since you got me to make the construction for them, and I never judged it to be difficult. I would prefer you to have proposed the construction by trisecting the angle, a method which, if I am not mistaken, I gave you at the same time as the other; for it is a little less easy, and Mr. Mydorge admitted to me that he had not been able to demonstrate it. But I should like it even better if they would practice attempting Pappus's proposition, for it is said that Mr. Mydorge has put a solution to it in his *Conics*; but those who, like me, have examined it a little closely, cannot easily be persuaded of this. I do not think that they could persuade Mr. Golius of it either. He told me he had once proposed it to Mr. Mydorge, as you may easily ascertain, if you wish to write to him about it.

11.A6 Descartes's solution to the Pappus problem

This led me to try to find out whether, by my own method, I could go as far as they had gone.

First, I discovered that if the question be proposed for only three, four, or five lines, the required points can be found by elementary geometry, that is, by the use of the ruler and compasses only, and the application of those principles that I have already explained, except in the case of five parallel lines. In this case, and in the cases where there are six, seven, eight, or nine given lines, the required points can always be found by means of the geometry of solid loci, that is, by using some one of the three conic sections. Here, again, there is an exception in the case of nine parallel lines. For this and the cases of ten, eleven, twelve, or thirteen given lines, the required points may be found by means of a curve of degree next higher than that of the conic sections. Again, the case of thirteen parallel lines must be excluded, for which, as well as for the cases of fourteen, fifteen, sixteen, and seventeen lines, a curve of degree next higher than the preceding must be used; and so on indefinitely.

Next, I have found that when only three or four lines are given, the required points lie not only all on one of the conic sections but sometimes on the circumference of a circle or even on a straight line.

When there are five, six, seven, or eight lines, the required points lie on a curve of degree next higher than the conic sections, and it is impossible to imagine such a curve that may not satisfy the conditions of the problem; but the required points may possibly lie on a conic section, a circle, or a straight line. If there are nine, ten, eleven, or twelve lines, the required curve is only one degree higher than the preceding, but any such curve may meet the requirements, and so on to infinity.

Finally, the first and simplest curve after the conic sections is the one generated by the intersection of a parabola with a straight line in a way to be described presently.

I believe that I have in this way completely accomplished what Pappus tells us the ancients sought to do, and I will try to give the demonstration in a few words, for I am already wearied by so much writing.

Let $AB, AD, EF, GH, \ldots$ be any number of straight lines given in position, and let it be required to find a point C, from which straight lines $CB, CD, CF, CH, \ldots$ can be drawn, making given angles $CBA, CDA, CFE, CHG, \ldots$ respectively, with the given lines, and such that the product of certain of them is equal to the product of the rest, or at least such that these two products shall have a given ratio, for this condition does not make the problem any more difficult.

First, I suppose the thing done, and since so many lines are confusing, I may simplify matters by considering one of the given lines and one of those to be drawn (as, for example, AB and BC) as the principal lines, to which I shall try to refer all the others. Call the segment of the line AB between A and B, x, and call BC, y. Produce all the other given lines to meet these two (also produced if necessary) provided none is parallel to either of the principal lines. Thus, in the figure, the given lines cut AB in the points A, E, G, and cut BC in the points R, S, T.

Now, since all the angles of the triangle ARB are known, the ratio between the sides AB and BR is known. If we let $AB{:}BR = z{:}b$, since $AB = x$, we have $RB = \frac{bx}{z}$; and since B lies between C and R, we have $CR = y + \frac{bx}{z}$. (When R lies between C and B, CR is equal to $y - \frac{bx}{z}$, and when C lies between B and R, CR is equal to $-y + \frac{bx}{z}$.) Again, the three angles of the triangle DRC are known, and therefore the ratio between the sides CR and CD is determined. Calling this ratio $z{:}c$, since $CR = y + \frac{bx}{z}$, we have $CD = \frac{cy}{z} + \frac{bcx}{z^2}$. Then, since the lines AB, AD, and EF are given in position, the distance from A to E is known. If we call this distance k, then $EB = k + x$; although $EB = k - x$ when B lies between E and A, and $E = -k + x$ when E lies between A and B. Now the angles of the triangle ESB being given, the ratio of BE to BS is known. We may call this ratio $z{:}d$. Then $BS = \frac{dk + dx}{z}$ and $CS = \frac{zy + dk + dx}{z}$. When S lies between B and C we have $CS = \frac{zy - dk - dx}{z}$, and when C lies between B and S we have $CS = \frac{-zy + dk + dx}{z}$. The angles of the triangle FSC are known, and hence, also the ratio of CS to CF, or $z{:}e$. Therefore, $CF = \frac{ezy + dek + dex}{z^2}$. Likewise, AG or l is given, and $BG = l - x$. Also, in triangle BGT, the ratio of BG to BT, or $z{:}f$, is known. Therefore, $BT = \frac{fl - fx}{z}$ and $CT = \frac{zy + fl - fx}{z}$. In triangle TCH, the ratio of TC to CH, or $z{:}g$, is known, whence $CH = \frac{gzy + fgl - fgx}{z^2}$.

And thus you see that, no matter how many lines are given in position, the length of any such line through C making given angles with these lines can always be expressed by three terms, one of which consists of the unknown quantity y multiplied or divided by some known quantity; another consisting of the unknown quantity x multiplied or divided by some other known quantity; and the third consisting of a known quantity. An exception must be made in the case where the given lines are parallel either to AB (when the term containing x vanishes), or to CB (when the term containing y vanishes). This case is too simple to require further explanation. The signs of the terms may be either $+$ or $-$ in every conceivable combination.

You also see that in the product of any number of these lines the degree of any term containing x or y will not be greater than the number of lines (expressed by means of x and y) whose product is found. Thus, no term will be of degree higher than the second if two lines be multiplied together, nor of degree higher than the third, if there be three lines, and so on to infinity.

Furthermore, to determine the point C, but one condition is needed, namely, that the product of a certain number of lines shall be equal to, or (what is quite as simple), shall bear a given ratio to the product of certain other lines. Since this condition can be expressed by a single equation in two unknown quantities, we may give any value we please to either x or y and find the value of the other from this equation. It is obvious that when not more than five lines are given, the quantity x, which is not used to express the first of the lines can never be of degree higher than the second.

Assigning a value to y, we have $x^2 = \pm ax \pm b^2$, and therefore x can be found with ruler and compasses, by a method already explained. If then we should take successively an infinite number of different values for the line y, we should obtain an infinite number of values for the line x, and therefore an infinity of different points, such as C, by means of which the required curve could be drawn.

11.A7 'Geometric' curves

[1] Probably the real explanation of the refusal of ancient geometers to accept curves more complex than the conic sections lies in the fact that the first curves to which their attention was attracted happened to be the spiral, the quadratrix, and similar curves, which really do belong only to mechanics, and are not among those curves that I think should be included here, since they must be conceived of as described by two separate movements whose relation does not admit of exact determination. [...]
[2] Consider the lines AB, AD, AF, and so forth, which we may suppose to be described by means of the instrument YZ. This instrument consists of several rulers hinged together in such a way that YZ being placed along the line AN the angle XYZ can be increased or decreased in size, and when its sides are together the points B, C, D, E, F, G, H, all coincide with A; but as the size of the angle is increased, the ruler BC, fastened at right angles to XY at the point B, pushes toward Z the ruler CD which slides along YZ always at right angles. In like manner, CD pushes DE which slides along YX always parallel to BC; DE pushes EF; EF pushes FG; FG pushes GH, and so on. Thus we may imagine an infinity of rulers, each pushing another, half of them making equal angles with YX and the rest with YZ.

Now as the angle XYZ is increased the point B describes the curve AB, which is a circle; while the intersections of the other rulers, namely, the points D, F, H describe other curves, AD, AF, AH, of which the latter are more complex than the first and this more complex than the circle. Nevertheless I see no reason why the description of the first cannot be conceived as clearly and distinctly as that of the circle, or at least as that of the conic sections; or why that of the second, third, or any other that can be thus described, cannot be as clearly conceived of as the first: and therefore I see no reason why they should not be used in the same way in the solution of geometric problems.

I could give here several other ways of tracing and conceiving a series of curved lines, each curve more complex than any preceding one, but I think the best way to group together all such curves and then classify them in order, is by recognizing the fact that all points of those curves which we may call 'geometric', that is, those which admit of precise and exact measurement, must bear a definite relation to all points of a straight line, and that this relation must be expressed by means of a single equation. If this equation contains no term of higher degree than the rectangle of two unknown quantities, or the square of one, the curve belongs to the first and simplest class, which contains only the circle, the parabola, the hyperbola, and the ellipse; but when the equation contains one or more terms of the third or fourth degree in one or both of the two unknown quantities (for it requires two unknown quantities to express the relation between two points) the curve belongs to the second class; and if the equation contains a term of the fifth or sixth degree in either or both of the unknown quantities the curve belongs to the third class, and so on indefinitely. [...]

[3] But the curve AD is of the second class, while it is possible to find two mean proportionals by the use of the conic sections, which are curves of the first class. Again, four or six mean proportionals can be found by curves of lower classes than AF and AH respectively. It would therefore be a geometric error to use these curves. On the other hand, it would be a blunder to try vainly to construct a problem by means of a class of lines simpler than its nature allows.

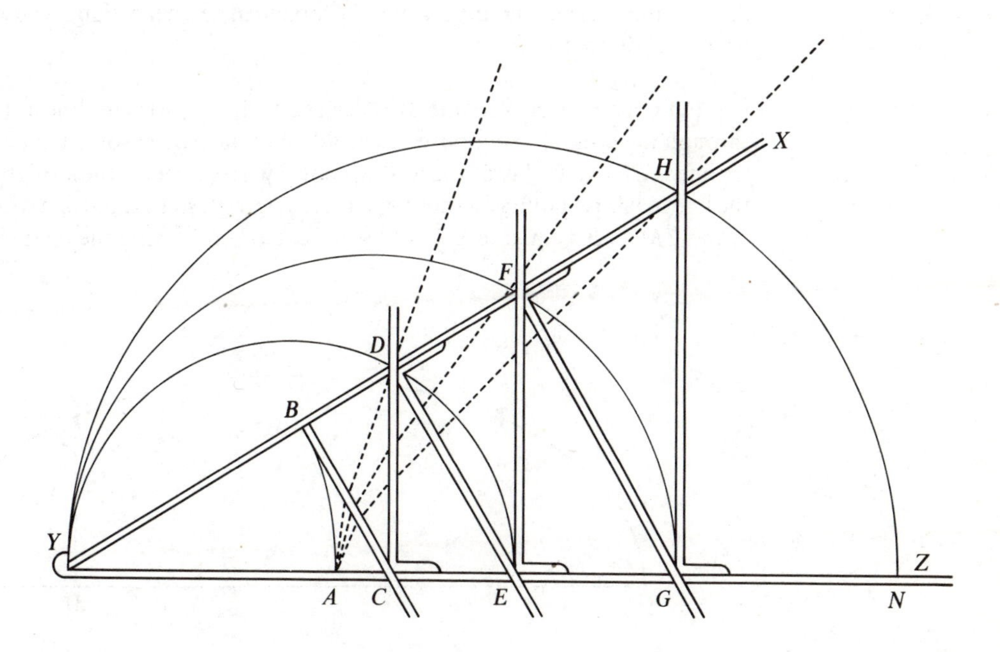

11.A8 Permissible and impermissible methods in geometry

But the fact that this method of tracing a curve by determining a number of its points taken at random applies only to curves that can be generated by a regular and continuous motion does not justify its exclusion from geometry. Nor should we reject the method in which a string or loop of thread is used to determine the equality or difference of two or more straight lines drawn from each point of the required curve to certain other points, or making fixed angles with certain other lines. We have used this method in *La Dioptrique* in the discussion of the ellipse and the hyperbola.

On the other hand, geometry should not include lines that are like strings, in that they are sometimes straight and sometimes curved, since the ratios between straight and curved lines are not known, and I believe cannot be discovered by human minds, and therefore no conclusion based upon such ratios can be accepted as rigorous and exact. Nevertheless, since strings can be used in these constructions only to determine lines whose lengths are known, they need not be wholly excluded.

11.A9 The method of normals

[1] The angle formed by two intersecting curves can be as easily measured as the angle between two straight lines, provided that a straight line can be drawn making right angles with one of these curves at its point of intersection with the other. This is my reason for believing that I shall have given here a sufficient introduction to the study of curves when I have given a general method of drawing a straight line making right angles with a curve at an arbitrarily chosen point upon it. And I dare say that this is not only the most useful and most general problem in geometry that I know, but even that I have ever desired to know.

[2] Let CE be the curved line. It is desired to draw a straight line at right angles to it, through the point C. I suppose the problem to have been solved, and that the sought-for line is CP, which I prolong to the point P where it meets the straight line GA. (GA is the line to whose points all those of CE are referred; so that putting MA or CB equal to y, and CM or BA equal to x, I have some equation showing the relation between x and

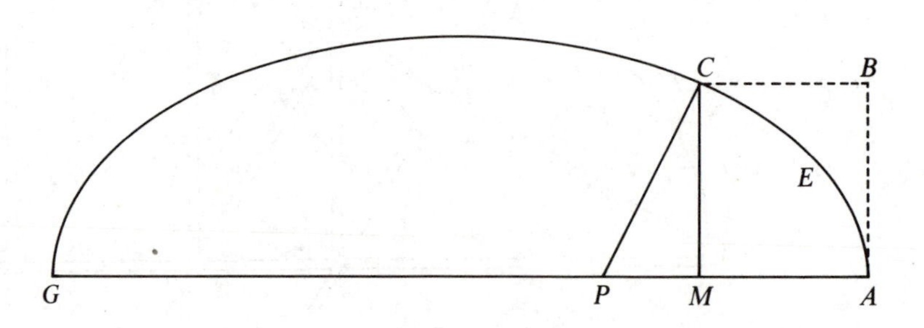

y.) Then I put $PC = s$, and $PA = v$, whence $PM = v - y$. Since the triangle PMC is right-angled, the square on the hypotenuse s^2 is equal to $x^2 + v^2 - 2vy + y^2$, the sum of the squares on the two sides. That is to say, $x = \sqrt{s^2 - v^2 + 2vy - y^2}$ or equally $y = v + \sqrt{s^2 - x^2}$. By this means I can get rid of one of the two unknown quantities x or y from the equation relating the points of the curve CE to those of the straight line GA. This is easily done by putting throughout $\sqrt{s^2 - v^2 + 2vy - y^2}$ in place of x, the square of this in the place of x^2, its cube in place of x^3, and so on. That is if it's x I want to get rid of; or if it's y, I put in its place $x + \sqrt{s^2 - x^2}$, and its square or cube, etc., in place of y^2, y^3 etc. After this process there always remains an equation in only one unknown quantity, x or y.

[3] For example, if CE is an Ellipse, MA the segment of its diameter on which CM is ordinate, and which has r for its *latus rectum* and q its major axis then by Book I Proposition 13 of Apollonius we have $x^2 = ry - ry^2/q$. Getting rid of x^2 from this gives $s^2 - v^2 + 2vy - y^2 = ry - ry^2/q$, or

$$y^2 + \frac{qry - 2qvy + qv^2 - qs^2}{q - r}$$

equals nothing. For it is better here to consider the whole together in this way, than as one part equal to the other. [...]

[4] Such an equation having been found it is to be used, not to determine x, y, or z, which are known, since the point C is given, but to find v or s, which determine the required point P. With this in view, observe that if the point P fulfills the required conditions, the circle about P as centre and passing through the point C will touch but not cut the curve CE; but if this point P be ever so little nearer to or farther from A than

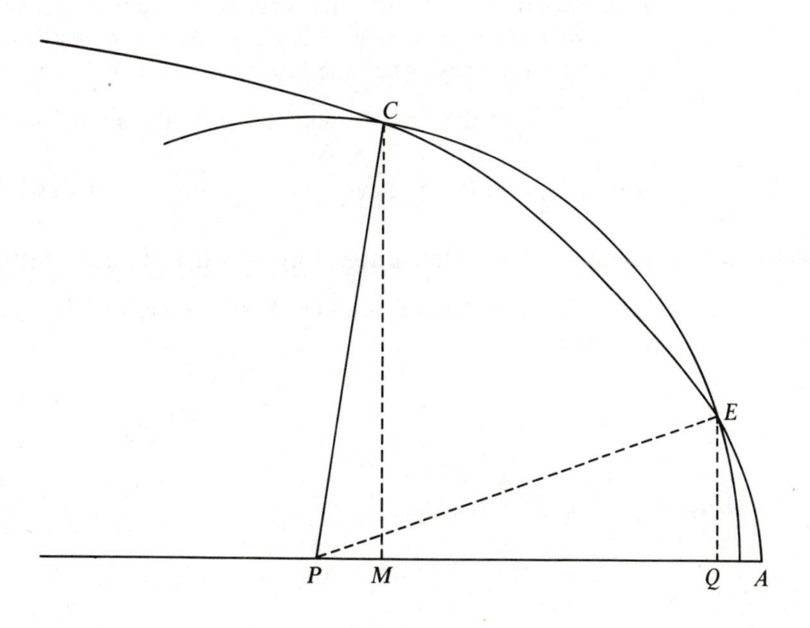

it should be, this circle must cut the curve not only at C but also in another point. Now if this circle cuts CE, the equation involving x and y as unknown quantities (supposing PA and PC known) must have two unequal roots. Suppose, for example, that the circle cuts the curve in the points C and E. Draw EQ parallel to CM. Then x and y may be used to represent EQ and QA respectively in just the same way as they were used to represent CM and MA; since PE is equal to PC (being radii of the same circle), if we seek EQ and QA (supposing PE and PA given) we shall get the same equation that we should obtain by seeking CM and MA (supposing PC and PA given). It follows that the value of x, or y, or any other such quantity, will be two-fold in this equation, that is, the equation will have two unequal roots. If the value of x be required, one of these roots will be CM and the other EQ; while if y be required, one root will be MA and the other QA. It is true that if E is not on the same side of the curve as C, only one of these will be a true root, the other being drawn in the opposite direction, or less than nothing. The nearer together the points C and E are taken however, the less difference there is between the roots; and when the points coincide, the roots are exactly equal, that is to say, the circle through C will touch the curve CE at the point C without cutting it.

[5] Furthermore, it is to be observed that when an equation has two equal roots, its left-hand member must be similar in form to the expression obtained by multiplying by itself the difference between the unknown quantity and a known quantity equal to it; and then, if the resulting expression is not of as high a degree as the original equation, multiplying it by another expression which will make it of the same degree. This last step makes the two expressions correspond term by term.

[6] For example, I say that the first equation found in the present discussion, namely

$$y^2 + \frac{qry - 2qvy + qv^2 - qs^2}{q - r},$$

must be of the same form as the expression obtained by making $e = y$ and multiplying $y - e$ by itself, that is, as $y^2 - 2ey + e^2$. We may then compare the two expressions term by term, thus: Since the first term, y^2, is the same in each, the second term, $\dfrac{qry - 2qvy}{q - r}$, of the first is equal to $-2ey$, the second term of the second; whence, solving for v, or PA, we have $v = e - \dfrac{r}{q}e + \dfrac{1}{2}r$; or, since we have assumed e equal to y, $v = y - \dfrac{r}{q}y + \dfrac{1}{2}r$. In the same way, we can find s from the third term, $e^2 = \dfrac{qv^2 - qs^2}{q - r}$; but since v completely determines P, which is all that is required, it is not necessary to go further.

11.A10 H. J. M. Bos on Descartes's *Geometry*

Descartes's view can be summarized as follows: Construction of problems by ruler and compass is certainly simpler than, and therefore preferable to, construction by means of the intersection of conics or more complex curves. In the construction of problems one should always use the simplest possible curves. But this does not imply that more complex curves are necessarily less geometrical than the straight line and the circle, or that constructions by means of these curves are less geometrical than constructions by ruler and compass. There is a collection of curves of ever increasing complexity (circles, conics, conchoids, etc.) which are in principle acceptable in geometrical constructions. If a problem can be constructed by the intersection of two such curves and it cannot be constructed by simpler curves, then that construction is the right one to choose and it is no less geometrical a construction than one by ruler and compass.

This vision of the geometrical procedure of constructing problems determined a programme in three parts. First Descartes had to determine which curves were acceptable as genuinely geometrical means for the construction of problems. Secondly, he had to make it clear on which criteria some curves would be considered simpler than others: this would lead to a classification in order of simplicity within the collection of geometrically acceptable curves. Finally, a method had to be devised for finding the simplest possible curves by which each problem could be constructed. This is essentially the programme which Descartes worked out in his *Géométrie*.

The first point of the programme—differentiating between the curves which are acceptable in geometry and those which are not—caused Descartes (and his successors) the greatest number of conceptual problems. Basically Descartes took as geometrical curves those 'which can be described by some regular motion'. But this is not a very clear criterion. Also Descartes wished to include in the collection of geometrically acceptable curves all curves that may occur as locus solutions of problems such as the problem of Pappus. This meant that in fact—although Descartes never explicitly said so—he wanted to regard all algebraic curves as geometrical. But to do so he would have to prove that all algebraic curves could be traced by continuous and geometrically acceptable motions, or that they could be traced by other means which were just as geometrical as the tracing by continuous motion.

Algebra, in the sense of the existence of an algebraic equation of the curve, was the essential criterion in the first part of the programme. But the algebra had to remain implicit. Descartes could not simply take as 'geometrical' all curves that admit an algebraic equation, because obviously that is not a geometrical criterion; if he were to adopt this criterion, Descartes could no longer claim that he was doing geometry.

[...] As we have seen, the whole structure of his *Géométrie* depended on the conception of construction by the intersection of geometrical curves. For Descartes, these intersections were actually found or constructed only if the curves could be traced by continuous motion. In that case one can conceive clearly and distinctly that the intersections are found. If he were to renounce his criterion of tracing by continuous motion and at the same time keep to his programme of construction by the intersection of curves, he would have to state as an axiom that for all curves having an algebraic equation the intersections are given or constructible.

It is evident that Descartes could not do this. An axiom which states that the intersections of curves are constructible is by no means clearly and distinctly evident, so it would not satisfy Descartes's criterion for accepting a statement as a basis for further argument.

Moreover, by adopting this approach Descartes could no longer claim that he was doing geometry; he would be doing some kind of algebra. But that would mean giving up the principal aim of his work; to bring order into the science of geometry.

Finally, the whole structure of the *Géométrie*, which was based on finding the simplest constructing curves for a given problem, would lose much of its meaning. If the intersections of all algebraic curves are by axiom constructible, there is no evident reason for finding the simplest curves for a given problem, and hence there is not much point in finding constructions for roots of equations. The roots of an equation $x^n + ax^{n-1} + \cdots = 0$ are the intersections of the curve $y = x^n + ax^{n-1} + \cdots$ with the straight line $y = 0$; thus they are already given as intersections of curves with algebraic equations.

We see that Descartes could not give up his definition of geometrical curves by continuous motion because then he would have lost the claim of doing geometry and hence the rationale of the whole structure of his work would have been destroyed.

11.B Responses to Descartes's *Geometry*

Descartes's *Geometry* provoked a flurry of activity when it came out, not least because it was so very difficult. One of the first to study it was Florimond Debeaune (1601–1652), who contributed the problem which carried his name for well over a century (see 11.B1); it asks for a curve given a property of its tangent at every point. As such it was the first inverse tangent problem to be raised, and so was destined to make a decisive contribution to the later study of differential equations (see 13.A2 and 13.B2).

Descartes's methods made great progress when the general study of curves and equations was at stake, for it was hard to imagine doing without them. But they were often avoided when the subject was conic sections, for then there were powerful, traditional methods which, moreover, were visibly geometric and did not seem to require dilution with algebraic analysis (for Newton's strictures on this point see 12.D3). Philippe de la Hire (1640–1718) was one who felt this tension acutely. His first book, an independent rediscovery of Girard Desargues's ideas about a projective approach to the theory of conics (see 11.D), was a failure, but when he tried again in 1679 he still gave a traditional sort of account. He defined them via their focal properties, rather than as sections of a cone, but worked with them via their so-called symptoms—i.e. as Apollonius had given them. His preface (see 11.B2–11.B3) makes clear how orthodox he was deliberately being in many ways, apparently successfully, for his book was translated into English by B. Robinson in 1704. However, when he turned to higher plane curves in the second part of his book, la Hire felt that it was necessary to use analysis (i.e. Cartesian algebra) in order to describe them. His distinction between determinate problems, or polynomial equations, and indeterminate problems, or loci, shows the two sides Descartes's work had in this period, and in the third part of his book la Hire went into more detail about the solution of equations which, following Descartes, he discussed in terms of the mutual intersection of curves of lower degree.

Hendrik van Heuraet's rectification of curves (11.B4), like William Neil's a little earlier, is remarkable for eroding a distinction which Descartes had insisted upon without making it clear, the distinction we express in the dichotomy of algebraic versus transcendental curves. By rectifying a family of algebraic curves, which Descartes had insisted was impossible, Heuraet helped to make Cartesian methods apply where their inventor had said they could not.

Like Heuraet's discovery, Frans Hudde's rules (11.B5) were published in van Schooten's important second edition of Descartes's *Geometry*. By helping to systematize the finding of tangents, they were to be a significant step on the way to the discovery of the algorithms of the calculus.

11.B1 Florimond Debeaune's inverse tangent problem

Florimond Debeaune was an early convert to Descartes's way of doing geometry, and in the course of teaching it to himself posed various questions to the mathematicians of the day. This one, which he was unable to solve himself, became deservedly famous, for it is the first problem which asks you to find a curve given a property of its tangents, whence it is called an inverse tangent problem. Descartes was unable to solve the problem when he heard of it, although, like Roberval, he showed that the curve had an asymptote; it is in fact a logarithmic curve and so a mechanical curve in Descartes's sense. Following the lead of Newton and Leibniz, inverse tangent problems became formulated in the calculus as differential equations.

(*a*) *Debeaune to Gilles Personne de Roberval*

Let there be a curve AXE whose vertex is A, axis AYZ, and the property of this curve is that, having taken any point on it you wish, say X, from which the line XY is drawn as a perpendicular ordinate to the axis, and having taken the tangent GXN through the same point X, and extended the perpendicular XZ to it at X until it meets the axis, there will be the same ratio of ZY to YX as a given line, like AB, has to the line $YX-AY$. [See figure overleaf.]

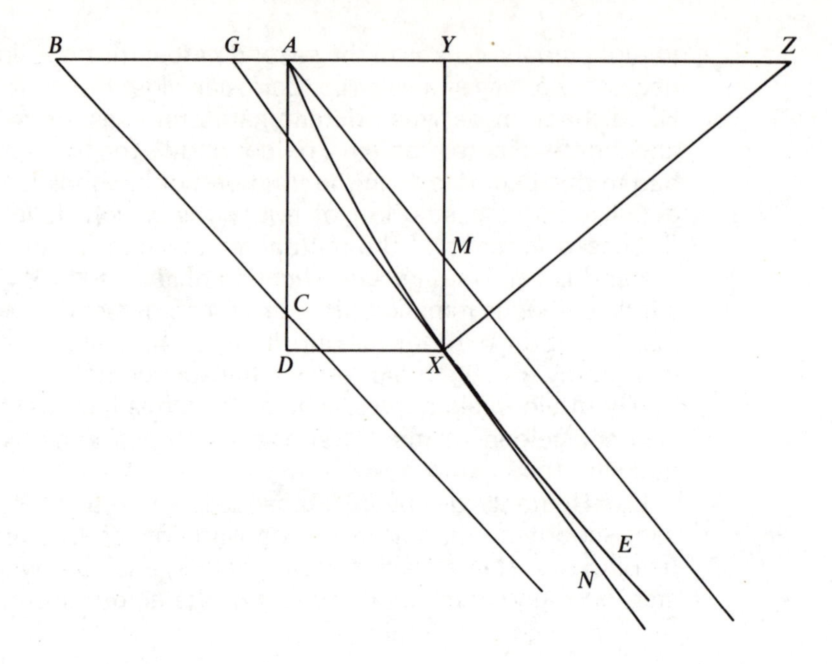

(*b*) *Debeaune to Marin Mersenne*

As for my curves, I don't pretend to prove by their means that a quadruple weight is necessary to raise the sound of a string by an octave, for I have a clear and simple proof of that, as of several other questions you have asked me. But I do need the curves, first of all to prove that a heavy body when suspended makes its oscillations through small arcs in the same time as through large ones, and likewise that the strings of a lute, or something similar, when taut make their oscillations in the same time when they are large and when they are small. For, while these are things about which one can experiment, and for the strings of the lute they are quite obvious because their sound has the same tone when their emotion is great and when it is small, nonetheless the thing being proved geometrically it would be of no little help with other speculations.

11.B2 Philippe de la Hire on conic sections

Some years ago I published a *Treatise of Conick Sections* [*Nouvelle Méthode, etc.*] in a new Method, and Demonstrated their Principal Properties from the Cone: But such as had not been sufficiently enured to Demonstrations about the Intersections of Planes and Solids, found it difficult to understand them, tho' they were very Simple and Plain when once comprehended. This put me upon seeking out another Method, whereby, without making use of the Cone, and only by drawing Curve lines upon a Plane, I might demonstrate the same Properties of the Conick Sections: And after having tried that Method as being the most Simple and easie of all others I laid it aside at last, not being able to Surmount at that time all the difficulties which occurred. But I was satisfied with having reduced the Conick Sections to a Plain, which I called Plane Conicks; and I applied to those Plain Sections the same Demonstrations which I had before made for the Solid. And I can say, that That Work had the good fortune to have the Approbation of many Learned Geometers.

But altho' it be very Advantagious to please the Learned, yet we ought by no means to make that the sole Object and Ultimate End of our Studies, and to neglect the Instruction of those that are desirous to understand things of this Nature; and I think they ought to be contented when they are shewed several ways of doing the same thing; for then every one may make choice of that which he likes best. [...]

This Book contains all the Elementary Properties of the Conick Sections, and there is no more Geometry required, than to understand thro'ly the first Six Books of Euclide's *Elements*; or the Substance of what is contained in them, as it hath been demonstrated divers ways by many Learned Geometers.

You will find in this Treatise, the most useful Part of the first Book of Apollonius; The Properties of the Assymptotes, which he delivers in his Second Book, and the first Propositions of the third, and those about the foci: which are the most considerable.

As to the Proportions which I make use of, I cite only Composition, Division and Equality of Ratio's; because I suppose my Reader knows how to change Alternatively, or to Invert or Convert the Terms of four proportionals, etc.

11.B3 Philippe de la Hire on the algebraic approach

I have tried to give a treatise on lines as simple and short as seemed to me to be possible, but having considered that it would be necessary to explain not only what mathematicians understand by lines but also how they come to be useful in geometry, I thought that it would be necessary to start from the beginning, and to that end I have added an initial chapter where I make clear in what way one can use analysis to solve problems in geometry, and I am persuaded that I have said enough to provide instruction to those who know the first rules of this art, which I don't claim to explain here. The examples which I give are very appropriate for training those who are not familiar with their resolution and they have been chosen so that each has some detail which can illustrate the general rule in similar situations. I divide problems into two kinds, determinate and indeterminate; the determinate are those which have only a certain number of different solutions, as for example when it is required to cut a straight line segment in two parts so that their rectangle shall equal a given square; one sees easily that one can find two points on this line which shall be as required, but if the side of the given square is equal to the half of the given line there will only be a single solution to this problem, because the point which divides the line into two equal pieces will be the division which is required; and finally if the side of the given square is greater than the half of the given line the question is impossible; and so it is with others. But if the problem is indeterminate it can have an infinity of different solutions, as for example one can find an infinity of straight lines of different lengths whose rectangles taken two by two shall be equal to a given square, and similarly with others.

11.B4 Hendrik van Heuraet on the rectification of curves

Let there be given two curves, e.g. *ABCDE*, *GHIKL*, and the straight line *AF* such that (if from an arbitrary point *M* on line *AF* the perpendicular *MI* is drawn, cutting the given curves in *C* and *I*, while *CQ* is taken perpendicular to the curve *ABCDE*) *MC* is to *CQ* as a certain given segment Σ to *MI*: then the superficies *AGHIKLF* will be equal to the rectangle contained by the given segment Σ and a segment equal to the curve *ABCDE*.

Let the line *AF* be divided in as many parts as you like, for example in the points *O*, *M* and *P*. And let the perpendiculars *OH*, *MI*, *PK* be drawn, cutting the curve *ABCDE* in the points *B*, *C* and *D*, and the curve *GHIKL* in the points *H*, *I* and *K*. And let through the points *A*, *B*, *C*, *D*, and *E* tangents be drawn, meeting each other in *R*, *S*, *T*, and *V*. And let through these points the lines *Ra*, *Yb*, *Zc*, *ed* be drawn, perpendicular to *AF*. And let through the points *G*, *H*, *I*, *K* and *L* lines be drawn parallel to *AF*, cutting *Ra* in *f* and *a*, *Yb* in *g* and *b*, *Zc* in *h* and *c*, *ed* in *i* and *d*. And finally let from *S SX* be drawn parallel to the line *AF*, and let the tangent *TS* be produced to *N*. Because of the right angle *NCQ*, *CM* will be to *CQ* as *MN* to *NC*. But *MN* is to *NC* as *SX* to *ST*. So *SX* will be to *ST* as *CM* to *CQ*. And because *CM* is to *CQ* as Σ to *MI*, there will also hold that *SX* is to *ST* as Σ to *MI*, and therefore the rectangle contained by *SX* or *YZ* and *MI* or *Yb* will be equal to the rectangle contained by *S* and Σ. In the same way one proves the rectangle *ce* to be equal to the rectangle contained by *TV* and Σ, and □*dF* ∞ □*VE*, Σ and □*aY* ∞ □ contained by *RS* and Σ. Therefore these rectangles taken together will be equal to the rectangle contained by Σ and another segment, equal to the tangents taken together. Because this is true for any number of rectangles and tangents, and the figure consisting of the parallelograms will finally become the superficies *AGHIKLF* if their number is increased to infinity, and because similarly the tangents will finally become the curve *ABCDE*, it is clear that the superficies *AGHIKLF* is equal to the rectangle contained by Σ and another segment equal to the curve *ABCDE*. Which had to be proved.

However, how the length of the given curve can be investigated using this, will reveal itself in the following examples.

Let first the curve *ABCDE* be such that, if on line *AF* an arbitrary point *M* is taken, if further the perpendicular *MC* is drawn, if *AM* is called *x* and *MC* is called *y*, *yy* always is ∞ $\frac{x^3}{a}$. Further putting *AQ* ∞ *s*, *CQ* ∞ *v* and *MI* ∞ *z*; there will hold *QM* ∞ $s - x$, and its square ∞ $ss - 2sx + xx$. And if the square of *MC*, that is *yy* or $\frac{x^3}{a}$, is added to it, one will find $ss - 2sx + xx + \frac{x^3}{a}$ ∞ *vv*. Because of two equal roots, multiply according to Hudde's method with $\underline{0 \quad 1 \quad 2 \quad 3 \quad 0}$,

and one will find $-2xs + 2xx + \frac{3x^3}{a}$ ∞ 0.

Therefore *AQ* or *s* ∞ $x + \frac{3xx}{2a}$, and if from this *AM* ∞ *x* is subtracted, there will remain *MQ* ∞ $\frac{3xx}{2a}$, the square of which is $\frac{9x^4}{4aa}$. To this add □*CM* or $\frac{x^3}{a}$, and there will appear

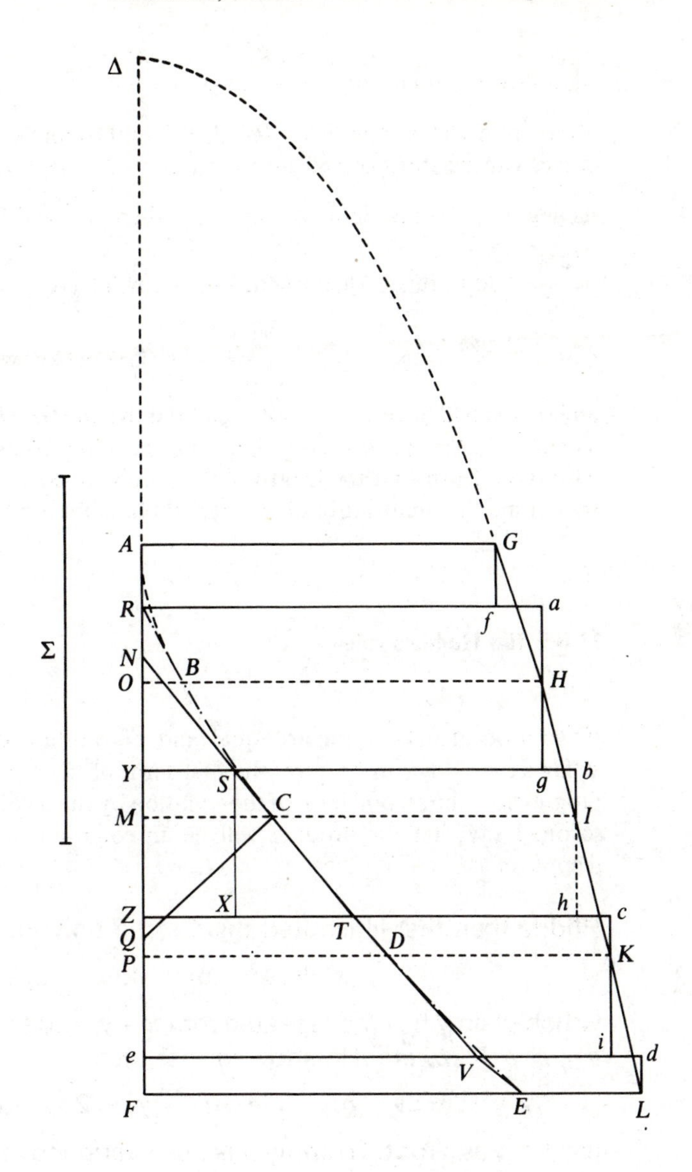

$\square CQ \propto \dfrac{9x^4}{4aa} + \dfrac{x^3}{a}$. Thus, as CM — i.e. $\sqrt{\dfrac{x^3}{a}}$ — will be to CQ — i.e. $\sqrt{\dfrac{9x^4}{4aa} + \dfrac{x^3}{a}}$ — so some known segment, put it $\frac{1}{3}a$ (for one may choose it arbitrary) to $MI \propto z$, and so you will have $z \propto \sqrt{\frac{1}{4}ax + \frac{1}{9}aa}$. And this argues that the line $GHIKL$ is a parabola, the vertex of which is in Δ, with $A\Delta \propto \frac{4}{9}a$, and with latus rectum $\propto \frac{1}{4}a$. And thus the length of the curve $ABCDE$ is $\sqrt{\dfrac{v^3}{a}} - \dfrac{8}{27}a$, if $\Delta F \propto v$. Similarly, if you put instead of $yy \propto \dfrac{x^3}{a}$

the following equation $y^4 \propto \frac{x^5}{a}$, or $y^6 \propto \frac{x^7}{a}$, or $y^8 \propto \frac{x^9}{a}$, and so on to infinity: you will always find such a superficies *AGHIKLF* that it can be squared, and therefore all these curves can be transformed into a straight line. If, however, *ABCDE* is a parabola, having axis *AG* and latus rectum $\propto a$: then you will find $MQ \propto \frac{2x^3}{aa}$ and its square $\propto \frac{4x^6}{a^4}$. Add to this *CM* squared, and you will have $\frac{4x^6}{a^4} + \frac{x^4}{aa}$ for $\square CQ$. From this: as CM—i.e. $\frac{xx}{a}$—is to CQ i.e. $\sqrt{\frac{4x^6}{a^4} + \frac{x^4}{aa}}$, so some known segment, put it a, to $MI \propto z$: and so you will have $z \propto \sqrt{4xx + aa}$, and the line *GHIKL* will be a hyperbola, having axis *AG*, centre *A*, latus rectum $\propto \frac{1}{2}a$ and latus transversum $\propto 2a$. And from this exactly we learn that the length of the parabolic curve cannot be found unless at the same time the quadrature of the hyperbola is found, and vice versa.

11.B5 Jan Hudde's rules

If two roots of an equation are equal, and it is multiplied by an arithmetic progression as far as you like; to be sure, the first term of the equation by the first term of the progression, the second term of the equation by the second term of the progression and so on, I say that the product will be an equation in which one of the said roots reappears.

Hudde then first illustrated this for the polynomial

$$(x^3 + px^2 + qx + r)\ (x^2 - 2yx + y^2),$$

which plainly has the repeated root $x = y$, and the arithmetic progression a, $a \pm b$, $a \pm 2b$, etc. He showed that the resulting polynomial,

$$x^5a + (-2y + p)\ x^4(a + b) + (y^2 - 2py + q)x^3(a + 2b) + \text{etc.},$$

has $x = y$ as a root. To apply this observation to maxima and minima, Hudde argued that the repeated root of the original polynomial is now common to two polynomials, and can be found by a technique like the Euclidean algorithm, which finds the common factors of numbers.

11.C Pierre de Fermat

Pierre de Fermat (1601–1665) was a solitary man, never straying far from his native Toulouse, and hardly ever meeting a fellow mathematician. He preferred writing to individuals to any form of publication, and even when he felt he had something to say he was content to ask Marin Mersenne or, later, his friend Pierre de Carcavy to distribute copies of his letters. As a result much of what he wrote circulated in manuscript, if at all, in his lifetime, and was only published posthumously by his son Samuel in 1679. However, a large number of his letters survive and give a good picture of him as a mathematician. We have included as 11.C1–11.C2 some of his investigations into tangents and maxima and minima, where, like Descartes, but in the formalism of Viète (which he had learned as a young man and preferred), he gave an algebraic treatment of geometrical problems. Necessarily, the validity of his method rests on an appeal to geometric intuition—in this case in his appeal to an obscure concept of *adequality* (approximate equality) which he claimed to have learned from his reading of Diophantus. Indeed, Fermat's whole approach to the study of curves is rooted in his involvement in the contemporary enthusiasm for restoring lost Greek texts, and one is struck by how unsure he was all his life about the possibility of going far beyond the classical masters.

In 11.C5 we present his striking discovery of an area of finite size with an infinite perimeter. This is also one of the first times an area was found without the use of infinitesimals or indivisibles but by means of strips of finite width.

These achievements notwithstanding, perhaps Fermat's most profound insights were into the theory of numbers, a branch of mathematics in which he was to be without a follower, let alone a peer, until Leonhard Euler came on the scene. Inspired by Claude Gaspard Bachet's edition of Diophantus' *Arithmetica*, with its many ingenious problems, Fermat took up the question of finding all the solutions to a given question, or of showing that none could exist. His remarkable invention is the *method of descent,* which can be used on occasion to show that no solutions exist, and it is in this way that he despatched his case of what is nowadays called 'Fermat's last theorem' (see 11.C8). As is clear from 11.C7, Fermat was well aware of the power of his method; sadly he failed to interest Christiaan Huygens in it. Fermat's 'little theorem' (see 11.C4), although not at all difficult to prove once you have thought of it, derived from his interest in prime numbers, and we have included in this connection one of his few mistakes, 11.C3 on 'Fermat primes'. Much more profound is his grasp of 'Pell's equation', $x^2 = Ay^2 + 1$, a glimpse of which is given in 11.C6. From his choice of numbers it is clear that Fermat was in possession of a general theory for dealing with problems of this kind, for the solutions are far too large to be found by trial and error. For example, the smallest values of x and y for which $x^2 = 109y^2 + 1$ has solutions in integers are $x = 158070671986249$ and $y = 15140424455100$. André Weil, in his book cited in 11.C8(b), has argued convincingly that Fermat was in possession of a general method adequate for the problems he treated successfully, and so may be said to have had a theory of numbers. The reader is referred to this book, and to Weil's interesting essays, for a thorough analysis of Fermat as a number theorist.

11.C1 On maxima and minima and on tangents

(a) Maxima and minima

The whole theory of evaluation of maxima and minima presupposes two unknown quantities and the following rule:

Let a be any unknown of the problem (which is in one, two, or three dimensions, depending on the formulation of the problem). Let us indicate the maximum or minimum by a in terms which could be of any degree. We shall now replace the original unknown a by $a + e$ and we shall express thus the maximum or minimum quantity in terms of a and e involving any degree. We shall *adequate*, to use Diophantus' term, the two expressions of the maximum or minimum quantity and we shall take out their common terms. Now it turns out that both sides will contain terms in e or its powers. We shall divide all terms by e, or by a higher power of e, so that e will be completely removed from at least one of the terms. We suppress then all the terms in which e or one of its powers will still appear, and we shall equate the others; or, if one of the expressions vanishes, we shall equate, which is the same thing, the positive and negative terms. The solution of this last equation will yield the value of a, which will lead to the maximum or minimum, by using again the original expression.

Here is an example:

To divide the segment AC at E so that AE × EC may be a maximum.

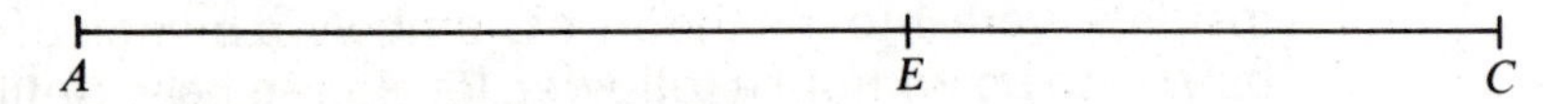

We write $AC = b$; let a be one of the segments, so that the other will be $b - a$, and the product, the maximum of which is to be found, will be $ba - a^2$. Let now $a + e$ be the first segment of b; the second will be $b - a - e$, and the product of the segments, $ba - a^2 + be - 2ae - e^2$; this must be adequated with the preceding: $ba - a^2$. Suppressing common terms: $be \sim 2ae + e^2$. Suppressing e: $b = 2a$. To solve the problem we must consequently take the half of b.

We can hardly expect a more general method.

(b) Tangents

We use the preceding method in order to find the tangent at a given point of a curve.

Let us consider, for example, the parabola BDN with vertex D and of diameter DC; let B be a point on it at which the line BE is to be drawn tangent to the parabola and intersecting the diameter at E.

We choose on the segment BE a point O at which we draw the ordinate OI; also we construct the ordinate BC of the point B. We have then: $CD/DI > BC^2/OI^2$, since the point O is exterior to the parabola. But $BC^2/OI^2 = CE^2/IE^2$, in view of the similarity of triangles. Hence $CD/DI > CE^2/IE^2$.

Now the point B is given, consequently the ordinate BC, consequently the point C, hence also CD. Let $CD = d$ be this given quantity. Put $CE = a$ and $CI = e$; we obtain

$$\frac{d}{d-e} > \frac{a^2}{a^2 + e^2 - 2ae}.$$

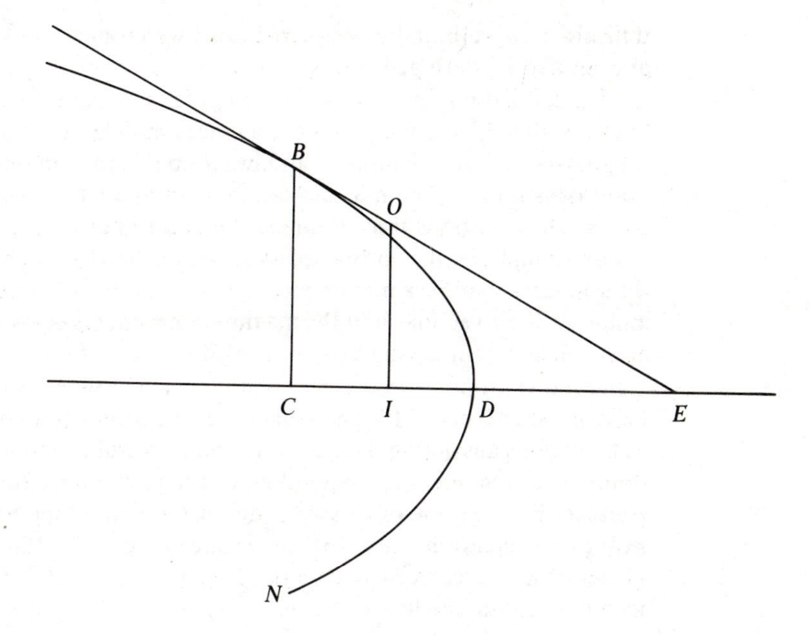

Removing the fractions:

$$da^2 + de^2 - 2dae > da^2 - a^2e.$$

Let us then adequate, following the preceding method; by taking out the common terms we find:

$$de^2 - 2dae \sim -a^2e,$$

or, which is the same,

$$de^2 + a^2e \sim 2dae.$$

Let us divide all terms by e:

$$de + a^2 \sim 2da.$$

On taking out de, there remains $a^2 = 2da$, consequently $a = 2d$.

Thus we have proved that CE is the double of CD—which is the result.

This method never fails and could be extended to a number of beautiful problems; with its aid, we have found the centres of gravity of figures bounded by straight lines or curves, as well as those of solids, and a number of other results which we may treat elsewhere if we have time to do so.

I have previously discussed at length with M. de Roberval the quadrature of areas bounded by curves and straight lines as well as the ratio that the solids which they generate have to the cones of the same base and the same height.

11.C2 A second method for finding maxima and minima

In studying the method of syncriseos and anastrophe of Viète, and carefully following its application to the study of the nature of correlative equations, it occurred to me to derive a process for finding maxima and minima and thus for resolving easily all the difficulties concerning limiting conditions which have caused so many problems for ancient and modern geometers.

Maxima and minima are in effect unique and singular, as Pappus said and as the ancients already knew, although Commandino claimed not to know what the term 'singular' signified in Pappus. It follows from this that on one side and the other of the point constituting the limit one can take an ambiguous equation, and that the two ambiguous equations thus obtained are accordingly correlative, equal and similar.

For example, let it be proposed to divide the line b in such a way that the product of the segments shall be a maximum. The point answering this question is evidently the middle of the given line, and the maximum product is equal to $b^2/4$; no other division of this line gives a product equal to $b^2/4$.

But if one proposes to divide the same line b in such a way that the product of the segments shall equal z'' (this area being besides supposed to be less than $b^2/4$) there will be two points answering the question, and they will be found situated on one side and the other of the point corresponding to the maximum product.

In fact let a be one of the segments of the line b, one will have $ba - a^2 = z''$; an ambiguous equation, since for the segment a one can take each of the two roots. Therefore let the correlative equation be $be - e^2 = z''$. Comparing the two equations according to the method of Viète:

$$ba - be = a^2 - e^2.$$

Dividing both sides by $a - e$, one obtains

$$b = a + e;$$

the lengths a and e will moreover be unequal.

If, in place of the area z'', one takes another greater value, although always less than $b^2/4$, the segments a and e will differ less from each other than the previous ones, the points of division approaching closer to the point constituting the maximum of the product. The more the product increases the more on the contrary diminishes the difference between a and e until it will vanish exactly at the division corresponding to the maximum product; in this case there will only be a unique and singular solution, the two quantities a and e becoming equal.

Now the method of Viète applied to the two correlative equations above leads to the equality $b = a + e$, therefore if $e = a$ (which will always happen at the point constituting the maximum or the minimum) one will have, in the case proposed, $b = 2a$, which is to say that if one takes the middle of the segment b, the product of the segments will be a maximum.

Let us take another example: to divide the segment b in such a way that the product of the square of one of the segments with the other shall be a maximum.

Let a be one of the segments; one must have $ba^2 - a^3$ maximum. The equal and similar correlative equation is $be^2 - e^3$. Comparing these two equations according to the method of Viète:

$$ba^2 - be^2 = a^3 - e^3;$$

dividing both sides by $a - e$ one obtains

$$ba + be = a^2 + ae + e^2,$$

which gives the form of the correlative equations.

To obtain the maximum, set $e = a$; one obtains

$$2ba = 3a^2 \qquad \text{or} \qquad 2b = 3a;$$

the problem is solved.

11.C3 Fermat to Bernard de Frenicle on 'Fermat primes'

But here is what most excites me: I am almost persuaded that all progressive numbers increased by unity, whose exponents are numbers in the double progression, are prime numbers, such as 3, 5, 17, 257, 65537, 4,294,967,297 and the following one of 20 letters 18,446,744,073,709,551,617, etc. I don't have an exact proof, but I have excluded such a great quantity of divisors by infallible demonstrations and I have shed so much light on the problem which will establish my idea, that it would be difficult for me to retract.

Fermat's conjecture is that numbers of the form $2^n + 1$, where n is of the form 2^k, are prime. The first five, up to $2^{2^4} + 1 = 65537$, are prime, but the next is not; Euler showed in 1732 that it is divisible by 641.

11.C4 Fermat to Marin Mersenne on his 'little theorem' for $a = 2$ (1640)

Here are three propositions I have found on which I hope to erect a great building.
The numbers which proceed from the double progression, reduced by unity, such as

1	2	3	4	5	6	7	8	9	10	11	12	13	
1	3	7	15	31	63	127	255	511	1023	2047	4095	8191	etc.

are called the radicals of perfect numbers because whenever they are prime they produce them. Let us put above them the numbers according to the natural progression 1, 2, 3, 4, 5 etc. which are called their exponents. This done I say that:

1 When the exponent of a radical number is compound, its radical is also compound. Thus, because 6 the exponent of 63 is compound I say that 63 is also compound.
2 When the exponent is a prime number, I say that its radical reduced by unity is measured by the double of the exponent. Thus because 7 the exponent of 127 is a prime number, I say that 126 is a multiple of 14.
3 When the exponent is a prime number, I say that its radical is not measured by any prime number except those which exceed by unity either a multiple of the double of the exponent or the double of the exponent. Thus, because 11, the exponent of 2047, is a prime number, I say that it cannot be measured except by a number which is greater by unity than 22, as 23, or by a number which is greater by unity than a multiple of 22; in fact 2047 is only measured by 23 or 89, from which, if you remove unity, 88 remains, a multiple of 22.

Here are three extremely beautiful propositions which I have found and proved, not without difficulty. I could call them the foundations of the invention of perfect numbers. I don't doubt that M. Frenicle got there earlier, but I have only begun and without doubt these propositions will pass as very lovely in the minds of those who have not become sufficiently hypercritical of these matters, and I would be very happy to have the opinion of M. Roberval.

11.C5 Fermat's evaluation of an 'infinite' area

[1] Archimedes did not employ geometric progressions except for the quadrature of the parabola; in comparing various quantities he restricted himself to arithmetic progressions. Was this because he found that the geometric progression was less suitable for the quadrature? Was it because the particular device that he used to square the parabola by this progression can only with difficulty be applied to other cases? Whatever the reason may be, I have recognized and proved that this progression is very useful for quadratures, and I am willing to present to modern mathematicians my invention which permits us to square, by a method absolutely similar, parabolas as well as hyperbolas.

The entire method is based on a well-known property of the geometric progression, namely the following theorem:

Given a geometric progression the terms of which decrease indefinitely, the difference between two consecutive terms of this progression is to the smaller of them as the greater one is to the sum of all following terms.

[2] This established, let us discuss first the quadrature of hyperbolas: I define hyperbolas as curves going to infinity, which, like *DSEF*, have the following property. Let *RA* and *AC* be asymptotes which may be extended indefinitely; let us draw parallel to the asymptotes any lines *EG*, *HI*, *NO*, *MP*, *RS*, etc. We shall then always have the same ratio between a given power of *AH* and the same power of *AG* on one side, and a power of *EG* (the same as or different from the preceding) and the same power of *HI* on the other. I mean by powers not only squares, cubes, fourth powers, etc., the exponents of which are 2, 3, 4, etc., but also simple roots the exponent of which is unity.

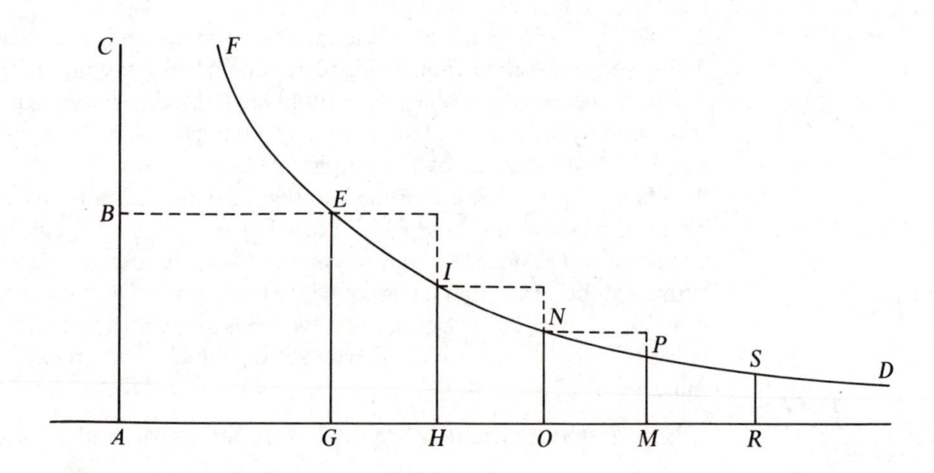

I say that all these infinite hyperbolas except the one of Apollonius, or the first, may be squared by the method of geometric progression according to a uniform and general procedure.

[3] Let us consider, for example, the hyperbolas the property of which is defined by the relations $AH^2/AG^2 = EG/HI$ and $AO^2/AH^2 = HI/NO$, etc. I say that the indefinite area which has for base EG and which is bounded on the one side by the curve ES and on the other side by the infinite asymptote GOR is equal to a certain rectilinear area.

Let us consider the terms of an indefinitely decreasing geometric progression; let AG be the first term, AH the second, AO the third, etc. Let us suppose that those terms are close enough to each other that following the method of Archimedes we could adequate according to Diophantus, that is, equate approximately the rectilinear parallelogram $GE \times GH$ and the general quadrilateral $GHIE$; in addition we shall suppose that the first intervals GH, HO, OM, etc. of the consecutive terms are sufficiently equal that we can easily employ Archimedes' method of exhaustion by circumscribed and inscribed polygons. It is enough to make this remark once and we do not need to repeat it and insist constantly upon a device well known to mathematicians.

[4] Now, since $AG/AH = AH/AO = AO/AM$, we have also $AG/AH = GH/HO = HO/OM$, for the intervals. But for the parallelograms,

$$\frac{EG \times GH}{HI \times HO} = \frac{HI \times HO}{ON \times OM}.$$

Indeed, the ratio $EG \times GH/HI \times HO$ of the parallelograms consists of the ratios EG/HI and GH/HO; but, as indicated, $GH/HO = AG/AH$; therefore, the ratio $EG \times GH/HI \times HO$ can be decomposed into the ratios EG/HI and AG/AH. On the other hand, by construction, $EG/HI = AH^2/AG^2$ or AO/AG, because of the proportionality of the terms; therefore, the ratio $GE \times GH/HI \times HO$ is decomposed into the ratios AO/AG and AG/GH; now AO/AH is decomposed into the same ratios; we find consequently for the ratio of the parallelograms: $EG \times GH/HI \times HO = AO/AH = AH/AG$.

Similarly we prove that $HI \times HO/NO \times MO = AO/AH$.

But the lines AO, AH, AG, which form the ratios of the parallelograms, define by their construction a geometric progression; hence the infinitely many parallelograms $EG \times GH, HI \times HO, NO \times OM$, etc., will form a geometric progression, the ratio of which will be AH/AG. Consequently, according to the basic theorem of our method, GH, the difference of two consecutive terms, will be to the smaller term AG as the first term of the progression, namely, the parallelogram $GE \times GH$, to the sum of all the other parallelograms in infinite number. According to the adequation of Archimedes, this sum is the infinite figure bounded by HI, the asymptote HR, and the infinitely extended curve IND.

Now if we multiply the two terms by EG we obtain $GH/AG = EG \times GH/EG \times AG$; here $EG \times GH$ is to the infinite area the base of which is HI as $EG \times GH$ is to $EG \times AG$. Therefore, the parallelogram $EG \times AG$, which is a given rectilinear area, is adequated to the said figure; if we add on both sides the parallelogram $EG \times GH$, which, because of infinite subdivisions, will vanish and will be reduced to nothing, we reach a conclusion that would be easy to confirm by a more lengthy proof carried out in the manner of Archimedes, namely, that for this kind of hyperbola the parallelogram

AE is equivalent to the area bounded by the base *EG*, the asymptote *GR*, and the curve *ED* infinitely extended.

It is not difficult to extend this idea to all the hyperbolas defined above except the one that has been indicated.

11.C6 Fermat's challenge concerning $x^2 = Ay^2 + 1$

Given any number which is not a square there are infinitely many squares which, multiplied by the given number and one added, give a square. *Example:* Given 3, a number which is not square, then it, multiplied by the square 1 and unity added, gives 4, which is square.

The same 3, multiplied by the square 16 and unity added makes 49, which is square. And, in place of 1 and 16 there can be found infinitely many squares with the same property but the general law is to be sought given any non-square number.

To look, for example, for a square which multiplied by 149, or 109, or 433 etc. and unity added gives a square.

11.C7 On problems in the theory of numbers: a letter to Christiaan Huygens

[1] For a long time I was unable to apply my method to affirmative questions, because the twists and turns to get there are much more difficult than those which served me for negative questions. So much so that when it occurred to me to prove that every prime number which is one more than a multiple of 4 is a sum of two squares, I found myself in a good deal of trouble. But finally a line of thought gone over many times showed me a light which did not fail, and affirmative questions surrendered to my method, with the help of some new principles which had to be joined with it of necessity. The progress in my thinking on these affirmative questions is this: if a prime number taken at one's discretion, which exceeds by one a multiple of four, is not a sum of two squares, there will be a prime number of the same kind, less than the given one, and then a third still less etc., descending infinitely this way until you arrive at the number 5 which is the smallest of all those of this kind, which it follows cannot be the sum of two squares, which it is nonetheless. From which one must infer from that deduction of an impossibility that all those of this kind are consequently a sum of two squares.

[2] There are infinitely many questions of this kind, but there are some others which demand new principles before the descent can be applied to them and the study of them is sometimes so difficult one cannot overcome without extreme effort. Such is the following question that Bachet said Diophantus had never been able to demonstrate, on the subject of which M. Descartes in one of his letters made the same declaration when he confessed that he found it so difficult that he could see no way of solving it.

Every number is a square, or a sum of two, three, or four squares.

I have finally organized this according to my method and shown that if a given number is not of this nature there will be a smaller which is also not, then a third less than the second, etc., to infinity, from which one infers that all numbers are of this nature.

[3] What I proposed to M. Frenicle and others is also of this great or even greater difficulty: Every non-square number is of such a nature that there are infinitely many squares which, multiplying the said number are one less than a square. I proved it by a descent applied in a quite particular manner.

I admit that M. Frenicle gave various particular solutions and M. Wallis also, but the general proof is found by a descent strictly and properly applied, which I will show them so that they can add the proof and general construction of the theorem and of the problem to the singular solutions which they gave.

[4] Finally I considered certain questions which, although negative, did not shrink from receiving very great difficulties, the way of applying the descent being quite as diverse as the preceding, as it will be easy to check. These are the following:

No cube is a sum of two cubes.

There is only one square in integers which, added to two, gives a cube. The said square is 25.

There are only two squares in integers which, added to 4, give a cube. The said squares are 4 and 121.

All the square powers of two, added to one, are prime numbers.

This last question is of a very subtle and ingenious study and, even though it is posed affirmatively, it is negative; for to say that a number is prime is to say that it cannot be divided by any number.

I put in this place the following question of which I have sent the proof to M. Frenicle after he told me and even showed me in his printed writings that he could not find it.

There are only the two numbers 1 and 7 which, being less by 1 than the double of a square make squares of the same kind, that is to say which are less by one than the double of a square.

11.C8 Fermat's last theorem

(a) On $x^4 - y^4 = z^2$

'Bachet: Find a triangle whose area is a given number.' The area of a triangle in numbers cannot be a square. I am going to give a proof of this theorem, which I have discovered; and I did not find it without painful and laborious thinking about it, but this kind of proof will lead to marvellous progress in the science of numbers.

If the area of a triangle was a square there would be two fourth powers whose difference was a square; it would follow equally that there would be two squares whose sum and difference would be squares. Consequently there would be a square number, the sum of a square and the double of a square, with the property that the sum of the two squares that make it up is likewise a square. But if a square number is the sum of a square and the double of a square its root is likewise the sum of a square and the double of a square, which I can prove without difficulty. One concludes from this that this root is the sum of the two sides of a right angle in a triangle of which one of the square components forms the base and the double of the other square the height.

This triangle will therefore be formed of two square numbers whose sum and difference are squares. But one will prove that the sum of these two squares is smaller than the first two of which one has likewise supposed that the sum and difference are squares. Therefore, if one has two squares whose sum and difference are squares one has at the same time, in integers, two squares enjoying the same property whose sum is less.

By the same reasoning, one has accordingly another sum smaller than that derived from the first, and continuing indefinitely one will always find smaller and smaller integers satisfying the same condition. But this is impossible because an integer being given there cannot be an infinity of integers which are smaller.

The margin is too narrow to receive the complete proof with all its developments.

In the same way I have discovered and proved that there cannot be a triangular number, except for one, which is also a fourth power.

(b) Commentary by André Weil

Fortunately, just for once, he had found room for this mystery in the margin of the very last proposition of Diophantus; this is how it goes.

Take a pythagorean triangle whose sides may be assumed mutually prime; then they can be written as $(2pq, p^2 - q^2, p^2 + q^2)$ where p, q are mutually prime, $p > q$, and $p - q$ is odd. Its area is $pq(p + q)(p - q)$, where each factor is prime to the other three; if this is a square, all the factors must be squares. Write $p = x^2, q = y^2, p + q = u^2$, $p - q = v^2$, where u, v must be odd and mutually prime. Then x, y and $z = uv$ are a solution of $x^4 - y^4 = z^2$; incidentally, v^2, x^2, u^2 are then three squares in an arithmetic progression whose difference is y^2. We have $u^2 = v^2 + 2y^2$; writing this as $2y^2 = (u + v)(u - v)$, and observing that the g.c.d. of $u + v$ and $u - v$ is 2, we see that one of them must be of the form $2r^2$ and the other of the form $4s^2$, so that we can write $u = r^2 + 2s^2$, $\pm v = r^2 - 2s^2$, $y = 2rs$, and consequently

$$x^2 = \tfrac{1}{2}(u^2 + v^2) = r^4 + 4s^4.$$

Thus $r^2, 2s^2$ and x are the sides of a pythagorean triangle whose area is $(rs)^2$ and whose hypotenuse is smaller than the hypotenuse $x^4 + y^4$ of the original triangle. This completes the proof 'by descent'.

11.D Girard Desargues

Girard Desargues (1591–1661) was born in Lyons, but not much is known about his life until his arrival in Paris in 1626. Eventually he gravitated to the scientific circles around Marin Mersenne, where he was about the only mathematician to earn the respect of Descartes. He seems to have had a lifelong interest in the practical arts, especially perspective as used by architects and stonemasons, and he published a short tract on this subject in 1636. Then he decided to investigate what his ideas about perspective might entail for theoretical geometry. The result was his *Rough Draft on Conics* of 1639, which he circulated in an edition of fifty copies in order to solicit a reaction from mathematicians (see 11.D1). An impression of Descartes's and Pascal's reactions is given in 11.D4 and 11.D5; in general, the reaction was poor, and by the time la Hire took up projective geometry (see 11.D7) it seems that Desargues's contribution was all but forgotten. Even his striking theorem on triangles in perspective (11.D6) seems not to have been known to la Hire.

Desargues's crucial idea was to isolate the properties of figures which remain invariant under projection, and to base the theory of conic sections on them. Since the property of being a conic is such an invariant, but all non-degenerate conics are projectively equivalent, this gives a remarkable degree of unity to the theory of conic sections, surpassing the classical treatment due to Apollonius. But the price is that Desargues had to devise new proofs, couched in projective terms, for almost the whole theory. First of all he introduced a line of points at infinity in the plane, with one infinite point on each line; this allowed him to speak of pencils of intersecting lines and pencils of parallel lines as projectively equivalent. In 11.D2 he states that the property of being six points in involution was a projective invariant. The special case when two pairs of points are allowed to coincide gave him the concept of four points in involution, thus: B,H; D,F; and C,G are six points in involution if $\frac{BD\cdot BF}{HD\cdot HF}=\frac{BC\cdot BG}{HC\cdot HG}$ and when $D=F$ and $C=G$ this gives $\frac{BD}{HD}=\frac{BC}{HC}$. This extract gives us a taste of Desargues's awful taste in neologisms; his proof of the claim is, however, a neat exercise in applying Menelaus's theorem.

11.D3 gives Desargues's celebrated theorem on six points in involution, together with his proof of it. As he rightly insists, it is enough to prove it when the conic (which he calls a 'section of a roll') is a circle, when the result can be found using simple Euclidean results, together with an earlier result of his that any line crosses the six lines of a complete quadrilateral in six points in involution (P,Q,I,G,H,K in the figure).

11.D1 Preface to *Rough Draft on Conics*

In this work there will be no difficulty in making the necessary distinction between the giving of names [definitions], propositions, demonstrations, when they follow on, and passages of other kinds; neither will it be difficult to select from among the figures the one which refers to the sentence one is reading, or to construct these figures from what is said in the text.

Everyone will form his own opinion both of what we deduce and of the manner in which we deduce it, and will see that our reason is trying to grasp, on the one hand, infinite quantities, and together with these, quantities so small that their opposing extremities coincide; [and] that these quantities are beyond our understanding, not only because they are unimaginably large or small, but also because ordinary reasoning leads us to deduce from their sizes properties which it cannot comprehend.

In this work every straight line is, if necessary, taken to be produced to infinity in both directions.

We indicate this infinite extension in both directions by means of a row of dots, lined up with the line, lengthening it in both directions.

To convey that several straight lines are either parallel to one another or are all directed towards the same point we say that these straight lines belong to the same ordinance, which will indicate that in the one case as well as in the other it is as if they all converged to the same place.

The place to which several lines are thus taken to converge, in the one case as well as in the other, we call the *butt* of the ordinance of the lines.

To convey that we are considering the case in which several lines are parallel to one another we often say that the lines belong to the same ordinance, whose butt is at an infinite distance along each of them in both directions.

To convey that we are considering the case in which all the lines are directed to the same point we say that all the lines belong to the same ordinance, whose butt is at a finite distance along each of them.

Thus any two lines in the same plane belong to the same ordinance, whose butt is at a finite or infinite distance.

In this work every plane is similarly taken to extend to infinity in all directions.

We indicate that a plane extends to infinity in all directions by means of a number of points scattered all round in this same plane.

11.D2 The invariance of six points in involution

When on a straight trunk *GH* there are three different pairs of knots *BH*, *DF*, *CG*, which are in involution, and through them there pass three pairs of branches springing from the trunk *FK*, *DK*; *BK*, *HK*; *CK*, *GK*, all belonging to the same ordinance, whose butt is *K*, these three pairs of branches all belonging to one ordinance are, taken together, called a *bough of a tree*, and on any other straight line *cb*, suitably drawn in their plane, each of them gives one of the three pairs of knots of an involution *gh*, *df*, *cg*.

11.D3 Desargues's involution theorem

When in a plane we have four points *B*, *C*, *D*, *E*, as marker posts paired three times among themselves, through which there pass three pairs of market lines *BCN*, *EDN*, *BEF*, *DCF*, *BDR*, *ECR*, each of these three pairs of marker lines and the curved edge of

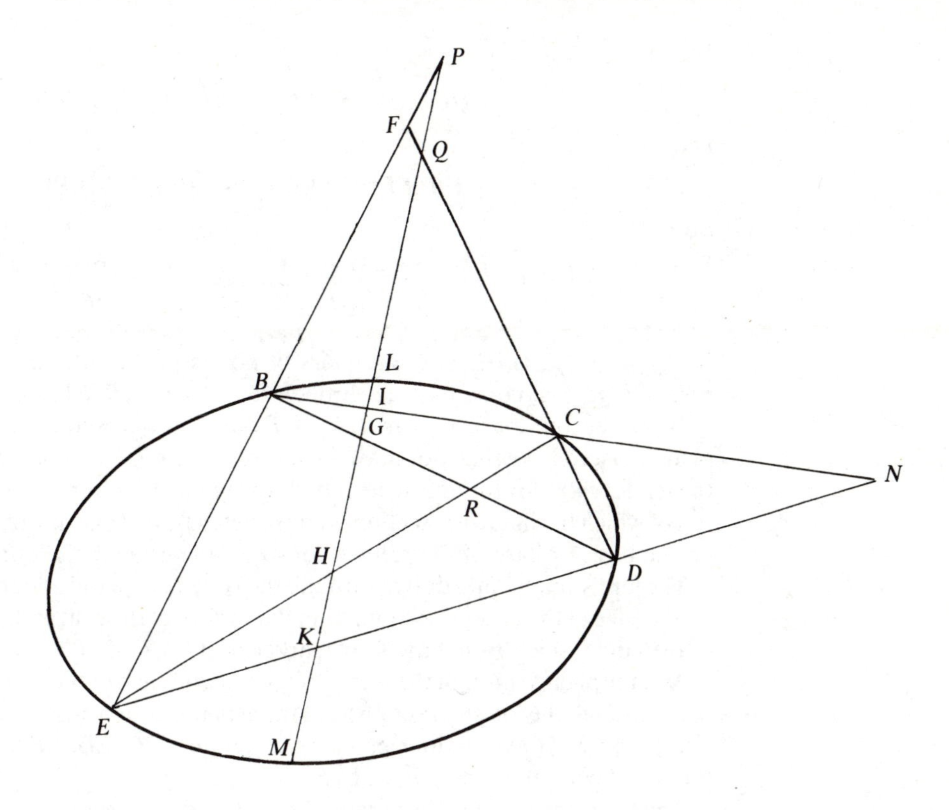

any section of a roll which passes through the four points B, C, D, E gives on any other straight line in their plane, such as a trunk I, G, K, one of the pairs of knots of an involution IK, PQ, GH and LM, and if the two marker lines of one of the pairs BCN, EDN are parallel to one another, the ratios between the rectangles of their related pairs of shoots springing from the trunk are the same as those between their twin rectangles, [that is] the rectangles of the shoots folded to the trunk and in the same order. [...]

When the curved edge of any section of a roll passes through the four points B, C, D, E, anyone who would like to seek a proof in the same words for every kind of section can do so; however, here is the proof in two instances, firstly for when it is the edge of a circle which passes through the points, and next for any one of the other kinds of section of a roll.

So, when the four marker posts B, C, D, E are on the edge of a circle which a seventh general line GPH meets in the points LM.

[Desargues's proof of this result, shorn of his terminology, goes like this: $QL \cdot QM = QC \cdot QD$, $PL \cdot PM = PB \cdot PE$, $FB \cdot FE = FC \cdot FD$. So

$$QL \cdot QM : PL \cdot PM = ((QC \cdot QD):(FC \cdot FD)) \cdot ((FC \cdot FD):(PB \cdot PE)).$$

But

$$QC \cdot QD : FC \cdot FD = (CQ : CF) \cdot (DQ : DF)$$

and

$$FB \cdot FE : PB \cdot PE = (BF : BP) \cdot (EF : EP).$$

So

$$QL \cdot QM : PL \cdot PM = (CQ : CF) \cdot (DQ : DF) \cdot (BF : BP) \cdot (EF : EP)$$
$$= (QI \cdot QK) : (PI \cdot PK) = (QG \cdot QH) : (PG : PH).$$

So L, M; P, Q; G, H; I, K are pairs of points in involution. He then turns to the case of a general conic passing through four points, as follows.]

When the four market posts B, C, D, E are on the curved edge of any other kind of section of a roll, we shall not draw so many figures for a mere sketch of a project, but if the reader cares for the amusement of drawing them for himself he will find that, if the roll of which the figure is a section is constructed from it [i.e. from the figure], and then on the basis or base of the roll we construct the general circle $BCDE$.

The four straight lines drawn through the vertex of the roll and the four marker posts on the edge of the general section, lie in the surface of the roll, and also give on the edge of the circle which is its base four marker posts B, C, D, E.

And the planes through the vertex of the roll and each of the marker lines of the three pairs through the four marker posts of the general section, also give, in the plane of the circular base of the roll, three pairs of marker lines BC, ED, BE, CD, BD, CE, passing through the marker posts B, C, D, E.

And the plane through the vertex of the roll and the seventh general line drawn in the plane of the section, gives, in the plane of the circular base of the roll, a corresponding seventh general line K, G, H, which meets the edge of the circle B, C, D, E, which is the base or basis of the roll, in two points LM, and this same line K, G, H also meets each of the marker lines of the three pairs in the plane of this circle, in points such as PQ, GH, IK.

And the straight lines drawn from the vertex of the roll through the points of LM, on its circular base, intersect the seventh, general, line in the plane of the section, in the same points as those given on this line by the edge of this general section.

And the straight lines drawn from the vertex of the roll through the points of each pair of knots QP, GH, IK on the seventh line GH, in the plane of the circular base of the roll, pass through the points given on the seventh line in the plane of the section by the three pairs of marker lines drawn in this plane.

Now it has been shown that the pairs of knots LM, QP, GH, IK in the plane of the circle are in involution with one another.

And the bough of the tree with three or four pairs of branches, all belonging to the same ordinance, whose butt is the vertex of the roll, gives on the seventh line in the plane of the section the same number of pairs of knots, also in involution. And consequently:

It will be understood that this proof is applicable on numerous occasions, and shows that each important straight line and point arises in the same way for every kind of section of a roll, and it is rare that a general line in the plane of a general section of a roll should have a significant property in relation to that section without a line corresponding to it in the plane of another section of the roll having its position and properties also given by a similar construction of the bough of an ordinance whose butt is at the vertex of the roll.

11.D4 Descartes to Desargues

The openness I have observed in your temperament, and my obligations to you, invite me to write to you freely what I can conjecture of the *Treatise on Conic Sections*, of which the Reverend Father Mersenne sent me the *Rough Draft*. You may have two designs, which are very good and very praiseworthy, but which do not both require the same course of action. One is to write for the learned, and to instruct them about some new properties of conics with which they are not yet familiar; the other is to write for people who are interested but not learned, and make this subject, which until now has been understood by very few people, but which is nevertheless very useful for Perspective, Architecture etc., accessible to the common people and easily understood by anyone who studies it from your book. If you have the first of these designs, it does not seem to me that you have any need to use new terms: for the learned, being already accustomed to the terms used by Apollonius, will not easily exchange them for others, even better ones, and thus your terms will only have the effect of making your proofs more difficult for them and discourage them from reading them. If you have the second design, your terms, being French, and showing wit and elegance in their invention, will certainly be better received than those of the Ancients by people who have no preconceived ideas; and they might even serve to attract some people to read your work, as they read works on Heraldry, Hunting, Architecture etc., without any wish to become hunters or architects but only to learn to talk about them correctly. But, if this is your intention, you must steel yourself to writing a thick book, and in it explain everything so fully, so clearly and so distinctly that these gentlemen, who cannot study a book without yawning and cannot exert their imagination to understand a proposition of Geometry, nor turn the page to look at the letters on a figure, will not find anything in your discourse which seems to them to be less easy of understanding than the description of an enchanted palace in a novel. And, to this end, it seems to me that, to make your proofs less heavy, it would not be out of the question to employ the terminology and style of calculation and of Arithmetic, as I did in my *Geometry*; for there are many more people who know what multiplication is than there are who know about compounding ratios, etc.

Concerning your treatment of parallel lines as meeting at a butt at infinite distance, so as to include them in the same category as lines which meet at a point, this is very good, provided you use it, as I am sure you do, rather as an aid to understanding what is difficult to see in one of the types, by comparing it with the other, where it is clear, and not conversely.

I have nothing to add on what you have written about the centre of gravity of a sphere: for I have already expressed my opinion sufficiently fully to the Reverend Father Mersenne and your comment at the end of your corrections shows that you have understood what I said. But I ask your pardon if I have allowed myself to be carried away by my enthusiasm in recounting my thoughts so freely, and I beg you to believe me.

11.D5 Pascal's hexagon

Definition I

When several straight lines meet at one point, or are all parallel to one another, all these lines are said to belong to the same order or ordinance, and the set of lines is called an order of lines or an ordinance of lines.

Definition II

By the terms conic section we mean the circumference of the circle, the ellipse, the hyperbola, the parabola and the pair of straight lines, since a cone, cut parallel to its base, or through its vertex or in the three other ways which give the ellipse, the hyperbola or the parabola, produces in the conical surface either the circumference, of a circle or a pair of straight lines, or the ellipse, or the parabola.

Lemma I

If in the plane M, S, Q we have two lines MK, MV through the point M and two lines SK, SV through the point S, and K is the point of intersection of the straight lines MK, SK, and V the point of intersection of the straight lines MV, SV and A the point of intersection of the lines MA, SA and μ the point of intersection of the straight lines MV, SK, and if through any two of the four points A, K, μ, V which do not lie on the line MS, such as the points K, V, there passes the circumference of a circle which cuts the straight lines MV, MK, SV, SK in the points O, P, Q, N, I say that the straight lines MS, NO, PQ belong to the same order.

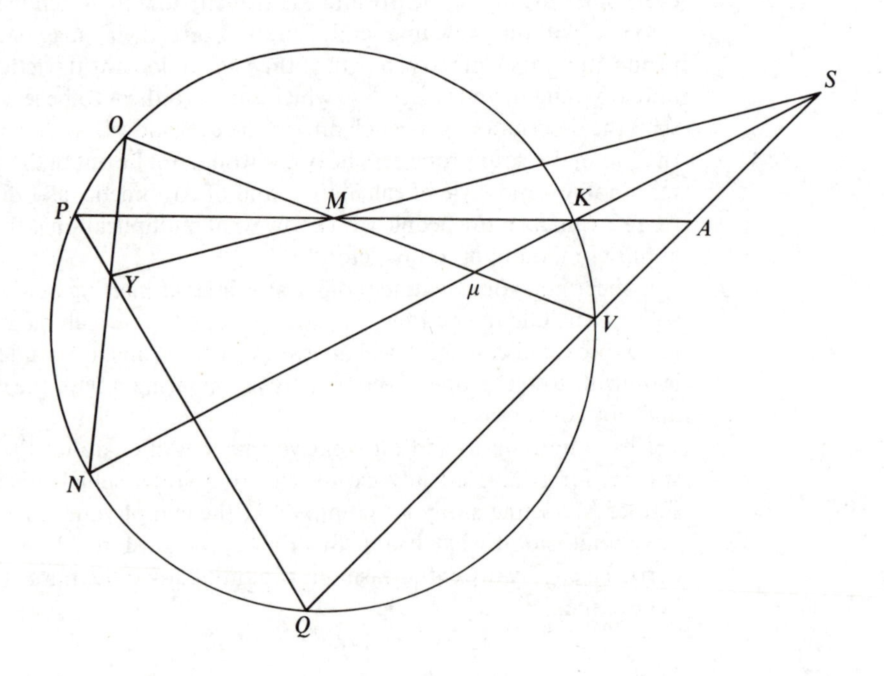

11.D6 Desargues's theorem on triangles in perspective

This proposition is Desargues's famous theorem on two triangles in perspective. Desargues gave two proofs of this theorem and its converse: the elegant one when the figure lies in three dimensions, and a proof based on Menelaus's theorem for the deeper two-dimensional case. It is a tribute to Desargues's obscure style that the line *cab* was omitted by Abraham Bosse, in whose book this, the first published statement of the theorem, appears.

When straight lines $HDa, HEb, cED, lga, lfb, Hlk, DgK, EfK$, [$cab$], which either lie in different planes or in the same one, cut one another in any order and at any angle in such points [as those implied in the lettering]; the points c, f, g lie on a straight line cfg. For, whatever form the figure takes, in every case: if the straight lines lie in different planes, the lines abc, lga, lfb lie in a plane; the lines DEc, DgK, KfE lie in another: and the points c, f, g lie in each of these two planes; consequently they lie on a straight line cfg. And if the same straight lines all lie in the same plane,

$$\frac{gD}{gK} = \frac{aD}{aH} \cdot \frac{lH}{lK} \quad \text{and} \quad \frac{fK}{fE} = \frac{lK}{lH} \cdot \frac{bH}{bE} \quad \text{and} \quad \frac{aD}{aH} = \frac{cD}{cE} \cdot \frac{bE}{bH}.$$

Therefore

$$\frac{cD}{cE} = \frac{gD}{gK} \cdot \frac{fK}{fE}.$$

Consequently c, g, f lie on a straight line.

11.D7 Philippe de la Hire's *Sectiones Conicae*

(*a*) *La Hire's theorem: Propositions 22 and 23 of Book II*

22 *If in the plane of the circle FNGM a line VAB is given which however far it is produced does not cut the circle, and from any of the points VAB on the line VB are drawn two lines touching the circle, such as VE, VD; AF, AG; BH, BI. I say that lines joining the two points of contact of lines drawn from the same point, as ED, FG, HI mutually intersect at the same point O inside the circle.*

Let M, C, N, A be the diameter through C the centre of the circle perpendicular to the line VB, and from the point A draw AF, AG, and let the join FG cut AM in O. By Proposition 21 AM is divided harmonically in the points $AMON$.

From the point V draw the line VO cutting the circle in K and L. By the preceding proposition this divides the line VL harmonically in the points $VKOL$, and by the corollary to Proposition 21 the line ED joining the points of contact of VE and VD meets the line VL at the point O inside the circle.

The same demonstrates that the line HI joining the points of contact H and I meets the others at the point O; which was to be shown.

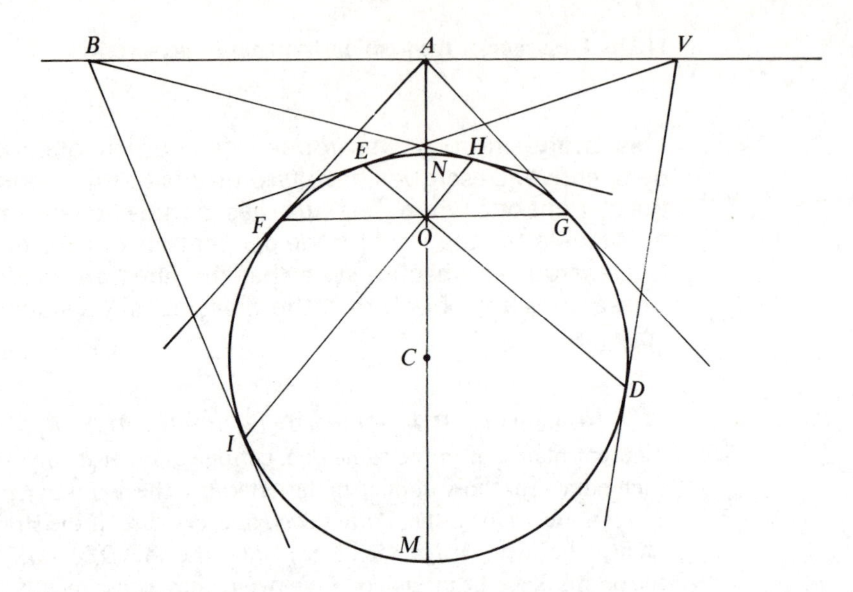

23 *In the plane of any conic section or section of opposites, if there is given a straight line VE which however far produced does not cut the section nor the section of opposites and does not pass through the centre, and if arbitrary points V, D etc. are taken on this line from which are drawn two lines VN, VL; DH, DM, etc. tangent to the same section or to the section of opposites, then, joining NL, MH etc., the points of contact of the tangents which are drawn from the same point: I say that such lines joining the contact points pass through one and the same point P in the section, and whatever line GPE is drawn through P is harmonically divided by the section or sections in GF and by the line first drawn at E or is bisected by such intersections if it is parallel to the line VE.*

And in the hyperbola or section of opposites, if the line first drawn VE cuts the asymptote OR at R and the line RT is tangent at T; I say that the line TP parallel to the asymptote OR drawn through T meets the other lines joining the other points of contact at the same point P; and the line PE meets the line VE at E equidistantly, bisected by the section cut at the point F. The same holds for the diameter of the parabola.

In Proposition 26 of Book I this proposition has been proved for the circle, from which the same will hold for all conic sections for this only concerns the harmonic division of lines arising from sections made by lines on a circular base howsoever divided; sprung from which harmonics, these lines are similarly divided, or only bisected, in prescribed plane sections. I shall *never* add *any*thing about asymptotes, nor about singular cases which can be deduced from the various harmonic divisions, enough having been said about them above, nor about many notable and unusual things.

(*b*) *M. Chasles on la Hire's* Sectiones Conicae

This work is divided into nine books. The first, which is fundamental to all the rest, successively treats the properties of the harmonic division of the line, harmonic pencils, and finally lines harmonically divided in a circle. So it contains some particular cases of the relationship of six points in involution, although this relationship is not given in all its generality. This book is an introduction from which is deduced, in the sequel, easy and general proofs of theorems which had cost the ancients a long and painful struggle. That is what the novelty and merit of de la Hire's method consists in.

With the exception of the locus to three and four lines, and the beautiful general theorems which form the basis of the works of Desargues and Pascal, all the other known properties of conics are brought together, for the first time, in de la Hire's treatise, and proved synthetically in a uniform and elegant way. Several are due to this geometer. Among them we cite above all the theory of poles, which consists of the following three theorems:

1 'If one rotates a transversal, which cuts a conic in two points, about a fixed point, the tangents at the points of intersection always meet on the same line' (Book I. 27, 28; Book II. 24, 27). And conversely 'If from each point on a line one draws the two tangents to a conic, then the line which joins the points of contact passes through a fixed point' (Book I. 26, 28; Book II. 23, 26).

This point was later called the *pole* of the line, and this line the *polar* of the point.

2 'If one takes several transversals through a fixed point which meet a conic, then the lines which join the pairs of intersection points of two arbitrary transversals meet on the polar of the fixed point' (Book I. 22, 33; Book II. 30).

3 Finally, 'The point where each transversal meets the polar of a fixed point will be the harmonic conjugate of the fixed point, with respect to the two points where this transversal meets the curve' (Book I. 21; Book II. 23, 26).

The last proposition was known to Apollonius. In de la Hire's treatise it is the fundamental proposition from which almost all the others are deduced. [...] So this proposition plays the same rôle in de la Hire's great treatise as the proposition about the *latus rectum* in Apollonius, and the theorem on six points in involution in the *Brouillon Project* of Desargues, and the mystic hexagram in the work of Pascal. It is easy to see that, of the three propositions we have stated, the first two are contained in the theorem on a quadrilateral inscribed in a conic, which we have said Pascal probably deduced from his hexagram, and the third is also a consequence of the same theorem by means of Proposition 131 of Book VII of the *Collectio*, which we have already cited when talking about Pappus. But Pascal's work never having been published, de la Hire has the merit of finding these beautiful propositions.

11.E Infinitesimals, Indivisibles, Areas and Tangents

There were several methods of trying to find areas in use in the mid-seventeenth century, and Gilles Personne de Roberval was as eclectic as anybody, for in his treatise on indivisibles (written in 1634 but only published in 1693) he also used infinitesimal methods. (The distinction is roughly that an indivisible in, say, an area problem is a line, something with no area, whereas an infinitesimal is an infinitely small strip and has some area, albeit infinitely little.) Roberval's discussion of the cycloid bears the weight of an early attempt to describe a curve as generated by two simultaneous motions, but once you interpret it as a set of instructions for an animation, it makes a fine moving picture, whether in your mind's eye or on a screen. His proof of the area property is a good illustration of indivisibles at their best, and the discussion of the tangency problem is likewise a good illustration of the power of the motion description of curves. Although this passage is a famous one, we have retranslated it (as 11.E1) to rescue it from a gratuitous transition to infinitesimal strips supplied by an earlier translator (E. Walker).

The practitioners of precalculus mathematics are often scolded for their lack of rigour—a somewhat whiggish position, to be sure—so it is interesting to see the robust attitude taken in this connection by Blaise Pascal in 11.E2, for he was no stranger to philosophical discussion. The crucial distinction to be made in the period is instead between the more geometric and the more algebraic methods, or the more particular versus the more general. Isaac Barrow's famous argument (11.E3), from a book that did not, in fact, sell very well in its day, is typical of how questions on areas and tangency were investigated before the subject was transformed by the algorithmically dominated approaches of Newton and Leibniz. What to them was to be a central result (the fundamental theorem of the calculus) was to Barrow only a rather good theorem.

11.E1 Gilles Personne de Roberval on the Cycloid

We suppose that the diameter *AB* of the circle *AEFGB* moves parallel to itself, as if carried by some other body, until it has arrived at *CD* and turned through a semi-circle. While it travelled, the point *A* at the extremity of the said diameter moves round the circumference of the circle *AEFGB*, and moves as far as the diameter, in such a way that when the diameter is at *CD* the point *A* has reached *B*, and the line *AC* is equal to the circumference *AGHB*. Now, the path of the diameter is divided into infinitely many parts equal amongst themselves and to each part of the circumference *AGB* which is also divided into infinitely many parts all equal to themselves and to the parts of *AC* run through by the diameter, as was said. And then I consider the path of the said point *A* carried by the two movements: the one of the diameter forwards, the other its proper motion along the circumference. To find the said path, I see that when it has reached *E* it has risen above its initial position which it has left; the height is marked by drawing the sine *E*1 through the point *E* to the diameter *AB*, and the versed sine *A*1 is the height of the said *A* when it has reached *E*. Likewise when it has reached *F*, I draw the sine *F*2 through the point *F* to *AB*, and *A*2 will be the height of *A* when it has reached *G*; and doing this at all the places on the circumference that *A* runs through I find all the heights and elevations above the end of the diameter *A*, which are *A*1, *A*2, *A*3, *A*4, *A*5, *A*6, *A*7; therefore to find the places which the said point *A* passes through, and to know the curve which is drawn by the two movements, I carry each of the heights on each of the diameters *M*, *N*, *O*, *P*, *Q*, *R*, *S*, *T* and I find that *M*1, *N*2, *O*3, *P*4, *Q*5, *R*6, *S*7 are the same as those taken on *AB*. Then I take the same sines *E*1, *F*2, *G*3 etc. and I carry them at the height found on this diameter and draw them towards the circle, and the ends of these sines form two curves of which one is *A*, 8, 9, 10, 11, 12, 13,

14, D and the other $A, 1, 2, 3, 4, 5, 6, 7, D$. I know how the curve $A89D$ is drawn, but to know which movements produce the other I say that while AB has run along the line AC the point A has risen up the line AB and marked all the points 1, 2, 3, 4, 5, 6, 7,—the first space while AB reached M, the second while AB reached N, and so always the one space is equal to the other until the diameter has arrived at CD, when the point A has risen to B. That is how the curve $A, 1, 2, 3, D$ is drawn. Now, these two curves enclose a space, being separated one from the other by all the sines and joining together again at the two ends A, D. Now, each part contained between these two curves is equal to each part of the area of the circle AEB contained in that circumference, for the ones and the others are made of equal lines, i.e. of height $A1$, $A2$ etc. and of sines $E1$, $F2$ etc., which are the same as those of the diameters M, N, O etc., and thus the figure $A, 4, D$, 12 is equal to the semi-circle AHB. Now, the line $A123D$ divides the parallelogram $ABCD$ in two equally, because the lines of one half are equal to the lines of the other half, and the line AC to the line BD, and consequently, according to Archimedes, the half is equal to the circle, to which on adding the semi-circle, i.e. the space between the two curves, one will have a circle and a half for the space $A89DC$; and doing the same for the other half, the whole figure of the cycloid will make three times the circle.

To find the tangent to the figure at a given point, I draw a tangent to the circle which passes through the said point, for each point of the circle moves along the tangent to the circle. I then consider the movement which we have given to our point by carrying it along the diameter moving parallel to itself. Drawing through the same point the curve of this movement, if I obtain a parallelogram (which will always have its four sides equal when the path of the point A on the circumference is equal to the path of the diameter AB on the line AC) and if at the same point I draw the diagonal, I have the tangent to the figure which has these two movements as its components, i.e. the circular and the direct. That is how one proceeds when one supposes the movements are equal. If one had supposed that, instead of being equal, the movements had been in some other ratio, the parallelogram would have been constructed with its sides in that ratio.

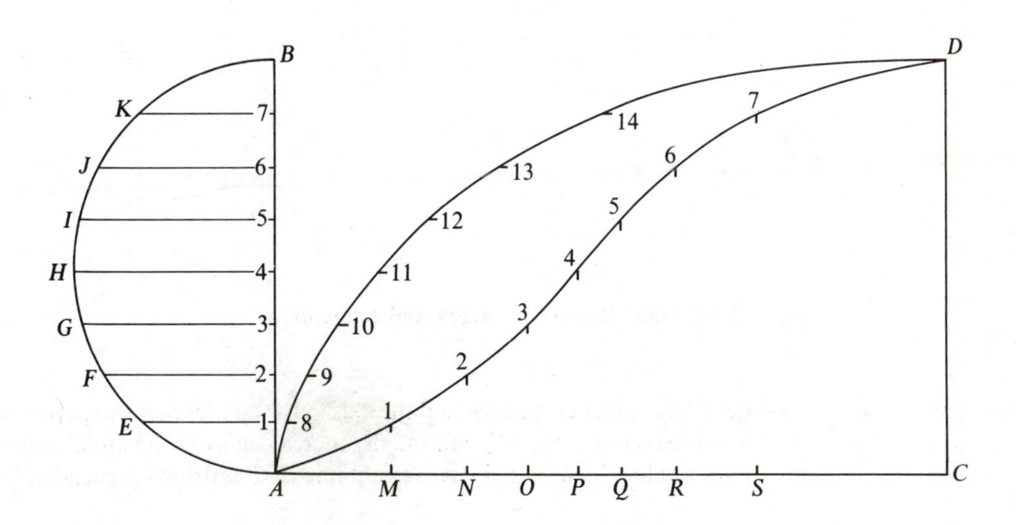

11.E2 Blaise Pascal to Pierre de Carcavy

I wanted to write this note to show that everything which is proved by the true rules of indivisibles will also be proved with the rigour and the manner of the ancients, and that therefore the methods differ, the one from the other, only in the way they are expressed: which cannot hurt reasonable people once one has alerted them to what that means.

And that is why I do not find any difficulty in what follows in using the language of indivisibles, the sum of lines or the sum of planes; and thus when for example I consider the diameter of a semicircle divided into an indefinite number of equal parts at the points Z, from which ordinates ZM are taken, I shall find no difficulty in using this expression, the sum of the ordinates, which seems not to be geometric to those who do not understand the doctrine of indivisibles and who imagine that it is to sin against geometry to express a plane by an indefinite number of lines; which only shows their lack of intelligence, for one understands nothing other by that than the sum of an indefinite number of rectangles made on each ordinate with each of the equal portions of the diameter, whose sum is certainly a plane which only differs from the space of the semi-circle by a quantity less than any given quantity.

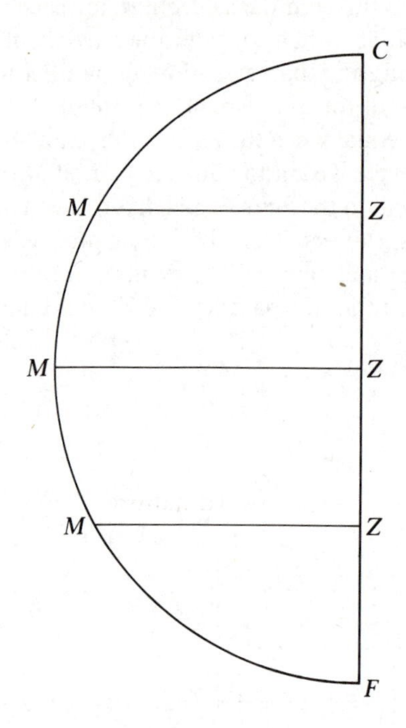

11.E3 Isaac Barrow on areas and tangents

Let ZGE be any curve of which the axis is VD and let there be perpendicular ordinates to this axis (VZ, PG, DE) continually increasing from the initial ordinate VZ; also let VIF be a line such that, if any straight line EDF is drawn perpendicular to VD, cutting the curves in the points E, F, and VD in D, the rectangle contained by DF and a given length R is equal to the intercepted space $VDEZ$; also let $DE:DF = R:DT$, and join [T and F]. Then TF will touch the curve VIF. For, if any point I is taken in the line VIF (first on the side of F towards V), and if through it IG is drawn parallel to VZ, and IL is parallel to VD, cutting the given lines as shown in the figure; then $LF:LK = DF:DT = DE:R$, or $R \times LF = LK \times DE$.

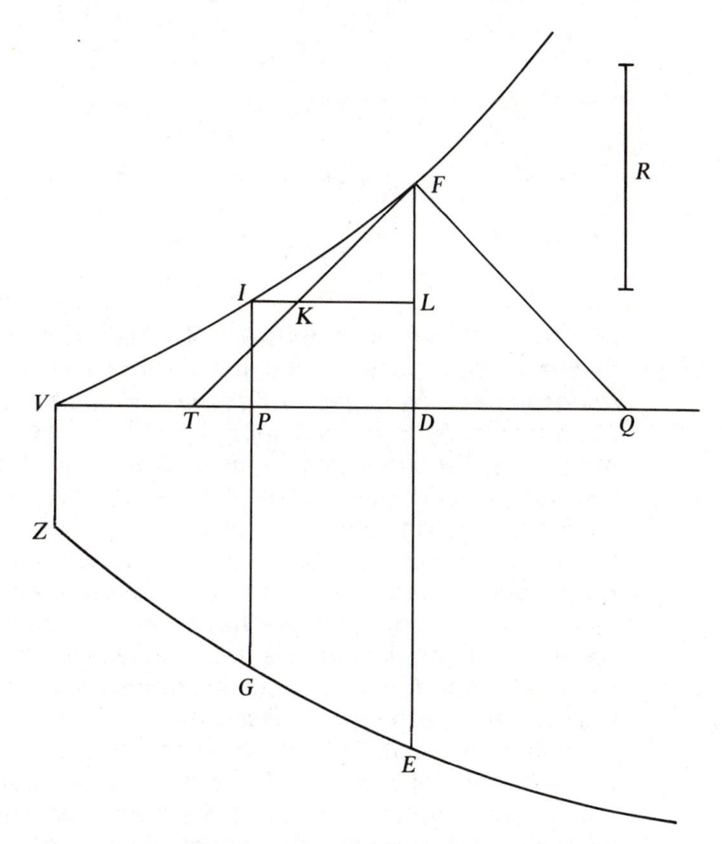

But, from the stated nature of the lines DF, LK, we have $R \times LF =$ area $PDEG$: therefore $LK \times DE =$ area $PDEG < DP \times DE$; hence $LK < DP < LI$.

Again, if the point I is taken on the other side of F, and the same construction is made as before, plainly it can be easily shown that $LK > DP > LI$.

From which it is quite clear that the whole of the line TKF lies within or below the curve $VIFI$.

Other things remaining the same, if the ordinates, VZ, PG, DE, continually decrease, the same conclusion is attained by similar argument; only one distinction occurs, namely, in this case, contrary to the other, the curve VIF is concave to the axis VD.

12 Isaac Newton

Sir Isaac Newton (1642–1727) did not dominate even English mathematical and scientific life until the successful publication of his *Philosophiae Naturalis Principia Mathematica* (*The Mathematical Principles of Natural Philosophy*) in 1687 (see 12.B). Until then the vast bulk of his discoveries lay in his desk drawers, known only in outline to a few friends and colleagues. Afterwards, when he left Cambridge and moved to London, where he was made director of the Royal Mint, he began to publish increasingly, but even so it has remained for modern scholars to print more of Newton's mathematical work than Newton ever did. So the picture of Newton that we have, and the nature of his influence, are necessarily complicated. His first love was for mathematics, and his initial years at Cambridge were spent mastering the literature; works by Oughtred, Wallis, and especially Descartes's *Geometry* and the numerous commentaries on it. But soon he left them behind, and in 1664 began to do his own original research. Our first selections show him at work in this period investigating curves in the Cartesian style, but insisting on the centrality of the problem of tangents (see 12.A1, 12.A3, 12.A5). His use of infinite series lent his work a generality which surpassed Descartes's (see 12.A2, 12.A3), but two other features of his thought are also particularly noteworthy: his emphasis on the tangent as the instantaneous direction of motion along the curve; and his discovery of a pattern in the results which yielded him an algorithm (see 12.A4, 12.A5). Soon he realized that quadrature problems were inverse to tangency problems, and he was then in possession of what can be called the Newtonian calculus.

This calculus makes certain kinds of problems easy which had been difficult, and suggested to Newton that it was now possible to tackle a much harder problem, the inverse tangent problem (raised in 12.A6), which he regarded as a generalization of the finding of areas. His formal rules for 'differentiation' or finding fluxions did not, of course, invert in any simple way, but he found, as his two letters to Leibniz make clear (see 12.C), that his method of infinite series was a great help here too, although it was not the only method.

But Newton did not only invent the calculus, and write the *Principia* and his *Opticks*; he was also a first-rate geometer. By this we do not only mean that his geometrical arguments in the *Principia* are skilful and elegant, as they indeed are, but that there and elsewhere he had dramatic new things to say. Another fruit of the 1660s that was not to see the light until his *Opticks* was published in 1704 was his remarkable classification of cubic curves (see 12.D2), which was both a *tour de force* of Cartesian algebraic methods, and the occasion for a lapidary statement of the power of projective geometry. The work was to provoke many eighteenth-century geometers, some of whom we shall look at in 14.D. In the *Principia* itself he not only naturally included many results about conics satisfying various conditions, but also presented his projective transformations of curves (see 12.D4); this, too, gradually brought forth a literature of its own. However, as he grew older, he developed increasingly firm views on the subject of geometry and its superiority over algebra; 12.D3 is just one of many similar passages on this theme.

12.A Newton's Invention of the Calculus

12.A1 Tangents by motion and by the *o*-method

Lemma

If two bodys A, B, move uniformely the $\genfrac{}{}{0pt}{}{\text{one}}{\text{other}}$ from $\genfrac{}{}{0pt}{}{a}{b}$ to $\genfrac{}{}{0pt}{}{c, d, e, f,}{g, h, k, l,}$ &c: in the same time. Then are the lines $\genfrac{}{}{0pt}{}{ac,}{bg,}$ & $\genfrac{}{}{0pt}{}{cd,}{gh,}$ & $\genfrac{}{}{0pt}{}{de,}{hk,}$ & $\genfrac{}{}{0pt}{}{ef,}{kl,}$ &c: as their velocitys $\genfrac{}{}{0pt}{}{p}{q}$. And though they move not uniformely yet are the infinitely little lines which each moment they describe, as

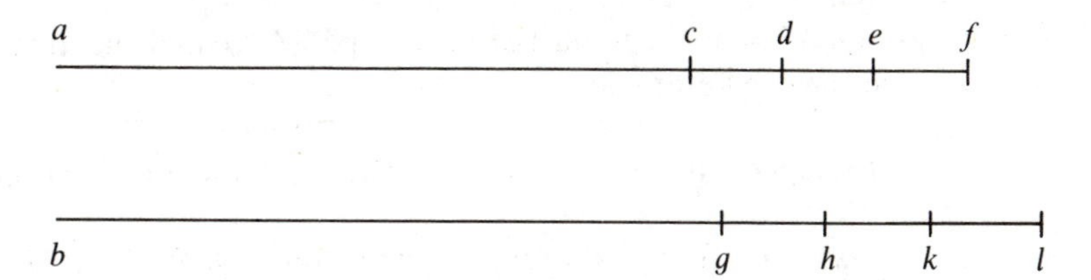

their velocitys which they have while they describe them. As if the body A with the velocity p describe the infinitely little line $(cd =)p \times o$ in one moment, in that moment the body B with the velocity q will describe the line $(gh =)q \times o$. For $p{:}q{::}po{:}qo$. Soe that if the described lines bee $(ac =)x$, & $(bg =)y$, in one moment, they will bee $(ad =)x + po$, & $(bh =)y + qo$ in the next.

Demonstration

Now if the equation expressing the relation twixt the lines x & y bee $x^3 - abx + a^3 - dyy = 0$. I may substitute $x + po$ & $y + qo$ into the place of x & y; because (by the lemma) they as well as x & y, doe signify the lines described by the bodys A & B. By doeing so there results

$$\begin{array}{l} x^3 + 3poxx + 3ppoox + p^3o^3 - dyy - 2dqoy - dqqoo = 0. \\ \qquad - abx \qquad - abpo \\ \qquad\qquad\qquad + a^3 \end{array}$$

But $x^3 - abx + a^3 - dyy = 0$ (by supp). Therefore there remaines onely

$$\begin{array}{l} 3poxx + 3ppoox + p^3o^3 - 2dqoy - dqqoo = 0. \\ \qquad\qquad - abpo \end{array}$$

Or dividing it by o tis

$$\begin{array}{l} 3px^2 + 3ppox + p^3oo - 2dqy - dqqoo = 0. \\ \qquad\qquad - abp \end{array}$$

Also those termes are infinitely little in which o is. Therefore omitting them there rests

$$3pxx - abp - 2dqy = 0.$$

The like may bee done in all other equations.

12.A2 Rules for finding areas

The general method which I had devised some time ago for measuring the quantity of curves by an infinite series of terms you have, in the following, rather briefly explained than narrowly demonstrated.

To the base AB of some curve AD let the ordinate BD be perpendicular and let AB be called x and BD y. Let again $a, b, c, \ldots$ be given quantities and m, n integers. Then

Rule 1
If $ax^{m/n} = y$, then will $[na/(m+n)]x^{(m+n)/n}$ equal the area ABD. The matter will be evident by example.

Example 1 If $x^2(= 1 \times x^{\frac{2}{1}}) = y$, that is, if $a = n = 1$ and $m = 2$, then $\frac{1}{3}x^3 = ABD$.

Example 2 If $4\sqrt{x}(= 4x^{\frac{1}{2}}) = y$, then $\frac{8}{3}x^{\frac{3}{2}}(= \frac{8}{3}\sqrt{x^3}) = ABD$.

Example 3 If $\sqrt[3]{x^5}(= x^{\frac{5}{3}}) = y$, then $\frac{3}{8}x^{\frac{8}{3}}(= \frac{3}{8}\sqrt[3]{x^8}) = ABD$.

Example 4 If $(1/x^2)(= x^{-2}) = y$, that is, if $a = n = 1$ and $m = -2$, then $([1/-1]x^{-\frac{1}{1}} =) - x^{-1}(= -[1/x]) = \alpha BD$ infinitely extended in the direction of α: the computation sets its sign negative because it lies on the further side of the line BD.

Example 5 If $2/3\sqrt{x^3}(= \frac{2}{3}x^{-\frac{3}{2}}) = y$, then $(2/-1)x^{-\frac{1}{2}} = -(2/\sqrt{x}) = BD\alpha$.

Example 6 If $(1/x)(= x^{-1}) = y$, then $(1/0)x^{\frac{0}{1}} = (1/0)x^0 = (1/0) \times 1 = \infty$, just as the area of the hyperbola is on each side of the line BD.

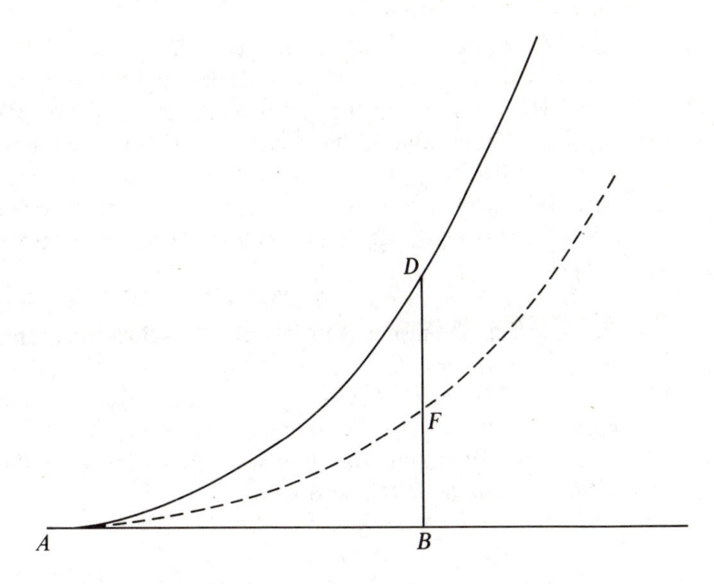

Rule 2
If the value of y is compounded of several terms of that kind the area also will be compounded of the areas which arise separately from each of those terms.

Let its first examples be these. If $x^2 + x^{\frac{3}{2}} = y$, then $\frac{1}{3}x^3 + \frac{2}{5}x^{\frac{5}{2}} = ABD$. For if there be always $BF = x^2$ and $FD = x^{\frac{3}{2}}$, then by the preceding rule $\frac{1}{3}x^3 =$ the surface AFB described by the line BF and $\frac{2}{5}x^{\frac{5}{2}} = AFD$ described by DF; and consequently $\frac{1}{3}x^3 + \frac{2}{5}x^{\frac{5}{2}} =$ the whole ABD. Thus if $x^2 - x^{\frac{3}{2}} = y$, then $\frac{1}{3}x^3 - \frac{2}{5}x^{\frac{5}{2}} = ABD$: and if $3x - 2x^2 + x^3 - 5x^4 = y$, then $\frac{3}{2}x^2 - \frac{2}{3}x^3 + \frac{1}{4}x^4 - x^5 = ABD$.

12.A3 The sine series and the cycloid

If it is desired to find the sine AB from the arc αD given, of the equation $z = x + \frac{1}{6}x^3 + \frac{3}{40}x^5 + \frac{5}{112}x^7 \ldots$ found above (supposing namely that $AB = x$, $\widehat{\alpha D} = z$ and $A\alpha = 1$) I extract the root, which will be $x = z - \frac{1}{6}z^3 + \frac{1}{120}z^5 - \frac{1}{5040}z^7 + \frac{1}{362880}z^9 \ldots$. If, moreover, you want the cosine $A\beta$ of that given arc, make $A\beta(= \sqrt{[1 - x^2]}) = 1 - \frac{1}{2}z^2 + \frac{1}{24}z^4 - \frac{1}{720}z^6 + \frac{1}{40320}z^8 - \frac{1}{3628800}z^{10} \ldots$.

Let it be noted here, by the way, that when you know 5 or 6 terms of those roots you will for the most part be able to prolong them at will by observing analogies. Thus you may prolong this $x = z + \frac{1}{2}z^2 + \frac{1}{6}z^3 + \frac{1}{24}z^4 + \frac{1}{120}z^5 \ldots$ by dividing the last term by these numbers in order $2, 3, 4, 5, 6, 7, \ldots$; and this $x = z - \frac{1}{6}z^3 + \frac{1}{120}z^5 - \frac{1}{5040}z^7 \ldots$ by these $2 \times 3, 4 \times 5, 6 \times 7, 8 \times 9, 10 \times 11, \ldots$; and this $x = 1 - \frac{1}{2}z^2 + \frac{1}{24}z^4 - \frac{1}{720}z^6 \ldots$ by these 1×2, 3×4, 5×6, 7×8, $9 \times 10, \ldots$; while this $z = x + \frac{1}{6}x^3 + \frac{3}{40}x^5 + \frac{5}{112}x^7 \ldots$ you may produce by multiplying by these $\frac{1 \times 1}{2 \times 3}, \frac{3 \times 3}{4 \times 5}, \frac{5 \times 5}{6 \times 7}, \frac{7 \times 7}{8 \times 9}, \ldots$. And so for others.

Let these remarks suffice for geometrical curves. But, indeed, if the curve is mechanical it yet by no means spurns our method. Take, for example, the cycloid

ADFG whose vertex is *A* and axis *AH* while *AKH* is the 'wheel' by which it is generated. And let the surface *ABD* be sought. Setting now $AB = x$, $BD = y$ (as above) and $AH = 1$, I seek in the first instance the length of *BD*. Precisely, by the nature of a cycloid, *KD* is equal to the arc *AK*, and therefore the whole line $BD = BK +$ the arc *AK*. But $BK(= \sqrt{[x - x^2]}) = x^{\frac{1}{2}} - \frac{1}{2}x^{\frac{3}{2}} - \frac{1}{8}x^{\frac{5}{2}} - \frac{1}{16}x^{\frac{7}{2}} \ldots$ and (from the preceding) the arc $AK = x^{\frac{1}{2}} + \frac{1}{6}x^{\frac{3}{2}} + \frac{3}{40}x^{\frac{5}{2}} + \frac{5}{112}x^{\frac{7}{2}} \ldots$, so that in consequence the whole line $BD = 2x^{\frac{1}{2}} - \frac{1}{3}x^{\frac{3}{2}} - \frac{1}{20}x^{\frac{5}{2}} - \frac{1}{56}x^{\frac{7}{2}} \ldots$. And (by Rule 2) the area $ABD = \frac{4}{3}x^{\frac{3}{2}} - \frac{2}{15}x^{\frac{5}{2}} - \frac{1}{70}x^{\frac{7}{2}} - \frac{1}{252}x^{\frac{9}{2}} \ldots$.

Or more briefly thus. Since the straight line *AK* is parallel to the tangent *TD*, *AB* will be to *BK* as the momentum of the line *AB* to the momentum of the line *BD*, that is, $x{:}\sqrt{[x - x^2]} = 1{:}x^{-1}\sqrt{[x - x^2]}$ or $x^{-\frac{1}{2}} - \frac{1}{2}x^{\frac{1}{2}} - \frac{1}{8}x^{\frac{3}{2}} - \frac{1}{16}x^{\frac{5}{2}} - \frac{5}{128}x^{\frac{7}{2}} \ldots$. Hence (by Rule 2) $BD = 2x^{\frac{1}{2}} - \frac{1}{3}x^{\frac{3}{2}} - \frac{1}{20}x^{\frac{5}{2}} - \frac{1}{56}x^{\frac{7}{2}} - \frac{5}{576}x^{\frac{9}{2}} \ldots$ and the surface $ABD = \frac{4}{3}x^{\frac{3}{2}} - \frac{2}{15}x^{\frac{5}{2}} - \frac{1}{70}x^{\frac{7}{2}} - \frac{1}{252}x^{\frac{9}{2}} - \frac{5}{3168}x^{\frac{11}{2}} \ldots$.

In a not dissimilar way you will (on setting *C* as the circle's centre and $CB = x$) obtain the area *CBDF*, and so on.

12.A4 Quadrature as the inverse of fluxions

Let any curve $AD\delta$ have base $AB = x$, perpendicular ordinate $BD = y$ and area $ABD = z$. Take $B\beta = o$, $BK = v$ and the rectangle $B\beta HK(ov)$ equal to the space $B\beta\delta D$. It is, therefore, $A\beta = x + o$ and $A\delta\beta = z + ov$. With these premisses, from any arbitrarily assumed relationship between *x* and *z* I seek *y* in the way you see following.

Take at will $\frac{2}{3}x^{\frac{3}{2}} = z$ or $\frac{4}{9}x^3 = z^2$. Then, when $x + o(A\beta)$ is substituted for *x* and $z + ov(A\delta\beta)$ for *z*, there arises (by the nature of the curve) $\frac{4}{9}(x^3 + 3x^2o + 3xo^2 + o^3) = z^2 + 2zov + o^2v^2$. On taking away equal quantities ($\frac{4}{9}x^3$ and z^2) and dividing the rest by *o*, there remains $\frac{4}{9}(3x^2 + 3xo + o^2) = 2zv + ov^2$. If we now suppose $B\beta$ to be infinitely small, that is, *o* to be zero, *v* and *y* will be equal and terms multiplied by *o* will vanish and there will consequently remain $\frac{4}{9} \times 3x^2 = 2zv$ or $\frac{2}{3}x^2(= zy) = \frac{2}{3}x^{\frac{3}{2}}y$, that is, $x^{\frac{1}{2}}(= x^2/x^{\frac{3}{2}}) = y$. Conversely therefore if $x^{\frac{1}{2}} = y$, then will $\frac{2}{3}x^{\frac{3}{2}} = z$.

Or in general if $[n/(m + n)]ax^{(m+n)/n} = z$, that is, by setting $na/(m + n) = c$ and $m + n = p$, if $cx^{p/n} = z$ or $c^nx^p = z^n$, then when $x + o$ is substituted for *x* and $z + ov$ (or, what is its equivalent, $z + oy$) for *z* there arises $c^n(x^p + pox^{p-1} \ldots) = z^n + noyz^{n-1} \ldots$, omitting the other terms, to be precise, which would ultimately vanish. Now, on taking away the equal terms c^nx^p and z^n and dividing the rest by *o*, there remains $c^npx^{p-1} = nyz^{n-1}(= nyz^n/z) = nyc^nx^p/cx^{p/n}$. That is, on dividing by c^nx^p, there will be $px^{-1} = ny/cx^{p/n}$ or $pcx^{(p-n)/n} = y$; in other words, by restoring $na/(m + n)$ for *o* and $m + n$ for *p*, that is, *m* for $p - n$ and *na* for *pc*, there will come $ax^{m/n} = y$. Conversely therefore if $ax^{m/n} = y$, then will $[n/(m + n)]ax^{(m+n)/n} = z$. As was to be proved.

Here in passing may be noted a method by which as many curves as you please whose areas are known may be found: namely, by assuming any equation at will for the relationship between the area *z* and from it in consequence seeking the ordinate *y*. So if you should suppose $\sqrt{[a^2 + x^2]} = z$, by computation you will find $x/\sqrt{[a^2 + x^2]} = y$. And similarly in other cases.

12.A5 Finding fluxions of fluent quantities

The moments of the fluent quantities (that is, their indefinitely small parts, by addition of which they increase during each infinitely small period of time) are as their speeds of flow. Wherefore if the moment of any particular one, say x, be expressed by the product of its speed $\dot{x}$ and an infinitely small quantity o (that is, by $\dot{x}o$), then the moments of the others, v, y, z, [...], will be expressed by $\dot{v}o, \dot{y}o, \dot{z}o$, [...] seeing that $\dot{v}o, \dot{x}o, \dot{y}o$ and $\dot{z}o$ are to one another as $\dot{v}, \dot{x}, \dot{y}$ and $\dot{z}$.

Now, since the moments (say, $\dot{x}o$ and $\dot{y}o$) of fluent quantities (x and y say) are the infinitely small additions by which those quantities increase during each infinitely small interval of time, it follows that those quantities x and y after any infinitely small interval of time will become $x + \dot{x}o$ and $y + \dot{y}o$. Consequently, an equation which expresses a relationship of fluent quantities without variance at all times will express that relationship equally between $x + \dot{x}o$ and $y + \dot{y}o$ as between x and y; and so $x + \dot{x}o$ and $y + \dot{y}o$ may be substituted in place of the latter quantities, x and y, in the said equation.

Let there be given, accordingly, any equation $x^3 - ax^2 + axy - y^3 = 0$ and substitute $x + \dot{x}o$ in place of x and $y + \dot{y}o$ in place of y: there will emerge

$$(x^3 + 3\dot{x}ox^2 + 3\dot{x}^2o^2x + \dot{x}^3o^3) - (ax^2 + 2a\dot{x}ox + a\dot{x}^2o^2)$$
$$+ (axy + a\dot{x}oy + a\dot{y}ox + a\dot{x}\dot{y}o^2) - (y^3 + 3\dot{y}oy^2 + 3\dot{y}^2o^2y + \dot{y}^3o^3) = 0.$$

Now by hypothesis $x^3 - ax^2 + axy - y^3 = 0$, and when these terms are erased and the rest divided by o there will remain

$$3\dot{x}x^2 + 3\dot{x}^2ox + \dot{x}^3o^2 - 2a\dot{x}x - a\dot{x}^2o + a\dot{x}y + a\dot{y}x + a\dot{x}\dot{y}o - 3\dot{y}y^2 - 3\dot{y}^2oy - \dot{y}^3o^2 = 0.$$

But further, since o is supposed to be infinitely small so that it be able to express the moments of quantities, terms which have it as a factor will be equivalent to nothing in respect of the others. I therefore cast them out and there remains $3\dot{x}x^2 - 2a\dot{x}x + a\dot{x}y + a\dot{y}x - 3\dot{y}y^2 = 0$.

It is accordingly to be observed that terms not multiplied by o will always vanish, as also those multiplied by o of more than one dimension; and that the remaining terms after division by o will always take on the form they should have according to the rule. This is what I wanted to show.

12.A6 Finding fluents from a fluxional relationship

Problem 1

When a fluent quantity is exhibited, the relationship of whose moments to those of some other fluent quantity is given, to find the relation of the quantities to one another.

Multiply the value of the ratio of the moments of the quantity sought to the moments of the exhibited quantity (so long as it is free from irrationals and not affected with some denominator of several terms) by the exhibited quantity, then divide each term individually by its own number of dimensions in this same quantity: what results will be the value of the quantity sought.

For instance, if x be exhibited and y sought, multiply the value of the ratio $\dot{y}/\dot{x}$, displayed in any given equation, by x and then divide each term by the number of its dimensions, setting the result equal to y.

Example If there be given $\dot{y}/\dot{x} = x/a$, I multiply x/a by x and there comes x^2/a. Here since x is of two dimensions I divide by 2 and there comes $x^2/2a$, which I set equal to y.

Problem 2

When an equation involving the fluxions of quantities is exhibited, to determine the relation of the quantities one to another.

Since this problem is the converse of the preceding, it ought to be resolved the contrary way: namely, by arranging the terms multiplied by $\dot{x}$ according to the dimensions of x and dividing by $\dot{x}/x$ and then by the number of dimensions or perhaps another arithmetic progression, by carrying out the same operation in the terms multiplied by $\dot{v}$, $\dot{y}$ or $\dot{z}$, and, with redundant terms rejected, setting the total of the resulting terms equal to nothing.

Example Thus when the equation $3\dot{x}x^2 - 2a\dot{x}x + a\dot{x}y - 3\dot{y}y^2 + a\dot{y}x = 0$ is exhibited, I operate in this manner:

I divide	$3\dot{x}x^2 - 2a\dot{x}x + a\dot{x}y$			I divide	$-3\dot{y}y^2 + a\dot{y}x$	
by $\frac{\dot{x}}{x}$, making	$3x^3$	$- 2ax^2$	$+ axy$.	by $\frac{\dot{y}}{y}$, making	$-3y^3$	$+ axy$.
Then I divide by	3.	2.	1.	Then by	3.	1.
making	x^3	$- ax^2$	$+ axy$	making	$- y^3$	$+ axy$.

The total $x^3 - ax^2 + axy - y^3 = 0$ will be the desired relationship of the quantities x and y. Here it should be noticed that, even though the term axy arose twice, I do not, however, put it twice into this total $x^3 - ax^2 + axy - y^3$ but lay aside one term as redundant. And so universally where some term arises twice (or more often if several fluent quantities are in question) I write it but once in the total of the terms.

There are other circumstances, too, which I leave to the notice of the skilled practitioner, for it would be superfluous to expend a lot of words on this topic since the Problem cannot always be resolved by this practice. I may add, however, that if, after the practised mathematician has obtained a relation between the fluent quantities by this method, regress may be had by Problem 1 to the exhibited equation which involves their fluxions, the procedure has been correctly carried out, but otherwise faultily so. So in the example proposed, where I obtained the equation $x^3 - ax^2 + axy - y^3 = 0$, if in turn the relation between $\dot{x}$ and [$\dot{y}$] be thence required with the help of the first Problem, the exhibited equation $3\dot{x}x^2 - 2a\dot{x}x + a\dot{x}y - 3\dot{y}y^2 + a\dot{y}x = 0$ will be got. This would consequently establish that the equation $x^3 - ax^2 + axy - y^3 = 0$ had been correctly found. But if the equation $\dot{x}x - \dot{x}y + \dot{y}a = 0$ were to be exhibited and thence by the method prescribed I were to elicit $\frac{1}{2}x^2 - xy + ay = 0$ for the relation between x and y, the procedure would be faulty since therefrom in turn by Problem 1 there would be produced $\dot{x}x - \dot{x}y - \dot{y}x + \dot{y}a = 0$, an equation different from that first exhibited.

12.B Newton's *Principia*

The *Principia* was a revolutionary work. Not just the first definitive theory of the heavens, but, as I. B. Cohen has argued, a dramatic new model of how to do science. For two-thirds of its length the work is overwhelmingly mathematical; only in the final book does Newton show how well his ideas can be made to fit with observation. The result is that his conception of gravity, although unintelligible to many a well-trained contemporary mind, emerges as a strikingly good theoretical construct, and the alternative hypothesis, that of planetary vortices, is, as Huygens was reluctantly to concede, 'utterly destroyed'. Not only did Newton thereby present a celestial physics in which the elementary modern ideas are all to be found for the first time, but also, by basing them on such a thoroughgoing mathematical analysis he was to establish the new mathematics in its central role in the new science—as Galileo, Kepler and Descartes had not. Contrary to Newton's oft-quoted remark about not making hypotheses, the whole of the *Principia* is a colossal piece of hypothesis testing; Newton's hostility was reserved for *ad hoc* hypotheses designed to shore up an idea which, like the vortex idea, could be shown to be in trouble when examined mathematically.

The famous laws of motion (12.B2) with which *Principia* opens themselves represent a breakthrough, for Newton here distinguishes between force and momentum in a way not commonly done before. In 12.B3 we see just a glimpse of Newton's wonderful ability to argue geometrically about momentary increments, now cast in the language of first and last ratios (further discussed in 12.B4). The pay-off is 12.B5, his law of areas which generalizes Kepler's to any central force. What Newton finally established (12.B6, 12.B7) is that if the planets are attracted to the sun by the force of gravity, then because each planet obeys Kepler's 3/2 power law, the gravitational force obeys an inverse square law. But Newton did more. In 12.B8 we see that he tried, with, as he admitted, only imperfect success, to calculate the motion of the moon. And though he did much better than his predecessors in this respect, here nonetheless he left much for his successors to do, too (see 14.B). In 12.B9, after a lengthy discussion of motion in a resisting medium, he delivers his destructive critique of the vortex theory. In 12.B11 he argues for the Earth being flatter at the poles, which indeed it is, but his underlying physical hypotheses about rotating masses of fluids were to wear less well.

Later mathematicians were also to read the *Principia* for clues to Newton's calculus. While it seems hard to reach a simple verdict on the question of how helpful the calculus was to Newton himself, certainly several passages in the book employ the concepts of first and last ratios (e.g. 12.B3, 12.B4) and were to be carefully studied for that reason by such as Pierre Varignon and Jean le Rond d'Alembert. But when, as an old man, Newton came to revise it for the second and third editions, he did not put it into calculus form. He did, however, take the opportunity to promote his claim to priority in the invention of the calculus, rewriting the scholium in

12.B9, and above all to deepen his attack on Descartes. 12.B12 and 12.B13 are typical of the careful way he now defended his concept of gravity. These passages, and the second half of 12.B10, are the only places where our selection, taken from the standard Motte-Cajori translation of the third edition, cannot be taken to represent Newton's views in 1687.

12.B1 Prefaces

(a) First edition

Since the ancients (as we are told by Pappus) esteemed the science of mechanics of greatest importance in the investigation of natural things, and the moderns, rejecting substantial forms and occult qualities, have endeavoured to subject the phenomena of nature to the laws of mathematics, I have in this treatise cultivated mathematics as far as it relates to philosophy. [. . .] But I consider philosophy rather than arts and write not concerning manual but natural powers, and consider chiefly those things which relate to gravity, levity, elastic force, the resistance of fluids, and the like forces, whether attractive or impulsive; and therefore I offer this work as the mathematical principles of philosophy, for the whole burden of philosophy seems to consist in this—from the phenomena of motions to investigate the forces of nature, and then from these forces to demonstrate the other phenomena; and to this end the general propositions in the first and second Books are directed. In the third Book I give an example of this in the explication of the System of the World; for by the propositions mathematically demonstrated in the former Books, in the third I derive from the celestial phenomena the forces of gravity with which bodies tend to the sun and the several planets. Then from these forces, by other propositions which are also mathematical, I deduce the motions of the planets, the comets, the moon, and the sea. I wish we could derive the rest of the phenomena of Nature by the same kind of reasoning from mechanical principles, for I am induced by many reasons to suspect that they may all depend upon certain forces by which the particles of bodies, by some causes hitherto unknown, are either mutually impelled towards one another, and cohere in regular figures, or are repelled and recede from one another. These forces being unknown, philosophers have hitherto attempted the search of Nature in vain; but I hope the principles here laid down will afford some light either to this or some truer method of philosophy.

In the publication of this work the most acute and universally learned Mr Edmund Halley not only assisted me in correcting the errors of the press and preparing the geometrical figures, but it was through his solicitations that it came to be published; for when he had obtained of me my demonstrations of the figure of the celestial orbits, he continually pressed me to communicate the same to the Royal Society, who afterwards, by their kind encouragement and entreaties, engaged me to think of publishing them. But after I had begun to consider the inequalities of the lunar motions, and had entered upon some other things relating to the laws and measures of gravity and other forces; and the figures that would be described by bodies attracted according to given laws; and the motion of several bodies moving among themselves; the motion of bodies in resisting mediums; the forces, densities, and motions, of mediums; the orbits of the comets, and such like, I deferred that publication till I had made a search into those matters, and could put forth the whole together. What relates to the lunar motions (being imperfect), I have put all together in the corollaries of Proposition LXVI, to avoid being obliged to propose and distinctly demonstrate the several things there contained in a method more prolix than the subject deserved and interrupt the series of the other propositions. Some things, found out after the rest, I chose to insert in places less suitable, rather than change the number of the propositions and the citations. I heartily beg that what I have here done may be read with forbearance; and that my labours in a subject so difficult may be examined, not so much with the view to censure, as to remedy their defects.

(*b*) *Second edition*

In this second edition of the *Principia* there are many emendations and some additions. In the second section of the first Book, the determination of forces, by which bodies may be made to revolve in given orbits, is illustrated and enlarged. In the seventh section of the second Book the theory of the resistances of fluids was more accurately investigated, and confirmed by new experiments. In the third Book the lunar theory and the precession of the equinoxes were more fully deduced from their principles; and the theory of the comets was confirmed by more examples of the calculation of their orbits, done also with greater accuracy.

12.B2 Axioms, or laws of motion

Law I

Every body continues in its state of rest, or of uniform motion in a right line, unless it is compelled to change that state by forces impressed upon it.

Projectiles continue in their motions, so far as they are not retarded by the resistance of the air, or impelled downwards by the force of gravity. A top, whose parts by their cohesion are continually drawn aside from rectilinear motions, does not cease its rotation, otherwise than as it is retarded by the air. The greater bodies of the planets and comets, meeting with less resistance in freer spaces, preserve their motions both progressive and circular for a much longer time.

Law II

The change of motion is proportional to the motive force impressed; and is made in the direction of the right line in which that force is impressed.

If any force generates a motion, a double force will generate double the motion, a triple force triple the motion, whether that force be impressed altogether and at once, or gradually and successively. And this motion (being always directed the same way with the generating force), if the body moved before, is added to or subtracted from the former motion, according as they directly conspire with or are directly contrary to each other; or obliquely joined, when they are oblique, so as to produce a new motion compounded from the determination of both.

Law III

To every action there is always opposed an equal reaction: or, the mutual actions of two bodies upon each other are always equal, and directed to contrary parts.

Whatever draws or presses another is as much drawn or pressed by that other. If you press a stone with your finger, the finger is also pressed by the stone. [. . .] If a body impinge upon another, and by its force change the motion of the other, that body also (because of the equality of the mutual pressure) will undergo an equal change, in its own motion, towards the contrary part. The changes made by these actions are equal, not in the velocities but in the motions of bodies; that is to say, if the bodies are not hindered by any other impediments. For, because the motions are equally changed, the changes of the velocities made towards contrary parts are inversely proportional to the bodies. This law takes place also in attractions, as will be proved in the next Scholium.

Corollary I

A body, acted on by two forces simultaneously, will describe the diagonal of a parallelogram in the same time as it would describe the sides by those forces separately.

If a body in a given time, by the force M impressed apart in the place A, should with an uniform motion be carried from A to B, and by the force N impressed apart in the same place, should be carried from A to C, let the parallelogram $ABCD$ be completed, and, by both forces acting together, it will in the same time be carried in the diagonal

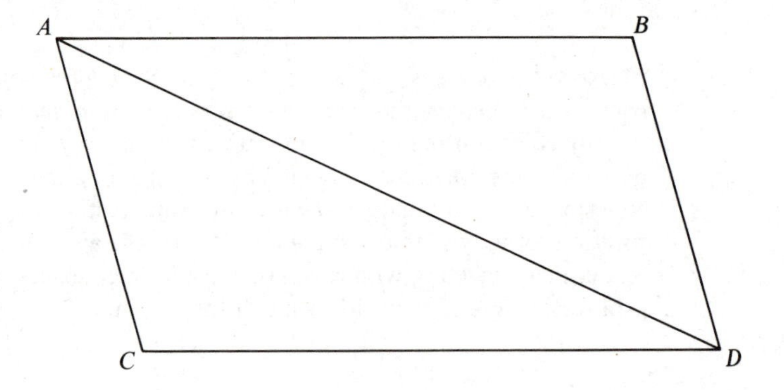

from A to D. For since the force N acts in the direction of the line AC, parallel to BD, this force (by the second Law) will not at all alter the velocity generated by the other force M, by which the body is carried towards the line BD. The body therefore will arrive at the line BD in the same time, whether the force N be impressed or not; and therefore at the end of that time it will be found somewhere in the line BD. By the same argument, at the end of the same time it will be found somewhere in the line CD. Therefore it will be found in the point D, where both lines meet. But it will move in a right line from A to D, by Law I.

12.B3 The method of first and last ratios

Lemma I

Quantities, and the ratios of quantities, which in any finite time converge continually to equality, and before the end of that time approach nearer to each other than by any given difference, become ultimately equal.

If you deny it, suppose them to be ultimately unequal, and let D be their ultimate difference. Therefore they cannot approach nearer to equality than by that given difference D; which is contrary to the supposition.

Lemma II

If in any figure AacE, terminated by the right lines Aa, AE, and the curve acE, there be inscribed any number of parallelograms Ab, Bc, Cd, &c., comprehended under equal bases AB, BC, CD, &c., and the sides, Bb, Cc, Dd, &c., parallel to one side Aa of the figure; and the parallelograms aKbl, bLcm, cMdn, &c., are completed: then if the breadth of those parallelograms be supposed to be diminished, and their number to be augmented in infinitum, *I say, that the ultimate ratios which the inscribed figure*

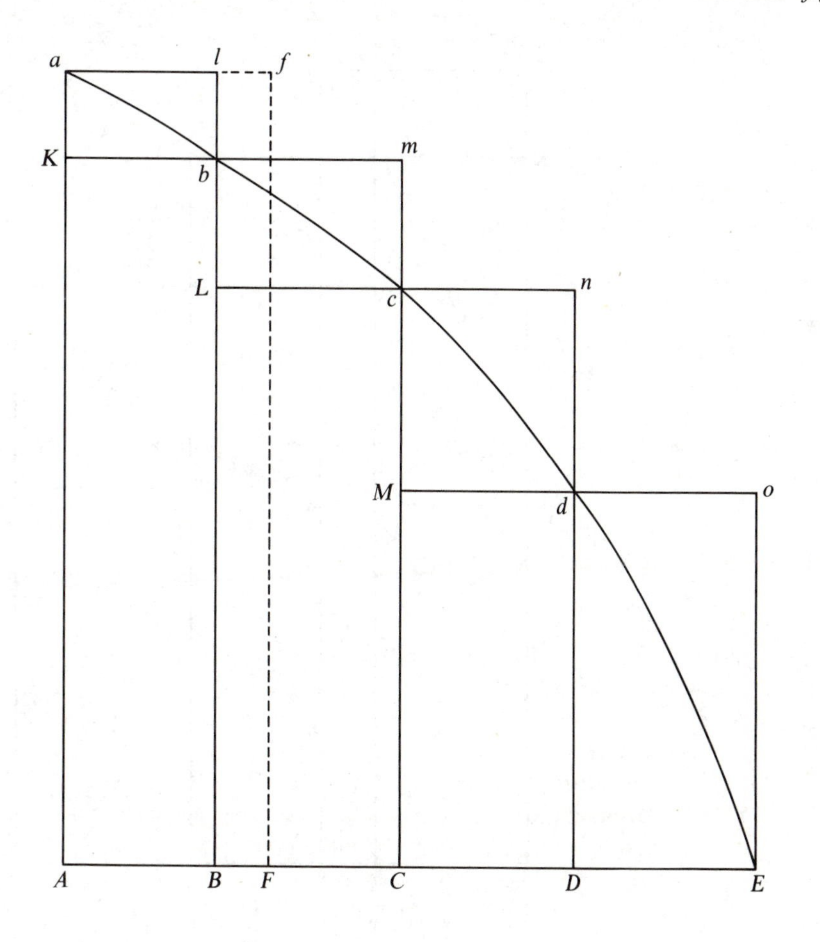

AKbLcMdD, the circumscribed figure AalbmcndoE, and curvilinear figure AabcdE, will have to one another, are ratios of equality.

For the difference of the inscribed and circumscribed figures is the sum of the parallelograms *Kl*, *Lm*, *Mn*, *Do*, that is (from the equality of all their bases), the rectangle under one of their bases *Kb* and the sum of their altitudes *Aa*, that is, the rectangle *ABla*. But this rectangle, because its breadth *AB* is supposed diminished *in infinitum*, becomes less than any given space. And therefore (by Lem. I) the figures inscribed and circumscribed become ultimately equal one to the other; and much more will the intermediate curvilinear figure be ultimately equal to either. Q.E.D.

Lemma III

The same ultimate ratios are also ratios of equality, when the breadths AB, BC, DC, &c., of the parallelograms are unequal, and are all diminished in infinitum.

For suppose *AF* equal to the greatest breadth, and complete the parallelogram *FAaf*. This parallelogram will be greater than the difference of the inscribed and circumscribed figures; but, because its breadth *AF* is diminished *in infinitum*, it will become less than any given rectangle. Q.E.D.

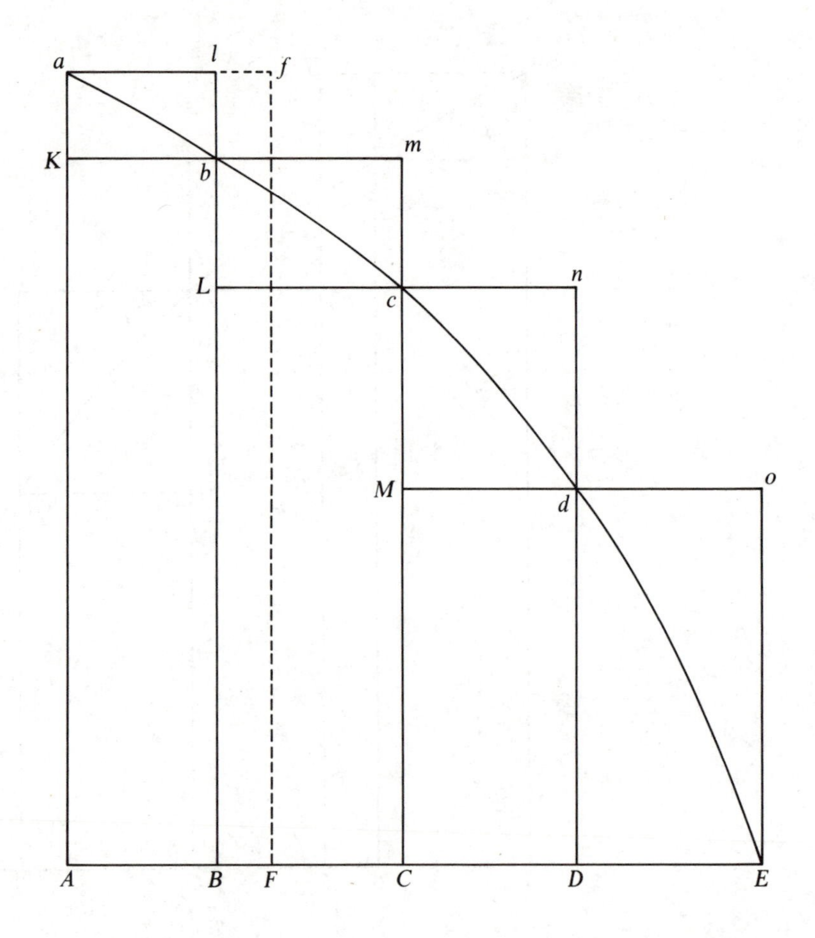

Corollary I Hence the ultimate sum of those evanescent parallelograms will in all parts coincide with the curvilinear figure.

Corollary II Much more will the rectilinear figure comprehended under the chords of the evanescent arcs *ab*, *bc*, *cd*, &c., ultimately coincide with the curvilinear figure.

Corollary III And also the circumscribed rectilinear figure comprehended under the tangents of the same arcs.

Corollary IV And therefore these ultimate figures (as to their perimeters *acE*) are not rectilinear, but curvilinear limits of rectilinear figures.

12.B4 The nature of first and last ratios

Those things which have been demonstrated of curved lines, and the surfaces which they comprehend, may be easily applied to the curved surfaces and contents of solids. These Lemmas are premised to avoid the tediousness of deducing involved demonstrations *ad absurdum*, according to the method of the ancient geometers. For demonstrations are shorter by the method of indivisibles; but because the hypothesis of indivisibles seems somewhat harsh, and therefore that method is reckoned less geometrical, I chose rather to reduce the demonstrations of the following Propositions to the first and last sums and ratios of nascent and evanescent quantities, that is, to the limits of those sums and ratios, and so to premise, as short as I could, the demonstrations of those limits. For hereby the same thing is performed as by the method of indivisibles; and now those principles being demonstrated, we may use them with greater safety. Therefore if hereafter I should happen to consider quantities as made up of particles, or should use little curved lines for right ones, I would not be understood to mean indivisibles, but evanescent divisible quantities; not the sums and ratios of determinate parts, but always the limits of sums and ratios; and that the force of such demonstrations always depends on the method laid down in the foregoing Lemmas.

Perhaps it may be objected, that there is no ultimate proportion of evanescent quantities; because the proportion, before the quantities have vanished, is not the ultimate, and when they are vanished, is none. But by the same argument it may be alleged that a body arriving at a certain place, and there stopping, has no ultimate velocity; because the velocity, before the body comes to the place, is not its ultimate velocity; when it has arrived, there is none. But the answer is easy; for by the ultimate velocity is meant that with which the body is moved, neither before it arrives at its last place and the motion ceases, nor after, but at the very instant it arrives; that is, that velocity with which the body arrives at its last place, and with which the motion ceases. And in like manner, by the ultimate ratio of evanescent quantities is to be understood the ratio of the quantities not before they vanish, nor afterwards, but with which they vanish. In like manner the first ratio of nascent quantities is that with which they begin to be. And the first or last sum is that with which they begin and cease to be (or to be augmented or diminished). There is a limit which the velocity at the end of the motion may attain, but not exceed. This is the ultimate velocity. And there is the like limit in all quantities and proportions that begin and cease to be. And since such limits are certain and definite, to determine the same is a problem strictly geometrical. But whatever is geometrical we may use in determining and demonstrating any other thing that is also geometrical.

It may also be objected, that if the ultimate ratios of evanescent quantities are given, their ultimate magnitudes will be also given: and so all quantities will consist of indivisibles, which is contrary to what Euclid has demonstrated concerning incommensurables, in the tenth Book of his *Elements*. But this objection is founded on a false supposition. For those ultimate ratios with which quantities vanish are not truly the ratios of ultimate quantities, but limits towards which the ratios of quantities decreasing without limit do always converge; and to which they approach nearer than by any given difference, but never go beyond, nor in effect attain to, till the quantities are diminished *in infinitum*. This thing will appear more evident in quantities infinitely great. If two quantities, whose difference is given, be augmented *in infinitum*, the ultimate ratio of these quantities will be given, namely, the ratio of equality; but it does not from thence follow, that the ultimate or greatest quantities themselves, whose ratio that is, will be given. Therefore if in what follows, for the sake of being more easily understood, I should happen to mention quantities as least, or evanescent, or ultimate, you are not to suppose that quantities of any determinate magnitude are meant, but such as are conceived to be always diminished without end.

12.B5 The determination of centripetal forces

Book I, Proposition I, Theorem I

The areas which revolving bodies describe by radii drawn to an immovable centre of force do lie in the same immovable planes, and are proportional to the times in which they are described.

For suppose the time to be divided into equal parts, and in the first part of that time let the body by its innate force describe the right line *AB*. In the second part of that time, the same would (by Law I), if not hindered, proceed directly to *c*, along the line *Bc* equal to *AB*; so that by the radii *AS*, *BS*, *cS*, drawn to the centre, the equal areas *ASB*, *BSc*, would be described. But when the body is arrived at *B*, suppose that a centripetal force acts at once with a great impulse, and, turning aside the body from the right line *Bc*, compels it afterwards to continue its motion along the right line *BC*. Draw *cC* parallel to *BS*, meeting *BC* in *C*; and at the end of the second part of the time, the body (by Cor. I of the Laws) will be found in *C*, in the same plane with the triangle *ASB*. Join *SC*, and, because *SB* and *Cc* are parallel, the triangle *SBC* will be equal to the triangle *SBc*, and therefore also to the triangle *SAB*. By the like argument, if the centripetal force acts successively in *C*, *D*, *E*, &c., and makes the body, in each single particle of time, to describe the right lines *CD*, *DE*, *EF*, &c., they will all lie in the same plane; and the triangle *SCD* will be equal to the triangle *SBC*, and *SDE* to *SCD*, and *SEF* to *SDE*. And therefore, in equal times, equal areas are described in one immovable plane: and, by composition, any sums *SADS*, *SAFS*, of those areas, are to each other as the times in which they are described. Now let the number of those triangles be augmented, and their breadth diminished *in infinitum*; and (by Cor. IV, Lem. III) their

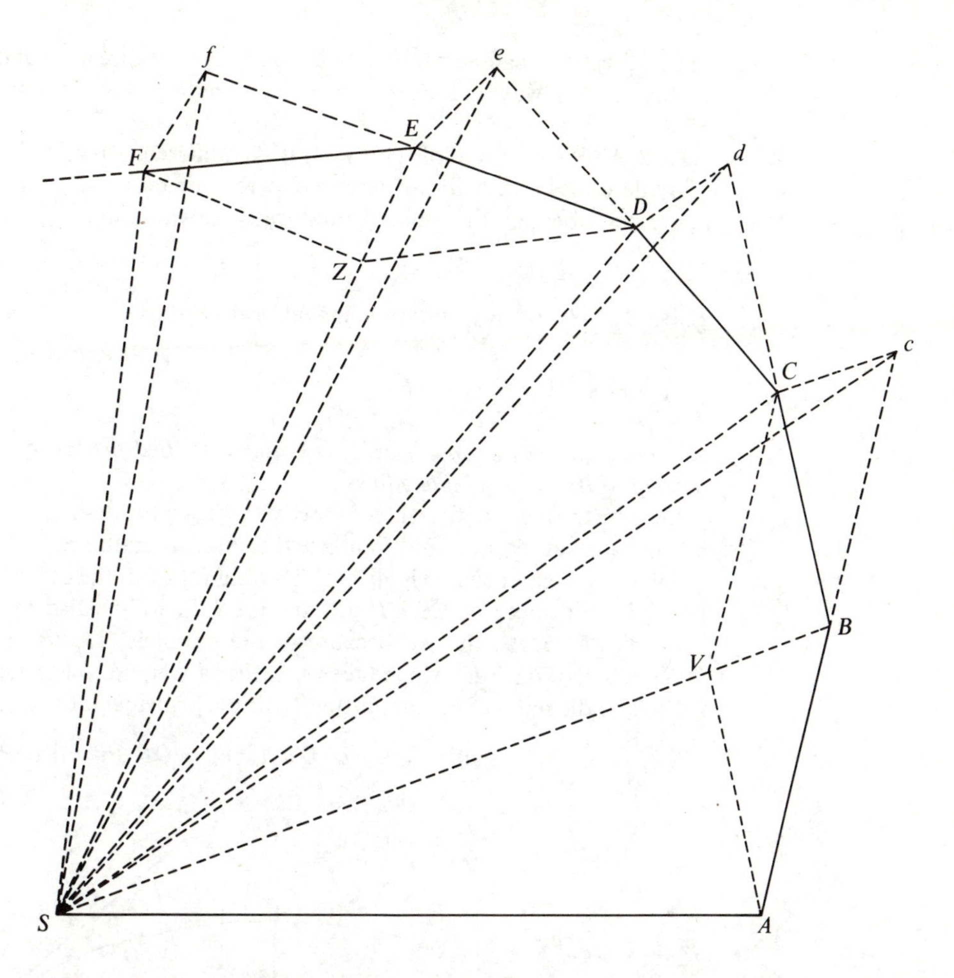

ultimate perimeter *ADF* will be a curved line: and therefore the centripetal force, by which the body is continually drawn back from the tangent of this curve, will act continually; and any described areas *SADS*, *SAFS*, which are always proportional to the times of description, will, in this case also, be proportional to those times. Q.E.D.

Proposition II, Theorem II

Every body that moves in any curved line described in a plane, and by a radius drawn to a point either immovable, or moving forwards with an uniform rectilinear motion, describes about that point areas proportional to the times, is urged by a centripetal force directed to that point.

Case 1 For every body that moves in a curved line is (by Law I) turned aside from its rectilinear course by the action of some force that impels it. And that force by which the body is turned off from its rectilinear course, and is made to describe, in equal times, the equal least triangles *SAB*, *SBC*, *SCD*, &c., about the immovable point *S* (by Euclid's *Elements*, I.40 and Law II), acts in the place *B*, according to the direction of a line parallel to *cC*, that is, in the direction of the line *BS*; and in the place *C*, according to the direction of a line parallel to dD, that is, in the direction of the line CS, &c.; and therefore acts always in the direction of lines tending to the immovable point S. Q.E.D.

Case 2 And (by Cor. V of the Laws) it is indifferent whether the surface in which a body describes a curvilinear figure be at rest, or moves together with the body, the figure described, and its point S, uniformly forwards in a right line.

12.B6 The law of force for an elliptical orbit

Proposition XI, Problem VI

If a body revolves in an ellipse; it is required to find the law of the centripetal force tending to the focus of the ellipse.

Let S be the focus of the ellipse. Draw SP cutting the diameter DK of the ellipse in E, and the ordinate Qv in x; and complete the parallelogram $QxPR$. It is evident that EP is equal to the greater semiaxis AC: for drawing HI from the other focus H of the ellipse parallel to EC, because CS, CH are equal, ES, EI will be also equal; so that EP is the half-sum of PS, PI, that is (because of the parallels HI, PR, and the equal angles IPR, HPZ), of PS, PH, which taken together are equal to the whole axis $2AC$. Draw QT perpendicular to SP, and putting L for the principal latus rectum of the ellipse (or for $\frac{2BC^2}{AC}$), we shall have $L \cdot QR : L \cdot Pv = QR : Pv = PE : PC = AC : PC$, also,

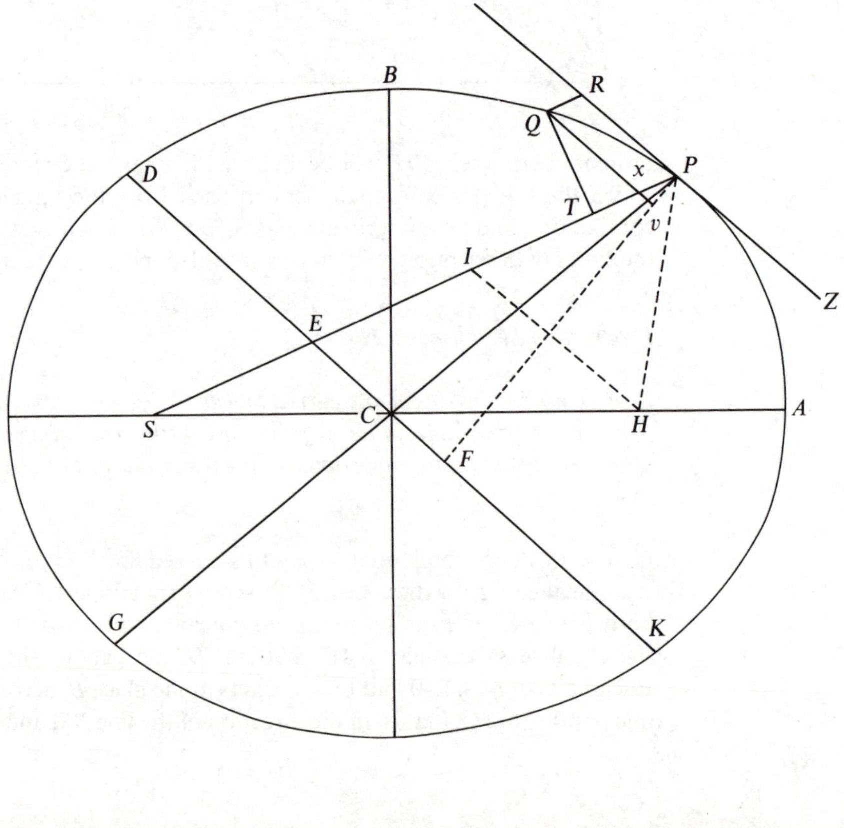

$L \cdot Pv:Gv \cdot Pv = L:Gv$, and, $Gv \cdot Pv:Qv^2 = PC^2:CD^2$. By Cor. II, Lem. VII, when the points P and Q coincide, $Qv^2 = Qx^2$, and Qx^2 or $Qv^2:QT^2 = EP^2:PF^2 = CA^2:PF^2$, and (by Lem. XII) $= CD^2:CB^2$. Multiplying together corresponding terms of the four proportions, and simplifying, we shall have

$$L \cdot QR:QT^2 = AC \cdot L \cdot PC \cdot CD^2:PC \cdot Gv \cdot CD^2 \cdot CB^2 = 2PC:Gv,$$

since $AC \cdot L = 2BC^2$. But the points Q and P coinciding, $2PC$ and Gv are equal. And therefore the quantities $L \cdot QR$ and QT^2, proportional to these, will be also equal. Let those equals be multiplied by $\frac{SP^2}{QR}$, and $L \cdot SP^2$ will become equal to $\frac{SP^2 \cdot QT^2}{QR}$. And therefore (by Cor. I and V, Prop. VI) the centripetal force is inversely as $L \cdot SP^2$, that is, inversely as the square of the distance SP. Q.E.I.

12.B7 Gravity obeys an inverse square law

Book III, Proposition II, Theorem II

That the forces by which the primary planets are continually drawn off from rectilinear motions, and retained in their proper orbits, tend to the sun; and are inversely as the squares of the distances of the places of those planets from the sun's centre.

The former part of the Proposition is manifest from Phen. V, and Prop. II, Book I; the latter from Phen. IV, and Cor. VI, Prop. IV, of the same Book. But this part of the Proposition is, with great accuracy, demonstrable from the quiescence of the aphelion points; for a very small aberration from the proportion according to the inverse square of the distances would (by Cor. I, Prop. XLV, Book I) produce a motion of the apsides sensible enough in every single revolution, and in many of them enormously great.

Book I, Proposition IV, Theorem IV

The centripetal forces of bodies, which by equable motions describe different circles, tend to the centres of the same circles; and are to each other as the squares of the arcs described in equal times divided respectively by the radii of the circles.

These forces tend to the centres of the circles (by Prop. II, and Cor. II, Prop. I), and are to one another as the versed sines of the least arcs described in equal times (by Cor. IV, Prop. I); that is, as the squares of the same arcs divided by the diameters of the circles (by Lem. VII); and therefore since those arcs are as arcs described in any equal times, and the diameters are as the radii, the forces will be as the squares of any arcs described in the same time divided by the radii of the circles. Q.E.D.

[...]

Corollary VI If the periodic times are as the $\frac{3}{2}$th powers of the radii, and therefore the velocities inversely as the square roots of the radii, the centripetal forces will be inversely as the squares of the radii; and conversely.

12.B8 Motion of the apsides

Proposition XLV, Problem XXXI

To find the motion of the apsides in orbits approaching very near to circles.

This problem is solved arithmetically by reducing the orbit, which a body revolving in a movable ellipse (as in Cor. II and III of the above Prop.) describes in a fixed plane, to the figure of the orbit whose apsides are required; and then seeking the apsides of the orbit which that body describes in a fixed plane. [...]

Example Taking m and n for any indices of the powers of the altitude, and b and c for any given numbers, suppose the centripetal force to be as $(bA^m + cA^n) \div A^3$, that is, as $[b(T-X)^m + c(T-X)^n] \div A^3$, or (by the method of converging series above mentioned) as

$$[bT^m + cT^n - mbXT^{m-1} - ncXT^{n-1} + \frac{mm-m}{2}bXXT^{m-2} + \frac{nn-n}{2} - cXXT^{n-2}, \text{\&c.}] \div A^3;$$

and comparing the terms of the numerators, there will arise,

$$RGG - RFF + TFF : bT^m + cT^n = -FF : -mbT^{m-1} - ncT^{n-1} + \frac{mm-m}{2}bXT^{m-2} + \frac{nn-n}{2}cXT^{n-2}, \text{\&c.}$$

And taking the last ratios that arise when the orbits come to a circular form, there will come forth $GG : bT^{m-1} + cT^{n-1} = FF : mbT^{m-1} + ncT^{n-1}$; and again, $GG : FF = bT^{m-1} + cT^{n-1} : mbT^{m-1} + ncT^{n-1}$. This proportion, by expressing the greatest altitude CV or T arithmetically by unity, becomes, $GG : FF = b + c : mb + nc = 1 : \frac{mb + nc}{b + c}$. Whence G becomes to F, that is, the angle VCp to the angle VCP, as 1 to $\sqrt{\frac{mb+nc}{b+c}}$. [...]

Corollary Hence also if a body, urged by a centripetal force which is inversely as the square of the altitude, revolves in an ellipse whose focus is in the centre of the forces; and a new and foreign force should be added to or subtracted from this centripetal force, the motion of the apsides arising from that foreign force may (by the Example) be known; and conversely: If the force with which the body revolves in the ellipse be as $\frac{1}{AA}$; and the foreign force as cA, and therefore the remaining force as $\frac{A - cA^4}{A^3}$; then (by the Example) b will be equal to 1, m equal to 1, and n equal to 4; and therefore the angle of revolution between the apsides is equal to $180° \sqrt{\frac{1-c}{1-4c}}$. Suppose that foreign force to be 357.45 times less than the other force with which the body revolves in the ellipse; that is, c to be $\frac{100}{35745}$, A or T being equal to 1; and then $180° \sqrt{\frac{1-c}{1-4c}}$ will be $180° \sqrt{\frac{35645}{35345}}$ or 180°.7623, that is, 180° 45′ 44″. Therefore the body, parting from the upper apse, will arrive at the lower apse with an angular motion of 180° 45′ 44″, and this angular motion being repeated, will return to the upper apse; and therefore the upper apse in each revolution will go forward 1° 31′ 28″. The apse of the moon is about twice as swift.

12.B9 Against vortices

Book II, Proposition LIII, Theorem XLI

Bodies carried about in a vortex, and returning in the same orbit, are of the same density with the vortex, and are moved according to the same law with the parts of the vortex, as to velocity and direction of motion.

For if any small part of the vortex, whose particles or physical points continue a given situation among themselves, be supposed to be congealed, this particle will move according to the same law as before, since no change is made either in its density, inertia, or figure. And again; if a congealed or solid part of the vortex be of the same density with the rest of the vortex, and be resolved into a fluid, this will move according to the same law as before, except so far as its particles, now become fluid, may be moved among themselves. Neglect, therefore, the motion of the particles among themselves as not at all concerning the progressive motion of the whole, and the motion of the whole will be the same as before. But this motion will be the same with the motion of other parts of the vortex at equal distances from the centre; because the solid, now resolved into a fluid, is become exactly like the other parts of the vortex. Therefore a solid, if it be of the same density with the matter of the vortex, will move with the same motion as the parts thereof, being relatively at rest in the matter that surrounds it. If it be more dense, it will endeavour more than before to recede from the centre; and therefore overcoming that force of the vortex, by which, being, as it were, kept in equilibrium, it was retained in its orbit, it will recede from the centre, and in its revolution describe a spiral, returning no longer into the same orbit. And, by the same argument, if it be more rare, it will approach to the centre. Therefore it can never continually go round in the same orbit, unless it be of the same density with the fluid. But we have shown in that case that it would revolve according to the same law with those parts of the fluid that are at the same or equal distances from the centre of the vortex.

Corollary I Therefore a solid revolving in a vortex, and continually going round in the same orbit, is relatively quiescent in the fluid that carries it.

Corollary II And if the vortex be of an uniform density, the same body may revolve at any distance from the centre of the vortex.

Scholium

Hence it is manifest that the planets are not carried round in corporeal vortices; for, according to the Copernican hypothesis, the planets going round the sun revolve in ellipses, having the sun in their common focus; and by radii drawn to the sun describe

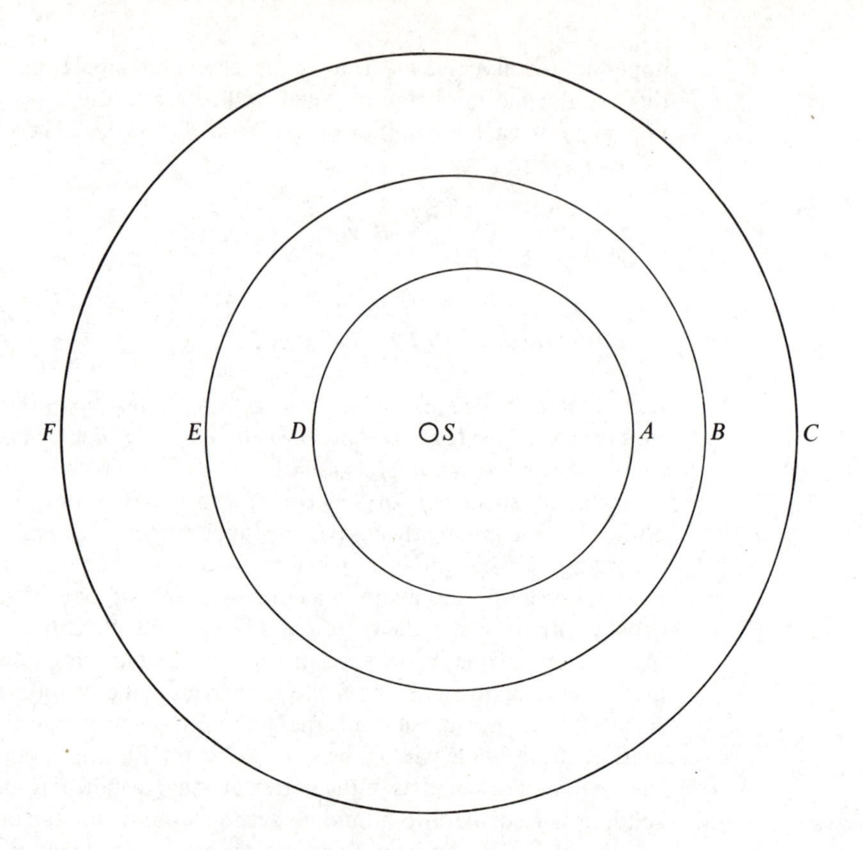

areas proportional to the times. But the parts of a vortex can never revolve with such a motion. For, let *AD*, *BE*, *CF* represent three orbits described about the sun *S*, of which let the outmost circle *CF* be concentric to the sun; let the aphelions of the two innermost be *A*, *B*; and their perihelions *D*, *E*. Hence a body revolving in the orb *CF*, describing, by a radius drawn to the sun, areas proportional to the times, will move with an uniform motion. And, according to the laws of astronomy, the body revolving in the orbit *BE* will move slower in its aphelion *B*, and swifter in its perihelion *E*; whereas, according to the laws of mechanics, the matter of the vortex ought to move more swiftly in the narrow space between *A* and *C* than in the wide space between *D* and *F*; that is, more swiftly in the aphelion than in the perihelion. Now these two conclusions contradict each other.

12.B10 Rules of reasoning in philosophy

If it universally appears, by experiments and astronomical observations, that all bodies about the earth gravitate towards the earth, and that in proportion to the quantity of matter which they severally contain; that the moon likewise, according to the quantity of its matter, gravitates towards the earth; that, on the other hand, our sea gravitates towards the moon; and all the planets one towards another; and the comets in like manner towards the sun; we must, in consequence of this rule, universally allow that all bodies whatsoever are endowed with a principle of mutual gravitation. For the argument from the appearances concludes with more force for the universal gravitation of all bodies than for their impenetrability; of which, among those in the celestial regions, we have no experiments, nor any manner of observation. Not that I affirm gravity to be essential to bodies: by their *vis insita* I mean nothing but their inertia. This is immutable. Their gravity is diminished as they recede from the earth.

Rule IV

In experimental philosophy we are to look upon propositions inferred by general induction from phenomena as accurately or very nearly true, notwithstanding any contrary hypotheses that may be imagined, till such time as other phenomena occur, by which they may either be made more accurate, or liable to exceptions.

This rule we must follow, that the argument of induction may not be evaded by hypotheses.

12.B11 The shape of the planets

Book III, Proposition XVIII, Theorem XVI

That the axes of the planets are less than the diameters drawn perpendicular to the axes.

The equal gravitation of the parts on all sides would give a spherical figure to the planets, if it was not for their diurnal revolution in a circle. By that circular motion it comes to pass that the parts receding from the axis endeavour to ascend about the equator; and therefore if the matter is in a fluid state, by its ascent towards the equator it will enlarge the diameters there, and by its descent towards the poles it will shorten the axis. So the diameter of Jupiter (by the concurring observations of astronomers) is found shorter between pole and pole than from east to west. And, by the same argument, if our earth was not higher about the equator than at the poles, the seas would subside about the poles, and, rising towards the equator, would lay all things there under water.

12.B12 General scholium

The hypothesis of vortices is pressed with many difficulties. That every planet by a radius drawn to the sun may describe areas proportional to the times of description, the periodic times of the several parts of the vortices should observe the square of their distances from the sun; but that the periodic times of the planets may obtain the $\frac{3}{2}$th power of their distances from the sun, the periodic times of the parts of the vortex ought to be as the $\frac{3}{2}$th power of their distances. That the smaller vortices may maintain their lesser revolutions about Saturn, Jupiter, and other planets, and swim quietly and undisturbed in the greater vortex of the sun, the periodic times of the parts of the sun's vortex should be equal; but the rotation of the sun and planets about their axes, which ought to correspond with the motions of their vortices, recede far from all these proportions. The motions of the comets are exceedingly regular, are governed by the same laws with the motions of the planets, and can by no means be accounted for by the hypothesis of vortices; for comets are carried with very eccentric motions through all parts of the heavens indifferently, with a freedom that is incompatible with the notion of a vortex.

12.B13 Gravity

Hitherto we have explained the phenomena of the heavens and of our sea by the power of gravity, but have not yet assigned the cause of this power. This is certain, that it must proceed from a cause that penetrates to the very centres of the sun and planets, without suffering the least diminution of its force; that operates not according to the quantity of the surfaces of the particles upon which it acts (as mechanical causes used to do), but according to the quantity of the solid matter which they contain, and propagates its virtue on all sides to immense distances, decreasing always as the inverse square of the distances. Gravitation towards the sun is made up out of the gravitations towards the several particles of which the body of the sun is composed; and in receding from the sun decreases accurately as the inverse square of the distances as far as the orbit of Saturn, as evidently appears from the quiescence of the aphelion of the planets; nay, and even to the remotest aphelion of the comets, if those aphelions are also quiescent. But hitherto I have not been able to discover the cause of those properties of gravity from phenomena, and I frame no hypotheses; for whatever is not deduced from the phenomena is to be called an hypothesis; and hypotheses, whether metaphysical or physical, whether of occult qualities or mechanical, have no place in experimental philosophy. In this philosophy particular propositions are inferred from the phenomena, and afterwards rendered general by induction. Thus it was that the impenetrability, the mobility, and the impulsive force of bodies, and the laws of motion and of gravitation, were discovered. And to us it is enough that gravity does really exist, and act according to the laws which we have explained, and abundantly serves to account for all the motions of the celestial bodies, and of our sea.

12.C Newton's Letters to Leibniz

12.C1 From the Epistola Prior (1676)

Though the modesty of Mr Leibniz, in the extracts from his letter which you have lately sent me, pays great tribute to our countrymen for a certain theory of infinite series, about which there now begins to be some talk, yet I have no doubt that he has discovered not only a method for reducing any quantities whatever to such series, as he asserts, but also various shortened forms, perhaps like our own, if not even better. Since, however, he very much wants to know what has been discovered in this subject by the English, and since I myself fell upon this theory some years ago, I have sent you some of those things which occurred to me in order to satisfy his wishes, at any rate in part.

Fractions are reduced to infinite series by division; and radical quantities by extraction of the roots, by carrying out those operations in the symbols just as they are commonly carried out in decimal numbers. These are the foundations of these reductions: but extractions of roots are much shortened by this theorem,

$$(P + PQ)^{m/n} = P^{m/n} + \frac{m}{n}AQ + \frac{m-n}{2n}BQ + \frac{m-2n}{3n}CQ + \frac{m-3n}{4n}DQ + \text{etc.}$$

where $P + PQ$ signifies the quantity whose root or even any power, or the root of a power, is to be found; P signifies the first term of that quantity, Q the remaining terms divided by the first, and m/n the numerical index of the power of $P + PQ$, whether that power is integral or (so to speak) fractional, whether positive or negative. For as analysts, instead of aa, aaa, etc., are accustomed to write a^2, a^3, etc., so instead of $\sqrt{a}, \sqrt{a^3}, \sqrt{c\text{:}\,a^5}$, etc. I write $a^{\frac{1}{2}}, a^{\frac{3}{2}}, a^{\frac{5}{3}}$, and instead of $1/a, 1/aa, 1/a^3$, I write a^{-1}, a^{-2}, a^{-3}. And so for $\dfrac{aa}{\sqrt{c\text{:}\,(a^3 + bbx)}}$ I write $aa(a^3 + bbx)^{-\frac{1}{3}}$, and for $\dfrac{aab}{\sqrt{c\text{:}\,\{(a^3 + bbx)(a^3 + bbx)\}}}$ I write $aab(a^3 + bbx)^{-\frac{2}{3}}$: in which last case, if $(a^3 + bbx)^{-\frac{2}{3}}$ is supposed to be $(P + PQ)^{m/n}$ in the Rule, then P will be equal to a^3, Q to bbx/a^3, m to -2, and n to 3. Finally, for the terms found in the quotient in the course of the working I employ A, B, C, D, etc., namely, A for the first term, $P^{m/n}$; B for the second term, $m/n\ AQ$; and so on. For the rest, the use of the rule will appear from the examples.

Example 1

$$\sqrt{(c^2 + x^2)} \text{ or } (c^2 + x^2)^{\frac{1}{2}} = c + \frac{x^2}{2c} - \frac{x^4}{8c^3} + \frac{x^6}{16c^5} - \frac{5x^8}{128c^7} + \frac{7x^{10}}{256[c]^9} + \text{etc.}$$

For in this case, $P = c^2$, $Q = x^2/c^2$, $m = 1$, $n = 2$, $A(= P^{m/n} = (cc)^{\frac{1}{2}}) = c$, $B(= (m/n)AQ) = x^2/2c$, $C\left(= \frac{m-n}{2n}BQ\right) = -\frac{x^4}{8c^3}$; and so on.

Example 2

$$\sqrt[5]{(c^5 + c^4x - x^5)},$$

i.e.

$$(c^5 + c^4x - x^5)^{\frac{1}{5}} = c + \frac{c^4x - x^5}{5c^4}[+]\frac{-2c^8x^2 + 4c^4x^6 - 2x^{10}}{25c^9} + \text{etc.}$$

as will be evident on substituting 1 for m, 5 for n, c^5 for P and $(c^4x - x)^5/c^5$ for Q, in the rule quoted above. Also $-x^5$ can be substituted for P and $(c^4x + c^5)/(-x^5)$ for Q. The result will then be

$$\sqrt[5]{(c^5 + c^4x - x^5)} = -x + \frac{c^4x + c^5}{5x^4} + \frac{2c^8x^2 + 4c^9x + [2]c^{10}}{25x^9} + \quad \text{etc.}$$

The first method is to be chosen if x is very small, the second if it is very large.

12.C2 From the Epistola Posterior (1676)

(a) I can hardly tell with what pleasure I have read the letters of those very distinguished men Leibniz and Tschirnhaus. Leibniz's method for obtaining convergent series is certainly very elegant, and it would have sufficiently revealed the genius of its author, even if he had written nothing else. But what he has scattered elsewhere throughout his letter is most worthy of his reputation—it leads us also to hope for very great things from him. The variety of ways by which the same goal is approached has given me the greater pleasure, because three methods of arriving at series of that kind had already become known to me, so that I could scarcely expect a new one to be communicated to us. One of mine I have described before; I now add another, namely, that by which I first chanced on these series—for I chanced on them before I knew the divisions and extractions of roots which I now use. And an explanation of this will serve to lay bare, what Leibniz desires from me, the basis of the theorem set forth near the beginning of the former letter.

At the beginning of my mathematical studies, when I had met with the works of our celebrated Wallis, on considering the series by the intercalation of which he himself exhibits the area of the circle and the hyperbola, the fact that, in the series of curves whose common base or axis is x and the ordinates

$$(1-x^2)^{\frac{0}{2}},\quad (1-x^2)^{\frac{1}{2}},\quad (1-x^2)^{\frac{2}{2}},\quad (1-x^2)^{\frac{3}{2}},\quad (1-x^2)^{\frac{4}{2}},\quad (1-x^2)^{\frac{5}{2}},\quad \text{etc.},$$

if the areas of every other of them, namely

$$x,\quad x-\tfrac{1}{3}x^3,\quad x-\tfrac{2}{3}x^3+\tfrac{1}{5}x^5,\quad x-\tfrac{3}{3}x^3+\tfrac{3}{5}x^5-\tfrac{1}{7}x^7,\quad \text{etc.}$$

could be interpolated, we should have the areas of the intermediate ones, of which the first $(1-x^2)^{\frac{1}{2}}$ is the circle: in order to interpolate these series I noted that in all of them the first term was x and the second terms $\frac{0}{3}x^3, \frac{1}{3}x^3, \frac{2}{3}x^3, \frac{3}{3}x^3$, etc., were in arithmetical progression, and hence that the first two terms of the series to be intercalated ought to be $x-\frac{1}{3}(\frac{1}{2}x^3), x-\frac{1}{3}(\frac{3}{2}x^3), x-\frac{1}{3}(\frac{5}{2}x^3)$, etc. To intercalate the rest I began to reflect that the denominators $1, 3, 5, 7$, etc. were in arithmetical progression, so that the numerical coefficients of the numerators only were still in need of investigation. But in the alternately given areas these were the figures of powers of the number 11, namely of these, $11^0, 11^1, 11^2, 11^3, 11^4$, that is, first 1; then 1, 1; thirdly, 1, 2, 1; fourthly 1, 3, 3, 1; fifthly 1, 4, 6, 4, 1, etc. And so I began to inquire how the remaining figures in these series could be derived from the first two given figures, and I found that on putting m for the second figure, the rest would be produced by continual multiplication of the terms of this series,

$$\frac{m-0}{1}\times\frac{m-1}{2}\times\frac{m-2}{3}\times\frac{m-3}{4}\times\frac{m-4}{5},\quad \text{etc.}$$

For example, let $m=4$, and $4\times\frac{1}{2}(m-1)$, that is 6 will be the third term, and $6\times\frac{1}{3}(m-2)$, that is 4 the fourth, and $4\times\frac{1}{4}(m-3)$, that is 1 the fifth, and $1\times\frac{1}{5}(m-4)$, that is 0 the sixth, at which term in this case the series stops. Accordingly, I applied this rule for interposing series among series, and since, for the circle, the second term was $\frac{1}{3}(\frac{1}{2}x^3)$, I put $m=\frac{1}{2}$, and the terms arising were

$$\frac{1}{2}\times\frac{\frac{1}{2}-1}{2}\quad \text{or}\quad -\tfrac{1}{8},\qquad -\frac{1}{8}\times\frac{\frac{1}{2}-2}{3}\quad \text{or}\quad +\tfrac{1}{16},\qquad \frac{1}{16}\times\frac{\frac{1}{2}-3}{4}\quad \text{or}\quad -\tfrac{5}{128},$$

and so to infinity. Whence I came to understand that the area of the circular segment which I wanted was

$$x - \frac{\frac{1}{2}x^3}{3} - \frac{\frac{1}{8}x^5}{5} - \frac{\frac{1}{16}x^7}{7} - \frac{\frac{5}{128}x^9}{9} \quad \text{etc.}$$

And by the same reasoning the areas of the remaining curves, which were to be inserted, were likewise obtained: as also the area of the hyperbola and of the other alternate curves in this series $(1 + x^2)^{\frac{0}{2}}$, $(1 + x^2)^{\frac{1}{2}}$, $(1 + x^2)^{\frac{2}{2}}$, $(1 + x^2)^{\frac{3}{2}}$, etc. And the same theory serves to intercalate other series, and that through intervals of two or more terms when they are absent at the same time. This was my first entry upon these studies, and it had certainly escaped my memory, had I not a few weeks ago cast my eye back on some notes.

But when I had learnt this, I immediately began to consider that the terms

$$(1 - x^2)^{\frac{0}{2}}, \quad (1 - x^2)^{\frac{2}{2}}, \quad (1 - x^2)^{\frac{4}{2}}, \quad (1 - x^2)^{\frac{6}{2}}, \quad \text{etc.,}$$

that is to say,

$$1, \quad 1 - x^2, \quad 1 - 2x^2 + x^4, \quad 1 - 3x^2 + 3x^4 - x^6, \quad \text{etc.}$$

could be interpolated in the same way as the areas generated by them: and that nothing else was required for this purpose but to omit the denominators 1, 3, 5, 7, etc., which are in the terms expressing the areas; this means that the coefficients of the terms of the quantity to be intercalated $(1 - x^2)^{\frac{1}{2}}$, or $(1 - x^2)^{\frac{3}{2}}$, or in general $(1 - x^2)^m$, arise by the continued multiplication of the terms of this series

$$m \times \frac{m-1}{2} \times \frac{m-2}{3} \times \frac{m-3}{4}, \quad \text{etc.,}$$

so that (for example)

$$(1 - x^2)^{\frac{1}{2}} \quad \text{was the value of} \quad 1 - \tfrac{1}{2}x^2 - \tfrac{1}{8}x^4 - \tfrac{1}{16}x^6 \quad \text{etc.,}$$

and

$$(1 - x^2)^{\frac{3}{2}} \quad \text{of} \quad 1 - \tfrac{3}{2}x^2 + \tfrac{3}{8}x^4 + \tfrac{1}{16}x^6, \quad \text{etc.,}$$

$$(1 - x^2)^{\frac{1}{3}} \quad \text{of} \quad 1 - \tfrac{1}{3}x^2 - \tfrac{1}{9}x^4 - \tfrac{5}{81}x^6, \quad \text{etc.}$$

So then the general reduction of radicals into infinite series by that rule, which I laid down at the beginning of my earlier letter became known to me, and that before I was acquainted with the extraction of roots. But once this was known, that other could not long remain hidden from me. For in order to test these processes, I multiplied

$$1 - \tfrac{1}{2}x^2 - \tfrac{1}{8}x^4 - \tfrac{1}{16}x^6, \quad \text{etc.}$$

into itself; and it became $1 - x^2$, the remaining terms vanishing by the continuation of the series to infinity. And even so $1 - \frac{1}{3}x^2 - \frac{1}{9}x^4 - \frac{5}{81}x^6$, etc. multiplied twice into itself also produced $1 - x^2$. And as this was not only sure proof of these conclusions so too it guided me to try whether, conversely, these series, which it thus affirmed to be roots of the quantity $1 - x^2$, might not be extracted out of it in an arithmetical manner. And the matter turned out well. This was the form of the working in square roots.

$$1 - x^2 \;(\; 1 - \tfrac{1}{2}x^2 - \tfrac{1}{8}x^4 - \tfrac{1}{16}x^6, \text{ etc.}$$

$$\begin{array}{l} 1 \\ \hline 0 - x^2 \\ \quad - x^2 + \tfrac{1}{4}x^4 \\ \hline \qquad - \tfrac{1}{4}x^4 \\ \qquad - \tfrac{1}{4}x^4 + \tfrac{1}{8}x^6 + \tfrac{1}{64}x^8 \\ \hline \qquad 0 \quad - \tfrac{1}{8}x^6 - \tfrac{1}{64}x^8. \end{array}$$

After getting this clear I have quite given up the interpolation of series, and have made use of these operations only, as giving more natural foundations. Nor was there any secret about reduction by division, an easier affair in any case.

(b) But in that treatise infinite series played no great part. Not a few other things I brought together, among them the method of drawing tangents which the very skilful Sluse communicated to you two or three years ago, about which you wrote back [to him] (on the suggestion of Collins) that the same method had been known to me also. We happened on it by different reasoning: for, as I work it, the matter needs no proof. Nobody, if he possessed my basis, could draw tangents any other way, unless he were deliberately wandering from the straight path. Indeed we do not here stick at equations in radicals involving one or each indefinite quantity, however complicated they may be; but without any reduction of such equations (which would generally render the work endless) the tangent is drawn directly. And the same is true in questions of maxima and minima, and in some others too, of which I am not now speaking. The foundation of these operations is evident enough, in fact; but because I cannot proceed with the explanation of it now, I have preferred to conceal it thus: *6accdæ13eff7i3l9n4o4qrr4s8t12vx*. On this foundation I have also tried to simplify the theories which concern the squaring of curves, and I have arrived at certain general Theorems. And, to be frank, here is the first Theorem.

For any curve let $dz^\theta \times (e + fz^\eta)^\lambda$ be the ordinate, standing normal at the end z of the abscissa or the base, where the letters d, e, f denote any given quantities, and θ, η, λ are the indices of the powers of the quantities to which they are attached. Put

$$\frac{\theta + 1}{\eta} = r, \quad \lambda + r = s, \quad \frac{d}{\eta f} \times (e + fz^\eta)^{\lambda + 1} = Q \quad \text{and} \quad r\eta - \eta = \pi,$$

then the area of the curve will be

$$Q \times \left\{ \frac{z^\pi}{s} - \frac{r-1}{s-1} \times \frac{eA}{fz^\eta} + \frac{r-2}{s-2} \times \frac{eB}{fz^\eta} - \frac{r-3}{s-3} \times \frac{eC}{fz^\eta} + \frac{r-4}{s-4} \times \frac{eD}{fz^\eta}, \text{ etc.} \right\}$$

the letters A, B, C, D, etc., denoting the terms immediately preceding; that is, A the term z^π/s, B the term $-((r-1)/(s-1)) \times (eA)/(fz^\eta)$, etc. This series, when r is a fraction or a negative number, is continued to infinity; but when r is positive and integral it is continued only to as many terms as there are units in r itself; and so it exhibits the geometrical squaring of the curve. I illustrate the fact by examples.

(c) When I said that almost all problems are soluble I wished to be understood to refer specially to those about which mathematicians have hitherto concerned themselves, or at least those in which mathematical arguments can gain some place. For of course one may imagine others so involved in complicated conditions that we do not succeed in understanding them well enough, and much less in bearing the burden of such long calculations as they require. Nevertheless—lest I seem to have said too much—inverse problems of tangents are within our power, and others more difficult than those, and to solve them I have used a twofold method of which one part is neater, the other more general. At present I have thought fit to register them both by transposed letters, lest, through others obtaining the same result, I should be compelled to change the plan in some respects.

5accdae10effh11i4l3m9n6oqqr8s11t9y3x:11ab3cdd10eaeg10ill4m7n6o3p3q6r5s11t8vx,

3acae4egh5i4l4m5n8oq4r3s6t4v, aaddaececceiijmmnnooprrrsssssttuu.

This inverse problem of tangents, when the tangent between the point of contact and the axis of the figure is of given length, does not demand these methods. Yet it is that mechanical curve the determination of which depends on the area of an hyperbola. The problem is also of the same kind, when the part of the axis between the tangent and the ordinate is given in length. But I should scarcely have reckoned these cases among the sports of nature. For when in the right-angled triangle, which is formed by that part of the axis, the tangent and the ordinate, the relation of any two sides is defined by any equation, the problem can be solved apart from my general method. But when a part of the axis ending at some point given in position enters the bracket, then the question is apt to work out differently.

The communication of the solution of affected equations by the method of Leibniz will be very agreeable; so too an explanation how he comports himself when the indices of the powers are fractional, as in this equation $20 + x^{\frac{3}{7}} - x^{\frac{6}{5}}y^{\frac{2}{3}} - y^{\frac{7}{11}} = 0$, or surds, as in $(x^{\sqrt{2}} + x^{\sqrt{7}})^{\sqrt[3]{\frac{2}{3}}} = y$, where $\sqrt{2}$ and $\sqrt{7}$ do not mean coefficients of x, but indices of powers or dignities of it, and $\sqrt[3]{\frac{2}{3}}$ means the power of the binomial $x^{\sqrt{2}} + x^{\sqrt{7}}$. The point, I think, is clear by my method, otherwise I should have described it. But a term must at last be set to this wordy letter. The letter of the most excellent Leibniz fully deserved of course that I should give it this more extended reply. And this time I wanted to write in greater detail because I did not believe that your more engaging pursuits should often be interrupted by me with this rather austere kind of writing.

The second anagram runs as follows:

'Una Methodus consistit in extractione fluentis quantitatis ex æquatione simul involvente fluxionem ejus: altera tantum in assumptione Seriei pro quantitate qualibet incognita ex qua cætera commode derivari possunt, & in collatione terminorum homologorum æquationis resultantis, ad eruendos terminos assumptæ seriei.' ('One method consists in extracting a fluent quantity from an equation at the same time involving its fluxion; but another by assuming a series for any unknown quantity whatever, from which the rest could conveniently be derived, and in collecting homologous terms of the resulting equation in order to elicit the terms of the assumed series.')

On counting the letters, one finds that there are two *i*s or *j*s too few and one *s* too many. The anagram was inaccurately transcribed in the copy which Leibniz received.

12.D Newton on Geometry

12.D1 On the locus to three or four lines

Descartes, in regard to his accomplishment of this problem, makes a great show as if he had achieved something so earnestly sought after by the Ancients and for whose sake he considers that Apollonius wrote his books on conics. With all respect to so great a man I should have believed that this topic remained not at all a mystery to the Ancients. For Pappus informs us of a method for drawing an ellipse through five given points and the reasoning is the same in the case of the other conics. And if the Ancients knew how to draw a conic through five given points, does any one not see that they found out the composition of the solid locus. To be sure, their method is more elegant by far than the Cartesian one. For he achieved the result by an algebraic calculus which, when transposed into words, following the practice of the Ancients in their writings, would prove to be so tedious and entangled as to provoke nausea, nor might it be understood. But they accomplished it by certain simple proportions, judging that nothing written in a different style was worthy to be read, and in consequence concealing the analysis by which they found their constructions. To reveal that this topic was no mystery to them, I shall attempt to restore their discovery by following in the steps of Pappus' problem. And to this end I propose these problems:

1. *To describe a conic through three given points A, B, C which shall have the given centre O.* [...]
2. *To describe a conic through the five given points A, B, C, D, E.* [...]

With these premissed nothing further remains to be done in composing the *locus solidus* but to look for five points through which the figure shall pass. In the Ancients' problem this is very easy. Let AT, ST, AG, SG be four straight lines given in position and to these from some common point C four more CB, CF, CD, CH are to be drawn at given angles subject to the condition that the product $CB \times CF$ of the first two shall bear a given ratio to the product $CD \times CH$ of the remainder. The curve in which the point C is perpetually found will pass through the four intersections A, G, S, T of the given lines, for when FC is nil, the product $FC \times CB$ will be nil, and so also the product $CD \times CH$ along with one of the lines CD, CH. If it is CD, the point C will fall at T; if CH, at S. And thus when CB is nil the point C will fall either at A or G. As a result the four points A, G, S, T through which the figure shall pass are given and it remains only to discover a fifth. Which is very easy. For through the point A let any straight line AC be drawn and look for the point C in it which is to satisfy the problem. Now because of the given angle DAC there is given the ratio of DC to AC and that of AC to BC, and hence that of DC to BC, and consequently also the ratio of CH to FC seeing that the ratio of $DC \times CH$ to $BC \times FC$ is given. Draw therefore the straight line SC defined by

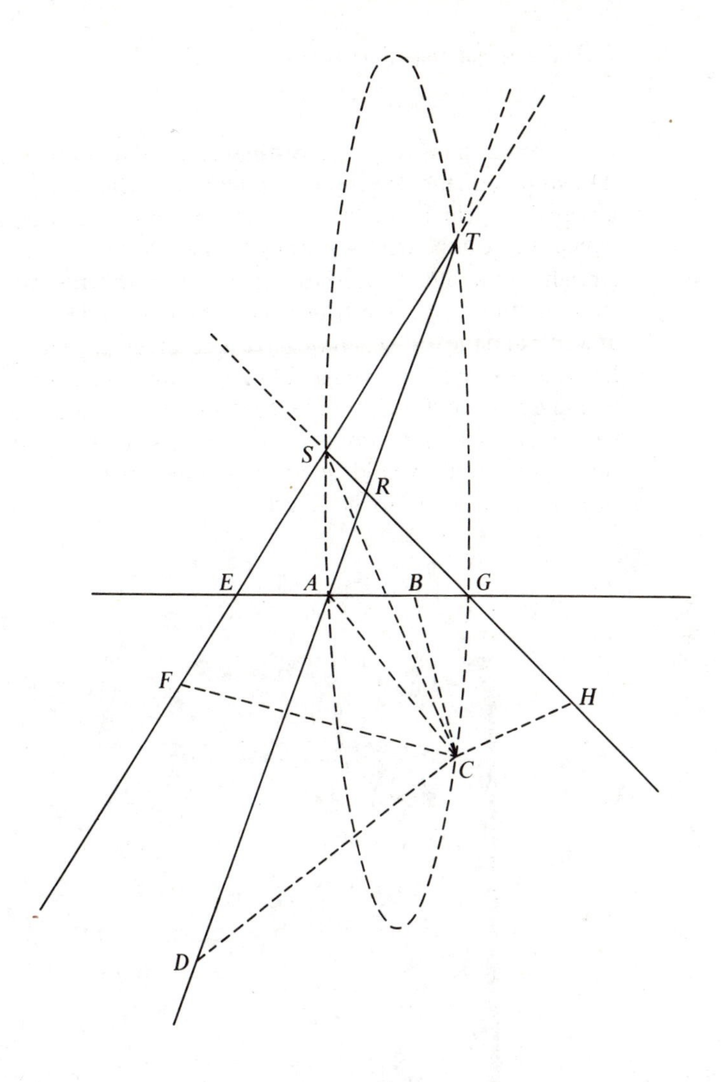

CH to FC being in that given ratio, and this will cut the line AC in the required point C. By the same precept countless points may be found, but when we have found some single one we shall have the five points which, by determining a conic in line with the preceding, satisfy the problem.

And this seems the most natural method of solving the problem, not merely because it is relatively simple but since the first part of the problem (in the form propounded by Descartes himself) is to find some point having the given condition, and thereafter, since there are an infinity of points of this sort, to determine the locus in which they are all found. What then is more natural than to reduce the difficulties of this latter part to those of the former by determining the locus from a few points after they are found? In consequence, since the Ancients did develop a procedure for constructing a conic through five given points, no one should have doubted that they composed solid loci by this means.

12.D2 The enumeration of cubics

Geometrical lines are best distinguished into orders according to the number of dimensions of the equation by which the relationship between their ordinates and abscissas is defined, or (what is the very same) according to the number of points in which they can be cut by a straight line. On this basis a line of first order will be the straight line alone, those of second or quadratic order will be conics and the circle, and those of third or cubic order the cubic parabola, Neilian parabola, cissoid of the ancients and the rest which we have here undertaken to enumerate. A curve of first kind however—seeing that the straight line is not to be counted among curves—is the same as a line of second order, and a curve of second kind the same as a line of third order; while a line of infinite order is one which a straight line can cut in an infinity of points, such as the spiral, cycloid, quadratrix and every line which is generated by an infinity of revolutions of a radius or wheel. [. . .]

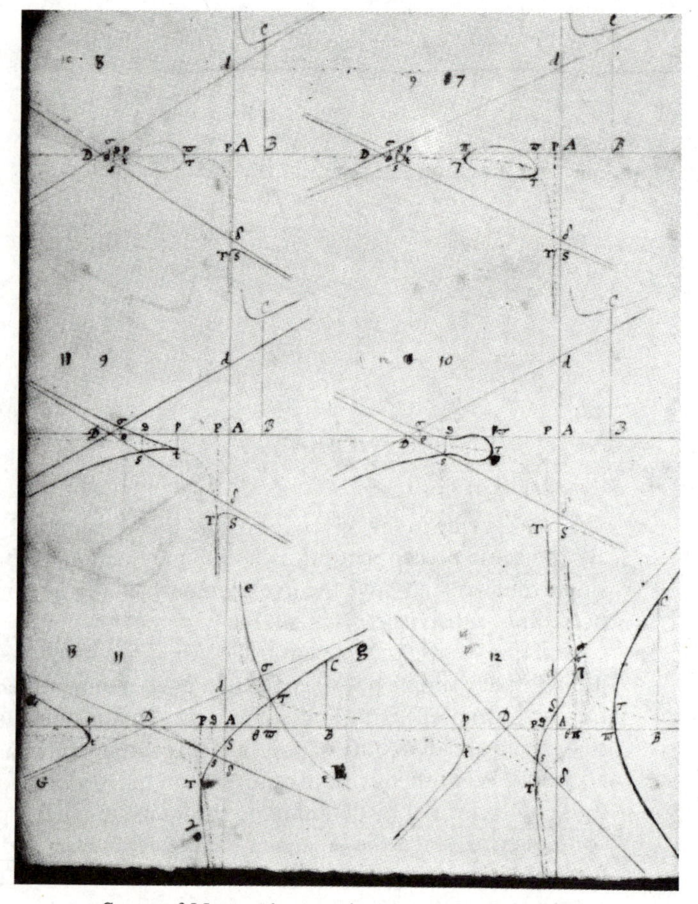

Some of Newton's own drawings of cubic curves.

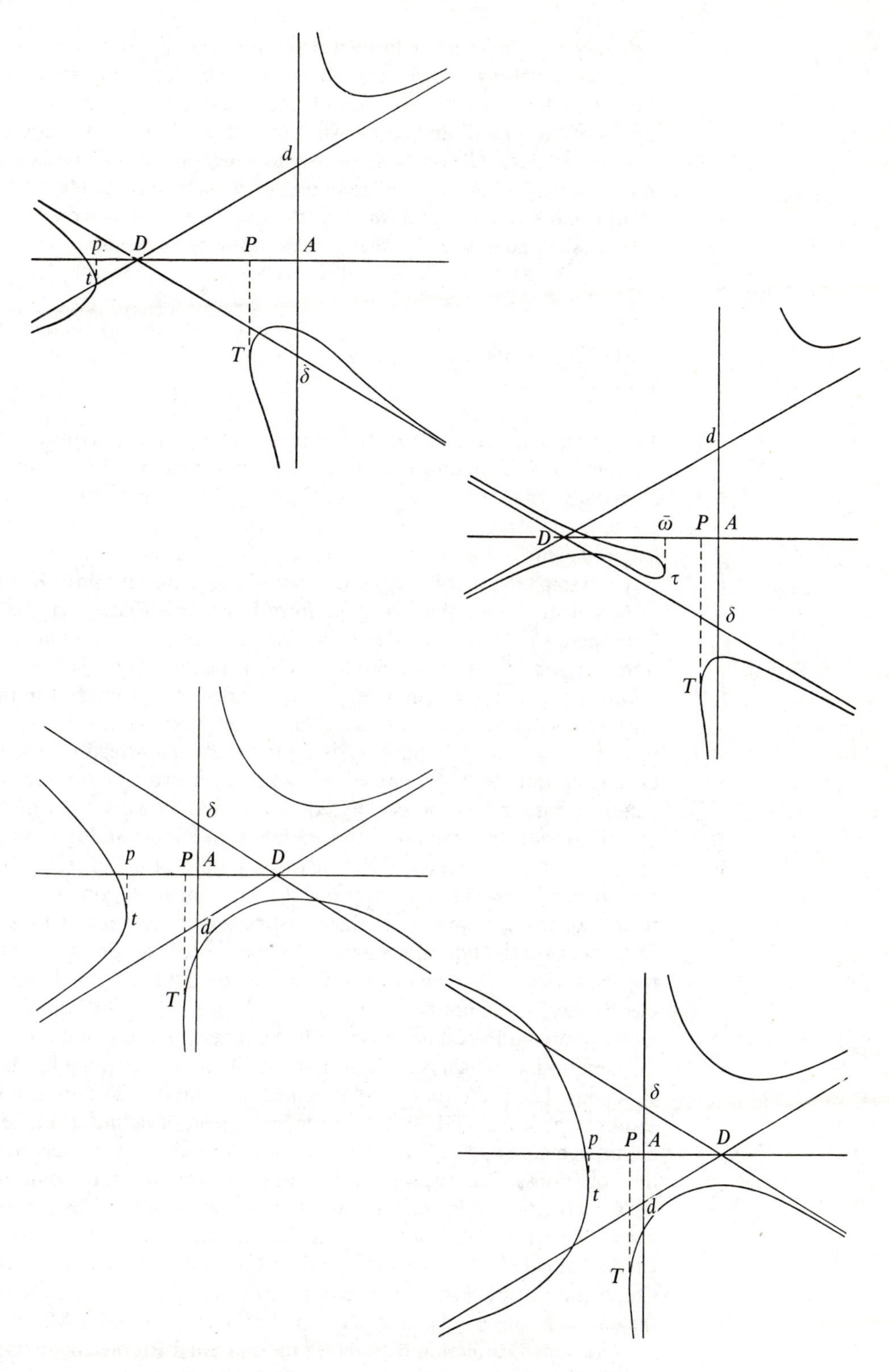

Cubic curves, from Newton's *Enumeration*.

If onto an infinite plane lit by a point-source of light there should be projected the shadows of figures, the shadows of conics will always be conics, those of curves of second kind will always be curves of second kind, those of curves of third kind always curves of third kind, and so on without end. And just as the circle by projecting its shadow generates all conics, so the five divergent parabolas by their shadows generate and exhibit all other curves of second kind; while in this manner certain simpler curves of other kinds can be found which by their shadows cast by a point-source of light onto a plane shall delineate all other curves of the same kinds.

12.D3 On geometry and algebra

The Antients, as we learn from Pappus, at first in vain endeavoured at the Trisection of an Angle, and the finding out of two mean Proportionals by a right Line and a Circle. Afterwards they began to consider several other Lines, as the Conchoid, and the Cissoid, and the Conick Sections, and by some of these to solve those Problems. At length, having more thoroughly examined the Matter and the Conick Sections being received into Geometry, they distinguished Problems into three Kinds; viz. Into *Plane ones*, which deriving their Original from Lines on a Plane, may be solved by a right Line and a Circle, into *Solid ones*, which were solved by Lines deriving their Original from the Consideration of a Solid, that is, of a Cone: And *Linear ones*, to the Solution of which were required Lines more compounded. And according to this distinction, we are not to solve Solid Problems by other Lines than the Conick Sections, especially if no other Lines but right ones, a Circle, and the Conick Sections, must be received into Geometry. But the Moderns advancing yet much farther, have received into Geometry all Lines that can be expressed by Equations, and have distinguished, according to the Dimensions of the Equations, those Lines into Kinds; and have made it a Law, that you are not to construct a Problem by a Line of a Superior Kind, that may be constructed by one of an inferior one. In the Contemplation of Lines, and finding out their Properties, I approve of their Distinction of them into Kinds, according to the Dimensions of the Equations by which they are defined. But it is not the Equation, but the Description that makes the Curve to be a Geometrical one. The Circle is a Geometrical Line, not because it may be expressed by an Equation, but because its Description is a Postulate. It is not the Simplicity of the Equation, but the Easiness of the Description, which is to determine the Choice of our Lines for the Construction of Problems. For the Equation that express a Parabola, is more simple than that that expresses a Circle, and yet the Circle, by reason of its more simple Construction, is admitted before it. The Circle and the Conick Sections, if you regard the Dimension of the Equations, are of the same Order, and yet the Circle is not numbered with them in the Construction of Problems, but, by reason of its simple Description, is depressed to a lower Order, viz. that of a right Line; so that it is not improper to construct that by a Circle that may be constructed by a right Line. But it is a Fault to construct that by the Conick Sections which may be constructed by a Circle. Either therefore you must fix the Law to be observed in a Circle from the Dimensions of Equations, and so take away as vitious the Distinction between Plane and Solid Problems; or else you must grant, that the Law is not so strictly to be observed in Lines of Superior Kinds, but that some by reason of their more simple Description, may be preferred to others of the same Order, and may be numbered with Lines of inferior Orders in the Construction of Problems. In Constructions that are equally Geometrical, the most simple are always to be preferred. This Law is beyond all Exception. But Algebraick Expressions add nothing to the Simplicity of the Construction. The bare Descriptions of the Lines only are here to be considered. These alone were considered by those Geometricians who joined a Circle with a right Line. And as these are easy or hard, the Construction becomes easy or hard. And therefore it is foreign to the Nature of the Thing, from anything else to establish Laws about Constructions. Either therefore let us, with the Antients, exclude all Lines besides a right Line, the Circle, and perhaps the Conick Sections, out of Geometry, or admit all, according to the Simplicity of the Description. If the Trochoid were admitted into Geometry, we might, by its Means, divide an Angle in any given Ratio. Would you therefore blame those who should make use of this Line to divide an Angle in the Ratio of one Number to another, and contend that this Line was not defined by an Equation, but that you must make use of such Lines as are defined by Equations? If therefore, when an Angle was to be divided, for Instance, into 10001 Parts, we should be obliged to bring a Curve defined by an Equation of above an hundred Dimensions to do the Business; which no Mortal could describe, much less understand; and should prefer this to the Trochoid, which is a Line well known, and described easily by the Motion of a Wheel or a Circle, who would not see the Absurdity? Either therefore the Trochoid is not to be admitted at all into Geometry, or else, in the Construction of Problems, it is to be preferred to all Lines of a more difficult Description. And there is the same Reason for other Curves. For which Reason we approve of the Trisections of an Angle by a Conchoid, which Archimedes in his Lemma's, and Pappus in his Collections, have preferred to the Inventions of all others in this Case; because we ought either to exclude all Lines, besides the Circle and right Line, out of Geometry, or admit them according to the Simplicity of their Descriptions, in which Case the Conchoid yields to none, except the Circle. Equations are Expressions of Arithmetical Computation, and properly have no place in Geometry, except as far as Quantities truly Geometrical (that is, Lines, Surfaces, Solids, and Proportions) may be said to be some equal to others. Multiplications, Divisions, and such sort of Computations, are newly received into Geometry, and that unwarily, and contrary to the first Design of this Science. For whosoever considers the Construction of Problems by a right Line and a Circle, found out by the first Geometricians, will easily perceive that Geometry was invented that we might expeditiously avoid, by drawing Lines, the Tediousness of Computation. Therefore these two Sciences ought not to be confounded. The Antients did so industriously distinguish them from one another, that they never introduced Arithmetical Terms into Geometry. And the Moderns, by confounding both, have lost the Simplicity in which all the Elegancy of Geometry consists. Wherefore that is *Arithmetically* more simple which is determined by the more simple Equations, but that is *Geometrically* more simple which is determined by the more simple drawing of Lines; and in Geometry, that ought to be reckoned best which is Geometrically most simple. Wherefore, I ought not to be blamed, if, with the Prince of Mathematicians, Archimedes, and other Antients, I make use of the Conchoid for the Construction of Solid Problems. But if any one thinks otherwise, let him know, that I am here sollicitous not for a Geometrical Construction, but any one whatever, by which I may the nearest Way find the Roots of the Equations in Numbers.

12.D4 Newton's projective transformation

To change figures into other figures of the same class.

Let it be required to transform any figure HGI. Draw at pleasure two parallel straight lines AO, BL intersecting any third one AB, given in position, in A and B, and from any point G of the figure to the straight line AB draw GD parallel to OA. Then from some point O given in the line OA draw to the point D the straight line OD meeting BL in d, and from the meeting-point erect a straight line dg containing any given angle with the straight line BL and having to Od the ratio which GD bears to OD: g will then be the point in the new figure hgi corresponding to the point G. By the same procedure individual points of the first figure will yield an equal number of points in the new figure. Conceive therefore, that the point G travels in a continuous movement through all points of the first figure, and the point g will travel with a similarly continuous motion through all the points of the new figure, describing it. For distinction's sake let us name DG the prime ordinate, dg the new ordinate; BD the prime abscissa, Bd the new abscissa; O the pole, OD the abscinding radius, OA the prime ordinate radius and Oa (by which the parallelogram $OABa$ is completed) the new ordinate radius.

I now assert that, if the point G traces a straight line given in position, the point g will also trace a straight line given in position; if the point G traces a conic, then the point g will also trace a conic (with conics I here count the circle); while if the point G traces a line of third analytic order, the point g will trace a line likewise of third order; and so with curves of higher orders: the two lines traced by the points G and g will be ever of the same analytic order. For, indeed, as ad is to OA, so are Od to OD, dg to DG and AB to AD; and consequently AD is equal to $OA \times AB/Oa$ and DG equal to $OA \times dg/ad$. Now, if the point G traces a straight line, and hence in any equation by which the relationship between the abscissa AD and ordinate DG is exhibited those indeterminates AD and DG rise but to a single dimension, on writing in this equation $OA \times AB/Oa$ in place of AD and $OA \times dg/ad$ in place of DG there will be produced a new equation in which the new abscissa ad and new ordinate dg will rise but to a single dimension, and which shall hence denote a straight line. But were AD and DG (or one or other of them) rising to two dimensions in the first equation, then ad and dg will rise likewise to two in the second equation. And so for three or more dimensions: the indeterminates ad, dg in the second equation and AD, DG in the first will ever rise to the same number of dimensions, and accordingly the lines which the points G and g trace are of the same analytic order.

I further assert that, should some straight line touch a curve in the first figure, after transformation this straight line will touch the curve in the new figure; and conversely so. For, if any two points of a curve approach one another and coalesce in the first figure, the same points will after transformation coalesce in the second figure, and hence the straight lines joining these points will come simultaneously to be tangent in each figure. Proofs of these assertions might be composed in a more geometrical style, but it is my intention to be brief.

12.E Newton's Image in English Poetry

Alexander Pope's *Epitaph* (12.E1) is well-known; but the tone of his reference to Newton in *An Essay on Man* (12.E2) is ambiguous—Pope felt that the virtual deification of Newton was somewhat excessive. William Wordsworth was a student at Cambridge in the late 1780s, sixty years after Newton's death. In his autobiographical poem *The Prelude* (12.E3) his image of Newton is one of chilling perplexity. William Blake was in no doubt, however, that Newton was foremost among those responsible for establishing the cruel, mechanistic world-view he so abhorred (12.E4).

12.E1 Alexander Pope, *Epitaph*

Intended for Sir ISAAC NEWTON,
In Westminster-Abbey.

ISAACUS NEWTONIUS
Quem Immortalem,
Testantur Tempus, Natura, Cælum:
Mortalem
Hoc Marmor fatetur.

NATURE, and Nature's Laws lay hid in Night.
God said, *Let Newton be!* and All was *Light*.

12.E2 Alexander Pope, *An Essay on Man*, Epistle II

Know then thyself, presume not God to scan;
The proper study of Mankind is Man.
Plac'd on this isthmus of a middle state,
A being darkly wise, and rudely great:
With too much knowledge for the Sceptic side,
With too much weakness for the Stoic's pride,
He hangs between; in doubt to act, or rest,
In doubt to deem himself a God, or Beast;
In doubt his Mind or Body to prefer,
Born but to die, and reas'ning but to err;
Alike in ignorance, his reason such,
Whether he thinks too little, or too much:
Chaos of Thought and Passion, all confus'd;
Still by himself abus'd, or disabus'd;

Created half to rise, and half to fall;
Great lord of all things, yet a prey to all;
Sole judge of Truth, in endless Error hurl'd:
The glory, jest, and riddle of the world!
  Go, wond'rous creature! mount where Science guides,
Go, measure earth, weigh air, and state the tides;
Instruct the planets in what orbs to run,
Correct old Time, and regulate the Sun;
Go, soar with Plato to th' empyreal sphere,
To the first good, first perfect, and first fair;
Or tread the mazy round his follow'rs trod,
And quitting sense call imitating God;
As Eastern priests in giddy circles run,
And turn their heads to imitate the Sun.
Go, teach Eternal Wisdom how to rule—
Then drop into thyself, and be a fool!
  Superior beings, when of late they saw
A mortal Man unfold all Nature's law,
Admir'd such wisdom in an earthly shape,
And shew'd a Newton as we shew an Ape.
  Could he, whose rules the rapid Comet bind,
Describe or fix one movement of his Mind?
Who saw its fires here rise, and there descend,
Explain his own beginning, or his end?
Alas what wonder! Man's superior part
Uncheck'd may rise, and climb from art to art:
But when his own great work is but begun,
What Reason weaves, by Passion is undone.

12.E3 William Wordsworth (1770–1850), *The Prelude*, Book III

Near me hung Trinity's loquacious clock,
Who never let the quarters, night or day,
Slip by him unproclaimed, and told the hours
Twice over with a male and female voice.
Her pealing organ was my neighbour too;
And from my pillow, looking forth by light
Of moon or favouring stars, I could behold
The antechapel where the statue stood
Of Newton with his prism and silent face,
The marble index of a mind for ever
Voyaging through strange seas of Thought, alone.

12.E4 William Blake, *Jerusalem*, Chapter 1

I see the Four-fold Man, The Humanity in deadly sleep
And its fallen Emanation, The Spectre & its cruel Shadow.
I see the Past, Present & Future existing all at once
Before me. O Divine Spirit, sustain me on thy wings,
That I may awake Albion from his long & cold repose;
For Bacon & Newton, sheath'd in dismal steel, their terrors hang
Like iron scourges over Albion: Reasonings like vast Serpents
Infold around my limbs, bruising my minute articulations.

I turn my eyes to the Schools & Universities of Europe
And there behold the Loom of Locke, whose Woof rages dire,
Wash'd by the Water-wheels of Newton: black the cloth
In heavy wreathes folds over every Nation: cruel Works
Of many Wheels I view, wheel without wheel, with cogs tyrannic
Moving by compulsion each other, not as those in Eden, which,
Wheel within Wheel, in freedom revolve in harmony & peace.

12.F Biographical and Historical Comments

Newton's death in 1727 was, as Voltaire suggested (12.F2), the occasion for national mourning. It also called forth the obituary by Bernard le Bovier de Fontenelle (12.F1) which was at once translated into English, and which contains a most instructive comparison between Newton and Descartes, the first of several (see also 14.D1 and 14.D2). Voltaire, who published some of his writings about Newton in Amsterdam in order to avoid persecution at home, was caught up in that rare thing, a French craze for all things English. He was also fortunate to have as a companion and a guide to the *Principia* Madame du Châtelet, a gifted mathematician who was to go on to produce the French translation of Newton's work, and to whom he composed several moving passages; his summary of the *Principia* (see 12.F3) he certainly learned from her. 12.F4 is taken from John Maynard Keynes's interesting essay, which R. S. Westfall has hailed in his own fine biography (*Never at Rest*, 1983) as the first recent work that 'broke decisively with the established pattern of apotheosizing Newton'. Finally, 12.F5 is an attempt to select from so much so thoughtfully written about Newton by the man—D. T. Whiteside—who has, more than anyone else, restored him for the modern reader, just one item which might hint at a distillation of a lifetime's work. Much more is, of course, to be found in Whiteside's monumental eight-volume edition of the *Mathematical Papers of Isaac Newton*.

12.F1 Bernard de Fontenelle's *Eulogy of Newton*

The motion of the Moon is the least regular of any of the Planets, the most exact tables are sometimes wrong, and she makes certain excursions which could not before be accounted for. Dr Halley, whose profound skill in mathematicks has not hindered his being a good Poet, says in the Latin verses prefixt to the *Principia*,

> *Discimus, hinc tandem qua causa argentea phoebe*
> *Passibus haud aequis graditur; cur sudita nulli*
> *Hactenus Astronomo numerorum frena recuset.*

That the Moon till then never submitted to the bridle of calculations, nor was ever broke by any Astronomer; but that at last she is subdued in this new System: All the irregularities of her course are there shewn to proceed from a necessity by which they are foretold. It is difficult to imagine that a System in which they take this form should be no more than a lucky conjecture; especially if we consider this but as a small part of a Theory, which with the same success comprehends an infinite number of other solutions. The ebbing and flowing of the Tyde so naturally shews itself to proceed from the operation of the Moon upon the Sea, combined with that of the Sun, that the admiration which this phenomenon used to raise in us seems to be lessened by it.

The second of these two great Theories, upon which the *Principia* chiefly runs, is that of the Resistance of mediums to motion, which must enter into the consideration of all the chief phenomena of Nature, such as the motions of the celestial bodies, of Light and Sound. Sir Isaac, according to his usual Method, lays his foundations in the most solid proofs of Geometry, he considers all the causes from which resistance can possibly arise; the density of the medium, the swift motion of the body moved, the magnitude of its superficies, and from thence he at last draws conclusions which destroy all the Vortices of Des Cartes, and overturn that immense celestial edifice, which we might have thought immoveable. If the Planets move round the Sun in a certain medium whatever it be, in an aetherial matter which fills up the whole, and which notwithstanding its being extreamly subtil, will yet cause resistance as is demonstrated, whence comes it then, that the motions of the Planets are not perpetually, nay instantly lessened? But besides this, how can Comets traverse those Vortices freely every way, sometimes with a tendency absolutely opposite to theirs, without receiving any sensible alteration in their motions, tho of never so long a continuance? Whence comes it that these immense torrents whirling round with almost incredible velocity, do not instantly destroy the particular motion of any body, which is but an atom in comparison of them, and why do they not force it to follow their course? The Celestial Bodies do then move in a vast vacuum, unless their exhalations and the rays of Light which together form a thousand different mixtures should mingle a small quantity of matter with the almost infinite immaterial spaces. Thus Attraction and Vacuum banished from Physicks by Des Cartes, and in all appearance for ever, are now brought back again by Sir Isaac Newton, armed with a power entirely new, of which they were thought incapable, and only perhaps a little disguised.

These two great men, whose Systems are so opposite, resembled each other in several respects, they were both Genius's of the first rank, both born with superior understandings, and fitted for the founding of Empires in Knowledge. Being excellent Geometricians, they both saw the necessity of introducing Geometry into Physicks;

For both founded their Physicks upon discoveries in Geometry, which may almost be said of none but themselves. But one of them taking a bold flight, thought at once to reach the Fountain of All things, and by clear and fundamental ideas to make himself master of the first principles; that he might have nothing more left to do, but to descend to the phenomena of Nature as to necessary consequences; the other more cautious, or rather more modest, began by taking hold of the known phenomena to climb to unknown principles; resolved to admit them only in such manner as they could be produced by a chain of consequences. The former sets out from what he clearly understands, to find out the causes of what he sees; the latter sets out from what he sees, in order to find out the cause, whether it be clear or obscure. The self-evident principles of the one do not always lead him to the causes of the phenomena as they are; and the phenomena do not always lead the other to principles sufficiently evident. The boundaries which stop'd two such men in their pursuits through different roads, were not the boundaries of Their Understanding, but of Human understanding it self.

12.F2 Voltaire on Descartes and Newton

A Frenchman arriving in London finds things very different, in natural science as in everything else. He has left the world full, he finds it empty. In Paris they see the universe as composed of vortices of subtle matter, in London they see nothing of the kind. For us it is the pressure of the moon that causes the tides of the sea; for the English it is the sea that gravitates towards the moon, so that when you think that the moon should give us a high tide, these gentlemen think you should have a low one. Unfortunately this cannot be verified, for to check this it would have been necessary to examine the moon and the tides at the first moment of creation.

Furthermore, you will note that the sun, which in France doesn't come into the picture at all, here plays its fair share. For your Cartesians everything is moved by an impulsion you don't really understand, for Mr Newton it is by gravitation, the cause of which is hardly better known. In Paris you see the earth shaped like a melon, in London it is flattened on two sides. For a Cartesian light exists in the air, for a Newtonian it comes from the sun in six and a half minutes. Your chemistry performs all its operations with acids, alkalis and subtle matter; gravitation dominates even English chemistry.

The very essence of things has totally changed. You fail to agree both on the definition of the soul and on that of matter. Descartes affirms that the soul is the same thing as thought, and Locke proves to him fairly satisfactorily the opposite.

Descartes also affirms that volume alone makes matter, Newton adds solidity. There you have some appalling clashes.

Non nostrum inter vos tantas componere lites.

This Newton, destroyer of the Cartesian system, died in March last year, 1727. He lived honoured by his compatriots and was buried like a king who had done well by his subjects.

People here have eagerly read and translated into English the *Eulogy of Newton* that M. de Fontenelle delivered in the Académie des Sciences. In England it was expected that the verdict of M. de Fontenelle would be a solemn declaration of the superiority of

English natural science. But when it was realized that he compared Descartes with Newton the whole Royal Society in London rose up in arms. Far from agreeing with this judgement they criticized the discourse. Several even (not the most scientific) were shocked by the comparison simply because Descartes was a Frenchman.

In a criticism made in London of the discourse of M. de Fontenelle, people have dared to assert that Descartes was not a great mathematician. People who talk like that can be reproached for beating their own nurse. Descartes covered as much ground from the point where he found mathematics to where he took it as Newton after him. He is the first to have found the way of expressing curves by algebraical equations. His mathematics, now common knowledge thanks to him, was in his time so profound that no professor dared undertake to explain it, and only Schooten in Holland and Fermat in France understood it.

He carried this spirit of mathematics and invention into dioptrics, which became in his hands quite a new art, and if he committed some errors it is because a man who discovers new territories cannot suddenly grasp every detail of them: those who come after him and make these lands fertile do at least owe their discovery to him. I will not deny that all the other works of Descartes are full of errors.

Mathematics was a guide that he himself had to some extent formed, and which would certainly have led him in his physical researches, but he finally abandoned this guide and gave himself up to a fixed system. Thereafter his philosophy was nothing more than an ingenious novel, at the best only plausible to ignoramuses. He was wrong about the nature of the soul, proofs of the existence of God, matter, the laws of dynamics, the nature of light; he accepted innate ideas, invented new elements, created a world and made man to his own specification, and it is said, rightly, that Descartes's man is only Descartes's man and far removed from true man.

12.F3 Voltaire on gravity as a physical truth

At last in 1672, Mr Richer, in a Voyage to Cayenna, near the Line, undertaken by Order of Lewis XIV under the Protection of Colbert, the Father of all Arts; Richer, I say, among many Observations, found that the Pendulum of his Clock no longer made its Vibrations so frequently as in the Latitude of Paris, and that it was absolutely necessary to shorten it by a Line, that is, eleventh Part of our Inch, and about a Quarter more.

Natural Philosophy and Geometry were not then, by far, so much cultivated as at present. Who could have believed, that from this Remark, so trifling in Appearance, that from the Difference of the eleventh of our Inch, or thereabouts, could have sprung the greatest of physical Truths? It was found, at first, that Gravity must needs be less under the Equator, than in the Latitude of France, since Gravity alone occasions the Vibration of a Pendulum.

In Consequence of this it was discovered, that, whereas the Gravity of Bodies is by so much the less powerful, as these Bodies are farther removed from the Centre of the Earth, the Region of the Equator must absolutely be much more elevated than that of France; and so must be farther removed from the Centre; and therefore, that the Earth could not be a Sphere. Many Philosophers, on occasion of these Discoveries, did what Men usually do, in Points concerning which it is requisite to change their Opinion; they opposed the new-discovered Truth.

12.F4 John Maynard Keynes on Newton, the man

It is with some diffidence that I try to speak to you in his own home of Newton *as he was himself*. I have long been a student of the records and had the intention to put my impressions into writing to be ready for Christmas Day 1942, the tercentenary of his birth. The war has deprived me both of leisure to treat adequately so great a theme and of opportunity to consult my library and my papers and to verify my impressions. So if the brief study which I shall lay before you to-day is more perfunctory than it should be, I hope you will excuse me.

One other preliminary matter. I believe that Newton was different from the conventional picture of him. But I do not believe he was less great. He was less ordinary, more extraordinary, than the nineteenth century cared to make him out. Geniuses *are* very peculiar. Let no one here suppose that my object to-day is to lessen, by describing, Cambridge's greatest son. I am trying rather to see him as his own friends and contemporaries saw him. And they without exception regarded him as one of the greatest of men.

In the eighteenth century and since, Newton came to be thought of as the first and greatest of the modern age of scientists, a rationalist, one who taught us to think on the lines of cold and untinctured reason.

I do not see him in this light. I do not think that any one who has pored over the contents of that box which he packed up when he finally left Cambridge in 1696 and which, though partly dispersed, have come down to us, can see him like that. Newton was not the first of the age of reason. He was the last of the magicians, the last of the Babylonians and Sumerians, the last great mind which looked out on the visible and intellectual world with the same eyes as those who began to build our intellectual inheritance rather less than 10,000 years ago. Isaac Newton, a posthumous child born with no father on Christmas Day, 1642, was the last wonder-child to whom the Magi could do sincere and appropriate homage. [...]

I believe that the clue to his mind is to be found in his unusual powers of continuous concentrated introspection. A case can be made out, as it also can with Descartes, for regarding him as an accomplished experimentalist. Nothing can be more charming than the tales of his mechanical contrivances when he was a boy. There are his telescopes and his optical experiments. These were essential accomplishments, part of his unequalled all-round technique, but not, I am sure, his *peculiar* gift, especially amongst his contemporaries. His peculiar gift was the power of holding continuously in his mind a purely mental problem until he had seen straight through it. I fancy his pre-eminence is due to his muscles of intuition being the strongest and most enduring with which a man has ever been gifted. Anyone who has ever attempted pure scientific or philosophical thought knows how one can hold a problem momentarily in one's mind and apply all one's powers of concentration to piercing through it, and how it will dissolve and escape and you find that what you are surveying is a blank. I believe that Newton could hold a problem in his mind for hours and days and weeks until it surrendered to him its secret. Then being a supreme mathematical technician he could dress it up, how you will, for purposes of exposition, but it was his intuition which was pre-eminently extraordinary—'so happy in his conjectures', said de Morgan, 'as to seem to know more than he could possibly have any means of proving'. The proofs, for what they are worth, were, as I have said, dressed up afterwards—they were not the instrument of discovery.

His experiments were always, I suspect, a means, not of discovery, but always of verifying what he knew already.

Why do I call him a magician? Because he looked on the whole universe and all that is in it *as a riddle*, as a secret which could be read by applying pure thought to certain evidence, certain mystic clues which God had laid about the world to allow a sort of philosopher's treasure hunt to the esoteric brotherhood. He believed that these clues were to be found partly in the evidence of the heavens and in the constitution of elements (and that is what gives the false suggestion of his being an experimental natural philosopher), but also partly in certain papers and traditions handed down by the brethren in an unbroken chain back to the original cryptic revelation in Babylonia. He regarded the universe as a cryptogram set by the Almighty—just as he himself wrapt the discovery of the calculus in a cryptogram when he communicated with Leibniz. By pure thought, by concentration of mind, the riddle, he believed, would be revealed to the initiate.

He *did* read the riddle of the heavens. And he believed that by the same powers of his introspective imagination he would read the riddle of the Godhead, the riddle of past and future events divinely foreordained, the riddle of the elements and their constitution from an original undifferentiated first matter, the riddle of health and of immortality. All would be revealed to him if only he could persevere to the end, uninterrupted, by himself, no one coming into the room, reading, copying, testing—all by himself, no interruption for God's sake, no disclosure, no discordant breakings in or criticism, with fear and shrinking as he assailed these half-ordained, half-forbidden things, creeping back into the bosom of the Godhead as into his mother's womb. 'Voyaging through strange seas of thought *alone*', not as Charles Lamb 'a fellow who believed nothing unless it was as clear as the three sides of a triangle'.

And so he continued for some twenty-five years. In 1687, when he was forty-five years old, the *Principia* was published.

And when the turn of his life came and he put his books of magic back into the box, it was easy for him to drop the seventeenth century behind him and to evolve into the eighteenth-century figure which is the traditional Newton.

12.F5 D. T. Whiteside on Newton, the mathematician

Never ever to forget that mathematics had for Newton, above and beyond its place as trusty tool-kit, an inner beauty and strength independent of all outward motivation and application. To those who are insensitive to the elegance and power of mathematics as an intellectual discipline in its own right, shorn of all footholds into the world at large, it is difficult to put over just why Newton should have interested himself (as he did at various times in his life) in such 'useless' topics as projective and transformational geometry or Diophantine number theory or the exact quadrature of algebraic multinomials; why it was important for him to take forty pages of his *Principia* to give solution of problems of circle-tangency and of anharmonic conic theory; why it mattered to him to seek numbers which can be expressed in two separate ways as the sum of two cubes (he, unlike G. H. Hardy two centuries later, would have needed no prompting from Ramanujan to appreciate the significance of 1729 in this respect); why he should have spent so much effort in determining the most general Cartesian equation of the quartic trefoil from the condition that this be invariant under rotations through 120° about its centre, cheerfully setting himself to reduce the 106 terms of the ensuing transformed equation; why he just *had* to go on deriving formula after formula, each more general than the last, for the squaring of curves by series 'wch brake off'. There you have a 'pure' mathematician, as the old phrase has it, at times wholly absorbed in his Cambridge ivory tower in elaborating theorems and properties and algorithms and elegances of construction for their own sake; and magnificently did he practise his talents and expertise. There was no more gifted, no more widely versed mathematician in all the world in his day; no one more adept in algebra, more skilled in geometry, more adroitly wise in the subtleties of infinitesimal variation. Set him against James Gregory, Leibniz, Huygens, the Bernoullis Jakob and Johann, and in his heyday he was their peer in the mathematical realms in which each of these showed themselves at their best. Lagrange's aphorism that Newton was lucky to be born at just the right time to be able to unlock the secrets of the universe is not without truth; and there were giants a-plenty around with shoulders broad enough to raise him on high to the keyhole. But do give full measure to the trained acuity, the deep-read learning coupled with originality of approach, and—let me say it—the fund of genius which allowed him to insert the key and fully turn it in his *Principia*, the *Philosophiæ Naturalis Principia Mathematica*.

In its manifestation, that genius was never the seemingly quicksilver thought and deft Pascalian insight of so many lesser-gifted men, but a deeper, firmer, more solid and enduring sort. As you all know, Newton himself in 1692 epitomised his intellectual achievements as 'due to nothing but industry & a patient thought': which is not to say that his mental stamina and stubborn sweat was but average and commonplace. The power of intense, unremitting concentration which he developed at Cambridge in the 1660s—by all account he could work a 16-hour day, 7 days a week for months on end—was to remain with him undimmed over the next quarter of a century, and was gradually to be underpinned by a massive weight of knowledge and technical expertise, in mathematics as in the other areas of his scientific studies. And behind the concentration—do I show my bias in thinking so?—was an earthy core of straight-thinking common sense. In all the thousands of sheets of his preserved mathematical papers I can detect no hint of any belief by him in number-mysticism, no trace of the extravagance of hermetic mathesis: no neo-Pythagorean arithmologies lurk there that I know of, no cabbalistic *gematria*, no cryptic magic squares or motifs in John Deeist style complexly intertwining pentacles and stellate heptagons. Elsewhere it may be so, but I need to be shown. Karin Figala has just lately directed our overdue attention to the three simplest non-trivial solutions of the recursion $a_{n+1} + 2a_{n-1} = 3a_n$ which Newton set to be basic in his schemes for determining the successive degrees of rarefaction in 'sulphurial' and 'mercurial' and other substances, adding in comment that 'In this way'—combining one step forward and two backward!—'generation of different degrees are connected in a "trinity"'. Maybe so in that shadowy no-man's-land between renaissance alchemy and modern chemistry. But had any one else put such an interpretation to him, Newton in his trenchant way would have despatched the metaphorical spirit with the same curt 'See the mystery!' with which he afterwards in a couple of lines surgically dismembered Leibniz' claim to 'harmonic' perfection in summing to a finite aggregate the infinite series of reciprocals of triangular and higher polygonal numbers (binomial coefficients to us). There is the no-nonsense, down-to-earth Newton as I have always met with him.

13 Leibniz and His Followers

13.A Leibniz's Invention of the Calculus

A key ingredient in the invention of the calculus by Gottfried Wilhelm Leibniz (1646–1716) was his interest in logic and language, for it led him to think deeply about the basic processes involved and to devise a notation which, by capturing an underlying unity, made his discoveries easy to use. In 13.A1, which is taken from J. M. Child's fascinating edition of the early manuscripts, we can see Leibniz working out these ideas. He began with area questions, using Bonaventura Cavalieri's notation omn. *l* for 'all the lines' of a figure, and then introduced the much more perspicacious symbol ∫, and noted some of its algorithmic properties. Then, already alert to an inverse relationship between summing and differencing, finding areas and tangents, he symbolized the inverse of ∫ by *d,* written as a reciprocal on dimensional grounds. Soon, however, practical convenience persuaded him to abandon such considerations, and *d* for differential entered mathematics.

Scarcely nine months later Leibniz was able to see how his calculus could tackle Debeaune's inverse tangent problem (see 11.B1), and, although he floundered a little, his evident pleasure at getting beyond the rather arrogant remarks of Descartes, which he went to the trouble of transcribing, is completely understandable (see 13.A2). But he was only to publish his results much later (see 13.A3). This paper is not altogether clear, and is marred by careless mistakes, but Leibniz was at pains to stress the generality of his methods and their routine character. What he presented is a way of solving geometrical problems without, one might say, the need for geometrical methods. The fruit of his meditations was a way of calculating with symbols—truly, a *calculus*—and its scope and power is illustrated by the range of problems Leibniz tackled with it at the end. In a few pages, the reign of geometry was challenged in a way not even Descartes could have contemplated.

13.A1 A notation for the calculus

To resume, $\frac{l}{a} = \frac{p}{\text{omn.}\,l} = y$, therefore $p = \frac{\overline{\text{omn.}\,l}}{a} l$. Hence, omn. $y\frac{l}{a}$ does not mean the same thing as omn. y into omn. l, nor yet y into omn. l; for, since $p = \frac{y}{a} l$ or $\frac{\overline{\text{omn.}\,l}}{a} l$, it means the same thing as omn. l multiplied by that one l that corresponds with a certain p; hence, omn. p = omn. $\frac{\overline{\text{omn.}\,l}}{a} l$. Now I have otherwise proved omn. $p = \frac{y^2}{2}$, i.e., $= \frac{\overline{\text{omn.}\,l}^2}{2}$; therefore we have a theorem that to me seems admirable, and one that will be of great service to this new calculus, namely, $\frac{\overline{\text{omn.}\,l}^2}{2} = \text{omn.}\,\overline{\overline{\text{omn.}\,l}\frac{l}{a}}$, whatever l may be; that is, if all the ls are multiplied by their last, and so on as often as it can be done, the sum of all these products will be equal to half the sum of the squares, of which the sides are the sum of the ls or all the ls. This is a very fine theorem, and one that is not at all obvious.

Another theorem of the same kind is: omn. $xl = x$ omn. l − omn. omn. l, where l is taken to be a term of a progression, and x is the number which expresses the position or order of the l corresponding to it; or x is the ordinal number and l is the ordered thing.

N.B. In these calculations a law governing things of the same kind can be noted; for, if omn. is prefixed to a number or ratio, or to something indefinitely small, then a line is produced, also if to a line, then a surface, or if to a surface, then a solid; and so on to infinity for higher dimensions.

It will be useful to write ∫ for omn., so that $\int l$ = omn. l, or the sum of the ls. Thus, $\frac{\int \bar{l}^2}{2} = \int \overline{\int \bar{l}\frac{l}{a}}$, and $\int \overline{xl} = x \int \bar{l} - \int\int l$.

From this it will appear that a law of things of the same kind should always be noted, as it is useful in obviating errors of calculation.

N.B. If $\int l$ is given analytically, then l is also given; therefore if $\int\int l$ is given, so also is l; but if l is given, $\int l$ is not given as well. In all cases $\int x = x^2/2$.

N.B. All these theorems are true for series in which the differences of the terms bear to the terms themselves a ratio that is less than any assignable quantity.

$$\int x^2 = \frac{x^3}{3}$$

Now note that if the terms are affected, the sum is also affected in the same way, such being a general rule; for example, $\int \overline{\frac{a}{b} l} = \frac{a}{b} \times \int \bar{l}$, that is to say, if $\frac{a}{b}$ is a constant term, it is to be multiplied by the maximum ordinal; but if it is not a constant term, then it is impossible to deal with it, unless it can be reduced to terms in l, or whenever it can be reduced to a common quantity, such as an ordinal. [...]

I propose to return to former considerations.

Given l, and its relation to x, to find $\int l$.

This is to be obtained from the contrary calculus, that is to say, suppose that $\int l = ya$. Let $l = ya/d$; then just as $\int$ will increase, so d will diminish the dimensions. But $\int$ means a sum, and d a difference. From the given y, we can always find y/d or l, that is, the difference of the ys. Hence one equation may be transformed into the other; just as from the equation $\int c\int \overline{l^2} = \dfrac{c\int \overline{l^3}}{3a^3}$, we can obtain the equation $c\int \overline{l^2} = \dfrac{c\int \overline{l^3}}{3a^3 d}$.

$$\text{N.B.}\quad \int\frac{x^3}{b} + \int\frac{x^2 a}{e} = \int\overline{\frac{x^3}{b} + \frac{x^2 a}{e}}.\ \text{And similarly,}\ \frac{x^3}{db} + \frac{x^2 a}{de} = \frac{\dfrac{x^3}{b} + \dfrac{x^2 a}{e}}{d}.$$

13.A2 Debeaune's inverse tangent problem

In the third volume of the *Correspondence* of Descartes, I see that he believed that Fermat's method of Maxima and Minima is not universal: for he thinks that it will not serve to find the tangent to a curve, of which the property is that the lines drawn from any point on it to four given points are together equal to a given straight line.

Descartes to Debeaune:

'I do not believe that it is in general possible to find the converse to my rule of tangents, nor of that which M. Fermat uses, although in many cases the application of his is more easy than mine; but one may deduce from it *a posteriori* theorems that apply to all curved lines that are expressed by an equation, in which one of the quantities, x or y, has no more than two dimensions, even if the other had a thousand. There is indeed another method that is more general and *a priori*, namely, by the intersection of two tangents, which should always intersect between the two points at which they touch the curve, as near one another as you can imagine; for in considering what the curve ought to be, in order that this intersection may occur between the two points, and not on this side or on that, the construction for it may be found. But there are so many different ways, and I have practised them so little, that I should not know how to give a fair account of them.'

Descartes speaks with a little too much presumption about posterity; he says that his rule for resolving in general all problems on solids has been without comparison the most difficult to find of all things which have been discovered in geometry up to the present, and one which will possibly remain so after centuries, 'unless I take upon myself the trouble of finding others' (as if several centuries would not be capable of producing a man able to do something that would be of greater moment).

The question of the four spheres is one that is easy to investigate for a man who knows the calculus. It is due to Descartes, but as it is given in the book, it appears to be very prolix.

The problem on the inverse method of tangents, which Descartes says he has solved: EAD is an angle of 45 degrees. ABO is a curve, BL a tangent to it; and BC, the ordinate, is to CL as N is to BJ. Then $CL = \dfrac{BC = ny}{BJ = y - x}$, $CL = t$, hence, $t = \dfrac{ny}{y-x}$, $\dfrac{n}{t} = \dfrac{y-x}{y} = 1 - \dfrac{x}{y}$, hence, $\dfrac{x}{y} = \dfrac{t-n}{t}$; but $\dfrac{t}{y} = \dfrac{\overline{dx}}{dy}$; therefore $\dfrac{\overline{dx}}{dy} = \dfrac{n}{y-x}$, or $\overline{dx}\, y - x\,\overline{dx} = \overline{dy}\, n$; hence $\int \overline{dx}\, y - \int x\,\overline{dx} = -n\int dy$.

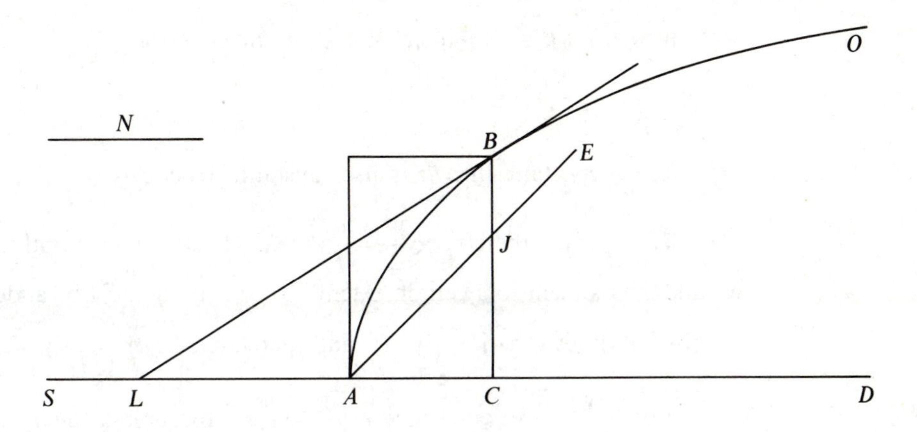

Now, $\int \overline{dy} = y$, and $\int \overline{x\,\overline{dx}} = x^2/2$, and $\int \overline{\overline{dx}\,y}$ is equal to the area $ACBA$, and the curve is sought in which the area $ACBA$ is equal to $(x^2/2) + ny = (AC^2/2) + nBC$.

Let this $x^2/2$, i.e., the triangle ACJ be cut off from the area, then the remainder $AJBA$ should be equal to the rectangle ny.

The line that Debeaune proposed to Descartes for investigation reduces to this, that if BC is an asymptote to the curve, BA the axis, A the vertex, AB, BC, fixed lines, for BAC is a right angle.

Let RX be an ordinate, XN a tangent, then RN is always to be constant and equal to BC; required the nature of the curve.

This is how I think it should be done.

Let PV be another ordinate, differing from the other one RX by a straight line VS, found by drawing XS parallel to RN; then the triangles SVX, RXN are similar, $RN = t = c$, a constant, $RX = y$, $SY = dy$, and therefore $\frac{\overline{dy}}{dx} = \frac{y}{t = c}$; hence $cy = \int \overline{y\,\overline{dx}}$ or $c\,\overline{dy} = y\,\overline{dx}$.

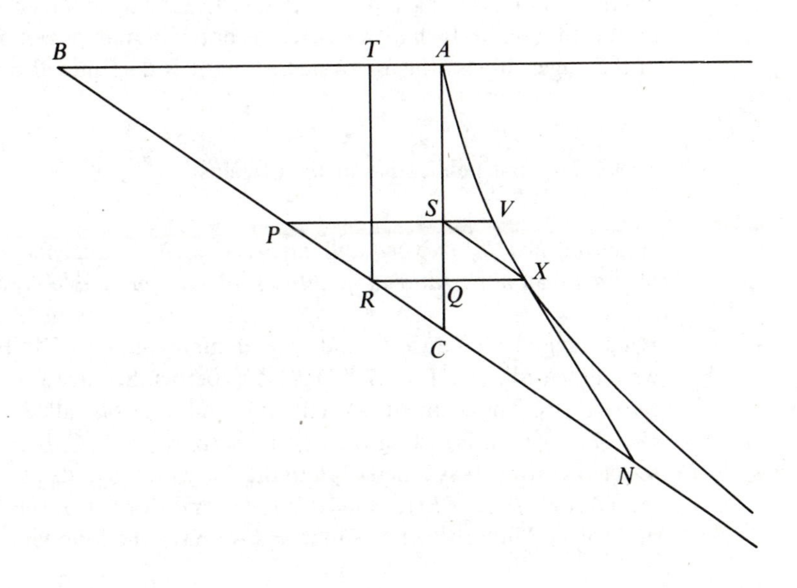

If AQ or $TR = z$, and $AC = f$, while $BC = a$; then, $\frac{AC}{BC} = \frac{f}{a} = \frac{TR}{BR} = \frac{z}{x}$; and thus $x = \frac{az}{f}$.

If $\overline{dx}$ is constant, then $\overline{dz}$ is also constant. Hence $c\, dy = \frac{a}{f} y\, \overline{dz}$, or $cy = \frac{a}{f} \int \overline{y\, dz}$, and $cy\, \overline{dy} = \frac{a}{f} y^2\, \overline{dz}$, therefore $c\frac{y^2}{2} = \frac{a}{f} \int y^2\, \overline{dz}$. Hence we have both the area of the figure and the moment to a certain extent (for something must be added on account of the obliquity); also $cz\, \overline{dy} = \frac{a}{f} yz\, \overline{dz}$, and therefore $c \int z\, \overline{dy} = \frac{a}{f} \int \overline{yz\, dz}$.

Also $\frac{c\, \overline{dy}}{y} = \frac{a}{f} dz$, and hence, $c \int \frac{\overline{dy}}{y} = \frac{a}{f} z$. Now, unless I am greatly mistaken, $\int \frac{\overline{dy}}{y}$ is in our power. The whole matter reduces to this, we must find the curve in which the ordinate is such that it is equal to the differences of the ordinates divided by the abscissae, and then find the quadrature of that figure. $d\overline{\sqrt{ay}} = \frac{1}{\sqrt{ay}}$.

Figures of this kind, in which the ordinates are dy/y, dy/y^2, dy/y^3, are to be sought in the same way as I have obtained those whose ordinates are $y\, \overline{dy}$, $y^2\, \overline{dy}$, etc. Now $w/a = \overline{dy}/y$, and since $\overline{dy}$ may be taken to be constant and equal to β, therefore the curve, in which $w/a = \overline{dy}/y$, will give $wy = \alpha\beta$, which would be a hyperbola. Hence the figure, in which $dy/y = z$, is a hyperbola, no matter how you express y, and if y is expressed by ϕ^2 we have $dy = 2\phi$, and $\frac{2\phi}{\phi^2} = \frac{2}{\phi}$. Now, $c \int \frac{dy}{y} = \frac{a}{f} z$, and therefore $\frac{fc}{a} \int \frac{dy}{y} = z$, which thus appertains to a logarithm.

Thus we have solved all the problems on the inverse method of tangents, which occur in Volume 3 of the *Correspondence* of Descartes, of which he solved one himself; but the solution is not given; the other he tried to solve but could not, stating that it was an irregular line, which in any case was not in human power, nay not within the power of the angels unless the art of describing it is determined by some other means.

13.A3 The first publication of the calculus

A new method for maxima and minima as well as tangents, which is neither impeded by fractional nor irrational quantities, and a remarkable type of calculus for them

Let an axis AX [Figure 1] and several curves such as VV, WW, YY, ZZ be given, of which the ordinates VX, WX, YX, ZX, perpendicular to the axis, are called v, w, y, z respectively. The segment AX, cut off from the axis is called x. Let the tangents be VB, WC, YD, ZE, intersecting the axis respectively at B, C, D, E. Now some straight line selected arbitrarily is called dx, and the line which is to dx as v (or w, or y, or z) is to XB (or XC, or XD, or XE) is called dv (or dw, or dy, or dz), or the difference of these v (or w, or y, or z). Under these assumptions we have the following rules of the calculus.

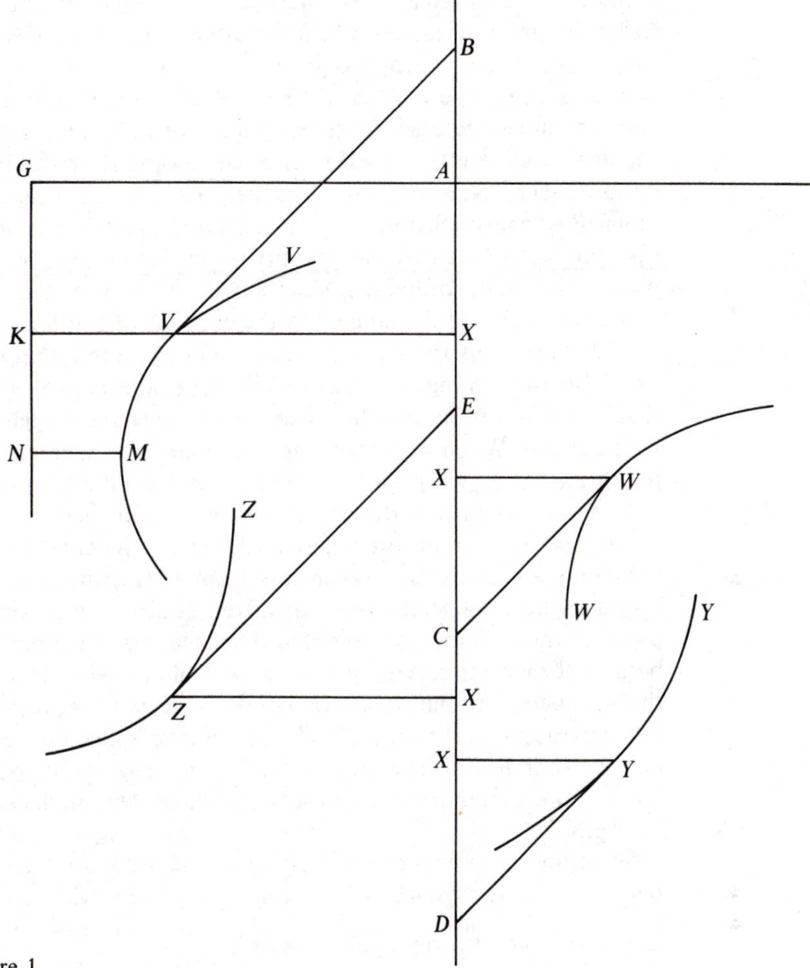

Figure 1

If a is a given constant, then $da = 0$, and $d(ax) = a\,dx$. If $y = v$ (that is, if the ordinate of any curve YY is equal to any corresponding ordinate of the curve VV), then $dy = dv$. Now *addition* and *subtraction*: if $z - y + w + z = v$, then $d(z - y + w + x) = dv = dz - dy + dw + dx$. Multiplication: $d(xv) = x\,dv + v\,dx$, or, setting $y = xv$, $dy = x\,dv + v\,dx$. It is indifferent whether we take a formula such as xv or its replacing letter such as y. It is to be noted that x and dx are treated in this calculus in the same way as y and dy, or any other indeterminate letter with its difference. It is also to be noted that we cannot always move backward from a differential equation without some caution, something which we shall discuss elsewhere.

Now *division*: $d\frac{v}{y}$ or $\left(\text{if } z = \frac{v}{y}\right) dz = \frac{\pm v\,dy \mp y\,dv}{yy}$.

The following should be kept well in mind about the *signs*. When in the calculus for a letter simply its differential is substituted, then the signs are preserved; for z we write dz, for $-z$ we write $-dz$, as appears from the previously given rule for addition and subtraction. However, when it comes to an explanation of the values, that is, when the relation of z to x is considered, then we can decide whether dz is a positive quantity or less than zero (or negative). When the latter occurs, then the tangent ZE is not directed toward A, but in the opposite direction, down from X. This happens when the ordinates z decrease with increasing x. And since the ordinates v sometimes increase and sometimes decrease, dv will sometimes be positive and sometimes be negative; in the first case the tangent VB is directed toward A, in the latter it is directed in the opposite sense. None of these cases happens in the intermediate position at M, at the moment when v neither increases nor decreases, but is stationary. Then $dv = 0$, and it does not matter whether the quantity is positive or negative, since $+0 = -0$. At this place v, that is, the ordinate LM, is *maximum* (or, when the convexity is turned to the axis, *minimum*), and the tangent to the curve at M is directed neither in the direction from X up to A, to approach the axis, nor down to the other side, but is parallel to the axis. When dv is infinite with respect to dx, then the tangent is perpendicular to the axis, that is, it is the ordinate itself. When $dv = dx$, then the tangent makes half a right angle with the axis. When with increasing ordinates v its increments or differences dv also increase (that is, when dv is positive, $d\,dv$, the difference of the differences, is also positive, and when dv is negative, $d\,dv$ is also negative), then the curve turns toward the axis its *concavity*, in the other case its *convexity*. Where the increment is maximum or minimum, or where the increments from decreasing turn into increasing, or the opposite, there is a *point of inflection*. Here concavity and convexity are interchanged, provided the ordinates too do not turn from increasing into decreasing or the opposite, because then the concavity or convexity would remain. However, it is impossible that the increments continue to increase or decrease, but the ordinates turn from increasing into decreasing, or the opposite. Hence a point of inflection occurs when $d\,dv = 0$ while neither v nor $dv = 0$. The problem of finding inflection therefore has not, like that of finding a maximum, two equal roots, but three. This all depends on the correct use of the signs.

Sometimes it is better to use *ambiguous signs*, as we have done with the division, before it is determined what the precise sign is. When with increasing x v/y increases (or decreases), then the ambiguous signs in $d\frac{v}{y} = \frac{\pm v\,dy \mp y\,dv}{yy}$ must be determined in such a way that this fraction is a positive (or negative) quantity. But $\mp$ means the opposite of $\pm$, so that when one is $+$ the other is $-$ or vice versa. There also may be several ambiguities in the same computation, which I distinguish by parentheses. For example, let $\frac{v}{y} + \frac{y}{z} + \frac{x}{v} = w$; then we must write

$$\frac{\pm v\,dy \mp y\,dv}{yy} + \frac{(\pm)y\,dz\,(\mp)\,z\,dy}{zz} + \frac{((\pm))x\,dv\,((\mp))\,v\,dx}{vv} = d[w],$$

so that the ambiguities in the different terms may not be confused. We must take notice that an ambiguous sign with itself gives $+$, with its opposite gives $-$, while with another ambiguous sign it forms a new ambiguity depending on both.

Powers. $dx^a = ax^{a-1}\,dx$; for example, $dx^3 = 3x^2dx$. $d\frac{1}{x^a} = -\frac{a\,dx}{x^{a[+]1}}$; for example, if $w = \frac{1}{x^3}$, then $dw = -\frac{3\,dx}{x^4}$.

Roots. $d\sqrt[b]{x^a} = \frac{a}{b} dx \sqrt[b]{x^{a-b}}$ (hence $d\sqrt[2]{y} = \frac{dy}{2\sqrt[2]{y}}$, for in this case $a = 1$, $b = 2$), therefore $\frac{a}{b}\sqrt[b]{x^{a-b}} = \frac{1}{2}\sqrt[2]{y^{-1}}$, but y^{-1} is the same as $\frac{1}{y}$; from the nature of the exponents in a geometric progression, and $\sqrt[2]{\frac{1}{y}} = \frac{1}{\sqrt[2]{y}}$, $d\frac{1}{\sqrt[b]{x^a}} = \frac{-a\,dx}{b\sqrt[b]{x^{a+b}}}$. The law for integral powers would have been sufficient to cover the case of fractions as well as roots, for a power becomes a fraction when the exponent is negative, and changes into a root when the exponent is fractional. However, I prefer to draw these conclusions myself rather than relegate their deduction to others, since they are quite general and occur often. In a matter that is already complicated in itself it is preferable to facilitate the operations.

Knowing thus the *Algorithm* (as I may say) of this calculus, which I call *differential calculus*, all other differential equations can be solved by a common method. We can find maxima and minima as well as tangents without the necessity of removing fractions, irrationals, and other restrictions, as had to be done according to the methods that have been published hitherto. The demonstration of all this will be easy to one who is experienced in these matters and who considers the fact, until now not sufficiently explored, that dx, dy, dv, dw, dz can be taken proportional to the momentary differences, that is, increments or decrements, of the corresponding x, y, v, w, z. To any given equation we can thus write its differential equation. This can be done by simply substituting for each *term* (that is, any part which through addition or subtraction contributes to the equation) its differential quantity. For any other quantity (not itself a term, but contributing to the formation of the term) we use its differential quantity, to form the differential quantity of the term itself, not by simple substitution, but according to the prescribed Algorithm. The methods published before have no such transition. They mostly use a line such as DX or of similar kind, but not the line dy which is the fourth proportional to DX, DY, dx—something quite confusing. From there they go on removing fractions and irrationals (in which undetermined quantities occur). It is clear that our method also covers transcendental curves—those that cannot be reduced by algebraic computation, or have no particular degree—and thus holds in a most general way without any particular and not always satisfied assumptions.

We have only to keep in mind that to find a *tangent* means to draw a line that connects two points of the curve at an infinitely small distance, or the continued side of a polygon with an infinite number of angles, which for us takes the place of the *curve*. This infinitely small distance can always be expressed by a known differential like dv, or by a relation to it, that is, by some known tangent. In particular, if y were a transcendental quantity, for instance the ordinate of a cycloid, and it entered into a computation in which z, the ordinate of another curve, were determined, and if we desired to know dz or by means of dz the tangent of this latter curve, then we should by all means determine dz by means of dy, since we have the tangent of the cycloid. The tangent to the cycloid itself, if we assume that we do not yet have it, could be found in a similar way from the given property of the tangent to the circle.

Now I shall propose an example of the calculus, in which I shall indicate division by $x:y$, which means the same as x divided by y, or $\frac{x}{y}$. Let the *first* or given equation be

$x{:}y + (a + bx)(c - xx){:}(ex + fxx)^2 + ax\sqrt{gg + yy} + yy{:}\sqrt{hh + lx + mxx} = 0$. It expresses the relation between x and y or between AX and XY, where a, b, c, e, f, g, h are given. We wish to draw from a point Y the line YD tangent to the curve, or to find the ratio of the line DX to the given line XY. We shall write for short $n = a + bx$, $p = c - xx$, $q = ex + fxx$, $r = gg + yy$, and $s = hh + lx + mxx$. We obtain $x{:}y + np{:}qq + ax\sqrt{r} + yy{:}\sqrt{s} = 0$, which we call the *second* equation. From our calculus it follows that

$$d(x{:}y) = (\pm x\,dy \mp y\,dx){:}yy,$$

and equally that

$$d(np{:}qq) = [(\pm)2np\,dq(\mp)q(n\,dp + p\,dn)]{:}q^3,$$

$$d(ax\sqrt{r}) = +ax\,dr{:}2\sqrt{r} + a\,dx\sqrt{r},$$

$$d(yy{:}\sqrt{s}) = ((\pm))yy\,ds\,((\mp))4ys\,dy{:}2s\sqrt{s}.$$

All these differential quantities from $d(x{:}y)$ to $d(yy{:}\sqrt{s})$ added together give 0, and thus produce a *third* equation, obtained from the terms of the second equation by substituting their differential quantities. Now $dn = b\,dx$ and $dp = -2x\,dx$, $d = e\,dx + 2fx\,dx$, $dr = 2y\,dy$, and $ds = l\,dx + 2mx\,dx$. When we substitute these values into the third equation we obtain a *fourth* equation, in which the only remaining differential quantities, namely dx, dy, are all outside of the denominators and without restrictions. Each term is multiplied either by dx or by dy, so that the law of homogeneity always holds with respect to these two quantities, however complicated the computation may be. From this we can always obtain the value of $dx{:}dy$, the ratio of dx to dy, or the ratio of the required DX to the given XY. In our case this ratio will be (if the fourth equation is changed into a proportionality):

$$\mp x{:}yy - axy{:}\sqrt{r}\,(\mp)2y{:}\sqrt{s}$$

divided by

$$\mp 1{:}y(\pm)(2npe + 2fx){:}q^3(\mp)(-2nx + pb){:}qq + a\sqrt{r}((\pm))yy(l + 2mx){:}2s\sqrt{s}.$$

Now x and y are given since point Y is given. Also given are the values of n, p, q, r, s expressed in x and y, which we wrote down above. Hence we have obtained what we required. Although this example is rather complicated we have presented it to show how the above-mentioned rules can be used even in a more difficult computation. Now it remains to show thair use in cases easier to grasp.

Let two points C and E [Figure 2] be given and a line SS in the same plane. It is required to find a point F on SS such that when E and C are connected with F the sum of the rectangle of CF and a given line h and the rectangle of FE and a given line r are as small as possible. In other words, if SS is a line separating two media, and h represents the density of the medium on the side of C (say water), r that of the medium on the side of E (say air), then we ask for the point F such that the path from C to E via F is the shortest possible. Let us assume that all such possible sums of rectangles, or all possible paths, are represented by the ordinates KV of curve VV perpendicular to the line GK [Figure 1]. We shall call these ordinates w. Then it is required to find their minimum NM. Since C and E [Figure 2] are given, their perpendiculars to SS are also given, namely CP (which we call c) and EQ (which we call e); moreover PQ (which we call p) is

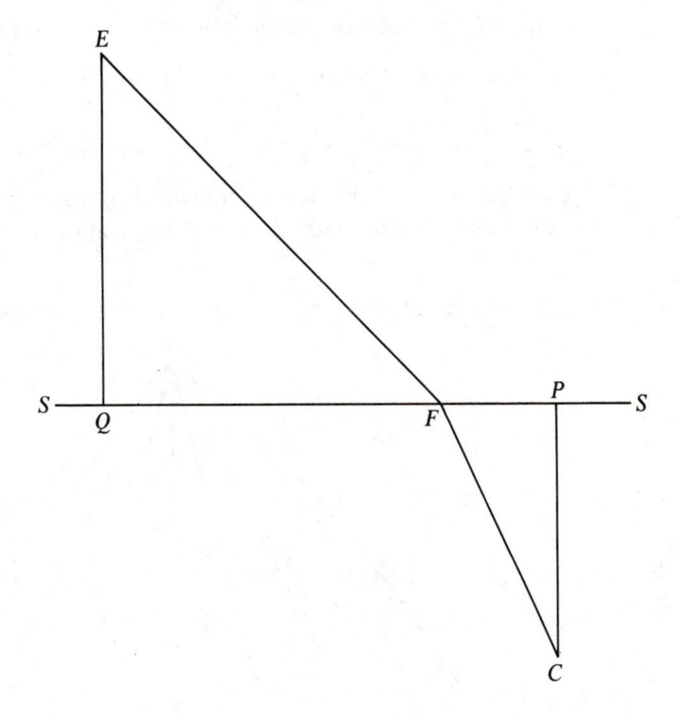

Figure 2

given. We denote $QF = GN$ (or AX) by x, CF by f, and EF by g. Then $FP = p - x$, $f = \sqrt{cc + pp - 2px + xx}$ or $= \sqrt{l}$ for short; $g = \sqrt{ee + xx}$ or $= \sqrt{m}$ for short. Hence

$$w = h\sqrt{l} + r\sqrt{m}.$$

The differential equation (since $dw = 0$ in the case of a minimum) is, according to our calculus,

$$0 = +h\,dl{:}2\sqrt{l} + r\,dm{:}2\sqrt{m}.$$

But $dl = -2(p - x)\,dx$, $dm = 2x\,dx$; hence

$$h(p - x){:}f = rx{:}g.$$

When we now apply this to dioptrics, and take f and g, that is, CF and EF, equal to each other (since the refraction at the point F is the same no matter how long the line CF may be), then $h(p - x) = rx$ or $h{:}r = x{:}(p - x)$, or $h{:}r = QF{:}FP$; hence the sines of the angles of incidence and of refraction, FP and QF, are in inverse ratio to r and h, the densities of the media in which the incidence and the refraction take place. However, this density is not to be understood with respect to us, but to the resistance which the light rays meet. Thus we have a demonstration of the computation exhibited elsewhere in these *Acta*, where we presented a general foundation of optics, catoptrics, and dioptrics. Other very learned men have sought in many devious ways what someone versed in this calculus can accomplish in these lines as by magic.

This I shall explain by still another example. Let *13* [Figure 3] be a curve of such a nature that, if we draw from one of its points, such as *3*, six lines *34, 35, 36, 37, 38, 39* to six fixed points *4, 5, 6, 7, 8, 9* on the axis, then their sum is equal to a given line. Let

T14526789 be the axis, *12* the abscissa, *23* the ordinate, and let the tangent *3T* be required. Then I claim that *T2* is to *23* as $\frac{23}{34}+\frac{23}{35}+\frac{23}{36}+\frac{23}{37}+\frac{23}{38}+\frac{23}{39}$ to $-\frac{24}{34}-\frac{25}{35}+\frac{26}{36}+\frac{27}{37}+\frac{28}{38}+\frac{29}{39}$. The same rule will hold if we increase the number of terms, taking not six but ten or more fixed points. If we wanted to solve this problem by the existing tangent methods, removing irrationals, then it would be a most tedious

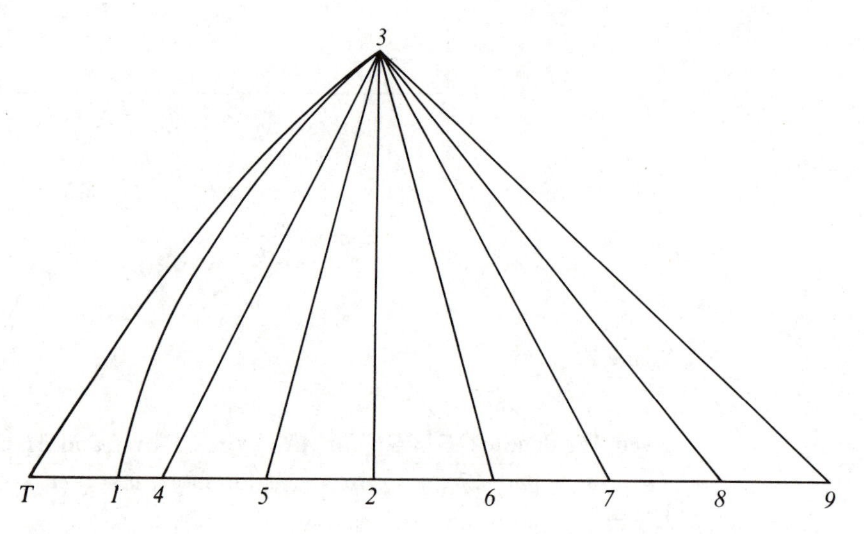

Figure 3

and sometimes insuperable task; in this case we would have to set up the condition that the rectangular planes and solids which can be constructed by means of all possible combinations of two or three of these lines are equal to a given quantity. In all these cases and even in more complicated ones our methods are of astonishing and unequaled facility.

And this is only the beginning of much more sublime Geometry, pertaining to even the most difficult and most beautiful problems of applied mathematics, which without our differential calculus or something similar no one could attack with any such ease. We shall add as appendix the solution of the problem which Debeaune proposed to Descartes and which he tried to solve in Volume 3 of the *Letters*, but without success. It is required to find a curve *WW* such that, its tangent *WC* being drawn to the axis, *XC* is always equal to a given constant line *a*. Then *XW* or *w* is to *XC* or *a* as *dw* is to *dx*. If *dx* (which can be chosen arbitrarily) is taken constant, hence always equal to, say, *b*, that is, *x* or *AX* increases uniformly, then $w = \frac{a}{b}dw$. Those ordinates *w* are therefore proportional to their *dw*, their increments or differences, and this means that if the *x* form an arithmetic progression, then the *w* form a geometric progression. In other words, if the *w* are numbers, the *x* will be logarithms, so that the curve *WW* is logarithmic.

13.B Johann Bernoulli and the Marquis de l'Hôpital

The published versions of Leibniz's calculus were not at all easy to understand—what, for example, could be the seemingly finite *dx* with which 13.A3 opens? The two brothers Jakob (1654–1705) and Johann (1667–1748) Bernoulli, however, took it up in 1687 and by 1690 were publishing articles using it. They also began a long correspondence with Leibniz about it, and gradually their work spread the news of the new method. In particular, Johann took it with him and literally sold it to the Marquis de l'Hôpital when he went to Paris in 1691 (see 13.B4). The first extract (13.B1) shows how Johann Bernoulli used the calculus to solve Debeaune's problem. It is interesting to see that he plainly felt that the full solution required him to give a geometric construction for the solution curve; the answer could not be left as the equation for the curve. The solution is taken from Bernoulli's lectures to l'Hôpital, which Bernoulli published more than fifty years after he had first given them. L'Hôpital had, of course, by then long since published his account of the differential calculus as he had learned it from Johann (a fact he acknowledged in what Johann thought was a rather off-hand manner in his preface: see 13.B5). The long gap may be a good measure of how much more difficult integration was (or is) felt to be.

13.B2 is Bernoulli's account of how to integrate rational functions. Not only does it testify to his growing mastery of the calculus, it also displays an implicit grasp of the relationships between the circular and exponential functions of real and imaginary quantities, and a confidence in the fundamental theorem of algebra. When Jean le Rond d'Alembert drew people's attention to the need to prove that theorem, in 1746, he was to cite Bernoulli's paper as a particularly important use of it. The unification of the elementary functions was to be one of Leonhard Euler's achievements (see 14.A4), but undoubtedly he was well aware of the insights of Bernoulli, his former professor, on this question.

13.B3 concerns another matter. Newton had shown convincingly that elliptical orbits under a force directed to a focus imply that the law of force is inverse square, and had gone on to prove the converse geometrically. Bernoulli was not satisfied with the latter proof, and gave one of his own, which stands here as evidence of a growing belief that if lesser mortals were ever to master the *Principia* they would need to avail themselves of the new calculus. As to which kind of calculus, the publications of Bernoulli and l'Hôpital were a great service to the Leibnizian formulation, whose foundations as given by l'Hôpital (13.B6) were truly infinitesimal; but it was Euler's reformulation which was to win the day. However, Newton's motion-theoretic ideas survived vigorously in dynamical problems, and one may also feel with Edmund Stone (13.B7) that one's working practice is hardly affected by mere questions of notation and names—whatever Stone's political purpose was in rewriting as he did. Rather, the decisive shift was to be away from geometry and into algebra, as we shall document in the next chapter.

13.B1 Bernoulli's lecture to l'Hôpital on the solution to Debeaune's problem

Another such example is the problem set to M. Descartes by M. Debeaune, the solution to which is not in his works but can be found in his *Letters* (Vol. III, No. 71). The solution of it does not appear to be very easy according to our method, indeed at first sight the problem appears impossible by this method [separation of variables]. But we shall see that by a change of variables it becomes easy to separate them, and that this problem can be solved completely once the quadrature of the hyperbola is given, for the curve is mechanical.

The problem goes like this: a line AC makes an angle of half a right angle with the axis AD, and E is a given constant line segment; what is the nature of the curve AB in which the ordinates BD are to the subtangents FD as the given E is to BC? *Solution*. Let $AD = x$, $DB = y$, $E = a$, suppose by hypothesis that $dy{:}dx = a{:}(y - x)$, then $adx = ydy - xdy$. From this equation the nature of the curve is to be found, either by integration or by rewriting y with dy on one side and x with dx on the other, for then two areas can be found and by comparing them the nature of the curve can be found. But the equation just found cannot be integrated, nor can x and dx be separated from y and dy; however, it can be changed into another by substituting the value of another variable. Therefore let $y - x = z$, $y = x + z$ and $dy = dz + dx$. The equation just found transforms into this: $adx = zdz + zdx$ or $adx - zdx = zdz$ and $dx = zdz{:}(a - z)$. Therefore these two variables separate, and we are led to the curve on multiplying by a, $adx = azdz{:}(a - z)$. And dropping normals GT and NH, $GN = GH = a$, and drawing through the points H and N HV and MR parallel to GT, $NR = NG$; erecting a perpendicular RS and asymptotes RM, RS and drawing a hyperbola LKG through G: then $GO = z$ & $GQ = x$, and $KO = az{:}(a - z)$, and because QI always equals a, the hyperbolic space KGO will equal the rectangle HQ & producing the lines IQ, KO, the point P where they meet draws out the curve GPW which satisfies the equation just found $adx = azdz{:}(a - z)$. Having constructed the curve AB there is then no more work; for, QP being produced to Z [Figure 1], as PZ shall equal the abscissa GQ the point Z will lie on the curve AB. Since $PZ = GQ = x = AD$ [Figure 2] and $QP = z$ [Figure 3], $QP + PZ$ will $= z + x = y = DB$ [Figure 2]. Q.E.I.

Corollary I NR [Figure 3] is asymptotic to GPW & $QP = BC$ [Figure 2]. This curve AB has its asymptote parallel to AC.

Corollary II The space $ADB = xy + ax - \frac{1}{2}yy$.

13.B2 Bernoulli on the integration of rational functions

Let the differential be $pdx{:}q$, of which p and q express rational quantities composed arbitrarily of a single variable x and constants; one seeks the integral or the algebraic sum or the means of reducing it to the quadrature of the hyperbola or the circle, the one or the other always being possible.

Let p be divided by q until the highest power of x in the remainder shall be less than q,

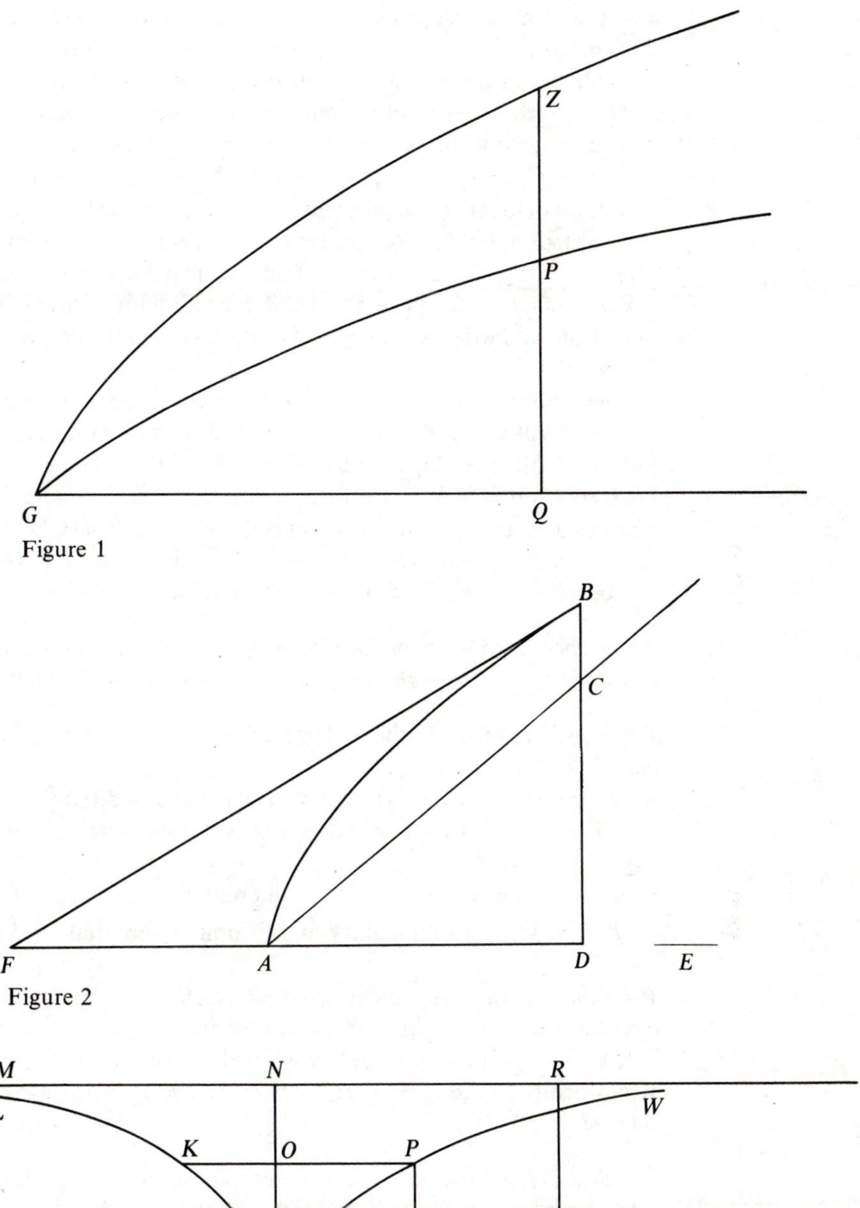

Figure 1

Figure 2

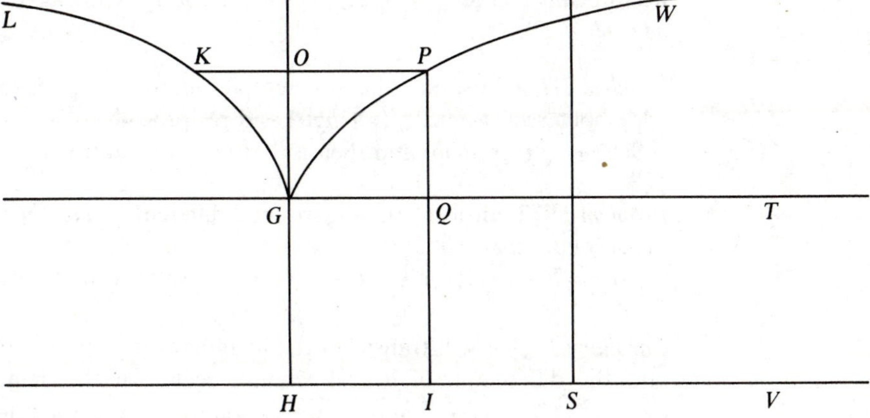

Figure 3

assuming the highest power of x in p is not already less than q, in which case there is no division to do. Then take the integral of the quotient of this division, which is always possible because this quotient (in respect of x) is always integral and rational.

Then for the integral of the remainder (which is properly the source of the difficulty) here is what one finds. Let the remainder be called r, and suppose that $rdx:q = adx:(x+f) + bdx:(x+g) + cdx:(x+h) +$ etc., that is to say, $rdx:q$ being equal to a collection of logarithmic differentials, that the highest power of x in q is unity. Here it is to be noted that a, b, c etc., and even f, g, h etc. are indeterminate constant quantities. [He goes on to explain how to find them by two methods. The second beings 'Let $rdx:q = sdx:t + adx:(x+f)$, taking for t a quantity involving x' and then argues by induction on the degree of t, which is one less than the degree of q. He then concludes as follows.]

One knows that $dx:(x+f)$, $dx:(x+g)$, $dx:(x+h)$ etc. are the differentials of logarithms of $x+f, x+g, x+h$ etc. And therefore that $\int(dx:(x+f))$, $\int(dx:(x+g))$, $\int(dx:(x+h))$ are logarithms themselves: in such a way that one will have $\int(dx:(x+f)) = l(x+f)$, and thus for the others, where l signifies logarithm as d signifies differential. Therefore $\int(adx:(x+f)) + \int(bdx:(x+g)) + \int(cdx:(x+h)) +$ etc. $= al(x+f) + bl(x+g) + cl(x+h) +$ etc. (by the nature of logarithms) $= \mathrm{Log}((x+f)^a \cdot (x+g)^b \cdot (x+h)^c$ etc.) Which was to be shown.

The abbreviated way of transforming compound differentials into simple ones and reciprocally; and even the imaginary simples into real compounds.

Problem I Transform the differential $adz:(bb-zz)$ into a logarithmic differential $adt:2bt$ and reciprocally.

Set $z = (t-1)b:(t+1)$, and you will have $adz:(bb-zz) = adt:2bt$. Reciprocally take $t = (+z+b):(-z+b)$ and you will have $adt:2bt = adz:(bb-zz)$.

Corollary One transforms the differential $adz:(bb+zz)$ in the same way into $-adt:2bt\sqrt{-1}$, an imaginary logarithmic differential, and reciprocally.

Problem II Transform the differential $adz:(bb+zz)$ into the differential of a sector or circular arc $-adt:2\sqrt{(t-bbtt)}$; and reciprocally.

Set $z = \sqrt{(1:t-bb)}$ and you will have $adz:(bb+zz) = -adt:2\sqrt{(t-bbtt)}$. Reciprocally take $t = 1:zz+bb$ and you will have $-adt:2\sqrt{(t-bbtt)} = adz:(bb+zz)$.

Problem III Transform the differential $adz:(bb-zz)$ into the differential of a hyperbolic sector $adt:2\sqrt{(t+bbtt)}$ and reciprocally.

Set $z = \sqrt{(1:t+bb)}$ and then $t = 1:(bb-zz)$; and you will have what is wanted.

Problem IV Transform the logarithmic differential $adt:2bt$ into the differential of a hyperbolic sector $adr:2\sqrt{(r+bbrr)}$.

Set $t = b+\sqrt{(1:r+bb)}:b-\sqrt{(1:r+bb)}$ and you will have what is wanted.

Corollary 1 One will transform the imaginary logarithmic differential $adt:2bt\sqrt{-1}$ into the differential of a real circular sector in the same way. For on setting $t = b\sqrt{-1}+\sqrt{(1:r-bb)}:b\sqrt{-1}-\sqrt{(1:r-bb)}$ one will have $adr:2\sqrt{(r-bbrr)}$.

Corollary 2 Then (Problem II) $\int (adz:(bb + zz))$ depends on the quadrature of the circle, and moreover $adz:(bb + zz)$ is $= \frac{1}{2}adz:(bb + bz\sqrt{-1}) + \frac{1}{2}adz:(bb - bz\sqrt{-1})$ which are two differentials of imaginary logarithms; one sees that imaginary logarithms can be taken for real circular sectors because the compensation which imaginary quantities make on being added together of destroying themselves in such a way that their sum is always real.

13.B3 Bernoulli on the inverse problem of central forces

The quadratures being supposed, and the centripetal law of forces being given arbitrarily in x and constants, to find the trajectory ABC which it makes a moving body describe.

Let $OA = a$, and from this radius the arc of the circle $AL = z$, $Ll = dz$, and in consequence $Nb = xdz:a$. Let also the time for Bb, proportional to $Nb \times BO$ (the double of triangle BOb), $= xxdz:a$. You know that this time, multiplied by the speed, i.e. [following the corollary to the preceding lemma] by $(ab - \int \phi dx)$ gives the space Bb. Therefore

$$xxdz\sqrt{(ab - \int\phi dx)}:a = Bb = \sqrt{(dx^2 + xxdz^2:aa)};$$

from which follows the equation

$$dz = aacdx:\sqrt{(abx^4 - x^4 \int \phi dx - aaccxx)}$$

which expresses the nature of the sought-for trajectory *ABC*, and in which equation *c* is an arbitrary constant put in to make everything homogeneous. *That is what was to be found.*

You will see, Monsieur, that I have reached at a stroke a differential equation of the first degree, in which there is no mixing up of indeterminates; and so the geometric construction can be easily deduced from it, the quadratures of the curved spaces being given, and even more conveniently than Mr Newton found it in his *Principia*.

Moreover, my equation displays whether the sought-for trajectory is algebraic or not, depending on what hypothesis is given for the force. For if the integral

$$aacdx:\sqrt{(abx^4 - x^4 \int \phi dx - aaccxx)}$$

is reducible to the arc of a circle whose radius stands to $OA(a)$ as number to number, then the sought-for curve will necessarily be algebraic. Thus the usual hypothesis of centripetal forces acting as the square of the reciprocal of the distance of the moving body from the centre, i.e. the hypothesis $\phi = aag:xx$, changes the preceding equation to

$$dz = aacdx:\sqrt{(abx^4 + aagx^3 - aaccxx)}$$
$$= aacdx:x\sqrt{(abxx + aagx - aacc)}$$

which can be reduced to such an arc of a circle; I see at once that your curve *ABC* must be algebraic on this hypothesis.

To see presently that this curve ABC, on this hypothesis, is always a conic section, as Mr Newton supposed, without proving it, there is much to say. Here is how I solve it.

First, to reduce this value of dz to an ordinary differential formula for a circular arc, let $x = aa{:}y$. The substitution of this value of x will give

$$aacdx{:}x\sqrt{(abxx + aagx - aacc)} = -acdy{:}\sqrt{(a^3b + aagy - ccyy)}$$

$$(\text{supposing } y = aag{:}2cc - t)$$

$$= acdt{:}\sqrt{(a^3b + a^4gg{:}4cc - cctt)}$$

$$(\text{supposing } cchh = ab^3 + a^4gg{:}4cc \text{ to abbreviate})$$

$$= adt{:}\sqrt{(hh - tt)}$$

$$= dz$$

and consequently $dz{:}a = dt{:}\sqrt{(hh - tt)} = (1{:}h) \times hdt{:}\sqrt{(hh - tt)}$ which is a differential of a circular arc (of which the radius $y = h$ and the sine $= t$) divided by its radius.

This being so, because the arc of a circle divided by its radius expresses the angle opposite to it, and, following that, the angle $LOl = dz{:}a$, one will also have $(1{:}h) \times hdt{:}\sqrt{(hh - tt)}$ for the quantity of the differential angle, which has radius $= h$ and sine $= t$. Therefore, because these two differential angles are equal, the integral angles will also be equal or, for greater generality, one exceeds the other by a constant angle. If therefore one draws a circle MST of radius $OM = h$, and an angle AOL ($\int dz{:}a$) either increased or decreased by a constant angle LOS to yield the angle MOS ($\int((1{:}h) \times hdt{:}\sqrt{(hh - tt)})$); it is obvious that the perpendicular SP on AO will $= t$, and from that (arguing backwards) one will find y, then x or OE, which will $= 2aacc{:}(aag - 2cct)$. Therefore drawing the radius OE, the arc EB which cuts OL at B, the point B will be one of those on the trajectory ABC, which I claim is a conic section: I prove it.

Every attentive mathematician will see that the constant angle LOS, which increases or decreases the angle AOL, doesn't change the nature of the curve ABC at all but only its position; by advancing or withdrawing it around the point O, all the points B advance or withdraw along their arcs EB just as if the whole plane of the curve ABC turned with it around this fixed centre O. However, to make the calculation easier, I am going to suppose that the angle AOL is neither increased or decreased i.e. that the angle MOS is equal to it.

Therefore erect the line $OQ = aag{:}2cc$ perpendicularly on AO at O; and draw an equilateral hyperbola VXZ between the asymptotes QO, QR with centre Q, of which the rectangle [of the coordinates] QYX or $QOZ = aa$; extend an arbitrary ordinate XY of the hyperbola until it meets the circle MST at S; take OS through this point S, and on it [prolonged, if necessary] take $OB = XY$. I say that the point B will be one of those on the trajectory ABC; because [by hypothesis] $OB = XY = QOZ{:}\ QY = aa{:}(aag{:}2cc - t) = 2aacc{:}(aag - 2cct)$; and that the present construction makes ABC a conic section, of which one will be convinced if one finds the equation which expresses the relationship of the coordinates OF, FB. For, calling them x and y, one will find the equation

$$(a^4gg - 4c^4hh)xx = 8aac^4hx - a^4ggyy + 4a^4c^4,$$

which describes a conic section: a parabola when $OQ(aa{:}2c) = OT(h)$; an ellipse when $OQ > OT$, and a hyperbola when $OQ < OT$. *Which was to be shown.*

13.B4 O. Spiess on Bernoulli's first meeting with l'Hôpital

Immediately on arriving in Paris in late Autumn 1691, Bernoulli visited Father Malebranche who once a week played host to the best known scholars of the city. Rather as an admission card, so to speak, he showed the famous philosophers, who were also good mathematicians, his construction of the catenary—on a single piece of paper—that had just been published a little earlier in the June volume of the *Acta Eruditorum* and which was his first important achievement. Because of this he was invited by Malebranche to take part regularly at his meetings. So the 24-year-old student appeared in the illustrious circle on the next occasion and was immediately introduced to the Marquis de l'Hôpital, to whom Malebranche had showed his paper a few days before.

'From the conversation' [Bernoulli wrote to his friend Montmort in 1718] 'which I had with M. le Marquis I knew right away that he was a good geometer for what was already known, but that he knew nothing at all of the differential calculus, of which he scarcely knew the name, and still less had he heard talk of the integral calculus which was only just being born, the little that there was of this calculus in the *Acta* of Leipzig having not yet reached him because of the war.' The Marquis, who found it difficult to see the conqueror of the catenary in the young man, examined him backwards and forwards 'but he saw soon enough that I was neither an adventurer nor the pretender that he believed I wanted to play at being. The conversation finally fell to the developed curve [evolute] or osculating circle, for the study of which he prided himself on an entirely particular rule drawn from M. Fermat's method of max. and min. To test him, I proposed an example of an algebraic curve (for this supposedly general rule only worked for algebraic curves and only gave the radius at the maximum).'

'Mr l'Hôpital took paper and ink, and began to calculate. After he had used up nearly an hour in scribbling over several pieces of paper he finally found the correct value of the radius at the maximum of the curve.'

Bernoulli then said to him that there was a formula that would find the radius of curvature of any curve at any point in a few minutes, and put forward a curve for which he could find the sought-for value at once 'which struck him so much with surprise that from that moment he became charmed with the new analysis of the infinitely small and excited with the desire to learn it from me'. Bernoulli visited the Marquis the very next day, who asked him to visit four times a week 'to explain to him on each occasion and then to deliver a lesson based on the paper which I had written at home the evening before'. And, what is of particular importance now 'one of my friends from Basel who was lodging with me had the kindness to copy each of the papers I was to take to M. le Marquis, so I have preserved them all'. [...]

The lessons in Paris went on from the end of 1691 to the end of July 1692, so for over half a year; then l'Hôpital took his young instructor to his estate in Oucques where Bernoulli presented his lectures in daily contact with the Marquis and his spirited wife. 'I didn't hesitate' he wrote in the letter to Montmort 'to give to M. l'Hôpital new memoirs always written in my own hand whenever I found appropriate material, and he furnished me himself with the occasion for all sorts of questions.'

13.B5 Preface to l'Hôpital's *Analyse des Infiniment Petits*

The Defect of this Method was supplied by that of Mr Leibnitz's (footnote by l'Hôpital *Acta Erudit. Lips. Ann., 1684, p. 467*) [footnote by Stone: or rather the great Sir Isaac Newton—see Commercium Epistolicum]. He began where Dr Barrow and others left off. His Calculus has carried him into Countries hitherto unknown; and he has made Discoveries by it astonishing the greatest Mathematicians of Europe. The Messieurs Bernoulli were the first who perceived the Beauty of the Method; and have carried it to such a length, as by its means to surmount Difficulties that were before thought insuperable.

I intended to have added another Section to shew the surprising use of this calculus in Physicks, and to what degree of Exactness it may bring the same; as likewise the use thereof in Mechanicks; But Sickness has prevented me herein. However, I hope to effect it hereafter, and present it the Publick with interest. And indeed the whole of the present Treatise is only the First Part of the Calculus of Mr Leibnitz, or the Direct Method, wherein we descend from Whole Magnitudes to their infinitely small Parts of what kind soever comparing them with each other, which is called the Calculus Differentialis: But the other Part, called the Calculus Integralis, [or Inverse Method of Fluxions] consists on ascending from these infinitely small Parts to the Magnitudes, or Wholes, whereof they are the Parts. This Inverse Method I also designed to publish but Mr Leibnitz's having wrote to me, that he was at work upon this subject, in order for a Treatise de Scientia Infiniti, I was unwilling to deprive the Publick of so fine a Piece, which must needs contain whatever is curious in the Inverse Method of Tangents, Rectifications of Curves, Quadratures, Investigation of Superficies of Solids, and their Solidities, Centres of Gravity, etc. Neither would I ever have published the present Treatise, had he not intreated me to it by Letter; as likewise because I believed it might prove a necessary Introduction to whatever should hereafter be discovered on the subject.

I must own myself very much obliged to the labours of Messieurs Bernoulli, but particularly to those of the present Professor at Groeningen, as having made free with their Discoveries as well as those of Mr Leibnitz: So that whatever they please to claim as their own I frankly return them.

I must here in justice own (as Mr Leibnitz himself has done, in *Journal des Sçavans* for August, 1694) that the learned Sir Isaac Newton likewise discovered something like the Calculus Differentialis, as appears by his excellent *Principia*, published first in the Year 1687 which almost wholly depends on the Use of the said Calculus. But the Method of Mr Leibnitz's is much more easy and expeditious, on account of the Notation he uses, not to mention the wonderful assistance it affords on many occasions.

13.B6 l'Hôpital on the foundations of the calculus

1. *Definition I Variable quantities are those that continually increase or decrease; and constant or standing quantities, are those that continue the same while others vary.*

As the ordinates and abscisses of a parabola are variable quantities, but the parameter is a constant or standing quantity.

Definition II The infinitely small part whereby a variable quantity is continually increased or decreased, is called the differential of that quantity.

For example: let there be any curve line *AMB* [Figure 1] whose axis or diameter is the line *AC*, and let the right line *PM* be an ordinate, and the right line *pm* another infinitely near to the former.

Now if you draw the right line *MR* parallel to *AC*, and the chords *AM*, *Am*; and about the centre *A* with the distance *AM*, you describe the small circular arch *MS*: then shall *Pp* be the differential of *PA*; *Rm* the differential of *Pm*; *Sm* the differential of *AM*; and *Mm* the differential of the arch *AM*. In like manner, the little triangle *MAm*, whose base is the arch *Mm*, shall be the differential of the segment *AM*; and the small space *MPpm* will be the differential of the space contained under the right lines *AP*, *PM*, and the arch *AM*.

Corollary It is manifest, that the differential of a constant quantity (which is always one of the initial letters a, b, c, etc. of the alphabet) is 0: or (which is all one) that constant quantities have no differentials.

Scholium The differential of a variable quantity is expressed by the note or characteristic *d*, and to avoid confusion this note *d* will have no other use in the sequence of this calculus. And [Figure 1 overleaf] if you call the variable quantities AP, x; PM, y; AM, z; the arch AM, u; the mixtlined space APM, s; and the segment AM, t: then will dx express the value of *Pp*, dy the value of *RM*, dz the value of *Sm*, du the value of the small arch *Mm*, ds the value of the little space *MPpm*, and du the value of the small mixtlined triangle *MAm*.

2. *Postulate I* Grant that two quantities, whose difference is an infinitely small quantity, may be taken (or used) indifferently for each other: or (which is the same thing) that a quantity, which is increased or decreased only by an infinitely small quantity, may be considered as remaining the same.

For example: grant that *Ap* may be taken for *AP*; *pm* for *PM*; the space *Apm* for *APM*; the small space *MPpm* for the small rectangle *MPpR*; the small sector *AMS* for the small triangle *AMm*; the angle *pAm* for the angle *PAM*, etc.

3. *Postulate II* Grant that a curve line may be considered as the assemblage of an infinite number of infinitely small right lines: or (which is the same thing) as a polygon of an infinite number of sides, each of an infinitely small length, which determine the curvature of the line by the angles they make with each other [Figure 2 overleaf].

For example: grant that the part *Mm* of the curve, and the circular arch *MS*, may be considered as straight lines, on account of their being infinitely small, so that the little triangle *mSM* may be looked upon as a right-lined triangle.

4. *Proposition I To find the differentials of simple quantities connected together with the signs + and −.*

It is required to find the differentials of $a + x + y - z$. If you suppose x to increase by an infinitely small part, viz. till it becomes $x + dx$; then will y become $y + dy$; and z,

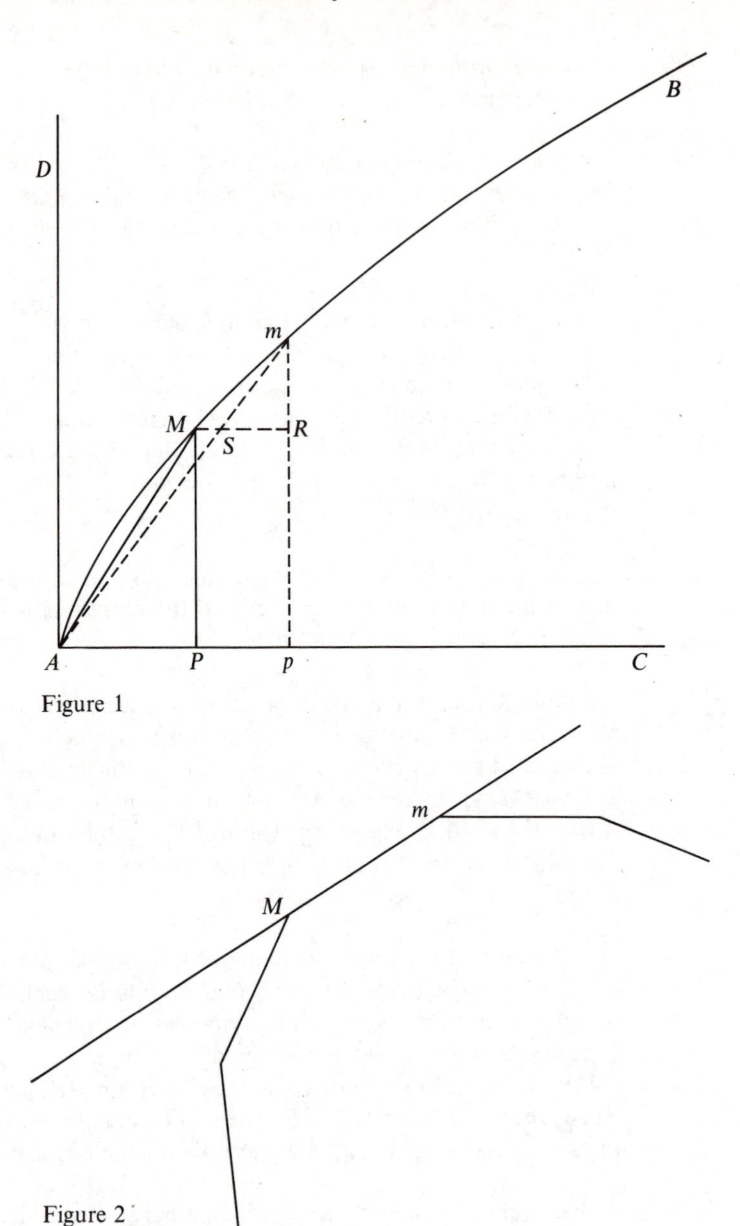

Figure 1

Figure 2

$z + dz$: and the constant quantity a will still be the same a. So that the given quantity $a + x + y - z$ will become $a + x + dx + y + dy - z - dz$; and the differential of it (which will be had in taking it from this last expression) will be $dx + dy - dz$; and so of others. From whence we have the following.

Rule I For finding the differentials of simple quantities connected together with the signs + and −.

Find the differential of each term of the quantity proposed; which connected together by the same respective signs will give another quantity, which will be the differential of that given.

5. *Proposition II To find the differentials of the product of several quantities multiplied, or drawn into each other.*

The differential of xy is $y\,dx + x\,dy$: for y becomes $y + dy$, when x becomes $x + dx$; and therefore xy then becomes $xy + y\,dx + x\,dy + dx\,dy$. Which is the product of $x + dx$ into $y + dy$, and the differential thereof will be $y\,dx + x\,dy + dx\,dy$, that is, $y\,dx + x\,dy$: because $dx\,dy$ is a quantity infinitely small, in respect of the other terms $y\,dx$ and $x\,dy$: For if, for example, you divide $y\,dx$ and $dx\,dy$ by dx, we shall have the quotients y and dx, the latter of which is infinitely less than the former.

Whence it follows, that the differential of the product of two quantities, is equal to the product of the differential of the first of those quantities into the second plus the product of the differential of the second into the first.

13.B7 Preface to the English edition of l'Hôpital's *Analyse des Infiniment Petits*

In this preface, Edmund Stone (the translator) first described his additions to l'Hôpital's work, and then briefly resumed the theory of fluxions and fluents 'according to the sense of the great Author and Inventor thereof, Sir Isaac Newton', in which, as he put it 'we are to consider Quantities not as made up of very small parts, but as described by a continued motion'. He then blithely continued as follows.

Upon this latter foundation is built the Calculus Differentialis, first published by Mr Leibnitz, in the Year 1684; having been since followed by almost all the Foreigners; who represent the first Increment, or Differential (as they call it) by the letter *d*, the second by *dd*, the third by *ddd*, etc.; the Fluents or Flowing Quantities, being called Integral. But since this method in the Practice thereof, does not differ from that of Fluxions, and an Increment or Differential may be taken for a Fluxion; out of regard to Sir Isaac Newton, who invented the same before the Year 1669, I have altered the Notation of our Author, and instead of d, dd, d^3 etc. put his Notation, viz. $\dot{x}$, $\ddot{x}$, $\dddot{x}$, etc. or some others of the final Letters of the Alphabet, pointed thus, and called the infinitely small Increment, or Differential of a Magnitude, the Fluxion of it.

14 Euler and His Contemporaries

14.A Euler on Analysis

The prodigious output of Leonhard Euler (1707–1783) contains many great achievements, among the most important of which must be the reformulation of the calculus around the ideas of function and variable, and his creation of a flexible and powerful algebraic language for handling functions. Two examples of this are illustrated. Following Dirk J. Struik, we present in 14.A2 Euler's reformulation of the theory of exponential and circular functions, adding to it (14.A3) Euler's later and deeper ideas about the logarithm. Euler here created a unified, functional and algebraic theory of the most important objects in mathematics. Our second example is Euler's mastery of the emerging theory of differential equations, which is how, post Bernoulli, inverse tangent problems came to be formulated. We look at his discovery of a simple uniform method for dealing with all linear, ordinary differential equations with constant coefficients, which he first presented to Johann Bernoulli in 1739 (14.A1). The example which Euler gave in 14.A1 (a) is not artificial; it is the equation for the transverse vibrations of a horizontal rod with one end fixed in a wall, which Daniel Bernoulli, Johann's son, had obtained and sent to Euler in 1735. At that time Euler could only solve it by the method of undetermined coefficients, which gives the answer in the unilluminating form of a power series (see J. T. Cannon, S. Dostrovsky *The Evolution of Dynamics*, Springer, 1981, Chapter 13). (It was in fact Daniel Bernoulli who was to complete Euler's method by dealing with the case when the auxilliary polynomial has repeated roots.)

14.A1 A general method for solving linear ordinary differential equations

(a) *Euler to Bernoulli*

I have recently found a remarkable way of integrating differential equations of higher degrees in one step, as soon as a finite [algebraic] equation has been obtained. Moreover this method extends to all equations which, on setting dx constant, are contained in this general form:

$$y + \frac{ady}{dx} + \frac{bddy}{dx^2} + \frac{cd^3y}{dx^3} + \frac{dd^4y}{dx^4} + \frac{ed^5y}{dx^5} + \text{etc.} = 0.$$

To find the integral of this equation I consider this equation or algebraic expression:

$$1 - ap + bp^2 - cp^3 + dp^4 - ep^5 + \text{etc.} = 0.$$

If possible this expression is resolved into simple real factors of the form $1 - \alpha p$: if, however, this cannot be done resolve it into factors of two dimensions of this form $1 - \alpha p + \beta pp$, which resolution can always be done in reals, for whatever form the equation may have it can always be put in the form of a product of factors either simple, $1 - \alpha p$, or of two dimensions $1 - \alpha p + \beta pp$, all real. This resolution being done, I say that the value of y is a finite expression in x and constants, obtained from all the members which have been factors of the algebraic expressions, and singular members supply singular terms of the integral. Certainly the simple factor $1 - \alpha p$ gives as member of the integral $Ce^{x/\alpha}$, and a composite factor $1 - \alpha p + \beta pp$ gives this member of the integral

$$e^{-\alpha x/2\beta}\left(C \sin A \, . \frac{x\sqrt{4\beta - \alpha\alpha}}{2\beta} + D \cos A \, . \frac{x\sqrt{4\beta - \alpha\alpha}}{2\beta}\right)$$

where for me sin A. and cos A. denote the sine and the cosine of arcs in a circle of radius $= 1$: however it is to be noticed that if the expression $1 - \alpha p + \beta pp$ cannot be resolved into simple real factors, when $4\beta > \alpha\alpha$, still the integrals are real. Let the following be taken as a suitable example

$$ydx^4 = K^4d^4y, \quad \text{or } y - \frac{K^4d^4y}{dx^4} = 0;$$

this gives rise to the algebraic expression $1 - K^4p^4$, whose real factors are these three $1 - Kp, 1 + Kp, 1 + K^2p^2$; and from these spring the integrals of the equation

$$y = Ce^{-x/K} + De^{x/K} + E \sin A \, . \frac{x}{K} + F \cos A \, . \frac{x}{K};$$

in which expression, because a four-fold integration has been done in one operation, there are four new constants as the nature of the integration demands. If it would please you, most excellent sir, I shall write down the method of proof on another occasion.

(*b*) *Bernoulli's reply to Euler*

Bernoulli observed that he seemed to remember finding something similar himself many years before. He then gave the following proof of Euler's theorem, in terms of the exponential function $y = n^{x/p}$, which we would write today as $y = e^{x/p}$.

What in fact belongs to your general formula

$$y + \frac{ady}{dx} + \frac{bddy}{dx^2} + \frac{cd^3y}{dx^3} + \frac{dd^4y}{dx^4} + \text{etc.} = 0$$

on setting dx constant, is that they can always be solved by logarithmic curves whose subtangent is to be found, as I shall now show. The general equation for such a curve is this: $y = n^{x/p}$, where p denotes the subtangent of the general logarithm and n denotes a number whose logarithm is unity, so $\ln = 1$. [...]. Therefore differentiating $n^{x/p}$ continually the values of dy, d^2y, d^3y, d^4y etc. are obtained as far as is necessary:

$$dy = \frac{dx}{p} \cdot n^{x/p}, \qquad ddy = \frac{dx^2}{pp} \cdot n^{x/p}, \qquad d^3y = \frac{dx^3}{p^3} \cdot n^{x/p}, \qquad d^4y = \frac{dx^4}{p^4} \cdot n^{x/p}, \quad \text{etc.}$$

Which values being substituted into your formula

$$y + \frac{ady}{dx} + \frac{bddy}{dx^2} + \text{etc.}$$

it becomes this:

$$n^{x/p}\left(1 + \frac{a}{p} + \frac{b}{p^2} + \frac{c}{p^3} + \frac{d}{p^4} + \text{etc.}\right) = 0.$$

Therefore I divide by $n^{x/p}$ and multiply by the highest dimension of this p to obtain an algebraic equation whose roots p give the subtangents of the logarithms we seek. The example of the differential equation of the fourth degree $ydx^4 = K^4d^4y$, or $y - \frac{K^4d^4y}{dx^4} = 0$ is most easily solved. With the letters, a, b, c missing and $d = -K^4$, we have this equation of dimension four, but without affects,

$$p^4 - K^4 = 0, \qquad \text{or} \qquad p = K.$$

So I say the logarithm whose subtangent is K, will satisfy the given equation $y - \frac{K^4d^4y}{dx^4} = 0$. While having in this way and for this example exhibited one such logarithm, you have indeed exhibited several curves

$$y = Ce^{-x/K} + De^{x/K} + E \sin A\frac{x}{K} + F \cos A\frac{x}{K}.$$

To do this, if $y + \frac{K^4d^4y}{dx^4} = 0$ is proposed, my logarithms will be impossible or imaginary; but it is also the same in your solution, allowed to be more general, for you must let K be impossible or non-real.

(*c*) *Euler's reply to Bernoulli*

Euler reminded Bernoulli that the solution must contain as many logarithms as the equation has real parameters. He then turned to the specific example of a fourth-order equation and the question of real and imaginary logarithms, as follows.

Being indeed led to this algebraic equation $p^4 + K^4 = 0$ which can be resolved into these two equations of two dimensions

$$p^2 + Kp\sqrt{2} + K^2 = 0 \quad \text{and} \quad p^2 - Kp\sqrt{2} + K^2 = 0,$$

whence I obtain the complete integral equation

$$y = Ce^{x/K\sqrt{2}} \sin A . \frac{x}{K\sqrt{2}} + De^{x/K\sqrt{2}} \cos A . \frac{x}{K\sqrt{2}} + Ee^{-x/K\sqrt{2}} \sin A . \frac{x}{K\sqrt{2}} + Fe^{-x/K\sqrt{2}} \cos A . \frac{x}{K\sqrt{2}},$$

this equation having four constants C, D, E and F it is obvious that this equation is the complete integral.

14.A2 Euler's unification of the theory of elementary functions

126. After logarithms and exponential quantities we shall investigate circular arcs and their sines and cosines, not only because they constitute another type of transcendental quantity, but also because they can be obtained from these very logarithms and exponentials when imaginary quantities are involved.

Let us therefore take the radius of the circle, or its sinus totus, = 1. Then it is obvious that the circumference of this circle cannot be exactly expressed in rational numbers; but it has been found that the semicircumference is by approximation = 3.14159.26535.89793... [127 decimal places are given] for which number I would write for short π, so that π is the semicircumference of the circle of which the radius = 1, or π is the length of the arc of 180 degrees.

127. If we denote by z an arbitrary arc of this circle, of which I always assume the radius = 1, then we usually consider of this arc mainly the sine and cosine. I shall denote the sine of the arc z in the future in this way $\sin A . z$, or only $\sin z$; and the cosine accordingly $\cos A . z$, or only $\cos z$. Hence we shall have, since π is the arc of 180°, $\sin 0 = 0$, $\cos 0 = 1$ and $\sin \frac{1}{2}\pi = 1$, $\cos \frac{1}{2}\pi = 0$.

[After a whole set of trigonometric formulas and identities, Euler continues as follows.]

132. Since $(\sin z)^2 + (\cos z)^2 = 1$, we shall have by factorization $(\cos z + i \sin z) \times (\cos z - i \sin z) = 1$, which factors, although imaginary, still are of great use in combining and multiplying sines and cosines.

[Now comes De Moivre's theorem (though the name is not mentioned), from which follows, in §133:]

$$\cos nz = \frac{(\cos z + i \sin z)^n + (\cos z - i \sin z)^n}{2}$$

and

$$\sin nz = \frac{(\cos z + i \sin z)^n - (\cos z - i \sin z)^n}{2i}.$$

When we develop these binomials in a series we shall get

$$\cos nz = (\cos z)^n - \frac{n(n-1)}{1.2}(\cos z)^{n-2}(\sin z)^2 + \text{etc.}$$

and

$$\sin nz = \frac{n}{1}(\cos z)^{n-1} \sin z - \frac{n(n-1)(n-2)}{1.2.3}(\cos z)^{n-3}(\sin z)^3 + \text{etc.}$$

134. Let the arc z be infinitely small; then we get $\sin z = z$ and $\cos z = 1$; let now n be an infinitely large number, while the arc nz is of finite magnitude.

Take $nz = v$; then since $\sin z = z = v/n$ we shall have

$$\cos v = 1 - \frac{v^2}{1.2} + \frac{v^4}{1.2.3.4} - \cdots + \text{etc.}$$

and

$$\sin v = v - \frac{v^3}{1.2.3} + \frac{v^5}{1.2.3.4.5} - \cdots + \text{etc.}$$

[...]

138. Let us now take in the formulas of §133 the arc z infinitely small and let n be an infinitely small number ε such that εz will take the finite value v. We thus have $\varepsilon z = v$ and $z = v/\varepsilon$, hence $\sin z = v/\varepsilon$ and $\cos z = 1$. After substituting these values we find

$$\cos v = \frac{\left(1 + \frac{vi}{\varepsilon}\right)^\varepsilon + \left(1 - \frac{vi}{\varepsilon}\right)^\varepsilon}{2},$$

$$\sin v = \frac{\left(1 + \frac{vi}{\varepsilon}\right)^\varepsilon - \left(1 - \frac{vi}{\varepsilon}\right)^\varepsilon}{2i}.$$

In the previous chapter we have seen that

$$\left(1 + \frac{z}{\varepsilon}\right)^\varepsilon = e^z,$$

where by e we denote the base of the hyperbolic logarithms; if we therefore write for z first iv, then $-iv$, we shall have

$$\cos v = \frac{e^{iv} - e^{-iv}}{2}$$

and

$$\sin v = \frac{e^{iv} - e^{-iv}}{2i}.$$

From these formulas we can see how the imaginary exponential quantities can be reduced to the sine and cosine of real arcs. Indeed, we have

$$e^{iv} = \cos v + i \sin v,$$

$$e^{-iv} = \cos v - i \sin v.$$

14.A3 Logarithms

Euler began by reviewing the correspondence between Leibniz and Johann Bernoulli on the logarithm of a negative number, recently published by Gabriel Cramer, and by observing a number of problems these two had encountered. He then proceeded as follows.

I therefore say that in order to make all these difficulties and contradictions disappear, that in virtue indeed of the given definition, there corresponds to each number an infinity of logarithms; I shall prove this in the following theorem.

Theorem There is always an infinity of logarithms which belong equally to each given number; or, if y denotes the logarithm of the number x, I say that y contains an infinity of different values.

Proof I restrict myself here to hyperbolic logarithms, because one knows that the logarithms of all other kinds have a constant ratio to them, so, when the hyperbolic logarithm of the number x is set $= y$, the tabular logarithm of this same number will $= 0{\cdot}4342944819\ldots y$.

Now, the basis of hyperbolic logarithms is that, if w signifies an infinitely small number, the logarithm of the number $1 + w$ will $= w$, or $l(1 + w) = w$. From this it follows that $l(1 + w)^2 = 2w$, $l(1 + w)^3 = 3w$, and in general $l(1 + w)^n = nw$. But, because w is an infinitely small number, it is evident that the number $(1 + w)^n$ cannot become equal to an arbitrary given number x, at least if the exponent n is not an infinite number. So let n be an infinitely large number and set $x = (1 + w)^n$ and the logarithm of x, which is set $= y$, will be $y = nw$. Therefore, to express y in terms of x, the first formula giving $1 + w = x^{1/n}$ and $w = x^{1/n} - 1$, this value being substituted for w in the other formula will produce $y = nx^{1/n} - n = lx$.

From which it is clear that the value of the formula $nx^{1/n} - n$ will approach the logarithm of x closer and closer as the number n becomes greater, and that if one puts an infinite number for n, this formula will give the true value of the logarithm of x. Now, just as it is certain that $x^{1/2}$ has two different values, $x^{1/3}$ three, $x^{1/4}$ four and so on, equally it is certain that $x^{1/n}$ must have an infinity of different values because n is an infinite number. Consequently this infinity of different values of $x^{1/n}$ will also produce an infinity of different values of lx, so that the number x must have an infinity of logarithms. Q.E.D.

From this it follows that the logarithm of $+1$ is not only $=0$ but there is also an infinity of other quantities each equally the logarithm of $+1$. However, one readily understands that all these other logarithms, apart from the first 0, are imaginary quantities so that in calculation one is right to regard only 0 as the logarithm of $+1$, exactly as when the cube root of $+1$ is concerned one only uses $+1$ although these imaginary quantities $\frac{-1+\sqrt{-3}}{2}$ and $\frac{-1-\sqrt{-3}}{2}$ are equally cube roots of 1. But when one wants to compare the logarithm of 1 with the logarithms of -1, or of $\sqrt{-1}$, which are all, as I will show in the sequel, imaginary, it is necessary to consider the logarithm of 1 in its entirety, and then all the difficulties and contradictions reported on above will disappear of their own accord. For, let $\alpha, \beta, \gamma, \delta, \varepsilon, \xi$, etc. be the imaginary logarithms of unity, which answer to it as well as 0, and one readily understands that it can be that $2l(-1) = l(+1)$, although all the logarithms of -1 are imaginary; for, to satisfy the equation $2l(-1) = l(+1)$ it suffices that the double of all the logarithms of -1 are found among the imaginary logarithms of $+1$. Likewise, because $4l(\sqrt{-1}) = l(+1)$, each logarithm of $\sqrt{-1}$ multiplied by 4 must occur in the series α, β, γ, δ, ε, ξ, etc. So the equalities $2l(-1) = (+1)$ and $4l(\sqrt{-1}) = l(+1)$ can be maintained, without being obliged to suppose that either $l(-1) = 0$ or $l(\sqrt{-1}) = 0$, as M. Bernoulli had claimed. But all this will be as clear as daylight when I shall have actually determined all the logarithms of any given number, which will be the object of the following problems.

Problem 1 To find all the logarithms corresponding to an arbitrary given positive number $+a$.

14.A4 The algebraic theory of conics

The properties, which we drew out in the preceding chapter, occur in like manner in all the lines which belong to the second order; and we have not mentioned any difference by which such lines are distinguished from one another. Yet although all lines of the second order generally enjoy these qualities which I have expounded, they nevertheless differ a great deal from one another in form. For this reason it is fitting for the lines contained in this order to be divided into types, so that the different forms, which appear in this order may more easily be distinguished and the properties which coincide to such an extent in individual types, may be disclosed.

Nevertheless, we produced a general equation for lines of the second order, by changing the axis and the origin of the abscissas to such a degree, that all the lines of the second order may be contained in this equation.

$$yy = \alpha + \beta x + \gamma xx,$$

in which x and y denote orthogonal co-ordinates. If therefore for any abscissa x, the ordinate y assumes two values, one positive one negative, that axis, in which the abscissas x are contained, will divide the curve into two similar and equal parts; moreover that axis will be the diameter of the orthogonal curve, and every line of the second order will have an orthogonal diameter, upon which I add the abscissas, just like an axis.

Therefore three constant quantities enter this equation, α, β, and γ, which may vary among themselves in infinite ways, and innumerable variations in curved lines arise, which will however differ from one another in shape, either greatly or slightly. For in the first place the same shape results an infinite number of times from the proposed equation $yy = \alpha + \beta x + \gamma xx$, the origin of the abscissas varying on the axis.

This happens when the abscissa x either increases or decreases by a given amount. Then each figure is also included in the equation under a different size, so that infinite curved lines appear, which differ from one another in quantity to such an extent, that circles of different radii are sketched. From this it is clear that not every variation of the letters, α, β, and γ produces different types and kinds of lines of the second order.

However the greatest difference in the curved lines which are included in the equation $yy = \alpha + \beta x + \gamma xx$ is produced by the character of the coefficient γ, depending on whether it has a positive or a negative value. For if γ has a positive value, assume that the abscissa x is infinite in which case the term γxx turns out to be infinitely greater than the remainder $\alpha + \beta x$ and for that reason the expression $\alpha + \beta x + \gamma xx$ acquires a positive value. The ordinate y will likewise acquire two infinitely large values, one positive, one negative, because the same thing happens, if $x = -\infty$. In this case, however, the expression $\alpha + \beta x + \gamma xx$ assumes an infinitely great positive value. On account of this, if y becomes a positive quantity, the curve will have four branches stretching out to infinity, two corresponding to the abscissa $x = +\infty$ and two corresponding to the abscissa $x = -\infty$. Therefore these curves which have four branches stretching out to infinity, are thought to constitute one type of lines of the second order, and are called 'hyperbolas'.

If, however, the coefficient γ has a negative value, then if $x = +\infty$ or $x = -\infty$, the expression $\alpha + \beta x + \gamma xx$ will have a negative value and therefore the ordinate y becomes imaginary. Therefore nowhere in these curves will an abscissa or an ordinate be able to be infinite. For that reason no part of the curve will be able to extend to infinity, but the whole curve will be contained in a finite and limited space. So this type of lines of the second order acquires the name of 'ellipses', on account of the fact that their character is contained in this equation $yy = \alpha + \beta x + \gamma xx$, if γ is a negative quantity.

Therefore if the value of γ produces such a different character of lines of the second order, depending on whether it is positive or negative, that on this account two different sorts are rightly created: if $\gamma = 0$, a value which is midway between affirmative and negative numbers, the curve resulting from this also constitutes a certain type midway between hyperbolas and ellipses, called the 'parabola', which therefore expresses its nature by the equation $yy = \alpha + \beta x$.

14.A5 The theory of elimination

In the preceding piece I recorded without proof this theorem, that two algebraic curves, of which one is of order m and the other of order n, can cut in mn points. The truth of this proposition is known to all mathematicians, but one must admit that one nowhere finds a sufficiently rigorous demonstration of it. There are general truths which our mind is ready to admit as soon as it sees the justice in some particular cases; it is amongst this kind of truth that one can rightly place the proposition which I have just mentioned, because one not only finds it true in some or several cases, but also in an infinity of different cases. However, one readily admits that all these infinite proofs are incapable of sheltering the proposition from all the objections an adversary could raise, and a rigorous demonstration is absolutely necessary to reduce him to silence.

Before undertaking a proof of this proposition it is necessary to fix its meaning precisely. First of all it is to be remarked that the number of intersections of two curves, one of order m the other of order n, is not necessarily $= mn$, but very often it can be less. Thus it can happen that two straight lines do not cut at all, when they are parallel, and a straight line only cuts a parabola in one point, and that two conic sections only cut each other in two points or not at all. So the meaning of our proposition is that the number of intersections can never be greater than mn, while it can often be less; and then one says either that several intersections have gone off to infinity or that they have become imaginary. So if one takes account of the intersections at infinity and the imaginary as well as the real ones, one can then say that the number of intersections is always $= mn$.

14.B Euler and Others on the Motion of the Moon

Before Continental mathematicians could come to terms with Newton's *Principia* they had, of course, to unlearn what Descartes had said about celestial mechanics. This seems to have been a generational process, chiefly brought about in France by Alexis-Claude Clairaut (1713–1765), Pierre de Maupertuis (1698–1759) and Jean d'Alembert (1717–1783). The crucial events were to be the discovery of the shape of the Earth and the elucidation of the motion of the moon. For the first, Maupertuis makes clear in 14.B1 just what a victory this was for Newtonianism. But in fact this was to be only the start of an immense study of hydrodynamics, by Clairaut, Colin MacLaurin and Euler among others; for, as Clifford Truesdell argued in the first volume of his *Archive*, Newton's theory here rested on some pretty unsatisfactory hypotheses.

For the second, it was Newton himself who pointed out exactly how his theory of gravity failed for the moon. For a while in the 1740s, as the Euler–Clairaut correspondence (14.B2) makes clear, all those most competent to judge were willing to entertain the idea that Newtonian gravitation might after all have to be abandoned, and it is interesting to see what old ideas they were willing to put in its place. D'Alembert, not represented here, advocated a second force; Clairaut wanted to introduce a weak inverse fourth-power term; and Euler even wanted to bring back vortices. The letters give a vivid impression of just how much was invested in these investigations, which Clairaut was the first to carry through successfully (14.B3). Later Euler was to work out a theory of the moon which in turn enabled the astronomer Tobias Mayer to produce tables of the moon which satisfied the needs of navigators, thus consummating the hopes raised by Maupertuis.

The letters also capture something of the flavour of academic life in the period; most mathematicians were employed in learned academies and research was rewarded by the competitions the academies ran regularly on questions of note, such as the motion of the moon or the motion of Saturn. These competitions focused the small communities' attention on particular topics, and the best entries were then published in due course, which might well mean three to five years after they were written. Entries were supposed to be anonymous and identified only by a motto, with what success you can judge from 14.B2(a).

14.B1 Pierre de Maupertuis on the Figure of the Earth

In the 1730s the French mounted two expeditions, one to Lapland and one to Peru, to determine the shape of the Earth. Their method was to time the oscillations of a standard pendulum, which, according to Newton's theory, would beat faster the nearer it was to the Earth's centre. Maupertuis and Clairaut went on the expedition to Lapland and, despite the many hardships they had to endure, they were able to report first. Our next extract is taken from Maupertuis's own account, which was almost immediately translated into English because of the importance of his findings.

It is well known of what different Opinions the Learned have been these 50 Years past, with relation to the Figure of the Earth; some hold it to be that of a Spheroid flattened towards the Poles; others that it is a Spheroid prominent in that Direction. This Question, for its Curiousity only, might well merit the consideration of Philosophers and Mathematicians: But the Advantages arising from the Discovery of the Earth's true Figure, go beyond mere speculation; they are real, and of very great importance.

[Maupertuis points out that were the Earth to be a perfect sphere it would be enough to determine any one degree of a meridian for distances to become known.] But why should the Earth be a sphere? In an age when nothing less than the utmost Precision in all Science is insisted on, it was not to be supposed that the Proofs the Ancients had given of its spherical Form could pass. Even the Reasonings of the most celebrated Mathematicians, who gave it the Figure of a flat Spheroid, were not thought entirely satisfying; because they seemed still be be connected with some Hypotheses, although these Hypotheses were such as one cannot help admitting. [...]

It is evident in general that Sir Isaac Newton's Figure of a flat Spheroid, and Mr Cassini's of a long one, will give very different Distances of Places that have the same Longitude and Latitude [...]. In a course of 100 Degrees Longitude, there might be a Mistake of more than two Degrees, if sailing really upon Sir Isaac Newton's Earth one should imagine himself to be upon Mr Cassini's. And how many Ships have perished by smaller Mistakes? [...]

This Determination would likewise be exceedingly useful in that important problem, *To Find the Parallax of the Moon*; which would greatly contribute to the compleating of a Theory of this Satellite of our Earth; upon which the best Astronomers have always most reckoned for the discovery of the Longitudes at Sea. [...]

[At the end of his book Maupertius gave these conclusions.] In fine, All the Experiments which the Academicians, sent by the KING to Peru, have made, either at S. Domingo or the Equator, conspire with ours, to make the Increase of Gravitation towards the Pole, greater than according to Sir Isaac Newton's Table, and by consequence the Earth flatter than he has made it. All of them fall so wide of Mr Huygens's Theory (*Discours de la Cause de la Pesanteur*) which makes it still less, that his Theory must itself be wide of the Truth.

14.B2 Correspondence between Euler and Alexis-Claude Clairaut

(*a*) *Clairaut to Euler, 11 September 1747*

I have just read with great avidity your piece on Saturn, which I admit frankly I recognized as yours at a glance. It is true there is no great merit in having deciphered it, because you have written your work in your own hand, and this is not what you have done most regularly between us. Anyway I don't reproach you for having done this because I am deeply convinced that I would have recognized it anyway, and besides I don't think myself susceptible of being influenced by any prejudices.

To return therefore to your piece, although I have not yet finished it I can't stop myself from speaking to you about it and asking you about the sample I have sent you.

I was delighted to see that, like me, you have thought about Newtonian attraction. It seems to me to have been shown to be insufficient to explain the phenomena, but the distinctive character which you give it for the moon does not appear to me to be as striking as that which I have noticed. Instead of seeing what the distance must be, I have examined what it entails for the motion of the apogee. And finding it to be scarcely half what it is in nature seemed to me to furnish the most complete proof of the insufficiency of the law of squares. It is true that on adding some other term one feels that the theory will better accord with the phenomena. But it seems to me that this term must be such that at the distances of Mercury, Venus, the Earth and Mars it must be almost insensible, in view of the extreme smallness of the motion of the apsides. And if, as it seems initially from your work, the law of squares is palpably in error at the distance of Saturn and Jupiter it would still be necessary to add terms which were significant only at that distance. I confess that the whole of gravitation seems to me to be only a speculative hypothesis.

And I couldn't restrain myself enough on your conjecture on what could influence the shape of Jupiter at distances which are so enormous with respect to its diameter. As I have not yet finished my calculations on Saturn, I have no feelings on the point; I think pending something better that the general agreement of the system requires that the law be for all nature as $\frac{1}{\text{dist}^2}$ + a small function of distances detectable for small distances like the moon and almost zero for great distances [...].

It seems to me, and I am not a candidate for the prize, much more important to know if Newtonian attraction holds or not than to treat simply of Saturn. And in seeing if the square law of attraction must suffer some correction which can only be for small distances it seems to me to be necessary to begin by finishing the theory of the moon. I have, however, had enough patience to finish the quadrature of two curves which have given me the variation of Saturn quite exactly. But it would be necessary to have a greater number of such for the other equations of its motion; and I shall not be able to have them finished in the prescribed time. I will speak to you longer at another time on my work and on yours, which pleases me so much, and meanwhile I ask you to tell me if you are thinking like me about the movement of the apogee of the moon and on the true law of gravitation.

(b) Euler to Clairaut, 30 September 1747

Mr Bradley has communicated a great number of observations to me, from which I have already determined these coefficients so that calculation never differs by more than 5′ from observation: and as these errors cannot be attributed to the observations I don't doubt that a certain alteration in the forces, which one supposes in the theory, must be the cause of them. This circumstance suggests to me that vortices, or some other material cause of these forces is very probable because it is then easy to conceive that these forces must be altered when they are transmitted by some other vortex. Thus I suspect that the force of the sun on the moon is considerably altered at opposition, because it then passes through the vortex of the earth. Likewise I think that the force of the sun on the outer planets is altered because of its passage through the atmospheres or vortices of the lower ones; and for the same reason the force of Jupiter on Saturn at opposition must be considerably altered. From this one will easily conceive that the outer planets, even without regard for their mutual interaction, are subject to much greater alterations than the lower ones; and this explanation seems to be more probable than the one you suspect, that the forces are as $\frac{1}{\text{dist}^2}$ plus a small function of distance quite palpable at small distances; for although the moon seems to confirm it, the regular movement of Mercury, however, appears to me to reverse the explanation.

[Euler then concluded with some further remarks about Saturn.]

(c) Clairaut to Euler, 7 December 1747

I saw with much pleasure in your last that you, like me, think about the motion of the apogee of the moon. [... But] I don't yet see the necessity of employing vortices to repair the law of squared distances. It seems to me that you are close enough to the observations with your addition to your piece on Saturn to believe that if you had better observations theory would be completely in agreement with astronomy at this point. As to the other planets, Mercury, Venus, etc., it is certain they fit very well with my law of attraction. If the term to add to $1/dd$ is, for example, $ll/357dd$ (l being the distance of the moon from the earth and d an arbitrary distance) then at the distance of Mercury from the sun one finds a motion of the aphelion so small that it is not for certain a lapse in the observations; and at the distance of the other planets it is even less palpable. Moreover I don't think that the term $l^2/357d^4$ can be true. It must be a function and not a power. For the $1/d^4$ law of attraction gives far too strong an attraction to bodies on or near to the earth. [...]

And if you have invoked the help of vortices to repair Newton's law, you therefore think that their effect is nothing other than to produce a movement of the apsides which seems bizarre to me.

Concerning vortices or material causes, recently your Opuscules fell into my hands where I saw several very curious pieces, among which some where you treated the retardation of orbits produced by the resistance of the medium where the planets are. You show for example that they don't give a motion of the apsides at all, which could even diminish the hope that the motion of the apsides which it is necessary to add to that which the moon must have in virtue of the sole law of squares must be attributed to vortices. I have also seen in the same work the tables which you give for the moon, which I don't understand at all. [...]

If you have sent me an example of the Opuscules I have just mentioned, and if you could send it to arrive at Geneva, I would not be at all embarrassed to have it here, just as your other works which might already have appeared since your comets and isoperimeters. People are speaking here of a treatise of yours on the analysis of the infinitely small, which can't fail to arouse my curiosity considerably. It is perhaps rash of me to ask you for your papers, especially after I have given you mine, for it might seem that I believe I am entitled to them and I don't think that at all. I only want to show you how flattered I would be to have them.

14.B3 Clairaut on the system of the world according to the principles of universal gravitation

(*a*) *November 1747*

The famous book *The Mathematical Principles of Natural Philosophy* has been the occasion of a great revolution in Physics. The method which Mr Newton, its illustrious author, has followed to derive facts from their causes, has shed the light of mathematics on a science which up till then had been in the shadows of conjectures and hypotheses.

But if it is just to recognize all that is due to that great man, one knows that one can't help but admit that the way in which he expounded his discoveries has considerably delayed the use to which they could be put; I don't speak at all here of the way in which he hid his method of fluxions, the key to all his scholarly research, because after that method had been ripped out of him it has become so familiar that one has forgotten all the wrong he did by not communicating it. But is it not right to reproach him for another wrong which without doubt has struck all those who have studied his book with a true desire to understand it? Namely, that in most of the difficult places he employed too few words to explain his principles [...].

[Clairaut went on to berate those who, having only a partial understanding of the *Principia* either embrace it or reject it on facile grounds, before pointing out that Newton's theory establishes the truth of Kepler's laws for the sun, the principal planets and their satellites. He then turned to the problem raised by the motion of the moon.]

After having treated this calculation with all the exactness it demands, I was quite astonished to find that it made the motion of the apogee at least twice as slow as that which it has from observation, that is to say the period of the apogee which follows from the inverse square law of attraction would be about 18 years, instead of being a little less than 9 as it is really.

A result so contrary to the principles of Mr Newton inclined me initially to abandon attraction entirely, but then noticing the quantity of phenomena which agreed with it, the observation of Kepler's laws of which I spoke above, the movement of the nodes of the moon which I calculated separately and found to conform with what Astronomy teaches, the tides, whose theory has been verified by the most capable mathematicians, and finally several other questions equally favourable to attraction, it appeared to me as difficult to reject it as to accept it. A supposition which only led to vague results could agree with Nature in some phenomena without being more solidly established, but when it gives numbers to these phenomena which agree with those provided by observation, the probability acquires a great degree of force: there must therefore be some way of reconciling the arguments which seemed both opposed and favourable to attraction.

What seemed to me to be most simple and appropriate to serve as a solution, was that attraction holds in Nature but it follows a different law to that established by Mr Newton. This idea which springs initially to mind at the same time overcomes a difficulty which seems to destroy it; without doubt the moon obeys another law of attraction than the square of the distances, but on the contrary do not the principal planets demand this law in consequence of the observation of Kepler's laws? However, it is easy to reply to this difficulty by noticing that there are an infinity of laws of attraction which differ quite markedly from the law of squares for small distances and which disagree very little for great ones; one can imagine these laws easily with the familiar analysis. One can take for example an analytic quantity which expresses the relation of attraction to distance as composed of two terms, one having the square of the distance as divisor, the other the fourth power [...].

[In fact, at the end of the article, Clairaut considers a law of the form $f(r) = \frac{\alpha}{r^2} + \frac{\beta}{r^3}$.]

(b) 17 May 1749

My present point is solely to bring to the notice of geometers who are interested in this question that having considered it anew from a viewpoint which no one had previously thought of I have been led to reconcile observations on the motion of the apogee of the moon with the theory of attraction without supposing any other attractive force than one proportional to the inverse square of the distance.

14.B4 Euler to Clairaut, 2 June 1750

You have given me infinite pleasure in sending me the pages of your memoirs for the year 1745, which contain the basis of your theory of the moon, for that volume has not yet reached us. I have examined your memoir with the greatest enthusiasm, and have been considerably enlightened by it, having learned from your second lemma how to eliminate the element of time from two differentio-differential equations which the theory of the moon leads to; this affords the most considerable advantages in the analysis of these equations. I have been particularly delighted by this discovery because I have sought for a long time in vain to reach such a point, and finally I gave up this line of enquiry, thinking that the elimination of the time would not be useful enough, seeing that this same time is still contained in the terms which contain the angle between the sun and the moon.

[Euler then gave a brief exposition of the difficulties he was still encountering, before concluding thus.]

From this I still find the same annual movement of the apogee as with my previous methods, so that I must admit I'm not yet sufficiently clear about your important discovery; at least I'm quite certain that if there is only this equation to solve, $ddx + xd\phi^2 = Axd\phi^2 + Bd\phi^2 \cos 2\eta$, the movement of the apogee which will result can only be about half the true value, and I don't understand how the other terms can influence the motion of the apogee. For this reason I still await the publication of your entire method with the greatest impatience.

14.C Euler's Later Work

In 1747 the learned world was startled by Jean d'Alembert's successful analysis of the vibrations of an elastic string via the wave equation, which he introduced and which may be called, with pardonable exaggeration, the first partial differential equation to appear in mathematics. A simple idea: but it spawned functions of unimaginable generality, and Euler appreciated this better than did d'Alembert, because Euler was always looking for the 'right' general formulation of any problem (see 14.C3). It was, for example, in this spirit that he was the first to reformulate Newton's laws of motion as being generally true for a dynamical system and to express them in the now standard form of second-order differential equations (see 14.C1).

Sadly, our selection cannot do justice to Euler as a number theorist. We give just one example, from a delightful letter to Christian Goldbach, of Euler as an 'algebraic' number theorist and display nothing of his mastery of power series, the zeta function, and much else (for which see *Leonhard Euler: Beiträge zu Leben und Werk,* Birkhäuser, 1983). But a vivid impression of Euler in this connection is given by André Weil in 14.C2(b).

'Read Euler' said Pierre Simon Laplace, 'he is our master in everything'. In 14.C4, as a measure of that esteem, is an extract from the *Elogium of Euler* by the distinguished Secretary of the Parisian Académie des Sciences, Nicolas Condorcet. His account of Euler, here translated by Henry Hunter in 1795, is a fine example. What no extract can make plain is that Condorcet preferred Euler's calculus and applied mathematics (including navigation and, regrettably, gunnery) to his number theory. Indeed Condorcet goes as far as to speak of number theory as being attractive to some only because it is difficult—thus illustrating in his own case the inevitable way in which historians' tastes influence what they write.

14.C1 A general principle of mechanics

(a) Euler

Let there be an infinitely small body, or one whose entire mass is concentrated at a single point, this mass being $= M$; suppose that this body has received an arbitrary motion and that it is affected by arbitrary forces. To determine the motion of this body one only has to know the distance of this body from an arbitrary fixed and immobile plane. At the present instant let the distance of the body from this plane $= x$; one decomposes all the forces which act on the body along directions which are either parallel to the plane or perpendicular to it, and lets P be the force which results from this decomposition along the direction perpendicular to the plane and which attempts in consequence either to bring the body towards the plane or to move it away. Let the element of time be dt, let $x + dx$ be the distance of the body from the plane taking this element dt as constant, then will

$$2Mddx = \pm Pdt^2,$$

according as the force P tends either to move the body away or bring it towards the plane. It is this formula alone which contains all the principles of mechanics.

The better to understand the force of this formula, it is necessary to explain in which units the various quantities M, P, x, and t are written. First it is to be observed that M, denoting the mass of the body, expresses at the same time the weight that this body would have near the surface of the Earth, in such a way that the force P being also reduced to that of a weight, the letters M and P would contain homogeneous quantities. Thus the speed of the body as it moves away from the plane would be $\frac{dx}{dt}$; if we suppose that this speed would be equal to that which a heavy body would acquire in falling from a height V then it is necessary to take $\frac{dx^2}{dt^2} = V$ where the element of time $dt = \frac{dx}{\sqrt{V}}$; whence one knows the relationship between the time t and the space V.

As this formula only expresses the separation or approach of the body to a fixed arbitrary plane, to find the true position of the body at each instant, one will only have to refer it at the same time to three fixed planes perpendicular to each other. Therefore as x denotes the distance from one of the planes, let y and z be its distances from the two other planes, and having decomposed all the forces which act on the body along the directions perpendicular to these three planes let P be the force perpendicular to the first, Q to the second, R to the third. Let us suppose that all these forces tend to move the body away from the three planes, for if they tend to bring it in one will only have to deal with negative forces. This done, the motion of the body will be contained in the three following formulae:

$$\text{(I) } 2Mddx = Pdt^2, \quad \text{(II) } 2Mddy = Qdt^2, \quad (III)\ 2Mddz = Rdt^2.$$

If the body is not acted on by any force, so that $P = 0, Q = 0, R = 0$, the three formulae obtained, because dt is constant, reduce to the integration of these:

$$Mdx = Adt, \quad Mdy = Bdt, \quad \text{and} \quad Mdz = Cdt,$$

from which one sees at once that in this case the body will move along a straight line, and consequently these formulae contain in themselves the first law of motion, in virtue of which a body being at rest will stay there or, if moving, will continue uniformly in the same direction, at least if it is not acted on by any force from outside. But it is clear that our formulae are not at all restricted to this great law but contain besides the laws according to which arbitrary forces act on bodies. Consequently the principle which I have established contains on its own all the principles which can lead to the knowledge of the motion of all bodies of whatever kind.

(b) *Clifford Truesdell on Euler*

We may justly wonder that it took more than sixty years for so simple an extension of Newton's ideas, but the literature of mechanics does not permit us to doubt that it did. As often happens in the history of science, the simple ideas are the hardest to achieve; simplicity does not come of itself but must be created. Euler's researches had moved slowly closer to the general principle of linear momentum, and d'Alembert's work on the string, following upon John Bernoulli's formulation of hydraulics, must have made it finally obvious to him. Euler's *Discovery of a New Principle of Mechanics* sets down as the axiom which 'includes all the laws of mechanics' the momentum principle in the now familiar form

$$(1) \qquad M\frac{d^2x}{dt^2} = F_x, \quad M\frac{d^2y}{dt^2} = F_y, \quad M\frac{d^2z}{dt^2} = F_z$$

where F is the total static force acting upon the body of mass M, and where it is stated explicitly that in a continuous body M and F are to be replaced by differential elements dM and dF.

The expression of the laws of motion in *rectangular Cartesian co-ordinates* is also of the greatest importance. Today this possibility is so obvious that many scientists seem to believe that Newton himself used Cartesian co-ordinates, but of course this is not so, and Lagrange, in 1788 still fairly close to the discovery, after describing the intrinsic resolution always used by Newton wrote 'Nevertheless, it is much simpler to refer the motion of the body to directions fixed in space'. After stating (1) in words, he added, 'this manner of determining the motion of a body impelled by arbitrary accelerating forces is by virtue of its simplicity preferable to all others. It seems that Maclaurin was the first to employ it, in his *Treatise of Fluxions*, printed in 1742; now it is universally adopted.' The attribution to Maclaurin is false, however; the method was first used by John Bernoulli and was developed in Euler's papers on special systems having many degrees of freedom. The importance of the use of Cartesian co-ordinates lies deeper than in mere simplicity; in these co-ordinates the addition of vectors located at different points is so natural as to become customary at once, and the possibility of performing this addition lies at the heart of the classical conception of space-time.

14.C2 Fermat's last theorem and the theory of numbers

(*a*) *Euler to Christian Goldbach*

There's another very lovely theorem in Fermat whose proof he says he has found. Namely, on being prompted by the problem in Diophantus, find two squares whose sum is a square, he says that it is impossible to find two cubes whose sum is a cube, and two fourth powers whose sum is a fourth power, and more generally that this formula $a^n + b^n = c^n$ is impossible when $n > 2$. Now I have found valid proofs that $a^3 + b^3 \neq c^3$ and $a^4 + b^4 \neq c^4$, where $\neq$ denotes cannot equal. But the proofs in the two cases are so different from one another that I do not see any possibility of deriving a general proof from them that $a^n + b^n \neq c^n$ if $n > 2$. Yet one sees quite clearly as if through a trellice that the larger n is, the more impossible the formula must be. Meanwhile I haven't yet been able to prove that the sum of two fifth powers cannot be a fifth power. To all appearances the proof just depends on a brainwave, and until one has it all one's thinking might as well be in vain. But since the equation $aa + bb = cc$ is possible, and so also is this possible, $a^3 + b^3 + c^3 = d^3$, it seems to follow that this, $a^4 + b^4 + c^4 + d^4 = e^4$, is possible, but up till now I have been able to find no case of it. But there can be five specified fourth powers whose sum is a fourth power.

(*b*) *André Weil on Euler*

[1] One must realize that Euler had absolutely nothing to start from except Fermat's mysterious-looking statements. Frequently Fermat states flatly 'I have proved this', 'I have proved that', but then he seems to say the same about 'Fermat's equation'

$$x^n = y^n + z^n$$

(more about this later). There were among Fermat's statements, along with the impossibility of that equation, also the fact that every prime of the form $p = 4n + 1$ can be written as $x^2 + y^2$, and similar statements about conditions for a prime to be of the form $x^2 + 3y^2, x^2 + 2y^2$, and so on, and a statement about every integer being a sum of four squares. Euler was fascinated by such statements; but he first had to reconstruct for himself all the most basic theorems in number-theory. For instance, there was what is now known as the 'little theorem' of Fermat: if p is a prime, then (in modern notation) $x^{p-1} \equiv 1$ modulo p for every integer x, not a multiple of p. For a man who took up Fermat at that time, one statement might well seem just as mysterious as the other, in spite of the ease with which one can verify many of them empirically up to large values. Euler had to reconstruct everything from scratch, all the things that are now to be found in all elementary textbooks, and which now look so simple on the basis of two concepts, the group concept and the concept of a prime ideal. It took him some time. To begin with, he didn't know that the integers prime to any modulus n make up a group modulo n; of course he didn't have the concept, but also, at first, the existence of an inverse was not immediately obvious. Also there are the facts involved in the statement, which to us looks so elementary, that given a field (e.g. the prime field of integers modulo a prime) any equation in one unknown has at most as many roots in that field as its degree indicates. This was not proved by Euler and Lagrange until about 1760, about thirty years after Euler had started working on number-theory and when he was working on far more difficult questions. He had no way of knowing which questions were simple and which ones were not so. For instance, the fact that all primes $p = 4n + 1$ are of the form $x^2 + y^2$ looked neither more nor less difficult to him than the assertion that an equation of the fifth degree (modulo p, i.e. over a prime field) has at most five roots. In fact, he would have considered the former question as easier because it involves only squares and the other involves fifth powers; following Diophantus and Fermat, Euler took the degree as the first element in the classification of problems; of course, he could guess that there are other aspects, but he could not be sure.

So he had to reconstruct everything from scratch as I said. It is actually very fascinating to see in his correspondence with Goldbach how his ideas developed, how he solved one problem after the other. He solves some question, modulo something else—sometimes he explains 'if I could prove this then I could prove that', and Goldbach has some remarks to make about it. Goldbach really took an interest even though he does not seem ever to have contributed anything of real value. As a correspondent, however, he was invaluable to Euler for many years. Later Lagrange appeared on the scene and started corresponding with Euler; he, of course, was a first-rate mathematician, and Euler realized this immediately.

For many years Euler worked on pure number-theory, taking as his starting point only Fermat's work. One main topic was about writing integers and particularly primes as sums of squares. Take e.g. Fermat's assertion that any prime p of the form

$p = 4n + 1$ can be written as $p = x^2 + y^2$. Euler proves it in his correspondence to Goldbach in the year 1749; he says 'at last now I have the valid and complete proof for this'. [...]

[2] Many people think that one great difference between mathematics and physics is that in physics there are theoretical physicists and experimentalists and that a similar distinction does not occur in mathematics. This is not true at all. In mathematics just as in physics the same distinction can be made, although it is not always so clear-cut. As in physics the theoreticians think the experimentalists are there only to get the evidence for their theories while the experimentalists are just as firmly convinced that theoreticians exist only to supply them with nice topics for experiments. To experiment in mathematics means trying to deal with specific cases, sometimes numerical cases. For instance an experiment may consist in verifying a statement like Goldbach's conjecture for all integers up to 1000, or (if you have a big computer) up to one hundred billion. In other words, an experiment consists in treating rigorously a number of special cases until this may be regarded as good evidence for a general statement. There are many ways of making experiments, some of which may involve no little theoretical knowledge; for instance nowadays there are people who are greatly interested in $GL(n)$ and who make experiments by taking first $n = 1$ (which is already non-trivial for many problems) and then $n = 2$ (which may be quite hard). Consequently the first-rate mathematician must have some strength on both sides, but still there is a distinction between temperament. Now Fermat was clearly a theoretician. He was interested in general methods and general principles and not really in special cases; this appears in all his work, in analysis as well as in number-theory. Euler, on the other hand, was basically an experimentalist. He was very happy when he could conjecture a general law, and he was willing to spend a great deal of time to prove it; but if, instead of a proof, he had merely some really convincing experimental evidence, that pleased him almost as well. Therefore he tends to branch out in all possible directions, whereas Fermat, being a theoretician, always speaks about 'his methods', thus giving us fair indications about the range of his number-theoretic interests.

14.C3 The motion of a vibrating string

(a) Euler

[Euler first showed that the equation of motion of the string can be written as

$$\frac{ddy}{dt^2} = cc\frac{ddy}{dx^2},$$

the first time the equation was written in this form, and then he factorized it to obtain a system of two equations equivalent to it,

$$\frac{dy}{dt} = k\frac{dy}{dx},$$

where $k = \pm c$. He showed that these equations had solutions which were functions of $x + kt$, and then continued as follows.]

27 From this we conclude reciprocally that every function of $x + kt$, on being

substituted for y, satisfies the condition expressed by the formula

$$\frac{dy}{dt} = k\frac{dy}{dx}.$$

This is also clear in its own right, for denoting an arbitrary function of the quantity $x + kt$ by $\phi(x + kt)$ and its complete differential by $(dx + kdt)\phi'(x + kt)$, where $\phi'(x + kt)$ denotes another finite function of $x + kt$ which depends on the nature of the function $\phi(x + kt)$; if we set $y = \phi(x + kt)$, we shall have $dy = (dx + kdt)\phi'(x + kt)$, and following our way of writing

$$\frac{dy}{dt} = k\phi'(x + kt) \quad \text{and} \quad \frac{dy}{dx} = \phi'(x + kt),$$

whence it is evident that

$$\frac{dy}{dt} = k\frac{dy}{dx}.$$

From this there follows a general construction for the formula

$$\frac{dy}{dt} = k\frac{dy}{dx},$$

which gives us for y an arbitrary function in finite quantities of the quantity $x + kt$. [...]

28 Therefore taking ϕ and ψ as arbitrary functions, either of $y = \phi(x + ct)$ and $y = \psi(x - ct)$ satisfy the equation which gives the motion of the string. [...]

29 To explain this better, it is necessary to remark that every function of whatever nature can always be represented by a curved line whose ordinates express a certain function of the abscissa. So having constructed in Figure 1 a curve ES for the function ϕ and a curve FT for the function ψ, if we take in the one the abscissa $EQ = x + ct$, and in the other the abscissa $FR = x - ct$, the ordinates will be $QS = \phi(x + ct)$ and $RT = \psi(x - ct)$ and the sum of these two ordinates, or an arbitrary multiple of them, will always provide us with a suitable value of y which satisfies the equation. [...]

Figure 1

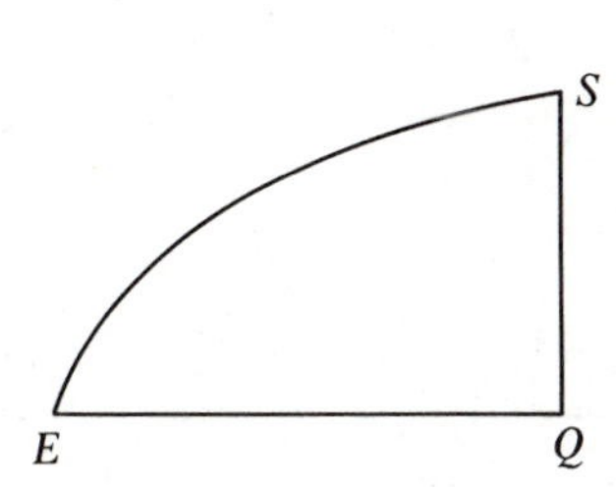

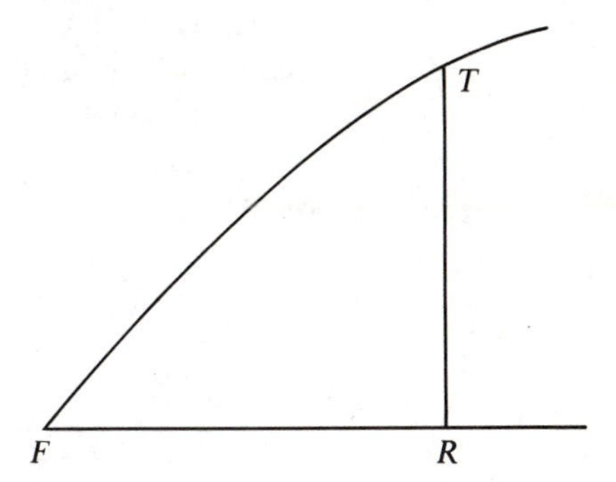

30 Now [...] stopping solely with the equation it is important to note that the two curves ES and FT are absolutely arbitrary, and that they can be taken at will; [...] But, and this is the main thing, these two curves generated in the way shown are equally

satisfactory, whether they are expressed by some equation or whether they are traced in any fashion, in such a way as not to be subject to any equation. The reader is asked to reflect carefully on this circumstance, which is the basis of the universality of my solution contested by M. d'Alembert.

[Euler then considered the extension of the curve which gives the initial shape of the string, and concluded that he needed to find ϕ and ψ for large values of t. He went on as follows.]

37 The different parts of this curve are thus not joined to each other through any law of continuity, and it is only by the description that they are joined together. For this reason it is impossible that all this curve should be comprised in any equation, unless perchance the [initial] figure be such that its natural continuation entails all these repeated parts; and this is the case when the figure is Taylor's sine curve or a mixture of such curves according to Mr Bernoulli. This is also, according to all appearances, the reason that Messrs Bernoulli and d'Alembert have believed the problem soluble in these cases only. But the manner in which I have just carried out the solution shows that it is not necessary for the directing curve to be expressed by any equation, and the shape of the curve is itself enough to let us infer the motion of the string, without subjecting it to calculation. I will make it plain also that the motion is not the less regular than if the initial shape were a sine curve, and thus the regularity of the motion cannot be alleged in favour of the sine curves to the exclusion of all others, as Mr Bernoulli seems to claim.

[Euler then showed (in 40) that the time for the string to make a complete vibration is known, and that each piece of the string completes its motion in the same amount of time, which depends only on the physical properties of the string. This time is a/c, where a is the length of the string. However, he went on, the string may beat faster than this.]

41 [...] the time of vibration can be reduced to a half, or a third, or any aliquot part. This depends on the initial shape which is given to the string, and when it has only one bulge the time of vibration will without doubt be $= a/c$ seconds. But if the initial figure of the curve has two equal bulges AEC, CFB as in Figure 2, and the part AEC is equal and similar to the part BFC, one sees by our construction that the point C will always remain at rest, and that the motion of the string will be the same as that of a string of half the length and an equal tension. Now it is necessary, to produce vibrations that are twice as rapid, that the node C is precisely in the middle, and that the two bulges AEC and BFC are equal and similar to one another, for without this condition the point C will not remain fixed.

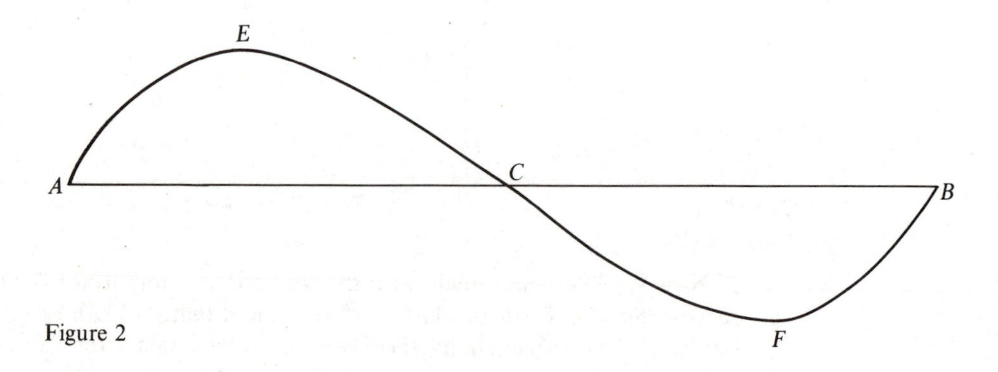

Figure 2

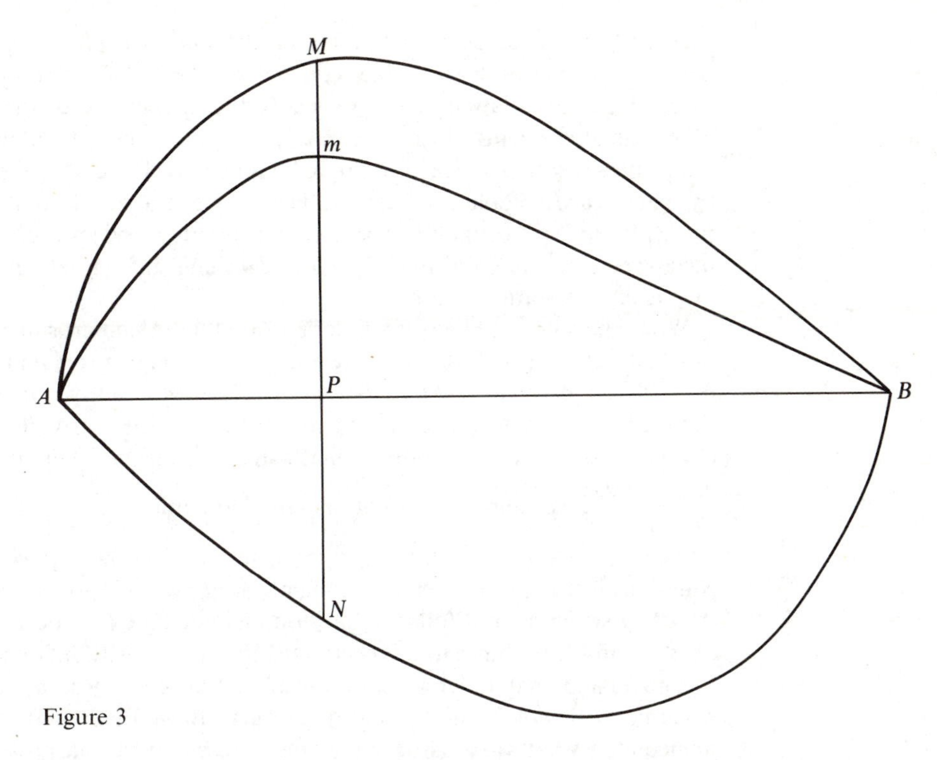

Figure 3

42 What Mr Bernoulli has observed about a mixture of two or more sine curves is equally applicable to all other curves imaginable. For, whatever the initial shape of the string in Figure 3 it can serve as the axis of another figure similar to that in Figure 2; and that can serve again as the axis of other similar figures with several bulges. In this case the total sound of the string will be a mixture of several sounds, of which one will be the octave, others the twelfth, the fifteenth, the seventeenth, and so on. This mixture of sounds made by one string simultaneously, which Mr Bernoulli was the first to explain so happily, is therefore not so essential an effect of the combination of Taylor's sine curves that it cannot equally well be produced by a similar combination of arbitrary curves. And when one thinks of the way in which one is accustomed to strike the strings, it is very probable that they never take the figure of sine curves, nor that they approach them more and more: for all the phenomena that Mr Bernoulli describes can equally well be brought about by an arbitrary figure.

(b) Clifford Truesdell on d'Alembert's and Euler's attitudes to the nature of solutions to the wave equation

To clarify d'Alembert's viewpoint it thus remains only to explain why he requires $z = \Phi(u)$ to be an 'equation'. He himself, while never giving any reason, shows by his obstinate repetitions from now on until the end of his life that he regards it as entirely obvious that 'mechanical' functions are to be exiled from mathematics, or at least from mathematical physics. This is a consequence of Leibniz's *law of continuity* as it was widely interpreted in the eighteenth century: Only 'continuous' functions occur in the solution of physical problems. While nowadays this seems a merely arbitrary prejudice, we must bear in mind that the majority of the geometers and more particularly the physicists of the day shared it. E.g. John Bernoulli and d'Alembert invoked Leibniz's law in order to justify the application of the laws of physics to infinitesimal elements. Less obvious, perhaps, is the advantage of the resultant uniqueness theorem, indeed not proved but nevertheless correctly believed at the time, by which each soluble physical problem has but a single solution, determinate in principle up to a singularity resulting from its very nature, and indeed such a metaphysics would furnish a basis for regarding differential equations as a correct means of formulating natural laws. [...]

While the difference between Euler's view and d'Alembert's might seem a matter of pure mathematics, in fact it is the very opposite. Today it is plain that *the phenomenon of wave motion contradicts Leibniz's law*. This was surely not obvious to Newton despite his enormous physical insight, nor to any other early physicist; rather, it is a discovery of Euler, by purely mathematical means. The differential equation $\frac{ddy}{dt^2} = cc\frac{ddy}{dx^2}$ certainly has solutions that are not analytic; d'Alembert's formula $y = \phi(x + ct) + \psi(x - ct)$, as Euler interprets it, gives them at will. If [the differential equation] is the entire statement of the physical principle governing the motion of the vibrating string, then it follows that non-analytic functions occur in the solutions of physical problems. Since to this everyone today agrees without question, it is now hard to understand that Euler's refutation of Leibniz's law was *the greatest advance in scientific methodology* in the entire century. Both Euler and d'Alembert realized immediately what was at issue in the otherwise rather tedious problem of the vibrating string. This is the only scientific reason for the sharpness of the controversy that Euler and d'Alembert were to carry on until their deaths at the end of the century.

14.C4 Nicolas Condorcet's *Elogium of Euler*

The Queen of Prussia, however, could extract from Euler monosyllables only: she taxed him with a timidity and reserve, which the cordiality of his reception could not possibly have inspired: 'Why, then, will you not talk to me', said the Queen? 'Because, Madam,' he replied, 'I have just come from a country, where people are hanged, if they talk'.

Feeling myself now called upon to give some account of Euler's immense scientific labours, I shrink from the impossibility of following him in detail, of conveying any thing like an accurate idea of that multiplicity of discoveries, of new methods of investigation, of ingenious views, diffused over more than thirty separate publications, and over near seven hundred memoirs, of which about two hundred, deposited in the Academy of Petersburg, previous to his death, are destined to enrich, in their order, the future collections published by that learned body.

But in particular character seems, to me, to distinguish Euler from the other illustrious men who, in pursuing the same career, have attained a glory which his has not eclipsed; that character is, his having embraced the mathematical sciences in their universality; his having brought to perfection, one after another, the different parts; and, enriching the whole by important discoveries, his having produced a very beneficial revolution in the manner of treating them. I imagined, therefore, that in sketching a methodical representation of the different branches of these sciences, in pointing out the progress of each, and the happy improvements to be ascribed to the genius of Euler, I should give, at least as far as my ability permits, a juster idea of this wonderful man, who, by uniting so many extraordinary talents, has presented a phenomenon, if the expression may be allowed, of which the history of science has, hitherto, furnished no example.

Algebra had long been a science of very limited use and application. The mode of considering the idea of magnitude, only in the highest degree of abstraction of which the human mind is susceptible; its rigorously separating from that idea every thing which, by employing imagination, might give some support; or repose, to the understanding; finally, the extreme generality of the signs which this science makes use of, render it, in some measure, too foreign to our nature, too remote from ordinary conception, to admit of the mind's taking extraordinary pleasure in it, and of easily acquiring a habit of tracing its operations. The algebraic method is apt to discourage even persons the most disposed to abstract speculation. If the object of pursuit be even a little complicated, we are forced to lose sight of it entirely, and to confine our whole attention to dry algebraic characters; the road is safe and sure, but the point which is aimed at, and that from whence we took our departure, equally vanish from the eye of the geometrician; and it required no slight degree of courage, to venture out of sight of land, without any other pilot than a recently discovered science. Accordingly, on examining the works of the great geometricians of the last age, even of those to whom algebra is indebted for the most important discoveries, we shall see how little they were accustomed to handle this very weapon, which has been brought to such a state of perfection; and it is impossible to refuse to Euler, the praise of having effected a revolution which renders algebraic analysis a mode of calculation luminous, universal, of general application, and of easy acquisition.

Thus, at certain epochs, when, after strenuous exertions, the mathematical sciences seemed to have exhausted all the resources of genius, and to have reached the *ne plus ultra* of their career; all at once a new method of calculation is introduced, and the face of the science is totally changed. We find it immediately, and with inconceivable rapidity, enriching the sphere of knowledge, by a solution of an incredible number of important problems, which geometricians had not dared to attempt, intimated by the difficulty, not to say the physical impossibility, of pursuing calculation to real issue. Justice would, perhaps, demand, in favour of the man who invented and introduced these methods, and who first taught their use and application, a share in the glory of all those who have practised them with success; he has, at least, claims upon their gratitude, which cannot be contested without a crime.

We have seen in the elogium of Daniel Bernoulli, that he had divided with Euler alone, the glory of having carried off thirteen prizes, proposed by the Academy of Sciences: They often contended for the same object, and occupied the same ground; and the honour of triumph over a competitor was, likewise, divided between them; but this rivalship never encroached on the expressions of reciprocal esteem, nor cooled the ardour of mutual friendship. On examining the subjects, for which the one or the other obtained the victory, we find that success depended, principally, on the character of talent peculiar to each. When the question required address, in the manner of taking it up, a dexterous application of experiment, or new and ingenious physical views, Daniel Bernoulli had the advantage: but did it present difficulties, which profound and accurate calculation could resolve; was it necessary to create a new method of analysis, victory declared for Euler. Were any one so presumptuous, as pretend to judge between them, he would find that he had to pronounce, not between two men, but between minds of a different genius, between two methods of employing genius.

I should have conveyed but a very imperfect idea of Euler's fertility of invention, unless I added to this faint sketch of his labours, that there are very few subjects of importance, once treated by him, that he did not retrace; nay, so far as to recompose his first work several times over. Sometimes he substituted a direct, and analytical method, in place of one more indirect; sometimes he extended his first solution, to cases which had, at first escaped him; adding, almost always, new examples, which he knew how to select with singular skill, among those which presented, or some useful observation, or curious remark.

The intention, merely, of giving to one of his productions a form more methodical, of rendering it somewhat more luminous, of bestowing on it a higher degree of simplicity, was, to him, motive sufficient for engaging in labours incredible. Never did geometrician write so much, and no one ever carried his work to such a height of perfection. When he published a memoir on a new subject, he simply explained the track which he pursued; he pointed out to his pupils its intricacies and aberrations, and having, with scrupulous exactness, made them accompany the progress of his own mind, in his first essays, he showed them, afterwards, how he had been enabled to trace a simpler path. It is evident, that he preferred the instruction of his disciples, to the silly satisfaction of dazzling them by his own superiority; and that he did not believe he had done enough for science, unless he added, to the new truths with which he was enriching it, a candid exposition of the ideas which led to discovery. [...]

The government of Russia had never treated Euler as a stranger. Notwithstanding his absence, part of his salary was always regularly paid; and in 1766, the Empress having given him an invitation to return to Petersburg, he complied.

In 1735, the exertion occasioned by an astronomical calculation, for which other academicians demanded several months, but completed by him in a few days, brought on an indisposition, which issued in the loss of one of his eyes. He had reason to apprehend a total loss of sight, if he continued to expose himself in a climate, the influence of which was unfavourable to his constitution. The interest of his family got the better of this apprehension; and if we reflect that, to Euler, study was an exclusive passion we shall readily conclude, that few examples of paternal tenderness have more completely demonstrated, that it is the most powerful, and the sweetest of all our affections.

A few years after, he was overtaken by the calamity which he foresaw and dreaded: but happily for himself, and for the sciences, he preserved still the faculty of distinguishing large characters traced on a slate with chalk. His sons, his pupils, copied his calculations; wrote, as he dictated, the rest of his memoirs; and if we may form a judgement of these from their number, and, frequently, from the genius transfused through them, it will appear abundantly credible, that from the absence still more absolute of all distraction, and from the new energy which this constrained recollection gave to all his faculty, he gained more, both as to facility and means of labour, than he lost by a diminution of sight.

Besides, Euler, by the nature of his genius, and his habits of life, had, even involuntarily, laid up for himself extraordinary supplies. On examining those great analytical formulas, so rare before his time, but so frequent in his works, the combination and display of which unite so much simplicity and elegance, whose very form pleases the eye as well as the mind, it will be evident, that they are not the result of a calculation traced on paper, but that, produced entirely in the head, they are the creation of an imagination equally vigorous and active.

There exist in analysis, and Euler greatly multiplied their number, formulas of a common, and almost daily application; he had them always present to his mind, knew them by heart, repeated them in conversation; and Mr d'Alembert, when he saw him at Berlin, was astonished at an effort of memory, which demonstrated, that Euler possessed, at once, a strength and a clearness of recollection almost incredible. At length, his facility of calculation by the head was carried to such a degree, as would exceed all belief, had not the history of his labours accustomed us to prodigies. He has been known, in the view of exercising his little grandson in the extraction of the square and cube roots, to have formed to himself the table of the first six powers of all numbers, from 1 to 100, and to have preserved it exactly in his memory. Two of his pupils had calculated as far as to the seventeenth term of a convergent series, abundantly complicated; their results, though formed after a written calculation, differed one unit at the fiftieth figure: they communicated this difference to their master: Euler went over the whole calculation in his head, and his decision was found to be the true one.

He had retained all his facility of thought, and apparently, all his mental vigour: no decay seemed to threaten the sciences with the sudden loss of their great ornament. On the 7th of September, 1783, after amusing himself with calculating on a slate, the laws of the ascending motion of air-balloons, the recent discovery of which was then making a noise all over Europe, he dined with Mr Lexell and his famly, talked of Herschell's planet, and of the calculations which determine its orbit. A little after he called his grand-child, and fell a playing with him as he drank tea, when suddenly, the pipe, which he held in his hand, dropped from it, and he ceased to calculate and to breathe.

14.D Some of Euler's Contemporaries

The two passages we have chosen from Jean-Paul de Gua's book of 1740 (14.D1) and Gabriel Cramer's of 1750 (14.D2) speak for themselves, both as brief histories of the algebraic theory of curves and as aesthetic statements about the nature of geometry. Condorcet, in his *Elogium of de Gua,* protested that no one was as single-minded as de Gua had advocated one should be, but arguably the powerful methods of the calculus and algebra were squeezing geometry into a corner, and it was appropriate for de Gua to take a stand on the matter. Indeed, it remains a live matter of debate what constitutes a truly geometric argument.

A few remarks might usefully be made about the technicalities of de Gua's and Cramer's work. An algebraic curve is, in their view, a real locus represented by a polynomial equation in two variables: $f(x, y) = 0$. Standard methods enabled one to rewrite this locally (i.e. for a small range of xs) in the form $y =$ power-series in x, provided that in that range the curve did not intersect itself. If it did, the power-series method did not know which branch of the curve to choose, but to solve that problem one then had recourse to the method of Newton's parallelogram. While Cramer's book was undoubtedly the definitive geometry text of the second half of the eighteenth century, it was in fact de Gua who saw deepest into the subject. His transformations amount to introducing a third variable, z, and then homogenizing the equation. This would, for example, replace the equation $x^3 + y^3 = 3xy$ by $x^3 + y^3 = 3xyz$. One can then select, say, x and z as coordinates and rewrite the original equation as $x^3 + 1 = 3xz$. In this way one gets a look at the curve 'at infinity'. By treating points at infinity on a par with finite points, in a strictly projective spirit, de Gua was able to prove his results about the sorts of infinite branches a (possibly singular) curve of a given order could have.

De Gua was, with Jean d'Alembert and Denis Diderot, one of the founders of the famous *Encyclopédie,* although he soon withdrew. D'Alembert filled pages of this work with his important essays on mathematics; we present here (14.D3) his marvellously Cartesian account of why mathematics works so well in science—Cartesian because of its emphasis on the power of theory, which was ever d'Alembert's belief even if he felt constrained on this occasion to sound Newtonian, even Baconian, as well (see also 18.A3).

Then we turn to Euler's greatest immediate successor, the Italian resident in Paris Joseph Louis Lagrange (1736–1813). The short extracts from his long article on the question of the solvability of polynomial equations by algebraic formulae ('by radicals', as it was called) give a hint of his novel approach, which was to consider the roots as abstract, formal objects (see 14.D4). He hoped in this way to explain why polynomials of low degree were solvable by radicals (because there were many symmetric expressions in the roots) but polynomial equations of higher degree were not. But to prove this claim on his approach it was necessary to show that not enough symmetric expressions could be found. Lagrange was intimidated by this exhausting work. But it could in fact be done, as Paolo Ruffini (more or less) and Nils Hendrik Abel were to show much later. In their work the study of permutations was carried further, a study first made explicit by Augustin Louis Cauchy. If, then, in this respect Lagrange's work was to have profound indirect consequences, his attention to Euler's theory of numbers was to have a much more direct effect. He summarized some of his earliest findings in the lengthy additions he made to Euler's *Algebra* (see 14.D5), which not only gave rigorous proofs in support of some of Euler's claims, but also gave them a coherence which enables one to speak for the first time of a true theory of numbers. This theory involves the essential use of 'numbers' of the form $a + b\sqrt{N}$, which is not only an interesting extension of the number concept, but also another significant use of complex numbers (when N is negative) at a time when historians have sometimes wanted to suggest that complex numbers were not well understood.

Our last contemporary of Euler is Johann Heinrich Lambert (1728–1777), who once described himself, reasonably enough, as the fourth-best mathematician of his day, behind Euler, Lagrange and d'Alembert. Lambert's strength was in setting up subjects clearly—for his remarks about non-Euclidean geometry see 16.A5—and here we give an extract from his important book on map-making (see 14.D6). As its translator W. R. Tobler points out, not only did Lambert go on to invent several map projections that are still in use today, but also his book helped direct the attention of mathematicians back to this useful application of their art. Soon after it appeared, Euler filled out the argument sketched by Lambert at the end of our extract, which shows that no map of the Earth on a flat piece of paper can get everything right. Perhaps the most decisive consummation of this line of research was to be Carl Friedrich Gauss's book of 1827 on differential geometry, from which we quote below (15.A5).

14.D1 Jean-Paul de Gua on the use of algebra in geometry

I will admit, however, that by the very nature of Descartes' analysis, the use I can make of it in the research I speak of is necessarily restricted. Should differential expressions enter into the given ratios of a problem, or should similar expressions arise, as the solution of the problem may determine—for example if some property of a mechanical curve is concerned, or if one must assign a value to an arc or to the area of even a geometric curve—these two cases, which are the only ones where the illustrious Mr Newton employed the calculus of the infinitely small, demand in effect that one necessarily has recourse to differential equations and one thereby supplements the range of ordinary equations. I have avoided speaking of more than ordinary analysis, because it would be improper to upbraid a method which does not give one the means of knowing what nature does not allow it to reveal.

But it is not thus with the relationships that the parameters of a geometrical curve can have with the distances from their origin to the points of inflexion, or cusps, or more generally to their singular or multiple points or even to the asymptotes of their different infinite branches; no more is it thus with the relationship of the abscissas with the subtangents of the corresponding points, simple or multiple, ordinary or singular; nor finally with those that the parameters of the curve and the sine of the angle of its coordinate axes have to the sine of the angles that the ultimate directions of the infinite branches make with the coordinate axes. On the contrary, in all the relationships of this kind one of the terms is necessarily expressible algebraically in terms of the other, and the determination of these relationships is consequently one of the most natural objects of the analysis of Descartes. To think otherwise would be to misunderstand completely the extent of the method of indeterminates, the most important discovery made in mathematics by that profound philosopher, the ornament of our nation and his century, who was the first to dare to open to the ordinary run of men the way to truth almost unknown and little taken until then, to whom thereby all the sciences are simultaneously obliged, all of whose ideas have been fertile and even his faults have been those of a great man.

[De Gua then observed that it was his reading of Newton's 'Enumeration of Curves of the Third Order' which inspired him to take up his own work, which he described as follows.]

In the second section I expound my principles at length, and describe in detail the different transformations which I have thought of for recognizing the principal properties of geometric curves. Then I expound a singular analogy between the result of most of my transformations and that of ordinary differentiation, and this analogy has enabled me to abbreviate and to facilitate considerably the practice of my own rules.

With their help I explain how to discover everything about the infinite branches which a geometric curve can present, whether the ultimate direction of these branches must be the same as their axes, whether they must be different, whether in this case—supposing the branches hyperbolic—the asymptotes of these branches must pass through the origin, whether they are more or less separated from it. I likewise explain how to discover in every respect all the simple or multiple, ordinary or singular points which can arise on a geometric curve, whether these points must, as one supposes initially, be placed at the origin itself, whether they arise as vertices of curves, whether they can be situated on curves at any place one wants to imagine. Finally I make an exact enumeration of the different kinds of infinite branches of which geometric curves of the first five orders are susceptible.

Later on I establish singular analogies between the methods provided by the consideration of the equations of curves for elucidating on the one hand the

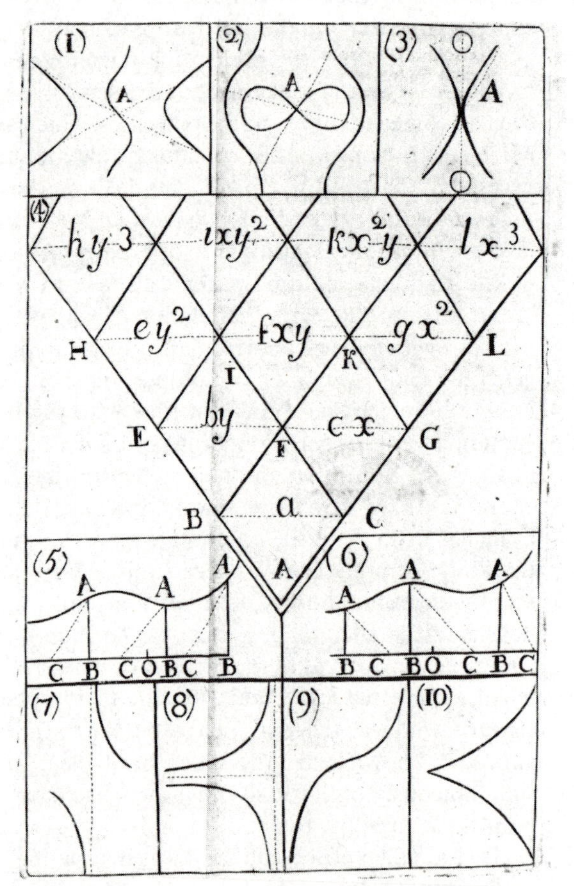

From de Gua's *Usages de l'Analyse*, showing his triangle for making coordinate transformations, and, in (8), (9) and (10), three different projections of the same cubic curve.

different points on the curves and on the other hand the different systems of infinite branches in mutually parallel directions. I find the source of these analogies in the theory of shadows or projections of curves, and I take the occasion to treat this last matter much more generally than any one has done before.

In other places I establish by my methods different properties of general curves and of curves of the third order in particular, properties whose discovery is due to Mr Newton, and which since he discovered them either have not been proved at all or only by means less direct and less simple than those which I use.

14.D2 Gabriel Cramer on the theory of algebraic curves

It was almost the same in every branch of mathematics until the invention of algebra: ingenious methods for reducing the problems to the most simple and easy calculations that the given question would admit. This universal mathematical key has opened the gate to many to whom it would always have been shut without its help. One can say that this discovery has produced a veritable revolution in all the sciences which depend on calculation.

It therefore seems to be a whim and a sort of caprice, to despise so useful a method and to glory in only using the geometrical analysis of the ancients. It has, I admit, the merit over algebra of being more apparent to the senses, and has a certain elegance which is infinitely pleasing, but it is nothing like so easy and so universal. So prefer it, if you wish, but do not exclude the other method. Mathematical truths are not so easy to find that it is worthwhile closing any of the routes that can lead to them.

Above all it is in the theory of curves that one feels most forcefully the utility of a method as general as that of algebra. Descartes, whose inventive mind shines no less brilliantly in geometry than in philosophy, had no sooner introduced the way of expressing the nature of curves by algebraic equations than this theory changed its face. Discoveries multiplied with extraordinary ease: each line of the calculus gave birth to new Theorems.

In this way, art made up for genius, and genius helped by such a helpful art had a success it could never have obtained with its own forces. For what is admirable here is that one does not discover by means of algebra some property of a particular curve without also finding out similar, or analogous, properties of an infinity of other curves.

Let us add that algebra alone furnishes the means to distribute curves into orders, classes, genera, and types: which, as in an arsenal where the arms are well arranged, enables one to choose without hesitation those which can assist in the resolution of a given problem

It is to the illustrious Mr Newton above all that geometry is indebted for this distribution. His 'Enumeration of Curves of the Third Order' is an excellent model for everything that needs to be done of this kind, and a convincing proof that that great man had penetrated to the bottom of what in the theory of curves is the most delightful and the most interesting.

It is unfortunate that Mr Newton contented himself with displaying his discoveries without appending their proofs and that he preferred the pleasure of being admired to that of instruction.

It is not only that in an attentive reading of the treatise one can find some traces of his method: one discovers that the guiding principles have been the doctrine of infinite series, which is almost entirely his, and the use of the analytic parallelogram, which he invented; one can even see that at some points he has not followed his guides with the exactitude one admires in his other works.

These slight inadvertencies did not escape Mr Stirling, who developed the principles and the method of Mr Newton in the excellent Commentary which he has given us on his book. One sees there that almost nothing is lacking for Mr Stirling to give a complete theory of curves, and that he would have left only a few things to say if he had not attached himself too scrupulously to be able to depart from his author.

Mr Nicole, in the *Mémoires de l'Académie Royale des Sciences*, gave an explanation as neat as it is exact of the principles that Mr Newton could have followed in his 'Enumeration of Curves of the Third Order': it is a great shame that so happy a beginning has not had a sequel. What could one not achieve with the genius and knowledge of Mr Nicole?

I would have made great use of the *Introductio in Analysin Infinitorum* of Mr Euler if that book had been better known to me. Its object being almost the same as mine it is not surprising if we are often together in our conclusions. But the difference of methods is as great as can be when one works on the same subject. I say nothing at all to prefer the route I have taken over that of Mr Euler, but only to advise the reader of this diversity. Here is an outline of the order I have thought it right to follow.

[Cramer now described his book in some detail; the following passages are but a selection.]

One goes in some detail into the manner in which an algebraic curve is represented by its equation; one indicates the method for deciding if a given equation expresses a single curve or an assemblage of several. One speaks of branches of curves, finite or infinite, real or imaginary. One examines in what sense the path of an algebraic curve is continuous, although a curve can be made up of separate parts. Finally one describes the method of describing a curve by assigning the position of an infinity of its points, not only when the equation can be resolved, but also in several other cases. It must be admitted that a general method for this is lacking; if Algebra could provide it one would have, by that alone, all that is necessary for the understanding of curves.

The following chapters discuss what is most remarkable in the path of a curve. These are the infinite branches and its singular points. It is by the infinite branches that one divides curves of each order into their genera, and by the singular points that one divides the curves of each genus into types. [...]

The eighth chapter is used to determine the number, nature, and position of the infinite branches a curve can have whose equation is given. [...]

In the ninth chapter one establishes the general division of curves based on the number, nature, and position of their infinite branches. One shows that there are only three curves of the second order: the ellipse, which includes the circle; the hyperbola and the parabola. This yields, in a few words, what is called the construction of geometric curves. One reduces curves of the third order to four classes, which are subdivided into fourteen genera in conformity with what Mr Newton has established in his 'Enumeration of Curves of the Third Order', of which this article can be regarded as a little commentary. There are nine classes of curves of the fourth order and each subdivides into various genera, but the enumeration of them is almost impossible.

The following chapters are devoted to the examination of the singular points of a curve. These points are the multiple points or points of inflexion. Multiple points are double, triple, or quadruple, etc. Inflexion points have one inflexion, or are simple, or double, or triple, etc. Inflexions of odd degree are visible, those of even degree are invisible and are only made manifest by the calculus; they are called serpentine.

14.D3 Jean d'Alembert on algebra, geometry and mechanics

In d'Alembert's view 'nothing is more indisputable than the evidence of the senses': these tell us of the existence of bodies, whose basic property is their impenetrability, occupying penetrable space, whose basic property is extension. He then argues as follows, in a manner the translator R. N. Schwab has pointed out is derived from Newton's *Arithmetica Universalis.*

Since our examination of shaped extension presents us with a large number of possible combinations, it is necessary to invent some means of achieving those combinations more easily; and since they consist chiefly in calculating and relating the different parts of which we conceive the geometric bodies to be formed, this investigation soon brings us to Arithmetic or the science of numbers. This [science] is simply the art of finding a short way of expressing a unique relationship which results from the comparison of several others. The different ways of comparing these relationships [numbers] give the different rules of Arithmetic.

Moreover, if we reflect upon these rules we almost inevitably perceive certain principles or general properties of the relationships, by means of which we can, expressing these relationships in a universal way, discover the different combinations that can be made of them. The results of these combinations reduced to a general form will in fact simply be arithmetical calculations, indicated and represented by the simplest and shortest expression consistent with their generality. The science or the art of thus denoting relationships is called Algebra. Thus, although no calculation proper is possible except by numbers, nor any magnitude measurable except by extension (for without space we could not measure time exactly), we arrive through the continual generalization of our ideas at that principal part of mathematics, and of all the natural sciences, called the Science of Magnitudes in general. It is the foundation of all possible discoveries concerning quantity, that is to say, concerning everything that is susceptible to augmentation or diminution.

This science is the farthest outpost to which the contemplation of the properties of matter can lead us, and we would not be able to go further without leaving the material universe altogether. But such is the progress of the mind in its investigations that after having generalized its perception to the point where it can no longer break them up further into their constituent elements, it retraces its steps, reconstitutes anew its perceptions themselves, and, little by little and by degrees, produces from them the concrete beings that are the immediate and direct objects of our sensations. These beings, which are immediately relative to our needs, are also those which it is most important for us to study. Mathematical abstractions help us in gaining this knowledge, but they are useful only insofar as we do not limit ourselves to them.

That is why, having so to speak exhausted the properties of shaped extension through geometric speculations, we begin by restoring to it impenetrability, which constitutes physical body and was the last sensible quality of which we had divested it. The restoration of impenetrability brings with it the consideration of the action of bodies on one another, for bodies act only insofar as they are impenetrable. It is thence that the laws of equilibrium and movement, which are the object of Mechanics, are deduced. We extend our investigations even to the movement of bodies animated by unknown driving forces or causes, provided the law whereby these causes act is known or supposed to be known.

Having at last made a complete return to the corporeal world, we soon perceive the use we can make of Geometry and Mechanics for acquiring the most varied and profound knowledge about the properties of bodies. It is approximately in this way that all the so-called physico-mathematical sciences were born. We can put at their head Astronomy, the study of which, next to the study of ourselves, is most worthy of our application because of the magnificent spectacle which it presents to us. Joining observation to calculation and elucidating the one by the other, this science determines with an admirable precision the distances and the most complicated movements of the heavenly bodies; it points out the very forces by which these movements are produced or altered. Thus it may justly be regarded as the most sublime and the most reliable application of Geometry and Mechanics in combination, and its progress may be considered the most incontestable monument of the success to which the human mind can rise by its efforts. [...]

With respect to the mathematical sciences, which constitute the second of the limits of which we have spoken, their nature and their number should not overawe us. It is principally to the simplicity of their object that they owe their certitude. Indeed, one must confess that, since all the parts of mathmatics do not have an equally simple aim, so also certainty, which is founded, properly speaking, on necessarily true and self-evident principles, does not belong equally or in the same way to all these parts. Several among them, supported by physical principles (that is, by truths of experience or by simple hypotheses), have, in a manner of speaking, only a certitude of experience or even pure supposition. To be specific, only those that deal with the calculation of magnitudes and with the general properties of extension, that is, Algebra, Geometry, and Mechanics, can be regarded as stamped by the seal of evidence. Indeed, there is a sort of gradation and shading, so to speak, to be observed in the enlightenment which these sciences bestow upon our minds. The broader the object they embrace and the more it is considered in a general and abstract manner, the more also their principles are exempt from obscurities. It is for this reason that Geometry is simpler than Mechanics, and both are less simple than Algebra. This will not be a paradox at all for whose who have studied these sciences philosophically. The most abstract notions, those that the common run of men regard as the most inaccessible, are often the ones which bring with them a greater illumination. Our ideas become increasingly obscure as we examine more and more sensible properties in an object. Impenetrability, added to the idea of extension, seems to offer us an additional mystery; the nature of movement is an enigma for the philosophers; the metaphysical principle of the laws of percussion is no less concealed from them. In a word, the more they delve into their conception of matter and of the properties that represent it, the more this idea becomes obscure and seems to be trying to elude them.

14.D4 Joseph Louis Lagrange on solvability by radicals

[1] With regard to the solution of literal equations there has scarcely been any advance since the time of Cardano, who was the first to publish one for the equations of the third and fourth degrees. The first success of the Italian analysts in this subject seems to have been the end of the discoveries one can make there; at least it is certain that all the attempts which have been made to push back the limits of this part of algebra have still only served to find new methods for equations of the third and fourth degrees which do not appear to be applicable in general to equations of higher degree.

I propose in this memoir to examine the different methods which have been found up till now for the algebraic solution of equations, to reduce them to general principles, and to examine *a priori* why these methods succeed with the third and fourth degrees but fail for higher degrees.

This examination will have a two-fold advantage: on the one hand it will serve to shed greater light on the known solutions to the third and fourth degrees; on the other hand it will be useful to those who wish to occupy themselves with the solution of higher degrees, in providing them with different views of this object and above all in saving them a great number of steps and useless attempts. [...]

[2] We conclude our analysis of the methods which concern the solution of equations of the fourth degree here. Not only have we related these methods to one another and show their interconnections and their mutual dependence, but we have also, and this is the principal point, given the *a priori* reason why they lead, some to resolvents of the third degree, others to resolvents of the sixth, but which can be reduced to the third. One has seen how this derives in general from the fact that the roots of these resolvents are functions of quantities x', x'', x''', x^{iv}, which, on making all the possible permutations of these four quantities, only receive three different values, like the function $x'x'' + x'''x^{\text{iv}}$, or six values of which two are equal and of opposite sign, like the function $x' - x'' - x''' - x^{\text{iv}}$, or even six values which, on dividing them into three pairs and taking the sum or the product of the values of each pair, the three sums or the three products are always the same, whatever permutation one makes of the quantities x', x'', x''', x^{iv},[...]. It is precisely the existence of such functions on which the solution of equations of the fourth degree depends. [...]

[3] It follows from these reflections that it is very doubtful if the methods of which we have been speaking can give a complete solution of equations of the fifth degree, and still more so to those of higher degrees. And this uncertainty, coupled with the length of the calculations which these methods involve, must repel in advance all those who would seek to use them to solve one of the most famous and important problems in algebra. Also we observe that the authors of these methods have themselves been content to apply them to the third and fourth degrees and that no one has yet undertaken to push their work further.

It would therefore be very desirable if one could judge *a priori* the success that one can expect in applying these methods to degrees higher than the fourth. We are going to try and give the means for this by an analysis similar to that which has served us up till now in respect of the known methods for the solutions of equations of the third and fourth degree.

14.D5 Joseph Louis Lagrange's additions to Euler's *Algebra*

The geometricians of the last century paid great attention to the Indeterminate Analysis, or what is commonly called the *Diophantine Algebra*; but Bachet and Fermat alone can properly be said to have added any thing to what Diophantus himself has left us on that subject.

To the former we particularly owe a complete method of resolving, in integer numbers, all indeterminate problems of the first degree: the latter is the author of some methods for the resolution of indeterminate equations, which exceed the second degree; of the singular method, by which we demonstrate that it is impossible for the sum, or the difference of two biquadrates to be a square; of the solution of a great number of very difficult problems; and of several admirable theorems respecting integer numbers, which he left without demonstration, but of which the greater part has since been demonstrated by M. Euler in the Petersburg *Commentaries*.

In the present century, this branch of analysis has been almost entirely neglected; and, except M. Euler, I know no person who has applied to it: but the beautiful and numerous discoveries, which that great mathematician has made in it, sufficiently compensate for the indifference which mathematical authors appear to have hitherto entertained for such researches. The *Commentaries* of Petersburg are full of the labours of M. Euler on this subject, and the preceding Work is a new service, which he has rendered to the admirers of the *Diophantine Algebra*. Before the publication of it, there was no work in which this science was treated methodically, and which enumerated and explained the principal rules hitherto known for the solution of indeterminate problems. The preceding Treatise unites both these advantages: but, in order to make it still more complete, I have thought it necessary to make several Additions to it, of which I shall now give a short account.

The theory of Continued Fractions is one of the most useful in arithmetic, as it serves to resolve problems with facility, which, without its aid, would be almost unmanageable; but it is of still greater utility in the solution of indeterminate problems, when integer numbers only are sought. This consideration has induced me to explain the theory of them, at sufficient length to make it understood. As it is not to be found in the chief works on arithmetic and algebra, it must be little known to mathematicians; and I shall be happy, if I can contribute to render it more familiar to them. At the end of this theory, which occupies the first Chapter, follow several curious and entirely new problems, depending on the truth of the same theory; but which I have thought proper to treat in a distinct manner, in order that the solution of them may become more interesting. Among these will be particularly remarked a very simple and easy method of reducing the roots of equations of the second degree to Continued Fractions, and a rigid demonstration, that those fractions must necessarily be always periodical.

The other Additions chiefly relate to the resolution of indeterminate equations of the first and second degree; for these I give new and general methods, both for the case in which the numbers are only required to be rational, and for that in which the numbers sought are required to be integer; and I consider some other important matters relating to the same subject.

The last Chapter contains researches on the functions, which have this property, that the product of two or more similar functions is always a similar function. I give a general method for finding such functions, and shew their use in the resolution of different indeterminate problems, to which the usual methods could not be applied.

Such are the principal objects of these Additions, which might have been made much more extensive, had it not been for exceeding proper bounds; I hope, however, that the subjects here treated will merit the attention of mathematicians, and revive a taste for this branch of algebra, which appears to me very worthy of exercising their skill.

Chapter VIII
Remarks on Equations of the form $p^2 = Aq^2 + 1$, *and on the common method of resolving them in Whole Numbers.*

The method of Chap VII of the preceding Treatise, for resolving equations of this kind, is the same that Wallis gives in his *Algebra* (Chap. XCVIII), and ascribes to Lord Brouncker. We find it, also in the *Algebra* of Ozanam, who gives the honour of it to M. Fermat. Whoever was the inventor of this method, it is at least certain, that M. Fermat was the author of the problem which is the subject of it. He had proposed it as a challenge to all the English mathematicians, as we learn from the *Commercium Epistolicum* of Wallis; which led Lord Brouncker to the invention of the method in question. But it does not appear that this author was fully apprised of the importance of the problem which he resolved. We find nothing on the subject, even in the writings of Fermat, which we possess, nor in any of the works of the last century, which treat of the Indeterminate Analysis. It is natural to suppose that Fermat, who was particularly engaged in the theory of integer numbers, concerning which he has left us some very excellent theorems, had been led to the problem in question by his researches on the general resolution of equations of the form,

$$x^2 = Ay^2 + B,$$

to which all quadratic equations of two unknown quantities are reducible. However, we are indebted to Euler alone for the remark, that this problem is necessary for finding all the possible solutions of such equations.

The method which I have pursued for demonstrating this proposition is somewhat different from that of M. Euler; but it is, if I am not mistaken, more direct and more general. For, on the one hand, the method of M. Euler naturally leads to fractional expressions, where it is required to avoid them; and, on the other, it does not appear very evidently, that the suppositions, which are made in order to remove the fractions, are the only ones that could have taken place. Indeed, we have elsewhere shewn, that the finding of one solution of the equation $x^2 = Ay^2 + B$, is not always sufficient to enable us to deduce others from it, by means of the equation $p^2 = Aq^2 + 1$; and that, frequently, at least when B is not a prime number, there may be values of x and y, which cannot be contained in the general expressions of M. Euler.

With regard to the manner of resolving equations of the form $p^2 = Aq^2 + 1$, I think that of Chap. VII, however ingenious it may be, is still far from being perfect. For, in the first place, it does not shew that every equation of this kind is always resolvable in whose numbers, when A is a positive number not a square. Secondly, it is not demonstrated, that it must always lead to the solution sought for. Wallis, indeed, has professed to prove the former of these propositions; but his demonstration, if I may presume to say so, is a mere *petitio principii* (see Chap. XCIX). Mine, I believe, is the first rigid demonstration that has appeared. It is in the *Mélanges de Turin*, Vol. IV; but it is very long, and very indirect: that of Art. 37 is founded on the true principles of the subject, and leaves, I think, nothing to wish for. It enables us, also to appreciate that of Chap. VII, and to perceive the inconveniences into which it might lead, if followed without precaution. This is what we shall now discuss. [...]

I believe I had, at the same time with M. Euler, the idea of employing the irrational, and even imaginary factors of formulae of the second degree, in finding the conditions, which render those formulae equal to squares, or to any powers. On this subject, I read a Memoir to the Academy in 1768, which has not been printed; but of which I have given a summary at the end of my researches *On Indeterminate Problems*, which are to be found in the volume for the year 1767, printed in 1769, before even the German edition of M. Euler's *Algebra*.

In the place now quoted, I have shewn how the same method may be extended to formulae of higher dimensions than the second; and I have by these means given the solution of some equations, which it would perhaps have been extremely difficult to resolve in any other way. It is here intended to generalize this method still more, as it seems to deserve the attention of mathematicians, from its novelty and singularity.

14.D6 Johann Heinrich Lambert on the making of maps

1. Several conditions which an adequate map should satisfy have been recognized. It should: (1) not disturb the shape of countries; (2) Countries should maintain their true relative sizes; (3) The distance of every place from every other place should also be proportional to the true distance; (4) Places lying on a straight line on the earth's surface, that is, on a spherical great circle, should also lie on a straight line on the map; (5) The geographical latitude and longitude of places should be easily found on the map; and so on. In summary, a map should bear the same relation to countries, to hemispheres, or even to the entire earth, as do engineering drawings to a house, yard, garden, field or forest. This would work if the surface of the earth were a flat surface. But it is a spherical surface, and all of the requirements cannot be satisfied simultaneously, and it is therefore necessary to emphasize one or several especially valuable requirements at the expense of others. [...]
5. The notion that a terrestrial map should be viewed as a plan view of the earth's surface has not been binding, and perspective sketches have been employed instead of plan views. Here the earth's surface is drawn as it would be seen by the eye from a particular viewing point. [...]
6. [...] It is possible to choose innumerable points as the position of the eye from which to view the earth perspectively, but three advantageous points are emphasized. In one instance the eye is placed infinitely far from the globe, and this yields [...] orthographic projection. In a second instance the point is taken somewhere on the surface of the earth, and this method of projection is called stereographic, presumably because of the lack of a more specific expression. Finally, the eye is taken at the midpoint of the earth, and this method of projection, since no other name is known to me, will be called the central projection.

7. The central projection has the advantage that all spherical great circles are represented on it by straight lines. The small circles are in exceptional circumstances shown as circles but are otherwise always represented by conic sections. This method of composition therefore has the advantage that all places which lie on a straight line are on a great circle on the surface of the earth. I am not aware of any maps drawn in this manner, unless one includes those drawn on sundials where this construction occurs. Charts of the heavens, on the other hand, are advantageously drawn in this manner.
8. The three perspective compositions cited therefore have their advantages and disadvantages, and none satisfies all of the conditions cited in (1). In particular, the condition that the sizes of the countries maintain their true relations, is not found in any, and the conditions concerning measurement of separation of places either require restrictions or special constructions. [...]

Maps to determine the distances of places

12. Let there be three places A, P, B on the surface of the sphere. Their distances are $AP = \xi$, $BP = \eta$, $AB = \zeta$, and the angle $APB = \lambda$.

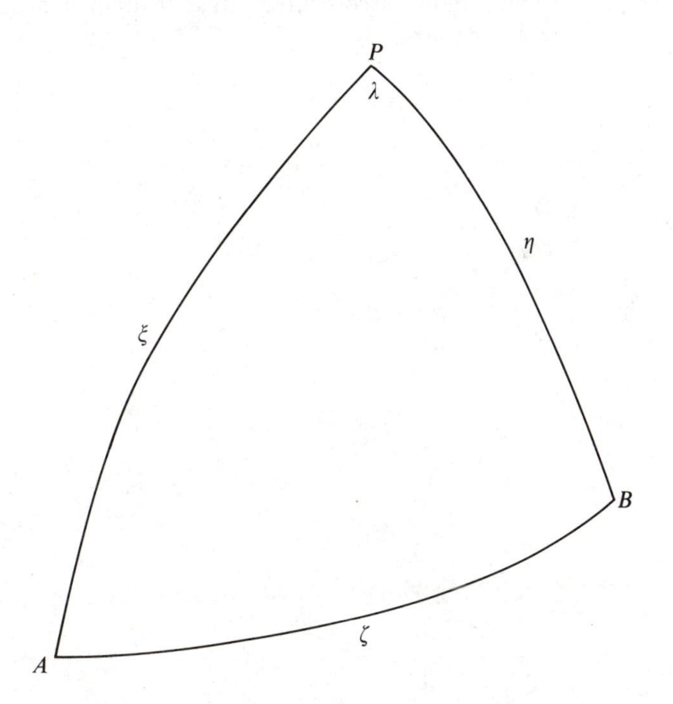

Here P can be the pole, and then one sees easily that λ is the difference in longitude of the two places, A, B.
13. Now let these places on the to-be-composed map be a, p, b. Let their distances be $ap = x$, $bp = y$, $ab = z$. By this I mean the lengths of the straight lines ap, bp, ab.

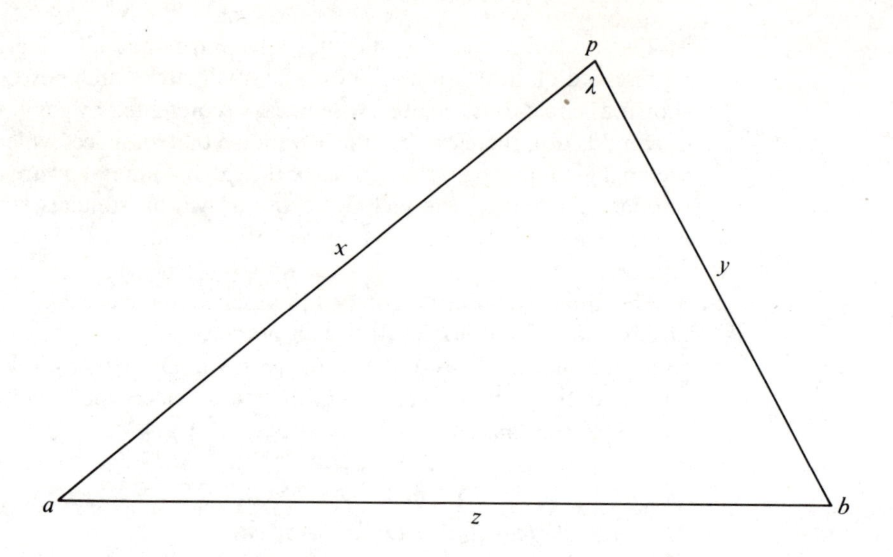

14. The three angles A, P, B on the spherical surface are together greater than 180 degrees, whereas the three plane angles a, p, b always equal 180 degrees. Therefore they cannot readily be compared with each other, nor can they be put into a proper relation with each other.

15 Gauss, and the Origins of Structural Algebra

Mathematics in the early nineteenth century was dominated by the aloof figure of Carl Friedrich Gauss (1777–1855). He had established himself as the leading number theorist and algebraist of his day, as perhaps the leading theoretical astronomer, and as a profound innovator in the theory of statistics. Much more than this lay unpublished in his notebooks. A positive treasure trove is alluded to by our first document (15.A1), which records the forty-nine items he confided to his mathematical diary in 1796, the first year in which it was kept. Not all of these items are comprehensible, but these cryptic notes seem to have been intended to provide Gauss with psychological reassurance; later on he laments that even he cannot always reconstruct their original meaning. Even so, the first entry records Gauss's celebrated discovery that the regular 17-gon is constructible by ruler and compass (15.A3(a)—all the more remarkable when one sees (as in 15.A3(b)) how deeply rooted this already was in his new theory of cyclotomy, a matter mentioned again and again in the diary. (Cyclotomy is concerned with complex algebraic numbers obtained from the roots of $x^n - 1 = 0$.) The second entry is also dramatic, being a crucial lemma on the way to the theorem of quadratic reciprocity, for which he finds another proof in June. By September his theory of elliptic functions is underway.

The second extract (15.A2) shows how novel, and how deep, were his criticisms of the work of his predecessors, who had been somewhat pragmatic on the question of the existence of mathematical objects. Gauss was much more precise. (For Gauss's proof of the fundamental theorem of algebra, the reader may consult Struik's *Source Book*.)

Gauss's first masterpiece was *Disquisitiones Arithmeticae* of 1801, from which we have taken his lucid and elementary introduction to modular arithmetic. Notice though that, curiously, Gauss stopped short of spelling out that it was now possible to calculate with residue classes as if they were essentially numbers themselves. This is the more interesting when one recalls that Gauss did exactly that with his remarkable theory of the composition of quadratic forms, which is the major theme of the book. Other important chapters in this work reported on his understanding of quadratic reciprocity (see 15.A4 for a special case) and of cyclotomy.

Typical of Gauss's insistence on working through a subject until he reached what he regarded as a truly basic organizing idea was his discovery of the intrinsic curvature of a surface (15.A5). This was the start of a wholesale reformulation of differential geometry, and led, in the hands of Bernhard Riemann later (see 16.C2), to a revolution in our ideas about geometry and physical space.

We have selected a few items from Gauss's voluminous correspondence, for this was one way in which he was willing to communicate with the scholarly world. To Sophie Germain (15.B1) he revealed himself pleasantly as one capable of appreciating good mathematics done by a woman and recognizing the difficulties women had to overcome in society before they could do it. We also see a nice illustration of his astonishing mastery of arithmetic. To his life-long friend Friedrich Wilhelm Bessel he divulged many of his doubts about the *a priori* validity of Euclidean geometry (15.B2). In the end he came to believe, correctly, that it could not be necessarily true, but the first thorough descriptions of a non-Euclidean geometry are due to Nicolai Lobachevskii in Russia and Janos Bolyai in Hungary (see Chapter 16). It is interesting to see that Bessel, like Gauss a professional astronomer, shared his doubts about the nature of geometry.

From a growing list of papers on number theory we have selected the one in which Adrien Marie Legendre stated the theorem of quadratic reciprocity in 1785, and observed some reservations about his proof (15.C1). Legendre, like Gauss, was well aware of the surprising connections between seemingly different questions about numbers; it was in exploring these connections that the systematic theory of numbers was created. Thus Gauss, as we can see, looked for proofs of quadratic reciprocity which would generalize to higher reciprocity laws. It was this question which animated a German mathematician of the next generation, E. E. Kummer (1810–1893) and led, as H. M. Edwards has argued (15.C2), to his invention of ideal complex numbers. It was formerly believed that this work derived from an interest in Fermat's last theorem, but Edwards argues convincingly for a richer and different association, which goes back to Gauss's work. Nonetheless, Fermat's last theorem enjoyed a revival of interest in the first half of the nineteenth century, with important papers by Germain and Dirichlet. 15.C2 gives a vivid picture of these debates.

Gauss was an increasingly conservative figure in a world polarized by the French Revolution and the Napoleonic wars. No mathematician could have been further from him than the ardent revolutionary Evariste Galois (1812–1832), but oddly enough they shared a radical attitude to mathematical arguments. Galois's famous letter to Auguste Chevalier (15.D1) is well-known, but the prefaces partially reprinted in 15.D2 perhaps give a more vivid picture of a young man shockingly estranged from, or beyond, the orthodoxies of his day. Galois's ideas took a generation to reach an audience, not least because the Académie des Sciences lost one of his manuscripts. But finally Joseph Liouville saw to it that some of Galois's papers were published, by which time Augustin Louis Cauchy had taken up the study of permutation groups (15.D3). Our last extract (15.D4) is taken from the preface to Camille Jordan's important book on the subject, written in 1870, and shows very clearly how aware he was of the steps on the road from eighteenth-century formal algebra à la Lagrange to the structural algebra with its Gaussian and Galoisian emphasis on the importance of concepts as opposed to computation.

15.A Gauss's Mathematical Writings

15.A1 Gauss's mathematical diary for 1796

[1] The principles upon which the division of the circle depend, and geometrical divisibility of the same into seventeen parts, etc. March 30 Brunswick.

[2] Furnished with a proof that in case of prime numbers not all numbers below them can be quadratic residues. April 8 Brunswick.

[3] The formulae for the cosines of submultiples of angles of a circumference will admit no more general expression except into two periods.
April 12 Brunswick.

[4] An extension of the rules for residues to residues and magnitudes which are not prime. April 29 Göttingen.

[5] Numbers which can be divided variously into two primes. May 14 Göttingen.

[6] The coefficients of equations are given easily as sums of powers of the roots.
May 23 Göttingen.

[7] The transformation of the series $1 - 2 + 8 - 64 + \ldots$ into the continued fraction

$$\cfrac{1}{1+\cfrac{2}{1+\cfrac{2}{1+\cfrac{8}{1+\cfrac{12}{1+\cfrac{32}{1+\cfrac{56}{1+128\ldots}}}}}}}$$

$$1 - 1 + 1\cdot3 - 1\cdot3\cdot7 + 1\cdot3\cdot7\cdot15 + . = \cfrac{1}{1+\cfrac{1}{1+\cfrac{2}{1+\cfrac{6}{1+\cfrac{12}{1+28\ldots}}}}}$$

and others. May 24 Göttingen.

[8] The simple scale in series which are recurrent in various ways is a similar function of the second order of the composite of the scales. May 26.

[9] A comparison of the infinities contained in prime and compound numbers.
May 31 Göttingen.

[10] A scale where the terms of the series are products or even arbitrary functions of the terms of arbitrarily many series. June 3 Göttingen.

[11] A formula for the sum of factors of an arbitrary compound number: general term

$$\frac{a^{n+1}-1}{a-1}.$$

June 5 Göttingen.

[12] The sum of the periods when all numbers less than a [certain] modulus are taken as elements: general term $[(n+1)a - na]a^{n-1}$. June 5 Göttingen.

[13] Laws of distributions. June 19 Göttingen.

[14] The sum to infinity of factors $= \frac{\pi^2}{6} \cdot$ sum of the numbers. June 20 Göttingen.

[15] I have begun to think of the multiplicative combination (of the forms of divisors of quadratic forms). June 22 Göttingen.

[16] A new proof of the golden theorem all at once, from scratch, different, and not a little elegant. June 27

[17] Any partition of a number a into three $\square$ gives a form separable into three $\square$. July 3

[17a] The sum of three squares in continued proportion can never be a prime: a clear new example which seems to agree with this. Be bold! July 9

[18] EUREKA. number $= \triangle + \triangle + \triangle$. July 10 Göttingen.

[19] Euler's determination of the forms in which composite numbers are contained more than once. [July Göttingen]

[20] The principles for compounding scales of series recurrent in various ways. July 16 Göttingen.

[21] Euler's method for demonstrating the relation between rectangles under line segments which cut each other in conic sections applied to all curves. July 31 Göttingen.

[22] $a^{2n+1(p)} \equiv 1$ can always be solved. August 3 Göttingen.

[23] I have seen exactly how the rationale for the golden theorem ought to be examined more thoroughly and preparing for this I am ready to extend my endeavours beyond the quadratic equations. The discovery of formulae which are always divisible by primes: $\sqrt[n]{1}$ (numerical). August 13 Göttingen

[24] On the way developed $(a + b\sqrt{-1})^{m+n\sqrt{-1}}$. August 14.

[25] Right now at the intellectual summit of the matter. It remains to furnish the details. August 16 Göttingen.

[26] $(a^p) \equiv (a) \bmod p$, a the root of an equation which is irrational in any way whatever. [August] 18.

[27] If P, Q are algebraic functions of an indeterminate quantity which are incommensurable. One is given $tP + uQ = 1$ in algebra as in number theory. [August] 19 Göttingen.

[28] The sums of powers of the roots of a given equation are expressed by a very simple law in terms of the coefficients of the equation (with other geometric matters in the Exercitiones). [August] 21 Göttingen.

[29] The summation of the infinite series

$$1 + \frac{x^n}{1\ldots n} + \frac{x^{2n}}{1\ldots 2n} \text{ etc.}$$

same day August 21.

[30] Certain small points aside, I have happily attained the goal namely if $p^n \equiv 1 \pmod{\pi}$ then $x^\pi - 1$ is composed of factors not exceeding degree n and therefore a sum of conditionally solvable equations; from this I have deduced two proofs of the golden theorem. September 2 Göttingen.

[31] The number of different fractions whose denominators do not exceed a certain bound compared to the number of all fractions whose numerators or

denominators are different and less than the same bound when taken to infinity is $6:\pi^2$. September 6.

[32] If $\int^{[x]} \frac{dt}{\sqrt{(1-t^3)}}$ is denoted $\Pi(x) = z$, and $x = \Phi(z)$ then

$$\Phi(z) = z - \frac{1}{8}z^4 - \frac{1}{112}z^7 - \frac{1}{1792}z^{10} + \frac{3z^{13}}{1792\cdot 52} - \frac{3\cdot 185z^{16}}{1792\cdot 52\cdot 14\cdot 15\cdot 16}\cdots$$

September 9.

[33] If $\Phi\left(\int \frac{dt}{\sqrt{(1-t^n)}}\right) = x$ then

$$\Phi{:}z = z - \frac{1\cdot z^n}{2\cdot n+1}A + \frac{n-1\cdot z^n}{4\cdot 2n+1}B - \frac{nn-n-1[z^n]}{2\cdot n+1\cdot 3n+1}C\ldots.$$

[34] An easy method for obtaining an equation in y from an equation in x, if given

$$x^n + ax^{n-1} + bx^{n-2}\ldots = y.$$

September 14.

[35] To convert fractions whose denominator contains irrational quantities (of any kind?) into others freed of this inconvenience. September 16.

[36] The coefficients of the auxiliary equation for the elimination are determined from the roots of the given equation. Same day.

[37] A new method by means of which it will be possible to investigate, and perhaps try to invent, the universal solution of equations. Namely by transforming into another whose roots are $\alpha\rho' + \beta\rho'' + \gamma\rho''' + \ldots$ where $\sqrt[n]{1} = \alpha, \beta, \gamma$ etc. and the number n denotes the degree of the equation. September 17.

[38] It seems to me the roots of an equation $x^n - 1$ $[=0]$ can be obtained from equations having common roots, so that principally one ought to solve such equations as enjoy rational coefficients. September 29 Brunswick.

[39] The equation of the third degree is this:

$$x^3 + xx - nx + \frac{n^2 - 3n - 1 - mp}{3} = 0$$

where $3n + 1 = p$ and m is the number of cubic residues omitting similarities. From this it follows that if $n = 3k$ then $m + 1 = 3l$, if $n = 3k \pm 1$, then $m = 3l$. Or

$$z^3 - 3pz + pp - 8p - 9pm = 0.$$

By these means m is completely determined, $m + 1$ is always $\square + 3\square$.

October 1 Brunswick.

[40] It is not possible to produce zero as a sum of integer multiples of the roots of the equation $x^p - 1 = 0$. ⊙ October 9 Brunswick.

[41] Obtained certain things concerning the multipliers of equations for the elimination of certain terms, which promise brightly.

⊙ October 16 Brunswick.

[42] Detected a law: and when it is proved a system will have been led to perfection.

October 18 Brunswick.

[43] Conquered GEGAN. October 21 Brunswick.

[44] An elegant interpolation formula. November 25 Göttingen.

[45] I have begun to convert the expression

$$1 - \frac{1}{2^\omega} + \frac{1}{3^\omega}$$

into a power series in which ω increases. November 26 Göttingen.
[46] Trigonometric formulae expressed in series. By December.
[47] Most general differentiations. December 23.
[48] A parabolic curve is capable of quadrature, given arbitrarily many points on it. December 26.
[49] I have discovered a true proof of a theorem of Lagrange. December 27.

15.A2 Critiques of attempts on the fundamental theorem of algebra

Gauss began by summarizing d'Alembert's *Recherches sur le Calcul Intégral* as asserting that a function X of x which takes the value 0 either for $x = 0$ or $x = \infty$ and which varies by infinitely small amounts as x does can always be written as a power series in infinitely small values of x. The convergence of this series guarantees that it takes infinitely small positive or negative values for values of x which are either real or of the form $p + q\sqrt{-1}$. He then continued as follows.

From this it follows that if X is a function of x which may obtain a real value V from the values of real v, and which may also obtain real values either greater or smaller than this by an infinitely small quantity, then it can receive a value which is smaller or greater then V by a finite amount, by assigning to x a value which is expressed in the form $p + q\sqrt{-1}$. This can be derived with no difficulty from the preceding if for X one can conceivably substitute $V + Y$, and for x, $v + y$. Finally, d'Alembert maintains that if one supposes that X can span the whole of some interval between two real values R and S (i.e. it becomes equal to R, to S, and to all the real values in between) by attributing to x values always of the form $p + q\sqrt{-1}$, then the function X can increase or diminish like a real finite quantity (as $S > R$ or $S < R$) while x remains of the form $p + q\sqrt{-1}$. [Gauss sketched d'Alembert's proof of this, and continued.] Now, if X is supposed to designate a function such as $x^m + Ax^{m-1} + Bx^{m-2} + \text{etc.} + M$, it is perceived without difficulty that real values can be given to x such that X may span a whole interval between two real values. Therefore some such value expressed in the form $p + q\sqrt{-1}$ could also be obtained for which X becomes $= 0$. Q.E.D.

Here are the basic points which seem able to be brought against d'Alembert's demonstration.

1. D'Alembert raises no doubts about the *existence* of values of x, to which given values of X may correspond, but supposes their existence, and investigates only the *form* of the values. Although this objection is indeed very serious in itself,

nevertheless it refers here only to the form of the expression, which can be corrected so easily that the objection altogether loses its force.

2. The assertion, that ω can always be expressed through such a series as he proposes, is certainly false, if X is meant to designate any transcendental function (as d'Alembert indicates in several places). This is clear if, for example, $X = e^{1/x}$, or $x = \frac{1}{\log X}$. However, if we confine the demonstration to that case, where X is an algebraic function of x (which in the present affair is sufficient), the proposition is certainly true. But d'Alembert has brought forward nothing to confirm his supposition. [...]
3. He uses infinitely small quantities more freely than one can concede from a scrupulous analysis is consistent with the rigours of geometry, at least in our age (where those things are rightly censured); and he also did not explain clearly enough the leap from an infinitely small value of Ω to a finite one.

Moreover, because this whole argument seems too uncertain to permit any rigorous conclusion to be drawn from it, I observe that there certainly are series which, no matter how small a value may be granted to the quantity of which it proceeds in powers, nevertheless always diverge, so that if it continues uninterrupted for long enough, you can reach values which are greater than any given quantity. This happens when the terms of the series constitute a hypergeometric progression. For this reason it would have been necessary to show that such a hypergeometric series cannot arise in the present instance. [...]

Euler tacitly supposes that the equation $X = 0$ has $2m$ roots, of which he determines the sum to be $= 0$ because the second term in X is missing. What I think of this licence I have already declared in art. 3. The proposition that the sum of all the roots of an equation is equal to the first coefficient with the sign changed, does not seem applicable to other equations unless they have roots; now although it ought to be proved by this same demonstration that the equation $X = 0$ really does have roots, it does not seem permissible to suppose the existence of these. No doubt those people who have not yet penetrated the fallacy of this expression will reply, 'Here it has not been demonstrated that the equation $X = 0$ can be satisfied (for this expression means that the equation has roots) but it has only been demonstrated that the equation can be satisfied by values of x of the form $a + b\sqrt{-1}$; and indeed that is taken as axiomatic'. But although types of quantities other than real and imaginary $a + b\sqrt{-1}$ cannot be conceived of, it does not seem sufficiently clear how the proposition awaiting demonstration differs from that supposed as axiomatic. [...] Therefore that axiom can have no other meaning than this: Any equation can be satisfied *either* by the real value of an unknown, *or* by an imaginary value expressed in the form $a + b\sqrt{-1}$, or perhaps by a value in some other form which we do not know, or by a value which is not totally contained in any form. But how such quantities which are shadowy and inconceivable can be added or multiplied is certainly not understood with the clarity which is required in mathematics. [...]

Finally, Lagrange has dealt with our theorem in the commentary *Sur la Forme des Racines Imaginaires des Equations*, 1772. This great geometer handed his work to the printers when he was worn out with completing Euler's first demonstration [...]. However, he does not touch upon the third objection at all, for all his investigation is built upon the supposition that an equation of the mth degree does in fact have roots.

15.A3 The constructability of the regular 17-gon

(a) The discovery of the constructability

It is well known to every beginner in Geometry that various regular polygons can be constructed geometrically, namely the triangle, pentagon, 15-gon, and those which arise from these by repeatedly doubling the number of sides. One had already got this far in Euclid's time, and it seems that one has persuaded oneself ever since that the domain of elementary geometry could not be extended; at least I do not know of any successful attempts to enlarge its boundaries on this side.

It seems to me then to be all the more remarkable that *besides the usual polygons there is a collection of others which are constructable geometrically, e.g. the 17-gon.* This discovery is properly only a corollary of a not quite completed discovery of greater extent which will be laid before the public as soon as it is completed.

(b) The theory underlying the constructability

Nevertheless none of these equations is so tractable and so suitable for our purposes as $x^n - 1 = 0$. Its roots are intimately connected with the roots of the above. That is, if for brevity we write i for the imaginary quantity $\sqrt{-1}$, the roots of the equation $x^n - 1 = 0$ will be

$$\cos\frac{kP}{n} + i\sin\frac{kP}{n} = r$$

where for k we should take all the numbers $0, 1, 2, \ldots, n-1$. Therefore since $1/r = \cos kP/n - i\sin kP/n$ the roots of equation I will be $[r - (1/r)]/2i$ or $i(1 - r^2)/2r$; the roots of equation II, $[r + (1/r)]/2 = (1 + r^2)/2r$; finally the roots of equation III, $i(1 - r^2)/(1 + r^2)$. For this reason we build our investigation on a consideration of the equation $x^n - 1 = 0$, and presume that n is an odd prime number.

[When Gauss considered the 17-gon, whose vertices are the complex roots of $x^{17} - 1$, he labelled the 16 roots other than $x = 1$ $[1], [2], \ldots, [16]$, and then calculated them by working through a chain of four quadratics. His results, expressed numerically, are as follows.]

If we then compute the remaining roots we will obtain the following numerical values, where the upper sign is to be taken for the first root, the lower sign for the second:

$$\begin{aligned}
&[1],[16]\ldots\ 0.9324722294 \pm 0.3612416662i\\
&[2],[15]\ldots\ 0.7390089172 \pm 0.6736956436i\\
&[3],[14]\ldots\ 0.4457383558 \pm 0.8951632914i\\
&[4],[13]\ldots\ 0.0922683595 \pm 0.9957341763i\\
&[5],[12]\ldots -0.2736629901 \pm 0.9618256432i\\
&[6],[11]\ldots -0.6026346364 \pm 0.7980172273i\\
&[7],[10]\ldots -0.8502171357 \pm 0.5264321629i\\
&[8],[\ 9]\ldots -0.9829730997 \pm 0.1837495178i
\end{aligned}$$

[...]

Thus by the preceding discussions we have reduced the division of the circle into n parts, if n is a prime number, to the solution of as many equations as there are factors in the number $n-1$. The degree of the equations is determined by the size of the factors. Whenever therefore $n-1$ is a power of the number 2, which happens when the value of n is 3, 5, 17, 257, 65537, etc. the sectioning of the circle is reduced to quadratic equations only, and the trigonometric functions of the angles P/n, $2P/n$, etc. can be expressed by square roots which are more or less complicated (according to the size of n). Thus in these cases the division of the circle into n parts or the inscription of a regular polygon of n sides can be accomplished by geometric constructions. Thus, e.g., for $n = 17$, by articles 354, 361 we get the following expression for the cosine of the angle $P/17$:

$$-\tfrac{1}{16} + \tfrac{1}{16}\sqrt{17} + \tfrac{1}{16}\sqrt{[34 - 2\sqrt{17}]} + \tfrac{1}{8}\sqrt{[17 + 3\sqrt{17} - \sqrt{(34 - 2\sqrt{17})} - 2\sqrt{(34 + 2\sqrt{17})}]}$$

The cosine of multiples of this angle will have a similar form, but the sine will have one more radical sign. It is certainly astonishing that although the geometric divisibility of the circle into three and five parts was already known in Euclid's time, nothing was added to this discovery for 2000 years. And all geometers had asserted that, except for those sections and the ones that derive directly from them (that is, division into 15, $3\cdot 2^{\mu}$, $5\cdot 2^{\mu}$, and 2^{μ} parts), there are no others that can be effected by geometric constructions. But it is easy to show that if the prime number $n = 2^m + 1$, the exponent m can have no other prime factors except 2, and so it is equal to 1 or 2 or a higher power of the number 2. For if m were divisible by any odd number ζ (greater than unity) so that $m = \zeta\eta$, then $2^m + 1$ would be divisible by $2^{\eta} + 1$ and so necessarily composite. All values of n, therefore, that can be reduced to quadratic equations, are contained in the form $2^{2^v} + 1$. Thus the five numbers 3, 5, 17, 257, 65537 result from letting $v = 0, 1, 2, 3, 4$ or $m = 1, 2, 4, 8, 16$. But the geometric division of the circle cannot be accomplished for *all* numbers contained in the formula but only for those that are prime. Fermat was misled by his induction and affirmed that all numbers contained in this form are necessarily prime, but the distinguished Euler first noticed that this rule is erroneous for $v = 5$ or $m = 32$, since the number $2^{32} + 1 = 4294967297$ involves the factor 641.

Whenever $n-1$ implies prime factors other than 2, we are always led to equations of higher degree, namely, to one or more cubic equations when 3 appears once or several times among the prime factors of $n-1$, to equations of the fifth degree when $n-1$ is divisible by 5, etc. *We can show with all rigour that these higher-degree equations cannot be avoided in any way nor can they be reduced to lower-degree equations.* The limits of the present work exclude this demonstration here, but we issue this warning lest anyone attempt to achieve geometric constructions for sections other than the ones suggested by our theory (e.g. sections into 7, 11, 13, 19, etc. parts) and so spend his time uselessly.

15.A4 The charms of number theory

The fundamental theorem on quadratic residues is one of the most beautiful truths of higher Arithmetic, [and] was indeed easily found by induction [inspection from particular cases], but it was far more difficult to prove it. In this kind of research it is usually seen that the proofs of the simplest truths, which the researcher has so to speak discovered for himself in an inductive way, lie hidden very deeply and can finally be brought to the light of day only after many unsuccessful attempts and then in a way very different from how one originally sought them. Moreover, it appears it is then quite often the case that as soon as one way has been found, several other ways open up which lead to the same goal, some shorter and more direct, others coming in almost from the side and proceeding on quite different principles, and one scarcely conjectured any connection between these and the previous researches. Such a wonderful connection between widely separated truths gives these researches not only a certain particular charm, but also deserves to be diligently studied and clarified, because it is not seldom that new techniques and advances of the science can be made on this account.

15.A5 Curvature and the differential geometry of surfaces

Although geometers have given much attention to general investigations of curved surfaces and their results cover a significant portion of the domain of higher geometry, this subject is still so far from being exhausted, that it can well be said that, up to this time, but a small portion of an exceedingly fruitful field has been cultivated. Through the solution of the problem, to find all representations of a given surface upon another in which the smallest elements remain unchanged, the author sought some years ago to give a new phase to this study. The purpose of the present discussion is further to open up other new points of view and to develop some of the new truths which thus become accessible. We shall here give an account of those things which can be made intelligible in a few words. But we wish to remark at the outset that the new theorems as well as the presentations of new ideas, if the greatest generality is to be attained, are still partly in need of some limitations or closer determinations, which must be omitted here.

In researches in which an infinity of directions of straight lines in space is concerned, it is advantageous to represent these directions by means of those points upon a fixed sphere, which are the end points of the radii drawn parallel to the lines. The centre and the radius of this *auxiliary sphere* are here quite arbitrary. The radius may be taken equal to unity. This procedure agrees fundamentally with that which is constantly employed in astronomy, where all directions are referred to a fictitious celestial sphere of infinite radius. Spherical trigonometry and certain other theorems, to which the author has added a new one of frequent application, then serve for the solution of the problems which the comparison of the various directions involved can present.

If we represent the direction of the normal at each point of the curved surface by the corresponding point of the sphere, determined as above indicated, namely, in this way, to every point on the surface, let a point on the sphere correspond; then, generally speaking, to every line on the curved surface will correspond a line on the sphere, and to every part of the former surface will correspond a part of the latter. The less this part differs from a plane, the smaller will be the corresponding part on the sphere. It is, therefore, a very natural idea to use as the measure of the total curvature, which is to be assigned to a part of the curved surface, the area of the corresponding part of the sphere. For this reason the author calls this area the *integral curvature* of the corresponding part of the curved surface. [...]

The solution of the problem, to find the measure of curvature at any point of a curved surface, appears in different forms according to the manner in which the nature of the curved surface is given. When the points in space, in general, are distinguished by three rectangular coordinates, the simplest method is to express one coordinate as a function of the other two. In this way we obtain the simplest expression for the measure of curvature. But, at the same time, there arises a remarkable relation between this measure of curvature and the curvatures of the curves formed by the intersections of the curved surface with planes normal to it. Euler, as is well known, first showed that two of these cutting planes which intersect each other at right angles have this property, that in one is found the greatest and in the other the smallest radius of curvature; or, more correctly, that in them the two extreme curvatures are found. It will follow then from the above mentioned expression for the measure of curvature that this will be equal to a fraction whose numerator is unity and whose denominator is the product of the extreme radii of curvature. The expression for the measure of curvature will be less simple, if the nature of the curved surface is determined by an equation in x, y, z. And it will become still more complex, if the nature of the curved surface is given so that x, y, z are expressed in the form of functions of two new variables p, q. In this last case the expression involves fifteen elements, namely, the partial differential coefficients of the first and second orders of x, y, z with respect to p and q. But it is less important in itself than for the reason that it facilitates the transition to another expression, which must be classed with the most remarkable theorems of this study. If the nature of the curved surface be expressed by this method, the general expression for any linear element upon it, or for $\sqrt{(dx^2 + dy^2 + dz^2)}$, has the form $\sqrt{(Edp^2 + 2Fdp \cdot dq + Gdq^2)}$, where E, F, G are again functions of p and q. The new expression for the measure of curvature mentioned above contains merely these magnitudes and their partial differential coefficients of the first and second order. Therefore we notice that, in order to determine the measure of curvature, it is necessary to know only the general expression for a linear element; the expressions for the coordinates x, y, z are not required. A direct result from this is the remarkable theorem: If a curved surface, or a part of it, can be developed upon another surface, the measure of curvature at every point remains unchanged after the development. In particular, it follows from this further: Upon a curved surface that can be developed upon a plane, the measure of curvature is everywhere equal to zero. From this we derive at once the characteristic equation of surfaces developable upon a plane, namely,

$$\frac{\partial^2 z}{\partial x^2} \cdot \frac{\partial^2 z}{\partial y^2} - \left(\frac{\partial^2 z}{\partial x \cdot \partial y}\right)^2 = 0,$$

when z is regarded as a function of x and y. This equation has been known for some time, but according to the author's judgement it has not been established previously with the necessary rigour. [...]

If upon a curved surface a system of infinitely many shortest lines of equal lengths be drawn from one initial point, then will the line going through the end points of these shortest lines cut each of them at right angles. If at every point of an arbitrary line on a curved surface shortest lines of equal lengths be drawn at right angles to this line, then will all these shortest lines be perpendicular also to the line which joins their other end points. Both these theorems, of which the latter can be regarded as a generalization of the former, will be demonstrated both analytically and by simple geometrical considerations. *The excess of the sum of the angles of a triangle formed by shortest lines over two right angles is equal to the total curvature of the triangle.* [...]

Evidently we can express this important theorem thus also: the excess over the two right angles of the angles of a triangle formed by shortest lines is to eight right angles as the part of the surface of the auxiliary sphere, which corresponds to its as its integral curvature, is to the whole surface of the sphere. In general, the excess over $2n - 4$ right angles of the angles of a polygon of n sides, if these are shortest lines, will be equal to the integral curvature of the polygon.

15.B Gauss's Correspondence

15.B1 Three letters between Gauss and Sophie Germain

(*a*) *Germain to Gauss, 21 November 1804*

Germain corresponded with Gauss on questions concerning number theory using the pseudonym Antoine Le Blanc.

Your *Disquisitiones Arithmeticae* have been the object of my admiration and my study for a long time. The last chapter of this book contains, amongst other remarkable things, the beautiful theorem about the equation $4\frac{(x^n - 1)}{x - 1} = y^2 \pm nz^2$; I believe that it can be generalized. [She went on to describe how, and continued as follows.]

I add to this art some other considerations which relate to the famous equation of Fermat $x^n + y^n = z^n$ whose impossibility in integers has still only been proved for $n = 3$ and $n = 4$; I think I have been able to prove it for $n = p - 1$, p being a prime number of the form $8k + 7$. I shall take the liberty of submitting this attempt to your judgement, persuaded that you will not disdain to help with your advice an enthusiastic amateur in the science which you have cultivated with such brilliant success....

(*b*) *Germain to Gauss, 20 February 1807*

Germain used her influence to help ensure that Gauss came to no harm when the Napoleonic War brought danger to Brunswick, but her emissary inadvertently revealed her true name. She then wrote to explain.

The consideration due to superior men will explain the care I have taken to ask General Pernety to make it known to whomever he thought appropriate that you have the right to the esteem of any enlightened government.

In describing the honourable mission I charged him with, M. Pernety informed me that he had made known to you my name. This has led me to confess that I am not as completely unknown to you as you might believe, but that fearing the ridicule attached to a female scientist, I have previously taken the name of M. LeBlanc in communicating to you those notes that, no doubt, do not deserve the indulgence with which you have responded.

The appreciation I owe you for the encouragement you have given me, in showing me that you count me among the lovers of sublime arithmetic whose mysteries you have developed, was my particular motivation for finding out news of you at a time when the troubles of the war caused me to fear for your safety; and I have learned with complete satisfaction that you have remained in your house as undisturbed as circumstance would permit. I hope, however, that these events will not keep you too long from your astronomical and especially your arithmetical researches, because this part of science has a particular attraction for me, and I always admire with new pleasure the linkages between the truths exposed in your book. Unfortunately, the ability to think with force is an attribute reserved for a few privileged minds, and I am sure that I will not encounter any of the developments that you deduce, seemingly so effortlessly, from those that you have already made known.

I include with my letter a note intended to show you that I have maintained an appetite for analysis that the reading of your work has inspired, and that has continually provided me with the confidence to send you my feeble attempts, without any other recommendation to you than the goodwill accorded by scientists to admirers of their work.

I hope that the information that I have today confided to you will not deprive me of the honour you have accorded me under a borrowed name, and that you will devote a few minutes to write me news of yourself.

(*c*) *Gauss to Germain, 30 April 1807*

Your letter of February 20, which did not arrive until March 12, was for me the source of as much pleasure as surprise. How pleasant and heartwarming to acquire a friend so flattering and precious. The lively interest that you have taken in me during this war deserves the most sincere appreciation. Your letter to General Pernety would have been most useful to me, if I had needed special protection on the part of the French government.

Happily, the events and consequences of war have not affected me so much up until now, although I am convinced that they will have a large influence on the future course of my life. But how can I describe my astonishment and admiration on seeing my esteemed correspondent M. LeBlanc metamorphosed into this celebrated person, yielding a copy so brilliant it is hard to believe? The taste for the abstract sciences in general and, above all, for the mysteries of numbers, is very rare: this is not surprising, since the charms of this sublime science in all their beauty reveal themselves only to those who have the courage to fathom them. But when a woman, because of her sex, our customs and prejudices, encounters infinitely more obstacles than men in familiarizing herself with their knotty problems, yet overcomes these fetters and penetrates that which is most hidden, she doubtless has the most noble courage, extraordinary talent, and superior genius. Nothing could prove to me in a more flattering and less equivocal way that the attractions of that science, which have added so much joy to my life, are not chimerical, than the favour with which you have honoured it.

The scientific notes with which your letters are so richly filled have given me a thousand pleasures. I have studied them with attention, and I admire the ease with which you penetrate all branches of arithmetic, and the wisdom with which you generalize and perfect. I ask you to take it as a proof of my attention if I dare add a remark to your last letter. It seems to me that the inverse proposition 'If the sum of the nth powers of two numbers is of the form $hh + nff$, then the sum of the numbers themselves will be of the same form' is put a little too strongly. Here is an example of where this rule fails:

$$\begin{aligned} 15^{11} + 8^{11} &= 8649755859375 + 8589934592 \\ &= 8658345793967 \\ &= (1595826)^2 + 11(745391)^2 \end{aligned}$$

Nevertheless $15 + 8 = 23$ cannot be reduced to the form $xx + 11yy$.

15.B2 Three letters between Gauss and Friedrich Wilhelm Bessel

(a) Gauss to Bessel, 27 January 1829

Also, another theme which is almost 40 years old with me that I have been thinking about now and then in a few free hours; I mean the foundations of geometry. I don't know if I've told you my opinions about this. I have consolidated some things further here too, and my opinion that we cannot establish geometry completely *a priori* is, if possible, much firmer. Meanwhile I will still not get round to it for some time and work up my very extensive researches for publication, and perhaps they will never appear in my lifetime, for I fear the howl of the Boeotians if I speak my opinion out loud. But it is strange that apart from the well-known holes in Euclid's geometry which up to now one has gratuitously sought to fill, and never will fill, still there is another omission in it which to my knowledge no one has pointed out, and is in no way easy (although possible) to put right. This is the definition of the plane as a surface in which the line joining any two points lies entirely. This definition entails more, as it is necessary to determine the surface and tacitly involves a theorem which first must be proved.

(b) Bessel to Gauss, 10 February 1829

I would be very sorry if you allowed yourself to be stopped by 'the howl of the Boeotians' from setting down your geometric opinions. From what Lambert has said and what Schweikart has been saying, it has become clear to me that our geometry is incomplete and should receive a correction, which is hypothetical and vanishes if the sum of the angles in a plane triangle is = 180°. This will be the true geometry, Euclid's the practical, at least for figures on the Earth.

(c) Gauss to Bessel, 9 April 1830

The readiness that you have shown me to engage with my opinions about geometry has been a real delight to me, especially since so few have an open mind about it. It is my deepest conviction that the study of space in our science *a priori* has a quite different position to the study of quantity; our knowledge of it differs from that complete conviction of its necessity (thus also from its absolute truth) which is proper to the latter; we must clothe ourselves in meekness, for, if number is entirely a product of our minds, space has a reality outside of our minds and we cannot prescribe its laws completely *a priori*.

15.C Two Number Theorists

15.C1 Adrien Marie Legendre on quadratic reciprocity

(a) Legendre's discovery of quadratic reciprocity

It is a matter for considerable regret that Fermat, who cultivated the theory of numbers with so much success, did not leave us with the proofs of the theorems he discovered. In truth, Messrs Euler and de la Grange, who have not disdained this kind of research, have proved most of these theorems, and have even substituted extensive theories for the isolated propositions of Fermat. But there are several which have resisted their efforts, either because Fermat did not really have a solid proof of them, which is difficlt to believe, or because the means to discover them is still quite unknown. Amongst the propositions not yet proved it is most necessary to note these two: every number is the sum of at most three triangular numbers; every prime number of the form $8n - 1$ is of the form $p^2 + q^2 + 2r^2$ or, which comes to the same thing, its double is a sum of three squares.

[After reviewing the state of the art, Legendre says that Lagrange's work has inspired him to prove 'some very general propositions about prime numbers which appear to advance this part of analysis and to merit the attention of mathematicians'. He then stated eight results, in which the letters A, a stand for prime numbers of the form $4n + 1$ and B, b for prime numbers of the form $4n - 1$. These results between them form the celebrated theorem of quadratic reciprocity. The equalities are to be understood with respect to the prime number in the exponent, and as he pointed out arise when one tries to take square roots in Fermat's little theorem: if $a^{b-1} \equiv 1 \pmod b$ what is $a^{(b-1)/2} \pmod b$? His answers are as follows.]

1: If $b^{(a-1)/2} = 1$, then $a^{(b-1)/2} = 1$.

2: If $a^{(b-1)/2} = -1$, then $b^{(a-1)/2} = -1$.

3: If $a^{(A-1)/2} = 1$, then $A^{(a-1)/2} = 1$.

4: If $a^{(A-1)/2} = -1$, then $A^{(a-1)/2} = -1$.

5: If $a^{(b-1)/2} = 1$, then $b^{(a-1)/2} = 1$.

6: If $b^{(a-1)/2} = -1$, then $a^{(b-1)/2} = -1$.

7: If $b^{(B-1)/2} = 1,\quad$ then $B^{(b-1)/2} = -1.$

8: If $b^{(B-1)/2} = -1,\quad$ then $B^{(b-1)/2} = 1.$

The theorems thus described are of great generality, but can all be included in the following statement: c and d being prime numbers, the expressions $c^{(d-1)/2}$ and $d^{(c-1)/2}$ are only of different signs when c and d are both of the form $4n - 1$; in all other cases these expressions are of the same sign.

[Legendre then set about proving this theorem, using his own result about sums of three squares. He concluded his paper in the following way.]

It will perhaps be necessary to prove rigorously something which we have assumed at several places in this article, i.e. that there is an infinity of prime numbers contained in any arithmetic progression whose first term and increment are relatively prime, or, which comes to the same thing, are of the form $2mx + \mu$ where $2m$ and μ have no common divisor. This proposition is quite difficult to prove, however one can assure oneself that it is true by comparing this arithmetic progression with the ordinary progression 1, 3, 5, 7 etc. If one takes a considerable number of terms in these progressions, the same in both, and arranges them, for example, in such a way that the greatest terms shall be equal and in the same place on both sides; one will see that in omitting from each side the multiples of 3, 5, 7 etc. up to a certain prime number p, the same number of terms must remain on each side or else there will be fewer remaining in the progression 1, 3, 5, 7, etc. But as prime numbers necessarily remain in this one, they must also remain in the other. I content myself with outlining the means for proving this theorem, which it would be too long to give in detail, and besides this memoir has already exceeded the ordinary limits.

(*b*) *Gauss's commentary on Legendre's discovery*

The illustrious Legendre presented his demonstration again in his excellent work *Essai d'une Théorie des Nombres* but in such a way as to change nothing essential. So this method is still subject to all [my earlier objections]. It is true that the theorem (on which one supposition is based) which states that any arithmetic progresion $l, l + k, l + 2k$, etc. contains prime numbers if k and l do not have a common divisor, is given more fully here, but it does not yet seem to satisfy geometric rigour. But even if this theorem were fully demonstrated, the second supposition remains (that there are prime numbers of the form $4n + 3$ for which a given positive prime number of the form $4n + 1$ is a quadratic nonresidue) and I do not know whether this can be proven *rigorously* unless the fundamental theorem is *presumed*. But it must be remarked that Legendre did not tacitly assume this last supposition, nor did he ignore it.

15.C2 E. E. Kummer: ideal numbers and Fermat's last theorem

On 1 March 1847 Gabriel Lamé startled the Paris Académie by outlining a 'proof' of Fermat's last theorem; which Joseph Liouville at once criticized, on the grounds that Lamé imputed to ideal numbers the property of unique factorization, which was known to hold for ordinary integers but required proof in the general case. Augustin Louis Cauchy, however, indicated that he was also working in the direction taken by Lamé, but was not yet secure in all the details. Things then heated up considerably, as H. M. Edwards now describes.

In the March 22nd proceedings it is recorded that *both* Cauchy and Lamé deposited 'secret packets' with the Academy. The depositing of secret packets was an institution of the Academy which allowed members to go on record as having been in possession of certain ideas at a certain time—without revealing them—in case a priority dispute later developed. In view of the circumstances of March 1847, there is little doubt what the subject of these two packets was. As it turned out, however, there was no priority dispute whatever on the subject of unique factorization and Fermat's Last Theorem.

In the following weeks, Lamé and Cauchy each published notices in the proceedings of the Academy, notices that are annoyingly vague and incomplete and inconclusive. Then, on May 24, Liouville read into the proceedings a letter from Kummer in Breslau which ended, or should have ended, the entire discussion. Kummer wrote to Liouville to tell him that his questioning of Lamé's implicit use of unique factorization had been quite correct. Kummer not only asserted that unique factorization *fails*, he also included with his letter a copy of a memoir he had *published three years earlier* in which he had demonstrated the failure of unique factorization in cases where Lamé had been asserting it was valid. However, he went on to say, the theory of factorization can be 'saved' by introducing a new kind of complex numbers which he called 'ideal complex numbers'; these results he had *published one year earlier in the proceedings of the Berlin Academy* in resumé form and a complete exposition of them was soon to appear in Crelle's Journal. He had for a long time been occupied with the application of his new theory to Fermat's Last Theorem and said he had succeeded in reducing its proof for a given n to the testing of two conditions on n. For the details of this application and of the two conditions, he refers to the notice he had published that same month in the proceedings of the Berlin Academy (15 April 1847). There he in fact stated the two conditions in full and said that he 'had reason to believe' that $n = 37$ did not satisfy them.

The reaction of the learned gentlemen of Paris to this devastating news is not recorded. Lamé simply fell silent. Cauchy, possibly because he had a harder head than Lamé or possibly because he had invested less in the success of unique factorization, continued to publish his vague and inconclusive articles for several more weeks. In his only direct reference to Kummer he said, 'What little [Liouville] has said [about Kummer's work] persuades me that the conclusions which Mr Kummer has reached are, at least in part, those to which I find myself led by the above considerations. If Mr Kummer has taken the question a few steps further, if in fact he has succeeded in removing all the obstacles, I would be the first to applaud his efforts; for what we should desire the most is that the works of all the friends of science should come together to make known and to propagate the truth.' He then proceeded to ignore—rather than to propagate—Kummer's work and to pursue his own ideas with only an occasional promise that he would eventually relate his statements to Kummer's work, a promise he never fulfilled. By the end of the summer, he too fell silent on the subject of Fermat's Last Theorem. (Cauchy was not the silent type, however, and he merely began producing a torrent of papers on mathematical astronomy.) This left the field to Kummer, to whom, after all, it had already belonged for three years.

It is widely believed that Kummer was led to his 'ideal complex numbers' by his interest in Fermat's Last Theorem, but this belief is surely mistaken. Kummer's use of the letter λ (lambda) to represent a prime number, his use of the letter α to denote a 'λth root of unity'—that is, a solution of $\alpha^\lambda = 1$—and his study of the factorization of prime numbers $p \equiv 1 \bmod \lambda$ into 'complex numbers composed of λth roots of unity' all derive directly from a paper of Jacobi which is concerned with *higher reciprocity laws*. Kummer's 1844 memoir was addressed by the University of Breslau to the University of Königsberg in honour of its jubilee celebration, and the memoir was definitely meant as a tribute to Jacobi, who for many years was a professor at Königsberg. It is true that Kummer had studied Fermat's Last Theorem in the 1830s and in all probability he was aware all along that his factorization theory would have implications for Fermat's Last Theorem, but the subject of Jacobi's interest, namely, higher reciprocity laws, was surely more important to him, both at the time he was doing the work and after. At the same time that he was demolishing Lamé's attempted proof and replacing it with his own partial proof, he referred to Fermat's Last Theorem as 'a curiosity of number theory rather than a major item', and later, when he published his version of the higher reciprocity law in the form of an unproved conjecture, he referred to the higher reciprocity laws as 'the principal subject and the pinnacle of contemporary number theory'.

There is even an often told story that Kummer, like Lamé, believed he had proved Fermat's Last Theorem until he was told—by Dirichlet in this story—that his argument depended on the unproved assumption of unique factorization into primes. Although this story does not necessarily conflict with the fact that Kummer's primary interest was in the higher reciprocity laws, there are other reasons to doubt its authenticity. It first appeared in a memorial lecture on Kummer given by Hensel in 1910 and, although Hensel describes his sources as unimpeachable and gives their names, the story is being told at third hand over 65 years later. Moreover, the person who told it to Hensel was not apparently a mathematician and it is very easy to imagine how the story could have grown out of a misunderstanding of known events. Hensel's story would be confirmed if the 'draft ready for publication' which Kummer is supposed to have completed and sent to Dirichlet could be found, but unless this happens the story should be regarded with great scepticism. Kummer seems unlikely to have assumed the validity of unique factorization and even more unlikely to have assumed it unwittingly in a paper he intended to publish.

15.D Galois Theory

15.D1 Evariste Galois's letter to Auguste Chevalier, 29 May 1832

My dear friend, I have done several new things in analysis.

Some concern the theory of equations; others, integral functions.

In the theory of equations I have found out in which cases the equations are solvable by radicals, which has given me the occasion to deepen the theory and to describe all the transformations admitted by an equation, even when it is not solvable by radicals.

One could make three memoirs of all that.

The first is written, and despite what Poisson has said about it, I hold it aloft with the corrections that I have made.

The second contains quite curious applications of the theory of equations. Here is a summary of the most important things.

(1) Following propositions II and III of the first memoir one sees a great difference between adjoining to an equation one or all the roots of an auxiliary equation.

In both cases the group of the equation is partitioned by the adjunction into groups such that one passes from one to the other by means of the same substitution; but it is only in the second case that it is certain that these groups have the same substitutions. This is called a 'proper decomposition'.

In other words, when a group G contains another group H, then the group G can be divided into groups that are obtained by performing the same substitution on the permutations of H, so that

$$G = H + HS + HS' + \cdots$$

It can also be divided into groups with the same substitutions so that

$$G = H + TH + T'H + \cdots$$

These two kinds of decomposition do not ordinarily coincide. When they do, the decomposition is said to be 'proper'.

It is easy to see that when the group of an equation does not admit any proper decomposition then one can transform it all one wants; the groups of the transformed equations will always have the same number of permutations.

If on the contrary, the new group admits a proper decomposition, so that it is divided into M groups of N permutations, then one can solve the given equation by means of two equations, one having a group of M permutations, and the other one of N permutations.

Thus, when one has exhausted in the group of an equation all the possibilities of proper decomposition, then one has groups that one can transform but that always have the same number of permutations.

If each of these groups has a prime number of permutations, then the equation will be solvable by radicals; otherwise, not.

The smallest number of permutations an indecomposable group can have, when this number is not prime, is $5 . 4 . 3$.

(2) The simplest decompositions are those which arise by Gauss's method.

As these decompositions are obvious even in the actual form of the group of the equation, it is pointless to spend a long time on them [...].

You will print this letter in the *Revue Encyclopedique*.

I have often in my life dared to advance propositions I was not sure of; but everything I have written here has been in my head for over a year, and it is too much in my interest not to deceive myself so that someone could suspect me of stating theorems of which I didn't have a complete proof.

You will publicly beg Jacobi or Gauss to give their opinion not of the truth but of the importance of the theorems.

After this, there will, I hope, be people who will find it to their advantage to decipher all this mess.

15.D2 An unpublished preface by Galois

First of all, the second page of this work is not encumbered with names, forenames, qualities, titles, and elegies of some miserly prince whose purse will be opened with the fumes of incense—with the threat of being closed when the censer-bearer is empty. You will not see, in characters three times larger than the text, a respectful homage to someone of high position in the sciences, to a wise protector—something indispensable (I was going to say inevitable) to someone twenty years old who wants to write. I do not say to anyone that I owe to their advice and their encouragement everything that is good in my work. I do not say it because that would be to lie. If I had to address anything to the great in the world or the great in science (at the present time the distinction between these two classes of people is imperceptible) I swear that it would not be in thanks. I owe to the ones that I have published the first of these two memoirs so late, to the others that I have written it all in prison, a place it would be wrong to consider a place of meditation, and where I am often amazed at my self-restraint and keeping my mouth shut in the face of my stupid ill-natured critics [Zoiles]; I think I may use the word Zoiles without fear of being immodest when my adversaries are in my mind base. It is not my subject to say how and why I was detained in prison, but I must say how my manuscripts have been lost most often in the cartons of Messieurs the members of the Institute, although in truth I cannot imagine such thoughtlessness on the part of men who have the death of Abel on their consciences. For me, who does not want to be compared with that illustrious mathematician, it suffices to say that my memoir on the theory of equations was deposited, in substance, with the Academy of Sciences in February 1830, that extracts from it had been sent in 1829, that no report on it was then issued and it has been impossible to recover the manuscript [...].

In the third place, the first memoir is not innocent of the master's eye: an extract was sent to the Academy of Sciences in 1831 and submitted to the inspection of M. Poisson, who reported to a session that he could not understand it at all. Which to my eyes, obsessed as they are with the amour-propre of the author, simply proves that M. Poisson did not want to or could not understand it, but which will certainly prove in the eyes of the public that my work is of no importance [...].

To begin with, long algebraic calculations have been less necessary to the progress of mathematics: the most simple theorems have been translated into the language of analysis. It is not only that since Euler this extremely terse language has become indispensable to the new extent that that great mathematician gave to science. Since Euler, calculations have become more and more necessary but more and more difficult, in proportion as they apply to the more advanced objects of science. Since the turn of the century the algorithm has attained such a degree of complexity that any progress has become impossible by that means, without the elegance that modern geometers have given to their research and by means of which the mind understands a great number of operations promptly and at a glance.

It is evident that elegance so praised and so valued has no other purpose.

From the agreed fact that the efforts of the most advanced mathematicians have elegance as their objective, one can therefore conclude with certainty that it has become more and more necessary to embrace several operations at once, because the mind can no longer stop to look at details.

Now I think that the simplifications produced by the elegance of the calculations (intellectual simplifications, understand; of material they are not) have their limits; I think that the moment will arrive when the algebraic transformations foreseen by the analysts will find neither time nor place to be produced; at such a point it will have to be enough to have foreseen them. I don't want to say that there is nothing new in analysis without this help, but I think that one day everything will be exhausted without it.

Jump on calculations with both feet; group the operations, classify them according to their difficulty and not according to their form; such, according to me, is the task of future geometers; such is the path I have embarked upon in this work.

Do not confuse the opinion I express here with the affectation of certain people for avoiding in appearance every kind of calculation, translating into very long phrases what can be expressed very briefly by algebra and so adding to the length of the operations the length of a language which was not made to express them. These people are a hundred years out of date. Here there is nothing like that; here one makes an analysis of analysis; here the most elevated calculations done up till now are considered as particular cases which it is useful, indispensable, to treat but which it would be fatal not to abandon for greater researches. There will be time to carry out the calculations envisioned by this high analysis and classified according to their difficulty but not specified by their form, when the details of a question reclaim them.

The general thesis which I advance can only be understood when one reads my work attentively, which is only an application of it: not that this theoretical point of view preceded the application; but I asked myself, my book finished, what made it so strange to most of my readers, and analysing myself I thought I observed a tendency in my mind to avoid calculations in the subjects I treated, and which, I recognized, is an insurmountable difficulty to those who would want to proceed generally with the material that I treated.

15.D3 Augustin Louis Cauchy on the theory of permutations

For more than thirty years I have been concerned with the theory of permutations, and especially with the problem of the number of values that can be taken on by functions. Lately, as I shall explain in a subsequent instalment, M. Bertrand has added a few new theorems to those previously established and to those obtained by myself. But as regards Lagrange's theorem, according to which the number of values taken on by a function of n letters divides the product $1 . 2 \ldots n$, one has until now obtained almost exclusively theorems concerning the impossibility of obtaining functions that take on certain prescribed numbers of values. In a new work I have attacked directly the following two questions: 1° What is the number of values that can be assumed by a function of n letters? and 2° How can one effectively form functions that take on the permissible numbers of values? Incidentally, research in this area has led me to new formulas in the theory of sequences that are not without interest. I plan to publish in *Exercises d'Analyse et de Physique Mathématique* the results of my work, including all the details that seem to me to be useful. I shall only ask the Academy for permission to publish, in *Compte Rendus*, abstracts indicating some of the most remarkable propositions at which I have arrived.

[Cauchy then proceeded to explain what he meant by the number of values a function could take as its independent variables are permuted, and gave the following simple example.]

So, for example, if one has $\Omega = x + y$, the two values that the function Ω can take, i.e. $x + y$ and $y + x$, will be equal to one another whatever values are attributed to x and y. But if one has $\Omega = x + 2y$, the two values of the function, i.e. $x + 2y$ and $y + 2x$, will be distinct values which cannot be called equal values for most often they are unequal and will only become equal in the particular case where one has $y = x$.

[Cauchy then considered the permutations themselves.]

One calls a permutation or substitution an operation which consists in moving the variables and substituting them for each other in a given value of the function Ω, or in the corresponding arrangement. To indicate this substitution we shall write the new arrangement which is produced above the first and enclose the system of the two arrangements in parentheses. So, for example, given the function $\Omega = x + 2y + 3z$, where the variables x, y, z occupy the first second and third positions respectively, and consequently follow one another in the order indicated by the arrangement xyz, if one interchanges the variables y, z which occupy the last two places one obtains a new value Ω' for Ω, which will be distinct from the first, and determined by the formula $\Omega' = x + 2z + 3y$. Moreover the new arrangement corresponding to this new value will be xzy, and the substitution by which one passes from the first value to the second will be found to be represented by the notation $\begin{pmatrix} xzy \\ xyz \end{pmatrix}$, which is a sufficient indication of the way in which the variables have been moved.

[Cauchy soon proceeded to consider how permutations can be combined.]

The *product* of a given arrangement xyz by a substitution $\begin{pmatrix} xzy \\ xyz \end{pmatrix}$ will be the new arrangement xzy which one obtains by applying this substitution to the given arrangement.

The *product* of two substitutions will be the new substitution which always furnishes the result to which the application of the first two, operating one after the other on an arbitrary arrangement, would lead. The two given substitutions shall be the two *factors* of the product. The product of an arrangement by a substitution or of a substitution by another will be indicated by one of the notations which serves to indicate the product of two quantities, the multiplicand being placed, following custom, on the right of the multiplier. So one finds for example

$$\begin{pmatrix} xzy \\ xyz \end{pmatrix} xyz = xzy, \quad \text{and} \quad \begin{pmatrix} yxuz \\ xyzu \end{pmatrix} = \begin{pmatrix} yx \\ xy \end{pmatrix}\begin{pmatrix} uz \\ zu \end{pmatrix}.$$

There is more: one can, in the second term of the last equation, exchange the two factors with each other without inconvenience, in such a way that one still has

$$\begin{pmatrix} yxuz \\ xyzu \end{pmatrix} = \begin{pmatrix} uz \\ zu \end{pmatrix} \cdot \begin{pmatrix} yx \\ xy \end{pmatrix}.$$

But this exchange is not always possible, and often the product of two substitutions will vary when one exchanges the two factors with each other. So, in particular, one will find

$$\begin{pmatrix} yx \\ xy \end{pmatrix}\begin{pmatrix} zy \\ yz \end{pmatrix} = \begin{pmatrix} yzx \\ xyz \end{pmatrix} \quad \text{and} \quad \begin{pmatrix} zy \\ yz \end{pmatrix}\begin{pmatrix} yx \\ xy \end{pmatrix} = \begin{pmatrix} zxy \\ xyz \end{pmatrix}.$$

We will say that the two substitutions are permutable with each other when their product is independent of the order in which the two factors occur.

15.D4 Camille Jordan on the background to his work on the theory of groups

These beautiful results [of Lagrange and Abel] were however only the prelude to a much greater discovery. It was reserved for Galois to put the theory of equations on its definitive footing and to show that to each equation there corresponds a group of substitutions, in which are reflected its essential characters, notably those that have to do with its solution by other auxiliary equations [e.g. by radicals]. According to this principle, given any algebraic equation it suffices to know one of its characteristic properties to determine its group, whence, reciprocally one can deduce its other properties.

From this high point of view the problem of solution by radicals which not long ago seemed to be the sole object of the theory of equations, only appears as the first link in a long chain of questions. [...]

[Three ideas set the theory off:] that of primitivity, which can be found already sketched out in the works of Gauss and Abel; that of transitivity, which belongs to Cauchy; finally, the distinction between simple and compound groups. This last concept, the most important of the three, is due to Galois.

The point of this work is to develop Galois's methods and shape them into a body of theory, by showing how easily they permit the solution of all the principal problems in the theory of equations.

16 Non-Euclidean Geometry

16.A Seventeenth- and Eighteenth-century Developments

Every lively mathematical culture has been critical of Euclid's assumption of the parallel postulate, and, when mathematics revived in the West, European mathematicians turned their attentions to what Henry Savile called a 'blot on Euclid'. Their first remarks were not as profound as those of their Arab predecessors; typically they tended to assume that the curve equidistant from a straight line is itself straight, which is actually logically equivalent to the parallel postulate. With John Wallis (16.A1), who was explicitly aware of some of the Arab work, the investigations took a new turn. Wallis was clear that his own analysis only revealed that (in the presence of the other axioms of Euclid) the parallel postulate was equivalent to postulating the existence of similar, but non-congruent figures. While this seems to have struck him as a more natural assumption to make, he knew he had not proved that the postulate was true. Many later workers were not so clear.

For example, Gerolamo Saccheri (1667–1733) gave a most thorough and careful account of the problem in the last year of his life, only to force it into a logical contradiction at the end by means of an invalid argument. Saccheri's line of attack was particularly neat. He showed that there are at most three geometries possible which are consistent with all of Euclid's *Elements* except the parallel postulate, and that these are distinguished by the angle sums of figures in them:

(1) the angle sum of every triangle is less than two right angles (Hypothesis of the Acute Angle = HAA);
(2) the angle sum of every triangle is exactly two right angles (HRA);
(3) the angle sum of every triangle is greater than two right angles (HOA).

He then aimed to show that hypotheses (1) and (3) each, separately, destroyed themselves. Indeed, as he showed, the third one does, but the first one is viable, and leads eventually to the creation of a new geometry, non-Euclidean geometry. Here, however, in his desire to vindicate Euclid, Saccheri committed the mistake alluded to before, allowing himself to invoke points at infinity which do not belong to the geometry. Saccheri (in 16.A2) shows how carefully he set out to avoid any tacit assumptions such as had bedevilled his predecessors. The reference to triangles 'in every way restricted' means that he argued first about small figures and only later about figures in which lines have been extended indefinitely. 16.A2 also presents Saccheri's account of a crucial moment in the whole subsequent story, his division of lines through a point into three kinds, depending on their relationship with a given line. We do not give his final, flawed, argument, but we have included his statement of this incorrect theorem because the phrase 'repugnant to the nature of the straight line' makes clear what it is easy to forget: Saccheri was, like Euclid (arguably), doing mathematical physics, not abstract axiomatics. A line was something with a physical significance to him.

From time to time philosophers interested themselves in this problem, most famously Immanuel Kant (1724–1804). Ever since his independent but nearly simultaneous realization with Lambert of the nature of the Milky Way, they had stood high in each other's esteem, as the letter in 16.A3 shows. Nonetheless the differences between their views casts a sharp light on the distinction between the philosophical and the critical, mathematical mind. Lambert was no mean philosopher, and was the author of two long works which were well received in their day, but his metaphysical stance is naive by comparison with Kant's mature epistemological position (see 16.A4). Yet Kant got very close to saying space *is* Euclidean, or at least that we have willy-nilly an *a priori* intuition of it as such which limits our mathematical investigations, while Lambert, doggedly pursuing geometrical arguments, got very close to discovering a new geometry altogether.

Lambert's approach was modelled on that of Saccheri, whose work he knew although he does not cite him by name. We give a lengthy extract in 16.A5 from his study of a geometry based on the HAA, in which he establishes several surprising results without thereby assuming that he has thus destroyed the hypothesis. These include his observation that on the HAA there would be an absolute measure of length—which causes him to depart from his philosophical mentor Christian Wolff—and that there is an intimate relationship between the angle sum and the area of figures, a theorem first rigorously proved by Gauss in 1831. Indeed Lambert may be compared in a sense with Moses, for he saw more of the promised land of the new geometry than any one before him, and knew that he had not proved it self-contradictory, but yet felt compelled to deny its ultimate validity. So, he said, he might 'almost' conclude that it was the geometry on an imaginary sphere, but he did not, lacking not just the tools to explain what such a thing might be, but perhaps the willingness to do so. The phrase remains a mere analogy in his work, based on the fact that triangles on a sphere have area = R^2 (angle sum $- \pi$), where R is the radius of the sphere; whereas on the HAA one may set area = $K^2(\pi -$ angle sum), for some constant of proportionality K; putting $K^2 = -R^2$, and so $K = iR$, suggests that the HAA can refer to triangles on a sphere of imaginary radius.

Legendre made several assaults—all of them flawed—on the parallel postulate in the course of reviving a critical geometric spirit in France. We present two of his attempts in 16.A6. The first is quite classical, and the second the one he always came back to when, as happened, for example, with the earlier proof, his tacit assumptions were pointed out to him. He held this ambivalent position because of his pedagogic aims, which were to present geometry as far as possible as an elementary, rigorous subject in the style of Euclid. At this he was successful, and the various editions of his *Eléments de Géométrie* were the standard texts for a generation. But when an elementary proof could not be given, he would reluctantly fall back on the second one, stored away in an appendix so that its unfamiliar function-theoretic approach would not hinder beginners.

16.A1 John Wallis's lecture on the parallel postulate

John Wallis gave a lecture on this topic at Oxford, where he was Savilian Professor of Mathematics, on the evening of 11 July 1663. He had been inspired by Nasr-Eddin's attempt on it, which he referred to in his lecture, to examine the question himself, and his analysis is remarkable both for its originality and its caution. Indeed, his view of the matter was to be much more profound than many a later writer's.

All his theorems concern a figure in which two lines, AB and CD, meet a line AC in such a way that the interior angles BAC and DCA sum to less than two right angles. His aim was to show that this means that if the lines are extended far enough then they must meet, thus proving the parallel postulate. To this end he proved a number of lemmas which we will merely state before he established his main result.

Lemma If AC is moved along $\alpha\gamma$ until α coincides with C and AB with $\alpha\beta$, and the angle BAC is never altered, then the line $\alpha\beta$, i.e. AB in its new position, lies beyond DC.
[His proof proceeded by looking at the angles involved.]

Lemma In its motion the line $\alpha\beta$ cuts CD before α reaches C.
[For, he remarked, $\alpha\beta$ must cross CD before α can reach C.]

Finally I shall assume that *to every figure there is always a similar one of arbitrary size.*

Indeed this seems (since magnitudes can always be divided and multiplied without restriction) to follow from the nature of relationships between magnitudes, since every figure (while it retains its own form) can always be made smaller or larger without restriction. In fact all geometers have made this assumption (without remarking on it explicitly or perhaps even noticing it themselves), as did Euclid. For when he showed that a circle with given centre and radius can always be drawn, he assumes that there is

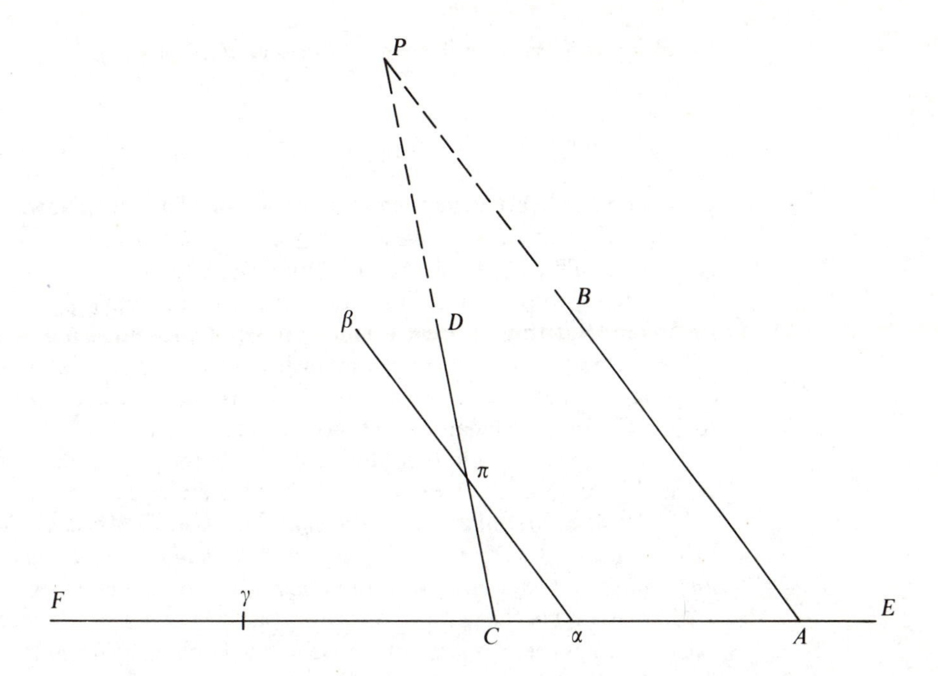

a circle of arbitrary size or with arbitrary radius, and when he assumes that something is possible he demands that one can construct it. Now indeed it is no cheap wish that one (without the necessary fore-knowledge) should be able to drag a similar figure to a given one but of arbitrary size. But it is as practicable to make this assumption for an arbitrary figure as for circles.

[After a little more in this vein Wallis turned to his proof of his main result.]

One thinks of the segment AC that lies between the two lines as moving along the unbounded line ACF. The line AB, that stands on AC, moves without altering the angle BAC until it is at $\alpha\beta$ which cuts the line CD at a point π. Then since $\pi C\alpha$ is a triangle and there are similar triangles of arbitrary size, one can draw a triangle on the segment CA that is similar to triangle $\pi C\alpha$ with base $C\alpha$. One thinks of this as done, and lets PCA be this triangle. [...]

Since PCA is a triangle the lines CP and AP meet at P, and since the triangle PCA is similar to the triangle $\pi C\alpha$ by construction, corresponding angles are equal. In particular angle PCA equals angle $\pi C\alpha$, which is angle DCA, and so the line CP lies along the line CD. Indeed if the line CD lay on one side or the other then the angle PCA would be greater or smaller than the angle DCA, while we have proved that they are equal.

Likewise the angle PAC is equal to the angle $\pi\alpha C$. But the same angle $\pi\alpha C$, i.e. angle $\beta\alpha F$, is equal to angle BAF or BAC, and therefore the angle BAC is equal to the angle PAC. Therefore the line AP lies along the line AB (for if it lay on one side or the other then the angles BAC and PAC would be different, whose equality is proved).

Therefore the line AP agrees with the line AB. Likewise CP and CD lie along a line. But, as was already shown, AP and CP meet at P, and so AB and CD must meet, and indeed at the same point P, i.e. on the side of the line EAF where lie the two angles whose sum is less than two right angles. As was to be shown.

16.A2 From Gerolamo Saccheri's *Euclides Vindicatus*

Preface to the reader

Of all who have learned mathematics, none can fail to know how great is the excellence and worth of Euclid's *Elements*. As erudite witnesses here I summon Archimedes, Apollonius, Theodosius, and others almost innumerable, writers on mathematics even to our times, who use Euclid's *Elements* as foundations long established and wholly unshaken. But this so great celebrity has not prevented many, ancients as well as moderns, and among them distinguished geometers, maintaining they had found certain blemishes in these most beauteous nor ever sufficiently praised *Elements*. Three such flecks they designate, which now I name.

The first pertains to the definition of parallels and with it the axiom which in Clavius is the thirteenth of the First Book, where Euclid says:

If a straight line falling on two straight lines lying in the same plane, make with them two internal angles toward the same parts less than two right angles, these two straight lines infinitely produced toward those parts will meet each other.

No one doubts the truth of this proposition; but solely they accuse Euclid as to it, because he has used for it the name axiom, as if obviously from the right understanding of its terms alone came conviction. Whence not a few (withal retaining Euclid's definition of parallels) have attempted its demonstration from those propositions of Euclid's First Book alone which precede the twenty-ninth, wherein begins the use of the controverted proposition. [...]

And this is enough to indicate to the reader what will be the material of the First Book of this work of mine: for a more complete explication of all that has been said will be given in the scholia after the twenty-first proposition of this Book.

I divide this Book into two parts. In the First Part I will imitate the antique geometers, and not trouble myself about the nature or the name of that line which at all its points is equidistant from a certain line supposed straight; but merely undertake without any *petitio principii* clearly to demonstrate the disputed Euclidean axiom. Therefore never will I use from those prior propositions of Euclid's First Book, not merely the twenty-seventh or the twenty-eighth, but not even the sixteenth or the seventeenth, except where clearly it is question of a triangle every way restricted.

Then in the Second Part for a new confirmation of the same axiom, I shall demonstrate that the line which at all its points is equidistant from an assumed straight line can only be a straight line. But every one sees that on this occasion the very first principles of all geometry are to be subjected to a rigid examination. [...]

Proposition XXXII

Now I say there is (in the hypothesis of acute angle) a certain determinate acute angle BAX drawn under which AX only at an infinite distance meets BX, and thus is a limit in part from within, in part from without; on the one hand of all those which under lesser acute angles meet the aforesaid BX at a finite distance; on the other hand also of the others which under greater acute angles, even to a right angle inclusive, have a common perpendicular in two distinct points with BX.

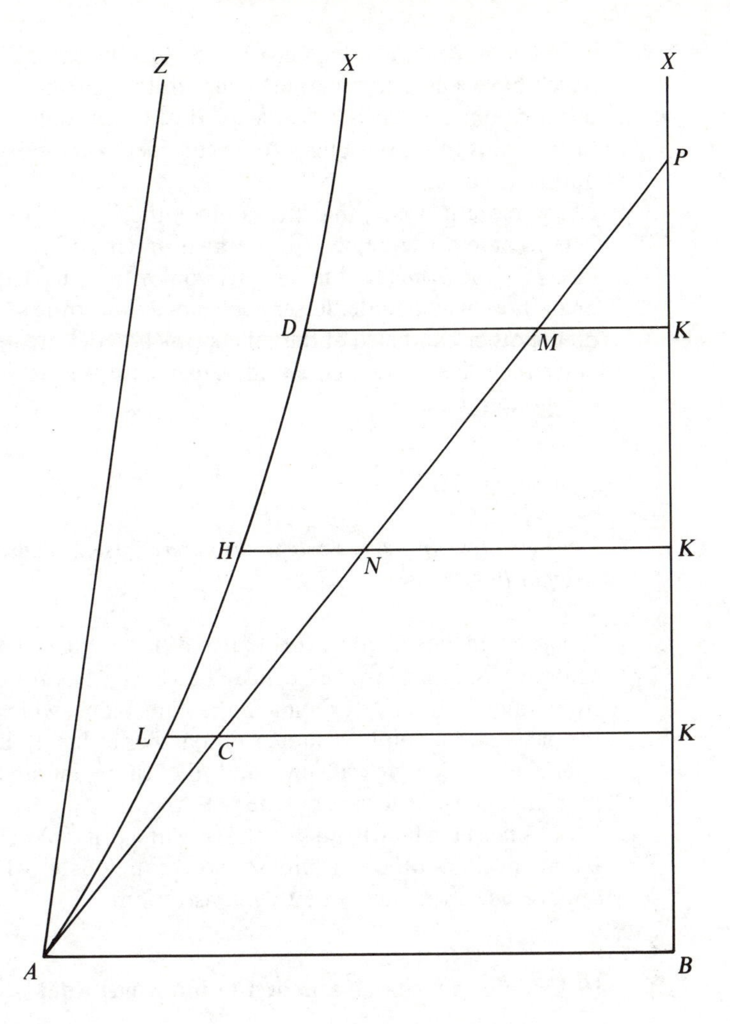

Proof First it holds (from Cor. II to P. XXIX) that no determinate acute angle will be the greatest of all drawn under which a straight from the poind *A* meets the aforesaid *BX* at a finite distance.

Secondly, it holds in like manner that (in the hypothesis of acute angle) no acute angle will be the least of all drawn, under which a straight has a common perpendicular in two distinct points with *BX*; since indeed (from what precedes) there can be no determinate limit, such that there cannot be found, under a lesser angle constituted at the point *A*, a common perpendicular in two distinct points, which is less than any assignable length *R*.

And hence follows thirdly, that (in this hypothesis) there must be a certain determinate acute angle *BAX*, drawn under which *AX* so approaches ever more to *BX*, that only at an infinite distance does it meet it.

But further that this *AX* is a limit in part from within in part from without of each of the aforesaid classes of straights is proved thus. First, it agrees with those straights which meet *BX* at a finite distance since it also finally meets: but it differs, because it meets only at an infinite distance.

But secondly it also agrees with, and at the same time differs from those straights which have a common perpendicular in two distinct points with *BX*; because it also has a common perpendicular with *BX*; but in one and the same point *X* infinitely distant. But this latter ought to be considered demonstrated in P. XXVIII, as I point out in its corollary.

Therefore it holds, that (in the hypothesis of acute angle) there will be a certain determinate acute angle *BAX*, drawn under which *AX* only at an infinite distance meets *BX*, and thus is a limit in part from within, in part from without; on the one hand of all those which under lesser acute angles meet the aforesaid *BX* at a finite distance; on the other hand also of the others which under greater acute angles, even to a right angle inclusive, have a common perpendicular in two distinct points with *BX*.

Quod erat etc.

Proposition XXXIII

The hypothesis of acute angle is absolutely false; because repugnant to the nature of the straight line.

Proof From the foregoing theorem may be established, that at length the hypothesis of acute angle inimical to the Euclidean geometry has as outcome that we must recognize two straights *AX*, *BX*, existing in the same plane, which produced *in infinitum* toward the parts of the points *X* must run together at length into one and the same straight line, truly receiving, at one and the same infinitely distant point a common perpendicular in the same plane with them.

But since I am here to go into the very first principles, I shall diligently take care, that I omit nothing objected almost too scrupulously, which indeed I recognize to be opportune to the most exact demonstration.

16.A3 Johann Heinrich Lambert to Immanuel Kant

You will gather easily enough how I conceive location and space. Ignoring the ambiguities of the words, I propose the analogy,

$$\text{Time} : \text{Duration} = \text{Location} : \text{Space}$$

The analogy is quite precise, except that space has three dimensions, duration only one, and besides this each of these concepts has something peculiar to it. Space, like duration, has absolute but also finite determinations. Space, like duration, has a reality peculiar to it, which we cannot explain or define by means of words that are used for other things, at least not without danger of being misleading. It is something simple and must be thought. The whole intellectual world is non-spatial; it does, however, have a counterpart of space, which is easily distinguishable from physical space. Perhaps this bears a still closer resemblance to it than merely a metaphoric one.

The theological difficulties that, especially since the time of Leibniz and Clarke, have made the theory of space a thorny problem have so far not confused me. I owe all my success to my preference for leaving undetermined various topics that are impervious to clarification. Besides, I did not want to peer at the succeeding parts of metaphysics when working on ontology. I won't complain if people want to regard time and space as mere pictures and appearances. For, in addition to the fact that constant appearance is for us truth, though the foundations are never discovered or only at some future time; it is also useful in ontology to take up concepts borrowed from appearance, since *the theory must finally be applied to phenomena again*. For that is also how the astronomers begins, with the phenomenon; deriving his theory of the construction of the world from phenomena, he applies it again to phenomena and their predictions in his *Ephemerides* [star calendars]. In metaphysics, where the problem of appearance is so essential, the method of the astronomer will surely be the safest. The metaphysician can take everything to be appearance, separate the empty from the real appearance, and draw true conclusions from the latter. If he is successful, he shall have few contradictions arising from the principles and win much favour. It only seems necessary to have time and patience. [...]

Our symbolic knowledge is a thing halfway between sensing and actual pure thinking. If we proceed correctly in the delineation of the simples and in the manner of our synthesizing, we thereby get reliable rules for constructing signs of things that are so highly synthesized that we need not review them again and can nevertheless be sure that the sign represents the truth. No one has yet formed himself a clear representation of all the members of an infinite series, and no one is going to do so in the future. But we are able to do arithmetic with such series, to give their sum, and so on, by virtue of the laws of *symbolic* knowledge. We thus extend ourselves far beyond the borders of our actual thinking. The sign $\sqrt{-1}$ represents an unthinkable non-thing. And yet it can be used very well in finding theorems. What are usually regarded as specimens of the pure understanding can be viewed most of the time as specimens of symbolic knowledge. This is what I said in No. 122 of my *Phaenomenology* with reference to question No. 119. And I have nothing against your making the claim quite general, in No. 10.

But I shall stop here and let you make whatever use you wish of what I have said. Please examine carefully the sentences I have underlined and, if you have time, let me know what you think of them. Never mind the postage. Till now I have not been able to deny all reality to time and space, or to consider them mere images and appearance. I think that every change would then have to be mere appearance too. And this would contradict one of my main principles (No. 54, *Phaenomenology*). If changes have reality, then I must grant it to time as well. Changes follow one another, begin, continue, cease, and so on, and all these expressions are temporal. If you can instruct me otherwise, I shall not expect to lose much. Time and space will be *real* appearances, and their foundation is an existent something that truly conforms to time and space just as precisely and constantly as the laws of geometry are precise and constant. The language of appearance will thus serve our purposes just as precisely as the unknown 'true' language. I must say, though, that an appearance that absolutely never deceives us could well be something more than mere appearance.

16.A4 Kant on our intuition of space

Philosophy, as well as mathematics, does indeed treat of quantities, for instance, of totality, infinity, etc. Mathematics also concerns itself with qualities, for instance, the difference between lines and surfaces, as spaces of different quality, and with the continuity of extension as one of its qualities. But although in such cases they have a common object, the mode in which reason handles that object is wholly different in philosophy and in mathematics. Philosophy confines itself to universal concepts; mathematics can achieve nothing by concepts alone but hastens at once to intuition, in which it considers the concept *in concreto*, though not empirically, but only in an intuition which it presents *a priori*, that is, which it has constructed, and in which whatever follows from the universal conditions of the construction must be universally valid of the object of the concept thus constructed.

Suppose a philosopher be given the concept of a triangle and he be left to find out, in his own way, what relation the sum of its angles bears to a right angle. He has nothing but the concept of a figure enclosed by three straight lines, and possessing three angles. However long he meditates on this concept, he will never produce anything new. He can analyse and clarify the concept of a straight line or of an angle or of the number three, but he can never arrive at any properties not already contained in these concepts. Now let the geometrician take up these questions. He at once begins by constructing a triangle. Since he knows that the sum of two right angles is exactly equal to the sum of all the adjacent angles which can be constructed from a single point on a straight line, he prolongs one side of his triangle and obtains two adjacent angles, which together are equal to two right angles. He then divides the external angle by drawing a line parallel to the opposite side of the triangle, and observes that he has thus obtained an external adjacent angle which is equal to an internal angle—and so on. In this fashion, through a chain of inferences guided throughout by intuition, he arrives at a fully evident and universally valid solution of the problem. [...]

Now what can be the reason of this radical difference in the fortunes of the philosopher and the mathematician, both of whom practise the art of reason, the one making his way by means of concepts, the other by means of intuitions which he exhibits *a priori* in accordance with concepts? The cause is evident from what has been said above, in our exposition of the fundamental transcendental doctrines.

It would therefore be quite futile for me to philosophize upon the triangle, that is, to think about it discursively. I should not be able to advance a single step beyond the mere definition, which was what I had to begin with. There is indeed a transcendental synthesis [framed] from concepts alone, a synthesis with which the philosopher is alone competent to deal; but it relates only to a thing in general, as defining the conditions under which the perception of it can belong to possible experience. But in mathematical problems there is no question of this, nor indeed of existence at all, but only of the properties of the objects in themselves, [that is to say], solely in so far as these properties are connected with the concept of the objects. [...]

All our knowledge relates, finally, to possible intuitions, for it is through them alone that an object is given. Now an *a priori* concept, that is, a concept which is not empirical, either already includes in itself a pure intuition (and if so, it can be constructed), or it includes nothing but the synthesis of possible intuitions which are not given *a priori*. In this latter case we can indeed make use of it in forming synthetic *a priori* judgements, but only discursively in accordance with concepts, never intuitively through the construction of the concept.

The only intuition that is given *a priori* is that of the mere form of appearances, space and time. A concept of space and time, as quanta, can be exhibited *a priori* in intuition, that is, constructed, either in respect to the quality (figure) of the quanta, or through number in their quantity only (the mere synthesis of the homogeneous manifold). But the matter of appearances, by which *things* are given us in space and time, can only be represented in perception, and therefore *a posteriori*. The only concept which represents *a priori* this empirical content of appearances is the concept of a *thing* in general, and the *a priori* synthetic knowledge of this thing in general can give us nothing more than the mere rule of the synthesis of that which perception may give *a posteriori*. It can never yield an *a priori* intuition of the real object, since this must necessarily be empirical.

Synthetic propositions in regard to *things* in general, the intuition of which does not admit of being given *a priori*, are transcendental. Transcendental propositions can never be given through construction of concepts, but only in accordance with concepts that are *a priori*.

16.A5 Lambert on the consequences of a non-Euclidean postulate

One easily sees that one can go much further with the third hypothesis [i.e. the HAA] in this way, and that similar theorems can be found to those derived on the second hypothesis [i.e. the HOA], albeit with quite opposite consequences. But I have principally sought such consequences of the third hypothesis to see if it did not condradict itself. From them all I saw that this hypothesis would not destroy itself at all easily. I will therefore adduce some such consequences without seeing how far they can be derived from the second hypothesis by making corresponding changes.

The most remarkable of such conclusions is that if the third hypothesis holds we would have an absolute measure of length for each line, for the content of each surface and each bodily space. Now this overturns a theorem that one can unhesitatingly count amongst the fundamentals of Geometry, and which up to now no one has doubted, namely that there is no such absolute measure. Indeed, Wolf makes it into a theorem which he derives from the definition of quantity, and which can be stated as follows: *quantitas dari sed non per se intelligi potest* [there can not be a quantity known in itself]. However, this theorem, like the definition, must be amended, because unquestionably there are quantities that are intelligible in themselves and have a definite unit. For lines, surfaces, and bodily spaces the same is also true; and I don't believe that one should introduce a definition into Geometry in order to put it right.

Now, in order to prove the first-mentioned corollary, let A, B, C, D, E be right angles, and, assuming the third hypothesis, let G, F, H, J be acute, and indeed $H < G$, and $J < H$, and likewise $F < G$ and $J < F$. Now I say that the angle G is the measure of the quadrilateral $ADGB$, if indeed $AB = AD$, and likewise the angle J shall be the measure of the quadrilateral $ACJE$, if $AC = AE$.

For, from considering the equality of the sides $AB = AD$ and the right angles A, B, D, the acute angle G can fit no other quadrilateral except those whose sides AB, AD have the absolute lengths of AB, AD. If, e.g., one takes the greater sides $AE = AC$ and puts right angles at E, C then on the third hypothesis the angle $J < G$. So G does not fit on J. If $AE = AC$ were to be taken smaller than $AD = AB$ then would J be $> G$ and so G would not fit on J in any case.

Therefore the angle G is the absolute measure of the quadrilateral $ADGB$. Since the angle has a measure intelligible in itself, if one took e.g. $AB = AD$ as a Paris foot and then the angle G was $80°$ this is only to say that if one should make the quadrilateral

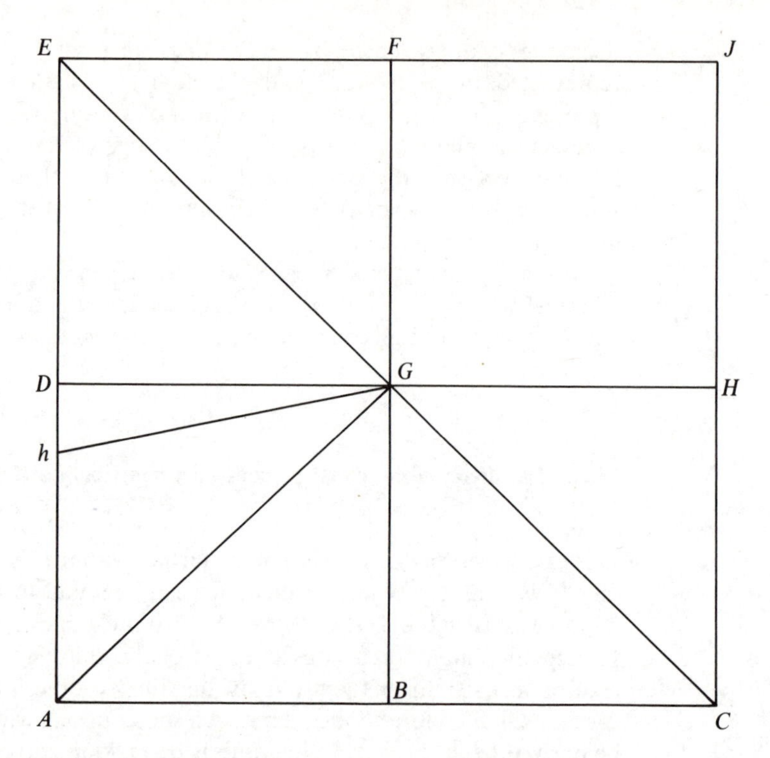

ADGB so big that the angle *G* was 80°: then one would have the absolute measure of a Paris foot on $AB = AD$.

This consequence is somewhat surprising, which inclines one to want the third hypothesis to be true! However, this advantage not withstanding, I still do not want it, because innumerable other inconveniences would thereby come about. Trigonometric tables would have to be infinitely extended; the similarity and proportionality of figures would entirely lapse; no figure could be presented except in its absolute size; Astronomy would be an evil task; etc.

But these are *argumenta ab amore et invidia ducta* [arguments drawn from love and hate] which Geometry, like all the sciences, must leave entirely on one side. I therefore return to the third hypothesis. According to it, it is not only the case that in every triangle the angle sum is less than 180°, as we have already seen, but also that the difference from 180° increases directly with the area of the triangle. So I want to say: if of two triangles one has a greater area than the other then the angle sum of the first triangle is smaller than that of the other.

I shall not prove this theorem completely here, as I state it, rather I shall give only so much of the proof as will enable the rest of it to be understood overall.

Let, e.g., the triangle *EFG* be put into triangle *ACB* so that the vertices of the former lie on the sides of the latter. In this way *EFG* is entirely inside *ABC*, so the area of the former is unquestionably smaller than the area of the latter. Now the angles sums are:

$$EFG + EGF + GEF = 180° - a$$

$$EGA + EAG + AEG = 180° - b$$

$$FGB + GBF + GFB = 180° - c$$

$$FCE + FEC + EFC = 180° - d$$

whereas

$$EGA + EGF + FGB = 180^\circ$$

$$AEG + GEF + FEC = 180^\circ$$

$$EFC + EFG + GFB = 180^\circ$$

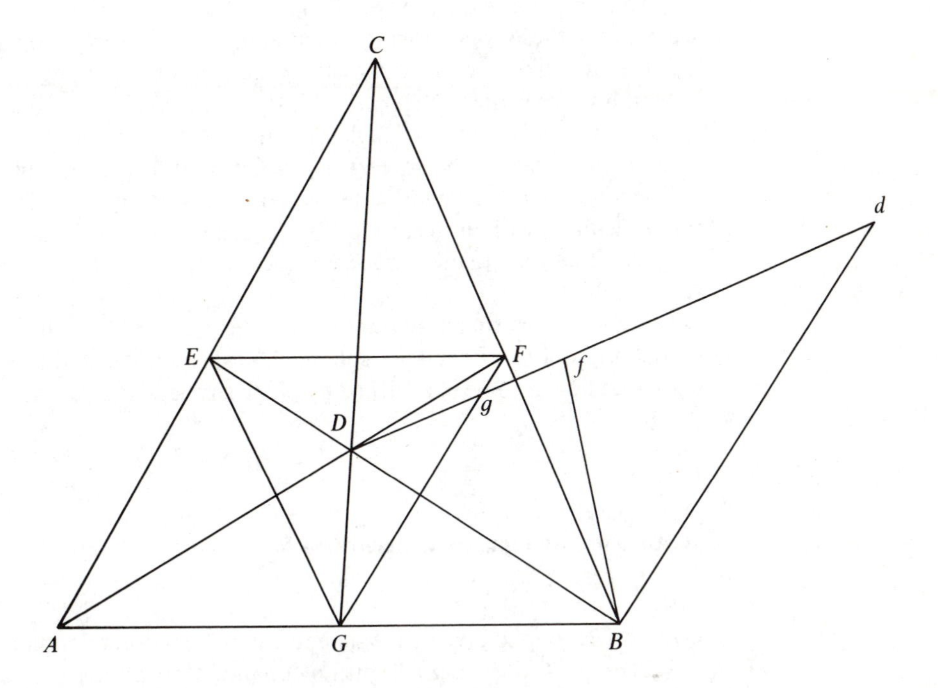

If one subtracts the sum of the last three equations from the sum of the first four, there remains

$$CAB + ABC + BCA = 180^\circ - a - b - c - d.$$

But since here not only a but also $a + b + c + d$ must be subtracted from 180°, one sees that in the triangle ABC all the defects a, b, c, d of the four triangles AEG, ECF, FBG, GEF occur together and therefore the sum of its three angles is exactly that much smaller than 180°.

If the smaller triangle does not quite lie inside the larger one then something lies outside which can be cut off and placed partially inside the larger triangle, and this process can be continued, cutting the smaller triangle into more and more pieces until it lies entirely inside the larger one. The uncovered space remaining in the larger one can be divided into triangles. The amount by which the sum of all the angles in these triangles is less than the same number of 180°s, will equal the amount by which the sum of the three angles of the greater of the given triangles is smaller than the sum of the three angles of the smaller of the given triangles.

If it were possible on the third hypothesis to cover with equal and similar triangles a larger triangle, then it would also be easy to deduce that for each triangle the excess of 180° over the sum of its three angles was proportional to the area of the triangle.

Meanwhile since this excess is connected with area it allows us to think of this proportionality in another way.

One takes, e.g., two triangles. If the first has twice as much area as the other, then the first can be divided up as one wishes until it can be made to lie doubly on the other. And if the angle sum of the smaller one falls short of 180° by $a°$, then the larger one will fall short by $2a°$.

I shall now make the following comments. Quite similar theorems are true on the second hypothesis, only now the angle sum of every triangle will be greater than 180°. The excess is likewise proportional to the area of the triangle.

Now it seems to me to be remarkable that the second hypothesis holds when one takes spherical instead of plane triangles, because for them not only is the sum of the angles greater than 180° but the excess is proportional to the area of the triangle.

It seems even more remarkable that what I am saying here about spherical triangles can be deduced without regard for the problem of parallel lines, and based on no other idea than that each plane surface passing through the centre of a sphere divides it into two equal parts.

I should almost therefore put forward the proposal that the third hypothesis holds on the surface of an imaginary sphere. At least there must always be something which does not allow it to be overturned by plane surfaces as easily as the second hypothesis can be.

16.A6 Two attempts by Legendre on the parallel postulate

(a) Let ABC be the given triangle and suppose, if possible, that the sum of its angles is $= 2P - Z$, P denoting a right angle, and Z an arbitrary quantity, such that one supposes the sum of the angles to be less than two right angles.

Let A be the smallest angle of triangle ABC and on the opposite side make angle $BCD = ABC$, and angle $CBD = ACB$; the triangles BCD, ABC will be equal, as they have an equal side BC adjacent to two pairs of equal angles. Through the point D draw an arbitrary line EF which meets the extended sides of the angle A.

Since the sum of the angles of each triangle ABC, BCD is $2P - Z$ and that of each triangle EBD, DCF cannot exceed $2P$ [by Legendre's immediately preceding theorem] it follows that the sum of the angles of the four triangles ABC, BCD, EBD, DCF cannot exceed $4P - 2Z + 4P$ or $8P - 2Z$. If one removes the angles at B, C, D, which make $6P$, from this sum, since the sum of the angles at each of B, C, D is $2P$ the remainder will be equal to the sum of the angles of triangle AEF. Therefore the sum of the angles of triangle AEF cannot exceed $8P - 2Z - 6P$ or $2P - 2Z$.

So, while it is necessary to add Z to the sum of the angles of triangle ABC to make two right angles, it is necessary to add at least $2Z$ to the angles of the triangle AEF to make it likewise up to two right angles.

By means of triangle DEF one similarly constructs a third triangle such that one must add at least $4Z$ to the sum of its three angles to make the total equal to two right angles; and by means of the third one constructs a fourth, to which it is necessary to add at least $8Z$ to the sum of its angles to make the total equal to two right angles, and so on.

Now, however small Z is with respect to the right angle P, the sequence $Z, 2Z, 4Z, 8Z$ etc, whose terms increase by doubling, eventually yields a term equal to or greater than $2P$. One is thereby led to a triangle to which it is necessary to add a quantity equal to or greater than $2P$ for the total to be only $2P$. This consequence is visibly absurd, therefore the hypothesis from which we began cannot be valid, i.e. it is impossible that the sum of the angles of triangle ABC can be smaller than two right angles.

(b) [The parallel] postulate has never hitherto been demonstrated in a way strictly geometrical, and independent of all considerations about infinity, a circumstance attributable, doubtless, to the importance of our common definition of a straight line, on which the whole of geometry hinges. But viewing the matter in a more abstract light, we are furnished by analysis with a very simple method of rigorously proving both this and the other fundamental properties of Geometry. We here propose to expound this method, with all the requisite minuteness, beginning with the theorem concerning the sum of the three angles of a triangle.

By superposition, it can be shown immediately, and without any preliminary propositions, that *two triangles are equal when they have two angles and an interadjacent side in each equal*. Let us call this side p, the two adjacent angles A and B, the third angle C. This third angle C therefore is entirely determined, when the angles A and B, with the side p, are known; [...] hence the angle C must be a determinate function of the three quantities A, B, p, which I shall express thus, $C = \phi:(A, B, p)$.

Let the right angle be equal to unity, then the angles A, B, C will be numbers included between 0 and 2; and since $C = \phi:(A, B, p)$, I assert, that the line p cannot enter into the function ϕ. For we have already seen that C must be entirely determined by the given quantities A, B, p alone, without any other line or angle whatever. But the line p is heterogeneous with the other numbers A, B, C; and if there existed any equation between A, B, C, p the value of p must be found from it in terms of A, B, C; whence it would follow that p is equal to a number, which is absurd; hence p cannot enter into the function ϕ, and we have simply $C = \phi:(A, B)$. (Against this demonstration it has been objected that if it were applied word for word to spherical triangles, we should find that two angles being known, are sufficient to determine the third, which is not the case in that species of triangle. The answer is, that in spherical triangles, there exists one element more than in plane triangles, the radius of the sphere, namely, which must not be omitted from our reasoning. Let r be the radius; instead of $C = \phi(A, B, p)$ we shall now have $C = \phi(A, B, p, r)$ or by the law of homogeneity, simply $C = \phi(A, B, p/r)$. But since the ratio p/r is a number, as well as A, B, C, there is nothing to hinder p/r from entering the function ϕ, and consequently we have no right to infer from it, that $C = \phi(A, B)$.)

This formula already proves that if two angles of one triangle are equal to two angles of another, the third angle of the former must also be equal to the third angle of the latter, and this granted, it is easy to arrive at the theorem we have in view.

First, let ABC be a triangle right angled at A, from the point A draw AB perpendicular to the hypoteneuse. The angles B and D of the triangle ABD are equal to the angles B and A of the triangle BAC; whence, from what has just been proved, the third angle BAD is equal to the third C. For a like reason the angle $DAC = B$, hence $BAD + DAC$ or $BAC = B + C$; but the angle BAC is right; hence *the two acute angles of a right angled triangle are together equal to a right angle*.

16.B Early Nineteenth-century Developments

One of the changes between the eighteenth and nineteenth centuries was one of attitude to the foundations of geometry. Gradually various brave spirits began to push into the new world and prove theorems on the basis of the HAA, or an equivalent postulate, without also hoping to find a contradiction. One such was the Professor of Law Ferdinand Karl Schweikart (1780–1859) (see 16.B1). But Gauss had other correspondents who looked to him for advice, such Wolfgang Bolyai (1775–1856), the father of Janos (also called Johann) (1802–1860). In due course, in 1831, Wolfgang sent Gauss his son's addition to Gauss's two-volume work on geometry, the short *Appendix* in which he described a non-Euclidean geometry (see the brief extract in 16.B5). Gauss's startling reply will be found in 16.B2. By now, unknown to all these parties, the new geometry had already been described elsewhere, by the Russian Nicolai Lobachevskii (1793–1856). 16.B3 contains lengthy extracts from his clearest presentation of his ideas, his little booklet of 1840.

Two features of the work of Bolyai and Lobachevskii are most worthy of notice. First, they *assumed* that the new geometry is possible—they did not prove that it is. In this they had an eighteenth-century confidence that the mathematical god would not let them down, which in turn meant that their work was not logically compelling, however seductive it might appear. This helps us understand how it could be that it was not accepted in their lifetimes. Second, they abandoned classical geometric arguments for function-theoretic, trigonometric ones. This was a remarkably shrewd move, whose full significance had to wait for a deeper grasp of the nature of differential geometry.

16.B1 Ferdinand Karl Schweikart's memorandum to Gauss

There are two kinds of geometry—a geometry in the strict sense—the Euclidean; and an astral geometry.

Triangles in the latter have the property that the sum of their three angles is not equal to two right angles.

This being assumed, we can prove rigorously:

(a) That the sum of the three angles of a triangle is less than two right angles;
(b) that the sum becomes ever less, the greater the area of the triangle;
(c) that the altitude of an isosceles right-angled triangle continually grows, as the sides increase, but it can never become greater than a certain length, which I call the *Constant*.

Squares have, therefore, the [form shown in the figure]. If this Constant were *for us* the Radius of the Earth, (so that every line drawn in the universe from one fixed star to another, distant 90° from the first, would be a tangent to the surface of the earth), it would be infinitely great in comparison with the spaces which occur in daily life.

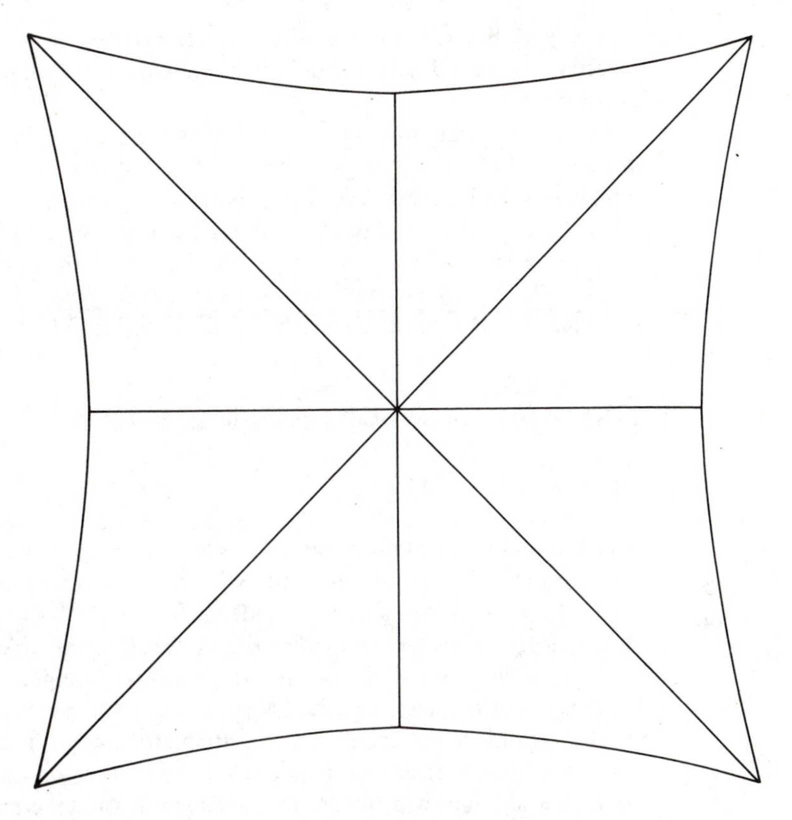

The Euclidean geometry holds only on the assumption that the Constant is infinite. Only in this case is it true that the angles of every triangle are equal to two right angles: and this can easily be proved, as soon as we admit that the Constant is infinite.

16.B2 Gauss on Janos Bolyai's *Appendix*

'If I commenced by saying that I *am unable to praise this work*, you would certainly be surprised for a moment. But I cannot say otherwise. To praise it, would be to praise myself. Indeed the whole contents of the work, the path taken by your son, the results to which he is led, coincide almost entirely with my meditations, which have occupied my mind partly for the last thirty or thirty-five years. So I remained quite stupefied. So far as my own work is concerned, of which up till now I have put little on paper, my intention was not to let it be published during my lifetime. Indeed the majority of people have not clear ideas upon the questions of which we are speaking, and I have found very few people who could regard with any special interest what I communicated to them on this subject. To be able to take such an interest it is first of all necessary to have devoted careful thought to the real nature of what is wanted and upon this matter almost all are most uncertain. On the other hand it was my idea to write down all this later so that at least it should not perish with me. It is therefore a pleasant surprise for me that I am spared this trouble, and I am very glad that it is just the son of my old friend, who takes the precedence of me in such a remarkable manner.'

Wolfgang Bolyai communicated this letter to his son, adding: 'Gauss's answer with regard to your work is very satisfactory and redounds to the honour of our country and of our nation.'

Altogether different was the effect Gauss's letter produced on Johann. He was both unable and unwilling to convince himself that others, earlier than and independent of him, had arrived at the *Non-Euclidean Geometry*. Further he suspected that his father had communicated his discoveries to Gauss before sending him the *Appendix* and that the latter wished to claim for himself the priority of the discovery. And although later he had to let himself be convinced that such a suspicion was unfounded, Johann always regarded the 'Prince of Geometers' with an unjustifiable aversion.

16.B3 Nicolai Lobachevskii's theory of parallels

(*a*) *Opening remarks*

In geometry I find certain imperfections which I hold to be the reason why this science, apart from transition into analytics, can as yet make no advance from that state in which it has come to us from Euclid.

As belonging to these imperfections, I consider the obscurity in the fundamental concepts of the geometrical magnitudes and in the manner and method of representing the measuring of these magnitudes, and finally the momentous gap in the theory of parallels, to fill which all efforts of mathematicians have been so far in vain.

For this theory Legendre's endeavours have done nothing, since he was forced to leave the only rigid way to turn into a side path and take refuge in auxiliary theorems which he illogically strove to exhibit as necessary axioms. My first essay on the foundations of geometry I published in the Kasan *Messenger* for the year 1829. In the hope of having satisfied all requirements, I undertook hereupon a treatment of the whole of this science, and published my work in separate parts in the *Gelehrten Schriften der Universität Kasan* for the years 1836, 1837, 1838, under the title 'New Elements of Geometry, with a Complete Theory of Parallels'. The extent of this work perhaps hindered my countrymen from following such a subject, which since Legendre had lost its interest. Yet I am of the opinion that the Theory of Parallels should not lose its claim to the attention of geometers, and therefore I aim to give here the substance of my investigations, remarking beforehand that contrary to the opinion of Legendre, all other imperfections—for example, the definition of a straight line—show themselves foreign here and without any real influence on the theory of parallels. [...]

All straight lines which in a plane go out from a point can, with reference to a given straight line in the same plane, be divided into two classes—into *cutting* and *not-cutting*.

The *boundary lines* of the one and the other class of those lines will be called *parallel to the given line*.

From the point A let fall upon the line BC the perpendicular AD, to which again draw the perpendicular AE.

In the right angle EAD either will all straight lines which go out from the point A meet the line DC, as for example AF, or some of them, like the perpendicular AE, will not meet the line DC. In the uncertainty whether the perpendicular AE is the only line which does not meet DC, we will assume it may be possible that there are still other lines,

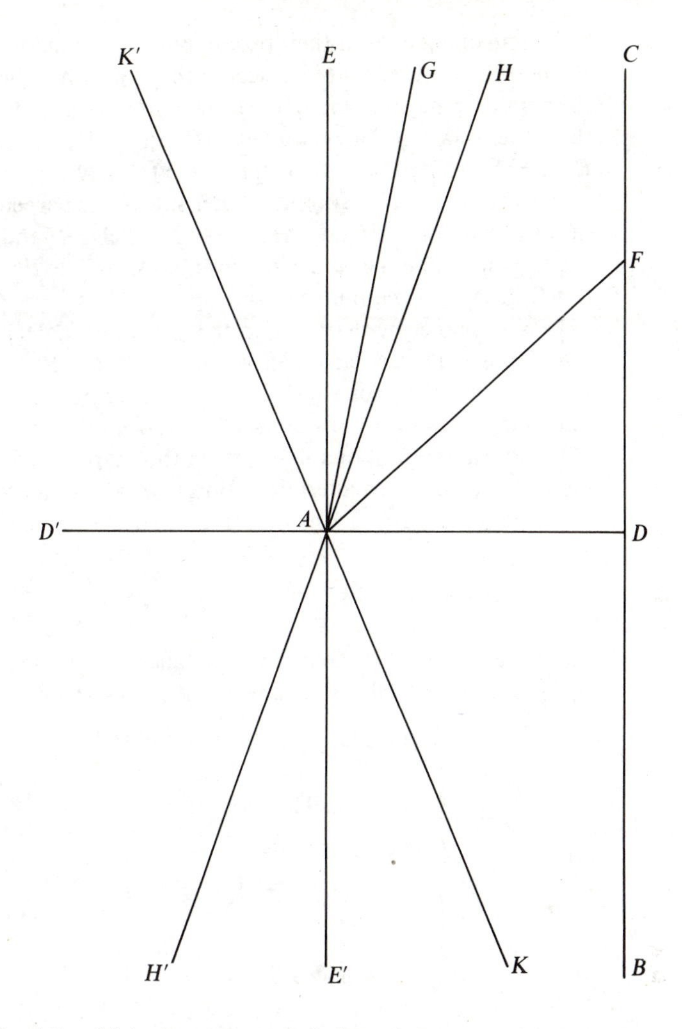

for example AG, which do not cut DC, how far soever they may be prolonged. In passing over from the cutting lines, as AF, to the not-cutting lines, as AG, we must come upon a line AH, parallel to DC, a boundary line, upon one side of which all lines AG are such as do not meet the line DC, while upon the other side every straight line AF cuts the line DC.

The angle HAD between the parallel HA and the perpendicular AD is called the parallel angle (angle of parallelism), which we will here designate by $\Pi(p)$ for $AD = p$.

If $\Pi(p)$ is a right angle, so will the prolongation AE' of the perpendicular AE likewise be parallel to the prolongation DB of the line DC, in addition to which we remark that in regard to the four right angles, which are made at the point A by the perpendiculars AE and AD, and their prolongations AE' and AD', every straight line which goes out from the point A, either itself or at least its prolongation, lies in one of the two right angles which are turned toward BC, so that except the parallel EE' all others, if they are sufficiently produced both ways, must intersect the line BC.

If $\Pi(p) < \frac{1}{2}\pi$, then upon the other side of AD, making the same angle $DAK = \Pi(p)$ will lie also a line AK, parallel to the prolongation DB of the line DC, so that under this assumption we must also make a distinction of *sides in parallelism*.

All remaining lines or their prolongations within the two right angles turned toward BC pertain to those that intersect, if they lie within the angle $HAK = 2\Pi(p)$ between the parallels; they pertain on the other hand to the non-intersecting AG, if they lie upon the other sides of the parallels AH and AK, in the opening of the two angles $EAH = \frac{1}{2}\pi - \Pi(p)$, $E'AK = \frac{1}{2}\pi - \Pi(p)$, between the parallels and EE' the perpendicular to AD. Upon the other side of the perpendicular EE' will in like manner the prolongations AH' and AK' of the parallels AH and AK likewise be parallel to BC; the remaining lines pertain, if in the angle $K'AH'$, to the intersecting, but if in the angles $K'AE$, $H'AE'$ to the non-intersecting.

In accordance with this, for the assumption $\Pi(p) = \frac{1}{2}\pi$, the lines can be only intersecting or parallel; but if we assume that $\Pi(p) < \frac{1}{2}\pi$, then we must allow two parallels, one on the one and one on the other side; in addition we must distinguish the remaining lines into non-intersecting and intersecting.

For both assumptions it serves as the mark of parallelism that the line becomes intersecting for the smallest deviation toward the side where lies the parallel, so that if AH is parallel to DC, every line AF cuts DC, how small soever the angle HAF may be.

(*b*) *Concluding remarks*

All four equations for the interdependence of the sides a, b, c, and the opposite angles A, B, C, in the rectilineal triangle will therefore be:

$$(1)\quad \begin{cases} \sin A \tan \Pi(a) = \sin B \tan \Pi(b), \\ \cos A \cos \Pi(b) \cos \Pi(c) + \dfrac{\sin \Pi(b) \sin \Pi(c)}{\sin \Pi(a)} = 1, \\ \cot A \sin C \sin \Pi(b) + \cos C = \dfrac{\cos \Pi(b)}{\cos \Pi(a)}, \\ \cos A + \cos B \cos C = \dfrac{\sin B \sin C}{\sin \Pi(a)}. \end{cases}$$

If the sides a, b, c, of the triangle are very small, we may content ourselves with the approximate determinations.

$$\cot \Pi(a) = a,$$
$$\sin \Pi(a) = 1 - \tfrac{1}{2}a^2$$
$$\cos \Pi(a) = a,$$

and in like manner also for the other sides b and c.

The equations (1) pass over for such triangles into the following:

$$b \sin A = a \sin B,$$
$$a^2 = b^2 + c^2 - 2bc \cos A,$$
$$a \sin (A + C) = b \sin A,$$
$$\cos A + \cos(B + C) = 0.$$

Of these equations the first two are assumed in the ordinary geometry; the last two lead, with the help of the first, to the conclusion

$$A + B + C = \pi.$$

Therefore the imaginary geometry passes over into the ordinary, when we suppose that the sides of a rectilineal triangle are very small.

I have, in the scientific bulletins of the University of Kasan, published certain researches in regard to the measurement of curved lines, of plane figures, of the surfaces and the volumes of solids, as well as in relation to the application of imaginary geometry to analysis.

The equations (1) attain for themselves already a sufficient foundation for considering the assumption of imaginary geometry as possible. Hence there is no means, other than astronomical observations, to use for judging of the exactitude which pertains to the calculations of the ordinary geometry.

This exactitude is very far-reaching, as I have shown in one of my investigations, so that, for example, in triangles whose sides are attainable for our measurement, the sum of the three angles is not indeed different from the two right angles by the hundredth part of a second.

In addition, it is worthy of notice that the four equations (1) of plane geometry pass over into the equations for spherical triangles, if we put $a\sqrt{-1}, b\sqrt{-1}, c\sqrt{-1}$, instead of the sides a, b, c; with this change, however, we must also put

$$\sin \Pi(a) = \frac{1}{\cos(a)},$$

$$\cos \Pi(a) = (\sqrt{-1}) \tan a,$$

$$\tan \Pi(a) = \frac{1}{\sin a(\sqrt{-1})},$$

and similarly also for the sides b and c.

In this manner we pass over from equations (1) to the following:

$$\sin A \sin b = \sin B \sin a,$$

$$\cos a = \cos b \cos c + \sin b \sin c \cos A,$$

$$\cot A \sin C + \cos C \cos b = \sin b \cot a,$$

$$\cos A = \cos a \sin B \sin C - \cos B \cos C.$$

16.B4 Correspondence between Wolfgang and Janos Bolyai

The Hungarian mathematician Wolfgang Bolyai wrote to his son, Johann:

> 'It is unbelievable that this stubborn darkness, this eternal eclipse, this flaw in geometry, this eternal cloud on virgin truth can be endured.'

At the same time the father is horrified by the thought that his son is attracted by the problem of parallels. Wolfgang Bolyai writes:

> 'You must not attempt this approach to parallels. I know this way to its very end. I have traversed this bottomless night, which extinguished all light and joy of my life. I

entreat you, leave the science of parallels alone. . . . I thought I would sacrifice myself for the sake of the truth. I was ready to become a martyr who would remove the flaw from geometry and return it purified to mankind. I accomplished monstrous, enormous labours; my creations are far better than those of others and yet I have not achieved complete satisfaction. For here it is true that *si paullum a summo discessit, vergit ad imum*. I turned back when I saw that no man can reach the bottom of this night. I turned back unconsoled, pitying myself and all mankind.'

And yet again:

'I admit that I expect little from the deviation of your lines. It seems to me that I have been in these regions; that I have travelled past all reefs of this infernal Dead Sea and have always come back with broken mast and torn sail. The ruin of my disposition and my fall date to this time. I thoughtlessly risked my life and happiness—*aut Caesar aut nihil*.'

In 1823 Johann Bolyai could tell his father, who had tried so hard to make him give up his interest in the problem, that he was succeeding:

'I am resolved to publish a work on parallels as soon as I can put it in order, complete it, and the opportunity arises. I have not yet made the discovery but the path which I have followed is almost certain to lead to my goal, provided this goal is possible. I do not yet have it but I have found things so magnificent that I was astounded. It would be an eternal pity if these things were lost as you, my dear father, are bound to admit when you see them. All I can say now is that I have created a new and different world out of nothing. All that I have sent you thus far is like a house of cards compared with a tower.'

His father advised him to publish his results as soon as posssible. Johann Bolyai's comment follows:

'He advised me that, if I was really successful, I should speadily make a public announcement and that for two reasons. One reason is that the idea might easily pass to someone else who would then publish it. Another reason—and one that seems valid enough—is that when the time is ripe for certain things, these things appear in different places in the manner of violets coming to light in early spring. And since scientific striving is like a war of which one does not know when it will be replaced by peace one must, if possible, win; for here preeminence comes to him who is first.'

16.B5 Janos Bolyai's *The Science Absolute of Space*

If the ray AM is not cut by the ray BN, situated in the same plane, but is cut by every ray BP comprised in the angle ABN, we will call ray BN *parallel* to ray AM; this is designated by $BN \parallel AM$.

It is evident that *there is one such ray BN, and only one*, passing through any point B (taken outside of the straight AM), and that the sum of the angles BAM, ABN can not exceed a st. $\angle$; for in moving BC around B until $BAM + ABC =$ st. $\angle$, somewhere ray

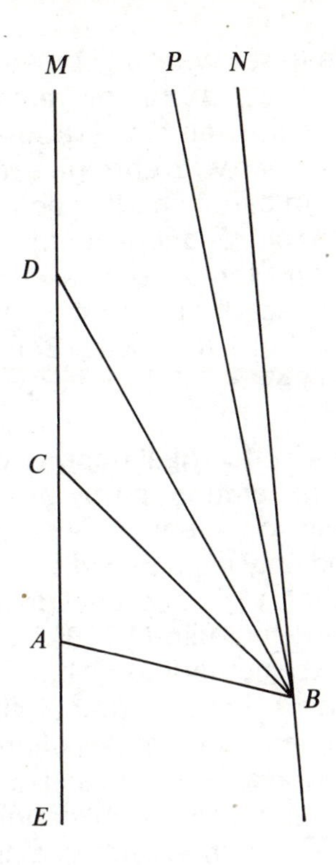

BC *first* does not cut ray AM, and it is then $BC \| AM$. It is clear that $BN \| EM$, wherever the point E be taken on the straight AM (supposing in all such cases $AM > AE$).

If while the point C goes away to infinity on ray AM, always $CD = CB$, we will have constantly $CDB = (CBD < NBC)$; but $NBC = 0$; and so also $ADB = 0$. [...]

The system of geometry resting on the hypothesis of the truth of Euclid's Axiom XI is called Σ; and the system founded on the contrary hypothesis is called S. All things which are not expressly said to be in Σ or in S, it is understood are enunciated absolutely, that is, are asserted true whether Σ or S is reality. [...]

It remains, finally (that the thing may be completed in every respect), to demonstrate the impossibility (apart from any supposition), of deciding *a priori*, whether Σ, or some S (and which one) exists. This, however, is reserved for a more suitable occasion.

16.C Later Nineteenth-century Developments

The immediate reaction of the scholarly community to the work of Bolyai and Lobachevskii can be described as malign neglect. In 16.C1 we reprint an extract from the standard history of the subject, in which Roberto Bonola gives some of the reasons why the tide began to turn in the 1860s. One does not have to agree with Bonola's arguments completely (see 16.C5) to appreciate the force of his points. But the crucial development was the creation of a new kind of differential geometry which made it possible to say what a geometry was, and so create many new geometries. The architects of this movement were Bernhard Riemann (1826–1866) and Eugenio Beltrami (1835–1900), who both saw that a geometry is simply a space with enough extra structure to be able to measure things like lengths. In particular, this ability can be obtained whether or not the new space has any relationship to a Euclidean space, and so Euclidean physical space is no longer special. In this way they were able to give the first rigorous account of the non-Euclidean geometry of Bolyai and Lobachevskii (this was Beltrami's achievement—see 16.C3) and to reformulate the whole concept of geometry (which is what Riemann did—see 16.C2).

General differential geometry was not to be developed in a big way for another generation, partly because the mathematical world c. 1870 was struggling to accept the existence of three geometries where previously there had only been Euclid's: non-Euclidean geometry, projective geometry (see Chapter 17) and Euclidean geometry. The way forward proposed in 1871 by Felix Klein (1849–1925) was particularly elegant. By accepting Riemann's methodology but recasting it in the language of projective geometry, he showed that all these geometries fitted into a hierarchy and were, indeed, all examples of projective geometry. In 16.C4 we give one of his first expressions of this idea, which became influential at the end of the century with the republication in several languages of his so-called *Erlanger Programm*. But it seems clear that the latter was not well known in the 1870s, which is why we have chosen to select instead from an item that was published in the then-recently founded *Mathematische Annalen*.

16.C1 Roberto Bonola on the spread of non-Euclidean geometry

The works of Lobachevskii and Bolyai did not receive on their publication the welcome which so many centuries of slow and continual preparation seemed to promise. However this ought not to surprise us. The history of scientific discovery teaches that every radical change in its separate departments does not suddenly alter the convictions and the presuppositions upon which investigators and teachers have for a considerable time based the presentation of their subjects.

In our case the acceptance of the non-Euclidean Geometry was delayed by special reasons, such as the difficulty of mastering Lobachevskii's works, written as they were in Russian, the fact that the names of the two discoverers were new to the scientific world, and the Kantian conception of space which was then in the ascendant.

Lobachevskii's French and German writings helped to drive away the darkness in which the new theories were hidden in the first years; more than all availed the constant and indefatigable labours of certain geometers, whose names are now associated with the spread and triumph of non-Euclidean Geometry. We would mention particularly: C. L. Gerling: (1788–1864), R. Baltzer (1818–1887) and Fr. Schmidt (1827–1901), in Germany; J. Hoüel (1823–1886), G. Battaglini (1826–1894), E. Beltrami (1835–1900), and A. Forti, in France and Italy.

From 1816 Gerling kept up a correspondence upon parallels with Gauss, and in 1819 he sent him Schweikart's memorandum on *Astralgeometrie*. Also he had heard from Gauss himself (1832), and in terms which could not help exciting his natural curiosity, of a *kleine Schrift* on non-Euclidean Geometry written by a young Austrian officer, son of W. Bolyai. The bibliographical notes he received later from Gauss (1844) on the works of Lobachevskii and Bolyai induced Gerling to procure for himself the *Geometrischen Untersuchungen* and the *Appendix*, and thus to rescue them from the oblivion into which they seemed plunged.

The correspondence between Gauss and Schumacher, published between 1860 and 1863, the numerous references to the work of Lobachevskii and Bolyai, and the attempts of Legendre to introduce even into the elementary text books a rigorous treatment of the theory of parallels, led Baltzer, in the second edition of his *Elemente der Mathematik* (1867), to substitute, for the Euclidean definition of parallels one derived from the new conception of space. Following Lobachevskii he placed the equation $A + B + C = 180°$, which characterises the Euclidean triangle, among the experimental results. To justify this innovation, Baltzer did nod fail to insert a brief reference to the possibility of a more general geometry than the ordinary one, founded on the hypothesis of two parallels. He also gave suitable prominence to the names of its founders. At the same time he called the attention of Hoüel, whose interest in the question of elementary geometry was well known to scientific men, to the Non-Euclidean geometry, and requested him to translate the *Geometrischen Untersuchungen* and the *Appendix* into French.

The French translation of this little book by Lobachevskii appeared in 1866 and was accompanied by some extracts from the correspondence between Gauss and Schumacher. That the views of Lobachevskii, Bolyai, and Gauss were thus brought together was extremely fortunate, since the name of Gauss and his approval of the discoveries of the two geometers, then obscure and unknown, helped to bring credit and consideration to the new doctrines in the most efficacious and certain manner.

The French translation of the *Appendix* appeared in 1867. It was preceded by a *Notice sur la vie et les travaux des deux mathématiciens hongrois W. et J. Bolyai de Bolya*, written by the architect Fr. Schmidt at the invitation of Hoüel, and was supplemented by some remarks by W. Bolyai, taken from Vol. I of the *Tentamen* and from a short analysis, also by Wolfgang, of the Principles of Arithmetic and Geometry.

16.C2 Bernhard Riemann on the hypotheses which lie at the basis of geometry

(*a*) *Plan of the investigation*

It is known that geometry assumes, as things given, both the notion of space and the first principles of constructions in space. She gives definitions of them which are merely nominal, while the true determinations appear in the form of axioms. The relation of these assumptions remains consequently in darkness; we neither perceive whether and how far their connection is necessary, nor, *a priori*, whether it is possible.

From Euclid to Legendre (to name the most famous of modern reforming geometers) this darkness was cleared up neither by mathematicians nor by such philosophers as concerned themselves with it. The reason of this is doubtless that the general notion of multiply extended magnitudes (in which space-magnitudes are included) remained entirely unworked. I have in the first place, therefore, set myself the task of constructing the notion of a multiply extended magnitude out of general notions of magnitude. It will follow from this that a multiply extended magnitude is capable of different measure-relations, and consequently that space is only a particular case of a triply extended magnitude. But hence flows as a necessary consequence that the propositions of geometry cannot be derived from general notions of magnitude, but that the properties which distinguish space from other conceivable triply extended magnitudes are only to be deduced from experience. Thus arises the problem, to discover the simplest matters of fact from which the measure-relations of space may be determined; a problem which from the nature of the case is not completely determinate, since there may be several systems of matters of fact which suffice to determine the measure-relations of space—the most important system for our present purpose being that which Euclid has laid down as a foundation. These matters of fact are—like all matters of fact—not necessary, but only of empirical certainty; they are hypotheses. We may therefore investigate their probability, which within the limits of observation is of course very great, and inquire about the justice of their extension beyond the limits of observation, on the side both of the infinitely great and of the infinitely small.

Notion of an n-*ply extended magnitude*

In proceeding to attempt the solution of the first of these problems, the development of the notion of a multiply extended magnitude, I think I may the more claim indulgent criticism in that I am not practised in such undertakings of a philosophical nature where the difficulty lies more in the notions themselves than in the construction; and that besides some very short hints on the matter given by Privy Councillor Gauss in his second memoir on Biquadratic Residues, in the *Göttingen Gelehrte Anzeige*, and in his Jubilee-book, and some philosophical researches of Herbart, I could make use of no previous labours.

(*b*) *On Euclidean, non-Euclidean and spherical geometry*

Riemann only briefly alluded to non-Euclidean geometry, which appears, without being named, as the geometry on a surface of constant negative curvature. Although we can infer that Riemann knew this by reading between the lines of the next extract, the first to spell out the details and so conclude two thousand years of investigation into Euclidean geometry was Beltrami in 1868 (see 16.C3). The extract here begins just after Riemann has explained that the surfaces of constant positive curvature are spheres.

The surfaces with less positive curvature are obtained from spheres of larger radii, by cutting out the lune bounded by two great half-circles and bringing the section-lines together. The surface with curvature zero will be a cylinder standing on the equator; the surfaces with negative curvature will touch the cylinder externally and be formed like the inner portion (towards the axis) of the surface of a ring. If we regard these surfaces as *locus in quo* for surface-regions moving in them, as Space is *locus in quo* for bodies, the surface-regions can be moved in all these surfaces without stretching. The surfaces with positive curvature can always be so formed that surface-regions may also be moved arbitrarily about upon them without *bending*, namely (they may be formed) into sphere-surfaces; but not those with negative curvature. Besides this independence of surface-regions from position, there is in surfaces of zero curvature also an independence of *direction* from position, which in the former surfaces does not exist.

Application to space

By means of these inquiries into the determination of the measure-relations of an *n-fold* extent the conditions may be declared which are necessary and sufficient to determine the metric properties of space, if we assume the independence of line-length from position and expressibility of the line-element as the square root of a quadric differential, that is to say, flatness in the smallest parts.

First, they may be expressed thus: that the curvature at each point is zero in three surface-directions; and thence the metric properties of space are determined if the sum of the angles of a triangle is always equal to two right angles.

Secondly, if we assume with Euclid not merely an existence of lines independent of position, but of bodies also, it follows that the curvature is everywhere constant; and then the sum of the angles is determined in all triangles when it is known in one.

16.C3 Eugenio Beltrami on the interpretation of non-Euclidean geometry

In recent times the mathematical public has begun to take an interest in some new concepts which seem destined, if they prevail, to profoundly change the whole complexion of classical geometry.

These concepts are not particularly recent. The master Gauss grasped them at the beginning of his scientific career, and although his writings do not contain an explicit exposition, his letters confirm that he had always cultivated them and attest his full support for the doctrine of Lobachevskii.

Such attempts at radical innovation in basic principles are encountered not infrequently in the history of ideas. Today they are a natural result of the critical spirit which accompanies all scientific investigation. When these attempts are presented as the fruits of conscientious and sincere investigations, and when they receive the support of a powerful, undisputed authority, it is the duty of men of science to discuss them calmly, avoiding equally both enthusiasm and disapproval. Moreover, in the science of mathematics, the triumph of new concepts cannot negate the truth already gained: it can only change the context or interpretation of the reasoning, and increase or diminish its value and frequency of use. A critique of principles cannot damage the solidity of the scientific edifice without leading to the discovery of a stronger foundation.

In this spirit we have sought, to the extent of our ability, to convince ourselves of the results of Lobachevskii's doctrine; then, following the tradition of scientific research, we have tried to find a real substrate for this doctrine, rather than admit the necessity for a new order of entities and concepts. We believe we have attained this goal for the planar part of the doctrine, but we believe that it is impossible to proceed further.

The present work is intended primarily to develop the first of these theses; the second is simply summarized briefly at the end, in order to allow the most straightforward judgement of the inherent significance of the proposed interpretation.

To avoid frequent interruptions to the exposition, we have postponed some necessary analytic results until a special note at the end.

The fundamental principle of proofs in elementary geometry is the *superimposability of equal figures.*

This principle is applicable not only to the plane, but to all surfaces on which there are equal figures in different positions, that is to say, to all surfaces of which any portion can be mapped onto any other by simple flexion. One sees that the rigidity of the surface on which the figure lies is not an essential condition for the application of the principle, so that, e.g., it does not affect the accuracy of proofs in euclidean plane geometry if the figure lies on the surface of a cylinder or cone rather than on a plane.

The surfaces with the above property, by a celebrated theorem of Gauss, are those which have a constant product of principal curvatures over all points, so that the spherical curvature is constant. The surfaces which do not satisfy the principle of superposition for figures traced on them therefore have a structure which varies with position.

The most essential figure in elementary geometry is the straight line. Its specific character is that of being completely determined by two points, so that two lines which pass through the same two points coincide throughout their extension.

These reflections are the starting point of our present research. We have begun by noting that the conclusion of a proof necessarily embraces all situations in which the hypotheses of the proof are satisfied. If the proof is stated in terms of a particular category of entities, without actually using any properties which differentiate them from a more extensive category, then it is clear that the conclusion of the proof acquires a generality much greater than that originally sought. It may very well happen that there are consequences seemingly incompatible with the nature of the entities originally contemplated, inasmuch as a property which holds generally for a given category of entities may be modified or disappear entirely for some particular ones. The apparent incongruence of the results of this kind of investigation, which the mind cannot reconcile, are due to an initially insufficient consciousness of the generality of the investigation.

This being understood at the outset, we consider which proofs in planimetry depend only on the principle of superposition and the postulate of the line, which are exactly those of non-euclidean planimetry. The conclusions of such proofs hold unconditionally wherever this principle and postulate are satisfied. Such situations are necessarily covered by the doctrine of surfaces of constant curvature, but they may not extend to surfaces with exceptional points. The principle of superposition in fact does not suffer from exceptions, but we have seen that the line postulate (for geodesics) meets with exceptions on the sphere, and consequently on all surfaces of positive curvature. Are there also exceptions on surfaces of constant negtive curvature? That is to say, can there be two points on such a surface which do not determine a unique geodesic?

This question is apparently still open. If it could be proved that such exceptions are impossible, it would become evident *a priori* that the theorems of non-euclidean planimetry hold unconditionally on all surfaces of constant negative curvature. In that case certain results which seem incompatible with the properties of the plane become interpretable on such surfaces, and receive a completely satisfactory explanation. At the present time we can explain the passage from euclidean to non-euclidean planimetry in terms of the difference between the surfaces of zero curvature and the surfaces of constant negative curvature.

Such are the considerations which have guided the following research.

16.C4 Felix Klein on non-Euclidean and projective geometry

The present discussion relates to the so-called non-Euclidean geometry of Gauss, Lobachevskii, Bolyai and the related considerations which Riemann and Helmholtz have put forward concerning the foundations of our geometric ideas. Nothing in it will pursue the philosophical speculations which have been attached to the works mentioned; the purpose is much more to put the mathematical results of those works, insofar as they relate to the theory of parallels, in a new, intuitive way and to make them accessible to a clear and general understanding. The route to this leads through projective geometry, whose independence from the question of the theory of parallels will be explained. One can now, after the start made by Cayley, construct a general projective metric which belongs on an arbitrarily chosen surface of the second degree taken as the so-called fundamental surface. This projective metric yields, according to the way in which the surface of the second degree is used, a picture for the different parallel theories of the above works. But it is not only a picture of them, it also shows their inner meaning.

The different theories of parallels

The eleventh axiom of Euclid [i.e. Postulate 5 in 3.B1(a)] is, as is well known, equivalent to the theorem that the sum of the angles in a triangle is equal to two right angles. Now it occurred to Legendre to prove that the angle sum of a triangle cannot be greater than two right angles; he showed further that if in one triangle the angle sum is two right angles then the same is the case in every triangle. But he did not succeed in showing that the angle sum cannot possibly be less than two right angles.

A similar reflection seems to have been the starting point for Gauss' researches into this question. Gauss took the view that it would in fact be impossible to prove the theorem that the angle sum would be two right angles, and rather that, as a consequence, one could construct a geometry in which the angle sum was less. Gauss called this geometry non-Euclidean; he occupied himself a lot with it, but sadly, apart from a few remarks, published nothing about it. In this non-Euclidean geometry a certain constant enters which is characteristic of the space metric. If one gives it an infinite value one obtains the usual Euclidean geometry. But if this constant has a finite value one obtains a different geometry in which, for example, the following holds: the angle sum of a triangle is less than two right angles and indeed the more so the greater is the area of the triangle. For a triangle whose vertices are infinitely far away the angle sum is zero. Through a point not on the line one can draw two parallels to the line, i.e.

lines which cut the line at the infinitely distant points one on the one side, one on the other. The lines through the given point which lie between the two parallels do not cut the given line at all.

Precisely this non-Euclidean geometry is the one discovered by Lobachevskii, Professor of Mathematics at the University of Kasan and, a year later, by the Hungarian mathematician J. Bolyai, and made the subject of detailed publications. However, these works remained quite unknown until one was made to take notice of them through the publication of the edition of letters between Gauss and Schumacher, published in 1862. Since then the view has spread that the theory of parallels is completely sorted out, i.e. that it is known to be truly undecidable.

[Klein then alluded to spherical geometry and discussions of it by Riemann and Helmholtz, before continuing as follows.]

In a similar way a geometry based on this representation will stand alongside the usual Euclidean geometry, as will the above-mentioned geometry of Gauss, Lobachevskii, and Bolyai. While in the latter, lines have two infinitely distant points, these lines have no (or rather two imaginary) infinitely distant points. Between the two Euclidean geometry stands as a limiting case; it ascribes to lines two coincident infinitely distant points.

A useful figure of speech in modern geometry is the following: the three geometries will be referred to respectively as hyperbolic, elliptic, or parabolic, according as the two infinitely distant points on the lines are real, imaginary, or coincident.

16.C5 J. J. Gray on four questions about the history of non-Euclidean geometry

Bonola (*Non-Euclidean Geometry*, 1912) divided the post-1600 history into four periods: one for forerunners, one for Gauss, Schweikart, and Taurinus, one for Bolyai and Lobachevskii, and one on later developments. Coolidge's account *A History of Geometrical Methods* (1940) likewise distinguished the period up to Bolyai and Lobachevskii from what he called the modern study of the problem by Beltrami, Riemann, and others. Kline devotes Chapter 36 of his *Mathematical Thought from Ancient to Modern Times*(1972) to the work done before 1840–50, and Chapter 38 to a further discussion of non-Euclidean geometry in the various contexts of projective geometry, metric geometry, models and consistency, and the question of reality. The intermediate chapter is devoted to a discussion of the differential geometry of Gauss and Riemann. The extent of agreement between these historical accounts greatly exceeds the extent of disagreement, and it will be convenient to regard them as providing what I may call the standard account. As such, the standard account gives a very good treatment of certain aspects of the subject. The division into periods coincides not only with the evident chronological divisions but with differences in mathematical methods: in the eighteenth century Saccheri and Lambert used classical geometry; in the early nineteenth century Bolyai and Lobachevskii used analysis; in the mid-nineteenth century Riemann and Beltrami turned to the techniques of differential geometry. It is also the case that before, roughly, 1800, mathematicians hoped to show that Euclidean geometry was the only possible geometry of space, whereas afterwards they sought to establish the possible validity of other geometries.

However, the standard account is open to several criticisms.

(1) It is well over 100 years from Saccheri's *Euclides Vindicatus* to Beltrami's *Saggio*, but the standard account does not explain why the development took so long.

(2) It does not explain, or adequately discuss, the choice of methods used at various times but subordinates it to a compilation of results. In Bonola's account, for instance, analytic methods appear unheralded in the discussion of Gauss, Schweikart, and Taurinus. This results in a failure to appreciate what it was that the mathematicians were intending to do.

(3) The exact nature of the accomplishment of J. Bolyai and Lobachevskii is not fully discussed. For, once it is admitted that it is not logically conclusive it must be asked why it has been found so compelling. This is the problem which makes it very hard to say who invented, or discovered, non-Euclidean geometry.

(4) Finally, it does not explain why it is that spherical geometry, well known throughout the entire period, did not at once settle the question, yet this geometry is now given almost immediately in modern textbooks as an example of a non-Euclidean geometry. This criticism bears with particular weight upon that part of the standard account which sees the problem as one in foundations: the existence or non-existence of a geometry differing from Euclid's in respect of the parallel postulate. If spherical geometry is such a good example of a geometry based on the HOA then the complete confidence of everyone from 1733 to 1854 in the impossibility of such a geometry is a historical problem which the standard account does not resolve. To say that spherical geometry is irrelevant because it flouts the 'Archimedean postulate' (that lines can be indefinitely extended) is no answer, because no one argued that way at the time.

It will be seen that these criticisms highlight one pervasive failing in the standard account; its tendency to see the subject as a prolonged attempt to answer one question (is the parallel postulate necessarily true?). The problem is essentially, therefore, regarded as one in foundations, and is resolved when the (negative) answer is given. Indeed, the standard account frequently ends with references to the logical independence of the postulate from the rest of geometry. This is partly the result of a historical accident. During the first half of this century mathematicians expended considerable effort in providing axiomatic foundations for various parts of their subject, and undoubtedly one of the reasons for this was the discovery of non-Euclidean geometry. However, much of mathematics was in need of the clarity that an axiomatization can bring, and if anything it was projective geometry that stood in greater need of an axiomatization to emancipate it fully from Euclidean and metric concepts. The coincidence of the problem of non-Euclidean geometry from the standpoint of, say, Proclus to Saccheri and, again to take examples, Pasch and Hilbert at the turn of this century diverted historians away from other, perhaps more important, matters. In general, historians of mathematics have been reluctant to discuss the changing nature of mathematical problems, preferring to produce a linear compilation of results.

The criticisms of the standard account can, I suggest, only be met by an account which dwells more on the mathematical methods and intentions of the actors in the historical drama, such as this book may provide and which I shall now summarize briefly.

16.D Influences on Literature

Non-Euclidean geometry became a fashionable topic of conversation at the end of the nineteenth century, perhaps because, like the theory of relativity later, it says something about the physical space in which we live and move and have our being. Writing in the late 1870s, Fyodor Dostoevsky evidently felt that his readers would understand Ivan Karamazov's metaphorical use of it (see 16.D1), even if Ivan is, arguably, mathematically confused. Ivan speculates that physical space might contain parallel lines that meet, or, to paraphrase him and make him more accurate, that physical space might be projective, although it could be that Ivan is thinking of Klein's elliptic (or spherical) geometry. It would seem that he does not have the generally accepted rival to Euclid's geometry in mind, the non-Euclidean or hyperbolic geometry with its many parallels. But then he admits that his puny Euclidean mind is incapable of grasping the infinite, which strikingly brings together the traditional position of the devout with Immanuel Kant's strictures about the limitations on human understanding.

While Dostoevsky's presentation neatly serves his reactionary purpose, the use Gabriel García Márquez has for geometry is suitably progressive (see 16.D2). In his novel *One Hundred Years of Solitude* he begins with a world in which ice, like many things, has no name, thus inverting St John's dictum that 'in the beginning was the word'. Here, materialist fashion, in the beginning are things. Knowledge is brought by travelling gypsies who, more Baconically than Pharaonically, wander the world with the wherewithal for scientific experiments. In another inversion, but one perhaps characteristic of the way mathematics intervenes in science, the central person in this part of the story, José Arcadio Buendía, works out for himself from the atlases that the gypsy Melquíades has given him that the world is round like an orange. This is as vivid an illustration as one could wish for of Gauss's claim (see 15.A5) that the curvature of a surface is intrinsic and can be detected by measurements taken just in the surface itself, without any need for there to be an ambient space from which the curvature could be seen. In just such a way, and with the right atlases, we might cry out that the world of non-Euclidean geometry is curved like a saddle!

16.D1 Fyodor Dostoevsky, from *The Brothers Karamazov*

'Joking? They said I was joking at the elder's yesterday. You see, my dear chap, there was an old sinner in the eighteenth century who delivered himself of the statement that if there were no God, it would have been necessary to invent him: *S'il n'existait pas Dieu il faudrait l'inventer*. And, to be sure, man has invented God. And what is so strange, and what would be so marvellous, is not that God actually exists, but that such an idea—the idea of the necessity of God—should have entered the head of such a savage and vicious animal as man—so holy it is, so moving and so wise and so much does it rebound to man's honour. So far as I'm concerned, I made up my mind long ago not to speculate whether man has created God or God has created man. Nor, of course, am I going to analyse all the modern axioms laid down by the Russian boys on that subject, all of them based on European hypotheses; for what is only an hypothesis there, becomes at once an axiom with a Russian boy, and not only with the boys but, I suppose, also with their professors, for Russian professors are quite often just the same Russian boys. And for this reason I'm going to disregard all the hypotheses. For what is it you and I are trying to do now? What I'm trying to do is to attempt to explain to you as quickly as possible the most important thing about me, that is to say, what sort of man I am, what I believe in and what I hope for—that's it, isn't it? And that's why I declare that I accept God plainly and simply. But there's this that has to be said: if God really exists and if he really has created the world, then, as we all know, he created it in accordance with the Euclidean geometry, and he created the human mind with the conception of only the three dimensions of space. And yet there have been and there still are mathematicians and philosophers, some of them indeed men of extraordinary genius, who doubt whether the whole universe, or, to put it more widely, all existence, was created only according to Euclidean geometry and they even dare to dream that two parallel lines which, according to Euclid, can never meet on earth, may meet somewhere in infinity. I, my dear chap, have come to the conclusion that if I can't understand even that, then how can I be expected to understand about God? I humbly admit that I have no abilities for settling such questions. I have a Euclidean, an earthly mind, and so how can I be expected to solve problems which are not of this world! And I advise you too, Alyosha, my friend, never to think about it, and least of all about whether there is a God or not. All these are problems which are entirely unsuitable to a mind created with the idea of only three dimensions. And so I accept God, and I accept him not only without reluctance, but, what's more, I accept his divine wisdom and his purpose—which are completely beyond our comprehension. I believe in the underlying order and meaning of life. I believe in the eternal harmony into which we are all supposed to merge one day. I believe in the Word to which the universe is striving and which itself was "with God" and which was God, and, well, so on and so forth, *ad infinitum*. Many words have been bandied about on this subject. So it would seem I'm on the right path—or am I? Anyway, you'd be surprised to learn, I think, that in the final result I refuse to accept this world of God's, and though I know that it exists, I absolutely refuse to admit its existence. Please understand, it is not God that I do not accept, but the world he has created. I do not accept God's world and I refuse to accept it. Let me put it another way: I'm convinced like a child that the wounds will heal and their traces will fade away, that all the offensive and comical spectacle of human contradictions will vanish like a pitiful mirage, like a horrible and odious invention of the feeble and infinitely puny Euclidean mind of man, and that in the world's finale, at the moment of eternal harmony, something so precious will happen and come to pass that it will suffice for all hearts, that it will allay all bitter resentments, that it will atone for all men's crimes, all the blood they have shed. It will suffice not only for the forgiveness but also for the justification of everything that has ever happened to men. Well, let it, let it all be and come to pass, but I don't accept it and I won't accept it! Let even the parallel lines meet and let me see them meet, myself—I shall see and I shall say that they've met, but I still won't accept it. That is the heart of the matter, so far as I'm concerned, Alyosha. That is where I stand. I'm telling you this in all seriousness. I deliberately began our talk as stupidly as I could, but I finished it with my confession, because that's all you want. You didn't want to hear about God, but only to find out what your beloved brother lived by. And I've told you.'

Ivan concluded his long tirade suddenly with a sort of special and unexpected feeling.

16.D2 Gabriel García Márquez, from *One Hundred Years of Solitude*

In spite of the fact that a trip to the capital was little less than impossible at that time, José Arcadio Buendía promised to undertake it as soon as the government ordered him to so that he could put on some practical demonstrations of his invention for the military authorities and could train himself in the complicated art of solar war. For several years he waited for an answer. Finally, tired of waiting, he bemoaned to Melquíades the failure of his project and the gypsy then gave him a convincing proof of his honesty: he gave him back the doubloons in exchange for the magnifying glass, and he left him in addition some Portuguese maps and several instruments of navigation. In his own handwriting he set down a concise synthesis of the studies by Monk Hermann, which he left José Arcadio so that he would be able to make use of the astrolabe, the compass, and the sextant, José Arcadio Buendía spent the long months of the rainy season shut up in a small room that he had built in the rear of the house so that no one would disturb his experiments. Having completely abandoned his domestic obligations, he spent entire nights in the courtyard watching the course of the stars and he almost contracted sunstroke from trying to establish an exact method to ascertain noon. When he became an expert in the use and manipulation of his instruments, he conceived a notion of space that allowed him to navigate across unknown seas, to visit uninhabited territories, and to establish relations with splendid beings without having to leave his study. This was the period in which he acquired the habit of talking to himself, of walking through the house without paying attention to anyone, as Úrsula and the children broke their backs in the garden, growing banana and caladium, cassava and yams, ahuyama roots and eggplants. Suddenly, without warning, his feverish activity was interrupted and was replaced by a kind of fascination. He spent several days as if he were bewitched, softly repeating to himself a string of fearful conjectures without giving credit to his own understanding. Finally, one Tuesday in December, at lunchtime, all at once he released the whole weight of his torment. The children would remember for the rest of their lives the august solemnity with which their father, devastated by his prolonged vigil and by the wrath of his imagination, revealed his discovery to them:

'The earth is round, like an orange.'

Úrsula lost her patience. 'If you have to go crazy, please go crazy all by yourself!' she shouted. 'But don't try to put your gypsy ideas into the heads of the children.' José Arcadio Buendía, impassive, did not let himself be frightened by the desperation of his wife, who, in a seizure of rage, smashed the astrolabe against the floor. He built another one, he gathered the men of the village in his little room, and he demonstrated to them, with theories that none of them could understand, the possibility of returning to where one had set out by consistently sailing east. The whole village was convinced that José Arcadio Buendía had lost his reason, when Melquíades returned to set things straight. He gave public praise to the intelligence of a man who from pure astronomical speculation had evolved a theory that had already been proved in practice, although unknown in Macondo until then, and as a proof of his admiration he made him a gift that was to have a profound influence on the future of the village: the laboratory of an alchemist.

17 Projective Geometry in the Nineteenth Century

Jean Victor Poncelet (1788–1867) was a pupil of Gaspard Monge at the Ecole Polytechnique, where he was exposed, as the passages by Michel Chasles in 17.A2 make clear, to the gifted teaching of an inspiring geometer. To keep up his spirits when a prisoner of war in Russia in 1813, Poncelet pushed Monge's ideas about the need for a kind of geometry which could match the generality of algebra. The result was his rediscovery of projective geometry, in which the sought-for generality was guaranteed by admitting a *principle of continuity,* whereby Poncelet hoped to establish a useful criterion for two different figures to be equivalent. Poncelet was clear that he was proselytizing for a revitalized geometry, and defended himself against the criticisms of Cauchy, who argued that Poncelet's principle could only be used if suitably restricted, much as a power-series only makes sense within its circle of convergence. Our extracts from the Preface to his important *Traité* of 1822 (see 17.A1) were literally written around Cauchy's critique, which he reprinted in his own work. It must be said, though, that his principle is not easy to understand and seems to lead him into outright error, until one appreciates Poncelet's radical conception of geometric figures and especially of ideal points. Later mathematicians were to prefer a blend of geometry and algebra, and Poncelet's ideas languished in his own country. As M. Paul has shown (see 17.A4), once Monge lost his influential position in the Ecole Polytechnique the cause of geometry was increasingly doomed. His disciples could not keep it alive; Poncelet himself turned directly to the applications Monge had hoped to encourage and devoted himself to the analysis of machines, and by 1837 Chasles had to admit that projective geometry had migrated to Germany and was now written in a language he could not speak.

Prominent in this new generation of German geometers was Julius Plücker (1801–1868), one of whose earliest discoveries is reprinted as 17.B2. It arose out of his interest in a disagreement between Poncelet and Joseph Diaz Gergonne, which is partially given in 17.A3. Gergonne, seeking to exploit the concept of duality in projective geometry, had wrongly claimed that the dual of a curve is another of the same order.

Poncelet pointed out the error but also noticed that his correction raised a second problem which he was unable to resolve. In seeing how to put it right, Plücker was led to discover his lovely result about the twenty-eight bitangents to a plane quartic, which continue to be fruitful objects of mathematical research. It marked the beginning of a rich period in the study of algebraic curves, in which curves were defined algebraically and studied projectively, a study that was soon extended to algebraic surfaces. Here perhaps the most dramatic discovery, made by Arthur Cayley and George Salmon in 1849, was that the general cubic surface has exactly twenty-seven straight lines on it. This discovery provoked a huge literature describing how the curves are related; discoveries such as Carl Friedrich Geiser's that these twenty-seven lines are close cousins of the twenty-eight bitangents; and a number of fine visual aids like the beautiful plaster of Paris models of the cubic surface and its lines, one of which is depicted stereoscopically in 17.B4.

In 17.B3 we present some passages from Alfred Clebsch's interesting obituary of Plücker. Clebsch was then the leader of the new school of German geometers (sadly he died two years later, aged only 39) and his comments are not only mathematically pertinent, but also historically rich. Clebsch rightly insisted on the importance of August Ferdinand Möbius's contribution to projective geometry, and 17.B1 gives some selections from Möbius's book of 1827. It is a remarkably algebraic work, and one may suppose that it was Möbius's intention to establish just what could be done without resorting to the difficult intuitions of Poncelet. Even when, as in 17.B1(b), Möbius describes various transformations in visual language, he still does not provide a picture. It is surely not the geometry Monge wanted.

To help you with this chapter here are some technicalities about duality, which it is easiest to present the way Möbius did, algebraically. In homogeneous coordinates the equation of a line in projective space is of the form

$$ax + by + cz = 0.$$

The so-called coordinates of the line are then the homogeneous coordinates $[a, b, c]$. In this way, every line of one projective plane can be thought of as a point in another projective plane, called later the dual space. To pass from a curve to its dual curve, one associates to each point on the curve its tangent, which one then thinks of as a point in the dual space. Because a bitangent is precisely a line tangent at two points, it gives rise to a double point on the dual curve. A cusp likewise gives rise to a point of inflection, as Plücker described.

Developments in France

17.A1 Jean Victor Poncelet on a general synthetic method in geometry

Poncelet began by praising analytic geometry or algebraic analysis for its general and uniform methods. Then he went on as follows.

[1] In ordinary geometry, which one often calls synthetic, the principles are quite otherwise, the development is more timid or more severe. The figure is described, one never loses sight of it, one always reasons with quantities and forms that are real and existing, and one never draws consequences which cannot be depicted in the imagination or before one's eyes by sensible objects. One stops when those objects cease to have a positive, absolute existence, a physical existence. Rigour is even pushed to the point of not admitting the consequences of an argument, established for a certain general disposition of the objects of a figure, for another equally general disposition of those objects which has every possible analogy with the first. In a word, in this restrained geometry one is forced to reproduce the entire series of primitive arguments from the moment where a line and a point have passed from the right to the left of one another, etc.

Now here precisely is in fact the weakness; here is what so strongly puts it below the new geometry, especially analytic geometry. If it was possible to apply implicit reasoning having abstracted from the figure, if only it was possible to apply the consequences of that kind of reasoning, this state of things would not exist, and ordinary geometry, without needing to employ the calculus and the signs of algebra, would rise to become in all respects the rival of analytic geometry, even if, as we have said already, it was not possible to conserve the explicit form of the reasoning.

Let us consider an arbitrary figure in a general position and indeterminate in some way, taken from all those that one can consider without breaking the laws, the conditions, the relationships which exist between the diverse parts of the system. Let us suppose, having been given this, that one finds one or more relations or properties, be they metric or descriptive, belong to the figure by drawing on ordinary explicit reasoning, that is to say by the development of an argument that in certain cases is the only one one regards as rigorous. Is it not evident that if, keeping the same given things, one can vary the primitive figure by insensible degrees by imposing on certain parts of the figure a continuous but otherwise arbitrary movement, is it not evident that the properties and relations found for the first system, remain applicable to successive states of the system, provided always that one has regard for certain particular modifications that may intervene, as when certain quantities vanish or change their sense or sign, etc., modifications which it will always be easy to recognize *a priori* and by infallible rules? [...]

Now this principle, regarded as an axiom by the wisest mathematicians, one can call the *principle* or *law of continuity* for mathematical relationships involving abstract and depicted magnitudes.

In the last analysis the principle of continuity has been admitted in its full extent and without any restriction by different geometers, who have employed it either overtly or tacitly, because without it they would be plunged into all the metaphysical consideradions of *imaginaries* which have always been driven back from the narrow sanctuary of rational geometry. Its explicit use in this science is almost always limited to real states of a system which is transformed by insensible degrees. And even there it gives rise to the *infinitely little* and the *infinitely great* which geometers still seek, in our day, to banish from the domain of the exact sciences. [...]

However, it will still not be difficult to establish this principle in an entirely direct and rigorous manner, with the aid of a calculus just like algebra the certainty of which is not the least to be doubted in our time, thanks to two centuries of efforts and success!

[2] In any case will it be necessary, and will one not immediately admit the principle of continuity in its full extent into rational geometry, as one does at once in algebraic calculus and then in the application of calculus to geometry, if it is not a means of proof but rather as a means of discovery or invention? Is it not at least as necessary to point out the resources employed at various times by men of genius for discovering the truth, as the feeble efforts they have then been obliged to use to prove them according to intellectual taste, either timid or less capable of bringing them home?

Finally, what harm can result, above all if one is restrained in one's conclusions, if one never uses half-truths, if one never admits analogy or induction, which are often deceptive, and which it is not necessary to confound with the principle of continuity? In fact, analogy and induction conclude from the particular to the general, from a series of isolated facts not necessarily related, in a word discontinuous, to a general and constant fact. The law of continuity, on the contrary, starts from a general state and some sort of indeterminacy of the system (that is to say that the conditions which govern it are never replaced by still more general ones) and they remain in a series of similar states going from one to the other by insensible gradations. One insists, besides, that the objects to which it is applied are by their nature continuous or submit to laws which can be regarded as such. Certain objects can even change their position by a series of variations undergone in the system, others can move away to infinity or approach to insensible distances, etc.; the general relations survive all the modifications without ceasing to apply to the system.

The only difficulty consists, as we have seen, in understanding fully what one wants to convey with the word *general* or *indeterminate* or *particular* state of a system. Now in each case the distinction is easy. For example, a line which meets another in a plane, is in a general state by comparison with the case where it becomes perpendicular or parallel to that line. Similarly a line (straight or curved) which meets another in a plane, is in a general or indeterminate state with regard to that other and the same thing takes place even when it ceases to meet it, proved that the two states do not suppose any particular relation of size or position beween these lines. The contrary will evidently hold when they become tangents, or asymptotic, or parallel etc.; they will then be in a particular state with regard to the primitive state.

17.A2 Michel Chasles

(a) *On descriptive geometry*

In recent times, after a rest of almost a century, pure Geometry has been enriched by a new doctrine, *descriptive geometry*, which is the necessary complement to the analytic geometry of Descartes and which, like it, must have immense results and mark a new era in the history of geometry.

This science is due to the creative genius of Monge.

It embraces two objects.

The first is to represent all bodies of a definite form on a plane area, and thus to transform into plane constructions graphical operations which it would be impossible to execute in space.

The second is to deduce from this representation of the bodies their mathematical relationships resulting from their forms and their relative positions.

This beautiful creation, which was initially intended for practical geometry and the arts which depend on it, really constitutes a *general theory*, because it reduces to a small number of abstract and invariant principles and to easy and always correct constructions, all the geometric operations which can be involved in stone cutting, carpentry, perspective, fortifications, gnomonics, etc, and which apparently can only be executed by mutually incoherent processes, which are uncertain and often scarcely rigorous.

But besides the importance due to its first intention, which gives a character of rationality and precision to all the constructive arts, descriptive geometry has another great importance due to the real services which it renders rational geometry, in several ways, and to the mathematical sciences in general. [For this reason] geometry thus reaches a state where it can most easily lend its generality and its intuitive evidence to mechanics and the physico-mathematical sciences.

(*b*) *On Gaspard Monge and his school*

Monge gave us, in his *Traité de Géométrie Descriptive*, the first examples of the utility of the intimate and systematic alliance between figures in three dimensions and plane figures. It is by such considerations that he proved, with rare elegance and perfect evidence, the beautiful theorems which constitute the theory of poles of curves of the second degree; the properties of centres of similitude of three circles taken two by two whose centres lie three by three on a line, and various other figures of plane geometry.

Since then, the pupils of Monge have cultivated this truly new kind of geometry with success, so that one has often, and with reason, given them the name of the school of Monge, and it consists, as we shall say, in introducing into plane geometry considerations of the geometry of three dimensions.

[Although declining to give details for reasons of space, Chasles in a footnote specifically mentioned the following as members of the school.]

Brianchon, who in a memoir of 1810 presented new and extensive reflexions on the subject to which, Poncelet tells us, he owes his own initial idea about the numerous beautiful geometrical researches contained in his *Traité des Proprietés Projectives* [and] Gergonne who performed the useful service of writing his own works, always imprinted with profound philosophical insight, and of founding his *Annales de Mathématiques* for the productions of former pupils of the Ecole Polytechnique.

(*c*) *On Monge's work*

Ancient geometry is a slew of figures. The reason is simple. Because one then lacked general and abstract principles, each question could only be treated in a concrete state, with the figure which was the object in question, and about which the only way forward could be to discover the elements necessary for the proof or the sought-for solution. [...]

This lack in ancient geometry was one of the relative advantages of analytic geometry where it was avoided in a most happy manner. So one could ask accordingly if there was not also a way of reasoning without the continual assistance of figures in pure and speculative geometry, for even when their construction is easy they are a real inconvenience and quite tire the spirit and exhaust the mind.

The writings of Monge and teaching of this illustrious master, whose style has been preserved by one of his most famous disciples (Arago), have resolved this question. They have taught us that it is enough, now that the elements of science have been created and are very extensive, to introduce into our language and our ideas of geometry, some general principles and transformations analogous to those of analysis, which, in making a truth known to us in its primitive purity and in all its facets, provide easy and fertile deductions by means of which one arrives naturally at one's goal. Such is the spirit of Monge's doctrines; and although his descriptive geometry, which provides us with examples of it, by its very nature makes essential use of figures, it is only in its effective and mechanical applications, where it plays the role of a tool, that it operates like this; but no one more than Monge could conceive of and do geometry without figures. It is a tradition in the Ecole Polytechnique that Monge knew to an extraordinary degree how to conceive of the most complicated forms in space, to penetrate to their general relation and their most hidden properties with no other help than his hands, whose movements accorded most admirably with his words, sometimes difficult but always with a true eloquence appropriate to the subject, in the neatness and precision, the richness and profundity of ideas.

17.A3 Joseph Diaz Gergonne on the principle of duality

(a) Duality

We have observed, not long ago, that at the point which mathematics has reached today, and encumbered as we are with theorems of which even the most intrepid memory cannot flatter itself it retains the statements, it would perhaps be less useful to science to seek new truths than to attempt to reduce the truths already discovered to a small number of guiding principles. In any case a science perhaps recommends itself less by the multitude of propositions which make it up then by the manner in which these propositions are related and connected to one another. Now, in each science there are certain elevated points of view where it is enough to stand there to embrace a great number of truths at a glance—which, from a less favourable position, one could believe were independent of one another—and which one realizes accordingly are derived from a common principle, often even incomparably more easy to establish than the particular truths of which it is the abridged expression.

It is with a view to confirming these considerations with some quite remarkable examples, on common points and common tangents to plane curves [etc.], that we are proposing to establish here a small number of general theorems offering an infinity of corollaries, among which we restrict ourselves to pointing out the simplest or those most worthy of notice [...].

Since we are not concerned at all here with metrical relations, all our theorems are double. To make this correspondence easier to grasp we place in two columns, the ones facing the others, the theorems which correspond to each other, as we have already done several times.

Properties of algebraic curves lying in a plane
Let there be a plane figure composed in any way one wishes of points, lines, and curves. Let us conceive that having drawn arbitrarily in the plane of this figure, an arbitrary

curve of the second order, one then constructs in the same plane another figure, all of whose points and all of whose lines are the poles and polars of all of the lines and all the points of the first, with respect to this curve of the second order considered as directrix. The two figures thus drawn are said to be polar reciprocals the one of the other, since the first can be derived from the second just as it was supposed to be the origin of the other. Now, in consequence of the properties of poles and polars, well-known today, here are the principal relations which are found to exist between these two figures.

1. If there is a system of a certain number of points on a line, then in the other figure there will be a system of exactly as many lines meeting in a point.
2. If there is a system of points lying on the same curve, then in the other figure there will be a system of exactly as many tangents to a curve of the same order.

1. If there is a system of a certain number of lines meeting in a point, then in the other figure there will be a system of exactly as many points lying on a line.
2. If there is a system of tangents to the same curve, then in the other figure there will be a system of exactly as many points lying on a curve of the same order.

(b) *A correction*

Unfortunately for Gergonne, both of the statements numbered 2 above are false, as Poncelet speedily pointed out. Gergonne reprinted Poncelet's criticisms in the next edition of his own *Annales,* and replied with an article of his own, from which the following is taken.

We shall presently need two words, one to state that a curve is such that a line cuts it in *m* points (or that a curved surface is pierced by a line in *m* points) and the other to state that a curve is such that one can draw *m* tangents to it from a given point in its plane (or that a curved surface is such that one can find *m* tangent planes to it through a given line). But in order not to introduce new words here for which the repugnance of the public, however ill-founded it might be, would nonetheless be invincible, we shall adopt the word *degree* for the first case and the word *class* for the second, i.e. we introduce the following definitions:

Definition I. A plane curve is said to be of the *m*th degree when it has *m* real or ideal intersections with a given line.

Definition I. A plane curve is said to be of the *m*th class when *m* real or ideal tangents can be drawn to it from a given point in its plane.

[...]

That is what we should have said [in] our 17th volume, but the use of the word order, which was entirely misplaced on that ocasion, led us to commit an error, as we have already noted above, and we owe a sincere obligation to M. Poncelet, whose doubts, however vaguely expressed, made us re-examine our work and made us feel the necessity of putting it right.

However, the corrections to be introduced are neither very numerous nor very difficult. First of all in the entire memoir one may regard everything in the left hand column as exact since everything is deduced independently of the principle of duality by a very simple and rigorous analysis. It will be the same with the columns on the right (and they form the greater number) which related only to curves and curved surfaces of the second order, because these lines and surfaces are both of the second degree and the second class. But the entire correction of the memoir, according to the ideas which have been omitted, can perhaps be summed up in these few words: Replace the word order by the word degree in the column on the left and by the word class in the column on the right, understanding by these latter words what has been explained above.

17.A4 M. Paul on students' studies at the Ecole Polytechnique

In the tables that follow, analysis includes differential and integral calculus as well as algebra; applications of geometry or analytic geometry refer to Mongean descriptive geometry in three dimensions, and mechanics means Lagrangean mechanics. The tables vividly document Paul's argument that the Ecole Polytechnique was speedily moved away from Monge's ideal of a utilitarian, geometric school and back towards the tradition of theoretical, research-oriented mathematics; a move directed by Pierre Laplace from 1806 on. Incidentally, the limited interest in chemistry is typical of the period. Mathematics was easily the preferred science of the French Revolution and the subsequent Napoleonic period. The figures are for the amount of time each student was to spend on each subject.

First year of study

	1801	1806	1812
Analysis	16%	29%	25%
Mechanics	10%	17%	18%
Geometry	40%	26%	23%
Chemistry	10%	9%	12%
	76%	81%	78%

Second year of study

	1801	1806	1812
Analysis	11%	18%	20%
Mechanics	12%	22%	25%
Geometry	4%	4%	2%
Chemistry	20%	9%	11%
Architecture	7%	13%	9%
Other applications	29%	14%	13%
	83%	80%	80%

17.B Developments in Germany

17.B1 August Ferdinand Möbius

(a) *Barycentric coordinates*

The remark that suitable weights can be attached to any three points in the plane so that a given fourth point can be considered as their centre of gravity, and that on the other hand these three weights stand in ratios to one another so the four points can be determined uniquely, leads me further to introduce a new method for determining the position of points in the plane. The three points which serve to determine all others I call the fundamental points, the lines joining them and the triangle made by them, the F-triangle; but the ratios which must exist between the weights of the F-points or its coefficients, which from now on I call the weights, are the coordinates of the point. I proceed to determine a point in space in the same way, where four F-points are necessary, and there are six F-lines and an F-pyramid.

The F-lines determined in this way are evidently the same as the axes in the usual method of parallel coordinates [...].

A principal object of study in analytic geometry is the properties of curves and surfaces. To be able to handle these with the barycentric calculus, it is necessary to take the coefficients of the F-points no longer as constants but rather as variables. If the coefficients of the three F-points in the plane, or the four F-points of space, are functions of a single variable quantity then all the centres of gravity I obtain, as I allow the variable to take all possible values successively, describe a curve in the plane or in space, and indeed a straight line if the coefficients are linear functions of the variable, a line of the second order if they are quadratic functions, etc. I undertake a more precise study of curves of the second order or conic sections, and I hope that the new analytic expressions for them and the manner in which some known and more new properties of these curves will be developed, will strike the reader as not disagreeable.

But at the same time I was persuaded to omit more of the same relations between figures and thereby to begin the second part of my book, which deals with geometric transformations, a study which in the sense used here embraces all of geometry, but which can be one of the most difficult if it is taken up exhaustively and in more complete generality. At present only the most simple kinds of transformation will be considered, namely those which are applied in elementary geometry.

[After reviewing the concepts of equality, similarity, and affinity, Möbius continued as follows.]

Still other problems lead to the more general transformations of collineation. The representation of these different kinds of problems and the solution of the novel ones amongst them by means of the baryc. calculus is the principal aim of the second part and at the same time is what in my book that I might most commend to the attention of the reader.

With the last mentioned transformation of collineation I was guided by the perspective representation of a plane figure. It is well known that one has often applied the theory of perspective projection to reduce difficult theorems and problems in geometry to simpler ones. (Here among others belongs Newton's generation of curves by shadows and the same geometer's problem: to transform figures into others of the same kind, *Principia*, Book I, Lemma 22.)

Another definition that can arise for the collineation belongs to the equality of a certain kind of compound ratio, here called the cross-ratio. The theory of this ratio, not considered hitherto so far as I know, is given in a special chapter. But to put the connection of this second definition with the first in its proper light researches have still seemed necessary, which are collected in the chapter on geometric nets.

Instruction in the theory of collineations is closely connected with the property of conics, well known to me, whereby to each point in the plane of such a curve there corresponds a line, and conversely. But just before I sent this account and some of its explanations to the press I learnt that the same material had already been dealt with in greater generality by French mathematicians. Without having been able to obtain a more detailed report, I have tried in the last chapter of my book to give an account at the same level of generality by means of the baryc. calculus.

(*b*) *Geometric transformations*

The simple means which we use to construct a system of collinear points in collineation with a given system, and whereby any specified point of the latter is produced by drawing a line, gives rise to the question if there is not just as simple a method which can be applied to the construction of collinear figures in the plane. One easily understands that if, from two planes which are in arbitrary positions with respect to one another, lines are drawn from the points of one plane through a fixed point, O, lying in neither of the planes, then the points in which these lines meet the other plane will form a system of points collinear with the first; for, from the manner of construction, each line a of the first plane will correspond with a line a' in the other, indeed the one in which the plane through O and a will cut it. (If one puts one's eye at O then the one system appears as a perspective representation of the other. The picture, thus sketched in perspective on the plane, of a plane figure is therefore a collineation of the latter. If the eye is taken infinitely far away then the pictures are affine one to another. If the two planes are parallel then similarity occurs, and if at the same time the eye is either infinitely far away or is in the middle between the two planes—equality and similarity.)

17.B2 Julius Plücker on twenty-eight bitangents

The discovery of the *principle of reciprocity* (theory of reciprocal polars) or which is exactly the same thing that of *duality* has given birth to a slew of new questions of which, with M. Poncelet, the one I regard as fundamental has still not been resolved despite the efforts of the most accomplished geometers.

M. Gergonne, when he introduced the first of his two columns which have since become so well known, had supposed that the degree of a curve would be equal to that of its polar or, in other terms, that the number of points of intersection of an algebraic curve of arbitrary degree with a given line should in general be equal to the number of tangents to the same curve passing through a given point. Several years earlier M. Poncelet had already recognized the contrary, in stating that if the degree of the given curve is m then the number of its tangents through a given point is $m(m-1)$: a fact which is obvious and very easy to prove by analysis. Finally conceding this in spite of himself M. Gergonne guaranteed his double columns their validity by only replacing in one of them the word *degree* with the word *class*. In this nomenclature a given curve of degree m is of class $m(m-1)$ and its polar is of class m. This new classification of curves seems to me to be very important; in adopting it, it pleases me to notice that even the mistake of an intelligent man bears its fruits.

For curves of degree more than two, there is a problem that I am going to expound. The degree of a given curve being m, that of its polar will in general be $m(m-1)$. Reciprocally, the new curve has as its polar the given one. Therefore while in general a curve of degree $m(m-1)$ has for its polar a curve of degree $m(m-1)[m(m-1)-1]$, a particular curve of the same degree has for polar a curve of degree m only. Thus, for example, to a curve of degree 3 there corresponds a polar of degree 6, to a curve of degree 6 in general a curve of degree 30, but in the particular case in question this degree descends by 27 units and becomes 3 again. Likewise to a curve of degree 4 there corresponds a polar a curve of degree 12, to a curve of degree 12 in general a curve of degree 132; but in this particular case the degree, by shedding 128 units, becomes 4 again. What therefore is the nature of the particular curve which produces such an extraordinary reduction? Here is the difficulty to be resolved.

M. Poncelet attempted to give an explanation in the present journal. He found it in the double points of the polar curve of the given one, of which each, according to him, produces a reduction in the degree of the polar reciprocal of two units. I immediately indicated that this explanation was incomplete, seeing that a curve of degree $m(m-1)$ cannot have enough double points to perform the required reduction. I should have to add that since, in the case $m = 3$, the given curve cannot be touched by the same line in two different points, its polar cannot have any double points. Somewhat later, in the succession of his memoirs M. Poncelet confessed in speaking of the explanation in question 'to having only skated over this important and difficult matter'.

Then he continued thus: 'In fact, the resolution of these questions depends on perfecting another theory, that of the singular points of curves, which is not yet complete despite the studies of our leading geometers, and it also leads to the most arduous difficulties in the science of elimination. But, although the considerations of a purely intuitive geometry, supported by notions which derive from the admission of the principle of continuity, have permitted us to advance much further in the fundamentals of this question, we should only flatter ourselves if we believed we have cleared up the difficulties entirely and dissipated all the clouds in which it still remains enveloped.'

This is the state of the question to which I propose to write down the answer here.

A curve can be regarded as described by a point or as enveloped by a straight line. The first mode of generation corresponds to the manner generally used for representing a curve by an equation in two variables. In this generation it is a quite singular case if the generating point stops in order to drive off in the opposite direction; curves represented by general equations of an arbitrary degree in general do not admit cusps. Nor do they admit in general multiple or isolated points. But they can have a certain number of inflexion points which do not behave at all like singular points, because the movement of the generating point is not interrupted at such points. In the second mode of generation, although it belongs to no less easy an algorithm, there are in general no points of inflexion, because at such a point the movement of the generating line would stop and then set off in the opposite direction. No more do there exist in general multiple tangents; but there will in general be a certain number of double points and cusps. At these last points the movement of the generating line presents no singularity, it continues to move in the same direction there. The definition of tangent is not the same in the two modes of generation; in the first every straight line passing through two coincident points (consecutive or not) plays the role of tangent, in the second the tangent is the generating line in its different positions. From this it follows that to each double point of the given curve there corresponds a reduction of *two* units in the degree of its polar, as M. Poncelet was the first to remark, provided however that the double point is not a *cusp*; in this case on the contrary the reduction is by *three* units. After these preliminaries I pass to the statement of the solution.

Curves represented by a general equation of degree m have in general a number of inflexion points equal to $3m(m-2) = M$ and a number of bitangents equal to $\frac{1}{2}(m+3)m(m-2)(m-3) = N$. Thus polar curves of degree $m(m-1)$ therefore in general have M cusps and N double points properly speaking. The degree of their polars which, for curves of their degree, is generally $m(m-1)[m(m-1)-1]$, therefore reduces by $3M + 2N$ units and so is down to m again.

In the case $m = 3$ the reduction therefore belongs solely to the inflexion points of the given curve, which are 9 in number. Six of these 9 points are always imaginary.

In the case of $m = 4$ the reduction belongs to 24 inflexion points and 28 bitangents of the given curve. And so on.

I must content myself here with having stated the result to which attaches a slew of questions of the greatest interest in the general theory of curves. One will find all the developments desirable in this important subject in a work which will appear in a few months, in which I propose to expound a system of analytic geometry, starting from new points of view and occupying myself almost exclusively with curves of higher degrees.

17.B3 From Alfred Clebsch's obituary of Julius Plücker

When one distinguishes between different aspects of the national times then, in scientific as in political life, the year 1826 may become particularly significant in the development of German mathematics and be taken to mark the start of an epoch. If we look over the decade in which this year lies we find in Germany Gauss, in scientific communication mostly with astronomers and who had dispensed with strictly mathematical connections; we find Pfaff, whose beautiful studies on partial differential equations earn him an enduring consideration. But compared to the numerous participants of manifold power which mathematics attracted in France we see in Germany only the school of Combinatorics, whose purpose lies far from our present ones.

How this relationship suddenly altered in the years after 1826, when one sees a real development of mathematics in Germany and a steadily growing circle of followers, is one of the most remarkable occurrences in the history of science. And the new life did not arise suddenly and randomly. On the one hand Jacobi developed a new and fundamental viewpoint on analytic functions. The profound work of Jacobi and Abel, as presented to German mathematicians in the closest inter-relationship in the newly founded journal of Crelle soon, led to perspectives on infinite expressions which have been worked through in a long series of works. At the same time Dirichlet announced the first of his rigorous and penetrating works on problems in number theory, which were to become fundamental to function theory in another sense. But simultaneously in Möbius, Steiner, and Plücker there appeared three geometers of the greatest significance and innermost originality, who, advancing in different ways, are yet unified in their essential points of view, and whom one must thank in major part for the present form of our geometric understanding. [...]

No one has had more influence on the further development of geometry than Monge. Above all, he understood the importance of arousing an interest in geometry in wider circles; his researches on the application of geometry to topics in Analysis showed the fruitfulness of this long neglected discipline in seemingly quite different domains. His pupils, brimming over with a truly geometrical sensibility, dedicated themselves to making the new ideas secure in a principled way. While some of them turned to the applications of analysis (in Monge's sense) to certain metrical problems, others took up purely projective considerations. From this the organic character developed which distinguishes the new geometry and makes it appear so different from the geometry of antiquity.

It was Poncelet who first proposed a projective geometry which comes very close to the present form. He treated the ratio of lengths clearly as a thing in itself; a major principle, fundamental to the development of the science, that of duality, was introduced in rivalry and competition between Poncelet and Gergonne, illuminated, and given its true significance. Above all, most of those elements that later were to be made into the principal basis of geometry are to be found already in Poncelet, only not presented as principles; e.g. the idea of cross-ratio and transformation.

[Clebsch now goes into the duality controversy in some mathematical detail, before continuing as follows.]

I have described these researches [on duality] which took place around 1827–30 [...] without saying that Möbius had already anticipated part of them in his *Barycentric Calculus* of 1827. In this work, which can never be marvelled at enough, and in which a great number of the fundamental ideas of geometry were spoken of for the first time, Möbius already dealt with collineations and duality completely [...]. It seems that Plücker did not at first know of Möbius' work or at least that he did not immediately relate it to his own. Perhaps one may put it down to the modest form in which Möbius published his deep and new ideas, that their content and significance were usually only first recognized when other geometers were led to appreciate the compelling necessity for the advancement of science of the elements introduced by Möbius. So some of his basic ideas appeared simultaneously in Steiner's geometric form and spread from there, where they were more organically and systematically developed, to a wider circle. They were later spread still further when Chasles, in his *Aperçu Historique* (1837) presented the same ideas, but, unable to read German, he presented them as new [...].

Simultaneously with his line coordinates Plücker introduced another technique into Geometry, whose application was equally consequential. These were the so-called triangular coordinates (*Crelle's Journal*, Vol. 5, 1829). By means of these coordinates and the fundamental theorem of homogeneous functions Plücker could give definitive form, in particular, to the equations for tangents and their contact points. One can again designate Möbius as the first (indeed, before Plücker) to have this idea, for

Möbius's barycentric coordinates are basically no different. Only they were described by an idea taken from mechanics, and there are essentially no applications of the theory of homogeneous functions, because he did not deal with the equation of the curve but rather with parametric representations of its coordinates.

The connection between the theory of algebraic curves and the theory of homogeneous functions which the introduction of these coordinates established, was only broached by Plücker. Hesse took the first step that leads to the present new algebra and allows the theory of algebraic-geometric figures to appear as a chapter in this subject.

17.B4 Christian Wiener's stereoscopic picture of the twenty-seven lines on a cubic surface

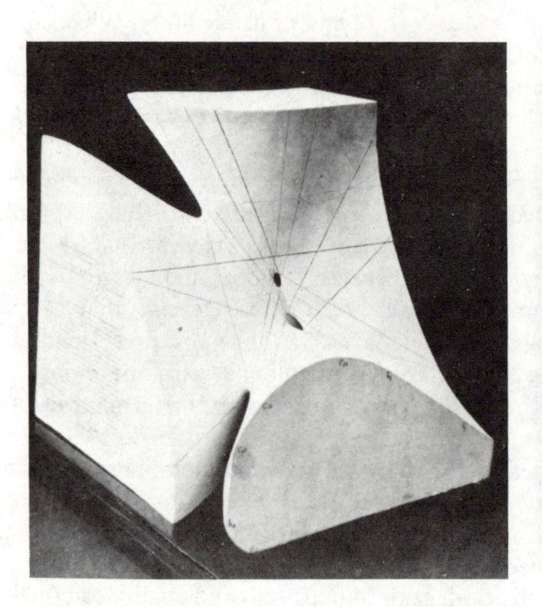

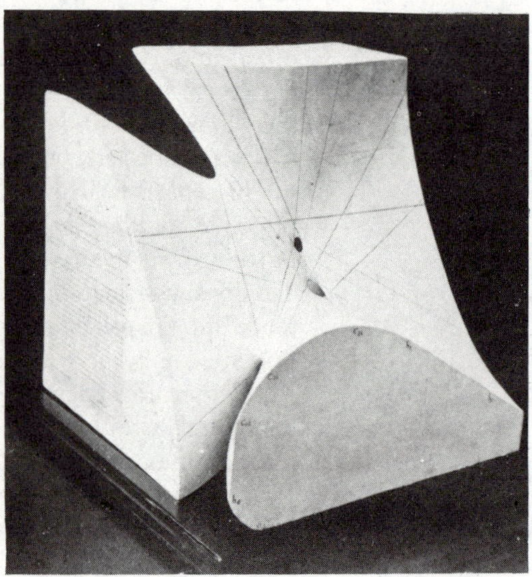

18 The Rigorization of the Calculus

The history of the calculus in the eighteenth century is not dominated by debates about the foundations of the calculus, a fact which seems to turn some historians quite schoolmasterly in their writings, but not only did progress in mathematics not seem then to depend on resolving any foundational questions, no one seemed to have enough of the answer for decisive progress to be possible. So what discussion there was is of a fitful character. The shrewd comments of George Berkeley (18.A1) reveal that he understood as well as anyone what the problems with the (Newtonian) calculus were, but his aim was only to show that mathematicians could presume to no exalted status as philosophers, not to put their house to rights. And although Colin MacLaurin (18.A2) showed that one could deal with all of Berkeley's criticisms case by case, it is equally clear that no one could be expected to write out all of their theorems concerning the calculus in the revived form of double *reductio ad absurdum.* So the question remained: what could a reformulation of the calculus consist of? In 18.A3 we give a glimpse of d'Alembert's ideas, which are based on an explicit but still naive appeal to the concept of *limit.* As d'Alembert makes clear, his presentation is heavily influenced by Newton's ideas, and his remarks about 0/0 demonstrate just how intuitive his ideas remained. 18.A3 shows that one could well be in possession of a partial solution but, for want of the rest, be unable to make significant progress.

Lagrange seems to have become interested in the foundations of the calculus when, following a prolonged depression in the mid-1780s, he lost his taste for research and turned instead to teaching. He organized a prize competition on the topic in 1786, and then, dissatisfied with the results, he started to develop his own ideas. His derivation of Taylor's theorem in this spirit, which gives a power-series expansion of an 'analytic' function, is given in 18.A4, and in 18.A5 we quote from a later reaffirmation of his views with the striking slogan 'algebra is only the theory of functions'.

Lagrange based his arguments on the power of formal algebra, but these could not suffice. There was also the sheer size of the problem. To rigorize the calculus it would not be enough to give it acceptable foundations; it would also be necessary to derive its results in a logical order. One of the most important results was the intermediate value theorem (see 18.B3), which asserts that if a continuous function is negative at one end of an interval and positive at the other, then it is zero somewhere in between. In 18.B1 we give a piece of Bernard Bolzano's proof of this result, which is needed even to prove results about polynomials like the fundamental theorem of algebra. Bolzano's proof highlights the need for a good (i.e. workable) definition of a real number. His view was that it was defined by any sensible limiting process—explicitly by means of the assumption of the existence of least upper bounds. So this was two problems: 'What is continuity?' and 'What is a real number?' In 18.B4 the historian J. V. Grabiner gives her estimation of the nature and significance of Augustin Louis Cauchy's achievement, and we have included her translation of Cauchy's proof of the intermediate value theorem in 18.B2. In 18.B3 we have given some of the definitions that he presented, which show how he built up a new calculus, the first to be adequately rigorous.

18.A Eighteenth-century Developments

18.A1 George Berkeley's criticisms of the calculus

Now the other method of obtaining a rule to find the fluxion of any power is as follows. Let the quantity x flow uniformly, and be it proposed to find the fluxion of x^n. In the same time that x by flowing becomes $x + o$, the power x^n becomes $\overline{x + o}|^n$, i.e. by the method of infinite series

$$x^n + nox^{n-1} + \frac{nn - n}{2} oox^{n-2} + \&c.,$$

and the increments

$$o \text{ and } nox^{n-1} + \frac{nn - n}{2} oox^{n-2} + \&c.$$

are one to another as

$$1 \text{ to } nx^{n-1} + \frac{nn - n}{2} ox^{n-2} + \&c.$$

Let now the increments vanish, and their last proportion will be 1 to nx^{n-1}. But it should seem that this reasoning is not fair or conclusive. For when it is said, let the increments vanish, i.e. let the increments be nothing, or let there be no increments, the former supposition that the increments were something, or that there were increments, is destroyed, and yet a consequence of that supposition, i.e. an expression got by virtue thereof, is retained. Which, by the foregoing lemma, is a false way of reasoning.

Certainly when we suppose the increments to vanish, we must suppose their proportions, their expressions, and every thing else derived from the supposition of their existence to vanish with them. [...]

It is curious to observe what subtilty and skill this great genius employs to struggle with an insuperable difficulty; and through what labyrinths he endeavours to escape the doctrine of infinitesimals; which as it intrudes upon him whether he will or no, so it is admitted and embraced by others without the least repugnance. Leibniz and his followers in their *calculus differentialis* making no manner of scruple, first to suppose, and secondly to reject, quantities infinitely small; with what clearness in the apprehension and justness in the reasoning, any thinking man, who is not prejudiced in favour of those things, may easily discern. The notion or idea of an *infinitesimal quantity*, as it is an object simply apprehended by the mind, hath been already considered. I shall now only observe as to the method of getting rid of such quantities, that it is done without the least ceremony. As in fluxions the point of first importance, and which paves the way to the rest, is to find the fluxion of a product of two indeterminate quantities, so in the *calculus differentialis* (which method is supposed to have been borrowed from the former with some small alterations) the main point is to obtain the difference of such product. Now the rule for this is got by rejecting the product or rectangle of the differences. And in general it is supposed that no quantity is bigger or lesser for the addition or subduction of its infinitesimal: and that consequently no error can arise from such rejection of infinitesimals.

[...]

Qu. 43 Whether an algebraist, fluxionist, geometrician, or demonstrator of any kind can expect indulgence for obscure principles or incorrect reasonings? And whether an algebraical note or species can at the end of a process be interpreted in a sense which could not have been substituted for it at the beginning? Or whether any particular supposition can come under a general case which doth not consist with the reasoning thereof?

Qu. 44 Whether the difference between a mere computer and a man of science be not, that the one computes on principles clearly conceived, and by rules evidently demonstrated, whereas the other doth not?

Qu. 45 Whether, although geometry be a science, and algebra allowed to be a science, and the analytical a most excellent method, in the application nevertheless of the analysis to geometry, men may not have admitted false principles and wrong methods of reasoning?

[...]

Qu. 59 If certain philosophical virtuosi of the present age have no religion, whether it can be said to be for want of faith?

Qu. 60 Whether it be not a juster way of reasoning, to recommend points of faith from their effects, than to demonstrate mathematical principles by their conclusions?

Qu. 61 Whether it be not less exceptionable to admit points above reason than contrary to reason?

Qu. 62 Whether mysteries may not with better right be allowed of in Divine Faith than in Human Science?

Qu. 63 Whether such mathematicians as cry out against mysteries have ever examined their own principles?

Qu. 64 Whether mathematicians, who are so delicate in religious points, are strictly scrupulous in their own science? Whether they do not submit to authority, take things upon trust, and believe points inconceivable? Whether they have not their mysteries, and what is more, their repugnancies and contradictions?

Qu. 65 Whether it might not become men who are puzzled and perplexed about their own principles, to judge warily, candidly and modestly concerning other matters?

Qu. 66 Whether the modern analytics do not furnish a strong *argumentum ad hominem* against the philomathematical infidels of these times?

Qu. 67 Whether it follows from the above-mentioned remarks, that accurate and just reasoning is the peculiar character of the present age? And whether the modern growth of infidelity can be ascribed to a distinction so truly valuable?

18.A2 Colin MacLaurin on rigorizing the fluxional calculus

The fluxion of the root A being supposed equal to a, the fluxion of the square AA will be equal to $2A \times a$.

Let the successive values of the root be $A-u$, A, $A+u$, and the corresponding values of the square be $AA - 2Au + uu$, AA, $AA + 2Au + uu$, which increase by the differences $2Au - uu$, $2Au + uu$, etc. and because those differences increase it follows from art. 704 that if the fluxion of A be represented by u, the fluxion of AA cannot be represented by a quantity that is greater than $2Au + uu$, or less than $2Au - uu$. This being premised, suppose, as in the proposition, that the fluxion A is equal to a; and if the fluxion of AA be not equal to $2Aa$, let it first be greater than $2Aa$ in any ratio, as that of $2A + o$ to $2A$, and consequently equal to $2Aa + oa$. Suppose now that u is any increment of A less than o; and because a is to u as $2Aa + oa$ to $2Au + ou$, it follows that if the fluxion of A should be represented by u, the fluxion of AA would be represented by $2Au + ou$, which is greater than $2Au + uu$. But it was shown from art. 704 that if the fluxion of A be represented by u the fluxion of AA cannot be represented by a quantity greater than $2Au + uu$. And these being contradictory, it follows that the fluxion of A being equal to a, the fluxion of AA cannot be greater than $2Aa$. If it can be less than $2Aa$, where the fluxion of A is supposed equal to a, let it be less in any ratio of $2A - o$ to $2A$, and therefore equal to $2Aa - oa$. Then because a is to u as $2Aa - oa$ is to $2Au - ou$, which is less than $2Au - uu$ (u being supposed less than o, as before) it follows that if the fluxion of A was represented by u, the fluxion of AA would be represented by a quantity less than $2Au - uu$, against what has been shown from art. 704. Therefore the fluxion of A being supposed equal to a, the fluxion of AA must be equal to $2Aa$.

18.A3 D'Alembert on differentials

Newton started out from another principle; and one can say that the metaphysics of this great mathematician on the calculus of fluxions is very exact and illuminating, even though he allowed us only an imperfect glimpse of his thoughts.

He never considered the *differential* calculus as the study of infinitely small quantities, but as the method of first and ultimate ratios, that is to say, the method of finding the limits of ratios. Thus this famous author has never differentiated quantities but only equations; in fact, every equation involves a relation between two variables and the differentiation of equations consists merely in finding the limit of the ratio of the finite differences of the two quantities contained in the equation. Let us illustrate this by an example which will yield the clearest idea as well as the most exact description of the method of the *differential* calculus.

Let AM be an ordinary parabola, the equation of which is $yy = ax$; here we assume that $AP = x$ and $PM = y$, and a is a parameter. Let us draw the tangent MQ to this parabola at the point M. Let us suppose that the problem is solved and let us take an ordinate pm at any finite distance from PM; furthermore, let us draw the line mMR

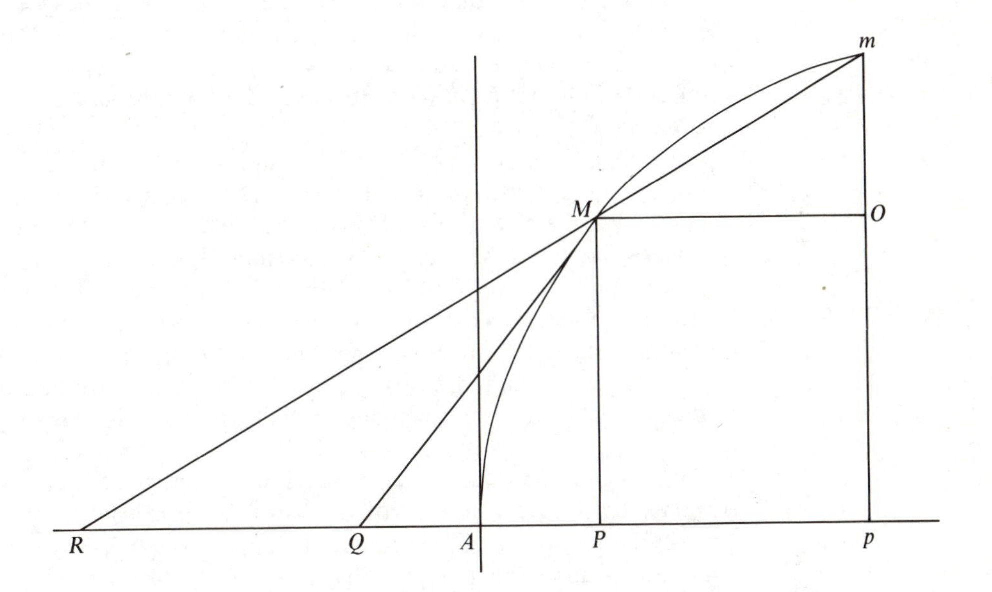

through the points M, m. It is evident, *first*, that the ratio MP/PQ of the ordinate to the subtangent is greater than the ratio MP/PR or mO/MO which is equal to it because of the similarity of the triangles MOm, MPR; *second*, that the closer the point m is to the point M, the closer will be the point R to the point Q, consequently the closer will be the ratio MP/PR or mO/MO to the ratio MP/PQ; finally, that the first of these ratios approaches the second one as closely as we please, since PR may differ as little as we please from PQ. Therefore, the ratio MP/PQ is the limit of the ratio of mO to OM. Thus, if we are able to represent the ratio mO/OM in algebraic form, then we shall have the algebraic expression of the ratio of MP to PQ and consequently the algebraic representation of the ratio of the ordinate to the subtangent, which will enable us to find this subtangent. Let now $MO = u$, $Om = z$; we shall have $ax = yy$, and $ax + au = yy + 2yz + zz$. Then in view of $ax = yy$ it follows that $au = 2yx + zz$ and $z/u = a/(2y + z)$.

This value $a/(2y + z)$ is, therefore, in general the ratio of mO to OM, wherever one may choose the point m. This ratio is always smaller than $a/2y$; but the smaller z is, the greater the ratio will be and, since one may choose z as small as one pleases, the ratio

$a/(2y + z)$ can be brought as close to the ratio $a/2y$ as we like. Consequently $a/2y$ is the limit of the ratio $a/(2y + z)$, that is to say, of the ratio mO/OM. Hence $a/2y$ is equal to the ratio MP/PQ, which we have found to be also the limit of the ratio of mO to OM, since two quantities that are the limits of the same quantity are necessarily equal to each other. To prove this, let X and Z be the limits of the same quantity Y. Then I say that $X = Z$; indeed, if they were to have the difference V, let $X = Z \pm V$: by hypothesis the quantity Y may approach X as closely as one may wish; that is to say, the difference between Y and X may be as small as one may wish. But, since Z differs from X by the quantity V, it follows that Y cannot approach Z closer than the quantity V and consequently Z would not be the limit of T, which is contrary to the hypothesis.

From this it follows that MP/PQ is equal to $a/2y$. Hence $PQ = 2yy/a = 2x$. Now, according to the method of the *differential* calculus, the ratio of MP to PQ is equal to that of dy to dx; and the equation $ax = yy$ yields $a\,dx = 2y\,dy$ and $dy/dx = a/2y$. So dy/dx is the limit of the ratio of z to u, and this limit is found by making $z = 0$ in the fraction $a/(2y + z)$.

But, one may say, is it not necessary also to make $z = 0$ and $u = 0$ in the fraction $z/u = a/(2y + z)$, which would yield $\frac{0}{0} = a/2y$? What does this mean? My answer is as follows. First, there is no absurdity involved; indeed $\frac{0}{0}$ may be equal to any quantity one may wish: thus it may be $= a/2y$. Secondly, although the limit of the ratio of z to u has been found when $z = 0$ and $u = 0$, this limit is in fact not the ratio of $z = 0$ to $u = 0$, because the latter one is not clearly defined; one does not know what is the ratio of two quantities that are both zero. This limit is the quantity to which the ratio z/u approaches more and more closely if we suppose z and u to be real and decreasing. Nothing is clearer than this; one may apply this idea to an infinity of other cases.

Following the method of differentiation (which opens the treatise on the quadrature of curves by the great mathematician Newton), instead of the equation $ax + au = yy + 2yz + zz$ we might write $ax + a0 = yy + 2y0 + 00$, thus, so to speak, considering z and u equal to zero; this would have yielded $\frac{0}{0} = a/2y$. What we have said above indicates both the advantage and the inconveniences of this notation: the advantage is that z, being equal to 0, disappears without any other assumption from the ratio $a/(2y + 0)$; the inconvenience is that the two terms of the ratio are supposed to be equal to zero, which at first glance does not present a very clear idea.

From all that has been said we see that the method of the *differential* calculus offers us exactly the same ratio that has been given by the preceding calculation. It will be the same with other more complicated examples. This should be sufficient to give beginners an understanding of the true metaphysics of the *differential* calculus. Once this is well understood, one will feel that the assumption made concerning infinitely small quantities serves only to abbreviate and simplify the reasoning; but that the *differential* calculus does not necessarily suppose the existence of those quantities; and that moreover this calculus merely consists in *algebraically determining the limit of a ratio, for which we already have the expression in terms of lines, and in equating those two expressions. This will provide us with one of the lines we are looking for.* This is perhaps the most precise and neatest possible definition of the *differential* calculus; but it can be understood only when one is well acquainted with this calculus, because often the true nature of a science can be understood only by those who have studied this science.

18.A4 Lagrange on derived functions

Now let us consider a function $f(x)$ of a variable x. If we replace x by $x + i$, i being any arbitrary quantity, it will become $f(x + i)$ and, by the theory of series, we can expand it in a series of the form

$$f(x) + pi + qi^2 + ri^3 + \cdots,$$

in which the quantities $p, q, r, \ldots$, the coefficients of the powers of i, will be new functions of x, which are derived from the primitive functions of x, and are independent of the quantity i.

But, in order to prove what we claim, we shall examine the actual form of the series representing the expansion of a function $f(x)$ when we substitute $x + i$ for x, which involves only positive integral powers of i.

This assumption is indeed fulfilled in the cases of various known functions; but nobody, to my knowledge, has tried to prove it *a priori*—which seems to me to be all the more necessary since there are particular cases in which it is not satisfied. On the other hand, the differential calculus makes definite use of this assumption, and the exceptional cases are precisely those in which objections have been made to the calculus.

I will first prove that in the series arising by the expansion of the function $f(x + i)$ no fractional power of i can occur except for particular values of x.

[Having accomplished this, Lagrange continues later as follows.]

We have seen that the expansion of $f(x + i)$ generates various other functions $p, q, r, \ldots$, all of them derived from the original function $f(x)$, and we have given the method for finding these functions in particular cases. But in order to establish a theory concerning these kinds of functions we must look for the general law of their derivation.

For this purpose, let us take once more the general formula $f(x + i) = f(x) + pi + qi^2 + ri^3 + \cdots$, and let us suppose that the undetermined quantity x is replaced by $x + o$, o being any arbitrary quantity independent of i. Then $f(x + i)$ will become $f(x + i + o)$, and it is clear that we shall obtain the same result by simply substituting $i + o$ for i in $f(x + i)$. The result must also be the same whether we replace the quantity i by $i + o$ or x by $x + o$ in the expansion $f(x)$.

The first substitution yields $f(x) + p(i + o) + q(i + o)^2 + r(i + o)^3 + \cdots$, or, expanding the powers of $i + o$ and writing out for the sake of simplicity no more than the first two terms of each power (since the comparison of these terms will be sufficient for our purpose):

$$f(x) + pi + qi^2 + ri^3 + si^4 + \cdots + po + 2qio + 3ri^3o + 4si^3o + \cdots.$$

In order to carry out the other substitution, we note that we obtain $f(x) + f'(x)o + \cdots$, $p + p'o + \cdots$, $q + q'o + \cdots$, $r + r'o + \cdots$ when we replace x by $x + o$ in the functions $f(x), p, q, r, \ldots$, respectively; here we retain in the expansion only the terms that include the first power of o. It is clear that the same expression will become $f(x) + pi + qi^2 + ri^3 + si^4 + \cdots + f'(x)o + p'io + q'i^2o + r'i^3o + \cdots$.

Since these two results must be identical whatever the values of i and o may be, comparison of the terms involving $o, io, i^2o, \ldots$, will give: $p = f'(x)$, $2q = p'$, $3r = q'$, $4s = r', \ldots$

Now it is clear that in the same way that $f'(x)$ is the first derived function of $f(x)$, p' is the first derived function of p, q' the first derived function of q, r' the first derived function of r, and so on. Therefore, if, for the sake of greater simplicity and uniformity, we denote by $f'(x)$ the first derived function of $f(x)$, by $f''(x)$ the first derived function of $f'(x)$, by $f'''(x)$ the first derived function of $f''(x)$, and so on, we have $p = f'(x)$, and hence $p' = f''(x)$; consequently $q = \frac{p'}{2} = \frac{f''(x)}{2}$, hence $q' = \frac{f'''(x)}{2}$; consequently $r = \frac{q'}{3} = \frac{f'''(x)}{2 \cdot 3}$, hence $r = \frac{f^{\text{iv}}(x)}{2 \cdot 3}$; consequently $s = \frac{r'}{4} = \frac{f^{\text{iv}}(x)}{2 \cdot 3 \cdot 4}$, hence $s' = \frac{f^{\text{v}}(x)}{2 \cdot 3 \cdot 4}$; and so on.

Then by substituting these values in the expansion of the function $f(x + i)$, we obtain

$$f(x+i) = f(x) + f'(x)i + \frac{f''(x)}{2}i^2 + \frac{f'''(x)}{2 \cdot 3}i^3 + \frac{f^{\text{iv}}(x)}{2 \cdot 3 \cdot 4}i^4 + \cdots.$$

This new expression has the advantage of showing how the terms of the series depend on each other and above all how we can form all the derived functions involved in the series provided that we know how to form the first derived function of any primitive function.

We shall call the function $f(x)$ the *primitive function* with respect to the functions $f'(x), f''(x), \ldots$ that are derived from it; these fuctions are called the *derived functions* with respect to the former one. Moreover, we shall call the first derived function $f'(x)$ the *first function*, the second derived function the *second function*, the third derived function the *third function*, and so on. In the same way, if y is supposed to be a function of x, we denote its derived function by $y', y'', y''', \ldots$, respectively, so that, y being the primitive function, y' will be its *first function*, y'' its *second function*, y''' its *third function*, and so on.

Consequently, if x is replaced by $x + i$, y will become

$$y + y'i + \frac{y''i^2}{2} + \frac{y'''i^3}{2 \cdot 3} + \cdots.$$

Thus, provided that we have a method of computing the first function of any primitive function, we can obtain, by merely repeating the same operation, all the derived functions, and consequently all the terms of the series that result from expanding the primitive function.

Finally, only a little knowledge of the differential calculus is necessary to recognize that the derived functions $y', y'', y''' \ldots$ of x coincide with the expressions

$$\frac{dy}{dx}, \frac{d^2y}{dx^2}, \frac{d^3y}{dx^3}, \quad \text{respectively.}$$

18.A5 Lagrange on algebra and the theory of functions

The development of functions, generally considered, gives rise to derived functions of different orders; and the algorithm for these functions once being found, one can consider them in themselves and independently of the series from which they have been obtained. So, a given function being regarded as *primitive* one can deduce from it by simple and uniform rules other functions which I call derived; and, being given arbitrary equations in several variables, one can pass successively from these to derived equations and return from these to primitive equations. These transformations correspond to differentiations and integrations; but in the theory of functions they only depend on purely algebraic operations based on the simple principles of the calculus.

Strictly speaking, in general algebra is only the theory of functions. In arithmetic, one seeks numbers from given conditions between these numbers and other numbers; and the numbers which one finds satisfy these conditions without preserving any trace of the operations which have been used to form them. In algebra, on the contrary, the quantities which one seeks must be functions of given quantities, that is to say, expressions which represent the different operations which it is necessary to perform on these quantities to obtain the values of the sought-for quantities.

In algebra, in the strict sense of the word, one only considers the primitive functions which are obtained from the ordinary algebraic operations; this is the first branch of the theory of functions. In the second branch one considers the derived functions, and it is this branch that we will designate simply by the name *theory of analytic functions* and which contains all that belongs to the new calculations.

18.B Augustin Louis Cauchy and Bernard Bolzano

The decisive move towards basing the calculus on something other than geometric intuition was the series of lectures by Cauchy on analysis given at the Ecole Polytechnique and published in his widely disseminated *Cours d'Analyse* of 1821. Although his work contained certain mistakes, and had gaps left for others to fill (notably the definition of real number: see 18.C2), it presented an acceptable set of definitions and a systematic derivation of the main results needed for a rigorous calculus. As a result it is possible to speak of a theory of analysis for the first time, as J. V. Grabiner argues in 18.B4.

Our first selections concentrate on the definitions Cauchy gave (see 18.B2). Notice that he gave a definition of the infinitely small—in strictly finite terms—but not of a real number. When he applied his definitions, the words were replaced by the ε and δ beloved of analysts ever since, and because of the inequalities that then appear his approach is often called the arithmetization of analysis. But it might be more accurate to speak of the metrization of analysis, because the central idea is that of the (arbitrary) closeness of points on a line. Cauchy's approach was to be followed many times by people interested in metric and, more generally, topological spaces.

Later we turn to Cauchy's use of these definitions to provide the crucial theorems of analysis. We have selected two: the intermediate value theorem and the fundamental theorem of the calculus (see 18.B3). The first result is the essential one for any study of continuous functions, as Bolzano had earlier observed, and is needed to complete the proofs of such results as the fundamental theorem of algebra. The second result is needed if, like Cauchy, you define integration independently of differentiation. In so doing Cauchy reversed a tendency that Newton had established, and set the pattern for all subsequent work. The third crucial result in elementary analysis is the mean value theorem for derivatives. We have omitted it for reasons of space, but the reader is invited to consult Grabiner's book, from which our translations have also been taken.

18.B1 Bolzano on the intermediate value theorem

There are two propositions in the theory of equations of which it could still be said, until recently, that a completely correct proof was unknown. One is the proposition: *that between any two values of the unknown quantity which give results of opposite sign there must always lie at least one real root of the equation*. The other is: *that every algebraic rational integral function of one variable quantity can be divided into real factors of first or second degree*. After several unsuccessful attempts by d'Alembert, Euler, de Foncenex, Lagrange, Laplace, Klügel, and others at proving the latter proposition Gauss finally supplied, last year, two proofs which leave very little to be desired. Indeed, this outstanding scholar had already presented us with a proof of this proposition in 1799, but it had, as he admitted, the defect that it proved a purely analytic truth on the basis of a *geometrical consideration*. But his two most recent proofs are quite free of this defect; the *trigonometric functions* which occur in them can, and must, be understood in a purely analytical sense.

The other proposition mentioned above is not one which so far has concerned scholars to any great extent. Nevertheless, we do find mathematicians of great repute concerned with the proposition, and already *different* kinds of proof have been attempted. To be convinced of this one need only compare the various treatments of the proposition which have been given by, for example, Kästner, Clairaut, Lacroix, Metternich, Klügel, Lagrange, Rösling, and several others.

However, a more careful examination very soon shows that none of these proofs can be viewed as adequate.

The most common kind of proof depends on a truth borrowed from *geometry*, namely, *that every continuous line of simple curvature of which the ordinates are first positive and then negative* (or conversely) *must necessarily intersect the x-axis somewhere at a point that lies in between those ordinates*. There is certainly no question concerning the *correctness*, nor indeed the *obviousness*, of this geometrical proposition. But it is clear that it is an intolerable offense against *correct method* to derive truths of *pure* (or general) mathematics (i.e., arithmetic, algebra, analysis) from considerations which belong to a merely *applied* (or special) part, namely, *geometry*. [...]

No less objectionable is the proof which some have constructed from the concept of the *continuity* of a function with the inclusion of the concepts of *time* and *motion*. 'If two functions fx and ϕx', they say, 'vary according to the law of continuity, and if for $x = \alpha$, $f\alpha < \phi\alpha$, but for $x = \beta$, $f\beta > \phi\beta$, then there must be some value u lying between α and β for which $fu = \phi u$. For if one imagines that the variable quantity x, in both these functions, gradually takes all values between α and β, and the same value is always taken by them both at the same moments, then at the *beginning* of this continuous change in x, $fx < \phi x$, and at the end, $fx > \phi x$. But since both functions, by virtue of their continuity, must first go through all intermediate values before they can reach a higher value, there must be some *intermediate moment* at which they are both equal to one another.' [...]

No one will deny that the concepts of *time* and *motion* are just as foreign to general mathematics as the concept of *space*. Nevertheless, if these two concepts were introduced here only for the sake for *clarification*, we would have nothing against them. For we are in no way party to such an exaggerated *purism*, which demands, in order to keep the science free from everything alien, that in its exposition one can never use an *expression* borrowed from another field, even if only in a metaphorical sense and with the purpose of describing a fact more briefly and clearly than could be done by a strictly literal description; nor even if it is just to avoid the jarring of the constant repetition of the same word, or so as to remember, by the mere name given to a thing, an example which would serve to confirm the assertion. Thus it may be noted that we do not regard *examples* and *applications* as detracting in the least from the perfection of a scientific exposition. On the other hand, we strictly require only this: that examples never be put forward instead of *proofs* and that the essence of a deduction never be based on the merely metaphorical use of phrases or on their related ideas, so that the deduction itself would become void as soon as these were changed.

In accord with these views, the inclusion of the concept of *time* in the above proof may still perhaps be excused, because no conclusion is based on phrases which contain it which would not also hold without it. But in no way can the last *illustration* about the motion of a body be viewed as anything more than a mere *example* which does not prove the proposition, but rather is only to be proved by it.

So let us drop this example and examine the rest of the reasoning. Let us *first* note that this is based on an incorrect concept of *continuity*. According to a *correct definition*, the expression *that a function fx varies according to the law of continuity for all values of x inside or outside certain limits* means just that: *if x is some such* value, the difference $f(x + \omega) - fx$ can be made smaller than any given quantity provided ω can be taken as small as we please. [...]

The following is a short summary of the method adopted.

The truth to be proved, that between any two values α and β which give results of opposite sign there always lies at least one real root, clearly rests on the more general truth that, if two continuous functions of x, fx and ϕx, have the property that for $x = \alpha$, $f\alpha < \phi\alpha$, and for $x = \beta$, $f\beta > \phi\beta$, there must always be some value of x lying between α and β for which $fx = \phi x$. However, if $f\alpha < \phi\alpha$, then by the law of continuity it is possible that $f(\alpha + i) < \phi(\alpha + i)$, if i is taken small enough. *The property of being smaller*, therefore, belongs to the function of i represented by the expression $f(\alpha + i)$, for all values smaller than a certain value. Nevertheless this property does not hold for *all* values of i without restriction, namely not for an $i = \beta - \alpha$, for $f\beta$ is already $> \phi\beta$. Now the *theorem* holds that whenever a certain property M belongs to all values of a variable quantity i which are smaller than a given value and yet not for *all values in general*, then there is always some *greatest* value u, for which it can be asserted that all $i < u$ possess property M. For this value of i itself $f(\alpha + u)$ cannot be $< \phi(\alpha + u)$, because then, by the law of continuity, $f(\alpha + u + \omega) < \phi(\alpha + u + \omega)$ if ω were taken small enough. And consequently it would not be true that u is the greatest of the values for which the assertion holds, that all lower values of i make $f(\alpha + i) < \phi(\alpha + i)$; for $u + \omega$ would be a still greater value for which this holds. But still less can it be true that

$f(\alpha + u) > \phi(\alpha + u)$, for then $f(\alpha + u - \omega) > \phi(\alpha + u - \omega)$ would also be true if ω were taken sufficiently small, and consequently, it would not be true that for all values of $i < u$, $f(\alpha + i) < \phi(\alpha + i)$. So therefore it must be that $f(\alpha + u) = \phi(\alpha + u)$; i.e., there is a value of x lying between α and β, namely, $\alpha + u$, for which the functions fx and ϕx are equal to one another. It is now only a question of the proof of the *theorem* mentioned. The theorem is proved by showing that those values of i of which it can be asserted that all smaller values possess property M and those of which this cannot be asserted can be brought as near one another as desired. Whence it follows, for anyone who has a correct concept of *quantity*, that the idea of a greatest value i of which it can be said that all below it possess property M is the idea of a real, i.e., *actual*, quantity.

18.B2 Cauchy's definitions

(*a*) *Limits and infinitesimals*

When the values successively attributed to the same variable approach a fixed value indefinitely, in such a way as to end up by differing from it as little as one could wish, this last value is called the *limit* of all the others. So, for example, an irrational number is the limit of the various fractions which provide values that approximate it more and more closely. In geometry, the surface of a circle is the limit to which the surfaces of inscribed polygons converge as the number of the sides steadily increases, etc.

When the successive numerical values of the same variable decrease indefinitely in such a way as to fall below any given number, this variable becomes what one calls an *infinitesimal* or an *infinitely small* quantity. A variable of this kind has zero for its limit.

When the successive numerical values of the same variables steadily increase in such a way as to exceed any given number, one says that this variable has *positive infinity* as its limit, indicated by the sign ∞, when a positive variable is concerned, and *negative infinity*, indicated by the sign $-\infty$, when a negative variable is concerned. The positive and negative infinities are jointly denoted by the name *infinite quantities*.

(*b*) *Continuous functions*

Among the objects which belong to the consideration of the infinitely small one must place notions relative to the continuity or discontinuity of functions. Let us first of all examine functions of a single variable from this point of view.

Let $f(x)$ be a function of the variable x and suppose that for each value of x between two given limits this function always takes a unique and finite value. If, having a value of x between these limits, one attributes to the variable x an infinitely small increase α, the function itself increases by the difference

$$f(x + \alpha) - f(x),$$

which depends simultaneously on the new variable α and the value of x. This done, the function $f(x)$ will be, between the two limits assigned to the variable x, a *continuous* function of this variable if, for each value of x intermediate between these limits the numerical value of the difference

$$f(x + \alpha) - f(x)$$

decreases indefinitely with α. In other words, *the function $f(x)$ will remain continuous with respect to x between the given limits if, between these limits an infinitely small increase in the variable always produces an infinitely small increase in the function itself.* One says furthermore that the function $f(x)$ is, in the neighbourhood of a particular value attributed to x, a continuous function of this variable, whenever it is continuous between two limits of x, however close, which contain that value of x.

(c) *Convergence*

A *sequence* is an infinite succession of quantities $u_0, u_1, u_2, u_3, \ldots$ which succeed each other according to some fixed law. These quantities themselves are the different *terms* of the sequence considered. Let

$$s_n = u_0 + u_1 + u_2 + \cdots + u_{n-1}$$

be the sum of the first n terms, where n is some integer. If the sum s_n tends to a certain limit s for increasing values of n, then the series is said to be *convergent*, and the limit in question is called the *sum* of the series. On the contrary, if the [partial] sum s_n approaches no fixed limit as n increases indefinitely, the series is *divergent* and has no sum. In either case, the term corresponding to the index n, namley u_n, is called the *general term*. It suffices to give this general term as a function of the index n in order for the sequence to be completely determined.

One of the simplest sequences is the geometric progression $1, x, x^2, x^3, \ldots$, whose general term is x^n, that is, the nth power of x. If one sums the first n terms of this sequence, one finds

$$1 + x + x^2 + \cdots + x^{n-1} = 1/(1 - x) - x^n/(1 - x)$$

and, since the magnitude of the fraction $x^n/(1 - x)$ either converges to zero for increasing values of n or increases beyond all limits, depending on whether one supposes the magnitude of x to be less than or greater than unity; one must conclude that under the first hypothesis the progression $1, x, x^2, x^3, \ldots$ defines a convergent series whose sum is $1/(1 - x)$, while under the second hypothesis the same progression defines a divergent series which has no sum.

By the principles established above, for the series

$$u_0 + u_1 + u_2 + \cdots + u_n + u_{n+1} + \cdots \tag{1}$$

to converge it is necessary and sufficient that the sum[s] $s_n = u_0 + u_1 + u_2 + \cdots + u_{n-1}$ converge to a fixed limit s as n increases; in other words, it is necessary and sufficient that for infinitely large values of n the sums $s_n, s_{n+1}, s_{n+2}, \ldots$ differ from the limit s, and hence from each other, by infinitesimal quantities. Besides, the successive differences between the first sum s_n and each of the following are respectively determined by the following equations:

$$s_{n+1} - s_n = u_n, \; s_{n+2} - s_n = u_n + u_{n+1}, \; s_{n+3} - s_n = u_n + u_{n+1} + u_{n+2}, \ldots.$$

Hence, for the series (1) to converge, it is necessary that the general term u_n decrease indefinitely as n increases; but this condition is not sufficient, and it must also be true that, for increasing values of n, the different sums $u_n + u_{n+1}$, $u_n + u_{n+1} + u_{n+2}, \ldots$, that is, the sums of quantities $u_n, u_{n+1}, u_{n+2}, \ldots$ taken in arbitrary number from the first will always have a magnitude which is less than any assignable limit. Conversely, when these various conditions are satisfied, the convergence of the series is assured.

(d) *Derivatives*

When the function $y = f(x)$ is continuous between two given limits of the variable x, and one assigns a value between these limits to the variable, an infinitesimal increment Δx of the variable produces an infinitesimal increment in the function itself. Consequently, if we then set $\Delta x = h$, the two terms of the *difference quotient*

$$\frac{\Delta y}{\Delta x} = \frac{f(x + h) - f(x)}{h}$$

will be infinitesimals. But whereas these terms tend to zero simultaneously, the ratio itself may converge to another limit, either positive or negative. This limit, when it exists, has a definite value for each particular value of x; but it varies with x. Thus, for example, if we take $f(x) = x^m$, m being a [positive] integer, the ratio of the infinitesimal differences will be

$$\frac{(x + h)^m - x^m}{h} = mx^{m-1} + \frac{m(m-1)}{1 \cdot 2} x^{m-2}h + \cdots + h^{m-1},$$

and it will have for [its] limit the quantity mx^{m-1}, that is to say, a new function of the variable x. The same will hold generally; only the form of the new function which serves as the limit of the ratio $[f(x + h) - f(x)]/h$ will depend upon the form of the given function $y = f(x)$. In order to indicate this dependence, we give to the new function the name derivative and we designate it, using a prime, by the notation y' or $f'(x)$.

Differentials of functions of a single variable

Let $y = f(x)$ remain a function of the independent variable x; let h be an infinitesimal and k a finite quantity. If we set $h = \alpha k$, α will also be an infinitesimal quantity, and we will have identically

$$\frac{f(x + h) - f(x)}{h} = \frac{f(x + \alpha k) - f(x)}{\alpha k},$$

whence one concludes that

$$\frac{f(x + \alpha k) - f(x)}{\alpha} = \frac{f(x + h) - f(x)}{h} k. \tag{1}$$

The limit toward which the left side of equation (1) converges as the variable α tends to zero, the quantity k remaining constant, is called the *differential* of the function $y = f(x)$. We indicate this differential by the symbol d, as follows:

$$dy \quad \text{or} \quad df(x).$$

It is easy to obtain its value when we know that of the derivative y' or $f'(x)$. Indeed, taking the limits of the two sides of equation (1), we shall find generally

$$df(x) = kf'(x). \tag{2}$$

In the special case where $f(x) = x$, equation (2) reduces to

$$dx = k.$$

(e) The definite integral

Suppose that the function $y = f(x)$ is continuous with respect to the variable x between the two finite limits $x = x_0$, $x = X$. We designate by $x_1, x_2, \ldots, x_{n-1}$ new values of x placed between these limits and suppose that they either always increase or always decrease between the first limit and the second. We can use these values to divide the difference $X - x_0$ into elements

$$x_1 - x_0, x_2 - x_1, x_3 - x_2, \ldots, X - x_{n-1},$$

which all have the same sign. Once this has been done, let us multiply each element by the value of $f(x)$ corresponding to the left-hand end point of that element: that is, the element $x_1 - x_0$ will be multiplied by $f(x_0)$, the element $x_2 - x_1$ by $f(x_1), \ldots$ and finally, the element $X - x_{n-1}$ by $f(x_{n-1})$; and let

$$S = (x_1 - x_0)f(x_0) + (x_2 - x_1)f(x_1) + \cdots + (X - x_{n-1})f(x_{n-1})$$

be the sum of the products so obtained. The quantity S clearly will depend upon

1st: the number n of elements into which we have divided the difference $X - x_0$;

2nd: the values of these elements and therefore the mode of division adopted.

It is important to observe that if the numerical values of these elements become very small and the number n very large, the mode of division will have only an insignificant effect on the value of S. This in fact can be proved as follows.

[After proving this result, Cauchy resumed as follows.]

Now suppose that we consider two separate modes of division of the difference $X - x_0$, in both of which the elements of the difference have very small numerical values. We can compare these two modes with a third mode, chosen so that each element, from either the first or second mode, is formed by bringing together several elements of the third mode. To satisfy this condition, it suffices for each of the values of x placed between the limits x_0 and X in the first two modes to be used in the third; and we can prove that we change the value of S very little in going from the first or the second mode to the third—and therefore, in going from the first to the second. Thus, when the elements of the difference $X - x_0$ become infinitely small, the mode of division has only an imperceptible effect on the value of S; and, if we let the numerical values of these elements decrease while their numbers increases, the value of S ultimately becomes, for all practical purposes, constant. Or, in other words, it ultimately reaches a certain limit that depends uniquely on the form of the function $f(x)$ and on the bounding values x_0, X of the variable x. This limit is what is called a *definite integral*.

18.B3 Cauchy on two important theorems of the calculus

(a) *The fundamental theorem of the calculus*

If in the definite integral $\int_{x_0}^{X} f(x)\,dx$ we let one of the two limits of integration, for instance X, vary, the integral itself will vary with that quantity. And if the limit X, now variable, is replaced by x, we obtain as a result a new function of x, which will be what is called an integral taken from the *origin* $x = x_0$. Let

(1) $$\mathscr{F}(x) = \int_{x_0}^{x} f(x)\,dx$$

be that new function. We derive from [the formula $\int_{x_0}^{X} f(x)\,dx = (X - x_0)f[x_0 + \theta(X - x_0)]$ that]

(2) $$\mathscr{F}(x) = (x - x_0)f[x_0 + \theta(x - x_0)],\ \mathscr{F}(x_0) = 0,$$

θ being a [nonnegative] number less than one. Also, from [the formula $\int_{x_0}^{X} f(x)\,dx = \int_{x_0}^{\xi} f(x)\,dx + \int_{\xi}^{X} f(x)\,dx$, where $x_0 \leqslant \xi \leqslant X$],

$$\int_{x_0}^{x+\alpha} f(x)\,dx - \int_{x_0}^{x} f(x)\,dx = \int_{x}^{x+\alpha} f(x)\,dx = \alpha f(x + \theta\alpha),$$

or

(3) $$\mathscr{F}(x + \alpha) - \mathscr{F}(x) = \alpha f(x + \theta\alpha).$$

It follows from equations (2) and (3) that if the function $f(x)$ is finite and continuous in the neighbourhood of some particular value of the variable x, the new function $\mathscr{F}(x)$ will not only be finite but also continuous in the neighbourhood of that value, since an infinitely small increment of x will correspond to an infinitely small increment in $\mathscr{F}(x)$. Thus if the function $f(x)$ remains finite and continuous from $x = x_0$ to $x = X$, the same will hold for the function $\mathscr{F}(x)$. In addition, if both members of formula (3) are divided by a, we may conclude by passing to the limits that

$$\mathscr{F}'(x) = f(x).$$

Thus the integral (1) considered as a function of x has as its derivative the function $f(x)$ under the integral sign $\int$.

(b) *The intermediate value theorem*

Let $f(x)$ be a real function of the variable x, continuous with respect to that variable between $x = x_0, x = X$. If the two quantities $f(x_0), f(X)$ have opposite sign, the equation

$$f(x) = 0$$

can be satisfied by one or more real values of x between x_0 and X.

Proof Let x_0 be the smaller of the two quantities x_0, X. We will set $X - x_0 = h$, and we designate by m any integer greater than one. Because one of the two quantities $f(x_0)$, $f(X)$ is positive, the other negative, it follows that if we form the sequence

$$f(x_0),\ f(x_0 + h/m),\ f(x_0 + 2h/m), \ldots, f(X - h/m),\ f(X),$$

and compare successively the first term in that sequence with the second, the second with the third, the third with the fourth, etc., we necessarily will finish by finding once—or more than once—two consecutive terms that are of opposite sign.

Let $f(x_1), f(X')$ be two such terms, where x_1 is the smaller of the two corresponding values of x. Clearly we will have $x_0 < x_1 < X' < X$ and $X' - x_1 = h/m = 1/m(X - x_0)$. [Cauchy uses the sign $<$ for *less than or equal to*.] Having determined x_1 and X' in the way just described, similarly we can place between these two new values of x two other values x_2, X'', which, when substituted in $f(x)$, give results of opposite sign, and which satisfy the conditions $x_1 < x_2 < X'' < X'$ and $X'' - x_2 = 1/m(X' - x_1) = 1/m^2(X - x_0)$. Continuing thus, we obtain, first, a series of increasing values of x, that is,

$$x_0, x_1, x_2, \ldots; \tag{1}$$

second, a series of decreasing values

$$X, X', X'', \ldots \tag{2}$$

whose terms, since they exceed those of the first series by quantities equal, respectively, to the products

$$(X - x_0),\ (1/m)(X - x_0),\ (1/m^2)(X - x_0), \ldots,$$

ultimately will differ from the values in the first series by as little as desired. We must conclude from this that the general terms of the series (1) and (2) converge toward a common limit. Let a be that limit. Since the function $f(x)$ is continuous between $x = x_0$ and $x = X$, the general terms of the following series,

$$f(x_0),\ f(x_1),\ f(x_2), \ldots, f(X), f(X'), f(X''), \ldots$$

both converge toward the common limit $f(a)$; and since they always remain of opposite sign when they approach this limit, it is clear that the quantity $f(a)$, which must be finite, cannot differ from zero. Therefore, the equation

$$f(x) = 0 \tag{3}$$

will be satisfied if we give the variable x the particular value a, which lies between x_0 and X. In other words, $x = a$ will be a *root* of equation (3).

18.B4 J. V. Grabiner on the significance of Cauchy

What is our estimate of Cauchy's achievement? Cauchy's work established a new way of looking at the concepts of the calculus. As a result, the subject was transformed from a collection of powerful methods and useful results into a mathematical discipline based on clear definitions and rigorous proofs. His views were less intuitive than the old ones, but they provide a new set of interesting questions. His definition of limit and elaboration of the associated method of proof by the inequalities are the basis for modern theories of continuity, convergence, derivative, and the integral. And many of the important consequences of these theories—in the study of convergence, existence proofs for the solution of differential equations, and the properties of definite integrals—were pioneered by Cauchy himself.

Moreover, Cauchy's rigorization of the calculus was much more than the sum of its separate parts. It was not merely that Cauchy gave this or that definition, proved particular existence theorems, or even presented the first reasonably acceptable proof of the fundamental theorem of calculus. He brought all these things together into a logically connected system of definitions, theorems, and proofs.

The implications of this achievement go beyond the calculus. In a very important sense, it may be said that Cauchy brought ancient and modern mathematics together. He cast his rigorous calculus in the deductive mould characteristic of ancient geometry. And unlike his predecessors, he did this successfully; that is, he, not only gave his work a Euclidean form but presented definitions that generally are adequate to support the desired results, proofs that basically are valid, and methods that were fruitful sources for later mathematical work. Cauchy, then, brought together three elements: the major *results* of analysis, most of which he could now prove; some fruitful *concepts* and *techniques* from algebra (particularly algebraic approximations) and analysis; and the *rigour* and *proof structure* of Greek geometry. For a long time Greek geometry had been considered the model for all of mathematics. If the origins of modern mathematics are traced to the Renaissance, then the rigour and structure characteristic of Greek geometry first effectively became part of modern mathematics only with Cauchy's work. Of course the late-nineteenth-century idea that mathematics is the science of abstract logical systems in general is absent from Cauchy's work. But Cauchy's rigorization of the calculus was an indispensable first step in that direction.

Cauchy left some unfinished business, as subsequent history shows. Some gaps in specific proofs had to be filled; some assumptions had to be proved, or at least explicitly stated; some crucial distinctions had yet to be made. But there is little in nineteenth-century analysis that was not marked, directly or indirectly, by his ideas. The basic logical structure Cauchy erected provides the framework in which we still think about rigorous calculus.

18.C Richard Dedekind and Georg Cantor

It was typical of Richard Dedekind's (1831–1916) modesty in mathematics that he did not publish his construction of the real numbers for fourteen years after he discovered it, but nonetheless it was an important event. For in his short paper (see 18.C1) he repaired the largest single omission in Cauchy's work, and sketched how it was now possible for the first time to give a rigorous account of the basic theorems about continuous functions. Dedekind's approach was to use the fact that the rational numbers are ordered to create sets of rationals which define new numbers, the *real* numbers. So it is rather like using families of parallel lines to define points at infinity in the projective plane. Indeed, throughout his life Dedekind was a leading exponent of the modern view concerning the existence of mathematical objects and hostile to the naïve attitudes of the eighteenth century.

Dedekind was a close friend of Georg Cantor (1845–1918) and one of the few who understood and sympathized with Cantor's emerging theory of sets. Our selection mostly concentrates on Cantor's ideas about the real numbers: how they might be defined via sequences of rational approximations to them (18.C2); and how many there are (18.C4). It seems that 18.C5 may be the first time his statement of the continuum hypothesis has been translated into English. His provocative ideas about this question only grew in importance after his death. In 18.C3 we give just a hint of his correspondence with Dedekind; Cantor's remarkable discovery that there is a one-to-one correspondence between the line and the plane destroyed any hope that there could be a simple set-theoretic characterization of the difference between the two.

18.C1 Dedekind on irrational numbers and the theorems of the calculus

(a) Introduction

My attention was first directed toward the considerations which form the subject of this pamphlet in the autumn of 1858. As professor in the Polytechnic School in Zürich I found myself for the first time obliged to lecture upon the elements of the differential calculus and felt more keenly than ever before the lack of a really scientific foundation for arithmetic. In discussing the notion of the approach of a variable magnitude to a fixed limiting value, and especially in proving the theorem that every magnitude which grows continually, but not beyond all limits, must certainly approach a limiting value, I had recourse to geometric evidences. Even now such resort to geometric intuition in a first presentation of the differential calculus, I regard as exceedingly useful, from the didactic standpoint, and indeed indispensable, if one does not wish to lose too much time. But that this form of introduction into the differential calculus can make no claim to being scientific, no one will deny. For myself this feeling of dissatisfaction was so overpowering that I made the fixed resolve to keep meditating on the question till I should find a purely arithmetic and perfectly rigorous foundation for the principles of infinitesimal analysis. The statement is so frequently made that the differential calculus deals with continuous magnitude, and yet an explanation of this continuity is nowhere given; even the most rigorous expositions of the differential calculus do not base their proofs upon continuity but, with more or less consciousness of the fact, they either appeal to geometric notions or those suggested by geometry, or depend upon theorems which are never established in a purely arithmetic manner. Among these, for example, belongs the above-mentioned theorem, and a more careful investigation convinced me that this theorem, or any one equivalent to it, can be regarded in some way as a sufficient basis for infinitesimal analysis. It then only remained to discover its true origin in the elements of arithmetic and thus at the same time to secure a real definition of the essence of continuity. I succeeded Nov. 24, 1858, and a few days afterward I communicated the results of my meditations to my dear friend Durège with whom I had a long and lively discussion. Later I explained these views of a scientific basis of arithmetic to a few of my pupils, and here in Braunschweig read a paper upon the subject before the scientific club of professors, but I could not make up my mind to its publication, because in the first place, the presentation did not seem altogether simple, and further, the theory itself had little promise. Nevertheless I had already half determined to select this theme as subject for this occasion, when a few days ago,

March 14, by the kindness of the author, the paper, *Die Elemente der Funktionenlehre* by E. Heine (*Crelle's Journal*, Vol. 74) came into my hands and confirmed me in my decision. In the main I fully agree with the substance of this memoir, and indeed I could hardly do otherwise, but I will frankly acknowledge that my own presentation seems to me to be simpler in form and to bring out the vital point more clearly. While writing this preface (March 20, 1872), I am just in receipt of the interesting paper *Ueber die Ausdehnung eines Satzes aus der Theorie der trigonometrischen Reihen*, by G. Cantor (*Math. Annalen*, Vol. 5), for which I owe the ingenious author my hearty thanks. As I find on a hasty perusal, the axiom given in Section II of that paper, aside from the form of presentation, agrees with what I designate in Section III as the essence of continuity. But what advantage will be gained by even a purely abstract definition of real numbers of a higher type, I am as yet unable to see, conceiving as I do of the domain of real number as complete in itself.

(*b*) *Continuity of the straight line*

Of the greatest importance, however, is the fact that in the straight line L there are infinitely many points which correspond to no rational number. If the point p corresponds to the rational number a, then, as is well known, the length op is commensurable with the invariable unit of measure used in the construction, i.e., there exists a third length, a so-called common measure, of which these two lengths are integral multiples. But the ancient Greeks already knew and had demonstrated that there are lengths incommensurable with a given unit of length, e.g., the diagonal of the square whose side is the unit of length. If we lay off such a length from the point o upon the line we obtain an end-point which corresponds to no rational number. Since further it can be easily shown that there are infinitely many lengths which are incommensurable with the unit of length, we may affirm: The straight line L is infinitely richer in point-individuals than the domain R of rational numbers in number-individuals.

If now, as is our desire, we try to follow up arithmetically all phenomena in the straight line, the domain of rational numbers is insufficient and it becomes absolutely necessary that the instrument R constructed by the creation of the rational numbers be essentially improved by the creation of new numbers such that the domain of numbers shall gain the same completeness, or as we may say at once, the same *continuity*, as the straight line.

The previous considerations are so familiar and well known to all that many will regard their repetition quite superfluous. Still I regarded this recapitulation as necessary to prepare properly for the main question. For, the way in which the irrational numbers are usually introduced is based directly upon the conception of extensive magnitudes—which itself is nowhere carefully defined—and explains number as the result of measuring such a magnitude by another of the same kind. Instead of this I demand that arithmetic shall be developed out of itself.

That such comparisons with non-arithmetic notions have furnished the immediate occasion for the extension of the number-concept may, in a general way, be granted (though this was certainly not the case in the introduction of complex numbers); but this surely is no sufficient ground for introducing these foreign notions into arithmetic, the science of numbers. Just as negative and fractional rational numbers are formed by a new creation, and as the laws of operating with these numbers must and can be reduced to the laws of operating with positive integers, so we must endeavour completely to define irrational numbers by means of the rational numbers alone. The question only remains how do to this.

(c) *Gaps*

The above comparison of the domain R of rational numbers with a straight line has led to the recognition of the existence of gaps, of a certain incompleteness or discontinuity of the former, while we ascribe to the straight line completeness, absence of gaps, or continuity. In what then does this continuity consist? Everything must depend on the answer to this question, and only through it shall we obtain a scientific basis for the investigation of *all* continuous domains. By vague remarks upon the unbroken connection in the smallest parts obviously nothing is gained; the problem is to indicate a precise characteristic of continuity that can serve as the basis for valid deductions. For a long time I pondered over this in vain, but finally I found what I was seeking. This discovery will, perhaps, be differently estimated by different people; the majority may find its substance very commonplace. It consists of the following. In the preceding section attention was called to the fact that every point p of the straight line produces a separation of the same into two portions such that every point of one portion lies to the left of every point of the other. I find the essence of continuity in the converse, i.e., in the following principle: *If all points of the straight line fall into two classes such that every point of the first class lies to the left of every point of the second class, then there exists one and only one point which produces this division of all points into two classes, this severing of the straight line into two portions.*

As already said I think I shall not err in assuming that every one will at once grant the truth of this statement; the majority of my readers will be very much disappointed in learning that by this commonplace remark the secret of continuity is to be revealed. To this I may say that I am glad if every one finds the above principle so obvious and so in harmony with his own ideas of a line; for I am utterly unable to adduce any proof of its correctness, nor has any one the power. The assumption of this property of the line is nothing else than an axiom by which we attribute to the line its continuity, by which we find continuity in the line. If space has at all a real existence it is *not* necessary for it to be continuous; many of its properties would remain the same even were it discontinuous. And if we knew for certain that space was discontinuous there would be nothing to prevent us, in case we so desired, from filling up its gaps, in thought, and thus making it continuous; this filling up would consist in a creation of new point-individuals and would have to be effected in accordance with the above principle.

(d) *Dedekind cuts*

From the last remarks it is sufficiently obvious how the discontinuous domain R of rational numbers may be rendered complete so as to form a continuous domain. [Earlier] it was pointed out that every rational number a effects a separation of the system R into two classes such that every number a_1 of the first class A_1 is less than every number a_2 of the second class A_2; the number a is either the greatest number of the class A_1 or the least number of the class A_2. If now any separation of the system R

into two classes A_1, A_2, is given which possesses only *this* characteristic property that every number a_1 in A_1 is less than every number a_2 in A_2, then for brevity we shall call such a separation a *cut* and designate it by (A_1, A_2). We can then say that every rational number a produces one cut or, strictly speaking, two cuts, which, however, we shall not look upon as essentially different; this cut possesses, *besides*, the property that either among the numbers of the first class there exists a greatest or among the numbers of the second class a least number. And conversely, if a cut possesses this property, then it is produced by this greatest or least rational number.

But it is easy to show that there exist infinitely many cuts not produced by rational numbers.

[Dedekind went on to call such cuts irrational numbers, and the set of all cuts he called the real numbers, denoted $\mathfrak{R}$.]

(e) Completeness

Besides these properties, however, the domain $\mathfrak{R}$ possesses also *continuity*; i.e., the following theorem is true: *If the system $\mathfrak{R}$ of all real numbers breaks up into two classes $\mathfrak{U}_1, \mathfrak{U}_2$ such that every number α_1 of the class $\mathfrak{U}_1$ is less than every number α_2 of the class $\mathfrak{U}_2$ then there exists one and only one number α by which this separation is produced.*

Proof By the separation or the cut of $\mathfrak{R}$ into $\mathfrak{U}_1$ and $\mathfrak{U}_2$ we obtain at the same time a cut (A_1, A_2) of the system R of all rational numbers which is defined by this that A_1 contains all rational numbers of the class $\mathfrak{U}_1$ and A_2 all other rational numbers, i.e., all rational numbers of the class $\mathfrak{U}_2$. Let α be the perfectly definite number which produces this cut (A_1, A_2). If β is any number different from α, there are always infinitely many rational numbers c lying between α and β. If $\beta < \alpha$, then $c < \alpha$; hence c belongs to the class A_1 and consequently also to the class $\mathfrak{U}_1$, and since at the same time $\beta < c$ then β also belongs to the same class $\mathfrak{U}_1$, because every number in $\mathfrak{U}_2$ is greater than every number c in $\mathfrak{U}_1$. But if $\beta > \alpha$, then is $c > \alpha$; hence c belongs to the class A_2 and consequently also to the class $\mathfrak{U}_2$, and since at the same time $\beta > c$, then β also belongs to the same class $\mathfrak{U}_2$, because every number in $\mathfrak{U}_1$ is less than every number c in $\mathfrak{U}_2$. Hence every number β different from α belongs to the class $\mathfrak{U}_1$ or to the class $\mathfrak{U}_2$ according as $\beta < \alpha$ or $\beta > \alpha$; consequently α itself is either the greatest number in $\mathfrak{U}_1$ or the least number in $\mathfrak{U}_2$, i.e., α is one and obviously the only number by which the separation of R into the classes $\mathfrak{U}_1, \mathfrak{U}_2$ is produced. Which was to be proved.

(f) Infinitesimal analysis

Here at the close we ought to explain the connection between the preceding investigations and certain fundamental theorems of infinitesimal analysis.

We say that a variable magnitude x which passes through successive definite numerical values approaches a fixed limiting value α when in the course of the process x lies finally between two numbers between which α itself lies, or, what amounts to the same, when the difference $x - \alpha$ taken absolutely becomes finally less than any given value different from zero.

One of the most important theorems may be stated in the following manner: *If a magnitude x grows continually but not beyond all limits it approaches a limiting value.*

I prove it in the following way. By hypothesis there exists one and hence there exist infinitely many numbers α_2 such that x remains continually $<\alpha_2$; I designate by $\mathfrak{U}_2$ the system of all these numbers α_2, by $\mathfrak{U}_1$ the system of all other numbers α_1; each of the latter possesses the property that in the course of the process x becomes finally $\geqq \alpha_1$, hence every number α_1 is less than every number α_2 and consequently there exists a number α which is either the greatest in $\mathfrak{U}_1$ or the least in $\mathfrak{U}_2$ [by the theorem above]. The former cannot be the case since x never ceases to grow, hence α is the least number in $\mathfrak{U}_2$. Whatever number α_1 be taken we shall have finally $\alpha_1 < x < \alpha$, i.e., x approaches the limiting value α.

18.C2 Cantor's definition of the real numbers

The rational numbers form the basis for the determination of the further idea of a numerical quantity; I will say that they form a domain A (and include the number zero in it). When I speak of a numerical quantity in the extended sense, it is immediately the case that it is presented in the form of a given infinite series of rational numbers

$$a_1, a_2, \ldots, a_n, \ldots \tag{1}$$

which have the property that the difference $a_{n+m} - a_n$ becomes infinitely small with increasing n, whatever the positive integer m may be, or in other words that for an arbitrary given (positive, rational) ε there is an integer n_0 so that $|a_{n+m} - a_n| < \varepsilon$, when $n \geqslant n_0$ and m is an arbitrary positive integer. I express the property of sequence (1) in the following words: *sequence (1) has a definite limit b.* [...]

If there is a second series

$$a'_1, a'_2, \ldots, a'_n, \ldots \tag{1'}$$

which has a definite limit b', one finds that the two sequences (1) and (1′) can be related to each other in one of the following three ways, which are mutually exclusive; either (i) $a_n - a'_n$ becomes infinitely small with increasing n, or (ii) $a_n - a'_n$ from a certain n onwards always remains greater than a positive (rational) quantity ε, or (iii) $a_n - a'_n$ from a certain n onwards always remains smaller than a certain negative (rational) quantity $-\varepsilon$.

When the first condition occurs, I set $b = b'$, if the second $b > b'$, if the third $b < b'$.

One likewise finds that the sequence (1), which has the limit b, can be related to a rational number a in only one of the following three ways. Either (i) $a_n - a$ becomes infinitely small with increasing n, or (ii) $a_n - a$ from a certain n onwards always remains greater than a positive (rational) quantity ε, or (iii) $a_n - a$ from a certain n onwards always remains smaller than a negative (rational) quantity $-\varepsilon$. According to which of these conditions exists we set, respectively, $b = a$, $b > a$, $b < a$. The totality of numerical quantities b may be denoted by B.

[Cantor next showed how one can add, subtract, multiply and divide the quantities in B. For example, the sum of (1) and (1′) above is the limit of $a_1 + a'_1, a_2 + a'_2, \ldots, a_n + a'_n, \ldots$. The set B forms the set of real numbers. However, although we would regard the set C formed from the limits of the set B as equivalent to B (because the real numbers form a complete set) Cantor said he found it 'essential to distinguish between them ... although they were in a certain sense mutually reciprocal'.]

18.C3 The correspondence between Cantor and Dedekind

Cantor developed many of his ideas in a long and fruitful correspondence with his friend Dedekind, who was one of the few people to understand him and sympathize with his aims at a time when, for example, most of the faculty at Berlin, Cantor's old university, were becoming increasingly hostile. Having discovered that the real numbers were uncountable (see 18.C4), it was naturally with Dedekind that Cantor discussed his next rather surprising observation.

Cantor to Dedekind, 5 January 1874

Is it possible to map uniquely a surface (suppose a square including its boundaries) onto a line (suppose a straight line including its endpoints), so that to each point of the surface corresponds one point of the line and reciprocally to each point of the line corresponds one point of the surface? It seems to me at the moment that the resolution of this question—however much one is attracted to the answer 'no' that one may feel proof is almost superfluous—has very great difficulties.

After his marriage in early 1874, Cantor appears to have set this problem aside for a while, but he wrote again to Dedekind in 1877 to say that 'although I have believed the opposite for a year' in fact the answer to the question is 'yes'. Cantor's proof was a slightly too naïve version of the now-standard argument in which the square is mapped to the line by sending $(0{\cdot}a_1a_2a_3a_4\ldots, 0{\cdot}b_1b_2b_3b_4\ldots)$ to $0{\cdot}a_1b_1a_2b_2a_3b_3\ldots$ As Dedekind pointed out, this argument falls foul of the fact that $0{\cdot}1999\ldots = 0{\cdot}2000\ldots$, but Cantor was able to reply with a more sophisticated, and convincing, proof. He then waited anxiously for Dedekind's reply.

Cantor to Dedekind, 29 June 1877

Your latest reply about our work was so unexpected and so novel that in a manner of speaking I will not be able to attain a certain composure until I have had from you, my very dear friend, a decision on its validity. As long as you have not confirmed it, I can only say: *I see it, but I don't believe it* [italics in French in the original]. [...] the distinction between domains of *different* dimensions must be sought for in quite another way than by the characteristic number of independent coordinates.

Dedekind to Cantor, 2 July 1877

I have checked over your proof once and can find no hole in it; I therefore believe that your interesting theorem is true and congratulate you on it. [...] But now I believe tentatively in the following theorem: given a unique and complete correspondence between the points of a continuous manifold A of dimension a on the one hand and the points of a continuous manifold B of dimension b on the other, then this *correspondence itself*, if a and b are *unequal, is necessarily thoroughly discontinuous.*

18.C4 Cantor on the uncountability of the real numbers

(a) The 1874 proof

By a real algebraic number will be understood in general a real numerical quantity ω which satisfies a non-identity equation of the form

$$a_0\omega^n + a_1\omega^{n-1} + \cdots + a_n = 0, \tag{1}$$

where $n, a_0, a_1, \ldots, a_n$ are integers; we may consider that the numbers n and a_0 are positive, the coefficients $a_0, a_1, \ldots, a_n$ have no common factors, and the equation (1) is irreducible.

[Cantor first shows that the algebraic numbers are countable, as follows, starting with his definition of their *height*, N]

$$N = n - 1 + |a_0| + |a_1| + \cdots + |a_n|.$$

The height N is then a definite positive integer for each real algebraic number ω, conversely for each value of N there is only a finite number of real algebraic numbers with height N; let the number of these be $\phi(N)$ then for example $\phi(1) = 1$, $\phi(2) = 2$, $\phi(3) = 4$. This implies that the totality of the collection (ω), i.e. of all real algebraic numbers, can be ordered in the following way: one takes as the first number the unique number ω_1 with height 1; setting it aside, as the index increases the $\phi(2) = 2$ real algebraic numbers with $\phi(2) = 2$ follow, denote them by ω_2 and ω_3; to these may be attached the $\phi(3) = 4$ numbers with height $N = 3$ whose index is still greater; quite generally one may count all the numbers in (ω) up to a certain height $N = N_1$ in this way, and the real algebraic numbers with height $N = N_1 + 1$ will then follow at a definite place, and the index will increase, and so one obtains the collection of all real algebraic numbers (ω) in the form:

$$\omega_1, \omega_2, \ldots, \omega_n, \ldots$$

and with respect to this ordering one can speak of the nth real algebraic number, and none of the collection (ω) is forgotten.

[Cantor now showed that the collection of real numbers was, however, uncountable, by showing that if they were countable, then given any listing of them we could find a real number which was not in the list. His proof of this was not his diagonal argument, but a clever use of the fact that between any two real numbers there is always a third. Putting his two results together, Cantor now deduced that there are infinitely many transcendental, i.e. non-algebraic numbers.]

(b) The 1891 proof using a diagonal argument

Namely let m and w be any two different characters, and we form the collection M of elements

$$E = (x_1, x_2, \ldots, x_n, \ldots)$$

which depends on infinitely many coordinates, each of which is either m or w. M is the set of all elements E. Amongst the elements of M are for example the following three

$$E' = (m, m, m, m, \ldots),$$
$$E'' = (w, w, w, w, \ldots),$$
$$E''' = (m, w, m, w, \ldots).$$

I now show that such a manifold M does not have the power of the series $1, 2, \ldots, n, \ldots$. This follows from the following theorem.

If $E_1, E_2, \ldots, E_n, \ldots$ is any simply infinite sequence of the manifold M, then there is always an element E_0, of M which does not agree with any E.

To prove this let

$$E_1 = (a_{11}, a_{12}, \ldots, a_{1n}, \ldots),$$
$$E_2 = (a_{21}, a_{22}, \ldots, a_{2n}, \ldots),$$
$$E_m = (a_{m1}, a_{m2}, \ldots, a_{mn}, \ldots).$$

Here each a_{mn} is a definite m or w. A sequence $b_1, b_2, \ldots, b_n$ will now be so defined for which each b is also equal to m or w and different from a. So if $a_{nn} = m$, then $b_n = w$, and if $a_{nn} = w$, then $b_n = m$. Then we consider the element

$$E_0 = (b_1, b_2, b_3, \ldots)$$

of M, and one sees without further ado that the equation

$$E_0 = E_m$$

can be satisfied by no integer value of k, since otherwise for that value of k and all integers n

$$b_n = a_{mn}$$

and also in particular

$$b_m = a_{mm}$$

which is excluded by the definition of b_m. It follows immediately from this theorem that the totality of all elements of M cannot be put in the form of a series: $E_1, E_2, \ldots, E_n, \ldots$ since otherwise we would have the contradiction that the thing E_0 would be an element of M and also not an element of M.

18.C5 Cantor's statement of the continuum hypothesis

Since we are led in this way to an extraordinary rich and broad domain of manifolds with the property that they can be put into a unique and complete correspondence with a line or a part of a line [...] the question arises of how the different parts of a continuous straight line, i.e. the thinkable different manifolds of points, relate to their respective powers. If we clothe this problem in its geometric dress and understand [...] by a *linear* manifold of real numbers any thinkable collection of infinitely many distinct real numbers, then one can ask into *how many* and what classes the linear manifolds fall, if manifolds of the same power are put into one and the same class and manifolds of different powers into different classes. By an inductive process, which we will not present here, we are led to the theorem that the number of classes which arise according to this principle of classification is equal to *two*.

19 The Mechanization of Calculation

Although there was quite an interest in devising mechanical aids to calculation in the seventeenth century (19.A), the figure of Charles Babbage in the nineteenth century (19.B) will always remain outstanding as a pioneer of mechanizing calculation. Babbage's ambitious engines were digital, like the modern computer, and the important stream of analogue machines, described by Samuel Lilley in 19.C, is no longer as familiar as once it was. We omit any extracts on the modern computer, which are well covered in the volume edited by Brian Randell, *The Origins of Digital Computers: selected papers* (Springer-Verlag, 1973). The final section (19.D) concerns the use of computers within pure mathematics: what status can be given to proofs heavily reliant on computers? This question arose with particular emphasis over the proof in 1976 of a long-standing conjecture, the four-colour theorem.

19.A Leibniz on Calculating Mechanics in the Seventeenth Century

When, several years ago, I saw for the first time an instrument which, when carried, automatically records the numbers of steps taken by a pedestrian, it occurred to me at once that the entire arithmetic could be subjected to a similar kind of machinery so that not only counting but also addition and subtraction, multiplication and division could be accomplished by a suitably arranged machine easily, promptly, and with sure results.

The calculating box of Pascal was not known to me at that time. I believe it has not gained sufficient publicity. When I noticed, however, the mere name of a calculating machine in the preface of his 'posthumous thoughts' (his arithmetical triangle I saw first in Paris) I immediately inquired about it in a letter to a Parisian friend. When I learned from him that such a machine exists I requested the most distinguished Carcavius by letter to give me an explanation of the work which it is capable of performing. He replied that addition and subtraction are accomplished by it directly, the other [operations] in a round-about way by repeating additions and subtractions and performing still another calculation. I wrote back that I venture to promise something more, namely, that multiplication could be performed by the machine as well as addition, and with greatest speed and accuracy.

He replied that this would be desirable and encouraged me to present my plans before the illustrious King's Academy of that place.

In the first place it should be understood that there are two parts of the machine, one designed for addition (subtraction) the other for multiplication (division) and that they should fit together.

The adding (subtracting) machine coincides completely with the calculating box of Pascal. Something, however, must be added for the sake of multiplication so that several and even all the wheels of addition could rotate without disturbing each other, and nevertheless any one of them should precede the other in such a manner that after a single complete turn unity would be transferred into the next following. If this is not performed by the calculating box of Pascal it may be added to it without difficulty.

The multiplying machine will consist of two rows of wheels, equal ones and unequal ones. Hence the whole machine will have three kinds of wheels: the wheels of addition, the wheels of the multiplicand and the wheels of the multiplier. [...]

It is sufficiently clear how many applications will be found for this machine, as the elimination of all errors and of almost all work from the calculations with numbers is of great utility to the government and science. It is well known with what enthusiasm the calculating rods of Napier were accepted, the use of which, however, in division is neither much quicker nor surer than the common calculation. For in his [Napier's] multiplication there is need of continual additions, but division is in no way faster than by the ordinary [method]. Hence the calculating rods soon fell into disuse. But in our [machine] there is no work when multiplying and very little when dividing.

Pascal's machine is an example of the most fortunate genius but while it facilitates only additions and subtractions, the difficulty of which is not very great in themselves, it commits the multiplication and division to a previous calculation so that it commended itself rather by refinement to the curious than as of practical use to people engaged in business affairs.

And now that we may give final praise to the machine we may say that it will be desirable to all who are engaged in computations which, it is well known, are the managers of financial affairs, the administrators of others' estates, merchants, surveyors, geographers, navigators, astronomers, and [those connected with] any of the crafts that use mathematics.

But limiting ourselves to scientific uses, the old geometric and astronomic tables could be corrected and new ones constructed by the help of which we could measure all kinds of curves and figures, whether composed or decomposed and unnamed, with no less certainty than we are now able to treat the angles according to the work of Regiomontanus and the circle according to that of Ludolphus of Cologne, in the same manner as straight lines. If this could take place at least for the curves and figures that are most important and used most often, then after the establishment of tables not only for lines and polygons but also for ellipses, parabolas, hyperbolas, and other figures of major importance, whether described by motion or by points, it could be assumed that geometry would then be perfect for practical use.

Furthermore, although optical demonstration or astronomical observation or the composition of motions will bring us new figures, it will be easy for anyone to construct tables for himself so that he may conduct his investigations with little toil and with great accuracy; for it is known from the failures [of those] who attempted the quadrature of the circle that arithmetic is the surest custodian of geometrical exactness. Hence it will pay to undertake the work of extending as far as possible the major Pythagorean tables; the table of squares, cubes, and other powers; and the tables of combinations, variations, and progressions of all kinds, so as to facilitate the labour.

Also the astronomers surely will not have to continue to exercise the patience which is required for computation. It is this that deters them from computing or correcting tables, from the construction of Ephemerides, from working on hypotheses, and from discussions of observations with each other. For it is unworthy of excellent men to lose hours like slaves in the labour of calculation, which could be safely relegated to anyone else if the machine were used.

What I have said about the construction and future use [of the machine] should be sufficient, and I believe will become absolutely clear to the observers [when completed].

19.B Charles Babbage

19.B1 Babbage on Gaspard de Prony

In the midst of that excitement which accompanied the Revolution of France and the succeeding wars, the ambition of the nation, unexhausted by its fatal passion for military renown, was at the same time directed to some of the nobler and more permanent triumphs, which mark the era of a people's greatness—and which receive the applause of posterity long after their conquests have been wrested from them, or even when their existence as a nation may be told only by the page of history. Amongst their enterprises of science, the French government was desirous of producing a series of mathematical tables, to facilitate the application of the decimal system which they had so recently adopted. They directed, therefore, their mathematicians to construct such tables, on the most extensive scale. Their most distinguished philosophers, responding fully to the call of their country, invented new methods for this laborious task; and a work, completely answering the large demands of the government, was produced in a remarkably short period of time. M. Prony, to whom the superintendence of this great undertaking was confided, in speaking of its commencement, observes: 'I shall carry it out with all the ardour of which I am capable, and will occupy myself first of all with a general plan of execution. The number of conditions I shall have to meet will necessitate the use of a great number of calculators, and this leads me straight away to think of applying the division of labour to the construction of these tables, which the commercial arts have drawn on so advantageously to join together the economy of expenditure and time for the improvement of manpower.' The circumstance which gave rise to this singular application of the principle of *the division of labour* is so interesting, that no apology is necessary for introducing it from a small pamphlet printed at Paris a few years since, when a proposition was made by the English to the French government, that the two countries should print these tables at their joint expense.

The origin of the idea is related in the following extract:

'The existence of the work which the British government wants to make the scholarly world enjoy, is probably due to a chapter in a justly celebrated English work [*An Enquiry into the Nature and Causes of the Wealth of Nations*, by Adam Smith].

Here is the anecdote: M. de Prony was committed, with government committees, to composing logarithmic and trigonometric tables for *the centesimal division of the circle, whose exactness was not only to leave nothing to be desired but which would provide the most vast and imposing monument of calculation which had ever been executed, or even conceived*. The logarithms of the numbers from 1 to 200,000 formed a necessary and demanding supplement to this work. It was easy for M. de Prony to convince himself that, even in association with three or four skilled colleagues, the greatest expectation of life he could have would not suffice for him to meet his commitments. He was wrestling with this troublesome thought when, standing in front of a bookshop, he saw the beautiful English edition of Smith printed in London in 1776; he opened the book at random and fell on the first chapter which treats of *the division of labour*, where the production of pins is given as an example. Scarcely had he turned over the first pages when, by a kind of inspiration, he conceived the expedient of *manufacturing* his logarithms as if they were pins. At that time he was giving classes at the Ecole Polytechnique on a part of analysis related to this type of work, *the method of differences*, and its applications to *interpolation*. He spent some days in the country and returned to Paris with the plan of *manufacture*, which has been followed in its execution. He put together two workshops, which performed the same calculations separately, and provided for mutual verification.'

The ancient methods of computing tables were altogether inapplicable to such a proceeding. M. Prony, therefore, wishing to avail himself of all the talent of his country in devising new methods, formed the first section of those who were to take part in this enterprise out of five or six of the most eminent mathematicians in France.

First section The duty of this first section was to investigate, amongst the various analytical expressions which could be found for the same function, that which was most readily adapted to simple numerical calculation by many individuals employed at the same time. This section had little or nothing to do with the actual numerical work. When its labours were concluded, the formulae on the use of which it had decided, were delivered to the second section.

Second section This section consisted of seven or eight persons of considerable acquaintance with mathematics: and their duty was to convert into numbers the formulae put into their hands by the first section—an operation of great labour; and then to deliver out these formulae to the members of the third section, and receive from them the finished calculations. The members of this second section had certain means of verifying the calculations without the necessity of repeating, or even of examining, the whole of the work done by the third section.

Third section The members of this section, whose number varied from sixty to eighty, received certain numbers from the second section, and, using nothing more than simple addition and subtraction, they returned to that section the tables in a finished state. It is remarkable that nine-tenths of this class had no knowledge of arithmetic beyond the two first rules which they were thus called up to exercise, and that these persons were usually found more correct in their calculations, than those who possessed a more extensive knowledge of the subject.

When it is stated that the tables thus computed occupy seventeen large folio volumes, some idea may perhaps be formed of the labour. From that part executed by the third class, which may almost be termed mechanical, requiring the least knowledge and by far the greatest exertions, the first class were entirely exempt. Such labour can always be purchased at an easy rate. The duties of the second class, although requiring considerable skill in arithmetical operations, were yet in some measure relieved by the higher interest naturally felt in those more difficult operations. The exertions of the first class are not likely to require, upon another occasion, so much skill and labour as they did upon the first attempt to introduce such a method; but when the completion of a calculating-engine shall have produced a substitute for the whole of the third section of computers, the attention of analysts will naturally be directed to simplifying its application, by a new discussion of the methods of converting analytical formulae into numbers.

The proceeding of M. Prony, in this celebrated system of calculation, much resembles that of a skilful person about to construct a cotton or silk-mill, or any similar establishment. Having, by his own genius, or through the aid of his friends, found that some improved machinery may be successfully applied to his pursuit, he makes drawings of his plans of the machinery, and may himself be considered as constituting the first section. He next requires the assistance of operative engineers capable of executing the machinery he has designed, some of whom should understand the nature of the processes to be carried on; and these constitute his second section. When a sufficient number of machines have been made, a multitude of other persons, possessed of a lower degree of skill, must be employed in using them; these form the third section; but their work, and the just performance of the machines, must be still superintended by the second class.

19.B2 Dionysius Lardner on the need for tables

Viewing the infinite extent and variety of the tables which have been calculated and printed, from the earliest periods of human civilization to the present time, we feel embarrassed with the difficulties of the task which we have imposed on ourselves:—that of attempting to convey to readers unaccustomed to such speculations, anything approaching to an adequate idea of them. These tables are connected with the various sciences, with almost every department of the useful arts, with commerce in all its relations; but, above all, with Astronomy and Navigation. So important have they been considered, that in many instances large sums have been appropriated by the most enlightened nations in the production of them; and yet so numerous and insurmountable have been the difficulties attending the attainment of this end, that after all, even navigators, putting aside every other department of art and science, have, until very recently, been scantily and imperfectly supplied with the tables indispensably necessary to determine their position at sea. [...]

Among the extensive classes of tables here enumerated, there are several which are in their nature permanent and unalterable, and would never require to be recomputed, if they once could be computed with perfect accuracy on accurate data; but the data on which such computations are conducted, can only be regarded as approximations to truth, within limits the extent of which must necessarily vary with our knowledge of astronomical science. It has accordingly happened, that one set of tables after another has been superseded with each advance of astronomical science. Some striking examples of this may not be uninstructive. In 1765, the Board of Longitude paid to the celebrated Euler the sum of 300*l*., for furnishing general formulae for the computation of lunar tables. Professor Mayer was employed to calculate the tables upon these formulae, and the sum of 3000*l*. was voted for them by the British Parliament, to his widow, after his decease. These tables had been used for the years, from 1766 to 1776, in computing the Nautical Almanac, when they were superseded by new and improved tables, composed by Mr Charles Mason, under the direction of Dr Maskelyne, from calculations made by order of the Board of Longitude, on the observations of Dr Bradley. A farther improvement was made by Mason in 1780; but a much more extensive improvement took place in the lunar calculations by the publication of the tables of the Moon, by M. Bürg, deduced from Laplace's theory, in 1806. Perfect, however, as Bürg's tables were considered, at the time of their publication, they were, within the short period of six years, superseded by a more accurate set of tables published by Burckhardt in 1812; and these also have been followed by the tables of Damoiseau. [. . .]

We have before us a catalogue of the tables contained in the library of one private individual, consisting of not less than one hundred and forty volumes. Among these there are no duplicate copies; and we observe that many of the most celebrated voluminous tabular works are not contained among them. They are confined exclusively to arithmetical and trigonometrical tables; and, consequently, the myriad of astronomical and nautical tables are totally excluded from them. Nevertheless, they contain an extent of printed surface covered with figures amounting to above sixteen thousand square feet. We have taken at random forty of these tables, and have found that the number of errors *acknowledged* in the respective errata, amounts to above *three thousand seven hundred*.

To be convinced of the necessity which has existed for accurate numerical tables, it will only be necessary to consider at what an immense expenditure of labour and of money even the imperfect ones which we possess have been produced.

19.B3 Anthony Hyman's commentary on the analytical engine

At the time and since his death most commentators have assumed that Babbage was trying to build an Analytical Engine and that he failed. In fact, as he was well aware, he did not have the necessary financial resources; but if he was not actually building one it is reasonable to ask what exactly he was doing with such enormous effort and at such high cost. The answer is clear: he was carrying out experiments and preparing designs. This is a common procedure in modern industrial laboratories, and if it was incomprehensible to Babbage's contemporaries it should not be difficult to comprehend today. Babbage may have entertained hopes that ultimately he would be able to build an Analytical Engine, but with the bitter experience of the difficulties he had faced, and indeed still faced, in securing government understanding and support for the relatively straightforward first Difference Engine, he had no illusions about the magnitude of the problems which would arise in financing one of his far more ambitious new engines. For the foreseeable future the intrinsic interest of the research would have to be its own justification.

The work proceeded at speed. In less than two years he had sketched out many of the salient features of the modern computer. A crucial step was the adoption of a punched-card system derived from the Jacquard loom. In this type of loom, and in an earlier loom made by another Frenchman, Falcon, the patterns woven were controlled by the patterns of holes in a set of punched cards strung together in the sequence in which they were to be used. In the loom it was necessary to push a set of wooden rods lifting the particular set of threads required to form the pattern designed. Falcon had used separated cards with holes punched in them to select the required sets of rods: a workman pressed the appropriate cards one after another on to the wooden levers. Then one of the greatest French engineers, Vaucanson, had suggested stringing the cards together and using an automatic system for feeding them in sequence.

Many different methods have been used through the ages for storing information. One technique which is quite familiar is the drum with little spikes raised on it, used for example in musical boxes. Vaucanson himself actually made a loom using punched cards moved by a perforated cylinder. He also made many beautiful and ingenious automata. But it was Jacquard, using the punched-card method, who designed the first really successful loom which automatically produced complex patterns and pictures woven in silk.

Babbage's plans separated the store, or storehouse, holding the numbers, from the mill which carried out numerical operations, such as addition, subtraction, multiplication, and division, using numbers brought from the store. The analogy was with the cotton mills: numbers were held in store, like materials in the storehouse, until they were required for processing in the mill or despatch to the customer. Following the introduction of punched cards early in 1836 four functional units familiar in the modern computer could soon be clearly distinguished: input/output system, mill, store, and control.

A great series of engineering drawings, about three hundred in all, among the finest ever made, came from Babbage's drawing office. His principal draftsman was Jarvis, who had earlier worked under Clement on the drawings for the first Difference Engine. An outstanding draftsman with far greater technical knowledge than Clement, Jarvis deserves an honoured place in the history of the computer. In November 1835 Babbage came to a permanent agreement with Jarvis, who had been offered employment abroad with a heavy penalty clause if he should break the contract. The pay was high, as the railway boom had begun and a skilled draftsman could command a high salary. Babbage consulted his mother who gave him encouragement and excellent advice: 'My dear son, you have advanced far in the accomplishment of a great object, which is worthy of your ambition. You are capable of completing it. My advice is—pursue it, even if it should oblige you to live on bread and cheese.' It never came to that, but Babbage paid Jarvis a guinea for an eight-hour day, a very high wage for the time.

Accompanying the drawings are hundreds of large sheets of the corresponding mechanical notations. Together they form a complete description of the Analytical Engines. The detailed study of these Engines is a task of considerable complexity. Each part had to carry out the required logical and arithmetical functions, but it had also to work mechanically. In addition the various parts had to be so grouped that interconnection between them could be made. Even in modern computers, where wires or other conductors can be placed and joined with relative ease, the design of the interconnection system has necessitated a great deal of attention and has often been a limiting factor. In Babbage's mechanical system it was, at the early stages of the Analytical Engines, a problem of the greatest difficulty. A general pattern of development in the work may be observed. As the designs proceeded Babbage was able gradually to simplify each part of the Engines. This in turn permitted him to add more functional units, thus increasing the overall power of the Engines. By repeated simplification and enhancement of computing power the work proceeded, with an interruption between 1848 and 1857, for most of the remainder of his life. Gradually he developed plans for Engines of great logical power and elegant simplicity (although the term 'simple' is used here in a purely relative sense).

The store of an Analytical Engine was formed of a set of axes on each of which was placed a pile of toothed wheels. Each wheel was free to rotate independently of the other wheels on the same axis. In the more straightforward designs each axis stored a number, while each wheel on the axis stored a single digit. The toothed wheels, which were called figure wheels, were marked on their edges with the figures 0 to 9. Only the top wheel on each axis had a special function, indicating whether the number stored on that axis was positive or negative. It was not necessary for Babbage to use the decimal system and he did indeed consider using other number bases, including 2, 3, 4, 5, and 100. Nowadays the use of number bases other than the base 10 of the ordinary decimal system is widely familiar from modern school mathematics. However, if Babbage had adopted a base of 2 (the binary system commonly used in modern computers) the use of storage space would have been inefficient as each figure wheel would only have held a single binary digit. Thus either the numbers which could be held on each axis would have been too small, or the number of discs inconveniently large. On the other hand a number base of 100 made the use of storage space very efficient, but was impractical because arithmetical calculation on the engine would have been too slow.

Babbage used his punched cards for three principal purposes. The numerical value of a constant, such as π, could be introduced into the engine by a *number-card*. A second type of card, called a *variable-card*, defined the axis on which the number was to be placed. The variable-cards could also order the number held on a particular axis in the store to be transferred to the mill when required; or alternatively to transfer a number from the mill to a designated axis. Cards of the third type, called *operation-cards*, were smaller than the other cards. The operation cards controlled the action of the mill. An operation card might order an addition, subtraction, multiplication, or division, or any other operation for which facilities were provided on the Anaytical Engine in use. Babbage considered many different arrangements of cards at different times and for varying purposes.

Eighteen thirty-six was Babbage's *annus mirabilis*. It is quite astonishing that within two years, working with Jarvis and two or three other assistants, Babbage should have advanced so far in so short a time. Sequences of punched cards gave instructions to an Engine. Instructions were then decoded by a fixed store, which in turn distributed the detailed instructions appropriate to each particular part of the Engine for the operation being carried out. Both the detailed operations and the order in which the cards themselves were called into action could be modified by the results of calculations previously carried out by the Engine. This is not yet the stored-program computer which plays so large a part in modern life, but rather a versatile and powerful calculator.

However Babbage's ideas did not end at that point. In 1836 he was already thinking of designing an Algebra or Formula Engine. Indeed the famous term Analytical Engine seems to have come into use at this time in contradistinction to the more general Algebra Engine. Whether that is indeed the way the terms were coined, Babbage's plans went far beyond the Analytical Engines which have become widely known. Sometimes his more general plans were thought of as a separate class of Algebra Engines, sometimes as further developments of the Analytical Engines. It is in these more general plans that one can find Babbage's closest approach to the formulation of the general concepts of the modern computer. [...]

It would not be appropriate to discuss here the technical details of the Analytical Engine, but we may glance very briefly at the course which developments had followed. The main components of the Engines were described by Babbage in a paper wrtiten in 1837–38, a year and a half after he had introduced a punched-card input system. The control system was most ingenious, using several barrels, or fixed stores, in conjunction. To alter the arithmetical operations performed by an Engine it was necessary to do little more than use different barrels, or what was in most cases the same thing, replace the rows of studs on each barrel. In modern terminology each row of studs was a word. Addition and subtraction took a few seconds, multiplication and division, some minutes. The other two operations principally considered were the extraction of square roots, and the method of finite differences on which the Difference Engines were based.

Babbage went to enormous lengths to make calculation as rapid as possible with his mechanical system, even developing a complex form of arithmetic with a carry at 5 as well as 0. He called this 'half-zero' carriage. He developed plans for a double engine, that could carry out operations simultaneously on two numbers each with half the number of digits which could be handled by the whole engine. Stepping numbers up or down the columns was carried out by pinions, but in 1843 he thought of the more elegant method of stepping by raising an entire column of wheels. The latter technique was later used in the 'minimal engine' on which he was working when he died. He planned a whole range of peripherals, including card reader and punch, line printer, and curve plotter. These could work locally or remotely (in the next room). He could effect several operations simultaneously, including transfer of numbers between store and mill while operations were being performed in the mill, thus taking one or two steps towards the modern technique of pipelining. At several stages he developed a 'great engine'—in modern terms a number-cruncher—and a simple engine. The achievement was matchless: the detailed drawings and notations reveal incomparable logical power combined with a mechanical ingenuity worthy of the great Da Vinci.

19.B4 Ada Lovelace on the analytical engine

(a) In studying the action of the Analytical Engine, we find that the peculiar and independent nature of the considerations which in all mathematical analysis belong to *operations*, as distinguished from *the objects operated upon* and from the *results* of the operations performed upon those objects, is very strikingly defined and separated.

It is well to draw attention to this point, not only because its full appreciation is essential to the attainment of any very just and adequate general comprehension of the powers and mode of action of the Analytical Engine, but also because it is one which is perhaps too little kept in view in the study of mathematical science in general. It is, however, impossible to confound it with other considerations, either when we trace the manner in which that engine attains its results, or when we prepare the data for its attainment of those results. It were much to be desired, that when mathematical processes pass through the human brain instead of through the medium of inanimate mechanism, it were equally a necessity of things that the reasonings connected with *operations* should hold the same just place as a clear and well-defined branch of the subject of analysis, a fundamental but yet independent ingredient in the science, which they must do in studying the engine. The confusion, the difficulties, the contradictions which, in consequence of a want of accurate distinctions in this particular, have up to even a recent period encumbered mathematics in all those branches involving the consideration of negative and impossible quantities, will at once occur to the reader who is at all versed in this science, and would alone suffice to justify dwelling somewhat on the point, in connexion with any subject so peculiarly fitted to give forcible illustration of it as the Analytical Engine. It may be desirable to explain, that by the word *operation*, we mean *any process which alters the mutual relation of two or more things*, be this relation of what kind it may. This is the most general definition, and would include all subjects in the universe. In abstract mathematics, of course operations alter those particular relations which are involved in the considerations of number and space, and the *results* of operations are those peculiar results which correspond to the nature of the subjects of operation. But the science of operations, as derived from mathematics more especially, is a science of itself, and has its own abstract truth and value; just as logic has its own peculiar truth and value, independently of the subjects to which we may apply its reasonings and processes. Those who are accustomed to some of the more modern views of the above subject, will know that a few fundamental relations being true, certain other combinations of relations must of necessity follow; combinations unlimited in variety and extent if the deductions from the primary relations be carried on far enough. They will also be aware that one main reason why the separate nature of the science of operations has been little felt, and in general little dwelt on, is the *shifting* meaning of many of the symbols used in mathematical notation. First, the symbols of *operation* are frequently *also* the symbols of the *results* of operations. We may say that these symbols are apt to have both a *retrospective* and a *prospective* signification. They may signify either relations that are the consequences of a series of processes already performed, or relations that are yet to be effected through certain processes. Secondly, figures, the symbols of *numerical magnitude*, are frequently *also* the symbols of *operations*, as when they are the indices of powers. Wherever terms have a shifting meaning, independent sets of considerations are liable to become complicated together, and reasonings and results are frequently falsified. Now in the Analytical Engine, the operations which come under the first of the above heads are ordered and combined by means of a notation and of a train of mechanism which belong exclusively to themselves; and with respect to the second head, whenever numbers meaning *operations* and not *quantities* (such as the indices of powers) are inscribed on any column or set of columns, those columns immediately act in a wholly separate and independent manner, becoming connected with the *operating mechanism* exclusively, and re-acting upon this. They never come into combination with numbers upon any other columns meaning *quantities*; though, of course, if there are numbers meaning *operations* upon n columns, these may *combine amongst each other*, and will often be required to do so, just as numbers meaning *quantities* combine with each other in any variety. It might have been arranged that all numbers meaning *operations* should have appeared on some separate portion of the engine from that which presents numerical *quantities*; but the present mode is in some cases more simple, and offers in reality quite as much distinctness when understood.

The operating mechanism can even be thrown into action independently of any object to operate upon (although of course no *result* could then be developed). Again, it might act upon other things besides *number*, were objects found whose mutual fundamental relations could be expressed by those of the abstract science of operations, and which should be also susceptible of adaptations to the action of the operating notation and mechanism of the engine. Supposing, for instance, that the fundamental relations of pitched sounds in the science of harmony and of musical composition were susceptible of such expression and adaptations, the engine might compose elaborate and scientific pieces of music of any degree of complexity or extent.

(b) There is no finite line of demarcation which limits the powers of the Analytical Engine. These powers are co-extensive with our knowledge of the laws of analysis itself, and need be bounded only by our acquaintance with the latter. Indeed we may consider the engine as the *material and mechanical representative* of analysis, and that our actual working powers in this department of human study will be enabled more effectually than heretofore to keep pace with our theoretical knowledge of its principles and laws, through the complete control which the engine gives us over the *executive manipulation* of algebraical and numerical symbols.

Those who view mathematical science, not merely as a vast body of abstract and immutable truths, whose intrinsic beauty, symmetry and logical completeness, when regarded in their connexion together as a whole, entitle them to a prominent place in the interest of all profound and logical minds, but as possessing a yet deeper interest for the human race, when it is remembered that this science constitutes the language through which alone we can adequately express the greater facts of the natural world, and those unceasing changes of mutual relationship which, visibly or invisibly, consciously or unconsciously to our immediate physical perceptions, are interminably going on in the agencies of the creation we live amidst: those who thus think on mathematical truth as the instrument through which the weak mind of man can most effectually read his Creator's works, will regard with especial interest all that can tend to facilitate the translation of its principles into explicit practical forms.

The distinctive characteristic of the Analytical Engine, and that which has rendered it possible to endow mechanism with such extensive faculties as bid fair to make this engine the executive right-hand of abstract algebra, is the introduction into it of the principle which Jacquard devised for regulating, by means of punched cards, the most complicated patterns in the fabrication of brocaded stuffs. It is in this that the distinction between the two engines lies. Nothing of the sort exists in the Difference Engine. We may say most aptly, that the Analytical Engine *weaves algebraical patterns* just as the Jacquard-loom weaves flowers and leaves. Here, it seems to us, resides much more of originality than the Difference Engine can be fairly entitled to claim. We do not wish to deny to this latter all such claims. We believe that it is the only proposal or attempt ever made to construct a calculating machine *founded on the principle of successive orders of differences*, and capable of *printing off its own results*; and that this engine surpasses its predecessors, both in the extent of the calculations which it can perform, in the facility, certainty and accuracy with which it can effect them, and in the absence of all necessity for the intervention of human intelligence *during the performance of its calculations.* Its nature is, however, limited to the strictly arithmetical, and it is far from being the first or only scheme for constructing *arithmetical* calculating machines with more or less of success.

The bounds of *arithmetic* were however outstepped the moment the idea of applying the cards had occurred; and the Analytical Engine does not occupy common ground with mere 'calculating machines'. It holds a position wholly its own; and the considerations it suggests are most interesting in their nature. In enabling mechanism to combine together *general* symbols in successions of unlimited variety and extent, a uniting link is established between the operations of matter and the abstract mental processes of the *most abstract* branch of mathematical science. A new, a vast, and a powerful language is developed for the future use of analysis, in which to wield its truths so that these may become of more speedy and accurate practical application for the purposes of mankind than the means hitherto in our possession have rendered possible. Thus not only the mental and the material, but the theoretical and the practical in the mathematical world, are brought into more intimate and effective connexion with each other. We are not aware of its being on record that anything partaking in the nature of what is so well designated the *Analytical* Engine has been hitherto proposed, or even thought of, as a practical possibility, any more than the idea of a thinking or of a reasoning machine.

(c) It is desirable to guard against the possibility of exaggerated ideas that might arise as to the powers of the Analytical Engine. In considering any new subject, there is frequently a tendency, first, to *overrate* what we find to be already interesting or remarkable; and, secondly, by a sort of natural reaction, to *undervalue* the true state of the case, when we do discover that our notions have surpassed those that were really tenable.

The Analytical Engine has no pretensions whatever to *originate* anything. It can do whatever we *know how to order it* to perform. It can *follow* analysis; but it has no power of *anticipating* any analytical relations or truths. Its province is to assist us in making *available* what we are already acquainted with. This it is calculated to effect primarily and chiefly of course, through its executive faculties; but it is likely to exert an *indirect* and reciprocal influence on science itself in another manner. For, in so distributing and combining the truths and the formulae of analysis, that they may become more easily and rapidly amenable to the mechanical combinations of the engine, the relations and the nature of many subjects in that science are necessarily thrown into new lights, and more profoundly investigated. This is a decidedly indirect, and a somewhat *speculative*, consequence of such an invention. It is however pretty evident, on general principles, that in devising for mathematical truths a new form in which to record and throw themselves out for actual use, views are likely to be induced, which should again react on the more theoretical phase of the subject. There are in all extensions of human power, or additions to human knowledge, various *collateral* influences, besides the main and primary object attained.

(d) We refer the reader to the *Edinburgh Review* of July 1834, for a very able account of the Difference Engine. The writer of the article we allude to has selected as his prominent matter for exposition, a wholly different view of the subject from that which

M. Menabrea has chosen. The former chiefly treats it under its mechanical aspect, entering but slightly into the mathematical principles of which that engine is the representative, but giving, in considerable length, many details of the mechanism and contrivances by means of which it tabulates the various orders of differences. M. Menabrea, on the contrary, exclusively developes the analytical view; taking it for granted that mechanism is able to perform certain processes, but without attempting to explain *how*; and devoting his whole attention to explanations and illustrations of the manner in which analytical laws can be so arranged and combined as to bring every branch of that vast subject within the grasp of the assumed powers of mechanism. It is obvious that, in the invention of a calculating engine, these two branches of the subject are equally essential fields of investigation, and that on their mutual adjustment, one to the other, must depend all success. They must be made to meet each other, so that the weak points in the powers of either department may be compensated by the strong points in those of the other. They are indissolubly connected, though so different in their intrinsic nature, that perhaps the same mind might not be likely to prove equally profound or successful in both.

19.C Samuel Lilley on Machinery in Mathematics

While the ordinary calculating machine was being perfected in France and later in Germany and the USA, the division between industry and established science in England was being healed. From about 1850 onwards, science of the new type, concerned with thermodynamics, electricity and the new sort of chemistry that industry can use, was forcing its way into the universities and the Royal Society. The biggest part in bringing it there was played by Kelvin, whose activities ranged from the very academic calculation of the future life of the universe by thermodynamics to the very practical business of transatlantic telegraph engineering, and who demonstrated the respectability of the new science in a way that the 19th century could well understand—by amassing a considerable fortune from the exploitation of it. It is not surprising that Kelvin, who thus united industrial science and technique with the academic tradition of mathematics (for he was, of course, a brilliant, Cambridge-trained mathematician) should have been the central figure in the new advance of mechanical methods in mathematics. It was a revolutionary advance, too, for it took us beyond mere arithmetical machines, and into the realm of machines for solving differential equations, harmonic analyses, integral equations and so on.

The uniting of mechanical techniques with mathematics was one of the main causes of the rise of the new machines. Another lay in the greatly increased complexity of the mathematical problems thrown up by the new science and the advancing industry—problems like those of telegraphy, of stresses in intricate frameworks and so on, discussed below. But another important cause is not quite so obvious.

The actual use of mathematics to solve problems about complicated machinery, or analogous problems in the electrical sphere, provided a new way of thinking for mathematicians. If a mathematical equation adequately represented the motion of the machine, then the motion of the machine equally represented the mathematical equation *and could therefore be used as a means of solving the equation*. That was a new mathematical situation.

The men of the 17th century had plenty of machines, but they did not apply much mathematics to them—the machinery was not complicated enough to require it. They had plenty of mathematics, but it was not primarily applied to machines. It was applied to things like the motion of the moon or the motion of a projectile under gravity. And you cannot reverse that application and use the observed motion of the moon or the projectile to solve mathematical equations, for you cannot control the conditions as you wish. So that it was virtually impossible for the idea of using a machine to solve, say, a differential equation to arise before the 19th century. [...]

By the 17th century men's knowledge of machines was sufficient to permit them to mechanise the familiar abacus. But, since they had not to any great extent applied mathematics to machinery, they could not use machinery to provide *fundamentally new* mathematical methods. Even today the vast majority of purely arithmetical machines remain merely mechanical versions of the abacus, though a few modern examples have gone a step further by incorporating an actual multiplication table in their mechanism. But meanwhile these *arithmetical* machines have been submerged in the far wider class of *mathematical* machines, whose history dates almost entirely from the latter part of the last century.

The men of the 19th century, unlike those of the 17th, were familiar with the application of mathematics to complicated mechanical problems of many types and so they could reverse the connection and use machines to produce fundamentally new mathematical methods. They did; and the mechanical methods which they introduced were in a broad sense the first advances in the basic processes of mathematics since the abacus, all that came between being merely a memorizing systematizing and short-cutting of the abacus process, whether it be by the algorithms, by tables of logarithms, or by reducing differential equations to difference equations. Formerly the object of mathematical science was to reduce each problem to one of arithmetic, essentially one of the abacus; now the object is to reduce it to a form in which some machine or other can complete the work, often without any intervention of arithmetic except in the notation used for recording the results. And this fundamental change was largely due to the fact that men were now so familiar with the application of mathematics to describe and predict the motion of machines, that they were at last able to appreciate the reverse connection, the use of an appropriate machine to represent and therefore solve a mathematical problem.

As the requirements of technology and science have grown, the number of these machines has also grown, so rapidly that we can do no more than mention a few of them and the problems with which they were connected.

Fundamental to a very large class of mathematical problems is the differential equation. A large number of instruments called 'integraphs' were produced to solve special differential equations. But these are of little importance beside Kelvin's proposal for the mechanical integration of any ordinary differential equation. The principle of the thing is simple. The essential element, the integrator, is a continuously variable gear, which was already familiar in many integraphs and planimeters. The big step forward is to realize that by suitable mechanical connections between such integrators a machine can be produced to solve any differential equation. Kelvin pointed this out in regard to linear equations in 1876.

But he failed to overcome the difficulty that the friction drive from the integrator is not sufficiently powerful to drive the further mechanical connections which are required. And this difficulty was not overcome till 1931, when Bush of the U.S.A. solved the problem by means of the 'torque amplifier'. The principle of the torque amplifier is again extremely simple; it is essentially the same as the capstan of a ship. The capstan is kept rotating, a rope is wound round it, and it is found that a small pull on one end of the rope produces a much stronger pull on the other end. Kelvin, in fact, might have produced a complete differential analyser (as Bush's machine is called) in 1876, if he had happened to mention his difficulty to the deck-hand of a tramp steamer, or one of the many other workmen who use the capstan principle. In that way is scientific knowledge retarded in a society in which scientists and deck-hands do not usually move in the same circles. And it is interesting to note that Bush, who actually solved the problem, is one of those who are particularly concerned about the social questions connected with science; his attitude in this may have some relation to his ability to synthesize mathematics with a common technique.

Wave motion problems probably rank next in importance to differential equations in late 19th century mathematics. They occur, for example, in telegraphy, which was then rapidly developing, and in the ever-important problem of forecasting the tides. The latter, of course, had been important for a long time, but singularly little progress has been made with it, largely because the earlier attempts were synthetic ones, attempts to produce general formulae for tidal phenomena. Gradual accumulation of evidence showed that the problem was too complicated for this method. Thus Kelvin was led to suggest in 1876 to the British Association committee on the subject the alternative *analytical* method: to analyse the tidal curve into simple harmonic components, extrapolate these into the future and then synthesize them to produce the future tidal curve.

These problems of telegraphy and tides, and many others lent great importance to the mathematics of harmonic analysis and synthesis. And there was a great crop of machines to avoid the tedium of arithmetical solution of such problems. [...]

Let us end with a word or two on the present position of mathematical machines.

There is every sign that a very thorough appreciation now exists of the power that mechanical methods can add to the mathematician's elbow. But it also seems that there is as yet no real appreciation of certain conditions that must be fulfilled if that power is to be fully used. Many of these machines cost thousands of pounds to build. Any scientist can afford a slide-rule; a few of them can afford a Brunsviga machine. But it takes a very large institution to buy a machine like the differential analyser, while the use of mathematical machines is not fully efficient till many such machines are gathered in one institution. So far such institutions as do possess collections of mathematical machines are either closed academic corporations, like the Massachusetts Institute of Technology, or commercial computing services, working to the orders of customers. The former, very naturally, tend to restrict the use of their machines to their own members; the latter apparently cannot accumulate sufficient capital to obtain the use of the more expensive and specialized machines.

It is clear that the fullest use of mathematical machines can only be made through the establishment of central organizations on a national (and perhaps even an international) scale, organizations which could be equipped with a full range of machines. In both fundamental science and technology, to say nothing of the vast statistical work of the social sciences, investigators are finding more and more that the mathematical problems they meet require mechanical aids far beyond the resources of the individual laboratory. Such problems they could send to a central institute for solution. The proposal to set up a Mathematics Division of the National Physical Laboratory, which would include this among its functions, is therefore to be welcomed. It is to be hoped that it will receive the very generous financial allocations necessary to make a success of such a venture. It is also to be hoped that its service will not, as at present seems to be intended, be confined to Government departments and industry—fundamental research, which is ultimately of even greater importance, has equal need of its aid, while provision must be made in this or some other organization for the biological and social sciences.

19.D Computer Proofs

19.D1 Letter from Augustus De Morgan to William Rowan Hamilton

A student of mine asked me today to give him a reason for a fact which I did not know was a fact—and do not yet. He says that if a figure be anyhow divided and the compartments differently coloured so that figures with any portion of common boundary *line* are differently coloured—four colours may be wanted, but not more—the following is the case in which four *are* wanted. Query cannot a necessity for five or more be invented. As far as I see at this moment if four *ultimate* compartments have each boundary line in common with one of the others, three of them inclose the fourth and prevent any fifth from connexion with it. If this be true, four colours will colour any possible map without any necessity for colour meeting colour except at a point.

Now it does seem that drawing three compartments with common boundary *ABC* two and two you cannot make a fourth take boundary from all, except by inclosing one. But it is tricky work, and I am not sure of the convolutions—what do you say? And has it, if true, been noticed? My pupil says he guessed it in colouring a map of

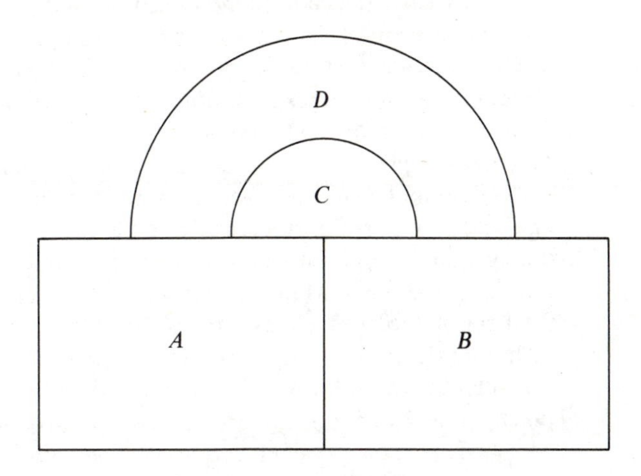

A, *B*, *C* and *D* are names of colours

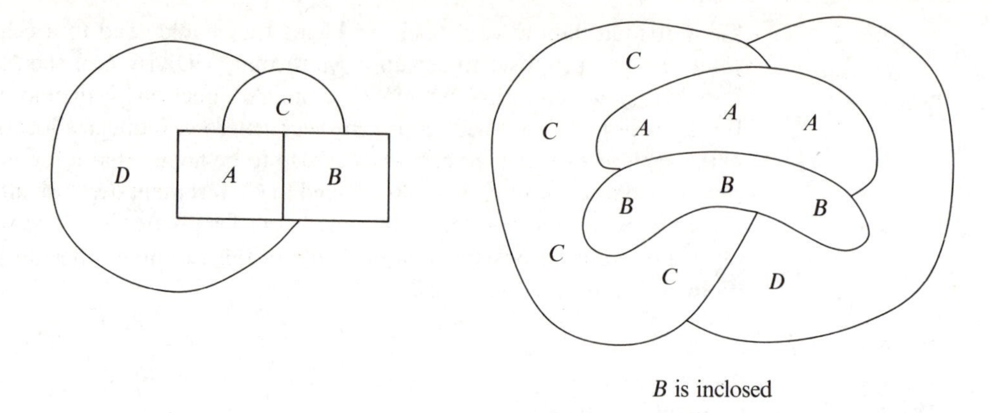

B is inclosed

England. The more I think of it, the more evident it seems. If you retort with some very simple case which makes me out a stupid animal, I think I must do as the Sphynx did...

19.D2 Donald J. Albers

In August of 1976, Wolfgang Haken announced a proof of the 125-year-old Four-Colour Conjecture to members of the American Mathematical Society and the Mathematical Association of America, who were assembled at the University of Toronto for their annual summer meeting. The elegant and old lecture hall was jammed with mathematicians anxious to hear Professor Haken give the proof. It seemed like the perfect setting to announce a great mathematical result. He proceeded to outline clearly the computer-assisted proof that he and his colleagues had devised. At the conclusion of his remarks, I had expected the audience to erupt with a great ovation. Instead, they responded with polite applause! I was puzzled by their cool reception and sought explanations from those who had heard his presentation. Mathematician after mathematician expressed uneasiness with a proof in which a computer had played a major role. They were bothered by the fact that more than 1000 hours of computer time had been expended in checking some 100,000 cases and often suggested (hoped?) that there might be an error buried in the hundreds of pages of computer print outs. Beyond that concern was the hope that a much shorter proof could be found. Yu. I. Manin's observation that 'a proof only becomes a proof after the social act of "accepting it as a proof"' fits the situation very well. [...]

Although there exist earlier cases of exhaustive computer checking, especially in some number-theoretic proofs, it seems that the computer-assisted work of Appel, Haken, and Koch on the well-known Four-Colour Problem may represent a watershed in the history of mathematics. Their work has been remarkably successful in forcing us to ask What is a Proof Today?

19.D3 F. F. Bonsall

It is natural and desirable that mathematicians should be attracted by famous unsolved problems and should make great efforts to solve them. Such efforts bring new and valuable methods into mathematics, but it is well to keep some sense of proportion. The problem itself may have very little intrinsic importance. If such a problem, the four colour problem, for example, is solved by some clever new idea, that is magnificent; but a solution by a cumulative application of existing methods may do nothing more than demonstrate the cleverness of the solver. It is worse still if the solution involves computer verification of special cases, and in my view such a solution does not belong to mathematical science at all.

It is no better to accept without verification the word of a computer than the word of another mathematician. In fact the tedium of routine tasks makes programming errors exceedingly probable. We cannot possibly achieve what I regard as the essential element of a proof—our own personal understanding—if part of the argument is hidden away in a box.

19.D4 Thomas Tymozco

People sometimes object to the preceding argument on the grounds that at least in some cases they can be more certain of a computer based result than a result of a fellow mathematician—so what's the big deal about computer-supplemented proofs? If we examine this objection a little more closely, we can clarify the philosophical significance of the Four-Colour Proof.

In brief, the objection is this: even though computers are fallible, on the mathematical matters at issue they are less fallible than mathematicians. So a mathematician might be more certain of a computer-supplemented result like the Four-Colour Theorem than a result such as 'all groups of odd order are solvable' which, while checked by mathematicians, has a very long and complex proof. In fact, most mathematicians borrow the complete proof of the latter theorem from Feit and Thompson. But they borrow only (the major) *part* of the Four-Colour Proof from Appel, Haken, and Koch—part is borrowed from a computer and this can be viewed as an *increase* in certainty. Although this objection is not as obvious as it might first appear, rather than question it I would like to point out an implicit premise in it—the premise that mathematicians are fallible.

Obviously the objection assumes mathematicians are fallible. If they are not, if mathematicians are infallible, then it is simply wrong to claim that computers are less fallible than mathematicians.

On the other hand it is easy to establish that the great tradition in philosophy of mathematics ignores this assumption and proceeds as if mathematicians are infallible. (To assume infallibility one need not explicitly assume that mathematicians are infallible, one need only refrain from introducing the idea of mathematical mistake!) By 'great tradition' I mean the rather coherent line of thought on mathematics developed by the Greeks (Euclid, Plato), advanced by modern philosophy beginning with Descartes and Leibniz through Kant, and dramatically advanced by the founders of modern mathematics such as Dedekind, Cantor, Frege, Russell, Hilbert, and Heyting. This tradition either ignores error or consigns it to nonmathematical causes (e.g., in Descartes' philosophy, mathematical error turns out to be a species of sin). In either case the effect is the same. Error is an accident, not part of the essence of mathematics: the mathematician *qua* mathematician is treated by philosophy as infallible. My claim is 'checkable'—check the writings of the tradition for discussions of mathematical error!

The upshot of this is that the assumption that mathematicians are fallible, for all its common sense appeal, is philosophically very significant. It is not part of traditional theories of mathematical knowledge. It introduces a 'new' topic to the philosophy of mathematics—error, uncertainty, fallibility. If philosophical reflection on the Four-Colour Proof leads us to articulate and develop the assumption of mathematical fallibility, then I would say that the Four-Colour Proof has proved its philosophical significance. But in fact it does more. The Four-Colour Proof provides a positive argument for the inclusion of fallibility in the philosophy of mathematics.

Suppose we tried denying that mathematicians were fallible, would this keep error out of mathematics? No, not if mathematicians rely on computers to check proofs! Even if mathematicians *qua* mathematicians are infallible, computers are not. An actual mathematician might be regarded as a rational mind (the source of knowledge) caged in a physical body (the source of error). But present computers are just physical structures. Mathematicians' reliance on computers brings with it, in principle, the fallibility inherent in any physical science. In practice, it brings with it the special fallibility of computer science (computers are not just Turing machines and to the extent to which computers approximate Turing machines, computer programs do not approximate Turing algorithms). Thus, if for no other reason, fallibility is present in mathematics when mathematicians rely on computers.

Of course once we admit error into mathematics, it seems silly to recognize it only for computers. After all we began with the intuition that on occasion computers can be less fallible than mathematicians. We are returned to the conclusion of our original investigation. In order to understand the place of the Four-Colour Proof in terms of mathematical knowledge, we have to reconstruct the theory of mathematical knowledge.

Sources

Chapter 1

1.A1 Aristotle, *Problems*, XV, 910^{b}23–911^{a}4; tr. T. L. Heath, *Mathematics in Aristotle*, Oxford, 1949, pp. 258–259.

1.A2 J. Leslie, *Philosophy of Arithmetic*, Edinburgh, 1820, pp. 1–3.

1.A3 K. Lovell, *The Growth of Basic Mathematical and Scientific Concepts in Children*, University of London Press, 1961, pp. 26–27.

1.A4 K. Menninger, *Number Words and Number Symbols*, MIT Press, 1969, p. 9.

1.A5 A Seidenberg, 'Ritual Origin of Counting', *Archive for History of Exact Sciences*, **2** (1962) pp. 8–10.

1.B1 J. de Heinzelin, 'Ishango', *Scientific American*, **206** (June 1962) pp. 109–111.

1.B2 A. Marshack, *The Roots of Civilization*, Weidenfeld and Nicolson, 1972, pp. 25–27, 31.

1.C1 A. Thom, *Megalithic Sites in Britain*, Oxford, 1967, pp. 34–35.

1.C2 S. Piggott, *Philosophical Transactions of the Royal Society A*, **276** (1974) pp. 275–276.

1.C3 E. W. MacKie, *Science and Society in Prehistoric Britain*, Elek, 1977, pp. 36–37.

1.C4 B. L. van der Waerden, *Geometry and Algebra in Ancient Civilisations*, Springer, 1983, pp. xi, 33, 35.

1.C5 W. R. Knorr, 'The Geometer and the Archaeoastronomers', *British Journal of the History of Science*, **18** (1985) pp. 202, 208–211.

1.D1 A. B. Chace (tr. and ed.), *The Rhind Mathematical Papyrus*, Mathematical Association of America, 1927.

1.D2 As 1.D1.

1.D3 A. Erman, *The Literature of the Ancient Egyptians*, E. P. Dutton, 1927, p. 223.

1.D4(a) Herodotus, *History*, II, 109; tr. A. D. Godley, Heinemann, 1920.

1.D4(b) Plato, *Phaedrus*, 274 cd; tr. H. Cary, Bell & Daldy, 1872.

1.D4(c) Aristotle, *Metaphysics*, 981^{b}20–25; as 1.A1, p. 195.

1.D4(d) Proclus, *On Euclid*, I; tr. I. Thomas, *Greek Mathematical Works I*, Heinemann, 1939, pp. 145–147.

1.D5 A. Gardiner, *Egyptian Grammar*, Oxford, 1957 (third edition), pp. 196–197.
1.D6 As 1.D1, p. 42–43.
1.D7 G. J. Toomer, 'Mathematics and Astronomy', in J. R. Harris (ed.), *The Legacy of Egypt*, Oxford, 1971, pp. 37–40, 45.
1.E1(a)–(e), (g) O. Neugebauer and A. Sachs, *Mathematical Cuneiform Texts*, American Oriental Society, American Schools of Oriental Research, 1945.
1.E1(f) F. Thureau-Dangin, *Textes Mathematiques Babyloniens*, Brill, 1938; tr. J. G. Fauvel.
1.E2 R. Creighton Buck, 'Sherlock Holmes in Babylon', *American Mathematical Monthly*, **87** (1980) pp. 338–345.
1.E3 J. Friberg, 'Methods and Traditions of Babylonian Mathematics: Plimpton 322, Pythagorean Triples, and the Babylonian Triangle Parameter Equations', *Historia Mathematica*, **8** (1981) p. 302.
1.E4(a) A. W. Sjoberg, 'In Praise of the Scribal Art', *Journal of Cuneiform Studies*, **24** (1971/2), p. 127.
1.E4(b) S. N. Kramer, 'Schooldays: a Sumerian Composition relating to the Education of a Scribe', *Journal of the American Oriental Society*, **65** (1949) p. 206.
1.E5 M. A. Powell Jr, 'The Antecedents of Old Babylonian Place Notation and the Early History of Babylonian Mathematics', *Historia Mathematica*, **3** (1976) pp. 432–434.
1.E6 J. Høyrup, *Influences of Institutionalised Mathematics Teaching on the Development and Organisation of Mathematical Thought in the Pre-modern Period*, Bielefeld, 1980, pp. 14–17.

Chapter 2

2.A1 Proclus, *A Commentary on the First Book of Euclid's Elements*; tr. G. R. Morrow, *Proclus: a Commentary on the First Book of Euclid's Elements*, Princeton University Press, 1970, pp. 52–57.
2.A2 W. Burkert, *Lore and Science in Ancient Pythagoreanism*, Harvard University Press, 1972, pp. 409–412.
2.B Simplicius, *Commentary on Aristotle's Physics;* tr. I. Thomas, *Selections Illustrating the History of Greek Mathematics*, I, Heinemann, 1939, pp. 239–249.
2.C1 Parmenides, *The Way of Truth*; tr. M. Schofield, from G. S. Kirk, J. E. Raven and M. Schofield, *The Presocratic Philosophers*, Cambridge, 1983, pp. 243, 245, 248–250.
2.C2 Aristophanes, *The Birds*; tr. D. Barrett, Penguin, 1978, pp. 187–188.
2.C3 Aristophanes, *The Clouds*; tr. J. Ferguson, *Socrates: a Source Book*, Macmillan, 1970, pp. 163–165.
2.D1 Archytas, *Fragment 1*; tr. A. C. Bowen, *Ancient Philosophy*, **2** (1982) p. 82.
2.D2 Plato, *Protagoras*, 318d–319a; tr. W. K. C. Guthrie, Penguin, 1956.
2.D3 As 2.A1, pp. 29–30.
2.D4 Nicomachus, *Introduction to Arithmetic*; tr. M. L. D'Ooge, Michigan, 1926.

2.D5 Boethius, *De Institutione Arithmetica*; tr. M. Masi, *Boethian Number Theory*, Rodopi, 1983, pp. 71, 73.

2.D6 Hrosvitha, *Pafnutius*; as 2.D5, p. 35.

2.D7 R. Bacon, *Opus Majus*, 4, III; tr. R. B. Burke, University of Pennsylvania Press, 1928.

2.D8 As 2.A2, pp. 406–407.

2.E1 Plato, *Meno*, 81e–86c; tr. W. K. C. Guthrie, Penguin, 1956.

2.E2 Plato, *Republic*, 521–531; tr. H. D. P. Lee, Penguin, 1955.

2.E3 Plato, *Theaetetus*, 147d–148b; tr. J. McDowell, Oxford, 1973.

2.E4 Plato, *Philebus*, 55c–57d; tr. R. Hackforth, Cambridge, 1945.

2.E5 Plato, *Timaeus*, 32–33, 53–56, 34–36; tr. H. D. P. Lee, Penguin, 1965.

2.E6 Plato, *Laws*, 817e–820d; as 2.B, pp. 21–27.

2.E7 Plato, *Epistle XIII*; tr. L. A. Post, Oxford, 1925.

2.E8 Aristoxenus, *Elements of Harmony*, II; as 2.B, pp. 389–391.

2.F1 Theon of Smyrna, *Mathematical Exposition*, as 2.B, p. 257.

2.F2 As 2.A1, p. 167.

2.F3 Eutocius, *Commentary on Archimedes' Sphere and Cylinder*; tr. I. E. Drabkin, in M. R. Cohen and I. E. Drabkin (eds), *A Source Book in Greek Science*, Harvard University Press, 1948, pp. 62–66.

2.F4 Eutocius, *Commentary on Archimedes' Sphere and Cylinder*; as 2.B, pp. 279–283.

2.G1 As 2.A1, p. 335.

2.G2(a) Aristotle, *Physics*, $185^{a}14$–17; as 1.A1, p. 94.

2.G2(b) Themistius, *Commentary on Aristotle's Physics*; as 2.B, pp. 235–237.

2.G2(c) Simplicius, *Commentary on Aristotle's Physics*; tr. A. Wasserstein, *Phronesis*, **4** (1959) p. 93.

2.G3(a) Aristotle, *Sophistical Refutations*, $171^{b}12$–$172^{a}7$; as 1.A1, p. 47.

2.G3(b) Alexander, *Commentary on Aristotle's Sophistical Refutations*; as 2.F3, p. 54.

2.G4 Pappus, *Mathematical Collection*, IV; as 2.B, pp. 337–347.

2.H All extracts from T. L. Heath, *Mathematics in Aristotle*, Oxford, 1949.

2.H1(a) Aristotle, *Posterior Analytics*, $76^{a}31$–$77^{a}4$.

2.H1(b) Aristotle, *Metaphysics*, $996^{b}26$–33.

2.H1(c) Aristotle, *Metaphysics*, $1006^{a}5$–15.

2.H2 Aristotle, *Nicomachean Ethics*, $1112^{b}11$–24.

2.H3(a) Aristotle, *Metaphysics*, $1061^{a}28$–$^{b}3$.

2.H3(b) Aristotle, *Physics*, $193^{b}22$–$194^{a}15$.

2.H3(c) Aristotle, *Posterior Analytics*, $78^{b}34$–$79^{a}6$.

2.H3(d) Aristotle, *Nicomachean Ethics*, $1142^{a}11$–20.

2.H4 Aristotle, *Metaphysics*, $985^{b}23$–$986^{a}3$.

2.H5 Aristotle, *Physics*, $206^{a}9$–$^{b}27$, $207^{a}31$–$^{b}34$.

2.H6(a) Aristotle, *Posterior Analytics*, $71^{a}1$–4.

2.H6(b) Aristotle, *Prior Analytics*, $41^{a}23$–7.

Chapter 3

All extracts from Euclid's *Elements* from T. L. Heath, *The Thirteen Books of Euclid's Elements*, Cambridge, 1925.

All extracts from Proclus from G. R. Morrow, *Proclus: a Commentary on the First Book of Euclid's Elements*, Princeton University Press, 1970.

3.A1(a) Proclus, pp. 60–61.
3.A1(b) Proclus, pp. 58–59.
3.B1(b) Proclus, pp. 77, 88, 94, 98–100, 107–108, 124–125, 150.
3.B2(b) Aristotle, *Prior Analytics*, 41[b]13–22; as 2.H, p. 23.
3.B2(c) Proclus, pp. 194–195.
3.B3(e) Proclus, pp. 251–252.
3.B4(b) Aristotle, *Metaphysics*, 1051[a]21–24; as 2.H.
3.B4(c) Proclus, p. 298.
3.B5(b) Proclus, pp. 332–333.
3.B5(f) Proclus, pp. 337–338.
3.D3(b) As 2.D4.
3.F1 A. Aaboe, *Episodes from the Early History of Mathematics*, Random House, 1964, pp. 35–37.
3.F2(a) J. Aubrey, *Brief Lives*; ed. O. L. Dick, Secker and Warburg, 1949.
3.F2(b) B. Russell, *The Autobiography of Bertrand Russell: 1872–1914*, Allen & Unwin, 1967, p. 36.
3.G1 B. L. van der Waerden, *Science Awakening*, Oxford, 1961, pp. 118–119.
3.G2 S. Unguru, 'On the Need to Rewrite the History of Greek Mathematics', *Archive for History of Exact Sciences*, **15** (1975) pp. 67–114.
3.G3 B. L. van der Waerden, 'Defence of a "Shocking" Point of View', *Archive for History of Exact Sciences*, **15** (1976), pp. 199–210.
3.G4 S. Unguru, 'History of Ancient Mathematics: some reflections on the present state of the art', *Isis*, **70** (1979) pp. 555–565.
3.G5 I. Mueller, *Philosophy of Mathematics and Deductive Structure in Euclid's Elements*, MIT Press, 1981, pp. 43–44, 50–52.
3.G6 J. L. Berggren, 'History of Greek Mathematics: a survey of recent research', *Historia Mathematics*, **11** (1984) pp. 394–410.

Chapter 4

4.A All extracts from T. L. Heath, *The Works of Archimedes*, Cambridge, 1897.
4.B1 Plutarch, *Life of Marcellus*, 14–19; tr. I. Scott-Kilvert, *Makers of Rome: Nine lives by Plutarch*, Penguin, 1965.
4.B2 Vitruvius, *The Ten Books on Architecture*; tr. M. H. Morgan, Harvard University Press, 1914, pp. 253–254.
4.B3 J. Wallis, *A Treatise of Algebra*, London, 1685, p. 3.
4.B4 T. L. Heath, *History of Greek Mathematics*, II, Oxford, 1921, pp. 25–27.
4.C All extracts from Diocles, *On Burning Mirrors*; tr. G. J. Toomer, *Diocles on Burning Mirrors*, Springer-Verlag, 1976.
4.D All extracts from Apollonius, *Conics*.
4.D1 tr. R. Catesby Taliaferro, in *Great Books of the Western World*, Encyclopaedia Britannica, 1952, p. 603.
4.D2 As 4.B4, pp. 130–132.
4.D3 As 4.D1, p. 604.

4.D4 As 4.D1, pp. 615–618.
4.D5 As 4.D1, pp. 659–660, 701–702.
4.D6(a)–(c) As 4.D1, pp. 710–711, 772–774.
4.D6(d) tr. (into French) P. ver Eecke, *Les Coniques d'Apollonius de Perge*, 1959, pp. 287–288; tr. (into English) J. J. Gray.
4.D7 As 4.D1, pp. 790–792.

Chapter 5

5.A1 As 2.A1, pp. 31–35.
5.A2 Pappus, *Mathematical Collection*, VII; tr. I. Thomas, *Selections Illustrating the History of Greek Mathematics*, II, Heinemann, 1939, pp. 615–621.
5.A3(a) As 4.B2, p. 198.
5.A3(b) Euclid, *Optics*; as 2.F3, pp. 257–260.
5.A3(c) Hero, *Catoptrics*, 4–5; as 2.F3, pp. 264–265.
5.A4(a) Archytas, *Fragment 2*; as 2.F3, pp. 6–7.
5.A4(b) As 2.D4.
5.A4(c) Aristotle, *Problems*, XIX, 41; tr. E. S. Forster, Oxford, 1927.
5.A4(d) Theon of Smyrna, *Mathematical Exposition*, II, 12; as 2.F3, pp. 294–295.
5.A5 Heron, *Metrica*, I, 8; as 5.A2, pp. 471–477.
5.A6 As 4.B2, pp. 252–253.
5.B1 Theon of Smyrna, *Mathematical Exposition*, I, 1–2.2; as 5.A2, pp. 401–403.
5.B2 As 2.A1, pp. 156–157.
5.B3 As 5.A2, pp. 597–601.
5.B4 Pappus, *Mathematical Collection*, IV; tr. I. E. Drabkin, as 2.F3, pp. 67–68.
5.B5 Pappus, *Mathematical Collection*, V; as 5.A2, pp. 589–593.
5.B6 As 2.A1, p. 343.
5.C1 As 2.A1, pp. 339–340.
5.C2 Heron, *Geometrica*; as 5.A2, pp. 503–509.
5.C3 tr. I. E. Drabkin, as 2.F3, pp. 27–29.
5.C4 *The Greek Anthology*, XIV; tr. W. R. Paton, London, 1918.
5.C5(a) As 1.D1.
5.C5(b) *The Oxford Dictionary of Quotations*, 1953, p. 366.
5.D All extracts from Diophantus, *Arithmetica*.
5.D1(a) tr. I. E. Drabkin, as 2.F3, p. 29.
5.D1(b) tr. T. L. Heath, *Diophantus of Alexandria*, Cambridge, 1910.
5.D2 As 5.D1(b).
5.D3 tr. I. Thomas, as 5.A2, pp. 551–553.
5.D4 As 5.D1(b).
5.D5 tr. J. Sesiano, *Books IV to VII of Diophantus' Arithmetica in the Arabic Translation Attributed to Qusta ibn Luqa*, Springer-Verlag, 1982.
5.D6 As 5.D5.
5.D7 As 5.D5.
5.D8 As 5.D1(b).
5.D9 As 5.D1(b).
5.D10 As 5.D1(b).

Chapter 6

6.A1 Gerard of Cremona; tr. M. Clagett, *Archimedes in the Middle Ages I: the Arabo-Latin Tradition*, University of Wisconsin Press, 1964, pp. 239–241, 353–355.

6.A2 Al-Sijzi; quoted in J. P. Hogendijk, 'Greek and Arabic Constructions of the Regular Heptagon', *Archive for History of Exact Sciences*, **30** (1984) pp. 305–308.

6.A3 Omar Khayyam, *A Paper of Omar Khayyam*; tr. A. R. Amir-Moez, *Scripta Mathematica*, **26** (1961) pp. 323–331.

6.B1(a) Al-Khwarizmi, *The Algebra of Mohammed Ben Musa*; ed. and tr. F. Rosen, John Murray, 1831, pp. 5–6, 8, 11.

6.B1(b) Al-Khwarizmi; ed. L. C. Karpinski, J. G. Winter, *Robert of Chester's Latin Translation of the Algebra of al-Khowarizmi*, University of Michigan Press, 1930, pp. 77–81; reprinted in R. Calinger, *Classics of Mathematics*, Moore, 1982, pp. 183–186.

6.B2 Abu-Kamil, *Algebra*, tr. and ed. M. Levey, University of Wisconsin Press, 1966, pp. 40, 52, 54.

6.B3 Omar Khayyam; quoted in H. J. J. Winter, W. Arafat, 'The Algebra of Omar Khayyam', *Journal of the Royal Asiatic Society of Bengal: Science*, **16** (1950) pp. 55–56.

6.C1 Al-Haytham; B. H. Sude, *Ibn al-Haytham's Commentary on the Premises of Euclid's Elements*, PhD thesis, Princeton University, 1974, pp. 94–96.

6.C2 Omar Khayyam, *Discussion of Difficulties in Euclid*; tr. A. R. Amir-Moez, *Scripta Mathematica*, **24** (1959) pp. 276–277.

6.C3 A. P. Youshkevitch, 1961; tr. M. Cazenove, K. Jaouiche, *Les Mathématiques Arabes*, Vrin, 1976, pp. 112–113; tr. (into English) J. J. Gray.

Chapter 7

7.A1(a)–(b), (d) Leonardo of Pisa (Fibonacci) *Liber Abaci*, 1202, 1228; in B. Boncompagni (ed.) *Scritti di Leonardo Pisano*, I, Rome, 1857–1862, pp. 138, 173, 182; tr. J. V. Field.

7.A1(c) Leonardo of Pisa (Fibonacci), *Letter to Master Theodorus*; in B. Boncompagni (ed.), *Scritti di Leonardo Pisano*, II, Rome, 1857–1862; pp. 247–252; tr. J. V. Field.

7.A2 Jordanus de Nemore, *De Numeris Datis*; tr. and ed. B. B. Hughes, University of California Press, 1981, pp. 127–132 (Publications of the Center for Medieval Studies, UCLA, 14).

7.A3 M. Biagio, *Chasi Exemplari alla Regola dell'Algibra, nella Trascelta a Cura di M° Benedetto*; ed. L. Pieraccini, University of Siena, 1983, p. 27 (Quaderni del Centro Studi della Matematica Medioevale 5); tr. F. R. Smith.

7.B1 J. Regiomontanus, *De Triangulis* Omnimodis; tr. and ed. B. B. Hughes, *Regiomontanus on Triangles*, University of Wisconsin Press, 1967, p. 59; in

D. J. Struik, *A Source Book in Mathematics, 1200–1800*, Harvard University Press, 1969, pp. 140–142.

7.B2 N. Chuquet; tr. and ed. H. G. Flegg, C. M. Hay, B. Moss, *Nicolas Chuquet, Renaissance Mathematician*, Reidel, 1985, pp. 144, 151–153.

7.B3 L. Pacioli, *Summa de Arithmetica, Geometria, Proportioni e Proportionalita*, Paganino de Paganini, 1494, Summary, pp. 111v, 144–146; tr. F. R. Smith.

Chapter 8

8.A1 N. Tartaglia, *Quesiti et Inventioni Diverse*, 1546; ed. A. Masotti, Brescia, 1959, pp. 106–108, 120–133; tr. F. R. Smith.

8.A2 As 8.A1, pp. 120–122; tr. F. R. Smith.

8.A3 A. Masotti (ed.), *Lodovico Ferrari e Niccolo Tartaglia, Cartelli di Sfida Matematica, 1547–48*, Brescia, 1972, pp. 67–68, 91, 113–114, 135–136; tr. F. R. Smith.

8.A4(a)–(b) G. Cardano, *Ars Magna: The Great Art, or the Rules of Algebra;* tr. R. B. McClenon, in D. E. Smith (ed.) *A Source Book in Mathematics*, Dover, 1959, pp. 203–206.

8.A4(c) G. Cardano, *De Vita Propria Liber*, 1575, first published Paris, 1643 tr. J. Stoner, *The Book of my Life*, Dover, 1962, pp. 142, 225–226.

8.A5 R. Bombelli, *L'Algebra*, Feltrinelli, 1966, pp. 8–9, 133–134; tr. F. R. Smith.

8.B1 J. Dee, 'Treatise of the Division of Superficies, etc.', in *Euclid's Elements of Geometry*, London, 1661, pp. 605–608.

8.B2 B. Baldi; from F. Ugolino, F.-L. Polidori (eds), *Versi e Prose di Bernardo Baldi*, Florence, 1859, pp. 528–531; tr. J. V. Field.

8.B3 P. Rose, *The Italian Renaissance of Mathematics*, Librarie Droz, 1975, pp. 169–171 (Studies on Humanists and Mathematicians from Petrarch to Galileo).

8.C1(a)–(c) S. Stevin, *L'Arithmétique*; in *The Principal Works of Simon Stevin*, IIB, 1955–1968, pp. 508–509, 581–582, 585–586, tr. C. M. Hay.

8.C1(d) S. Stevin, *De Thiende*; tr. D. J. Struik, in *The Principal Works of Simon Stevin*, IIA, pp. 403–407.

8.C2(a) F. Viète, *Introduction to the Analytic Art*, Preface; tr. J. W. Smith, in J. Klein, *Greek Mathematical Thought and the Origin of Algebra*, MIT Press, 1968.

8.C2(b)–(c) F. Viète, *Introduction to the Analytic Art*, Chapters 1, 7; tr. R. B. McClenon, in D. E. Smith (ed.), *A Source Book in Mathematics*, Dover, 1959, pp. 203–206.

Chapter 9

9.A1 R. Record, *The Ground of Artes: teaching the woorke and practise of Arithmetike, both in whole numbers and Fractions*, 1552 edition.

9.A2 R. Record, *The Pathway to Knowledge, containing the first principles of Geometrie, as they may most aptly be applied unto practise,...*, 1551.

9.A3 R. Record, *The Castle of Knowledge*, 1556.

9.A4 R. Record, *The Whetstone of Witte: the Seconde parte of Arithmetike, containyng the extraction of Rootes in diverse kindes, with the Arte of Cossike nombers,...*, 1557.

9.B1 J. Dee, 'Mathematical Praeface' to H. Billingsley's translation of Euclid's *Elements*, 1570.

9.B2 Euclid's *Elements*, I.9; tr. H. Billingsley, 1570.

9.B3(a) As 3.F2(a).

9.B3(b) E. G. R. Taylor, *The Mathematical Practitioners of Tudor and Stuart England*, Cambridge, 1954, pp. 170–171.

9.B3(c) F. A. Yates, *The Rosicrucian Enlightenment*, Paladin, 1975, pp. 22–23.

9.C1 R. Ascham, *The Schoolmaster*, 1570; Cassell, 1888, p. 23.

9.C2 W. Kempe, translator's dedication to P. Ramus, *The Art of Arithmeticke in Whole Numbers and Fractions,...*, 1592.

9.C3 G. Harvey, *Pierces Supererogation or a New Prayse of the Old Asse*, London, 1593, p. 190.

9.C4 T. Hylles, *The Arte of Vulgar Arithmeticke*, London, 1600.

9.C5 F. Bacon, *The Advancement of Learning*, 2, VIII.2, 1605.

9.D1 G. Chapman, *Achilles Shield*, London, 1598.

9.D2 T. Harriot; from R. C. H. Tanner, 'The Ordered Regiment of the Minus Sign: off-beat mathematics in Harriot's manuscripts', *Annals of Science*, **37** (1980) pp. 149–150.

9.D3 T. Harriot; from J. A. Lohne, 'A Survey of Harriot's Scientific Writings', *Archive for History of Exact Sciences*, **20** (1979) pp. 297–298; tr. J. G. Fauvel.

9.D4 W. Lower; from J. W. Shirley, *Thomas Harriot*, Oxford, 1983, p. 400.

9.D5 As 3.F2(a).

9.D6(a) J. Wallis, *Treatise of Algebra, both historical and practical,...*, London, 1685, Preface.

9.D6(b) J. Wallis, *De Algebra Tractatus*, 1693, p. 206; tr. F. Cajori, *William Oughtred*, Open Court, 1916, p. 70.

9.D6(c) As 9.D6(b), pp. 70–71.

9.D7(a) J. V. Pepper, '*Harriot's Earlier Work on Mathematical Navigation: theory and practice*' *in J. Shirley (ed.) Thomas Harriot; renaissance scientist*, Oxford, 1974, pp. 86–87.

9.D7(b) D. T. Whiteside, 'In Search of Thomas Harriot', *History of Science*, **13** (1975) pp. 61–62.

9.E1 J. Napier, *A Description of the Admirable Table of Logarithms*, preface; tr. E. Wright, London, 1616.

9.E2 H. Briggs, *Arithmetica Logarithmica*, 1624; preface; tr. C. Hutton, *Mathematical Tables*, London, 1785, p. 34.

9.E3 W. Lilly, *Mr Lilly's History of his Life and Times*, London, 1715, pp. 105–106.

9.E4 C. Hutton, *Mathematical Tables: ... to which is prefixed, a large and original history of the discoveries and writings relating to those subjects; ...*, London 1785; 1822 edition, pp. 49–54.

9.E5 J. Keil, *A Short Treatise of the Nature and Arithmetick of Logarithms*, 1733 edition, preface.

9.E6 E. Stone, *A New Mathematical Dictionary*, London, 1743.

9.F1 W. Oughtred, *The Key of the Mathematicks New Forged and Filed: . . .*, London, 1647.

9.F2 As 9.D6(a), pp. 67–69.

9.F3(a) J. Wallis, 1666; from S. P. Rigaud, *Correspondence of Scientific Men of the Seventeenth Century*, II, Oxford, 1842, p. 475.

9.F3(b) J. Wallis, 1666; as 9.F3(a) pp. 479–480.

9.F4 As 3.F2(a).

9.G All extracts as 3.F2(a).

9.H1 J. Pell; from P. J. Wallis, 'An Early Mathematical Manifesto—John Pell's Idea of Mathematics', *Durham Research Review*, **18** (1967) pp. 139–148.

9.H2(a)–(b) J. O. Halliwell (ed.), *A Collection of Letters Illustrative of the Progress of Science in England*, London, 1841, pp. 80, 86.

9.H2(c) H. Hervey, 'Hobbes and Descartes in the Light of some Unpublished Letters of the Correspondence between Sir Charles Cavendish and Dr. John Pell', *Osiris*, **10** (1952) pp. 67–90.

9.H3(a) T. Hobbes, *Six Lessons to the Professors of Mathematics*, 1656; in T. Hobbes, *Collected Works*, VII, 1839–45, pp. 315–316.

9.H3(b) J. Wallis, *Due Correction for Mr Hobbes, or Schoole Discipline, for not saying his Lessons right*, 1656, p. 50.

9.H4 J. Wallis; from C. J. Scriba, 'The Autobiography of John Wallis, F.R.S.', *Notes and Records of the Royal Society*, **25** (1970) pp. 26–27.

9.H5 S. Pepys; from R. Latham and W. Matthews (eds), *The Diary of Samuel Pepys*, Bell, 1970–1976; this selection in A. G. Howson, *A History of Mathematics Education in England*, Cambridge, 1982, p. 29.

Chapter 10

10.A1 J. Kepler, *Astronomia Nova*, Prague, 1609, Introduction; in J. Kepler, *Gesammelte Werke*, III, Munich, 1939; tr. A. R. Hall, T. Salusbury, in M. B. Hall (ed.), *Nature and Nature's Laws*, Macmillan, 1970, pp. 67–77.

10.A2 J. Kepler, *Harmonices Mundi*, Linz, 1619, Proem; from J. Kepler, *Gesammelte Werke*, VI, Munich, 1940; tr. C. G. Wallis, in *Great Books of the Western World*, XVI, Encyclopaedia Britannica, 1952.

10.A3(a) J. Kepler, *Harmonices Mundi*, Linz, 1619; tr. J. H. Walden, in H. Shapley, H. E. Howarth (eds), *A Source Book in Astronomy*, McGraw-Hill, 1929, pp. 30–41.

10.A3(b) J. Kepler, *Mysterium Cosmographicum*, Tübingen, 1596; from J. Kepler, *Gesammelte Werke*, I, Munich, 1938, facing p. 26.

10.A3(c) J. Kepler, *Harmonices Mundi*, Linz, 1619; from J. Kepler, *Gesammelte Werke*, VI, Munich, 1940, facing p. 79.

10.A4(a) J. Kepler, *Mysterium Cosmographicum*, Preface, Note; tr. A. Koestler, quoted in *The Sleepwalkers*, London, 1968, p. 268.

10.A4(b) J. Kepler, *Letter to Herwart*, 10 February 1605; tr. A. Koestler, quoted in *The Sleepwalkers*, London, 1968, p. 345.

10.A4(c) *Harmonices Mundi*, Linz, 1619, IV, Chapter I; tr. A. Koestler, quoted in *The Sleepwalkers*, London, 1968, p. 264.

10.B1(a) Galileo Galilei, *Il Saggiatore*, Rome 1623; tr. S. Drake, *Discoveries and Opinions of Galileo*, Doubleday Anchor, 1957, pp. 237–238.

10.B1(b)–(c) Galileo Galilei, *Dialogo sopra i Due Massimi Sistemi del Mondo*, Florence, 1632; tr. S. Drake, *Dialogue concerning the Two Chief World Systems*, University of California Press, 1967, pp. 103, 207–208.

10.B2 Galileo Galilei, *Discorsi e Dimonstrazione Matematiche intorno a Due Nuovo Science*, Leiden, 1638; tr. H. Crew, A. de Salvio, *Dialogues concerning Two New Sciences*, Macmillan, 1914; from Dover reprint, pp. 94–97.

10.B3 As 10.B2, pp. 161–162.

10.B4 As 10.B2, pp. 173–176.

10.B5 As 10.B2, pp. 248–250.

Chapter 11

11.A1 R. Descartes, *Discours de la Methode, etc.*, Leiden, 1637; tr. L. J. Lafleur, *Discourse on Method*, Bobbs-Merrill, 1964, pp. 15–16.

11.A2 R. Descartes, *La Géométrie*, Appendix; in tr. D. E. Smith, M. L. Latham, *The Geometry of René Descartes*, Dover, 1954, p. 5.

11.A3 As 11.A2, pp. 6–13.

11.A4 Pappus; tr. I. Thomas, *Greek Mathematical Works*, Loeb Heinemann, 1980, I, pp. 487–489, II, p. 603.

11.A5 R. Descartes, *Letter to Mersenne*; in M. Mersenne, *Correspondance*, II, Tannery, 1932, pp. 316–317; tr. G. Jagger.

11.A6 As 11.A2, pp. 25–37.

11.A7 As 11.A2, pp. 44–48, 156.

11.A8 As 11.A2, p. 91.

11.A9 As 11.A2, pp. 95–111; adapted by J. G. Fauvel.

11.A10 H. J. M. Bos, 'On the Representation of Curves in Descartes' *Géométrie*', *Archive for History of Exact Sciences*, **24** (1981) pp. 304–305, 331–332.

11.B1 F. Debeaune, *Letter to Roberval*, 1638, and *Letter to Mersenne*, 1639; in M. Mersenne, *Correspondence*, VIII, Tannery, 1963, pp. 142–143, 348; tr. J. J. Gray.

11.B2 P. de la Hire, *Les Lieux Géométriques*, Paris, 1679; tr. B. Robinson, *New Elements of Conic Sections*, London, 1704, Preface.

11.B3 P. de la Hire, *Les Lieux Géométriques*, Paris, 1679, Preface, pp. 179–181; tr. J. J. Gray.

11.B4 H. van Heuraet, 'De Transmutatione Curvarum Linearum in Rectas', in R. Descartes, *Geometria, etc.*, Amsterdam, 1659–1661 (2nd edition), pp. 517–520; tr. A. W. Grootendorst, J. A. van Maanen, 'On the Rectification of Curves', *Nieuw Archiv voor Wiskunde*, (3) **25** (1982) pp. 101–105.

11.B5 J. Hudde, 'Johannis Huddenii Epistola Secunda de Maximis et Minimis', in R. Descartes, *Geometria*, etc., Amsterdam, 1659–1661 (2nd edition), p. 507; tr. J. J. Gray.

11.C1 P. de Fermat; in P. Tannery, C. Henry (eds), *Pierre de Fermat: Oeuvres*, Paris, 1891–1922, I, pp. 133–135; tr. D. J. Struik, *A Source Book in Mathematics, 1200–1800*, Harvard University Press, 1969, pp. 223–224.

11.C2 As 11.C1, I, pp. 147–149; tr. J. J. Gray.
11.C3 As 11.C2, II, p. 206.
11.C4 As 11.C2, II, pp. 198–199.
11.C5 As 11.C1, I, pp. 255–259; tr. pp. 219–221.
11.C6 As 11.C2, II, p. 335.
11.C7 As 11.C2, II, pp. 432–434.
11.C8(a) As 11.C2, I, pp. 340–341.
11.C8(b) A. Weil, *Number Theory, an Approach Through History*, Birkhäuser, 1983, p. 77.
11.D1 G. Desargues, *Brouillon Project, etc.*, Paris, 1639; tr. J. V. Field, 'Rough Draft on Conics', in J. V. Field, J. J. Gray, *The Geometrical Work of Girard Desargues*, Springer, 1987, pp. 69–70.
11.D2 As 11.D1, p. 92.
11.D3 As 11.D1, p. 106–110.
11.D4 R. Descartes; tr. as 11.D1, pp. 175–177.
11.D5 B. Pascal; tr. as 11.D1, pp. 180–182.
11.D6 As 11.D1, pp. 161–162.
11.D7(a) P. de la Hire, *Sectiones Conicae*, Paris, 1685, pp. 12, 33–34; tr. J. J. Gray.
11.D7(b) M. Chasles, *Aperçu Historique sur l'Origine et le Développement des Methodes en Géométrie*, Bruxelles, 1837, pp. 122–123; tr. J. J. Gray.
11.E1 G. P. de Roberval, 'Traité des Indivisibles' in *Divers Ouvrages*, Paris, 1693, pp. 191–193; tr. J. J. Gray.
11.E2 B. Pascal, *Letter to Carcavy*, 1658; in L. Brunschvig, P. Boutroux (eds), *Oeuvres*, VIII, 1914, p. 352; tr. J. J. Gray.
11.E3 I. Barrow, *Lectiones Geometricae*, London, 1670, Lecture X; tr. D. J. Struik, in D. J. Struik, *A Source Book in Mathematics, 1200–1800*, Harvard University Press, 1969, pp. 255–256.

Chapter 12

All references to D. T. Whiteside (tr. and ed.), *The Mathematical Papers of Isaac Newton*, Cambridge University Press, 8 volumes, 1967 etc., are given as MPIN.

12.A1 I. Newton, *The October 1666 Tract on Fluxions*; MPIN, I, pp. 414–415.
12.A2 I. Newton, *De Analysi per Aequationes Infinitas*, 1669(?); MPIN, II, pp. 207–209.
12.A3 As 12.A2, pp. 237–239.
12.A4 As 12.A2, pp. 243–245.
12.A5 I. Newton, *The Method of Series and Fluxions*, 1671; MPIN, III, pp. 79–81.
12.A6 As 12.A5, pp. 83–87.
12.B1 I. Newton, *Mathematical Principles of Natural Philosophy and his System of the World*,1687; tr. A. Motte (1729), rev. F. Cajori (1934), University of California Press, 1962, pp. xvii–xix.
12.B2 As 12.B1, I, pp. 13–14.
12.B3 As 12.B1, I, pp. 29–30.
12.B4 As 12.B1, I, pp. 38–39.
12.B5 As 12.B1, I, pp. 40–42.

12.B6 As 12.B1, I, pp. 56–57.
12.B7 As 12.B1, III, p. 406 and I, pp. 45–46.
12.B8 As 12.B1, I, pp. 140–147.
12.B9 As 12.B1, II, pp. 394–396.
12.B10 As 12.B1, III, pp. 399–400.
12.B11 As 12.B1, III, p. 424.
12.B12 As 12.B1, II, p. 543.
12.B13 As 12.B1, II, pp. 546–547.
12.C1 I. Newton, *Epistola Prior*, 1676; in H. W. Turnbull (ed.), *The Mathematical Correspondence of Isaac Newton*, II, Cambridge, 1960, pp. 332–333.
12.C2 I. Newton, *Epistola Posterior*, 1676; as 12.C1, pp. 129–134.
12.D1 I. Newton, *Restoration of the Ancients' Solid Loci*, late 1670s; MPIN, IV, pp. 275–283.
12.D2 I. Newton, *An Improved Enumeration . . . of the General Cubic Curve*, 1695; MPIN, VII, pp. 589, 635.
12.D3 I. Newton; tr. J. Raphson, *Universal Arithmetic*, London, 1769 (2nd edition), pp. 465–470.
12.D4 I. Newton; MPIN, VI, pp. 269, 271.
12.E1 A. Pope, *Epitaphs*, 1730.
12.E2 A. Pope, *An Essay on Man*, 1732, Epistle II.
12.E3 W. Wordsworth, *The Prelude*, 1850 edition, Book III.
12.E4 W. Blake, *Jerusalem*, 1804, Chapter 1.
12.F1 B. de Fontenelle, *The Elogium of Sir Isaac Newton*, London, 1728, pp. 13–16.
12.F2 F. Voltaire, *Lettres Écrits de Londres sur les Anglois et Autre Sujets*, Basel, 1734; tr. L. W. Tancock, *Letters on England*, Penguin, 1984, pp. 68–71.
12.F3 F. Voltaire, *The Elements of Sir Isaac Newton's Philosophy*; tr. J. Hanna, 1738, pp. 166–167, 193–194, 200–201.
12.F4 J. M. Keynes, 'Newton, the Man', in *Newton Tercentenary Celebrations*, Cambridge, 1947, pp. 21–34.
12.F5 D. T. Whiteside, 'Newton, the Mathematician', in *Contemporary Newtonian Research,* Reidel, 1982, pp. 120–122, © Professor D. T. Whiteside, 1982.

Chapter 13

13.A1 G. W. Leibniz; in J. M. Child, *The Early Mathematical Manuscripts of Leibniz*, Open Court, 1920, pp. 80–84.
13.A2 G. W. Leibniz; as 13.A1, pp. 116–122.
13.A3 G. W. Leibniz, 'Nova Methodus pro Maximis et Minimis, etc.', *Acta Eruditorum,* **3** (1684) pp. 467–473; reprinted in G. W. Leibniz, *Mathematische Schriften*, 2.3, 1863; translation from D. J. Struik, *A Source Book in Mathematics, 1200–1800*, Harvard University Press, 1969, pp. 272–280.
13.B1 J. Bernoulli, *Lecture on Debeaune's Problem*; in N. Cramer (ed.), *Opera Omnia*, I, Lausanne and Geneva, 1742, pp. 423–424; tr. J. J. Gray.
13.B2 J. Bernoulli, *Solution of a Problem concerning the Integral Calculus*, Gröningen, 1702; as 13.B1, I, pp. 393–397.

13.B3 J. Bernoulli, *Problème Inverse des Forces Centrales*; as 13.B1, I, pp. 474–477.

13.B4 J. Bernoulli, *Letter to Montmort*, 1718; in O. Spiess (ed.), *Der Briefwechsel von Johann Bernoulli*, I, Birkhäuser, 1955, pp. 136–137; tr. J. J. Gray.

13.B5 G. de l'Hôpital, *Analyse des Infiniment Petits, pour l'Intelligences des Lignes Courbes*, Paris, 1696; tr. E. Stone, *The Method of Fluxions both Direct and Inverse*, London, 1730, Preface, pp. vii, x.

13.B6 As 13.B5, but with Stone's translation corrected by D. J. Struik, from D. J. Struik, *A Source Book in Mathematics, 1200–1800*, Harvard University Press, 1969, pp. 313–315.

13.B7 As 13.B5, translator's preface.

Chapter 14

14.A1(a) L. Euler, *Letter to Johann Bernoulli*, 15 September 1739; in G. Eneström (ed.), 'Der Briefwechsel zwischen Leonhard Euler and Johann I. Bernoulli', *Bibliotheca Mathematica*, **6** (1905) pp. 33–38; tr. J. J. Gray.

14.A1(b) J. Bernoulli, *Letter to Euler*, 9 December 1739; as 14.A1(a), pp. 39–43; tr. J. J. Gray.

14.A1(c) L. Euler, *Letter to Johann Bernoulli*, 19 January 1740; as 14.A1(a), pp. 43–52; tr. J. J. Gray.

14.A2 L. Euler, *Introductio in Analysin Infinitorum*, I, Lausanne, 1748, Chapter 8; in L. Euler, *Opera Omnia*, Leipzig Berlin Zurich Basel, 1911–present, (1) **8**, pp. 103–152; tr. in D. J. Struik, *A Source Book in Mathematics, 1200–1800*, Harvard University Press, 1969, pp. 346–351.

14.A3 L. Euler, 'De la Controverse entre Mrs Leibniz et Bernoulli sur les Logarithmes des Nombres Negatifs et Imaginaires', *Mem. Acad. Sci. Berlin*, 1749; as 14.A2, (1) **17**, pp. 195–232; tr. J. J. Gray.

14.A4 L. Euler, *Introductio in Analysin Infinitorum*, II; as 14.A2, (1) **9**, pp. 150–151; tr. G. Jagger.

14.A5 L. Euler, 'Demonstration sur la Nombre des Points ou Deux Lignes des Ordres Quelconques Peuvent se Couper', *Mem. Acad. Sci. Berlin*, 1748; as 14.A2, (1) **26**, pp. 46–47; tr. J. J. Gray.

14.B1 P. L. M. Maupertuis; tr. anon., *The Figure of the Earth, Determined from Observations*, 1738, pp. 3, 9, 11, 14, 224–225.

14.B2(a) A. C. Clairaut, *Letter to Euler*, 1747; as 14.A2, (4) **5**, pp. 173–175; tr. J. J. Gray.

14.B2(b) L. Euler, *Letter to Clairaut*, 1747; as 14.B2(a), pp. 175–177; tr. J. J. Gray.

14.B2(c) A. C. Clairaut, *Letter to Euler*, 1747; as 14.B2(a), pp. 177–179; tr. J. J. Gray.

14.B3 A. C. Clairaut, 'Du Système du Monde dans les Principes de la Gravitation Universelle', *Mém. Acad. Paris*, 1747/1749, pp. 329–364, 549; tr. J. J. Gray.

14.B4 L. Euler, *Letter to Clairaut*, 1750; as 14.B2(a), pp. 195–196.

14.C1(a) L. Euler, 'Découverte d'un Nouveau Principe de Mecanique', *Mem. Acad. Sci. Berlin*, 1750; as 14.A2, (2) **5**, pp. 81–108; tr. J. J. Gray.

14.C1(b) C. Truesdell, 'The Rational Mechanics of Flexible or Elastic Bodies 1638–1788', in L. Euler, *Opera Omnia*, (2) **11**, pp. 250–253.

14.C2(a) L. Euler, 'Letter to Goldbach' in P.-H. Fuss (ed.), *Correspondance Mathématique et Physique*, I, 1843, p. 618; tr. J. J. Gray.

14.C2(b) A. Weil, 'Two Lectures on Number Theory, Past and Present', *Oeuvres Scientifiques, Collected Papers*, III, Springer, 1980, pp. 282–284, 286–287.

14.C3(a) L. Euler, 'Remarques sur les Mémoires Précédens de M. Bernoulli', *Mem. Acad. Sci. Berlin*, 1753/1755, pp. 196–222; as 14.A2, (2) **10**, pp. 233–254; tr. J. J. Gray, inc. tr. by C. Truesdell from L. Euler, *Opera Omnia*, (2), **11**, p. 246.

14.C3(b) As 14.C1(b), pp. 244, 247–248.

14.C4 N. Condorcet, *Elogium of Euler*; tr. H. Hunter, in *Letters to a German Princess*, London, 1795, pp. xxv–lxiii.

14.D1 J. P. de Gua de Malves, *Usages de l'Analyse de Descartes, etc.*, Paris, 1740, preface and first plate of figures; tr. J. J. Gray.

14.D2 G. Cramer, *Introduction à l'Analyse des Lignes Courbes Algébriques*, Geneva, 1750, preface; tr. J. J. Gray.

14.D3 J. d'Alembert, *Discours Préliminaire*, 1751; tr. R. N. Schwab, *Preliminary Discourse to the Encyclopedia of 1751*, Bobbs-Merrill, 1963, pp. 19–22, 26–27.

14.D4 J. L. Lagrange, 'Réflexions sur la Résolution Algébriques des Équations', *Mem. Acad. Roy. Sci.* 1770/1771; in J. A. Serret, G. Darboux (eds), *Oeuvres*, Paris, 1867–1892, III, p. 206–207, 307; tr. J. J. Gray.

14.D5 J. L. Lagrange; in L. Euler, 'Elements of Algebra', in L. Euler, *Opera Omnia*, (1) **1**; tr. J. Hewlett, London, 1840 (5th edition); Springer reprint, 1985, pp. 463–465, 578–579, 583.

14.D6 J. H. Lambert, *Anmerkungen und Zusätze zur Entwerfung der Land- und Himmelscharten*, 1772; tr. W. R. Tobler, *Notes and Comments on the Compositions of Terrestrial and Celestial Maps*, University of Michigan Press, 1972, pp. 1–8 (Michigan Geographical Publication 8).

Chapter 15

15.A1 C. F. Gauss, *Tagebuch*, 1796; in C. F. Gauss, *Werke*, X.1, 1917, pp. 485–574; tr. J. J. Gray, 'Gauss's Mathematical Diary', *Expositiones Mathematicae*, **2** (1984) pp. 106–113.

15.A2 C. F. Gauss, *Demonstratio Nova Theorematis Omnem Functionem Algebraicam, etc.*, Göttingen, 1801; in C. F. Gauss, *Werke*, III, 1876 (2nd edition), pp. 3–64; tr. G. Jagger.

15.A3(a) C. F. Gauss, 'Neue Entdeckungen', *Intelligenzblatt der allgemeinen Literaturzeitung*, **66**, 1 June 1796, p. 554; in C. F. Gauss, *Werke*, X.1, 1917, p. 3; tr. J. J. Gray.

15.A3(b) C. F. Gauss, *Disquisitiones Arithmeticae*, Leipzig, 1801; in C. F. Gauss, *Werke*, I, 1863; tr. A. C. Clarke, *Disquisitiones Arithmeticae*, Yale University Press, 1966, paras 337, 364, 365.

15.A4 C. F. Gauss, 'Neue Beweise, etc.', *Commentationes soc. reg. Göttingen*, 1818; in C. F. Gauss, *Untersuchungen über höhere Arithmetik*, Chelsea, 1981 (2nd edition), p. 496; tr. J. J. Gray.

15.A5 C. F. Gauss, 'Disquisitiones Generales circa Superficies Curvas',

Commentationes soc. reg. Göttingen, 1828, Abstract; tr. A. Hiltebeitel, J. Morehead, reprinted in *astérisque*, **62** (1979) pp. 83–93.

15.B1(a) S. Germain, *Letter to Gauss*, 1804; in S. Germain, *Cinq Lettres*, Boncompagni, 1880; tr. J. J. Gray.

15.B1(b) S. Germain, *Letter to Gauss*, 1807; quoted in L. L. Bucciarelli, N. Dworsky, *Sophie Germain, an Essay in the History of the Theory of Elasticity*, Reidel, 1980, pp. 21–25.

15.B1(c) C. F. Gauss, *Letter to Germain*, 1807; as 15.B1(b), p. 25.

15.B2(a) C. F. Gauss, *Letter to Bessel*, 1829; in C. F. Gauss, *Werke*, VIII, 1900, pp. 200–201; tr. J. J. Gray.

15.B2(b) F. W. Bessel, *Letter to Gauss*, 1829; as 15.B2(a).

15.B2(c) C. F. Gauss, *Letter to Bessel*, 1829; as 15.B2(a).

15.C1(a) A. M. Legendre, 'Recherches d'Analyse Indéterminée', *Hist. Acad. Roy. des Sciences*, 1785/1788, pp. 513–517; tr. J. J. Gray.

15.C1(b) As 15.A3(b), Appendix.

15.C2 H. M. Edwards, *Fermat's Last Theorem*, Springer, 1977, pp. 79–81.

15.D1 E. Galois, *Letter to Chevalier*, 1832; in R. Bourgne, J.-P. Azra (eds), *Écrits et Mémoires Mathématiques d'Evariste Galois*, Paris, 1962, pp. 173–186; tr. L. Weisner, in D. E. Smith (ed.), *A Source Book in Mathematics*, Dover, 1959, pp. 278–285.

15.D2 E. Galois, *Preface*; as 15.D1, pp. 3–11; tr. J. J. Gray (includes some of a previous translation by H. M. Kline, 'Discussion on the Progress of Pure Analysis', *American Mathematical Monthly*, **85** (7) (1978) pp. 565–566).

15.D3 A. L. Cauchy, *Sur le Nombre des Valeurs, etc.*, 1845; in A. L. Cauchy, *Oeuvres*, (2) **1**, pp. 277–293; tr. J. J. Gray (includes a partial translation by A. Shenitzer in H. Wussing, *The Genesis of the Abstract Group Concept*, MIT Press, 1984, pp. 92–93).

15.D4 C. Jordan, *Traité des Substitutions et des Équations Algébriques*, Paris, 1870, Preface; tr. J. J. Gray.

Chapter 16

16.A1 J. Wallis, 'De Postulato Quinto et Definitione Quinta Lib. 6 Euclidis Disceptatio Geometrica', *Opera Mathematica*, **2** (1693) pp. 665–678; tr. J. J. Gray.

16.A2 G. Saccheri, *Euclides ab Omni Naevo Vindicatus*, 1733: tr. G. B. Halsted, *Euclid Freed from Every Flaw*, Open Court, 1920, pp. 3–9, 169–173.

16.A3 J. H. Lambert, *Letter to Kant*, 1770; in A. Zweig (ed.), *Kant's Philosophical Correspondence, 1759–1799*, University of Chicago Press, 1967, number 61.

16.A4 I. Kant, *Critik der reinen Vernunft*, Riga, 1781; tr. N. Kemp Smith, *Critique of Pure Reason*, Macmillan, 1970, pp. 578–581.

16.A5 J. H. Lambert, *Theorie der Parallellinien*, 1786; in F. Engel and P. Stäckel, *Theorie der Parallellinien von Euklid bis auf Gauss*, Teubner, 1899; this version from Johnson reprint, 1968, paras 79–82; tr. J. J. Gray.

16.A6 A. M. Legendre, *Éléments de Géométrie*, Paris, 1802 (4th edition), pp. 21–22; part tr. J. J. Gray; part tr. T. Carlyle, in A. M. Legendre, *Elements of*

Geometry and Trigonometry, Edinburgh, 1824, pp. 223–224.

16.B1 F. K. Schweikart, *Memorandum to Gauss*, 1818; in R. Bonola, *La Geometria Non-Euclidea*, 1906; tr. H. S. Carslaw, *Non-Euclidean Geometry*, Dover, 1955, p. 76.

16.B2 R. Bonola, *La Geometria Non-Euclidea*, 1906; tr. H. C. Carslaw, *Non-Euclidean Geometry*, Dover, 1955, p. 100.

16.B3 N. I. Lobachevskii, *Geometrische Untersuchungen*, 1840; tr. G. B. Halsted, *Theory of Parallels*, in 16.B2, Supplement, pp. 11–15, 44–45.

16.B4 H. Meschkowski, *Non-Euclidean Geometry*, Academic Press, New York, 1964, pp. 31–34.

16.B5 J. Bolyai, *The Science Absolute of Space*, 1831; in 16.B2, Supplement, paragraph 1 and final paragraph.

16.C1 As 16.B2, pp. 121–126.

16.C2 B. Riemann, 'Über die Hypothesen welche der Geometrie zugrunde liegen', *Abh. Göttingen*, **13** (1867); in B. Riemann, *Werke*, Leipzig, 1892 (2nd edition); Dover reprint, pp. 272–287; tr. W. K. Clifford, 'On the Hypotheses which lie at the Foundations of Geometry', in W. K. Clifford, *Mathematical Papers*, 1882, pp. 55–72; this version from Chelsea reprint, 1968, pp. 55–72.

16.C3 E. Beltrami, 'Saggio di Interpretazione della Geometria Non-Euclidea', *Giornale di Matematica*, **6** (1868), pp. 284–312; in E. Beltrami, *Opere Matematica*, I, pp. 374–405; tr. J. C. Stillwell, *Essay on the Interpretation of Non-Euclidean Geometry, 1868*, Monash University preprint, 1982, pp. 1–6 (Department of Mathematics Paper 4).

16.C4 F. Klein 'Über die so-genannte nicht-Euklidische Geometrie', *Mathematische Annallen*, **4** (1871); in B. Fricke, A. Ostrowski (eds), *Gesammelte Mathematische Abhanglungen*, I, Leipzig, 1921, pp. 243–247; tr. J. J. Gray.

16.C5 J. J. Gray, *Ideas of Space*, Oxford University Press, 1979, pp. 155–156.

16.D1 F. Dostoevsky, *The Brothers Karamazov*, 1880; tr. D. Magarshack, Penguin, 1958, pp. 274–275.

16.D2 G. G. Marquez, *One Hundred Years of Solitude*, 1967; tr. G. Rabassa, Picador, 1978, pp. 11–12.

Chapter 17

17.A1 J. V. Poncelet, *Traité des Propriétés Projectives des Figures*, 1822, Introduction, pp. xix–xxvii; tr. J. J. Gray.

17.A2 As 11.D7(b), pp. 189–190, 191, 208–209.

17.A3 J. D. Gergonne, 'Géométrie de Situation', *Annales des Mathématiques*, **18** (1827–28) pp. 150–152, 214–216; tr. J. J. Gray.

17.A4 M. Paul, *Gaspard Monges 'Geometrie Descriptive' und die Ecole Polytechnique*, Institut für Didaktik der Mathematik, 1980, pp. 123–124; tr. J. J. Gray.

17.B1 A. F. Möbius, *Der Barycentrische Calcul*, 1827, pp. v, vi–vii, x–xii, xiii–xiv, 320–321; tr. J. J. Gray.

17.B2 J. Plücker, 'Solution d'une Question Fondamentale concernant la Théorie

Générale des Courbes', *Journal für Mathematik*, **12** (1834) pp. 105–108; in A. Schoenflies, F. Pockels (eds), *Gesammelte Mathematische Abhandlungen*, I, Leipzig, 1895, pp. 298–301; tr. J. J. Gray.

17.B3 A. Clebsch, *J. Plücker, in Memoriam*; in A. Schoenflies, F. Pockels (eds), *Gesammelte Mathematische Abhandlungen*, I, Leipzig, 1895, pp. 1, 10, 14; tr. J. J. Gray.

17.B4 C. Wiener, *Stereoscopische Photographien des Modelles einer Fläche dritter Ordnung mit 27 reellen Geraden*, Leipzig, 1869.

Chapter 18

18.A1 G. Berkeley, *The Analyst*, London, 1734; in A. A. Luce, T. E. Jessup (eds), *The Works of George Berkeley, Bishop of Cloyne*, **4**, London, 1951, pp. 69–71, 72, 75–76, 100–102.

18.A2 C. MacLaurin, *Treatise on Fluxions*, II, Edinburgh, 1754, pp. 581–582.

18.A3 J. d'Alembert, 'Differentials', in *Encyclopédie*, **4**, 1754; tr. from D. J. Struik, *A Source Book in Mathematics, 1200–1800*, Harvard University Press, 1969, pp. 342–345.

18.A4 J. L. Lagrange, *Théorie des Fonctions Analytiques*, 1797; in J. A. Serret, G. Darboux (eds), *Oeuvres*, Paris, 1867–1892, IX, pp. 20ff.; tr. from D. J. Struik, *A Source Book in Mathematics, 1200–1800*, Harvard University Press, 1969, pp. 388–391.

18.A5 J. L. Lagrange, 'Discours sur l'Objet de la Théorie des Fonctions Analytiques', *Journal de l'École Polytechnique*, **2**; in J. A. Serret, G. Darboux (eds), *Oeuvres*, Paris, 1867–1892, VII, pp. 325–328; tr. J. J. Gray.

18.B1 B. Bolzano, *Rein analytischer Beweis, etc.*, Prague, 1817; tr. S. B. Russ, 'A Translation of Bolzano's Paper on the Intermediate Value Theorem', *Historia Mathematica*, **7** (2) (1980) pp. 156–185.

18.B2(a)–(b) A. L. Cauchy, *Oeuvres Complètes*, Académie des Sciences, 1882–1981, (2), III, pp. 19, 43; tr. J. J. Gray.

18.B2(c)–(d) As 18.B2(a), III, p. 114 and IV, pp. 22–23; tr. U. Merzbach, in G. Birkhoff, *A Source Book in Classical Analysis*, Harvard University Press, 1973, pp. 2–4.

18.B2(e) As 18.B2(a), IV, pp. 123–125; tr. J. V. Grabiner, *The Origins of Cauchy's Rigorous Calculus*, MIT Press, 1981, pp. 171–175.

18.B3 As 18.B2(e), IV, pp. 151–152; tr. pp. 167–168.

18.B4 J. V. Grabiner, *The Origins of Cauchy's Rigorous Calculus*, MIT Press, 1981, pp. 164–165.

18.C1 R. Dedekind, *Stetigkeit und irrationale Zahlen*, Braunschweig, 1872; in R. Fricke, E. Noether, O. Ore (eds), *Gesammelte Mathematische Werke*, 3, Braunschweig, 1932, pp. 315–334; tr. W. W. Beman, 'Continuity and Irrational Numbers', in R. Dedekind, *Essays on the Theory of Numbers*, Dover, 1963.

18.C2 G. Cantor, 'Über die Ausdehnung eines Satzes aus der Theorie der trigonometrischen Reihen', *Mathematische Annalen*, **5** (1872) pp. 123–124; in E. Zermelo (ed.), *Gesammelte Abhandlungen*, Berlin, 1932; this version from Hildesheim edition, 1962, pp. 92–93; tr. J. J. Gray.

18.C3 W. Purkert, H. J. Ilgauds, *Georg Cantor*, Teubner, 1985, pp. 32–35; tr. J. J. Gray.

18.C4(a) G. Cantor, 'Über eine Eigenschaft des Inbegriffes aller reellen algebraischen Zahlen', *Journal für die reine und angewandte Mathematik*, **77** (1874) pp. 258–259; as 18.C2, pp. 115–116.

18.C4(b) G. Cantor, 'Über eine elementare Frage der Mannigfaltigkeitslehre', *Jahresbericht der Deutschen Mathematiker-Vereinigung*, **1** (1891) pp. 75–78; as 18.C2, pp. 278–279.

18.C5 G. Cantor, 'Ein Beitrag zur Mannigfaltigkeitslehre', *Journal für die reine und angewandte Mathematik*, **77** (1874) p. 257; as 18.C2, p. 132.

Chapter 19

19.A1 G. W. Leibniz, *Machina Arithmetica in qua non Additio tantum Subtractio . . .;* tr. Mark Kormes, in D. E. Smith, *A Source Book in Mathematics*, McGraw-Hill, 1929, pp. 173–181.

19.B1 C. Babbage, *On the Economy of Machinery and Manufactures*, 1832, Chapter 19; part tr. J. J. Gray.

19.B2 D. Lardner, 'Babbage's Calculating Engine', *Edinburgh Review*, **120** (July 1834).

19.B3 A. Hyman, *Charles Babbage*, Oxford University Press, 1982, Chapter 12.

19.B4 Ada, Countess of Lovelace, 'Notes by the Translator' to L. F. Menabrea, 'Sketch of the Analytical Engine Invented by Charles Babbage', *Taylor's Scientific Memoirs*, III (1843).

19.C S. Lilley, 'Machinery in Mathematics', *Discovery*, **6** (1945) pp. 182–185.

19.D1 A. de Morgan, *Letter to William Rowan Hamilton*, 23 October 1852; in N. L. Biggs, E. K. Lloyd, R. J. Wilson, *Graph Theory 1736–1936*, Oxford, 1976, pp. 90–91.

19.D2 D. J. Albers, in *The Two-Year College Mathematics Journal*, **12** (1981) p. 82.

19.D3 F. F. Bonsall, 'A Down-to-earth View of Mathematics', *American Mathematical Monthly*, **89** (1982) p. 13.

19.D4 T. Tymoczko, 'Computers, Proofs and Mathematicians: a philosophical investigation of the four-colour proof', *Mathematics Magazine*, **53** (1980) pp. 137–138.

Name Index

Subject Index